£34.00

UCLA Symposia on Molecular and Cellular Biology, New Series

Series Editor
C. Fred Fox

RECENT TITLES

Volume 13
Molecular Biology of Host–Parasite Interactions, Nina Agabian and Harvey Eisen, *Editors*

Volume 14
Biosynthesis of the Photosynthetic Apparatus: Molecular Biology, Development and Regulation, J. Philip Thornber, L. Andrew Staehelin, and Richard B. Hallick, *Editors*

Volume 15
Protein Transport and Secretion, Dale L. Oxender, *Editor*

Volume 16
Acquired Immune Deficiency Syndrome, Michael S. Gottlieb and Jerome E. Groopman, *Editors*

Volume 17
Genes and Cancer, J. Michael Bishop, Janet D. Rowley, and Mel Greaves, *Editors*

Volume 18
Regulation of the Immune System, Harvey Cantor, Leonard Chess, and Eli Sercarz, *Editors*

Volume 19
Molecular Biology of Development, Eric H. Davidson and Richard A. Firtel, *Editors*

Volume 20
Genome Rearrangement, Melvin Simon and Ira Herskowitz, *Editors*

Volume 21
Herpesvirus, Fred Rapp, *Editor*

Volume 22
Cellular and Molecular Biology of Plant Stress, Joe L. Key and Tsune Kosuge, *Editors*

Volume 23
Membrane Receptors and Cellular Regulation, Michael P. Czech and C. Ronald Kahn, *Editors*

Volume 24
Neurobiology: Molecular Biological Approaches to Understanding Neuronal Function and Development, Paul O'Lague, *Editor*

Volume 25
Extracellular Matrix: Structure and Function, A. Hari Reddi, *Editor*

Volume 26
Nuclear Envelope Structure and RNA Maturation, Edward A. Smuckler and Gary A. Clawson, *Editors*

Volume 27
Monoclonal Antibodies and Cancer Therapy, Ralph A. Reisfeld and Stewart Sell, *Editors*

Volume 28
Leukemia: Recent Advances in Biology and Treatment, David W. Golde and Robert Peter Gale, *Editors*

Volume 29
Molecular Biology of Muscle Development, Charles Emerson, Donald Fischman, Bernardo Nadal-Ginard, and M.A.Q. Siddiqui, *Editors*

Volume 30
Sequence Specificity in Transcription and Translation, Richard Calendar and Larry Gold, *Editors*

Volume 31
Molecular Determinants of Animal Form, Gerald M. Edelman, *Editor*

Volume 32
Papillomaviruses: Molecular and Clinical Aspects, Peter M. Howley and Thomas R. Broker, *Editors*

Volume 33
Yeast Cell Biology, James Hicks, *Editor*

Volume 34
Molecular Genetics of Filamentous Fungi, William Timberlake, *Editor*

Volume 35
Plant Genetics, Michael Freeling, *Editor*

Volume 36
Options for the Control of Influenza, Alan P. Kendal and Peter A. Patriarca, *Editors*

Please contact the publisher for information about previous titles in this series.

UCLA Symposia Published Prior to 1983

(Numbers refer to the publishers listed below.)

1972
Membrane Research **(2)**

1973
Membranes **(1)**
Virus Research **(2)**

1974
Molecular Mechanisms for the Repair of DNA **(4)**
Membranes **(1)**
Assembly Mechanisms **(1)**
The Immune System: Genes, Receptors, Signals **(2)**
Mechanisms of Virus Disease **(3)**

1975
Energy Transducing Mechanisms **(1)**
Cell Surface Receptors **(1)**
Developmental Biology **(3)**
DNA Synthesis and Its Regulation **(3)**

1976
Cellular Neurobiology **(1)**
Cell Shape and Surface Architecture **(1)**
Animal Virology **(2)**
Molecular Mechanisms in the Control of Gene Expression **(2)**

1977
Cell Surface Carbohydrates and Biological Recognition **(1)**
Molecular Approaches to Eucaryotic Genetic Systems **(2)**
Molecular Human Cytogenetics **(2)**
Molecular Aspects of Membrane Transport **(1)**
Immune System: Genetics and Regulation **(2)**

1978
DNA Repair Mechanisms **(2)**
Transmembrane Signaling **(1)**
Hematopoietic Cell Differentiation **(2)**
Normal and Abnormal Red Cell Membranes **(1)**
Persistent Viruses **(2)**
Cell Reproduction: Daniel Mazia Dedicatory Volume **(2)**

1979
Covalent and Non-Covalent Modulation of Protein Function **(2)**
Eucaryotic Gene Regulation **(2)**
Biological Recognition and Assembly **(1)**
Extrachromosomal DNA **(2)**
Tumor Cell Surfaces and Malignancy **(1)**
T and B Lymphocytes: Recognition and Function **(2)**

1980
Biology of Bone Marrow Transplantation **(2)**
Membrane Transport and Neuroreceptors **(1)**
Control of Cellular Division and Development **(1)**
Animal Virus Genetics **(2)**
Mechanistic Studies of DNA Replication and Genetic Recombination **(2)**

1981
Immunoglobulin Idiotypes **(2)**
Initiation of DNA Replication **(2)**
Genetic Variation Among Influenza Viruses **(2)**
Developmental Biology Using Purified Genes **(2)**
Differentiation and Function of Hematopoietic Cell Surfaces **(1)**
Mechanisms of Chemical Carcinogenesis **(1)**
Cellular Recognition **(1)**

1982
B and T Cell Tumors **(2)**
Interferon **(2)**
Rational Basis for Chemotherapy **(1)**
Gene Regulation **(2)**
Tumor Viruses and Differentiation **(1)**
Evolution of Hormone-Receptor Systems **(1)**

Publishers

(1) Alan R. Liss, Inc.
41 East 11th Street
New York, NY 10003

(2) Academic Press, Inc.
Orlando, FL 32887

(3) W.A. Benjamin, Inc.
2725 Sand Hill Road
Menlo Park, CA 94025

(4) Plenum Publishing Corp.
233 Spring Street
New York, NY 10013

MOLECULAR BIOLOGY
OF MUSCLE DEVELOPMENT

MOLECULAR BIOLOGY OF MUSCLE DEVELOPMENT

Proceedings of a Roche–UCLA Symposium
Held in Park City, Utah
March 15–22, 1985

Editors

Charles Emerson
Department of Biology
University of Virginia
Charlottesville, Virginia

Bernardo Nadal-Ginard
Department of Cardiology
Children's Hospital
Department of Pediatrics
Harvard Medical School
Boston, Massachusetts

Donald Fischman
Departments of Cell Biology
and Anatomy
Cornell University
Medical College
New York, New York

M.A.Q. Siddiqui
Department of Biochemistry
Roche Institute of Molecular Biology
Nutley, New Jersey

Alan R. Liss, Inc. • New York

Address all Inquiries to the Publisher
Alan R. Liss, Inc., 41 East 11th Street, New York, NY 10003

Printed in the United States of America.

Library of Congress Cataloging-in-Publication Data

Roche-UCLA Symposium (1985 : Park City, Utah)
Molecular biology of muscle development.
(UCLA symposia on molecular and cellular biology; new ser., v. 29)
Includes bibliographies and index.
1. Muscles—Growth—Congresses. 2. Molecular biology—Congresses. 3. Cell physiology—Congresses.
I. Emerson, Charles. II. Hoffman-La Roche, Inc.
III. University of California, Los Angeles, IV. Title.
V. Series. [DNLM: 1. Molecular Biology—congresses.
2. Muscles—growth & development—congresses. 3. Muscular Dystrophy—congresses. W3 U17N new ser. v.29 /
WE 500 R673m 1985]
QP321.R7 1985 591.1′852 86-110
ISBN 0-8451-2628-8

Contents

V. GENE EXPRESSION

VI. MYOFIBRILLAR ASSEMBLY AND THE CYTOSKELETON

Contributors

Ueli Achtnich, Department of Cell Biology, Swiss Federal Institute of Technology, CH-8093 Zurich, Switzerland **[693]**

Janet Ajioka, Department of Molecular Biology, Pasteur Institute, Paris 75724, France **[237]**

Serge Alonso, Department of Molecular Biology, Pasteur Institute, Paris 75015, France **[487]**

N.L. Anderson, Molecular Anatomy Program, Division of Biological and Medical Research, Argonne National Laboratory, Argonne, IL 60459 **[445]**

Parker B. Antin, Department of Anatomy, School of Medicine, University of Pennsylvania, Philadelphia, PA 19104 **[709]**

Hans-Henning Arnold, Department of Toxicology, Medical School, University of Hamburg, D-2000 Hamburg, Federal Republic of Germany **[793]**

Saoko Atsumi, Division of Cell Biology, National Center for Nervous, Mental and Muscular Disorders, Kodaira, Tokyo 187, Japan, and Department of Anatomy, Yamanashi Medical College, Yamanashi Ken 409-38, Japan **[143]**

David Bader, Department of Cell Biology and Anatomy, Cornell University Medical College, New York, NY 10021 **[293, 825]**

Martin Bähler, Department of Cell Biology, Swiss Federal Institute of Technology, CH-8093 Zurich, Switzerland; present address: Laboratory of Molecular/Cellular Neuroscience, The Rockefeller University, New York, NY 10021-6399 **[683]**

William Bains, Department of Medicine, The MEDIGEN Project, Veterans Administration Medical Center, Palo Alto, CA 94305, and Department of Pharmacology, Stanford University Medical School, Palo Alto, CA 94305 **[507]**

Albert S. Baldwin, Center for Cancer Research and Department of Biology, Massachusetts Institute of Technology, Cambridge, MA 02139 **[1,21]**

R.J. Barstead, Department of Genetics, Washington University, St. Louis, MO 63110 **[605]**

Paul Barton, Department of Molecular Biology, Pasteur Institute, Paris 75015, France **[487]**

The number in brackets is the opening page number of the contributor's article.

G.M. Benian, Department of Genetics, Washington University, St. Louis, MO 63110 **[605]**

Derk Bergsma, Department of Cell Biology, Baylor College of Medicine, Houston, TX 77030 **[531]**

Corlee Bertelson, Division of Genetics, Children's Hospital, Boston, MA 02115; present address: Brandeis University, Waltham, MA 02254 **[865]**

I.M. Bird, Department of Biochemistry, University of Birmingham, Birmingham B15 2TT, United Kingdom; present address: Courtauld Institute of Biochemistry, The Middlesex Hospital Medical School, London W1P 7PN, United Kingdom **[423]**

Helen M. Blau, Department of Pharmacology, Stanford University Medical School, Palo Alto, CA 94305 **[105, 253, 507, 911]**

Christopher J. Brandl, Banting and Best Department of Medical Research, University of Toronto, C.H. Best Institute, Toronto, Ontario M5G 1L6, Canada **[199]**

Roger E. Breitbart, Department of Cardiology, Children's Hospital, and Department of Pediatrics, Harvard Medical School, Boston, MA 02115 **[387]**

Margaret Buckingham, Department of Molecular Biology, Pasteur Institute, Paris 75015, France **[487]**

Gabriele E. Bugaisky, Department of Medicine, University of Chicago, Chicago, IL 60637 **[363]**

Steve Burden, Department of Biology, Massachusetts Institute of Technology, Cambridge, MA 02139 **[189]**

Gillian S. Butler-Browne, Department of Molecular Biology, Pasteur Institute, Paris 75724, France **[237]**

Yassemi G. Capetanaki, Division of Biology, California Institute of Technology, Pasadena, CA 91125 **[749]**

Arnold I. Caplan, Department of Biology, Case Western Reserve University, Cleveland, OH 44106 **[117]**

David A. Carrino, Department of Biology, Case Western Reserve University, Cleveland, OH 44106 **[117]**

Lisbeth C. Cerny, Department of Cell Biology, Swiss Federal Institute of Technology, CH-8093 Zurich, Switzerland; present address: Department of Food Sciences and Technology, University of California, Davis, CA 95616 **[693]**

Choy-Pik Chiu, Department of Pharmacology, Stanford University School of Medicine, Stanford, CA 94305; present address: Laboratory of Molecular Carcinogenesis, Dana-Farber Cancer Institute, Boston, MA 02205 **[105]**

Jennifer Clyne, Department of Anatomy, University of California, San Francisco, CA 94143 **[547]**

Arlette Cohen, Department of Molecular Biology, Pasteur Institute, Paris 75015, France **[487]**

Reid S. Compton, Department of Biology, University of Virginia, Charlottesville, VA 22901; present address: Institute for Cancer Research, Fox Chase Cancer Center, Philadelphia, PA 19111 **[61]**

Thomas A. Cooper, Department of Anatomy, University of California, San Francisco, CA 94143 **[437]**

Christine B. Couch, Department of Neurology, Baylor College of Medicine, Houston, TX 77030 **[167]**

Michelle D. Coutu, Department of Biology, Johns Hopkins University, Baltimore, MD 21218 **[639]**

Leanne L. Cribbs, Department of Medicine, University of Chicago, Chicago, IL 60637 **[809]**

Michael T. Crow, Departments of Medicine and Oncology, Stanford University School of Medicine, Stanford, CA 94305; present address: Department of Biology, University of Houston, Houston, TX 77004 **[213]**

Douglas S. Darling, Department of Medicine, University of Chicago, Chicago, IL 60637 **[809]**

Shashiprabha Dasgupta, Department of Muscle Research, Boston Biomedical Research Institute and Department of Neurology, Harvard Medical School, Boston, MA 02114 **[591]**

Philippe Daubas, Department of Molecular Biology, Pasteur Institute, Paris 75015, France **[487]**

Norman Davidson, Department of Chemistry, California Institute of Technology, Pasadena, CA 91125 **[411]**

Amrut Deshpande, Department of Biochemistry, Roche Research Center, Roche Institute of Molecular Biology, Nutley, NJ 07110 **[773]**

G.K. Dhoot, Department of Immunology, University of Birmingham, Birmingham B15 2TT, United Kingdom **[423]**

Evelyn Dizon, Department of Medicine, University of Chicago, Chicago, IL 60637 **[809]**

Dennis Drayna, Howard Hughes Medical Institute and Department of Human Genetics, University of Utah School of Medicine, Salt Lake City, UT 84132 **[903]**

Catherine Duff, Department of Genetics, Hospital for Sick Children, Toronto, Ontario M5G 1X8, Canada **[887]**

A. Duran, Department of Anatomy, School of Medicine, University of Pennsylvania, Philadelphia, PA 19104 **[709]**

Anne H. Dutton, Department of Biology, University of California, San Diego, La Jolla, CA 92093 **[741]**

Marion S. Ecob, Muscular Dystrophy Research Laboratories, Newcastle General Hospital, Newcastle NE4 6BE, United Kingdom **[273]**

Brenda R. Eisenberg, Department of Medicine and Pharmacological and Physiological Sciences, University of Chicago, Chicago, IL 60637; present address: Department of Physiology, Rush Medical College, Chicago, IL 60612 **[841]**

Juanita Eldridge, Laboratory of Biochemistry, National Cancer Institute, National Institutes of Health, Bethesda, MD 20205 **[521]**

Mark S. Eller, Department of Muscle Research, Boston Biomedical Research Institute and Department of Neurology, Harvard Medical School, Boston, MA 02114 **[591]**

Charles P. Emerson, Jr., Department of Biology, University of Virginia, Charlottesville, VA 22901 **[xxix, 21]**

David J. Ende, Department of Medicine, University of Chicago, Chicago, IL 60637 **[809]**

Hans M. Eppenberger, Department of Cell Biology, Swiss Federal Institute of Technology, CH-8093 Zurich, Switzerland **[683, 693]**

Monika Eppenberger, Department of Cell Biology, Swiss Federal Institute of Technology, CH-8093 Zurich, Switzerland **[683]**

Henry F. Epstein, Department of Neurology, Baylor College of Medicine, Houston, TX 77030 **[653]**

James R. Eshleman, Department of Anatomy, School of Medicine, University of Pennsylvania, Philadelphia, PA 19104 **[709]**

Scott Falkenthal, Department of Genetics, Ohio State University, Columbus, OH 43210 **[411]**

Giorgio Filippi, Department of Medical Genetics, University of Trieste, Trieste, Italy **[911]**

Kenneth Fischbeck, Division of Genetics, Children's Hospital, Boston, MA 02115; present address: Department of Neurology, University of Pennsylvania, Philadelphia, PA 19104 **[865]**

Donald A. Fischman, Departments of Cell Biology and Anatomy, Cornell University Medical College, New York, NY 10021 **[xxix, 293, 309, 667]**

Marc Y. Fiszman, Department of Molecular Biology, Pasteur Institute, Paris 75724, France **[457]**

Irwin L. Flink, Departments of Internal Medicine and Pharmacology, University of Arizona, Tucson, AZ 85724 **[853]**

Suzanne Forry-Schaudies, Department of Anatomy, School of Medicine, University of Pennsylvania, Philadelphia, PA 19104 **[709]**

R. Francis, Department of Genetics, Washington University, St. Louis, MO 63110 **[605]**

Uta Francke, Department of Human Genetics, Yale University Medical School, New Haven, CT 06510 **[875]**

Eric A. Fyrberg, Department of Biology, Johns Hopkins University, Baltimore, MD 21218 **[639]**

Ian Garner, Department of Molecular Biology, Pasteur Institute, Paris 75015, France **[487]**

Sandra M. Gaston, Center for Cancer Research, Massachusetts Institute of Technology, Cambridge, MA 02139 **[189]**

Geraldine F. Gauthier, Department of Anatomy, University of Massachusetts Medical School, Worcester, MA 01605 **[225]**

C.S. Giometti, Molecular Anatomy Program, Division of Biological and Medical Research, Argonne National Laboratory, Argonne, IL 60459 **[445]**

Arlene Gonzalez-Sanchez, Department of Cell Biology, Cornell University Medical College, New York, NY 10021 **[825]**

Paula J. Grabowski, Center for Cancer Research and Department of Biology, Massachusetts Institute of Technology, Cambridge, MA 02139 **[1]**

N. Michael Green, National Institute for Medical Research, London NW7 1AA, United Kingdom **[199]**

David Greenberg, Department of Cell Biology, The Weizmann Institute of Science, Rehovot 76100, Israel **[559]**

Jim Grichnick, Department of Cell Biology, Baylor College of Medicine, Houston, TX 77030 **[531]**

Barbara K. Grove, Department of Cell Biology, Swiss Federal Institute of Technology, CH-8093 Zurich, Switzerland; present address: Department of Anatomy, University of Umea, Umea 90187, Sweden **[693]**

Stephen Grund, Department of Cardiology, Children's Hospital, Boston, MA 02115 **[345]**

James Gulick, Department of Biochemistry, University of Missouri-Columbia School of Medicine, Columbia, MO 65212 **[373]**

Peter Gunning, Department of Medicine, The MEDIGEN Project, Veterans Administration Medical Center, Palo Alto, CA 94305 **[507]**

Thomas A. Gustafson, Departments of Internal Medicine and Pharmacology, University of Arizona, Tucson, AZ 85724 **[853]**

Yasuko Hagiwara, Division of Cell Biology, National Center for Nervous, Mental and Muscular Disorders, Kodaira, Tokyo 187, Japan **[143]**

Yoshio Hamada, National Institute of Basic Biology, Okazaki, Aichi 444, Japan **[323]**

Kimi Hamilton, Department of Cell Biology, Cornell University Medical College, New York, NY 10021 **[825]**

John A. Hammer III, Laboratory of Cell Biology, National Heart, Lung and Blood Institute, National Institutes of Health, Bethesda, MD 20205 **[337]**

Edna Hardeman, Department of Pharmacology, Stanford University School of Medicine, Stanford, CA 94305 **[105, 507]**

Larry Hayward, Department of Cell Biology, Baylor College of Medicine, Houston, TX 77030 **[531]**

Ann W. Hobbs, Department of Anatomy, University of Massachusetts Medical School, Worcester, MA 01605 **[225]**

Howard Holtzer, Department of Anatomy, School of Medicine, University of Pennsylvania, Philadelphia, PA 19104 **[709]**

Thomas P. Horan, Department of Biochemistry, University of Missouri-Columbia School of Medicine, Columbia, MO 65212; present address: NPI, Inc., Salt Lake City, UT 84102 **[373]**

Hans Peter Hossle, Department of Cell Biology, Swiss Federal Institute of Technology, CH-8093 Zurich, Switzerland **[693]**

C. Houlker, Ludwig Institute for Cancer Research, MRC Centre, Cambridge CB2 2QH, United Kingdom **[445]**

Huey-Juang Hsu, Department of Medicine, University of Chicago, Chicago, IL 60637 **[809]**

Orest Hurko, Department of Neurology, Johns Hopkins University School of Medicine, Baltimore, MD 21205 **[921]**

Yuji Isobe, Department of Anatomy, School of Medicine, Chiba University, Chiba 280, Japan **[725]**

Seigo Izumo, Department of Cardiology, Children's Hospital, Beth Israel Hospital and Harvard Medical School, Boston, MA 02155 **[345]**

Smilja Jakovcic, Department of Medicine, University of Chicago, Chicago, IL 60637 **[363, 809]**

Chellah Jayabaskaran, Department of Muscle Research, Boston Biomedical Research Institute and Department of Neurology, Harvard Medical School, Boston, MA 02114 **[591]**

Elissavet Kardami, Department of Zoology, University of California, Berkeley, CA 94720 **[133, 153, 457]**

Christine C. Karlik, Department of Biology, Johns Hopkins University, Baltimore, MD 21218 **[639]**

Surender Singh Katoch, Institute of Basic Medical Sciences, University of Tsukuba, Niihari-gun, Ibaraki 305, Japan **[323]**

Maho Kawashima, Department of Biology, Chiba University, Chiba-shi, Chiba 260, Japan **[293]**

V. Kean, Department of Genetics, Hospital for Sick Children, Toronto, Ontario M5G 1X8, Canada **[887]**

Larry Kedes, Department of Medicine, The MEDIGEN Project, Veterans Administration Medical Center, Palo Alto, CA 94305 **[507]**

Alan M. Kelly, Department of Pathology, School of Veterinary Medicine, University of Pennsylvania, Philadelphia, PA 19104 **[77]**

Robert E. Kingston, Center for Cancer Research and Department of Biology, Massachusetts Institute of Technology, Cambridge, MA 02139; present address: Department of Molecular Biology and Genetics, Harvard Medical School, Massachusetts General Hospital, Boston, MA 02114 **[1]**

Shigekazu Kitani, Department of Biology, Chiba University, Chiba-shi, Chiba 260, Japan **[293]**

Maria M. Konarska, Center for Cancer Research and Department of Biology, Massachusetts Institute of Technology, Cambridge, MA 02139 **[1]**

Stephen F. Konieczny, Department of Biology, University of Virginia, Charlottesville, VA 22901 **[21]**

Irwin R. Konigsberg, Department of Biology, University of Virginia, Charlottesville, VA 22901 **[61, 283]**

Bozena Korczak, Banting and Best Department of Medical Research, University of Toronto, C.H. Best Institute, Toronto, Ontario M5G 1L6, Canada **[199]**

Edward D. Korn, Laboratory of Cell Biology, National Heart, Lung and Blood Institute, Bethesda, MD 20205 **[337]**

Keith Kropp, Department of Biochemistry, University of Missouri-Columbia School of Medicine, Columbia, MO 65212 **[373]**

Louis M. Kunkel, Division of Genetics, Children's Hospital, Boston, MA 02115 **[865]**

Richard Lappin, The Rockefeller University, New York, NY 10021 **[189]**

Christine Laurent, Department of Molecular Biology, Pasteur Institute, Paris 75724, France **[237]**

Elias Lazarides, Division of Biology, California Institute of Technology, Pasadena, CA 91125 **[749]**

Leslie A. Leinwand, Departments of Microbiology and Immunology and Genetics, Albert Einstein College of Medicine, New York, NY 10461 **[263]**

Marguerite Lemonnier, Department of Molecular Biology, Pasteur Institute, Paris 75724, France **[457]**

James E. Levin, Department of Molecular Genetics and Cell Biology, University of Chicago, Chicago, IL 60637 **[809]**

B.A. Levine, Inorganic Chemistry Laboratory, University of Oxford, Oxford OX1 3QR, United Kingdom **[423]**

Cairine Logan, Department of Genetics, Hospital for Sick Children, Toronto, Ontario M5G 1X8, Canada **[887]**

P. Lohse, Department of Toxicology, Medical School, University of Hamburg, D-2000 Hamburg, Federal Republic of Germany **[793]**

Susan Lowey, Rosenstiel Basic Medical Sciences Research Center, Brandeis University, Waltham, MA 02254 **[225]**

David H. MacLennan, Banting and Best Department of Medical Research, University of Toronto, C.H. Best Institute, Toronto, Ontario M5G 1L6, Canada **[199]**

A.R. MacLeod, Ludwig Institute for Cancer Research, MRC Centre, Cambridge CB2 2QH, United Kingdom **[445]**

James W. Mahaffey, Department of Biology, Johns Hopkins University, Baltimore, MD 21218; present address: Department of Biology, Indiana University, Bloomington, IN 47405 **[639]**

Vijak Mahdavi, Department of Cardiology, Children's Hospital, and Department of Paediatrics, Harvard Medical School, Boston, MA 02115 **[345, 387]**

Pamela A. Maher, Department of Biology, University of California, San Diego, La Jolla, CA 92093 **[741]**

Francis Manasek, Department of Medicine and Pharmacological and Physiological Sciences, University of Chicago, Chicago, IL 60637; present address: Department of Anatomy, University of Chicago, Chicago, IL 60637 **[841]**

Janet Mar, Department of Anatomy, University of California, San Francisco, CA 94143 **[547]**

Bruce E. Markham, Departments of Internal Medicine and Pharmacology, University of Arizona, Tucson, AZ 85724 **[853]**

Ichiro Maruyama, Laboratory of Molecular Biology, MRC Centre, Cambridge CB2 2QH, United Kingdom **[629]**

Tomoh Masaki, Institute of Basic Medical Sciences, University of Tsukuba, Niihari-gun, Ibaraki 305, Japan **[293, 323]**

Camillo Mastropaolo, Department of Child Neurology, University of Sassari, Sassari, Italy **[911]**

Mary Ann Matheson, Department of Cell Biology, Cornell University Medical College, New York, NY 10021 **[825]**

Sheila McClure, Department of Zoology, University of California, Berkeley, CA 94720 **[133]**

Danny McCormick, Department of Molecular Biology, Pasteur Institute, Paris 75724, France **[237]**

Linda McKee, Department of Neurology, Johns Hopkins University School of Medicine, Baltimore, MD 21205 **[921]**

Danielle Melloul, Department of Cell Biology, The Weizmann Institute of Science, Rehovot 76100, Israel **[559]**

Charmaine Mendola, Department of Biochemistry, Roche Research Center, Roche Institute of Molecular Biology, Nutley, NJ 07110 **[773]**

Peter Alan Merrifield, Department of Biology, University of Virginia, Charlottesville, VA 22901 **[61, 283]**

Julie Micou-Eastwood, Department of Zoology, University of California, Berkeley, CA 94720 **[153]**

William Middlesworth, Division of Genetics, Children's Hospital, Boston, MA 02115 **[865]**

Takashi Mikawa, National Institute of Physiology, Okazaki, Aichi 444, Japan **[323]**

David M. Miller, Department of Zoology, North Carolina State University, Raleigh, NC 27695 **[629]**

Jeffrey B. Miller, Departments of Medicine and Oncology, Stanford University School of Medicine, Stanford, CA 94305 **[213]**

Adrian J. Minty, Department of Medicine, The MEDIGEN Project, Veterans Administration Medical Center, Palo Alto, CA 94305 **[507]**

D.G. Moerman, Department of Genetics, Washington University, St. Louis, MO 63110 **[605]**

Anthony P. Monaco, Division of Genetics, Children's Hospital, Boston, MA 02115 **[865]**

Stephen E. Moore, Department of Neurochemistry, Institute of Neurology, Queen Square, London WC1N 38G, United Kingdom **[179]**

I. Mori, Department of Genetics, Washington University, St. Louis, MO 63110 **[605]**

Eugene Morkin, Departments of Internal Medicine and Pharmacology, University of Arizona, Tucson, AZ 85724 **[853]**

Paul Moss, Department of Zoology, University of California, Berkeley, CA 94720 **[153]**

Bernardo Nadal-Ginard, Department of Cardiology, Children's Hospital, and Department of Pediatrics, Harvard Medical School, Boston, MA 02115 **[xxix,345,387]**

Asish Nag, Department of Medicine and Pharmacological and Physiological Sciences, University of Chicago, Chicago, IL 60637; present address: Department of Biological Sciences, Oakland University, Rochester, MI 48063 **[841]**

Mark A. Nameroff, Department of Biological Structure, University of Washington School of Medicine, Seattle, WA 98195 **[35,47]**

Diarmuid Nicholson, Department of Biochemistry, Roche Research Center, Roche Institute of Molecular Biology, Nutley, NJ 07110 **[773]**

Uri Nudel, Department of Cell Biology, The Weizmann Institute of Science, Rehovot 76100, Israel **[559]**

Takashi Obinata, Department of Biology, Chiba University, Chiba-shi, Chiba 260, Japan **[293,309,323]**

Charles P. Ordahl, Department of Anatomy, University of California, San Francisco, CA 94143 **[437,547]**

Anthony J. Otsuka, Department of Genetics, University of California, Berkeley, CA 94720 **[619]**

Eijiro Ozawa, Division of Cell Biology, National Center for Nervous, Mental and Muscular Disorders, Kodaira, Tokyo 187, Japan **[143]**

Richard A. Padgett, Center for Cancer Research and Department of Biology, Massachusetts Institute of Technology, Cambridge, MA 02139 **[1]**

Joel D. Pardee, Departments of Cell Biology and Anatomy, Cornell University Medical College, New York, NY 10021 **[667]**

Bruce M. Paterson, Laboratory of Biochemistry, National Institutes of Health, Bethesda, MD 20205 **[337,521]**

Grace K. Pavlath, Department of Pharmacology, Stanford University School of Medicine, Stanford, CA 94305 **[105]**

Muthu Periasamy, Department of Cardiology, Children's Hospital, and Department of Pediatrics, Harvard Medical School, Boston, MA 02115 **[345,387]**

Jean-Claude Perriard, Department of Cell Biology, Swiss Federal Institute of Technology, CH-8093 Zurich, Switzerland **[693]**

Christian Pinset, Department of Molecular Biology, Pasteur Institute, Paris 75724, France **[237]**

LeBris Smith Quinn, Departments of Biological Structure and Pathology, University of Washington School of Medicine, Seattle, WA 98195; present address: Howard Hughes Medical Institute, University of Washington, Seattle, WA 98195 **[35]**

Peter N. Ray, Department of Genetics, Hospital for Sick Children, Toronto, Ontario M5G 1X8, Canada **[887]**

F.C. Reinach, Ludwig Institute for Cancer Research, MRC Centre, Cambridge CB2 2QH, United Kingdom **[445]**

Graham P. Riley, Department of Molecular Biology, Pasteur Institute, Paris 75724, France **[237]**

Antoniettina Rinaldi, Department of Biology, University of Cagliari, Cagliari, Italy **[911]**

Jeffrey Robbins, Department of Biochemistry, University of Missouri-Columbia School of Medicine, Columbia, MO 65212; present address: Department of Pharmacology and Cell Biophysics, University of Cincinnati Medical Center, Cincinnati, OH 45267 **[373]**

Phillips W. Robbins, Center for Cancer Research, Massachusetts Institute of Technology, Cambridge, MA 02139 **[189]**

Benôit Robert, Department of Molecular Biology, Pasteur Institute, Paris 75015, France **[487]**

Charles E. Rozek, Department of Biology, Case Western Reserve University, Cleveland, OH 44106 **[411]**

Neal A. Rubinstein, Department of Anatomy, University of Pennsylvania, Philadelphia, PA 19104 **[77]**

Nelson Ruiz-Opazo, Department of Cardiology, Children's Hospital, and Department of Pediatrics, Harvard Medical School, Boston, MA 02115 **[387]**

Anuradha D. Saad, Departments of Cell Biology and Anatomy, Cornell University Medical College, New York, NY 10021 **[309,667]**

Lino J. Saez, Departments of Microbiology and Immunology and Genetics, Albert Einstein College of Medicine, New York, NY 10461 **[263]**

Chandrika Saidapet, Department of Biochemistry, Roche Research Center, Roche Institute of Molecular Biology, Nutley, NJ 07110 **[773]**

Kayoko Saito, Division of Cell Biology, National Center for Nervous, Mental and Muscular Disorders, Kodaira, Tokyo 187, Japan **[143]**

Koji Saito, Division of Cell Biology, National Center for Nervous, Mental and Muscular Disorders, Kodaira, Tokyo 187, Japan **[143]**

O. Saitoh, Department of Biology, Chiba University, Chiba-shi, Chiba 260, Japan **[293]**

Satyapriya Sarkar, Department of Muscle Research, Boston Biomedical Research Institute and Department of Neurology, Harvard Medical School, Boston, MA 02114 **[591]**

Saverio Sartore, Rosenstiel Basic Medical Sciences Research Center, Brandeis University, Waltham, MA 02254; present address: Institute of General Pathology, University of Padua, Padua 35100, Italy **[225]**

B. Schäfer, Department of Cell Biology, Swiss Federal Institute of Technology, CH-8093 Zurich, Switzerland **[693]**

Dorothy A. Schafer, Departments of Medicine and Oncology, Stanford University School of Medicine, Stanford, CA 94305 **[213]**

Roy D. Schmickel, Department of Human Genetics, University of Pennsylvania Medical School, Philadelphia, PA 19104 **[887]**

Robert J. Schwartz, Department of Cell Biology, Baylor College of Medicine, Houston, TX 77030 **[531]**

Anne Seiler-Tuyns, Laboratory of Biochemistry, National Cancer Institute, National Institutes of Health, Bethesda, MD 20205 **[521]**

Moshe Shani, Department of Cell Biology, The Weizmann Institute of Science, Rehovot 76100, Israel **[575]**

Phillip A. Sharp, Center for Cancer Research and Department of Biology, Massachusetts Institute of Technology, Cambridge, MA 02139 **[1]**

Donald G. Sheer, Departments of Internal Medicine and Pharmacology, University of Arizona, Tucson, AZ 85724; present address: Clinical Endocrinology Branch, National Institute of Arthritis, Diabetes, Digestive and Kidney Diseases, National Institutes of Health, Bethesda, MD 20205 **[853]**

Yutaka Shimada, Department of Anatomy, School of Medicine, Chiba University, Chiba 280, Japan **[725]**

Akiko Shoji, Division of Cell Biology, National Center for Nervous, Mental and Muscular Disorders, Kodaira, Tokyo 187, Japan **[143]**

M.A.Q. Siddiqui, Department of Biochemistry, Roche Research Center, Roche Institute of Molecular Biology, Nutley, NJ 07110 **[xxix,773]**

Laura Silberstein, Department of Pharmacology, Stanford University School of Medicine, Stanford, CA 94305 **[253]**

Paul Simpson, Cardiovascular Research Institute, University of California, San Francisco, CA 94143 **[547]**

S.J. Singer, Department of Biology, University of California, San Diego, La Jolla, CA 92093 **[741]**

Achyut M. Sinha, Department of Medicine, University of Chicago, Chicago, IL 60637 **[809]**

Marcello Siniscalco, Sloan-Kettering Institute for Cancer Research, New York, NY 10021 **[911]**

L.B. Smillie, Ludwig Institute for Cancer Research, MRC Centre, Cambridge CB2 2QH, United Kingdom; present address: Department of Biochemistry, University of Alberta, Edmonton, Alberta T6G 2H7, Canada **[445]**

Dennis Spector, Department of Zoology, University of California, Berkeley, CA 94720 **[153]**

Hansell Stedman, Department of Muscle Research, Boston Biomedical Research Institute and Department of Neurology, Harvard Medical School, Boston, MA 02114 **[591]**

Joseph P. Stein, Department of Internal Medicine, Division of Endocrinology, University of Texas Medical School, Houston, TX 77030 **[473]**

Frank E. Stockdale, Departments of Medicine and Oncology, Stanford University School of Medicine, Stanford, CA 94305 **[213]**

Emanuel E. Strehler, Department of Cardiology, Children's Hospital, and Department of Pediatrics, Harvard Medical School, Boston, MA 02115 **[345,387]**

Marie-Antoinette Strehler, Department of Cardiology, Children's Hospital, Boston, MA 02115 **[345]**

Warren J. Strittmatter, Department of Neurology, Baylor College of Medicine, Houston, TX 77030 **[167]**

Richard C. Strohman, Department of Zoology, University of California, Berkeley, CA 94720 **[133,153]**

Reiko Suzuki, Institute of Basic Medical Sciences, University of Tsukuba, Niihari-gun, Ibaraki 305, Japan **[323]**

Lauren J. Sweeney, Department of Medicine and Pharmacological and Physiological Sciences, University of Chicago, Chicago, IL 60637; present address: Department of Anatomy, University of Chicago, Chicago, IL 60637 **[841]**

Herbert M. Swick, Department of Neurology, Johns Hopkins University School of Medicine, Baltimore, MD 21205; present address: Department of Pediatric Neurology, Medical College of Wisconsin, Milwaukee Children's Hospital, Milwaukee, WI 53223 **[921]**

James E. Sylvester, Department of Human Genetics, University of Pennsylvania Medical School, Philadelphia, PA 19104 **[887]**

Hiromi Takano-Ohmura, Tokyo Metropolitan Institute of Medical Sciences, Bunkyo-ku, Tokyo 113, Japan **[323]**

K. Talbot, Ludwig Institute for Cancer Research, MRC Centre, Cambridge CB2 2QH, United Kingdom **[445]**

John Tamkun, Center for Cancer Research, Massachusetts Institute of Technology, Cambridge, MA 02139 **[189]**

Takeshi Tanaka, Research Division, Saitama Red Cross Blood Center, Yono City, Saitama 338, Japan **[323]**

R.D. Taylor, Department of Biochemistry, University of Birmingham, Birmingham B15 2TT, United Kingdom **[423]**

M.W. Thompson, Department of Genetics, Hospital for Sick Children, Toronto, Ontario M5G 1X8, Canada **[887]**

Sohei Tokunaka, Department of Anatomy, School of Medicine, University of Pennsylvania, Philadelphia, PA 19104; present address: Asahikawa Medical College, Asahikawa 078-11, Japan **[709]**

Kiyotero T. Tokuyasu, Department of Biology, University of California, San Diego, La Jolla, CA 92093 **[741]**

Massimo Tondi, Department of Child Neurology, University of Sassari, Sassari, Italy **[911]**

Madeleine Toutant, Department of Molecular Biology, Pasteur Institute, Paris 75724, France **[237]**

Naoji Toyota, Department of Cell Biology, Cornell University Medical College, New York, NY 10021 **[825]**

Patrick K. Umeda, Department of Medicine, University of Chicago, Chicago, IL 60637 **[363,809]**

Nancy Valente, Department of Biochemistry, University of Missouri-Columbia School of Medicine, Columbia, MO 65212 **[373]**

Guillermina S. Waller, Rosenstiel Basic Medical Sciences Research Center, Brandeis University, Waltham, MA 02254 **[225]**

Theo Wallimann, Department of Cell Biology, Swiss Federal Institute of Technology, CH-8093 Zurich, Switzerland **[683]**

Frank S. Walsh, Department of Neurology, Johns Hopkins University School of Medicine, Baltimore, MD 21205; present address: Molecular Neurobiology Laboratory, Institute of Neurology, Queen Square, London WC1N 3BG, United Kingdom **[179,921]**

R.H. Waterston, Department of Genetics, Washington University, St. Louis, MO 63110 **[605]**

Simon C. Watkins, Department of Molecular Biology, Pasteur Institute, Paris 75724, France **[237]**

Cecelia Webster, Department of Pharmacology, Stanford University School of Medicine, Stanford, CA 94305 **[105,911]**

André Weydert, Department of Molecular Biology, Pasteur Institute, Paris 75015, France **[487]**

Robert G. Whalen, Department of Molecular Biology, Pasteur Institute, Paris 75724, France **[237,273]**

Raymond I. White, Howard Hughes Medical Institute and Department of Human Genetics, University of Utah School of Medicine, Salt Lake City, UT 84132 **[903]**

Karin Wiebauer, Department of Toxicology, Medical School, University of Hamburg, D-2000 Hamburg, Federal Republic of Germany **[793]**

David Wieczorek, Department of Cardiology, Children's Hospital, Boston, MA 02115 **[345]**

J.M. Wilkinson, Department of Biochemistry, Royal College of Surgeons, London WC2A 3PN, United Kingdom **[423]**

Barbara Winter, Department of Toxicology, Medical School, University of Hamburg, D-2000 Hamburg 13, Federal Republic of Germany **[793]**

R.G. Worton, Department of Genetics, Hospital for Sick Children, Toronto, Ontario M5G 1X8, Canada **[887]**

Woodring Erik Wright, Department of Cell Biology, The University of Texas Southwestern Medical School, Dallas, TX 75235 **[85]**

Zipora Yablonka-Reuveni, Department of Biological Structure, University of Washington School of Medicine, Seattle, WA 98195 **[47]**

David Yaffe, Department of Cell Biology, The Weizmann Institute of Science, Rehovot 76100, Israel **[559,575]**

Akira Yoshikawa, Research Division, Saitama Red Cross Blood Center, Yono City, Saitama 338, Japan **[323]**

Barry Zadeh, Department of Cell Biology, Cornell University Medical College, New York, NY 10021 **[825]**

Radovan Zak, Department of Medicine and Pharmacological and Physiological Sciences, University of Chicago, Chicago, IL 60637 **[841]**

Ana Maria Zarraga, Department of Biochemistry, Roche Research Center, Roche Institute of Molecular Biology, Nutley, NJ 07110 **[773]**

Stephen Zolnay, Department of Muscle Research, Boston Biomedical Research Institute and Department of Neurology, Harvard Medical School, Boston, MA 02114 **[591]**

Judith G.E.M. Zuurveld, Department of Neurology, Johns Hopkins University School of Medicine, Baltimore, MD 21205 **[921]**

Preface

This book represents the proceedings of a UCLA Symposium entitled: "The Molecular Biology of Muscle Development," held in Park City, Utah, March 15–22, 1985. The antecedents of the symposium date back to 1977 when David Yaffe organized an EMBO workshop on Muscle Development at Shoresh, Israel. Since that time, two additional EMBO workshops have been held, one in Israel (1980) and another in Switzerland (1983), the latter coorganized by Hans Eppenberger and Jean-Claude Perriard. Progress in the field of myogenesis has clearly been facilitated by the international cooperation, exchange of ideas and generous sharing of biological reagents which have resulted from the previous workshops held in Israel and Switzerland. As such, we would like to acknowledge our indebtedness to Drs. Yaffe, Eppenberger, and Perriard for setting these conferences in motion. All of us look forward to the next meeting tentatively scheduled for France in the autumn of 1986.

Progress in the field of myogenesis over the past eight years has been impressive: recombinant DNA probes are now available for most contractile protein mRNAs and their genes; assignment of gene number and chromosome distribution will soon be completed for many if not all of the myofibrillar proteins both in cardiac and skeletal muscle; quantitative analyses of transcriptional and posttranscription regulatory processes have been initiated; transfectional assays of gene expression in cell culture and germ line systems have been established, making possible analysis of the cis-acting and trans-acting regulatory mechanisms that control and coordinate muscle gene expression, during muscle differentiation and maturation; hybrid cell fusion models are available for isolating putative trans-acting regulatory molecules; techniques are at hand for precisely defining muscle lineages in the chicken, and preliminary studies indicate that muscle cell diversity arises early in development and independent of innervation; even the very early embryonic steps in determination of myogenic cell lineages are becoming accessible to experimentation using cell culture and molecular genetic methods; clues are emerging about the functional implications of extracellular matrix composition and regional complexity; genetic systems have been established in *Drosophila* and *Caenorhabditis* for analyzing myofibrillar assembly; more precise biophysical methods are being applied to the assembly and exchange properties of the contractile proteins, which in combination with site-directed mutagenesis, should clarify many of the posttranslational steps in myofibrillogenesis; and finally, molecular genetics has made enormous strides in pinpointing the locus for Duchenne muscular dystrophy. Accompanying these advances has been the emergence of several unanticipated puzzles. For example: Why are different isoforms of so many contractile proteins expressed at specific stages of development? Is there developmental significance to the expression of several "cardiac-type" isoforms in embryonic skeletal muscle? Why has posttranscriptional splicing been so widely adopted by striated muscle as a stratagem for increasing contractile protein diversity? These are only a few of the questions which could be cited; in fact, the application of molecular genetics has probably raised as many questions as the approach has solved. What should be emphasized, however, is that fundamental questions of cellular determination and differentiation now can be examined with techniques or experimental approaches currently available. Will more complex levels of cell, tissue and organ morphogenesis be equally amenable to cell biological and molecular genetic approaches? Progress in the next few years should tell.

We gratefully acknowledge Hoffmann-La Roche, Inc. for their sponsorship of this meeting. Additional financial support was provided by E.I. du Pont de Nemours and Company, American Cyanamid Company, Agricultural Research Division, a USPHS grant from the National Cancer Institute (R13 Am35521) and a grant from the Muscular Dystrophy Association of America (W850123).

Charles Emerson
Donald Fischman
Bernardo Nadal-Ginard
M.A.Q. Siddiqui

Molecular Biology of Muscle Development, pages 1–19

EXPRESSION OF mRNA IN EUKARYOTIC CELLS[1]

Phillip A. Sharp, Robert E. Kingston[2], Albert S. Baldwin, Richard A. Padgett, Maria M. Konarska and Paula J. Grabowski

Center for Cancer Research and Department of Biology
Massachusetts Institute of Technology
Cambridge, Massachusetts 02139

ABSTRACT

Trans-activation of transcription by viral oncogenes is probably due to an increase in the frequency with which RNA polymerase II interacts with some of the same DNA binding factors involved in basal transcription. Thus, promoters stimulated by the adenovirus 2 EIa activity do not appear to have a common specific regulatory sequence element. The recently determined pathway for splicing of mRNA precursors is consistent with a splicing mechanism

[1]This work was supported by grants from NIH (R01-GM32467 and P01-CA26717), from NSF (PCM-8200309) to P.A.S. and partially from an NIH Core grant P01-CA14051. Postdoctoral fellowship support was from the Jane Coffin Childs Memorial Fund (R.E.K. and M.M.K.), NIH (A.S.B.), the Myron Bantrell Foundation (R.A.P.), and the Helen Hay Whitney Foundation (P.J.G.)
[2]Present address: Department of Molecular Biology and Genetics, Harvard Medical School, Massachusetts General Hospital, Boston, MA 02114

involving RNA catalysis. The major quasi-stable RNA splicing intermediate contains two RNAs, the 5' exon and a lariat RNA composed of the intervening sequence and the 3' exon. These two RNAs are part of a complex that sediments at 60S. The bipartite nature of the intermediate in splicing suggests that exons from two distinct RNA molecules might be joined by trans-splicing. Trans-splicing can be detected with appropriate substrates in *in vitro* reactions.

INTRODUCTION

The synthesis of most mRNAs in mammalian cells involves three major steps: transcription, splicing and polyadenylation, in addition to transport of the processed RNA to the cytoplasm. All three steps are essential and each can be a point of regulation for the level of an mRNA. The role of these steps in mRNA synthesis *in vivo* has been documented by studies of the structure of specific genes and their mRNAs and by analysis of the expression of mRNAs from mutant genes. The molecular mechanisms by which these steps are regulated are poorly understood primarily because of the lack of an understanding of their biochemistry. The recent development of soluble *in vitro* systems which accurately duplicate these three steps in mRNA synthesis offers the possibility of a future understanding of gene regulation at a molecular level.

Is it necessary to understand the molecular nature of gene regulation to understand the basis of development in multicellular organisms? Development of a tissue or an organism is the manifestation of the controlled generation of a series of cellular states in a particular three dimensional pattern. Such cellular states are distinguishable by the subset of genes expressed. Identifying the factors that regulate the genes in a particular cellular state and in turn identifying what regulates the expression of the genes encoding these factors is an iterative process that will yield an understanding of some aspects of development. In mammalian systems, where genetics cannot easily be used to identify genes encoding factors that regulate gene expression in particular cell types, the success of this iterative approach will be highly dependent on the availability of both *in vivo* and *in vitro* systems for assaying gene expression.

Transcription Regulation by Trans-activation

Few cellular genes have been isolated from mammalian cells that are known to regulate the transcription of other genes. One of the first examples is the recent report of the cloning of the cDNA for a RNA polymerase III transcription factor which is necessary for transcription of 5S rRNA (1, 2). Examples of cellular genes encoding factors that mediate RNA polymerase II transcription will undoubtedly shortly follow. Several viral genes that regulate transcription of both viral and cellular promoters have been identified and studied. This stimulation is commonly referred to as trans-activation of transcription. A list would include the EIa gene of adenovirus, the immediate early genes of herpes simplex virus (HSV), the large T antigens of both SV40 and polyoma virus and more recently the X or LOR gene of the human retrovirus HTLV families (3). All of these viral genes encode proteins that are concentrated in the nucleus of cells. The specificity and mechanism by which these viral genes stimulate transcription is being studied in a number of laboratories. This work is the most advanced for the adenovirus EIa gene and suggests some interesting insights.

During the adenovirus lytic cycle, the 289 a.a. protein encoded by the EIa gene stimulates transcription of five other viral RNA polymerase II promoters (4, 5). The specificity of this stimulation has been studied by comparing the activity of promoter deletion mutants in the presence and absence of EIa. For the adenovirus, EIb and EII promoters, such analysis has failed to identify a separate regulatory sequence which specifically mediates EIa stimulation (6, 7, 8). Deletions which severely impair the basal promoter activity still respond to positive regulation by the EIa gene. However, the final level of transcription after EIa stimulation of mutants with low basal levels of transcription was also reduced. This strongly suggests that the same set of factors and DNA sequences which specify basal transcription of the promoter are also important in mediating transcription after stimulation by EIa.

A large number of promoters is stimulated by cotransfection with EIa: the rat preproinsulin promoter (9), the human β-globin promoter (10, 11), the human ε-globin promoter (12) and the SV40 early promoter (11). Some promoters are not stimulated such as the thymidine kinase promoter of HSV and promoters with strong enhancer elements (13, 14). Deletion studies of the β-globin and SV40 early promoters have also failed to identify a specific regulatory sequence related to EIa stimulation (11). These studies have shown that stimulation of the SV40 early promoter is dependent upon the presence of the GC repeat region, as is the basal level of transcription. In the case of the β-globin promoter, mutations in the TATA sequence of the promoter reduced both the basal and the EIa-stimulated level of transcription. However, a single base change at -87 in the β-globin promoter reduced the level of transcription of this promoter when expressed under the influence of an enhancer but did not reduce the level of transcription when stimulated by EIa. This suggests that the factor(s) recognizing the -87 region of the β-globin promoter is not critical for EIa stimulation but factor(s) recognizing the TATA region are important.

Model for Regulation by Trans-activation

Interpretation of these studies of trans-activation by EIa suggests some insights into regulation of transcription of promoters in mammalian cells. A typical eucaryotic promoter (see figure 1) is thought to be composed of three types of sequence elements: enhancer elements that stimulate transcription over considerable distances from positions either upstream or downstream of a promoter; upstream sequence elements that are located from -50 to -200 bp from the initiation site and which only stimulate transcription when positioned near and upstream of a promoter; and TATA-related sequences located approximately -30 N from the initiation site. Mutations affecting any one of these elements commonly reduce the rate of initiation of transcription and thus the processes mediated by factors which recognize these elements are potentially rate limiting. Each of these sequence elements must be recognized by DNA binding factors, probably proteins (DBP). DBPs that recognize

FIGURE 1. Model of a typical eucaryotic promoter and factors which mediate initiation of transcription.

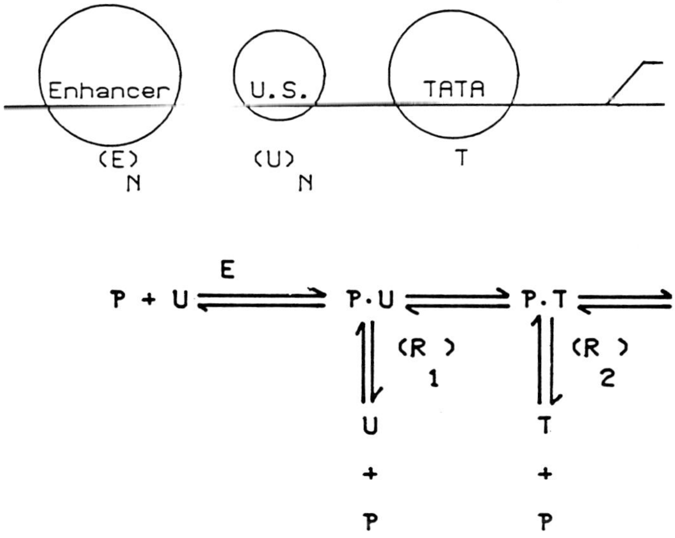

P free polymerase II (factors)
Pi initiated polymerase

enhancers (E-DBP) may be cell type specific, and each enhancer element could bind multiple factors. Also, many promoters do not have enhancer elements. All promoters probably have upstream sequence elements which are recognized by one or more U-DBPs. For a single promoter, these U-DBPs could be of a single type or of different types, the latter situation permits a wide range of specificities by combinatorial effects. There are probably many different types of U-DBPs within a cell. DNA binding factors which recognize the TATA region (T-DBP) are probably less diverse than the U set as the consensus sequence of this element is conserved for most promoters.

As shown in figure 1, initiation of transcription by polymerase II (Pi) could be pictured as resulting from interactions of RNA polymerase II with proteins bound to the TATA element (P.T). There are two paths to P.T by the polymerase, by interaction with proteins bound to U elements (P.U) or by direct binding (R2). For most promoters, the former path is the most efficient. Again there are two paths to the P.U state, direct binding (R1) or by the process pictured as being catalyzed by the enhancer element. The effect of the enhancer element could range from modification of the local chromatin structure, localization of the promoter to a particular structure in the nucleus, supercoiling of the DNA or entry sites for RNA polymerase II. Promoters without active enhancer elements depend upon direct interactions of RNA polymerase II with U-DBPs.

The data from EIa stimulation of promoters can be easily interpreted with this model. The EIa activity increases the rates of R1 and R2 for specific U-DBP and T-DBPs. Thus, the same sequence elements required for basal transcription are required for EIa stimulation. Some U-DBP and T-DBP are not recognized and thus promoters with these DBPs would not be stimulated by EIa. In general, the specificity of stimulation by a trans-activation activity of this type will reflect the number of U-DBPs and T-DBPs recognized. Finally, this model for EIa stimulation is consistent with the absence of stimulation of promoters which are strongly stimulated by an enhancer element (11). Under these conditions, the interactions of RNA polymerase II with U-DBPs might be so efficient that increases in the rates of R1 and R2 do not increase the total rate of initiation. In addition, it is possible that if the level of polymerase II and factors are limiting in the cell, then stimulation of R1 and R2 at promoters without enhancers would suppress the rate of initiation at promoters with enhancers due to a simple competition. Suppression of enhancer activity has been observed after transfection with EIa (13, 14). Whether simple competition or some other process is responsible is not clear.

The most interesting feature of the above model for transcription regulation is that it separates the degree of saturation by DNA binding proteins from regulation of transcription. Specific proteins would constantly remain bound to a promoter while the frequency with which polymerase II would interact with these protein-DNA complexes would be regulated. It has been a long standing observation that active promoters are denoted by nuclease sensitive sites. If this nuclease sensitivity indicates the presence of DNA binding proteins then the promoter is constantly adorned with these factors. However, initiation of polymerase II at most promoter sites is exceptionally infrequent. A typical mammalian gene is transcribed to yield 10 mRNAs per cell which have a steady state halflife of 12 hours. Thus, for this gene, polymerase II would initiate transcription approximately once every hour. Since a majority of the chromatin containing a specific active promoter is nuclease sensitive, this suggests that the binding of proteins to a promoter might not be rate limiting for initiation but that the utilization of this site by polymerase II might be.

In the above model, a promoter is best pictured as a DNA-protein complex. One might speculate that a few different DBPs might be synthesized in each differentiated cell type and that this would set up subgroups of promoters for subsequent regulation. This model would be consistent with the rapid regulation of transcription frequently observed in mammalian cells after addition of peptide hormones or proteins like interferon. Since protein-protein interactions are conjectured to be important in this model, the frequency of initiation at a promoter site could be regulated by protein modification and would not require de novo protein synthesis. In fact, regulation by mechanisms similar to that of EIa trans-activation might be responsible for many of the fast response transcription changes in mammalian cells, particularly, those related to control of cell growth.

Splicing of Messenger RNA Precursors

Elucidation of the molecular mechanism of splicing of mRNA precursors in eucaryotic cells is of interest for several reasons. It has been shown that differences in the specificity of splicing can process a particular precursor RNA into different mRNAs in different cell types (15, 16). Thus, the specificity of splicing can be regulated. In addition, one of the major puzzles in molecular biology is the evolutionary origin of introns. A RNA splicing process must have either co-evolved or predated the development of introns. Elucidation of the mechanism of splicing might suggest the origin of both splicing and introns.

We have extensively characterized the *in vitro* splicing of the first (L1) and second (L2) leader exons of the major late transcription unit of adenovirus (17). The *in vitro* splicing reaction contains either a whole cell extract (18) or nuclear extract of uninfected HeLa cells (19). The RNA substrate added to this reaction has an intervening sequence of 231 N and is synthesized *in vitro* with the bacterial phage RNA polymerase Sp6 primed with a cap (GpppG) dinucleotide (20). Thus, the *in vitro* substrate is capped, as are polymerase II transcribed precursors *in vivo*. Splicing in whole cell extracts is completely dependent upon the presence of a 5' terminal cap and the splicing of capped precursor is inhibited by addition of μM concentrations of cap analogs. It is interesting to speculate that recognition of the cap structure at the 5' ends of mRNA precursors may also be important for efficient splicing *in vivo*. Only polymerase II transcripts are capped and only polymerase II transcripts are spliced by the mRNA process.

Splicing of mRNA precursors in these reactions is dependent upon addition of both ATP and Mg^{++} (21, 22). The ATP cofactor may be required for a number of steps during the reaction. There is a kinetic lag between addition of substrate RNA and the linear kinetics of appearance of spliced RNA product. The events occurring during this lag require the presence of ATP (21). It is possible that the RNA is being assembled during the lag into a multicomponent complex which is competent for splicing. A candidate for one of the components of this complex is the small nuclear ribonucleoprotein particle containing U1 RNA (U1 snRNP; 23). There is a striking sequence complementarity between the 5' end of U1 RNA and the consensus sequence found at the 5' splice sites. The hypothesis that an association of the 5' splice sites and U1 snRNA is an essential step in splicing has been strengthened by the finding that: purified U1 snRNP will bind to and protect the 5' splice sites of an RNA from nuclease digestion (24); addition of antisera directed against U1 snRNP will inhibit the in vitro splicing reaction (25); and enzymatic removal of the 5' end of U1 RNA eliminates splicing activity (26).

RNA Intermediates and Products of Splicing

After the lag period, the first identifiable chemical change in the precursor RNA is cleavage at the 5' splice sites to produce a 5' exon with a 3' hydroxyl terminus and an RNA containing the entire intervening sequence and the 3' exon (27). This is shown in the second line of figure 2. This intermediate RNA has some peculiar features. First, it is topologically in the form of a lariat or tailed circle due to the fact that the 5' end of the intervening sequence has been covalently joined to a position near the 3' end of the intervening sequence (27-29). Second, it contains an unusual nucleotide structure which consists of an RNA branch where an adenosine residue has RNA chains linked to it by both 2'-5' and 3'-5' phosphodiester bonds (28, 30). As shown in figure 2, the phosphate group from the 5' splice sites is retained in the bond to the 2' position. Whether cleavage at the 5' splice sites and attachment of the 5' end of the intervening sequence to the branch point are concerted reactions is not known. The 2' hydroxyl of the adenosine residue could participate in a trans-esterification reaction at the 5' splice sites to generate the observed products or the reaction could proceed through several steps. What is

FIGURE 2. Schematic of intermediates and products of splicing process. Precursor RNA consists of 5' exon (L1, thick line), intervening sequences (thin line) and 3' exon (L2, thick line) from the major late transcript from adenovirus 2. The phosphates at the 5' and 3' splice sites are denoted by Ⓟ and ◇P, respectively. The adenosine at the branch site is boxed. The second line shows the complementarity between the sequences at the 5' splice site and branch site. The first covalent modification of the RNA is the formation of a phosphodiester bond between a 2' position on the ribose moiety of adenosine and the 5' phosphate at the 5' splice site (third line). Formation of this brach gives rise to the lariat RNA, intervening sequences - 3' exon, and the second RNA, 5' exon. This two part intermediate is reasonably stable in vitro but is ultimately resolved into the two products, the excised intervening sequences (IVS-A) and the spliced exons (L1-2, fourth line). The intervening sequences are released in a lariat form. The phosphate moieties at both splice sites are retained in the two reactions. The 3' termini of both the L1 exon and the excised intervening sequence are hydroxyl groups.

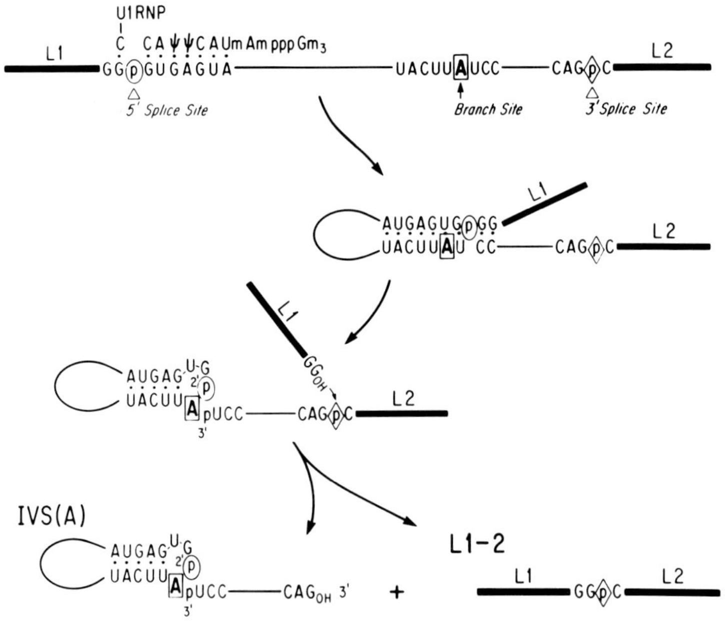

clear is that this reaction and the one discussed below produce no net change in the number of phosphodiester bonds and so the reactions could proceed in the absence of added energy.

The next reaction or series of reactions produces the spliced exons and a free intervening sequence still in a lariat form but with a shorter 3' tail due to the removal of the 3' exon. These products are shown in the last line of figure 2. The 3' end of the excised intervening sequence has a hydroxyl group due to the fact that the phosphate group at the 3' splice sites is retained in the newly formed phosphodiester bond in the spliced RNA (28). As in the case of the first reaction, the cleavage at the 3' splice sites and ligation of the 5' exon to the 3' exon could involve a concerted transesterification reaction or a series of separate reactions. Again, the products and reactants have the same number of phosphodiester bonds and thus, theoretically, the reaction could proceed in the absence of an energy cofactor. As mentioned previously, ATP is an essential cofactor in the splicing reaction. Neither the nonhydrolyzable α-β nor β-γ methylene analogs of ATP are active in the reaction. At the moment, it is difficult to determine whether this reflects an ATP dependent process for assembly of a ribonucleoprotein structure or for cleavage and ligation.

Location of Branch Sites

The significance of the sequence specificity and location of the branch site within introns of mammalian genes is unclear. In the two cases where the site is known, it is 24 and 36 nucleotides upstream of the 3' splice sites (29, 30). Tentative branch sites have been proposed on the basis of sequence homology between globin genes of different species. These sites are between 18 and 40 nucleotides upstream of the 3' splice sites (29). Studies *in vivo* of deletions within introns of yeast genes have identified a highly conserved sequence UACUAAC which is essential for splicing (31, 32). This sequence is located between 20 and 60 bases upstream of the 3' splice sites and the 3' most adenosine is the site of branch formation. Point mutations in this highly conserved sequence in yeast can inactivate splicing (33). Though branch sites in higher eukaryotes are obviously related in sequence to branch sites in yeast, their sequence is not as highly conserved. A search of a large number of mammalian introns for the analogy of the yeast sequence only yielded a limited consensus sequence $(CT\genfrac{}{}{0pt}{}{A}{C}A\genfrac{}{}{0pt}{}{T}{C})$ (34). Weiringa *et al.* (35) have constructed variants of the large intervening sequence of rabbit β-globin that have deletions of the normal branch site. These variants are accurately spliced both *in vivo* and *in vitro*. Characterization of the *in vitro* products have shown that the intervening sequences are excised as lariat RNAs with the branch site located 28N upstream of the 3' splice site. The rate of splicing of these variant RNAs *in vitro* was only 2 to 3 times slower than the w.t. RNA precursors. These results suggest that deletion of the normal branch site results in the utilization of a cryptic site. In situations where no acceptable cryptic site is available, the precursor is probably not spliced. Thus, branch formation is probably an essential part of the splicing mechanism and the specific sequences flanking the branch may influence the rate of the reaction.

The sequences containing the branch site of both the first intervening sequences of adenovirus and human beta globin show a striking complementarity to the sequences at the 5' splice site of their respective introns (see figure 2 for the adenovirus case). In one case, there are 10 consecutive base pairs of complementarity (counting G:U pairs) and in the other 9 of 10 bases of complementarity (29). A related but less extensive complementarity has previously been noted for the conserved UACUAAC sequences of yeast introns and the equally conserved sequences at the 5' splice site of yeast introns (32). It is possible that this complementarity could be important in specifying both the location of the branch site and the rate of splicing. If the 5' splice site and branch site were paired during branch formation, three to five nucleotides would have to remain single stranded to accommodate a loop for branch formation. It is interesting that the same set of sequences at the 5' splice site is complementary to sequences at both the 5' end of U1 RNA and the branch site. This may be fortuitous. Alternatively, U1 RNP recognition may be important in the selection of 5' splice sites and assembly of a ribonucleoprotein structure for splicing. Subsequently, the sequences at the 5' splice sites could be released from U1 RNP for interactions with sequences containing the branch site. This speculation aside, the role of complementarity between 5' splice sites and branch site will emerge from testing the effect of compensating base changes in these two sequences on the rate of the reaction.

Function of RNA Branch

It is somewhat surprising that RNA branches, i.e. 5' phosphodiester bonds to both the 2' and 3' positions of a ribose moiety, are not commonly found in nature. Synthetic nucleotide chemists have long been plagued by the reactivity of the 2' hydroxyl group of nucleotides. In fact, speculation concerning the existence of branched RNA was common in the early stages of molecular biology (36). RNA branches were first detected by Wallace and Edmonds (37) during an analysis of radioactively labeled nuclear polyA(+) RNA of HeLa cells. A nuclease resistant oligonucleotide was resolved with a net charge consistent with more 3' terminal phosphates than 5' terminal phosphates. Under the labeling protocol used, Wallace and Edmonds detected approximately one branch per 40,000 nucleotides in poly A(+) nuclear RNA and one branch per 10^6 nucleotides in poly A(+) cytoplasmic RNA. This low number suggests that branches are rapidly destroyed _in vivo_ and thus are not a common constituent of stable RNA. It is possible that under other cellular conditions, branches could be used to form networks of RNA or be important signals in RNA function or metabolism.

The function of the branch in the splicing process is unclear. _In vivo_, excised intervening sequences are rapidly degraded while the other product of the reaction, the joined exons are stable. Although these RNAs differ in several aspects, the presence of a branch in the excised intervening sequences could be a signal for a degradation pathway. The observed stability of the intervening sequences in vitro could be explained either by the absence of this degradation pathway or the failure of the system to release the intervening sequence from a protective complex. The most interesting possibility for the function of branch formation in mRNA splicing is that it is an integral part of the mechanism of splicing. As mentioned above, the phosphate moieties at both the 5' and 3' splice sites are retained in the branch of the excised intervening sequences and in the spliced product. As indicated in figure 2, in both branch formation and subsequent exon joining, a hydroxyl group is joined to the phosphate at the splice sites. These two reactions could occur by the same type of trans-esterification reaction responsible for the splicing of Tetrahymena ribosomal RNA precursor (38). The latter reaction is catalyzed by RNA structure in the intervening sequence and uses a 3' hydroxyl group on a guanosine nucleotide to initiate the cleavage reaction at the 5' splice site. If this model was extended to the splicing of mRNA precursors, then the RNA catalysis would have to be part of a ribonucleoprotein complex that recognized sequence and structure in the precursor RNA. This RNA catalysis would then promote cleavage at the 5' splice site by utilization of the 2' hydroxyl group of the branch site. The second reaction yielding the spliced exons could also be a trans-esterification reaction catalyzed by RNA where the 3' hydroxyl of the leader 1 exon is used to cleave the intervening sequence from the L2 exon. This speculation that the splicing of mRNA precursors may be RNA catalyzed is supported by the recent findings that intervening sequences of yeast mitochondrial RNA precursors are excised as lariats (L. Grivell, personal communication). These yeast introns have consensus sequences in common with the intron of Tetrahymena ribosomal RNA and thus are probably spliced by RNA catalysis (39-41).

Bipartite RNA Intermediate Suggests the Existence of a Complex and the Possibility of Trans-splicing

One of the more fascinating features of the splicing mechanism outlined in figure 2 is the involvement of a bipartite intermediate. This intermediate consisting of the 5' exon and the intervening sequence - 3' exon lariat RNA is reasonably stable in the in vitro reaction. Analysis of nuclear RNA from yeast and mammalian cells (42, 43) has detected equivalent intermediates. Thus, splicing in vivo probably occurs by way of intermediates where multiple RNA segments are subsequently joined to produce the final product. That this process occurs efficiently both in vivo and in vitro strongly suggests that these RNAs are held in juxtaposition by a complex RNP structure. At the moment, the components in this RNP complex are undefined. We have recently detected a 60S complex which contains the two RNAs of the bipartite intermediate discussed above. This complex was resolved by sedimenting the contents of an in vitro splicing reaction through a glycerol gradient. The two RNA species characteristic of the bipartite intermediate, L1 and IVS-L2, co-sediment in a 60S peak. Radioactive RNAs in this complex can be chased to spliced RNA products under conditions where exogenously added precursor RNA is not spliced. Thus, we believe this complex is a functional intermediate in the splicing process. We have not yet characterized this complex but it would not be surprising if the splicing substrate was complexed with both small nuclear ribonucleoprotein particles as well as the family of basic nuclear proteins found associated with nuclear RNA in eukaryotic cells (44, 45). One of the more important aspects of the analysis of splicing is the prospect that it will provide biochemical functions for these previously described cellular components.

A mechanism generating a bipartite splicing intermediate would offer the possibility of trans-splicing. Here, exon segments originating from two different RNAs would be joined by the splicing process. Trans-splicing must not occur frequently in cells since analysis of mRNAs has shown that all the exons found in a single RNA are also found in one covalent precursor RNA. However, it is possible that a low level of trans-splicing might occur and that these mRNAs are typically degraded. In particular biological systems, trans-splicing might be utilized as a frequent step in mRNA synthesis. Recent evidence has shown that many different mRNAs of trypanosomes have a common leader RNA

FIGURE 3. Schematic of substrate RNAs used to detect trans-splicing. Two independent substrate RNAs were prepared by transcription of two plasmids with bacterial phage Sp6 RNA polymerase. One plasmid contained the L1 exon, the 5' half of the intervening sequence and a 38 N segment of sequences downstream of the Sp6 promoter. The second plasmid had the Sp6 promoter inserted upstream of sequences complementary to the 38 N segment, the 3' half of the intervening sequence and the L2 exon. RNAs transcribed from these two plasmids had capped 5' termini. The structure of these two RNAs is shown in the top line. The structure of these RNAs after annealing is shown in the second line. This RNA substrate is spliced to yield the products shown on the bottom line. B.S. denotes the branch site.

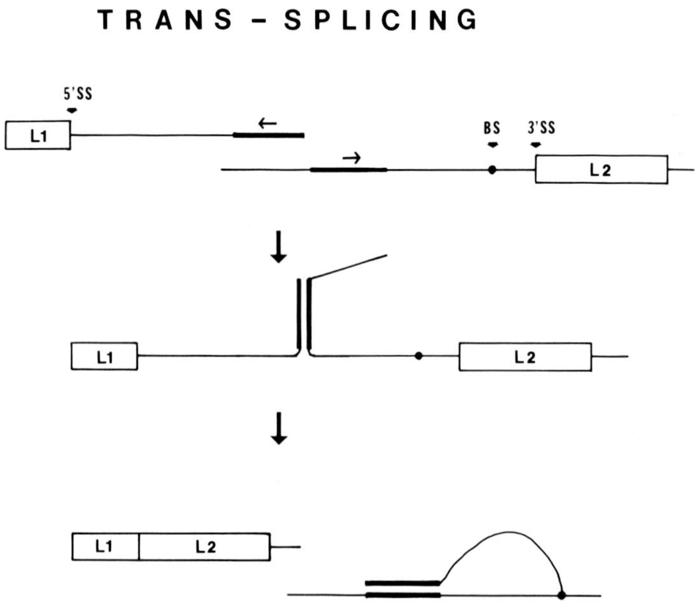

sequence at their 5' termini (46, 47). Sequences encoding this leader are located at a few sites in the genome, unlinked from the sequences coding for the rest of the mRNA. It is not clear how these two set of sequences are combined but splicing has been proposed as a possibility.

We have tested whether trans-splicing can be detected in *in vitro* reactions. As diagramed in figure 3, two substrate RNAs were prepared which contain (1) the L1 exon and part of the intervening sequence and (2) the intervening sequence and the L2 exon. A 38 N segment of complementary sequences was present near the 3' and 5' ends, respectively, of the intervening sequences in these two RNAs. Thus, the two RNA substrates can associate through duplex formation by these complementary sequences. This substrate with the two RNAs in an associated state are accurately spliced at about a third the efficiency as the covalent substrate. When the two RNAs are added in a nonassociated form, the two exons are spliced at about one tenth the efficiency of the covalent substrate. In this latter reaction, it is likely that the two RNAs anneal during the reaction. However, we have subsequently found that two similar RNA substrates containing the L1 and L2 exons and no complementary segments are spliced at a 1-5% efficiency in the nuclear extract reaction. We conclude from these results that the mechanism of mRNA splicing can join exons from two different RNAs and that the possibility of trans-splicing *in vivo* must be considered.

ACKNOWLEDGEMENTS

We thank Margarita Siafaca for preparation of this manuscript and Mary Esteve for technical assistance.

REFERENCES

1. Ginsberg, AM, King, BO, Roeder, RG (1984). Cell 39:479.
2. D. Brown, personal communication.
3. Kingston, RE, Baldwin, AS, Sharp, PA (1985). Cell (mini review), in press.

4. Jones, N, Shenk, T (1979). Proc. Natl. Acad. Sci. USA 76:3665.
5. Berk, AJ, Lee, F, Harrison, T, Williams, J, Sharp, PA (1979). Cell 17:935.
6. Bos, JL, ten Wolde-Kraamwinkel, HC (1983). EMBO J. 2:73.
7. Elkaim, R, Coding, C, Kedinger, C (1983). Nucleic Acids Res. 11:7105.
8. Kingston, RE, Kaufman, RJ, Sharp, PA (1984). Mol. Cell. Biol. 4:1970.
9. Gaynor, RB, Hillman, D, Berk, AJ (1984). Proc. Natl. Acad. Sci. USA 81:1193.
10. Treisman, R, Green, MR, Maniatis, T (1983). Proc. Natl. Acad. Sci. USA 80:7428.
11. Green, MR, Trisman, R, Maniatis, T (1983). Cell 35:137.
12. Allan, M, Zhu, J-d, Montague, P, Paul, J (1984). Cell 38:399.
13. Borrelli, E, Hen, R, Chambon, P (1984). Nature 312:608.
14. Velcich, A, Ziff, EB (1985). Cell 40:705.
15. Rozek, CE, Davidson, N (1983). Cell 32:23.
16. Nadal-Ginard, B, personal communication.
17. Berget, SM, Moore, C, Sharp, PA (1977). Proc. Natl. Acad. Sci. USA 74: 3171.
18. Manley, JL, Fire, A, Cano, A, Sharp, PA, Gefter, ML (1980). Proc. Natl. Acad. Sci. USA 77:3855.
19. Dignam, SD, Lebowitz, RM, Roeder, RG (1983). Nucl. Acids Res. 11:1475.
20. Konarska, MM, Padgett, RA, Sharp, PA (1984). Cell 38:731.
21. Hardy, SF, Grabowski, PJ, Padgett, RA, Sharp, PA (1984). Nature 308:375.
22. Hernandez, N, Keller, W (1983). Cell 35:89.
23. Lerner, MR, Boyle, JA, Mount, SM, Wolin, SL, Steitz, JA (1980). Nature 283:220.
24. Mount, SM, Pettersson, I, Hinterberger, M, Karmar, A, Steitz, JA (1983). Cell 33:509.
25. Padgett, RA, Mount, SM, Steitz, JA, Sharp, PA (1983). Cell 35:101.
26. Krämer, A, Keller, W, Appel, B, Lührmann, R (1984). Cell 38:299.
27. Grabowski, PJ, Padgett, RA, Sharp, PA (1984). Cell 37:415.

28. Padgett, RA, Konarska, MM, Grabowski, PJ, Hardy, SF, Sharp, PA (1984). Science 225:898.
29. Ruskin, B, Krainer, AR, Maniatis, T, Green, MR (1984). Cell 38:317.
30. Konarska, MM, Grabowski, PJ, Padgett, RA, Sharp, PA (1985). Nature 313:552.
31. Langford, CJ, Gallwitz, D (1983). Cell 33:519.
32. Pikielny, CW, Teem, JL, Rosbash, M (1983). Cell 34:395.
33. Langford, CJ, Klinz, F-J, Donath, C, Gallwitz, D (1984). Cell 36:645.
34. Keller, EB, and Noon, WA (1984). Proc. Natl. Acad. Sci. USA 81:7417.
35. Wieringa, B, Hofer, E, Weissmann, C (1984). Cell 37:915.
36. Allen, F (1954). Annu. Rev. Biochem. 23:99.
37. Wallace, JC, Edmonds, M (1983). Proc. Natl. Acad. Sci. USA 80:950.
38. Kruger, K, Grabowski, PJ, Zaug, AJ, Sands, J, Gottschling, DE, Cech, TR (1982). Cell 31:147.
39. Davies, RW, Waring, RB, Ray, JA, Brown, TA, Scazzocchio, C (1982). Nature 300:719.
40. Cech, TR, Tanner, NK, Tinoco, I, Weir, BR, Zuker, M, Perlman, PS (1983). Proc. Natl. Acad. Sci. USA 80:3903.
41. Burke, J, RajBhandary, UL (1982). Cell 31:509.
42. Domdey, H, Apostol, B, Lin, R, Newman, A, Brody, E, Abelson, J (1984). Cell 39:611.
43. Zeitlin, S, Efstratiadis, A (1984). Cell 39:589.
44. LeStourgen, WM, Lothstein, L, Walker, BJ, Beyer, AL (1981). In Busch, H (ed): "The Cell Nucleus: Nuclear Particles" New York: Academic Press, p 47.
45. Martin, TE, Okamura, CS (1981). In Busch, H (ed): "The Cell Nucleus: Nuclear Particles" New York: Academic Press, p 119.
46. Kooter, JM, De Longe, T, Borst, P (1984). EMBO J. 3:2387.
47. Milhausen, M, Nelson, R, Sather, T, Selkirk, S, Agabian, N (1984). Cell 38:721.

Molecular Biology of Muscle Development, pages 21–34

MYOGENIC DETERMINATION AND DIFFERENTIATION OF 10T1/2 CELL LINEAGES: EVIDENCE FOR A SIMPLE GENETIC REGULATORY SYSTEM[1]

Stephen F. Konieczny, Albert S. Baldwin[2]
and Charles P. Emerson, Jr.

Department of Biology, University of Virginia
Charlottesville, VA 22901

ABSTRACT We have investigated the molecular regulation of myogenic stem cell lineage determination and differentiation using the mouse embryonic cell line, C3H10T1/2. 5-Azacytidine converts 10T1/2 cells into distinct and genetically stable clonal lineages of myoblasts, chondroblasts or adipoblasts that can differentiate spontaneously under controlled culture conditions. Our results suggest that conversion of 10T1/2 cells to myogenic lineages is a result of the activation of a specific genetic locus that regulates myogenic linege determination and establishes a regulatory control system that mediates the activation of muscle-specific genes during myoblast differentiation. To test this model, a cloned quail troponin I gene was transfected into mouse 10T1/2 cells and its transcriptional expression examined during 10T1/2 conversion to myogenic lineages. The troponin I gene and endogenous mouse muscle genes are coactivated during myoblast differentiation regardless of whether the troponin I gene is introduced into 10T1/2 cells prior to or after myogenic determination. We propose that muscle genes are activated by the differentiation-specific expression of trans-acting signals that act on cis-acting regulatory sequences shared by muscle genes. The process of myogenic lineage determination

[1]This work was supported by research grants from the Muscular Dystrophy Association and the National Institutes of Health.

[2]Present address: Center for Cancer Research, Massachusetts Institute of Technology, Cambridge, MA 02139

establishes a regulatory control system that allows for the expression of these trans-acting muscle gene regulatory signals when myoblasts differentiate.

INTRODUCTION

Myoblasts are the embryonic stem cells that form skeletal muscle in vertebrates. Myoblast lineages originate in the somites and migrate into muscle-forming regions where they proliferate and then differentiate by fusion and activation of contractile protein genes. The molecular genetic mechanisms responsible for myoblast lineage determination previously have been inaccessible to direct analysis. However, it has been established that the process of myoblast determination produces cell lineages that, in clonal cell culture, retain their determined capacity to differentiate into skeletal muscle (1). Therefore, myoblast determination must imprint genetically stable, heritable changes in the myoblast genome that restrict the differentiation potential of myoblasts to muscle formation and to the transcriptional coactivation of a set of muscle-specific genes.

In this article we review our recent progress in the molecular analysis of myogenic determination in the embryonic mouse cell line, 10T1/2. We present evidence that the 10T1/2 cell is a multipotential mesodermal cell that offers a unique opportunity for molecular and genetic analysis of mesodermal stem cell lineage determination and of the determination-specific and differentiation-specific regulatory mechanisms that control muscle gene expression. Most significantly, our findings suggest that the myoblast determination and differentiation are under simple genetic control, involving a few regulatory genes that specify the determination of myoblast cell lineages and the coordinate activation of muscle genes during myoblast differentiation.

RESULTS

Clonal analysis of myoblast determination.

10T1/2 cells are an untransformed clonal cell line derived from early mouse embryos (2). Taylor and Jones made the remarkable discovery that brief treatment of proliferating 10T1/2 cells with the nucleotide analogue, 5-azacytidine, causes a high frequency conversion of 10T1/2 cells into differentiated muscle, fat and cartilage cells (3,4). Their studies also established that this mesodermal conversion phenomenon is a consequence of DNA hypomethylation, a process that also can lead to the activation of genes (5).

The earlier work of Taylor and Jones led us to a clonal cell lineage analysis of 5-azacytidine-induced, mesodermal conversion of 10T1/2 cells. Our findings, some of which have been published elsewhere in detail (6) provide direct evidence that 10T1/2 is a multipotential cell genetically blocked in the expression of three independent lineage pathways of mesodermal determination. 5-Azacytidine releases genetic blocks to these pathways and transforms 10T1/2 cells into one of three clonal stem cell lineages: myogenic, adipogenic or chondrogenic. These stem cells are proliferative, stably determined cell lineages that spontaneously differentiate in high density cultures.

The results of our clonal lineage studies, summarized in figure 1, establish that brief (24 hr) 5-azacytidine treatment converts 10T1/2 cells into clonal lineages of myoblasts, adipoblasts, or chondroblasts. The frequency of clonal conversion of 10T1/2 cells to these three new cell lineages is strikingly high. Almost 50% of all primary clones formed following 5-azacytidine treatment contain either muscle, fat or cartilage. Spontaneous conversion has not been detected by clonal analysis. Converted primary colonies always include parental 10T1/2 cells and usually only a single non-parental cell phenotype. The frequency of different phenotypes in primary clones differs somewhat, muscle-containing colonies being most prevalent (Figure 1). A low frequency of mixed colonies of muscle and fat also are observed.

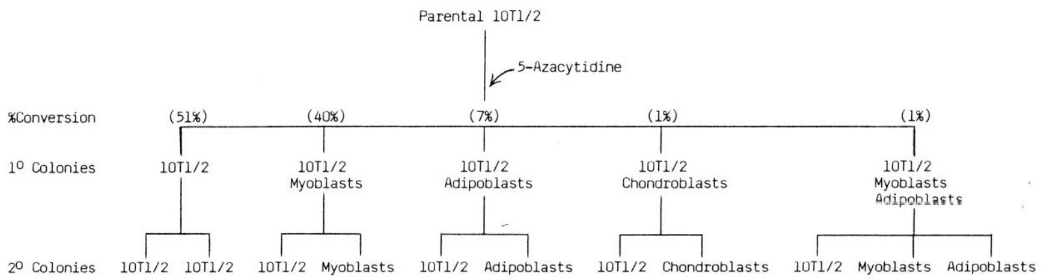

FIGURE 1. Clonal analysis of 5-azacytidine conversion of 10T1/2 cells. Parental 10T1/2 cells exposed to 5-azacytidine for 24 hr are plated at clonal density. The

approximate conversion frequencies derived from primary colonies are indicated. Subclones of individual primary colonies form secondary colonies, each containing a single cell type.

The cell phenotypes in secondary colonies derived by subcloning primary colonies, are informative. All phenotypes in primary colonies segregate clonally. For instance, subclones of a primary muscle colony include colonies of parental 10T1/2 cells (that retain their potential to undergo 5-azacytidine conversion to the mesodermal phenotypes) as well as colonies of myoblasts that proliferate and differentiate and that can be further subcloned as homogeneous cell lines. Subclones of mixed primary colonies of fat and muscle form three distinct secondary colony types: parental 10T1/2, muscle, and fat cell colonies. Thus, 5-azacytidine converts 10T1/2 cells to independently derived and genetically stable lineages of proliferative cells that can differentiate into muscle, or fat, or cartilage. The muscle clones resulting from 10T1/2 conversion are in all respects determined myoblasts. These cells have a bipolar morphology, distinct from parental 10T1/2 and typical of embryonic myoblasts, and synthesize a lineage-specific array of proteins distinct from parental 10T1/2 cells and the fat and cartilage cell lineages. Furthermore, these 10T1/2-derived myoblast lineages differentiate in high density cultures by fusing and activating contractile protein synthesis. 10T1/2 conversion accurately and remarkably reflects the developmental process of mesodermal lineage determination.

The clonal studies of 10T1/2 conversion provide an explanation for the action of 5-azacytidine, and this explanation leads to a simple genetic model for mesodermal determination that suggests an experimental test of this model. We have discussed this model in detail elsewhere (6), and therefore only briefly recount its basic features. 10T1/2 is a multipotential cell that is blocked from mesodermal conversion by the DNA methylation of three regulatory loci that are chromosomally unlinked. Demethylation is required for activation of these loci, which can act independently to convert 10T1/2 cells to new lineages. These loci may be methylated in the embryonic cell which gave rise to the 10T1/2 cell line, or these loci may have become methylated as a consequence of the establishment of this line. In any case, the 10T1/2 cell has lost the capacity in the culture environment to demethylate these loci and therefore is blocked in this multipotential state. Treatment during S phase with 5-azacytidine causes random hemimethylation of the 10T1/2 genome. In any cell, one or more of the methylated sites in these regulatory loci may become hemimethylated during semi-conservative DNA replication. Then, as chromosomes segregate and replicate during the subsequent two mitotic divisions, daughter cells inherit chromosomes having either completely unmethylated sites at various chromosomal loci or methylated sites identical to the parental 10T1/2 chromosome at these loci. Loci on different chromosomes will segregate randomly to these daughter cells. If demethylation is sufficient to activate a locus, some daughter cells in a primary clone will inherit the chromosome with an unmethylated locus and will convert to a stably inherited, new phenotype. In contrast, other daughter cells in this clone will inherit a chromosome that has the parental 10T1/2 methylation pattern.

This model of 10T1/2 conversion is consistent with our current understanding of the inheritance of methylated and hemimethylated DNA sites (7), and it explains the occurrence of parental 10T1/2 and converted lineages in primary clones and the segregation of stable phenotypes in mixed clones. Given that 5-azacytidine causes random hypomethylation in the 10T1/2 genome, as shown by other studies (8), the probablity that any one methylated site will become hypomethylated is a simple function of the overall extent of hypomethylation that occurs during 5-azacytidine treatment. Most significantly, the frequency of clonal conversion to any one phenotype should be directly proportional to the number of hypomethylation events required to activate a locus (or loci) within a cell. This reasoning led us to conclude that myogenic conversion is the result of only one or two hypomethylation events in a 10T1/2 cell and that adipogenic and chondrogenic conversion, which occur at lower frequencies, require substantially more hypomethylations, although these may be within one gene locus for each lineage (6).

The clonal analysis of 10T1/2 conversion suggests a simple experimental test of the idea that 10T1/2 conversion is directed by the activation of one or a few gene loci for each of the mesodermal phenotypes. If this model is correct, then transfected genomic DNA from a cell with an active, unmethylated determination locus should convert 10T1/2 cells to a mesodermal lineage (9). Our preliminary experiments establish the feasibility of such a genomic DNA transfection approach and support the idea that myogenic determination is controlled by a single genetic locus.

Genomic DNA isolated from a 10T1/2-derived myoblast cell line or from embryonic quail myoblasts was cotransfected into 10T1/2 cells with the neomycin-resistance plasmid, pKONEO. G418-resistant cells were selected and tested for muscle colony forming ability by fixation and staining of cultures at clonal density. In these experiments, we observed muscle colony formation at a frequency of $2x10^{-4}$ G418-resistant cells transfected with either 10T1/2-derived myoblast DNA or quail myoblast DNA. This frequency is close to that expected if myogenic conversion of 10T1/2 cells is under the control of a single gene or DNA segment. The number of myogenic colonies induced by genomic DNA transfection is still very small (4 colonies), but muscle colonies have not been observed to form spontaneously or in control cultures transfected with parental 10T1/2 DNA. The possibility remains, however, that the muscle colonies induced by myoblast DNA arise by a spontaneous or a non-specific, DNA-induced conversion process (Lassar and Weintraub, personal communication). A definitive test of this model requires recovery of myogenic colony-forming cells and DNA cloning of functional DNA segments that mediate myogenic conversion. Such experiments, now in progress, provide both a definitive test of a single gene model of myogenic determination and offer a means of characterizing the gene (or genes) that direct mesodermal determination.

Determination and coordinate muscle gene activation.

Conversion of 10T1/2 cells to myoblast cell lineages is accompanied by specific changes in gene expression, but these changes do not directly or immediately influence contractile protein synthesis. Rather, contractile protein synthesis is activated only when myoblasts differentiate in cultures at high cell density. This finding suggests that determination establishes a regulatory program that mediates the activation of muscle-specific genes, but does not act directly on these genes. We have used a gene transfection approach to examine the regulation of troponin I genes introduced into 10T1/2 cells and their myogenic lineages. Our findings support the view that myogenic determination is regulatory in nature. In these experiments, we compared the transcriptional expression of a transfected quail troponin I gene with that of the endogenous mouse alpha actin gene. The quail troponin I gene, described in detail elsewhere, includes 0.5 kb 5' flanking sequence and 1.5 kb 3' flanking sequence cloned in pBR322. This quail troponin I gene and pKONEO were cotransfected into 10T1/2 cells and into 10T1/2-derived myoblasts. 10T1/2 and myoblast cultures were then selected with G418, and 500 surviving clones from each group were pooled. Transfection conditions were adjusted such that the troponin I gene in the pooled clone population was integrated stably at an average copy number of 1-5 genes/cell. The expression of transfected troponin I and endogenous actin genes was assayed by hybridization of total cell RNA with end-labeled, exon-specific probes of these genes, followed by gel electrophoresis of S1 resistant fragments. These experiments, which will be described in detail elsewhere, show that both the transfected troponin I gene as well as the endogenous actin gene are not expressed in either 10T1/2 or in myoblast clones. However, when myoblasts are permitted to differentiate, the transfected troponin I gene and the endogenous actin gene are coordinately activated (Figure 2A) and express high and equivalent levels of transcripts. This troponin I gene also is not expressed when transfected directly into myoblasts, bypassing lineage determination, but its expression becomes fully activated once these converted myogenic cells undergo differentiation (Figure 2B). Additional S1 studies establish that the troponin I transcripts produced by these mouse muscle fibers have proper 5' and 3' termini and are processed. Nuclei transcription run-on experiments (10) also establish that the transfected troponin I gene is transcriptionally activated during the process of myoblast differentiation. Thus, myoblasts and 10T1/2 cells are equally non-permissive for the transcription of the transfected quail troponin I gene, and the endogenous mouse contractile protein genes, but both the transfected and the endogenous muscle genes are coactivated during differentiation. Furthermore, the transfected troponin I gene is regulated equivalently and appropriately regardless of whether this gene is transfected into 10T1/2 cells prior to or after myogenic lineage determination.

These gene transfection experiments lead to several important conclusions concerning the mechanism of muscle gene regulation. First, the troponin I gene and its flanking sequences must include evolutionally conserved cis-acting regulatory information that confers its quantitative, its temporal, and its lineage-specific

FIGURE 2A: S1 analysis of mouse alpha-actin and quail troponin I transcripts derived from pooled populations of stably transfected mouse myoblasts. Proliferating myoblasts (Day 2 and 4) do not express detectable levels of alpha-actin and troponin I whereas differentiated myofibers (Day 6 and 9) co-express the endogenous mouse alpha-actin and transfected quail troponin I genes to high levels.

FIGURE 2B: S1 analysis of mouse alpha-actin and quail troponin I transcripts derived from pooled populations of stably transfected 10T1/2 cells and 5-azacytidine converted myogenic lineages. 10T1/2 cells and converted myoblasts do not express alpha-actin or troponin I transcripts whereas differentiated myofiber cultures derived from this converted myoblast lineage co-express the transfected quail TnI and endogenous mouse alpha-actin genes.

---

activation in response to muscle differentiation. Second, the control mechanisms regulating troponin I gene activation function even though this gene is stably integrated at random sites in a mouse genome in the pooled population of transfected cells. This implies that the chromosomal domain of the troponin I gene does not influence its appropriate developmental regulation. Third, the process of myogenic determination is not required to prepare the troponin I gene for proper quantitative temporal or lineage-specific activation during differentiation. Furthermore, the process of lineage determination does not produce myoblasts that are in an intermediate regulatory state that is permissive for expression of either the transfected troponin I gene or the endogenous muscle genes. That is, these genes do not become partially activated in myoblasts followed by a quantitative activation during differentiation. Although determination may lead to a new regulatory state in myogenic lineages, such a hypothetical state is neither sufficient nor necessary for the proper activation of troponin I transcriptional activation when myoblasts differentiate. These results, therefore, support the simple conclusion that the process of differentiation, not determination, provides the regulatory signals for activation of troponin I transcription, and that the important role of determination is to establish the regulatory control system that allows for expression of these signals at the time of differentiation. This simple view of muscle gene regulation implies that muscle genes are not subject to multiple regulatory controls that act sequentially during development, but that these genes progress through cell lineage determination in a permissive state responsive to differentiation-specific regulatory signals. This conclusion is supported and also suggested by the findings that species-specific heterokaryons formed between differentiated muscle cells and non-muscle cells express muscle-specific contractile proteins directed by the non-muscle cell nucleus (11,12).

The molecular regulatory system that controls muscle gene activation during myoblast differentiation is currently unknown, but these troponin I transfection experiments and myoblast transfection studies with other muscle genes (13, this volume) establish that part of this system involves cis-acting regulatory elements closely associated with muscle genes. This regulation could be either negative or positive in its mode of action on muscle genes. That is, muscle genes may be actively repressed in myoblasts and activated by differentiation-specific signals that relieve this repression. Alternatively, muscle genes may be activated solely by differentiation-specific activators. Several possible models of muscle gene control are illustrated in figure 3.

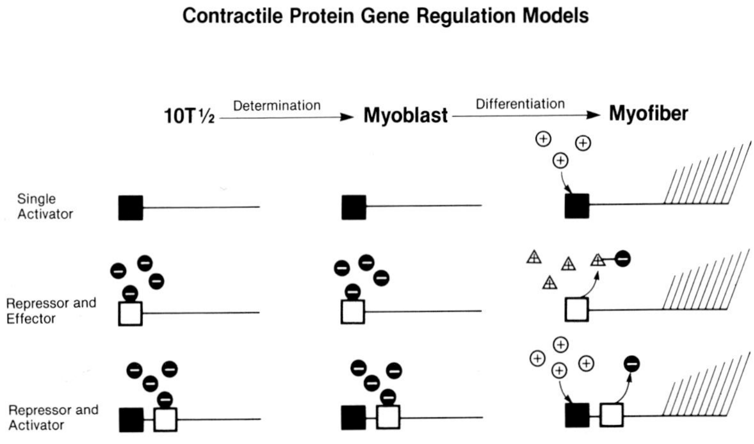

FIGURE 3: Proposed regulatory models for the transcriptional activation of contractile protein genes during myogenic differentiation. The two common features shared by these models are cis-acting regulatory elements closely associated with the muscle genes and the interaction of cell-type specific, trans-acting factors with these elements.

The cis-acting regulatory elements associated with these genes presumably interact with trans-acting regulatory signals. An important future goal will be to define the molecular components and interactions involved in the coordinate activation of muscle genes (14). Of particular interest is whether these genes are activated by a common cis-trans regulatory system or whether each gene has its distinct regulatory system that is coordinated at a step removed from transcription itself. If muscle genes are coordinately activated by a common cis-trans regulatory system, these genes should share homologous muscle-specific sequences, perhaps in their 5' flanking regions. Although muscle gene sequence data are relatively limited at this time, we have identified a 17 bp sequence, 320 bp upstream of troponin I transcription start site, that is homologous to a sequence in the upstream regions of alpha actin and other muscle genes (15-17) (Figure 4).

		5' 1	2	3 3'
Chick α-actin	-100	G C C C G A	c a c c	C A A A T A T
Rat α-actin	-100	G C C C a A	c a c c	C A A A T A T
Quail Troponin I	-329	G g a C G A	c c a g	C A A A T A T
Chick Myosin Light Chain 3	-256	G C C C G g	a c a a	g g A A T A T
Rat Cardiac Myosin Heavy Chain	-411	a g a C a g	g g g a	C A A A T A T

	1	2	3
Consensus Sequence	G C C C G A	c a c c	C A A A T A T
	a g a a g	a c a a	g g
		g g g g	

FIGURE 4: Proposed consensus sequence found in the 5' flanking regions of 5 contractile protein genes. The numbers to the left show the distance (bp) of these sequences to the transcription start sites.

This homologous sequence was previously identified as evolutionally conserved in the rat and chicken alpha actin genes (18). Computer homology analysis of mammalian gene library sequences shows that this homologous sequence is muscle-specific. These findings provide support for the idea that muscle genes are activated coordinately during differentiation by a shared cis-trans transcriptional control system. This hypothesis can now be tested directly by in vitro mutagenesis techniques.

DISCUSSION

Clonal cell lineage and gene transfection studies have established the 10T1/2 cell as a unique model system for analysis of the molecular genetic basis of myogenic cell lineage determination and contractile protein gene regulation. Our studies with the 10T1/2 cell line provide support for the view that both myogenic determination and muscle differentiation are under simple genetic control and establish new experimental approaches to explore the genetic control of both determination and differentiation. Identification of the gene or genes that specify cell lineage determination by genomic DNA transfection promises to provide insight into the fundamental, and previously unknown, mechanisms of cell lineage determination and the regulatory system that specifies muscle-specific genes for activation during myoblast differentiation. Further muscle gene transfection studies, in combination with in vitro mutagenesis and molecular techniques, will certainly identify the cis-acting regulatory sequences and the putative trans-acting signals that control and coordinate muscle gene activation. Such studies also should reveal whether muscle genes are co-regulated by a common transcriptional mechanism or by more complex mechanisms acting at levels distant from these genes. Most importantly, we can begin to understand, at a molecular genetic level, the developmental and genetic interrelationships between the processes of muscle cell lineage determination and myoblast differentiation. Such an understanding may provide a new developmental framework for viewing the process of cellular differentiation and, as suggested by our recent results, this framework may well be genetically simple and fundamental in nature.

ACKNOWLEDGMENTS

We would like to thank Irene Althaus, Sue Borzelleca, Jody Brock, Maggie Ober and Druen Robinson for excellent technical assistance. Some of the work presented here has been submitted for publication elsewhere (Konieczny and Emerson, submitted; Baldwin, Kittler and Emerson, submitted).

REFERENCES

1. Konigsberg IR (1963). Clonal analysis of myogenesis. Science 140:1273.
2. Reznikoff CA, Bankow DW, Heidelberger C (1973). Establishment and characterization of a cloned line of C3H mouse embryo cells sensitive to postconfluence inhibition of division. Cancer Res 33:3231.
3. Taylor SM, Jones PA (1979). Multiple new phenotypes induced in 10T1/2 and 3T3 cells treated with 5-azacytidine. Cell 17:771.
4. Taylor SM, Jones PA (1982). Changes in phenotypic expression in embryonic and adult cells treated with 5-azacytidine. J Cell Physiol 111:187.
5. Taylor JH (1984). DNA methylation and cellular differentiation. Cell Biol Monographs 11:1.
6. Konieczny SF, Emerson CP Jr (1984). 5-azacytidine induction of stable mesodermal stem cell lineages from 10T1/2 cells: evidence for regulatory genes controlling determination. Cell 38:791.
7. Razin A, Riggs AD (1980). DNA methylation and gene function. Science 210:604.
8. Hsiao WW, Gattoni-Celli S, Kirschmeier P, Weinstein IB (1984). Effects of 5-azacytidine on methylation and expression of specific DNA sequences in C3H10T1/2 cells. Mol Cell Biol 4:634.
9. Fassano O, Birnbaum D, Edlund L, Fogh J, Wigler M (1984). New human transforming genes detected by a tumorigenecity assay. Mol Cell Biol 4:1695.
10. Groudine M, Peretz M, Weintraub H (1981). Transcriptional regulation of hemoglobin switching in chicken embryos. Mol Cell Biol 1:281.
11. Blau HM, Chiu C-P, Webster C (1983). Cytoplasmic activation of human nuclear genes in stable heterokaryons. Cell 32:1171.
12. Wright WE (1984). Induction of muscle genes in neural cells. J Cell Biol 98:427.
13. Mellaul D, Aloni B, Calvo J, Yaffe D, Nudal U (1984). Developmentally regulated expression of chimeric genes containing muscle actin DNA sequences in transfected myogenic cells. EMBO J 3:983.
14. Devlin RB, Emerson CP Jr (1978). Coordinate regulation of contractile protein synthesis during myoblast differentiation. Cell 13:599.

15. Nabeshima Y-I, Fujii-Kuriyoma Y, Muramatsu M, Ogata K (1984). Alternative transcription and two modes of splicing result in two myosin light chains from one gene. Nature 308:333.
16. Mahdavi V, Chambers AP, Nadal-Ginard B (1984). Cardiac α- and ß-myosin heavy chain genes are organized in tandem. Proc Natl Acad Sci 81:2626.
17. Zakut R, Shani M, Gival D, Neuman S, Yaffe D, Nudel U (1982). Nucleotide sequence of the rat skeletal muscle actin gene. Nature 298:857.
18. Ordahl, CP, Cooper TA (1983). Strong homology in promoter and 3'-untranslated regions of chick and rat α-actin genes. Nature 303:348.

Molecular Biology of Muscle Development, pages 35–45

EVIDENCE FOR A MYOGENIC STEM CELL[1]

LeBris Smith Quinn[2] and Mark Nameroff

Departments of Biological Structure and Pathology
University of Washington, Seattle Washington 98195

ABSTRACT Recent evidence from both mass cultures and clonal analyses of primary myogenic cells indicates that a self-renewing stem cell compartment exists in the vertebrate skeletal muscle cell lineage. Myogenic stem cells generate committed muscle precursor cells by means of symmetric and asymmetric cell divisions. In the chick pectoral muscle, committed cells then undergo exactly four determined, symmetric divisions to generate cohorts of sixteen terminally differentiated muscle cells. This model of myogenesis contrasts sharply with the conventional model in which a homogeneous population of differentiation-competent muscle precursor cells is postulated. We propose that normal myogenic cells "count" a requisite number of symmetric cell divisions immediately prior to terminal differentiation. Environmental controls (e.g., ambient mitogen concentration) act upon cells in all compartments to regulate the rate of passage through the lineage. The commitment decision by a stem cell, however, may also be affected by the environment. These findings differ in many, but not all, respects from observations made on

[1]This work was supported by NIH Grant #AM-28154 to M.N. L.S.Q. was supported as a Postdoctoral Trainee on NIH Grants #HD-07152 to the University of Pennsylvania (Dr. Howard Holtzer) and #AG-00057 to Dr. George Martin, Dept. of Pathology, University of Washington.
[2]Present address: Howard Hughes Medical Institute, SL-15, University of Washington, Seattle, WA 98195.

myogenic cell lines, and are important in understanding the molecular and genetic bases of muscle differentiation.

INTRODUCTION

It has long been recognized that both intrinsic and extrinsic factors play a role in determining cell fate during development (1). Extrinsic factors could act in an instructive, permissive, or selective fashion. Distinguishing among these alternatives is important for investigating the molecular basis of development, since results obtained at the cell behavioral level heavily influence the design of biochemical and molecular biological experiments.

In this context, vertebrate myogenesis _in vitro_ has been considered a model system in which the role of extrinsic inductive factors in terminal differentiation has been well established. According to this paradigm, myogenic cells are considered to be a homogeneous population, each of which has two options: to keep cycling or to terminally differentiate. Ambient mitogen concentration is thought to determine the probability of choice between these options (2,3). The probability component of the model is introduced to account for the asynchrony which characterizes terminal muscle differentiation both _in vivo_ (4,5) and _in vitro_ (6). Much of the firm evidence in support of this hypothesis, however, has been derived from studies of myogenic cell lines (e.g., 7) which, by definition, have altered properties of growth control. Moreover, studies of primary myogenic cells, which are cited as evidence in favor of the induction model, are equally compatible with another interpretation; namely, that medium alterations act in a selective fashion on a heterogeneous population of myogenic cells (8).

A large body of literature also exists that demonstrates the importance of cell lineages in determination and differentiation (9). Much of the evidence in support of lineage-based mechanisms derives from studies of invertebrate development. However, many recent studies (e.g., 10) have pointed to the potential importance of intrinsic, lineage-based "programs" in vertebrate development. To investigate the role of cell lineage in vertebrate myogenesis, we have undertaken a series of cell cloning experiments using primary myogenic cells from the chick pectoral muscle. The use of cloning strategies has revealed heterogeneity in the division potential of individual myogenic cells prior to terminal differentiation. This has allowed us to trace the myogenic lineage back several generations from the terminal cell type. Additionally, the high plating efficiency and relatively low cell death rate in myogenic clones has permitted us to distinguish between selective and inductive effects of extrinsic factors on different compartments in the lineage. These investigations (6,8,11-13) have resulted in the model for the myogenic lineage illustrated in Figure 1. We found that the final four divisions in the chick myogenic lineage are symmetric and determined (i.e., modifiable in rate of occurrence but not in number by environmental alterations). These divisions are preceded by a stem cell compartment, the cells in which can undergo asymmetric divisions to produce another stem cell and a cell committed to the final four divisions of the lineage. Stem cells can also undergo symmetric divisions in which two stem cells or two committed cells are generated. The microenvironment in which the stem cell resides may well determine the probability of each of these types of division. Environmental factors can also influence the rate of progress of myogenic cells in all the lineage compartments by affecting the rate of cell division.

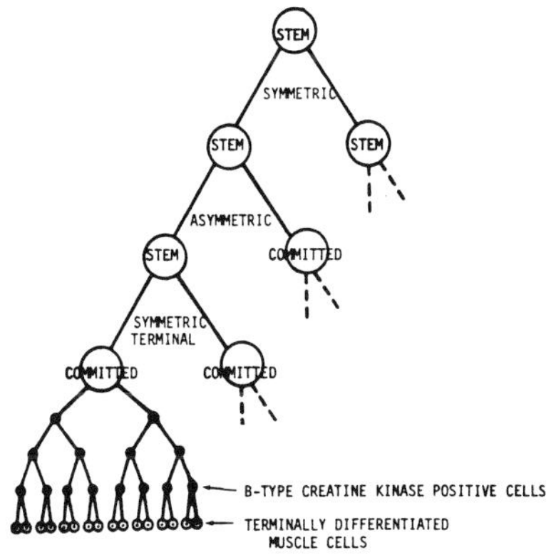

FIGURE 1. Model for the myogenic cell lineage.

This model provides a simple explanation for the protracted and asynchronous nature of muscle differentiation in vertebrate embryogenesis, in mass cultures, and even within single clones. Furthermore, it provides both a framework for resolving a long-standing debate concerning the mechanism by which extrinsic factors act to control populations of myogenic cells, and a new context in which myogenesis may be viewed.

METHODS

Myogenic cells from the pectoral muscles of chick embryos and post-hatch chickens were enzymatically liberated by standard methods (6). Clones were established and grown in fresh medium (Eagle's MEM, 5% embryo extract, 10% horse serum, and antibiotics) by plating 50 cells per 60 mm dish. The medium was changed every third day. The assay for terminal differentiation was peroxidase-antiperoxidase immunocytochemical labeling with a specific antiserum to M-type creatine kinase (6).

RESULTS AND DISCUSSION

Previous work from our laboratory (6,11,12) demonstrated that small myogenic clones tended to contain 1,2,4,8, or 16 terminally differentiated cells and no precursors. This pattern was not modifiable by growing the cells in mitogen-depleted ("conditioned") medium (8) and appeared despite mitotic asynchrony during the generation of individual clones (12). These findings suggested that the final stages of the chick myogenic lineage consisted of a series of at least four determined, symmetric cell divisions prior to the generation of post-mitotic, terminally differentiated muscle cells.

To determine how far back in the lineage the symmetry of replication extended, we analyzed large (17-300+ cells) myogenic clones which typically contained both cycling precursors and end-stage muscle cells. We counted the total number of cells and the number of differentiated cells (or nuclei, if the cells were fused) in 46 individual clones. We reasoned that if an asymmetric or stem cell-like division occurred farther back in the lineage, the pattern of terminally differentiated cells occurring in numbers which are powers of two should change, in the large clones, to a pattern of multiples of some power of two. The value of that power of two (4,5,6, etc.) would then indicate the number of generations back from the end-stage cells at which an asymmetric division occurred (i.e., the point or cell division at which sister cells with different replication potentials are generated).

The results of this analysis are illustrated in Figure 2, a histogram showing the divergence in the number of M-creatine kinase positive (MCK+) cells in each clone from the nearest multiple of 16. In 15 of the 46 clones examined so far, the number of MCK+ cells was an integer multiple of 16. (Values of the integer ranged from 1 x 16 to 9 x 16.) In all but 6 of the remaining cases, the number of MCK+ cells was just short of a multiple of 16 and fell away in an exponential fashion. In 2 cases, the number was equidistant (8), and in only 4 clones was the number of terminally differentiated cells "over" the nearest multiple of 16. This result is quite obviously non-random ($p < 0.001$, chi-squared test). It does not support the hypothesis that myogenic precur-

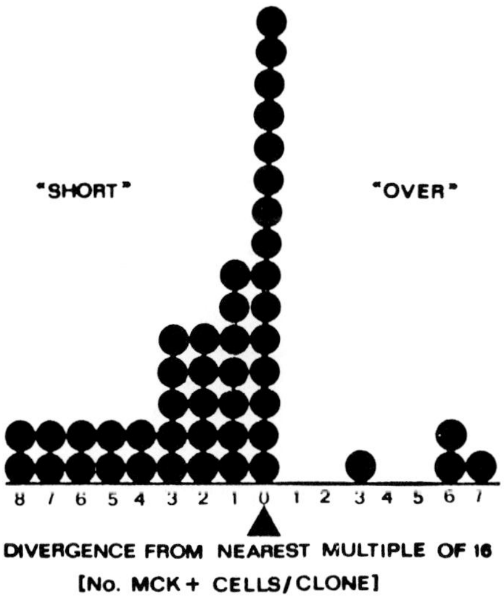

FIGURE 2. Clonal analysis of myogenic cells taken from chick pectoral muscles between day 10 of embryogenesis and 10 weeks post-hatching.

sor cells are a homogeneous population within which terminal differentiation occurs in a stochastic fashion. Such a model would predict a random uniform distribution of the data rather than the one-tailed exponential observed.

The simplest explanation for the data in figure 2 is that, after commitment to terminal differentiation, precursor cells "count" exactly four cell divisions prior to terminal differentiation (13). Cases in which the number of MCK+ cells was "short" of a multiple of 16 can be attributed to cell death or to mitotic asynchrony within cohorts of differentiating precursors; i.e., the clones were fixed before all the cells in a particular cohort had finished dividing. It was at first surprising to find <u>any</u> pattern in the number of MCK+ cells in the clones, since there was no way to predict the best time for fixation after several days in culture. However, in retrospect, it was apparent that mitotic asynchrony among different committed cohorts of cells would not have affected the data. Also, since terminally differentiated muscle cells are post-mitotic, a single cohort of cells whose members had gotten out of mitotic synchrony would be re-synchronized once all the cells had terminally differentiated. The one-sided appearance of the data in figure 2 is also strong support for a strict cell cycle-dependent timing mechanism for terminal differentiation. The basis for this interpretation is as follows: Since the clones were not fixed at any special time points, in most cases fixation would have occurred while at least one cohort of committed cells in each clone was traversing the final four divisions. Since mitotic asynchrony would arise over the course of those divisions, and since cell death also can occur, it is clear that many clones would exhibit numbers of MCK+ cells which were just below a power of 2. For example, a clone with 62 MCK+ cells (2 "short" in figure 2) might be composed of 3 "finished" cohorts (48 MCK+ cells) plus one almost finished cohort containing 14 MCK+ cells and one cell that had not yet undergone its final division to produce 2 MCK+ cells. Similarly, a clone with 31 MCK+ cells (1 "short" in figure 2) could be a case in which one MCK+ cell had died or detached. If mitotic asynchrony and cell death occur stochastically, the death of one or two cells (1 or 2 "short") would be expected to occur more often than the death of fifteen cells (1 "over"). The one-sided exponential curve we observed is exactly what would be expected given these considerations. Various other models (e.g., time-counting, factor dilution, microenvironmental) that, at first sight, might appear to explain the data in figure 2, can be envisioned. In these models a relatively loose, non-deterministic association exists between division and terminal differentiation, yet groups of MCK+ cells are generated in numbers centering around 16. For example, it could be argued that after the commitment step myogenic cells no longer synthesize a cytoplasmic regulatory factor. Through successive divisions (with a mean number of 4), the concentration of the factor would fall below some threshold and the cells would then undergo terminal differentiation. Since cytoplasmic partitioning of the factor would only be centered around a mean of one-half of the molecules to each sister cell in a division, it can be seen that in some cases a cell might require 5 divisions to reach the threshold, while its sister, which received less of the factor after cytokinesis, would require only 3 divisions. In these cases, the association of 4 divisions with terminal differentiation is correlative only. Such correlative models, however, would predict a two-sided distribution about groups of 16 terminally differentiated cells. Therefore, the one-sided distribution which was in fact observed is evidence for an extremely close linkage of the number of cell cycles and the process of myogenic differentiation.

In our analysis of the clones illustrated in Figure 2, we also found no consistent relationship between the number of MCK+ cells in the clones and the total number of cells in the clones. As shown in Figure 3, the percentage of MCK+ cells in each clone varied considerably,

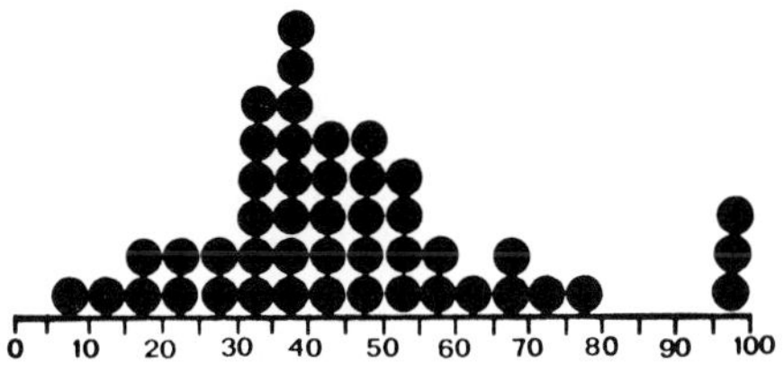

FIGURE 3. Percent differentiation data for myogenic clones fixed at days 4-9 after plating. Each circle represents one clone.

ranging from 8% (e.g., 32 MCK+ cells/398 total cells) to 100% (32 MCK+ cells/32 total cells). This variability indicates that myogenic stem cell divisions can be, but are not invariably, asymmetric. Cells in this compartment can undergo symmetric divisions in which 2 stem cells are generated or 2 committed cells are generated (symmetric terminal division). In the latter case, a 32-cell clone will be produced in which all the cells are post-mitotic and MCK+. These 3 types of division are schematized in Figure 1.

The question then arises as to what controls the direction (symmetric, asymmetric, or symmetric terminal) of myogenic stem cell divisions. The mean percent differentiation for the 46 clones in Figure 3 is 45.5% ± 20.6. This Figure represents data from clones derived from cells taken over a wide range of developmental stages (embryonic day 10 to 10 weeks post-hatching). If the percent differentiation data from each of these stages is grouped separately, we find that the mean percentage of MCK+ cells per clone remains constant at just below 50%. Preliminary data also indicate that serially passaged (mitotically "aged") stem cells likewise give rise to clones exhibiting a mean of 50% terminally differentiated cells. Although, under our culture conditions, stem cells appear to produce, on average, one committed cell at every division, the spread in percent terminal differentiation (Figure 3) suggests that the underlying mechanism of commitment is stochastic and not rigidly programmed. Percent differentiation data from stem cell-derived clones can be taken to reflect the value for stem cell commitment probability (about 0.5). These data therefore suggest a simple constant probability for stem cell commitment in culture. Myogenic stem cells may all be equivalent with regard to their tendency to give rise to committed precursor cells and may not acquire an increasing probability of commitment with each division as has been suggested in some other systems (14,15).

However, a separate line of investigation suggests that the probability of stem cell commitment is either different or can be altered under conditions not duplicated by our culture media (6). By cloning experiments, we estimated the number of myogenic stem cells and committed precursors present in the chick pectoral muscle at different stages of development (days 10 through 14 of embryogenesis). During this period of time, the total number of both types of cells increased, but the number of stem cells increased at a much greater rate. This result indicates that the stem cells must be reproducing themselves faster than they produce committed cells. Therefore, *in vivo*, the probability of commitment may be significantly less than the 0.5 exhibited by the same cells when tested *in vitro*. This difference is either due to the different microenvironment of the cells in the embryo or to effects of liberating the cells by trypsinization. Preliminary culture data indicate that culture in conditioned medium does not appreciably affect commitment probability of chick stem cells. The observation that stem cells increase as a proportion of the mononucleated cell population with age also explains the initially paradoxical result that mass cultures of myogenic cells from older embryos exhibit more protracted fusion kinetics than mass cultures from younger embryos (6). We also have observed (6,12) that, in the same culture conditions, mass cultures and clones (of the same initial cell populations) differentiate with virtually identical kinetics. This observation again suggests that proliferation potential prior to terminal differentiation is largely an intrinsic property of the cells and that the cell-cell interactions that occur in culture do not significantly affect stem cell commitment probability.

Some of the characteristics of the myogenic stem cell (microenvironmental control, stochastic commitment) are also exhibited by myogenic cell lines (16). In many cases, these lines undergo terminal differentiation in a highly asynchronous fashion, forming large colonies which contain both differentiated cells and precursors (16,17). It is possible that such lines represent populations of stem cells in which the tight linkage of commitment, cell division, and terminal differentiation has been disrupted. Alternatively, such lines may represent cells in other compartments of the lineage.

In summary, it is clear from both mass culture and clonal analyses that primary myogenic populations are comprised of a complex mixture of cell types, and that different controls may operate upon different cells within that population. We note that a specific biochemical marker, B-creatine kinase, has already been shown to be characteristic of one sub-class of precursor cells (18; see also Figure 1). The myogenic precursor population is thus both functionally and biochemically heterogeneous. The proportions of different cell types can change with both development *in vivo* and, by selective effects, in different culture conditions (6,8). We conclude that caution must be exercised in interpreting environmental effects on populations of cells as being inductive and/or instructive with regard to individual cell types within that population.

REFERENCES

1. Wilson EB (1940). "The cell in Development and Heredity." New York: Macmillan.
2. Konigsberg IR (1971). Diffusion-mediated control of myoblast fusion. Develop Biol 26:133.
3. O'Neill MC (1976). Population modeling in muscle cell culture: Comparisons with experiments. Develop Biol 53:190.
4. Holtzer H, Marshall JM, Finck H (1957). An analysis of myogenesis by the use of fluorescent anti-myosin. J Biophys Biochem Cytol 3:705.
5. Allbrook D, Han MD, Hellmuth AE (1971). Population of muscle satellite cells in relation to age and mitotic activity. Pathology 3:233.
6. Quinn LS, Nameroff M, Holtzer H (1984). Age-dependent changes in myogenic precursor cell compartment sizes. Evidence for the existence of a stem cell. Exp Cell Res 154:65.
7. Nadal-Ginard B (1978). Commitment, fusion, and biochemical differentiation of a myogenic cell line in the absence of DNA synthesis. Cell 15:855.
8. Quinn LS, Nameroff M (1983). Analysis of the myogenic lineage in chick embryos. IV. Effects of conditioned medium. Differentiation 24:124.
9. Gehring WJ (1978). In Gehring WJ (ed): "Genetic Mosaics and Cell Differentiation." New York: Springer-Verlag.
10. Wood WG, Bunch C, Kelly S, Gunn Y, Breckon G (1985). Control of hemoglobin switching by a developmental clock? Nature 313:320.
11. Kligman D, Nameroff M (1980). Analysis of the myogenic lineage in chick embryos. II. Evidence for a deterministic lineage in the final stages. Exp Cell Res 127:237.
12. Quinn LS, Nameroff M (1983). Analysis of the myogenic lineage in chick embryos. III. Quantitative evidence for discrete compartments of precursor cells. Differentiation 24:111.

13. Quinn LS, Holtzer H, Nameroff M (1985). Generation of chick skeletal muscle cells in groups of 16 from stem cells. Nature 313:692.
14. Holliday R, Huschtscha LI, Tarrant GM, Kirkwood TBL (1977). Testing the commitment theory of cellular aging. Science 198:366.
15. Prothero J, Gallant JA (1981). A model of clonal attenuation. Proc Nat Acad Sci US 78:333.
16. Wright WE (1984). Control of differentiation in heterokaryons and hybrids involving differentiation-defective myoblast variants. J Cell Biol 98:436.
17. Linkhart TA, Clegg CH, Hauschka SD (1981). Myogenic differentiation in permanent clonal mouse myoblast cell lines: Regulation by macromolecular growth factors in the culture medium. Develop Biol 86:191.
18. Robinson MM, Quinn LS, Nameroff M (1984). BB creatine kinase and myogenic differentiation. Immunocytochemical identification of a distinct precursor compartment in the chicken skeletal myogenic lineage. Differentiation 26:112.

Molecular Biology of Muscle Development, pages 47-60

IMMUNOCYTOCHEMICAL STUDIES ON THE EXPRESSION OF DESMIN BY DIVIDING CELLS FROM SKELETAL MUSCLE[1]

Z. Yablonka-Reuveni and M. Nameroff

Department of Biological Structure, School of Medicine, University of Washington, Seattle, Washington 98195

ABSTRACT Using a combination of [^{3}H]thymidine autoradiography and immunofluorescent staining with a desmin antibody, we examined mitotically active cells in primary cultures established from the breast muscles of 18-day-old chick embryos. In mass cultures about 5% of the [^{3}H]thymidine labeled cells were desmin positive. Skeletal myosin was not detected in the vast majority of the mitotically active, desmin positive cells observed in these cultures. In addition, some, but not all of the desmin positive, cycling cells expressed the BB isozyme of creatine kinase. In tissue sections from the breast muscles of 18-day-old embryos, the desmin antibody fluorescently stained muscle fibers and smooth muscle cells in blood vessel walls. To determine whether the [^{3}H]thymidine labeled, desmin positive cells were in the skeletal myogenic or another lineage, similar studies were carried out on myogenic clones established from the breast muscles. Again, desmin positive, cycling cells were observed in many of the clones which contained myotubes. These desmin positive cells may therefore reside in the myogenic lineage. Alternatively, these results could indicate that common stem cells may give rise to skeletal myogenic cells and to other cell types.

[1]This work was supported by NIH grant #AM-28154 and by a grant from the American Heart Association, Washington Affiliate.

INTRODUCTION

Desmin is the major subunit protein of muscle type intermediate filaments. The expression of this subunit by skeletal, cardiac and most types of smooth muscle is well documented in both embryonic and adult tissues (see Ref. 1 for review). In the skeletal muscle lineage, desmin expression has been thought to be specific to postmitotic myoblasts and myotubes. This conclusion has been drawn from studies on primary myogenic cultures using immunofluorescence techniques (2,3) and radiolabeling of proteins (3). The dividing cells that were examined by the fluorescence approach appeared to lack desmin (2). Similarily, using the biosynthetic approach, very low amounts of desmin were detected in extracts from myoblast cultures and were attributed to contaminating myotubes (3).

In contrast to the above-mentioned studies, we were able to detect dividing cells which expressed desmin using primary mass cultures from chicken embryonic breast muscles and employing a combination of [^{3}H]thymidine autoradiography and immunofluorescent staining with a desmin antibody. Although similar findings have been reported previously for primary myogenic cultures, the lineage identity of the cycling, desmin positive cells could not be determined (4). In the present study, we have detected the [^{3}H]thymidine labeled nuclei in desmin positive cells not only in mass cultures but also in individual myogenic clones. Some of our findings regarding these cells are described here. The fate of these desmin positive cells and their role, if any, in the myogenic lineage is still unknown.

METHODS

Myogenic cultures were established from the breast muscles of 18-day-old chick embryos as described previously (5) except that the tissue was exposed to 0.2% collagenase for 45 min prior to the trypsin digestion. To eliminate myofibril fragments which are present in muscle cell preparations from older embryos and adult chickens and prevent accurate cell counting, the muscle cell preparation was subjected to Percoll density centrifugation (6, and Yablonka-Reuveni, in preparation). Cells were grown in gelatin-coated, horse serum-treated tissue culture dishes (5). For mass cultures, cells were plated at

2×10^5 cells per 35 mm culture dish. For clonal cultures, cells were plated at 50-100 cells per 60 mm culture dish. Culture conditions and clone identification were as previously described (5,7).

A rabbit antiserum against chicken gizzard desmin was obtained from Dr. H. Holtzer. The preparation and characterization of this antibody have been reported (2,8). The IgG fraction of the desmin antiserum was obtained in our laboratory using protein A-sepharose (9). A guinea pig antiserum against skeletal muscle myosin heavy chain and a rabbit antiserum against BB-creatine kinase were prepared in our laboratory as described elsewhere (5,10). Single- and double-immunolabeling of cells in cultures were performed with the indirect immunofluorescence technique (10). In some experiments the desmin antibody was directly fluorescinated in the prescence of fluorescein isothiocyanate-celite (11) and direct:indirect immunostaining of cells was performed in sequence as previously described (3). Indirect immunofluorescent staining of tissue sections from the breast muscles of 18-day-old chick embryos was performed according to a recently described method (12, and Robinson et al., in preparation). Briefly, the tissue is fixed in a 4% solution of 1-ethyl-3-(3-dimethylaminopropyl)carbodiimide (CDI), infiltrated with sucrose and sectioned with a cryostat. Indirect immunoperoxidase staining was used to visualize specific proteins separated by SDS polyacrylamide gel electrophoresis (10).

Cultures exposed to $[^3H]$thymidine were processed for autoradiography subsequent to antibody staining (10).

RESULTS

Characterization of the desmin antibody.

The desmin antibody which we used in this study is well characterized (2,8). However, due to the nature of our studies it was very important to reensure that the antibody reacts only with desmin. Analysis of the antibody by the immunoblot technique (13) is shown in Figure 1. The only protein that reacted with the antibody had a molecular weight of 51,000 which is in agreement with molecular weights reported for desmin (1,8). In addition, the antibody did not react with partially purified

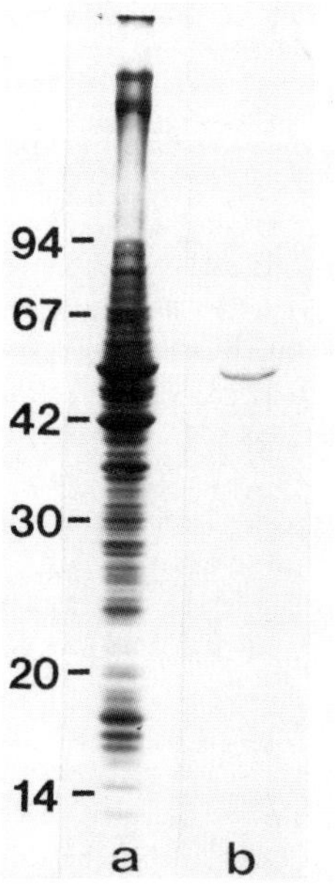

Figure 1. Immunoblot analysis of desmin antibody. a) Extract of a well fused muscle culture (4-day-old) was electrophoresced on a 10-20% SDS polyacrylamide gel and stained with Coomassie brilliant blue, b) Sample as in a) was further blotted onto nitrocellulose and the desmin band was visualized using the desmin antibody and the indirect immunoperoxidase technique.

vimentin, prepared from chicken erythrocyte ghosts according to published methods (14), and tested by the immunoblot technique (data not shown). As shown in Figure 2, when the anti-desmin was tested on a well fused myogenic culture (4-day-old), the myotubes reacted with the antibody whereas the majority of the mononucleated cells did not react. The desmin antibody was also tested on fixed sections from embryonic breast muscles as shown in Figure 3. Bundles of muscle fibers as well as smooth muscle cells in blood vessel walls were fluorescently stained. The connective tissue matrix in these muscle sections appeared to be lacking cells which were positive for desmin.

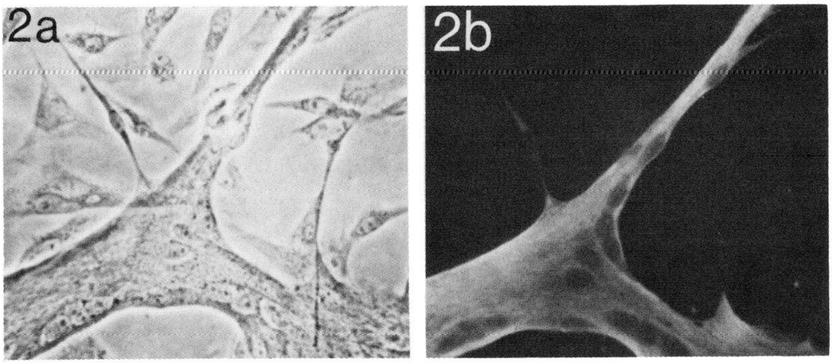

Figure 2. A phase and corresponding fluorescence micrographs of a 4-day primary muscle culture. a) Phase, b) Immunofluorescent staining with anti-desmin (0.1 mg/ml) indirectly labeled with fluorescein-conjugated goat anti-rabbit IgG (1:60). Micrographs were taken with a 40x objective and enlarged to an equivalent degree.

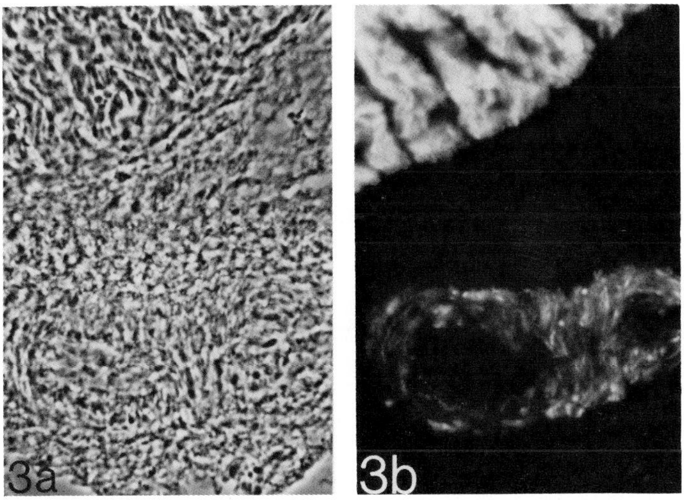

Figure 3. A phase and corresponding fluorescence micrographs of a tissue section from the breast muscles of an 18-day-old chicken embryo. a) Phase, b) Immunofluorescent staining with anti-desmin, details as in Figure 2.

Desmin expression in mass cultures.

To test the possible expression of desmin by mitotically active cells, primary myogenic cultures, 40 hr after plating, were pulsed for 2 hr with 1 μCi/ml [^{3}H]thymidine (specific activity 6.7 Ci/mmol) and immediately fixed and processed for immunofluorescent localization of desmin and skeletal myosin heavy chain, or desmin and BB-creatine kinase, and autoradiography. Selected microscopic fields are shown in Figure 4. The percentage of the total cells that incorporated the [^{3}H]thymidine was about 55%. Of these [^{3}H]thymidine labeled cells, about 5% were desmin positive. Desmin positive, postmitotic cells were also detected. Less than 0.1% of the [^{3}H]thymidine labeled cells were positive for both desmin and myosin (Figure 4, a-f). It has been suggested that the appearance of the BB-creatine kinase in cycling myogenic cells is a biochemical marker for the precursor compartment that gives rise to terminally differentiated myoblasts in the chicken skeletal myogenic lineage (10). Hence, the co-expression of desmin and BB-creatine kinase in dividing cells was investigated in order to determine whether the cycling, desmin positive cells are also in this lineage compartment (Figure 4, g-l). Desmin positive, cycling cells which are also BB-creatine kinase positive were detected. In addition, some of the desmin positive, mitotically active cells did

---

Figure 4. Micrographs of 2-day-old primary muscle cultures in which cycling cells were labeled with [^{3}H]thymidine. a) Phase, b,c) Double immunofluorescent staining of the same field with b) Anti-desmin (0.1 mg/ml) indirectly labeled with fluorescein-conjugated goat anti-rabbit (1:60), c) Anti-myosin (1:60) indirectly labeled with rhodamine-conjugated goat anti-guinea pig (1:60). d) Phase, e,f) The same field stained as described in b) and c), respectively. g) Phase, h,i) Double immunofluorescent staining of the same field with h) Anti-desmin (0.1 mg/ml) directly labeled with fluorescein, i) Anti-BB creatine kinase (1:50) indirectly labeled with rhodamine-conjugated goat anti-rabbit (1:60). j) Phase, k,l) The tively. All micrographs were taken with a 40x objective and enlarged to an equivalent degree. Arrows indicate [^{3}H]thymidine labeled nuclei.

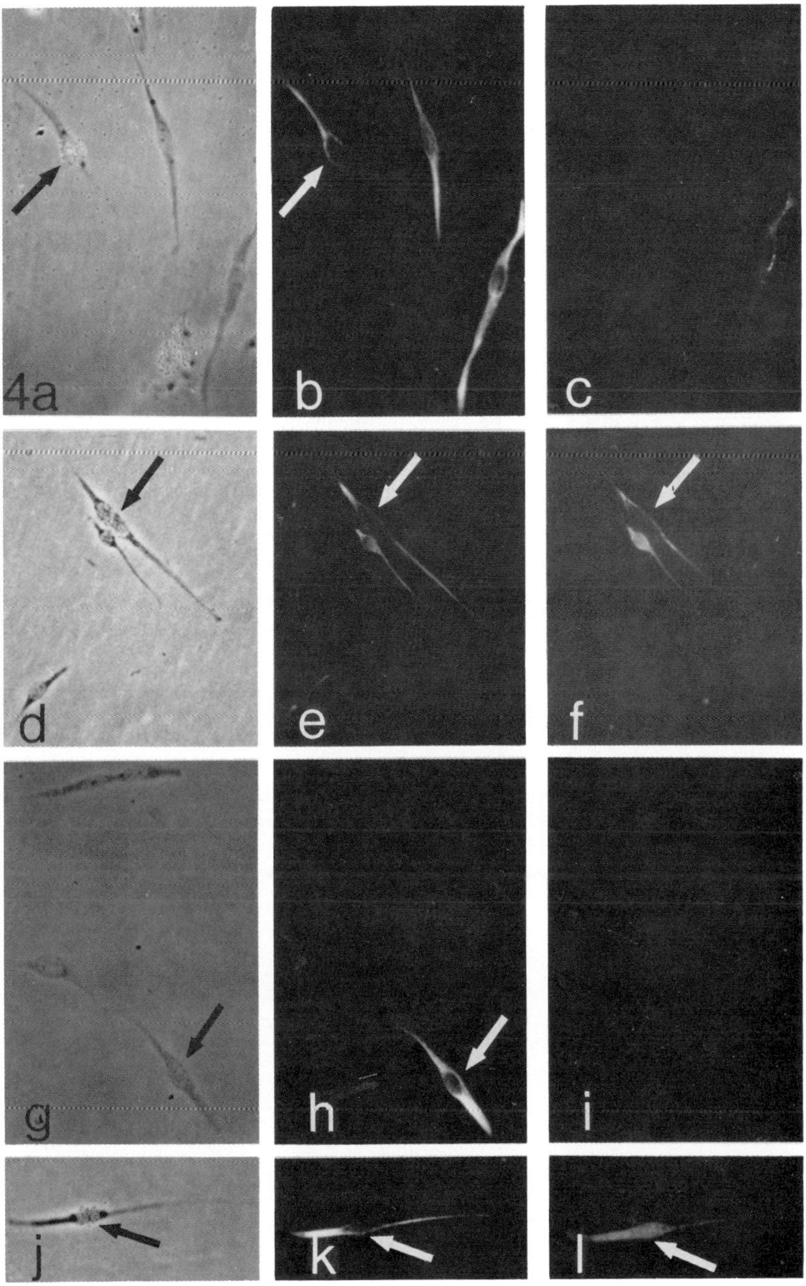
4a
b
c
d
e
f
g
h
i
j
k
l

not express the BB-creatine kinase. Also, not all of the BB-creatine kinase positive, cycling cells expressed desmin. Of the postmitotic cells, the majority, but not all, of the "myoblast-looking", desmin positive cells were BB-creatine kinase positive.

Desmin expression in myogenic clones.

The above experiments showed that there was a biochemical heterogeneity among the desmin positive, cycling cells in respect to BB-creatine kinase expression. One possible interpretation of these results was that some of the desmin positive, cycling cells, despite their skeletal myoblast appearance, are not residing in the skeletal myogenic lineage. For example, smooth muscle cells which express desmin and are present in the muscle tissue at this embryonic age (Figure 3) could be isolated and cultured along with the skeletal myogenic cells. The same argument is true for the BB-creatine kinase positive, cycling cells, especially since the BB-creatine kinase isozyme has been, biochemically, detected in smooth muscle and other tissues (15). Immunofluorescent studies on BB-creatine kinase in smooth muscle cells are lacking. To determine whether the desmin positive, mitotically active cells were, at least in part, progeny of skeletal muscle precursor cells we grew primary clones from cells prepared from the breast muscles of 18-day-old chicken embryos. After 8 days in culture the clonal cultures were pulsed with [^{3}H]thymidine and immediately processed for antibody staining and autoradiography as described above for mass cultures. About 90% of the clones contained myotubes and mononucleated cells of which about 50% were still mitotically active. These clones are referred to as myogenic clones. The remaining 10% of the clones contained only mononucleated cells which did not react with antibodies against muscle specific proteins. The myogenic clones contained many postmitotic

Figure 5. Phase and corresponding fluorescence micrographas of selected microscopic fields of 8-day-old myogenic clones in which cycling cells were labeled with [^{3}H]thymidine. a,c,e,g) Phase, b,d,f,h) Immunofluorescent staining with anti-desmin, details as in Figure 2. Arrows indicate [^{3}H]thymidine labeled nuclei.

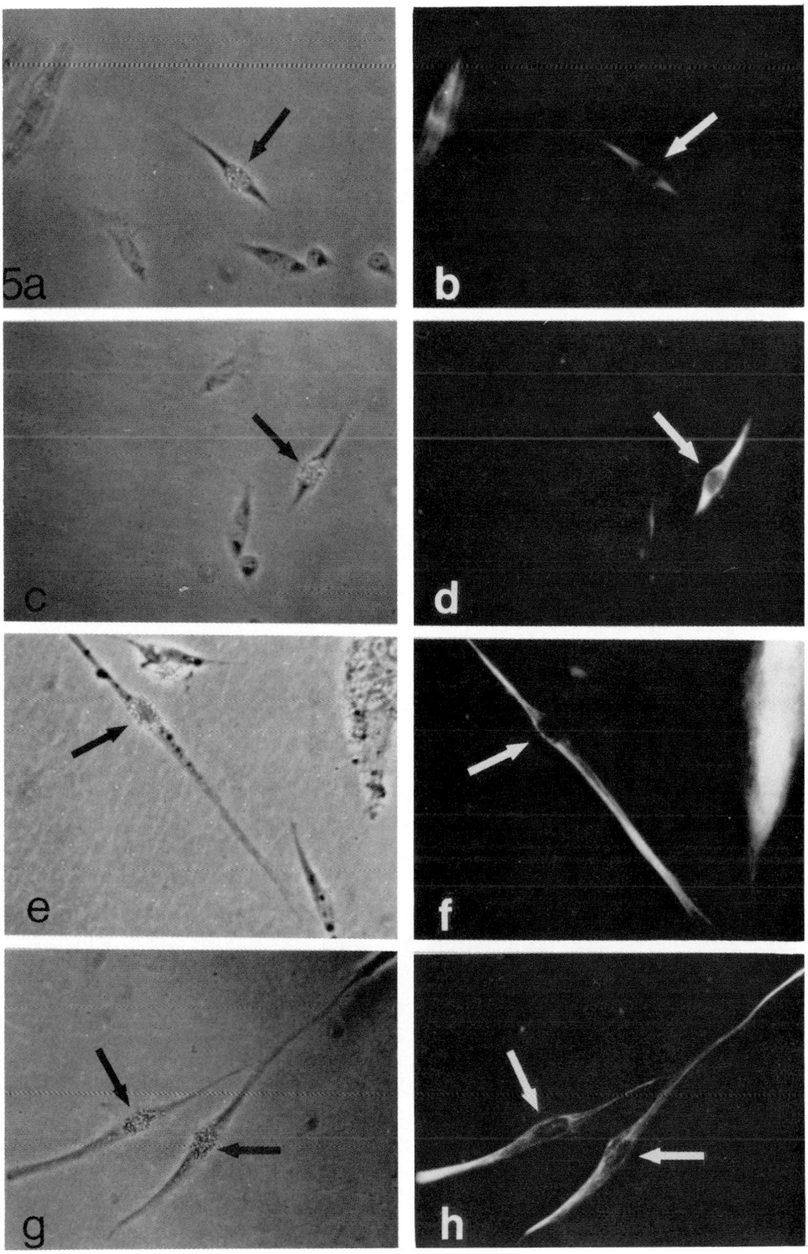
5a
b
c
d
e
f
g
h

cells which were desmin positive and myosin positive. In about 50% of the myogenic clones we detected 1 to 4 $[^3H]$thymidine labeled cells which also expressed desmin. Less than 0.1% of these desmin positive, cycling cells were positive for skeletal myosin. Figure 5 demonstrates several desmin positive, myosin negative, cycling cells from different myogenic clones. Whereas most of these cells resemble myoblasts (Figure 5 a-d), some show a very peculiar appearance (Figure 5, e-h).

DISCUSSION

In the skeletal muscle lineage, desmin expression has been thought to be specific to postmitotic myoblasts and myotubes (1-3). More recently, the existence of small numbers of desmin positive, cycling cells in primary myogenic cultures was described (4). However, the lineage identity of these cells was not determined. The present study detected mitotically active cells which express desmin in both primary mass cultures and primary myogenic clones from embryonic skeletal muscle.

The presence of $[^3]$thymidine labeled, desmin positive cells in myogenic mass culture is not sufficient to suggest that these cells are precursors of skeletal muscle cells. These cells can represent smooth muscle cells which were isolated from the muscle tissue along with the skeletal myogenic cells. However, the heterogeneity among mitotically active cells in myogenic cultures with respect to the expression of desmin and BB-creatine kinase demonstrates that these cells do not constitute a homogeneous cell population. This finding suggests, as previously discussed (10), that caution should be used when biochemical and molecular data from extracts of myogenic cultures are being interpreted.

The finding of desmin positive, mitotically active cells in individual myogenic clones clearly demonstrated that at least some of the cycling cells which express desmin are descendants of cells that give rise to terminally differentiated skeletal muscle cells. Hence, these desmin positive cells may represent cells that reside in the skeletal myogenic lineage. If the latter is true, several explanations can be suggested for the appearance of these desmin positive, cycling cells. These include: a) The expression of desmin by cycling cells may reflect an event which normally occurs in chicken skeletal muscle cells at a low rate where the synthesis of muscle specific proteins (e.g., desmin) and cell cycle withdrawal are separable. In fact, it has been shown that the addition of the tumor promotor PMA (phorbol 12-myristate 13-acetate) to muscle cultures causes an increase in the number of desmin positive, cycling cells. Hence, the frequency of this event can be manipulated. It has been suggested that the synthesis of muscle-specific proteins and cells cycle withdrawal are separable, non-interdependent events for quail myoblasts (16). Arguments against the separation of these two events as a regular pathway in chicken skeletal myogenesis were detailed in a recent publication from our laboratory (10). b) Desmin may be synthesized in all muscle precursor cells prior to the terminal M phase but accumulates to a degree that can be fluorescently detected only in some cells (i.e., cells that have prolonged G_2 phase, cells that can not enter or complete the final M phase). c) These cycling cells which express desmin are abnormal (i.e., dying cells) and might have initiated the differentiation program prior to its normal onset. Some of these suggestions have been also raised by others (4).

An alternative explanation to the appearance of cycling, desmin positive cells in skeletal myogenic clones is the existence of common stem cells which give rise to skeletal myogenic cells and to other cell types that express desmin, such as smooth muscle cells and cardiac cells. In a set of experiments, whose results are not shown here, we attempted to determine on primary myogenic mass cultures whether some of the desmin positive, cycling cells are actually smooth muscle cells. For this, we doubly labeled 2-day-old cultures with anti-desmin and a monoclonal antibody against smooth muscle-specific actin (actin was prepared from chicken gizzard, see Ref. 16 for details). None of the desmin positive, cycling cells reacted with the anti-smooth muscle-specific actin. It is noteworthy that some cells in the muscle culture (much less than 0.1% of the mononucleated cells) did react with the anti-smooth muscle-specific actin. These cells were spread out and almost unrecognizable when examined by phase microscopy, having the morphological appearance of smooth muscle cells in tissue culture. Fluorescence microscopy of these cells following staining with the anti-actin antibody, revealed a network of stress fibers. Surprisingly, these actin positive cells did not express desmin as judged by double immunofluorescence. We also detected in mass cultures, at low frequency, cells that are very spread out and flat and which reacted with the anti-desmin but were negative for smooth muscle actin. In view of the above findings, the fact that the desmin positive, cycling cells did not react with the anti-smooth muscle actin is not sufficient to rule out the idea that these desmin positive, mitotically active cells may be smooth muscle cells. However, none of the desmin positive, cycling cells in the myogenic clones resembled the spread out presumed smooth muscle cells.

To further investigate the nature of the desmin positive, cycling cells, we are currently studying their appearance in skeletal muscle mass cultures and clones from different embryonic ages and from adult skeletal muscle, while also investigating the existence of these cells in skeletal muscle tissue in vivo.

ACKNOWLEDGEMENTS

We are grateful to Dr. H. Holtzer for providing the desmin antiserum, to Dr. L.S. Quinn for isolating the IgG fraction from the desmin antiserum, to Dr. M.M. Robinson for providing the tissue sections and advice on the immunofluorescent staining of these sections, to Drs. A.M. Gown and D. Gordon for providing the smooth muscle actin antibody and to Mr. S.C. Braddy for excellent technical assistance.

REFERENCES

1. Lazarides E (1982). Intermediate filaments: A chemically heterogeneous, developmentally regulated class of proteins. Ann Rev Biochem 51: 219.
2. Bennett GS, Fellini SA, Holtzer H (1978). Immunofluorescent visualization of 100 Å filaments in different cultured chick embryo cell types. Differentiation 12: 71.
3. Gard DL, Lazarides E (1980). The synthesis and distribution of desmin and vimentin during myogenesis in vitro. Cell 19: 263.

4. Dlugosz AA, Tapscott SJ, Holtzer H (1983). Effects of Phorbol 12-Myristate 13-acetate on the differentiation program of embryonic chick skeletal myoblasts. Cancer Res 43: 2780.
5. Quinn LS, Nameroff M (1983). Analysis of the myogenic lineage in chick embryos. III. Quantitative evidence for discrete compartments of precursor cells. Differentiation 24: 111.
6. Yablonka-Reuveni Z, Nameroff M (1984). Isolation and characterization of different cell types from the breast muscles of chick embryos. Anat Rec 208: 198A.
7. Quinn LS, Nameroff M, Holtzer H (1984). Age-dependent changes in myogenic precursor cell compartment sizes. Evidence for the existence of a stem cell. Exp Cell Res 154: 65.
8. Fellini SA, Bennett GS, Toyama Y, Holtzer H (1978). Biochemical and immunological heterogeneity of 100 Å filament subunits from different chick cell types. Differentiation 12: 59.
9. Kraus JP, Rosenberg LE (1982). Purification of low-abundance messenger RNAs from rat liver by polysome immunoadsorption. Proc Natl Acad Sci USA 79: 4015.
10. Robinson MM, Quinn LS, Nameroff M (1984). BB creatine kinase and myogenic differentiation. Immunocytochemical identification of a distinct precursor compartment in the chicken skeletal myogenic lineage. Differentiation 26: 112.
11. Rinderknecht H (1962). Ultra-rapid fluorescent labelling of proteins. Nature 193: 167.
12. Robinson MM (1984). Skeletal myogenesis. An Immunocytochemical investigation of cell differentiation. Ph.D. thesis, University of Washington.
13. Towbin H, Staehelin T, Gordon J (1979). Electrophoretic transfer of proteins from polyacrylamide gels to nitrocellulose sheets: procedure and some applications. Proc Natl Acad Sci USA 76: 4350.
14. Granger BL, Repasky EA, Lazarides E (1982). Synemin and vimentin are components of intermediate filaments in avian erythrocytes. J Cell Biol 92: 299.
15. Rosenberg UB, Eppenberger HM, Perriard JC (1981). Occurence of heterogenous forms of the subunits of creatine kinase in various muscle and nonmuscle tissues and their behaviour during myogenesis. Eur J Biochem 116: 87.

16. Devlin BH, Konigsberg IR (1983). Reentry into the cell cycle of differentiated skeletal myocytes. Develop Biol 95: 175.

17. Gown AM, Vogel AM, Gordon D, Lu PL (1985). A smooth muscle-specific monoclonal antibody recognizes smooth muscle actin isozymes. J Cell Biol 100: in press.

Molecular Biology of Muscle Development, pages 61–75

MITOGENS INHIBIT CONTRACTILE PROTEIN SYNTHESIS IN THE ABSENCE OF CELL PROLIFERATION[1]

Reid S. Compton,[2] Peter A. Merrifield, and Irwin R. Konigsberg

Department of Biology, University of Virginia, Charlottesville, Virginia 22901

ABSTRACT Differentiated, fusion-blocked quail myocytes withdraw from the cell cycle and synthesize the normal complement of muscle-specific proteins when cultured in a differentiation-promoting (i.e., low mitogen) medium. The muscle phenotype is reversible under these conditions, however, since differentiated myocytes reenter the cell cycle and downregulate the synthesis of contractile proteins when refed with a growth-promoting (i.e., high mitogen) medium. Since dedifferentiation does not occur in totally fused myotubes, fusion is associated - at least temporally - with the terminal differentiation of muscle. We have also used a temperature sensitive mouse cell line to demonstrate that the mitogen-mediated downregulation of contractile protein synthesis can occur independently of reentry into S and can be mimicked by the addition of purified fibroblast growth factor (FGF) to differentiated myocytes. This myoblast cell line blocks in G_1 at the non-permissive temperature but does not synthesize contractile proteins in the presence of growth-promoting medium or FGF. This variant will synthesize contractile proteins when blocked at the temperature sensitive block point, however, if mitogen is removed. These results suggest that reentry into S is not required for the dedifferentiation of muscle and that the presence of growth factors directly

[1]This work was supported by an NIH grant (HD 07083) to IRK. RSC and PAM were supported by an NIH developmental biology training grant (HD 07192) and a NSERC of Canada postdoctoral fellowship, respectively.

[2]Author to whom correspondence should be addressed. Current address: Institute for Cancer Research, Fox Chase Cancer Center, 7701 Burholme Avenue, Philadelphia, PA 19111.

inhibits the expression of muscle-specific gene products in primary myogenic cells and established myogenic cell lines.

INTRODUCTION

Biologists have long been concerned with the timing mechanisms which control the expression of developmentally regulated events. During the development of embryonic skeletal muscle, determined but undifferentiated myoblasts proliferate for a period of time and then withdraw from the proliferative cycle, fusing to form multinucleated fibers in which a battery of muscle-specific contractile proteins and enzymes are coordinately activated. An understanding of the mechanisms which regulate the timing of this differentiative transition has been facilitated by the identification of culture conditions which promote this transition *in vitro*. One early model to explain the timing mechanism proposed that myoblasts were genetically programmed to undergo a limited number of cell divisions prior to differentiation and that a unique "quantal cell division" was necessary before myoblasts could differentiate (1). In contrast, evidence from our laboratory (2,3) and others (4) has demonstrated that myoblast differentiation is not rigidly determined, but can be regulated to a remarkable degree by extrinsic factors, most notably the level of mitogens in the medium.

We have recently shown that primary cultures of quail myoblasts withdraw from the cell cycle and synthesize myosin heavy chain (5) and creatine phosphokinase (6) when cultured in conditioned medium under fusion-impermissive (low Ca^{++}) conditions. Single cell analysis based on immunofluorescence has shown that greater than 95% of myocytes cultured under such conditions contain detectable levels of myosin after 48 hours (7) and that a significant proportion of these myocytes reenter the cell cycle when refed with growth-promoting medium (5). We also have shown that, in response to the mitogenic stimulus, differentiated myocytes downregulate the synthesis of myosin heavy chain (MHC) to levels observed in cultures of predominantly cycling myoblasts (5). These observations suggest that the muscle phenotype in fusion-blocked, primary myogenic cells *in vitro* is reversible and that the presence of mitogen in the cellular environment is somehow incompatible with the maintenance of muscle-specific contractile protein synthesis. This phenomenon has been further examined by studying the synthesis of a variety of muscle-specific contractile proteins in quail myocytes and in a temperature-sensitive mouse myoblast mutant in the presence and absence of mitogen. Our results support the observation that growth medium directly inhibits the synthesis of contractile proteins in fusion-blocked myocytes and suggest that purified mitogens like FGF can mimic this effect. In addition, our results suggest that downregulation of contractile protein synthesis can occur in stimulated myocytes without reentry into S phase. If the presence of trans-acting regulators or arrest in a putative G_0 phase of the cell cycle are necessary for muscle-specific gene expression in differentiating myoblasts, these requirements can be overridden by the presence of growth factor in the culture medium. This observation is important in interpreting experiments designed to identify positive regulators of muscle-specific gene expression.

MATERIALS AND METHODS

Primary cultures of differentiated, fusion-blocked quail myocytes were grown in 1-10 CM EGTA medium and refed with 10-15 MEM EGTA medium (7) in order to study mitogen-mediated reentry into the cell cycle and downregulation in primary cultures of myogenic cells. The mouse myoblast cell line used in similar studies (ts-36) was isolated from the tibialis anterior muscles of 12 week old, male BALB/c mice which had been injected with bupivacaine HCl. Cell lines were mutagenized with EMS, raised to the non permissive temperature (39.5^{o}C) and treated with BuDR and light to kill dividing cells and select for mutants which were temperature sensitive for growth. Ts-36 cells were grown on gelatin-coated culture dishes in F10 medium containing 3% embryo extract and 15% preselected horse serum. Growth medium (F10 3-15) was conditioned by exposing it to exponentially growing murine myoblasts (seeded at 165,000 cells/100 mm dish) for 96 hours. Fusion of mouse myoblasts was scored as previously described (6). Synthesis of muscle-specific contractile proteins in myogenic cells pulse-labelled with ^{35}S methionine was monitored by autoradiography or fluorography of two-dimensional polyacrylamide gels which contained a constant amount of labelled protein as described by Devlin and Emerson (8).

RESULTS

Downregulation of α Actin and Tropomyosin in Mitogen Stimulated Quail Myocytes.

During the normal differentiation of quail myoblasts _in vitro_, the synthesis of the major contractile proteins is highly coordinated. Since the mitogen-mediated downregulation of MHC synthesis which we previously observed resembles a reversal of normal differentiation, we were interested in knowing whether the synthesis of other contractile proteins is also inhibited by a mitogenic stimulus. Differentiated, fusion-blocked myocytes were obtained as previously described by culturing for 48 hrs in differentiation-promoting medium containing EGTA, and then refed with growth-promoting medium containing EGTA. At intervals during both differentiation and following the addition of growth medium, duplicate cultures were labelled for 2 hrs with ^{35}S methionine, extracted with lysis buffer and analyzed for contractile protein synthesis by two-dimensional gel electrophoresis and fluorography. α Actin and tropomyosin were identified on 2D gels by co-migration with purified samples on stained gels. 2D profiles of myoblasts labelled after 24 hrs in growth medium but before the addition of differentiation promoting medium (To) were also examined. The presence of small amounts of α actin in such myoblast cultures indicates that some cells have differentiated even at this early time (FIG. 1). Within 6 hrs of refeeding with growth medium, the relative amount of α actin synthesis decreases dramatically to levels similar to that observed in myoblast cultures (To). The amount of newly synthesized α actin (relative to ß and γ actin) remains low for up to 24 hours after the addition of growth medium. The relative amount of tropomyosin (Tm) synthesis also decreases in stimulated myocytes, but this downregulation occurs more slowly than is observed with MHC and α actin. Although substantial Tm synthesis is detectable in mitogen-stimulated myocytes up to 12 hrs, it is barely detectable at 24 hours after refeeding. Thus, while the downregulation of several contractile proteins is coordinately initiated following the addition of growth medium, this downregulation of synthesis proceeds at different rates for myosin heavy chain, tropomyosin and α actin.

This observation is not surprising. Although we still do not know the precise mechanism of the downregulation of contractile protein synthesis in stimulated quail myocytes, changes in the stability and/or the translatability of contractile protein mRNAs are most likely involved, since the half-lives of these messages normally exceed 6 hrs (9). Thus, in contrast to differentiating myoblasts where transcription seems to be the rate limiting step, downregulation may be the result of post-transcriptional as well as transcriptional events. In fact, where this question has been studied in a mutant myogenic cell line (10), both the cessation of contractile protein mRNA synthesis and the specific destabilization of cytoplasmic mRNA seem to play a role in downregulation. What is clear from our experiments is that the presence of mitogen in the culture medium is clearly incompatible with the synthesis of the major contractile proteins in differentiated, fusion-blocked myocytes.

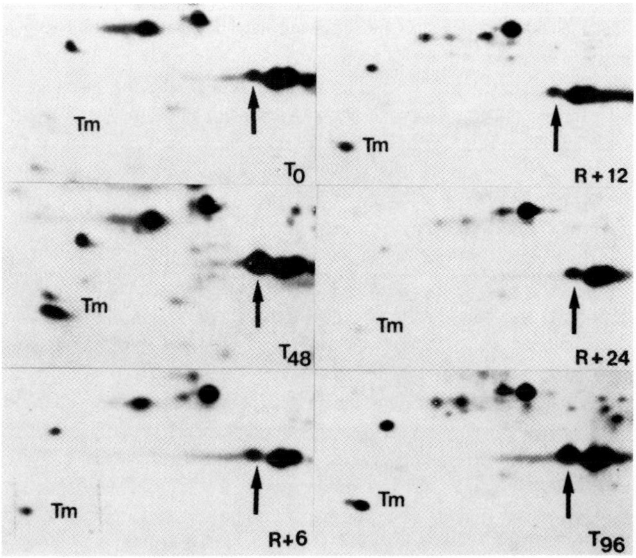

FIGURE 1 Two dimensional gel analysis of the relative synthesis of α actin (↑) and tropomyosin (Tm) subunits in differentiated myocytes using fluorography. Cultures were pulse-labelled with ^{35}S methionine at the start of culturing in 1-10 CM EGTA medium (T_0), just prior to refeeding with 10-15 MEM EGTA (T_{48}), or at 6,12 and 24 hrs after refeeding (R+6, R+12, R+24). While α actin and tropomyosin synthesis are downregulated in cells refed with growth medium, cells maintained in 1-10 CM EGTA for an additional 48 hrs maintain normal levels of synthesis (T_{96}).

Fused Myotubes Do Not Downregulate Contractile Protein Synthesis.

We attribute the small but significant amount of α actin synthesis detected in mitogen-stimulated myoblast cultures to the precocious differentiation of a small number of myoblasts. Significantly, we have observed that the level of α actin synthesis returns to this same baseline level following a mitogenic stimulus, suggesting that either a small subpopulation of muscle cells are unable to downregulate or that the inhibition of contractile protein synthesis per cell is not complete. We have previously observed some precocious fusion in myoblasts cultured in growth medium for 24 hours. Since we know that the level of EGTA used in these experiments to block fusion usually permits 5-10% of the cells to fuse, we suspected that the low levels of α actin synthesis observed in these experiments and the low level of myosin heavy chain synthesis previously observed (5) in stimulated myocytes were most probably restricted to a small population of fused myotubes in these cultures, which were refractory to the mitogenic stimulus. To test this possibility, myoblasts were cultured in low-growth medium for 5 days to ensure optimal fusion and then treated with BudR and light to kill any remaining myoblasts. These cultures of totally fused myotubes were then refed with growth medium, labelled with ^{35}S methionine and extracted at intervals for 2D gel analysis of contractile protein synthesis. The results of this analysis reveal that tropomyosin, actin and myosin light chains are synthesized at normal levels in fused myotubes under conditions which normally inhibit contractile protein synthesis in fusion-blocked, differentiated myocytes (FIG. 2). That myotubes do not respond when refed with high-growth medium demonstrates that growth medium does not inhibit synthesis generally but that downregulation is a precise mechanism restricted to fusion-blocked myocytes. The major difference between myocytes and myotubes is the ability of the former to reenter the cell cycle, since it is well established that myotubes are irreversibly withdrawn from the cell cycle.

If reentry into the cell cycle is necessary for downregulation, the critical transition probably occurs in G_1. We have previously shown, for example, that downregulation occurs before a significant number of stimulated myocytes have reentered S (11). Alternatively, one would anticipate that membrane-associated differences between myocytes and

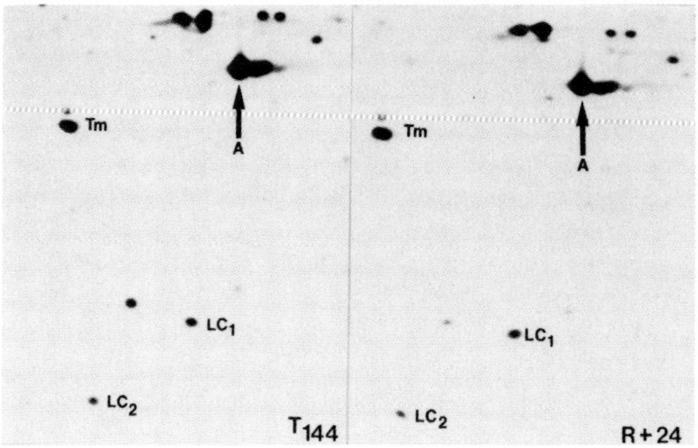

FIGURE 2. Two dimensional gel analysis of contractile protein synthesis in fused myotubes using fluorography. Cultures of totally fused myotubes were pulse-labelled with ^{35}S methionine either before (T_{144}) or 24 hrs after refeeding with high growth medium (R+24). The relative synthesis of α actin (A), tropomyosin (Tm), and myosin light chains 1 and 2 (LC_1, LC_2) in fused cells is unaffected by the addition of growth medium.

myotubes might exist, which might make myotubes refractile to the addition of mitogen. In order to better understand the mechanism by which mitogens regulate contractile protein synthesis and the actual component of growth medium which induces downregulation, we employed a mouse myogenic cell line which is temperature-sensitive for growth and which arrests in G_1 at the non-permissive temperature.

Characterization of a Mouse Myoblast Cell Line Which Is Temperature-Sensitive for Growth.

We have previously described the generation of a permanent myogenic cell line (4B3A) from the hindlimb musculature of postnatal BALB/c mice (11). We subsequently mutagenized and selected this line to obtain 8 myogenic cell lines which are temperature sensitive for growth. One of these lines, ts-36, has been characterized further.

At the permissive temperature (pt) in high growth medium the proliferative and differentiative behavior of ts-36 is indistinguishable from the wild type parental cell line,

4B3A, with the exception that ts-36 has a somewhat longer generation time (17 vs 13.5 hrs). In mass culture, the cells go through a period of rapid proliferation of approximately four days duration followed by a protraction of the average generation time as the cell density increases. At this time fusion of the mononucleated myoblasts is initiated concomitant with the synthesis of the muscle-specific contractile proteins (11).

At the non-permissive temperature (npt), ts-36 cells cease proliferating as assayed by DNA accumulation, cell number and reduced ^{3}H-Thymidine incorporation while wild type, parental cells continue cycling with a doubling time of approximately 15 hours. The inability of ts-36 cells to proliferate at the npt suggested that these cells might lose viability at the elevated temperature. Clonal analysis revealed, however, that ts-36 cells remain viable at the npt and capable of forming clones following a return to the permissive temperature (data not shown). The fact that the viability and cloning efficiency remain high in ts-36 cells maintained at the npt for extended periods of time is an indication that ts-36 is a G_1 mutant, since cells arrested by a ts mutation in any other phase of the cell cycle typically lose viability rapidly following a temperature shift. The kinetics of DNA synthesis after a shift-down to the npt and following return to the permissive temperature are also compatible with a block point several hours before the G_1/S interface (11).

Although ts-36 fuses normally at the permissive temperature, fusion is severly limited at the npt (FIG. 3). This is observed, both when the cells are allowed to condition their medium over the course of the experiment and when the cultures are refed daily with fresh medium. The low percent fusion observed with ts-36 at the npt is believed to be primarily a result of low cell density.

Thus, morphological criteria are poor indicators of differentiation in this line. A more sensitive approach is the analysis of newly synthesized proteins on polyacrylamide gels. 2D gel analysis and autoradiography of labelled cell extracts from cultures held at the npt for 96 hours in fresh medium do not reveal any of the muscle-specific proteins present in differentiated myotubes. Rather, that pattern matches completely the constellation of spots observed with proliferating cultures at the pt, suggesting that arrest of ts-36 myoblasts at the npt in G_1 is not sufficient to permit contractile protein synthesis. However, myocytes cultured for 96 hours at the npt in conditioned medium do

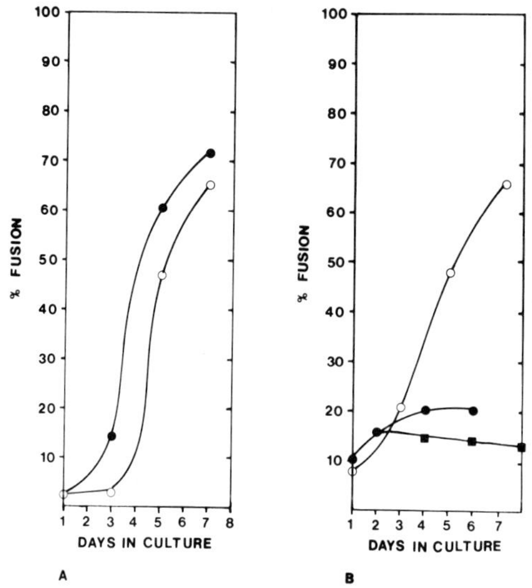

Figure 3. Fusion kinetics of the parental cell line (A) and the temperature sensitive line ts-36 (B) at the pt (37°) and the npt (39.5°C). Percent fusion was determined by scoring the number of mononucleated myoblasts and nuclei within myotubes on plates fixed and stained at intervals after refeeding and temperature shifting (if necessary) at day 1 after plating. ○ = cells cultured at 37°C, ● = cells cultured at 39.5°C, ■ = cells cultured at 39.5°C but refed daily. Fusion kinetics of ts-36 are comparable to that of the parental line at 37°C, but ts-36 fuses poorly at the npt.

differentiate normally as determined by the synthesis of muscle α actin, tropomyosin and the three myosin light chains. These observations are compatible with our thesis that growth factors directly inhibit the synthesis of contractile proteins.

Differentiated Myocytes of ts-36 Downregulate Contractile Protein Synthesis at the npt Following the Addition of Growth Medium or FGF.

In order to determine if murine myocytes would downregulate contractile protein synthesis following the addition of growth medium, cells which had been kept in conditioned medium at the npt for 72 hours and which had initiated the synthesis of muscle-specific contractile proteins were fed complete medium and labelled with ^{35}S methionine 24 hours later. Two dimensional gels reveal the abrupt termination of synthesis of the muscle-specific contractile proteins (FIG. 4). Significantly, this downregulation of contractile protein synthesis following refeeding with fresh medium occurs in the absence of cell proliferation. Complete growth medium thus appears to be incompatible with the maintenance as well as the initiation of the differentiated state.

These experiments indicate that myoblasts withdrawn from the cell cycle in G_1 by a ts mutation fail to differentiate in the presence of mitogen. Taken together, all of these observations suggested a dual role for the "mitogen" employed in these studies. To pursue this suggestion further under more defined conditions, we replaced the embryo extract normally used in our growth media with fibroblast growth factor (FGF). This purified mitogen, at nanogram levels (per ml), stimulates the proliferation of cell lines of murine myoblasts in culture almost to the extent stimulated by embryo extract and horse serum con-

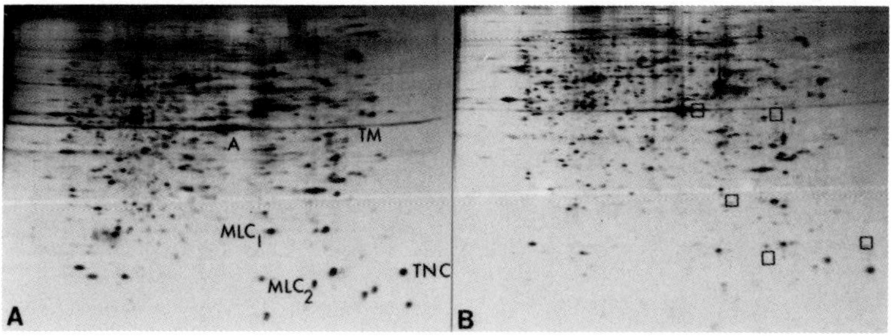

FIGURE 4. Two dimensional gel analysis of protein synthesis in ts-36 cultures allowed to differentiate for 72 hrs in conditioned medium at the npt and either maintained in conditioned medium (A) or refed with growth medium (B) while remaining at the npt. Cultures were pulse-labelled with ^{35}S methionine 24 hrs after refeeding and processed for PAGE and autoradiography. Although cells refed with growth medium at the npt do not reenter S, they do downregulate the synthesis of α actin (A), tropomyosin (TM), myosin light chains (MLC_1, MLC_2), and troponin C (TN C).

taining media (4,12).

The medium shift experiment described above was repeated except that after 72 hours in conditioned medium the cultures were fed conditioned medium supplemented with 100 ng/ml of FGF instead of complete growth medium. The results were the same as before; that is, the program of differentiation was terminated in the presence of the purified mitogen, even under conditions which did not stimulate proliferation. Subsequently, cells were cultured at the npt for 96 hours in conditioned medium supplemented with 100 ng/ml of FGF. The activation of differentiation previously observed in conditioned medium alone did not occur (data not shown). Taken together, these results suggest that both complete growth medium and conditioned medium supplemented with FGF prevent the initiation of the differentitative program. Both media also act in similar fashion to interrupt the program, once initiated. To this extent, FGF, a purified mitogen, mimics the effects of complete medium.

The specificity of the "FGF effect" was tested by maintaining cultures for 96 hours at the npt in conditioned medium supplemented with 100 ng/ml of epidermal growth factor (EGF). Like FGF, EGF is a low molecular weight polypeptide growth factor isolated from murine submaxillary glands. A series of recent studies has confirmed that murine myoblast membranes contain numerous EGF receptors capable of binding and internalizing mitogen (13). However, EGF has no mitogenic effect on murine myoblasts *in vitro*. Culturing growth arrested ts-36 cells with EGF containing medium did not prevent the initiation of the differentiation program. Obviously then, simply interacting with the cell surface and/or internalization of a hormone receptor complex is insufficient to elicit the downregulation of the differentiative program.

Undifferentiated ts-36 Cells Held at the npt Will Synthesize Contractile Proteins Following a Switch to Conditioned Medium Without Reentering S.

Some models attempt to explain the initiation of muscle differentiation by suggesting that the cell cycle immediately preceeding terminal differentiation (quantal cell cycle) is qualitatively different from other proliferative cycles and that myoblasts must pass through a unique S phase before becoming competent to differentiate (1). Other models reject this preprogrammed genetic view of muscle differentiation and propose instead that environmental cues recognized by proliferating cells are responsible for shunting development from proliferation into differentiation (2,3,4,15). The differentiation of ts-36 at the npt provides a unique system with which to test these alternative hypotheses.

Cells were cultured at the npt for 36 hours in F10 3-15, a time sufficient for all the cells to growth arrest. At that time one half of the cultures were refed with conditioned medium and the other half with complete medium. After an additional 72 hours at the npt, the cultures were labelled with ^{35}S methionine. Two-dimensional gels of whole cell extracts reveal that, as expected, cells kept continually in complete medium do not activate the synthesis of the muscle-specific contractile proteins. Those cells cultured in F10 3-15 and then switched to conditioned medium, however, do differentiate and initiate the synthesis of muscle-specific contractile proteins (FIG. 5). To determine whether cells in this latter group of cultures traverse S prior to differentiating, such cultures were pulse label-

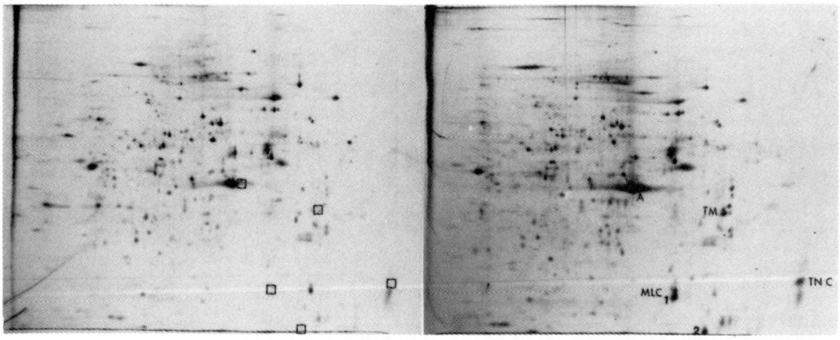

FIGURE 5. Two dimensional gel analysis of protein synthesis in ts-36 cultures before and after a medium switch from fresh to conditioned medium at the npt. Ts-36 cells were cultured in growth medium at the npt for 36 hrs and then refed with either fresh medium (A) or conditioned medium (B) for 72 hrs and pulse-labelled with ^{35}S methionine for 2 hrs. Although both sets of cultures remain blocked at the temperature block point following refeeding, only cultures refed with conditioned medium initiate the synthesis of muscle-specific contractile proteins (see FIG. 4). Muscle-specific gene expression thus occurs in the absence of a quantal mitosis.

led at intervals for 30 minutes with ^{3}H-TdR after the medium switch followed by autoradiography or scintillation counting of the TCA precipitable DNA. Ts-36 cells blocked at the npt in complete medium do not enter and pass through an S phase during the 24 hours subsequent to the medium switch (data not shown). These results demonstrate that myoblasts do not require passage through a special S phase prior to the acquisition of competence to differentiate, strengthening the argument that environmental cues play the major role in affecting the transition from the proliferative to the differentiated state.

DISCUSSION

Since the early 1970's we have proposed that the timing of the differentiative switch in myogenic cells _in vitro_ is regulated by the depletion of heterologous growth factors present in the horse serum and chick embryo extract used to prepare growth medium (1). This thesis is supported by the observation that the timing of differentiation can be manipulated by altering the level of these components, the density of cells or the volume of the medium. Since the depletion of growth factors is generally correlated with a protraction of and the eventual arrest in G_1, we have also postulated the existence of a differentiation-permissive subcompartment within G_1. Our results (5,7) and the results of others (10) are consistent with the idea that arrest in this subcompartment of G_1 is necessary for the initiation and the maintenance of muscle-specific gene expression. The results presented in this paper, however, suggest that while arrest in G_1 is necessary for differentiation this condition is not sufficient, since growth medium or purified growth factors like FGF actively inhibit differentiation even in the absence of cell proliferation. This is best illustrated by our observation that ts-36 cells arrested at a differentiation-permissive, ts block point in G_1 will downregulate contractile protein synthesis at the npt following a mitogenic stimulus. Our experiment in which the medium switch is conducted in inverse order serves to confirm our conclusion. Ts-36 cells blocked at the npt in growth medium will differentiate without reentering S when mitogen is removed. This experiment also serves to point out that, contrary to a mitotic clock model (1), myogenic cells can differentiate in the absence of DNA synthesis. Similar findings using different media shifts have recently been reported by Whalen and coworkers for the L_6 rat myoblast cell line (14). These results are also consistent with Linkhart's earlier observations that purified FGF can prevent differentiation and commitment in mouse cell lines in the absence of cell proliferation (15).

Since our results with ts-36 indicate that the downregulation of contractile protein synthesis in differentiated myocytes can occur without reentry into S, the failure of terminally differentiated quail myotubes to respond to a growth stimulus is somewhat puzzling. One explanation might be that the fusion process alters the muscle membrane so that myotubes can no longer respond to mitogen. In fact, Lim and Hauschka have demonstrated that the loss of at least one class of mitogen receptors - receptors for EGF - is temporally correlated with terminal differentiation (13). While EGF is not normally mitogenic for myogenic cells, these findings suggest that FGF receptors may be downregulated in a similar way in terminally differentiated myotubes.

While it is now clear that growth factors like FGF can inhibit contractile protein synthesis, little is known about the mechanism by which this is accomplished. It is still not known whether the FGF/receptor complex acts directly on muscle genes to regulate their transcription or whether the binding of FGF to the cell membrane simply sets off a cascade of cellular events which create a cellular environment incompatible with contractile protein synthesis. Elucidation of this mechanism should provide some useful insights into how muscle genes are regulated.

ACKNOWLEDGEMENTS

We are extremely grateful to Melanie Knutson for her dedicated help in completing many of the experiments described in this paper.

REFERENCES

1. Holtzer H (1978). Cell lineages, stem cells and the "quantal" cell cycle concept. In Lord, Potten, Cole (eds): "Stem Cells and Tissue Homeostasis," Cambridge University Press, p 1.
2. Buckley PA, Konigsberg IR (1974). Myogenic fusion and

the duration of the post-mitotic gap. Devel Biol 37:193.
3. Konigsberg IR (1971). Diffusion-mediated control of myoblast fusion. Devel Biol 26:133.
4. Linkhart TA, Clegg CH, Hauschka SD (1981). Myogenic differentiation in permanent clonal mouse myoblast cell lines: Regulation by macromolecular growth factors in the culture medium. Devel Biol 86:19.
5. Devlin BH, Merrifield PA, Konigsberg IR (1982). In Pearson, Epstein (eds): "Muscle Development: Molecular and Cellular Control," Cold Spring Harbor Laboratory, New York, p 355.
6. Sutherland WM, Konigsberg IR (1983). CPK accumulation in fusion-blocked quail myocytes. Devel Biol 99:287.
7. Devlin BH, Konigsberg IR (1983). Reentry into the cell cycle of differentiated skeletal myocytes. Devel Biol 94:175.
8. Devlin RB, Emerson Jr CP (1979). Coordinate regulation of contractile protein synthesis during myoblast differentiation. Cell 13:599.
9. Bowman LH, Emerson Jr CP (1980). Formation and stability of cytoplasmic mRNAs during myoblast differentiation: Pulse-chase and density labeling analyses. Devel Biol 80:146.
10. Nguyen HT, Medford RM, Nadal-Ginard B (1983). Reversibility of muscle differentiation in the absence of commitment: Analysis of a myogenic cell line temperature sensitive for commitment. Cell 34:281.
11. Merrifield PA, Compton RS, Konigsberg IR (1983). In Eppenberger HM, Perriard J-C (eds): "Molecular Aspects of Myogenesis and Myofibrillogenesis in Cell Culture: Normal and Diseased Muscle," Basel, New York: S. Karger, p 1.
12. Gospodarowicz D, Mescher AL (1977). A comparison of the response of cultured myoblasts and chondrocytes to fibroblast and epidermal growth factor. J Cell Physiol 93:117.
13. Lim RW, Hauschka SD (1984). A rapid decrease in EGF binding capacity accompanies the terminal differentiation of mouse myoblasts _in vitro_. J Cell Biol 98:739.
14. Pinset C, Whalen RG (1984). Manipulation of medium conditions and differentiation in the rat myogenic cell line L6. Devel Biol 102:269.
15. Linkhart TA, Clegg CH, Hauschka SD (1980). Control of mouse myoblast commitment to terminal differentiation by mitogens. J Supramol Struct 14:483.

Molecular Biology of Muscle Development, pages 77–84

MUSCLE HISTOGENESIS AND MUSCLE DIVERSITY

A.M. Kelly and N.A. Rubinstein
School of Veterinary Medicine and
School of Medicine
University of Pennsylvania
Philadelphia, Pennsylvania 19104

INTRODUCTION

That skeletal muscle is progressively built up from successive orders of cells has been recognized for over a century, but the morphogenic mechanisms by which this assembly is achieved have been the subject of controversy for many years and, as a result, the contributions of a number of early histologists have been overlooked. Bardeen (1) and Meves (2), in particular provided remarkably perceptive accounts of muscle histogenesis, proposing that fiber multi-nucleation came about by myoblasts "melting together" (Meves 1909) and that new generations of little fibers were formed as the result of proliferation of undifferentiated cells lying on the walls of primary myotubes (Fig. 1,2,3). Their model contradicted the more prevalent view of those times that muscle nuclei divided by amitotic division and new fibers were formed as yeast-like buds from the walls of primary myotubes which then separated by cell fission. The accuracy of these observations have been substantiated by more recent fine structural studies (3, 4).

In the present chapter, we wish to discuss some of the mechanisms which appear to be involved in this programmed pattern of muscle histogenesis and to consider some of its consequences in terms of fiber specialization. In particular, it is relevant to our discussion that the primary generations of myotubes serve as a scaffold upon which later generations develop.

RESULTS AND DISCUSSION

During early limb myogenesis in the rat the two premyo-

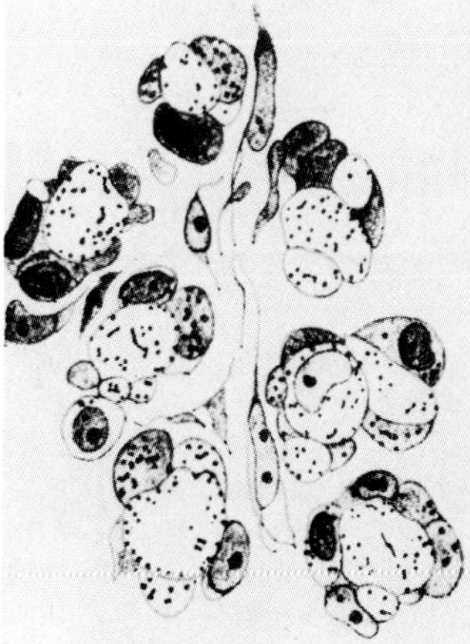

Fig. 1.

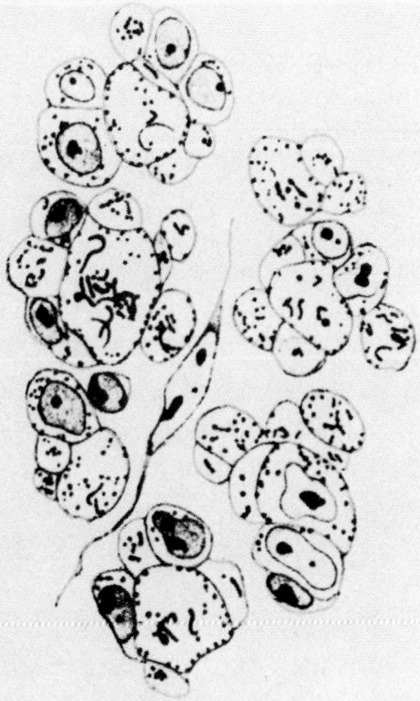

Fig. 2.

Cross section through leg muscle fibers of 10-day old chick embryo. The cells which ensheath muscle fibers are myoblasts. They arrange themselves in rows on muscle fibers and develop into young fibers. The young fibers at first lie in tight contact with the primary developed fibers and are united with them into a bundle (figure 1). Later, increasing in thickness, they become more and more independent (figure 2, 13-day old embryo). In this way the greater part of myoblasts are used up in a relatively short time. But one still sees some in figure 3 (15-day embryo) lying on single muscle fibers. These myoblasts can also change to myofibers.
Translated from text by Meves, 1909 (ref. 2).

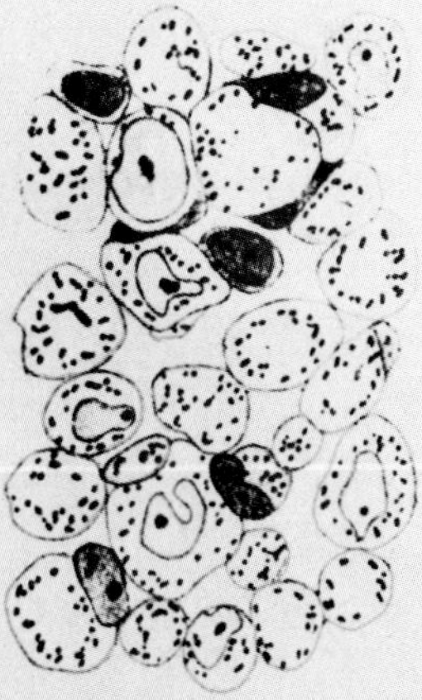

Fig. 3.

genic masses become progressively partitioned into specific muscle primordia at approximately 13 days gestation. This process involves selective association of groups of primary muscle cells and possibly includes several types of membrane interaction. One involves gap junctions (5), for myotubes within early primordia are known to be electrically coupled and to function as a syncitium (6). Secondly, myoblasts and myotubes are known to express cell adhesion molecules (CAM's) (7), and it seems likely that modulation of these proteins at specific cell surfaces contributes in a significant way to muscle tissue modelling. Finally, the surrounding connective tissues are thought to play an important role in the process of segregation (8), and these connective tissues almost certainly also provide the initial attachment sites upon which primary myogenic cells can elongate and differentiate. These early myogenic processes occur independent of the nerve which does not penetrate muscle primordia until later in development (9).

The contribution primary muscle cells make to the total fiber population varies among muscles and appears related to the pattern of phenotypic expression in adult muscle. In the rat diaphragm, which is predominantly a fast twitch muscle, primary cells constitute not more than 20 percent of the fibers (9). A similar figure applies in the fast twitch rat EDL (10). By contrast primary cells contribute approximately 50 percent of fibers in the slow twitch soleus and emerge later in fetal life than those in the EDL (10).

The affinity between primary myotubes is not persistent and in rat respiratory muscles they separate from one another between 16 & 17 days gestation and become independent units of contraction (3). It has been postulated that this separation coincides with a decline in capacity of primary cell membranes to cohere (3), possibly due to modulation of some cell aggregation factor such as CAM. Coincidently, new secondary generations of mononuclear cells intrude between neighboring primary cells. They then use the walls of primary myotubes as a scaffold upon which to elongate, fuse and differentiate into secondary myotubes. We presently know little of the membrane events which facilitate these side to side membrane interactions. These events are, however, clearly important in achieving the appropriate parallel arrangement of myotubes as muscle bundles are progressively assembled and provide sufficiently strong guidance to inhibit branching of myotubes as they elongate. It would be interesting to know whether fusion between primary and secondary generation cells is precluded in order to promote proliferation of new myotubes at this stage.

As a result of this pattern of development, rat intercostal muscles and diaphragm at 18 days gestation consist of groups of cells containing a well differentiated primary myotube surrounded by differentiated secondary myotubes and morphologically undifferentiated cells (3,9). Each cell cluster is ensheathed by a basal lamina the synthesis of which may play a role in altering expression of cell adhesion molecules. Subsequently, secondary generation myotubes separate from the walls of the primary myotubes to become independent units of contraction, each surrounded by a basal lamina. The most rapid rate of addition of secondary generation cells to the developing diaphragm is between 17 and 18 days gestation (9), and by birth, 70 percent of the adult number of fibers has been reached. A few fibers continue to be formed during the first two weeks post partum and the process then ceases despite continued proliferation of myoblasts on the walls of myofibers. Modulation of CAM's from an embryonic to adult form (7) offers an attractive way to account for this change and we suggest that, as the animal matures, the adhesive membrane qualities of fibers decline until they are no longer able to guide secondary cell differentiation. This change prevents formation of additional fibers. Remaining myoblasts then only have the option of fusing with a defined population of primary and secondary cells, contributing to their multinuclearity and progressive hypertrophy. The late phase of primary cell differentiation, comparatively small number of secondary fibers but large population of satellite cells in the rat soleus (13) is consistent with this interpretation. Whether satellite cells are predetermined and fuse selectively with one or other of the fiber lineages has yet to be determined.

Primary muscle cells provide not only the environment for secondary cell propagation, they also act as the target for initial neuromuscular synaptogenesis (11). The first synapses form at approximately the time primary myotubes separate from one another and this event represents the initial step in motor unit organization of muscle. Prior to this, primary cell development is not dependent upon neuromuscular contact (4,9), but the possibility of long range neurotrophic influences cannot be ruled out. Neural cell adhesion molecules (N.CAMS) may play a role in localizing early synapse formation (12), and possibly their transient expression is involved in the waxing and waning of polyneural innervation. The significance of these early synapses to muscle histogenesis has been elegantly demonstrated by Harris (9). Whereas primary generation cell development is autonomous, initiation of secondary myotube development is crucially dependent upon innervation and upon muscle activity. The nerve thus assumes control of myogenesis and through innervation of primary cells provides the permissive environment for secondary cell development. The nerve may also act as a mitogen, greatly increasing the number of secondary cells forming a muscle. Survival of secondary fibers depends upon availability of axon sprouts with which to synapse and their formation on the walls of innervated primary cells facilitates this (11).

Studies of skeletal muscle histogenesis thus demonstrate that new muscle cells continue to be formed over an extended period of fetal and neonatal life. During this time the environment in which new myofibers differentiate changes both with respect to motor activity and the hormonal milieu. That primary and secondary generation cells are intermingled in a mosaic comparable to the checkerboard distribution of specialized fibers in the adult suggests the successive orders of cells follow separate paths of specialization (12,14,15). We (10) have demonstrated immunochemically that many primary generation cells contain a form of slow muscle myosin as well as isozymes originally termed fast isozymes but now recognized as either embryonic or neonatal myosin. Whether the early slow myosin corresponds to adult slow twitch myosin or is a different isozyme, possibly a slow tonic myosin, is presently unclear. In the rat hind limb, fibers containing the early 'slow' isozyme have a distribution very comparable to that of Type I cells in the adult and it is proposed they are directly related one to another (16). The precocious differentiation of primary slow fibers implies a developmental strategy for maintaining isometric tension and supporting fetal posture. At least initially, slow myosin is not expressed by secondary generation cells; they contain embryonic and neonatal myosin and differentiate towards a fast phenotype (10).

Present immunochemical studies have been unable to demonstrate the presence of a 'slow' isozyme in the rat before 17 days gestation (10). Since this corresponds to the time synaptic contact is initiated on primary cells, the interpretation is that systhesis of this isozyme is neurally dependent. Studies of neonatally denervated rat muscles reveal a gradual elimination of slow myosin isozymes in the absence of the nerve and support this interpretation (7). The question, however, deserves further investigation for two reasons. First, many studies correlate the synthesis of slow muscle myosin with the imposition of tonic neural patterns of impulse activity and in the rat hind limb this does not emerge until after 12 days post partum when postural movements commence (18). Prior to this time all muscles are activated phasically. Secondly, in the chicken it has been shown that synthesis of a slow form of myosin is an intrinsic property of some primary generation cells and proceeds independent of the nerve (19).

As a result of the temporal sequence of muscle construction a fundamental pattern of specialization is established, primary cells predominantly, though not exclusively, develop as slow fibers and secondary cells as fast fibers. Positional cues also influence the pattern of phenotypic expression (16). As the animal matures this prototype model is modulated both by neural and by hormonal action. In the neonatal rat soleus, for example, there are equal proportions of primary generation slow and secondary generation fast fibers. In the maturing animal, changes occur in the properties of motoneurones innervating secondary cells and many are transformed from fast into slow fibers so that in the adult soleus 80 percent express the slow, Type I phenotype (22). The shift is regulated by the firing pattern of the motoneuron (22) and is related to circulating levels of thyroid hormone since thyroxine modulates expression of fast and slow myosin in the soleus (20,21). The soleus of the cat and guinea pig appear to follow the same developmental strategy (23) as both species have approximately equal proportions of large slow fibers and smaller, secondary generation fast fibers at birth. In contrast to the rat however, virtually all fast fibers in the cat and guinea pig soleus are transformed and in the adult only Type I fibers are present. The soleus of the adult mouse contains equal proportions of fast and slow fibers. This suggests to us that unlike the other species transformation of secondary cells in the mouse soleus does not proceed after birth. The high BMR of the mouse may contribute to this stability. Plasticity of the soleus thus appears to be mainly imparted by secondary generation cells.

In other muscles, particularly fast muscles, the primary slow fibers are modulated during development. In the extensor digitorum longus of some strains of mice (24), and in the guinea pig temporalis and masseter muscles (25), for example, there are small populations of primary slow fibers in the fetus and neonate but few, if any, in the adult (25). Apparently they are transformed into Type II, and most probably IIa fibers, as the animal matures. Since slow, Type I fibers emerge in some fast muscles as a functional adaptation to chronic overload (26) we are presently investigating the possibility that this represents a recepitulation of earlier programs of myosin synthesis. The primary and secondary motor units which are modulated during development may preferentially reexpress their initial program of specialization when the demands on the muscle change.

ACKNOWLEDGEMENTS

These studies were supported by USPHS grant #HL 15835 to the Pennslyvania Muscle Institute and by grants from the Muscular Dystrophy Association of America, Inc. We are grateful to the Historical Department, The College of Physicians Library, Philadelphia for permission to reproduce figures 1,2 and 3.

REFERENCES

1. Bardeen CR (1900). Development of musculature of the body wall in the pig. Johns Hopkins Hosp Rep 9:367.
2. Meves F (1909). Ueber Neubilding quergestrifter Muskelfasern nach Beobachtungen am Huhnerembryo. Anat Anz 34:161.
3. Kelly AM, Zacks SI (1969). The histogenesis of rat intercostal muscle. J Cell Biol 42:135.
4. Ontell M, Kozeka K (1984). The organogenesis of the murine striated muscle. A cytoarchitectural study. Am J Anat 171:133.
5. Rash JE, Staehlin LA (1974). Freeze-cleave demonstration of gap junctions between skeletal myogenic cells in vivo. Dev Biol. 36:455.
6. Dennis MJ, Ziskind-Conhaim L, Harris AJ (1980). Development of neuromuscular junctions in rat embryos. Dev Biol 81:266.
7. Edelman AM (1983). Cell adhesion molecules. Science 219:450.
8. Thiery JP (1984). Mechanisms of cell migration in the vertebrate embryo. Cell Different 15:1.
9. Harris AJ (1981). Embryonic growth and innervation of rat skeletal muscles. Proc Roy Soc B 293:20.
10. Rubinstein NA, Kelly AM (1981). Development of muscle fiber specialization in the rat hind limb. J Cell Biol 90:128.
11. Kelly AM, Zacks S (1969). The fine structure of motor end plate morphogenesis. J Cell Biol 42:154.
12. Grumet M, Rutishauser U, Edelman GM (1982). Neural cell adhesion molecule in an embryonic muscle cell mediates adhesion to nerve cells in vitro. Nature 295:693.

13. Kelly AM (1978). Satellite cells and myofiber growth in the rat soleus and extensor digitorum longus muscles. Dev Biol 65:1.
14. Kelly AM, Schotland D (1972). The evolution of the 'checkerboard' in a rat muscle. In Banker, BQ, Przybylski, RJ, van der Meulen, J, Victor, M (eds): "Research in Muscle Development and the Muscle Spindle," Excerpta Medica ICS No. 240, p. 32.
15. Ashmore CR, Robinson DW, Rattray P, Doerr L (1972). Biphasic development of muscle fibers in the fetal lamb. Dev Neurol 37:241.
16. Lyons G, Kelly AM, Rubinstein NA (1983). Myosin transitions in developing fast and slow muscles of the rat hind limb. Differentiation 25:168.
17. Rubinstein NA, Kelly AM (1978). Myogenic and neurogenic contributions to the development of fast and slow twitch muscle in rat. Dev Biol 62:473.
18. Navarette R, Vrbova G (1983). Changes of activity patterns in slow and fast muscles during posnatal development. Devel Brain Res 8:11.
19. Miller JB, Stockdale FE (1985). Diversity in myosin heavy chain expression in cultured myotubes. J Cellular Biochem Suppl 9B:40.
20. Ianuzzo D, Patel D, Chen V., O'Brien P, Williams C (1977). Thyroidal trophic influence on skeletal muscle myosin. Nature 270:74.
21. Winder W, Fitts R, Holloszy J, Kaiser K, Brooke M (1980). Effect of thyroid hormones on different types of skeletal muscle. In Pette, D (ed): "Plasticity of Muscle," Berlin: de Gruyter, p. 581.
22. Kugelberg E (1976). Adaptive transformation in rat soleus motor units during growth. J Neurol Sci 27:269.
23. Karpati G, Engel WK (1967). Neuronal trophic function. Arch Neurol 17:542.
24. Whalen RG, Johnstone D, Bryers PS, Butler-Browne GS, Ecob MS, Jaros E (1984). A developmentally regulated dissapearance of slow myosin in fast-type muscles of the mouse. FEBS Lett 177:510.
25. Kelly AM, Rubinstein NA. Unpublished observations.
26. Noble EG, Dabrowski BL, Ianuzzo CD (1983). Myosin transformation in hypertrophied rat muscle. Pflugers Arch 396:260.

Molecular Biology of Muscle Development, pages 85–103

SOMATIC CELL GENETIC ANALYSIS OF MYOGENESIS[1]

Woodring E. Wright

Department of Cell Biology, The University of Texas
Southwestern Medical School, Dallas, Texas 75235

ABSTRACT Cell hybridization studies involving differentiation-defective myoblasts led to the formulation of a molecular model for regulating the probability of the decision to differentiate. Heterokaryon experiments examining the inhibition of myogenesis by 5-bromodeoxyuridine (BUdR) suggested a way to exploit that model in order to isolate myoblasts overexpressing the responsible factors. The isolation of such BUdR resistant myoblasts has been accomplished, reinforcing our theory for the specificity of action of BUdR. The existence of such a cell line should permit the molecular cloning of the factors regulating myogenic differentiation.

INTRODUCTION

Although the factors regulating the transcription of structural genes are being intensively studied, little is known about the regulation of "upstream" events involved in initiating whole programs of differentiation such as myogenesis. We hoped that examining the interactions that occurred when cells of various differentiated states were fused together would provide sufficient information to permit a direct molecular approach to this problem. I would like to describe here a series of such studies and the model system that evolved from them. This model, in which myoblasts have become resistant to the differentiation-inhibiting effects of 5-bromodeoxyuridine (BUdR), presumably by amplifying the expression of the factors regulating the

1 This work was supported by grants from the National Institutes of Health and the Muscular Dystrophy Association.

decision to differentiate, should permit the cloning and isolation of the molecules involved in the decision to shift from the program of an undifferentiated myoblast to that of a differentiated myotube.

METHODS

L6 rat myoblasts (1) and primary chick myoblasts were cultured as described (2). Cell hybrids and heterokaryons were isolated using the complementarity between the lesions introduced by two different irreversible biochemical inhibitors (3). Although the experiments described here were performed using this selective system, we have now developed a much more efficient selection system that will be used in future experiments (4). Antimyosin immunoperoxidase staining (5) and the buoyant density analysis of DNA (6) were performed as described elsewhere.

RESULTS

Variable Expression of Regulatory Molecules

As is the case for most differentiating systems, when L6 rat myoblasts are kept in culture for a few months they begin to lose the ability to differentiate. It is then a simple matter to clone the cells and pick differentiation-defective variants that fail to form myotubes. In order to assess the kind of regulatory changes occurring in differentiation-defective cells, we formed cell hybrids between them and competent myoblasts. Individual hybrid clones were then plated at low density and grown for two weeks to allow large subclones to form, then stimulated to differentiate by feeding with 1% fetal calf serum containing 5μg/ml insulin. One week later the cells were fixed and immunoperoxidase stained with a monoclonal antimyosin heavy chain antibody.

Fig. 1 shows the frequency of myosin (+) cells for the different types of hybrids formed. Homohybrids of competent parental L6 rat myoblasts fused to themselves differentiated well, and about 65% of the cells became myosin (+). Homohybrids formed by fusing differentiation-defective cells to themselves only formed myosin (+) cells at very low frequencies. This approximately 0.05% frequency of differentiation is sufficiently low to explain why multinucleated myotube

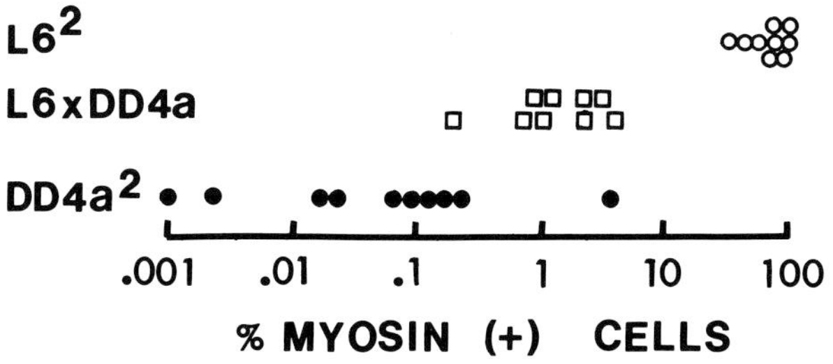

Figure 1. The frequency of differentiation of cell hybrids between differentiation competent (L6) and defective (DD4a) myoblasts (7).

formation was not observed in the differentiation-defective cells: the probability of three or four adjacent cells all making the decision to differentiate and being able to fuse together to form a myotube would be extremely low. The heterohybrids between competent and defective cells exhibited an intermediate probability of differentiation of about 2%.

Theoretical models for the way in which myoblasts regulate the decision to differentiate have to encompass two important conclusions drawn from the data in Fig. 1. Firstly, the mechanism must be able to regulate probabilities of differentiation covering at least four orders of magnitude and, secondly, it must explain how the interaction of the competent and defective phenotypes in a cell hybrid could generate a geometric rather than an arithmetic average of the two parental cells. One model that can explain these observations is illustrated in Fig. 2. Let us suppose that a threshold concentration of some regulatory factor has to be produced before an all-or-none decision to differentiate is made. If the capacity to synthesize this molecule in response to an environmental signal to differentiate follows a normal distribution and if the average concentration per cell is 0.36 standard deviations above threshold, then 65% of the population will exceed threshold and differentiate. Alternatively, if the mean concentration per cell is 3.3 standard deviations below threshold, then only 0.05% of the population would exceed threshold and differentiate. If cells from these two populations were fused and each genome in the hybrid continued to express its own setpoint (i.e., the production of the factor were regulated in cis), then

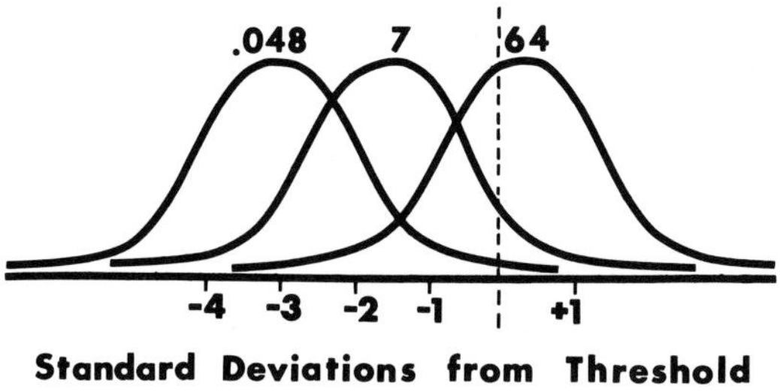

Figure 2. A molecular model for regulating the probability of differentiation (7). The number above each curve represents the proportion of the curve that exceeds the threshold (vertical dashed line).

the concentration of the factor would be the arithmetic average of that in the two parental lines. In this example, the average of 0.36 and -3.3 is -1.47. The middle curve in Fig. 2 shows that 7% of the cells in such a population would exceed threshold and differentiate. This value is very close to the behavior of the heterohybrid cells in Fig. 1, given the large number of assumptions made in the construction of this model.

The fact that it is very easy to isolate differentiation-defective myoblasts suggests that myoblasts producing decreased quantities of this factor are generated with a high frequency. Unless there is a directionality to the process producing these changes, one would predict that myoblasts overproducing such factors should also be generated with a high frequency. However, any myoblast significantly overproducing these factors would make the decision to differentiate, would become postmitotic, and would thus be lost from the population. A strategy for overcoming this problem was suggested by the following experiments.

Regulatory Factors and BUdR

As is the case for most differentiating systems, growing muscle cells in the presence of 5-bromodeoxyuridine (BUdR) inhibits differentiation (8,9). Although the mechanism for blocking differentiated but not "household" functions is unknown, there is substantial evidence that BUdR acts as a result of its substitution for thymidine in cellular DNA (10). In order to examine whether BUdR blocked the production of regulatory factors or the ability to respond to regulatory factors, we constructed heterokaryons between mononucleated differentiated chick skeletal myocytes and BUdR-blocked L6 rat myoblasts. We (3,5,11-13) and others (see Blau, this volume, also 14,15) have shown that differentiated muscle cells contain factors that can induce undifferentiated myoblasts or even non-muscle cells to synthesize muscle specific proteins. If BUdR blocked the production of regulatory factors but not the ability to respond to them and if the chick myocytes contained factors that activated differentiated structural protein genes directly and not just via some upstream events, then the factors provided by the differentiated chick myocytes should be able to induce myosin light chain one synthesis by the BUdR-blocked rat nuclei in the heterokaryons. Fig. 3 shows that this was not the case. Although both embryonic (e) and adult (a) rat myosin light chain one synthesis was induced in control heterokaryons formed by fusing chick myocytes to normal L6 myoblasts (Fig. 3A), only chick LC1 synthesis (c) was found when the chick myocytes were fused to BUdR-blocked L6 rat myoblasts (Fig. 3B) (16).

A variety of data suggested that the chick factors were present in limiting quantities. If BUdR acted by altering the binding of regulatory factors to BUdR-substituted DNA, then increasing the amount of chick factors might be able to overcome this inhibitory effect. Binucleated chick myocytes were produced by growing primary myoblast cultures in cytochalasin B (which inhibits cytokinesis) (17) for the

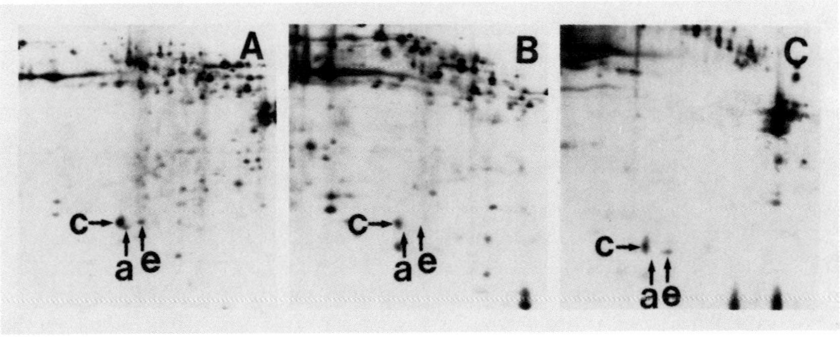

Figure 3. Synthesis of skeletal myosin light chain one in heterokayons (16). Mononucleated differentiated chick myocytes were fused to undifferentiated L6 rat myoblasts (A) or BUdR-blocked rat myoblasts (B). Binucleated chick myocytes were fused to BUdR-blocked rat myoblasts in (C).

last cell division before they differentiated. These binucleated chick myocytes were then fused to BUdR-blocked L6 myoblasts. Fig. 3C shows that these heterokaryons were able to synthesize rat myosin light chains.

Isolation of BUdR-resistant Myoblasts

The demonstration that increased amounts of regulatory factors could overcome the inhibition of differentiation produced by BUdR provided a strategy for isolating the myoblasts that were overproducing the factors regulating the decision to differentiate. If L6 cells were cloned in the presence of a minimally inhibitory concentration of BUdR, the first clones to differentiate would be those that were producing the highest levels of the regulatory factors. If those clones were picked and subjected to repeated cycles of selection in progressively higher concentrations of BUdR, then the levels of the regulatory molecules might be progressively amplified. The inhibitory effects of BUdR would thus prevent the overproducing cells from differentiating under growth conditions, while still permitting the cells to differentiate once the conditions of local confluency during clonal growth provided a stimulus to increase further the synthesis of the regulatory factors.

Approximately 10,000 cells were plated in a 150cm^2 petri dish and allowed to form colonies. The dish was scanned using a 4X objective on the day when differentiation was expected to begin, and the first 10-20 colonies to exhibit myotube formation were circled on the bottom of the dish using a fine tipped pen. These colonies were then trypsinized within glass cloning cylinders, pooled, and immediately replated at clonal density for the next cycle of selection. There were two major reasons for pooling groups of colonies rather than picking a single colony at each cycle. The most important one was that a variety of "random" events (the presence of a nearby colony, a colony that started from a clump of cells rather than a single cell, etc.) might cause a particular colony to differentiate earlier than others. Pooling groups of colonies minimized the chances of the selection procedure drifting off course as a result of picking a clone for the wrong reasons. A secondary reason was that by pooling many colonies sufficient cells were obtained so that the subsequent cycle could be initiated immediately without having to expand the number of cells prior to recloning.

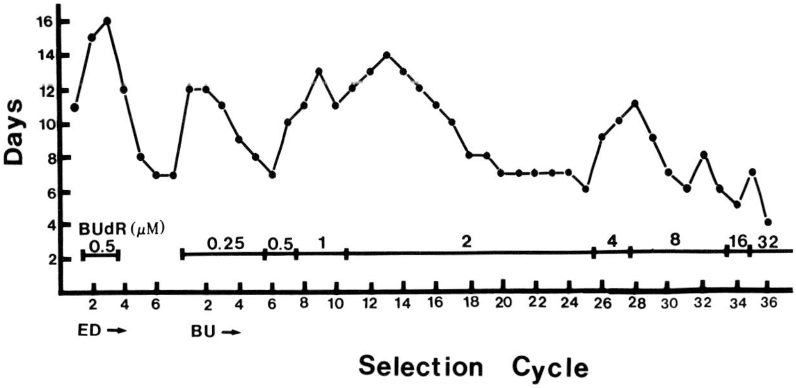

Figure 4. Selection history of the BUdR resistant myoblasts (18).

Fig. 4 summarizes the selection history of the amplified myoblasts. L6 myoblasts were first subjected to a single cycle of cloning in the absence of BUdR in order to obtain a well differentiating pool of cells for the selection protocol. These cells were then replated in 0.5μM BUdR, the 50% inhibitory concentration for L6 cells under mass culture conditions (6). This concentration of BUdR proved to be too high under clonal conditions. Although two cycles of selection were performed using 0.5μM BUdR, the "myotubes" that formed were so small that we were not confident that they actually represented differentiated structures rather than simply trinucleated cells. The cells were then subjected to four additional cycles of selection in the absence of BUdR, during which time cells with an increased capacity to differentiate under clonal conditions were isolated. These "Early Differentiating cycle 7" (ED7) cells then formed the starting material for reinitiating selection in the presence of BUdR. Myoblasts were maintained in 0.25μM BUdR during six cycles of selection until the time required for myotube formation dropped to 7 days. The concentration of BUdR was then doubled and the process repeated several times, first selecting for cells that could differentiate in higher concentrations of BUdR and then selecting for cells that differentiated at shorter times (smaller numbers of cells per colony). Individual colonies were picked and <u>not</u> pooled during the final cycle (cycle 36), so that during subsequent work a true clonal line of resistant cells would be available.

Cells can become resistant to the effects of BUdR for a variety of reasons, the most common being transport or incorporation mutations. The buoyant density of BU36b cells (BUdR; 36 cycles of selection, clone b) was determined to verify that the cells were incorporating BUdR into their DNA and were thus not resistant due to either of the above possibilities. The density shift of BU36b DNA was 0.0597g/cm^3, which corresponds to 65% of the thymidine residues being replaced by BUdR. This is about 70% of the level incorporated by control L6 myoblasts also grown in 32μM BUdR. Since the parental line was essentially totally inhibited from differentiating (under mass culture conditions) at a 20% level of BUdR for thymidine substitution (6), the slight reduction in incorporation does not explain the BUdR resistant phenotype of these cells.

If our model is correct and the cells have become resistant to BUdR because they have amplified the expression of the factor(s) regulating the decision to differentiate, then our model also predicts that the presence of BUdR might be necessary to prevent the cells from differentiating under growth conditions. Fig. 5 shows the local density (average colony size) at which cells differentiated when cloned in the presence versus the absence of BUdR. Parental L6

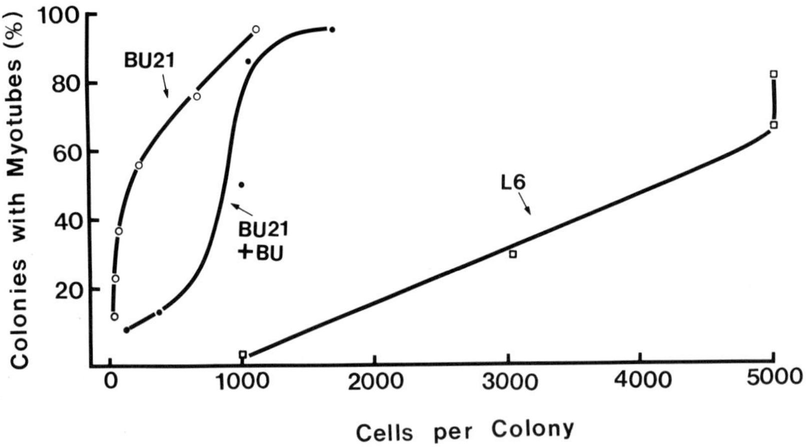

Figure 5. Differentiation of parental L6 and BUdR-resistant cycle 21 cells in the presence and absence of 2μM BUdR (18).

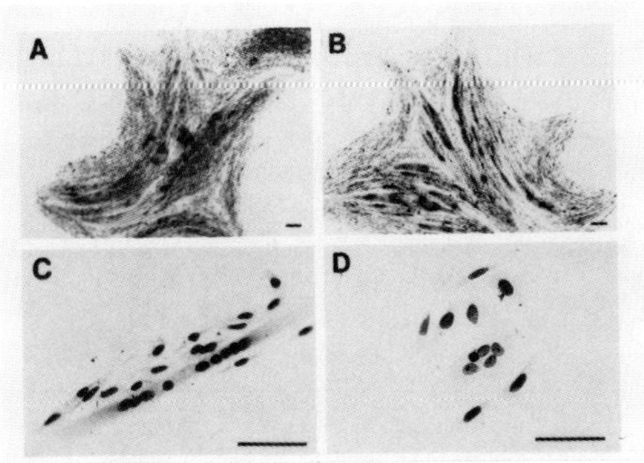

Figure 6. L6 (A,B) and BU cycle 21 (C,D) colonies fixed on the first day that myotube formation was observed (18). Bar=100μ.

myoblasts required approximately 3,000 cells per colony before they formed myotubes, and were totally prevented from differentiating in 2μM BUdR (7). BU21 colonies grown in the presence of 2μM BUdR contain myotubes at an average colony size of about 1,000 cells. However, if BU21 cells were cloned in the absence of BUdR, then 50% of the colonies contained myotubes when there were only about 100 cells per colony. Fig. 6 compares the appearance of colonies fixed and stained on the first day that myotubes appeared for the parental L6 myoblast line and BU21 cells grown in the absence of BUdR.

The model for the regulation of the decision to differentiate presented in Fig. 2 suggested that if the BUdR resistant cells are overproducing the molecules involved, then cell hybrids between them and the differentiation defective myoblast variants should exhibit much higher frequencies of differentiation than the original cross. Table 1 shows the results of that experiment. BU21 cells (resistant to 2μM BUdR) were fused to differentiation defective myoblasts, cloned, plated at low density in the absence of BUdR and allowed to grow for two weeks, stimulated to differentiate for one week, and then stained with an antimyosin antibody. In contrast to the 8% frequency of differentiation exhibited by the control set of hybrids (parental L6 myoblasts fused to defective cells), the hybrids between BUdR-resistant myoblasts and defective cells

TABLE 1
DIFFERENTIATION CAPACITY OF HYBRID CLONES

Cell Combination	% Differentiation[a]	Geometric Average
BU21 x DD4a	99 54 46 87 75 92 82 87 68 82 64 33 64 79 84 72 62 81	71
L6 x DD4a	8 5 8 4 4 27 16 6 12 10	8
ED7 x DD4a	2 13 7 13 38 3 17 28	11

[a]Each number represents the % myosin (+) cells observed in a different hybrid clone.

massively differentiated. In fact, most of these hybrids had differentiated after two weeks of clonal growth before being stimulated to differentiate with the additional week in 1% serum plus insulin. Their full ability to differentiate was thus not completely assessed. It is nonetheless abundantly clear that the hybrids were behaving as predicted by the model, as if the BU21 cells were resistant to BUdR because they were overproducing the factors that regulate the decision to differentiate.

One complication in the interpretation of this experiment results from a shift in karyotype that had occurred during the isolation of the BUdR resistant cells. Whereas the parental L6 cells had approximately 41 chromosomes, the BU21 cells had approximately 65 (no double minute chromosomes were observed and banding studies to identify homogeneous staining regions have not been done). Since gene dosage effects are known to have dramatic influences on the phenotype of cell hybrids (19), it is important to eliminate simple gene dosage effects as an explanation for the data in Table 1. This was partially accomplished by constructing a set of hybrids using ED7 cells (see Fig. 2). A major shift in karyotype had already occurred by the time ED7 cells were isolated, since they contained about 75 chromosomes. We have not determined whether this increase in chromosome number was simply a random shift or whether it was induced by the exposure to BUdR during cycles 2 and 3. However, since ED7 myoblasts were not resistant to BUdR and have even greater numbers of chromosomes than BU21 cells, they provided a useful cell line for examining whether the above results could be due to gross gene dosage effects. Table 1 shows that hybrids between ED7 cells and differentiation defective cells exhibited a frequency of differentiation (11%) that was no different from that of the L6xDD combination. The increased ability to differentiate in the BU21xDD hybrids thus reflects more than simple gene dosage effects.

DISCUSSION

Unambiguous proof of any molecular model for the regulation of cell differentiation requires that the molecules be isolated and shown to exhibit the predicted characteristics. Although the somatic cell genetic investigations described above do not meet this rigorous criterion, they nonetheless provide a number of important insights and working hypotheses that should greatly facilitate a direct molecular attack on the mechanisms involved. They provide models for the specificity of action of BUdR, the regulation of the probability of differentiation, a system for the molecular cloning of the molecules regulating the decision to commit to terminal myogenesis, a model for studying the number of steps that occur during terminal differentiation, and a cell line that should be useful in identifying a variety of other factors regulating differentiated functions.

The Mechanism for the Specificity of BUdR

The ability of BUdR to specifically inhibit differentiated functions has been a puzzle ever since its effects were discovered in the 1960's. Although melanin synthesis appears to be inhibited by a mechanism involving BUdR-induced deoxycytidine starvation (20), claims that BUdR inhibited myogenesis by a similar mechanism (21) have been disproved (6), and the bulk of the evidence favors a requirement that BUdR be incorporated into DNA before its effects are obtained. In bacterial systems, BUdR substitution in the DNA was found to increase the binding of the lac repressor and the catabolite activator proteins (22,23). However, an altered binding of regulatory proteins does not in and of itself explain why differentiated functions should be specifically inhibited. A rationale for this specificity comes from the growing body of evidence that household and differentiated functions are regulated in quite different fashions.

Most enzymatic pathways are composed of many steps, only a few of which are rate-limiting. Furthermore, the rate-limiting enzymes are generally controlled by feedback mechanisms, so that a reduction in their expression would lead to a signal to increase the levels of regulatory factors that control their expression. The consequences to such a pathway of a BUdR-induced alteration of the binding of regulatory proteins would be minimal. If the net effect of BUdR incorporation in the DNA were to reduce the levels of transcription (for example, by increasing the proportion of regulatory factors bound to nonspecific low affinity sites), then feedback mechanisms would operate to restore the rate-limiting enzymes to normal levels. Furthermore, since the non-rate-limiting enzymes were initially present in excess, small reductions in their levels would have little or no effect on the overall rate of the pathways. On the other hand, if the net effect were to increase transcription (for example, by increasing the binding of specific positive regulators more than the relevant negative factors), then the feedback mechanisms for the rate-limiting steps would intervene to restore the activity of the entire pathways toward normal levels. The combination of excess levels for most steps and feedback regulation for the limiting steps would thus make household functions resistant to the effects of BUdR.

In contrast, there is a growing body of evidence that the factors regulating the expression of differentiated functions are present in limiting quantities and are often subject to all-or-none kinetics. There are many examples in the cell hybrid literature of gene-dosage effects (e.g., 24-27). In these cases the typical result of the hybridization of two parental cell lines is the complete suppression of the differentiated phenotype of one of the parents. However, the result can be totally reversed in hybrids resulting from 2:1 fusions in favor of the normally suppressed parent, so that differentiated functions are then expressed. The implications of such gene dosage effects are that the molecules regulating the phenomenon are present in limiting quantities and are subject to threshold effects, since a simple two-fold change in the contribution of one parent can totally change the outcome of the experiment. Our data on the behavior of hybrids between competent and defective myoblasts (Fig. 1 and Table 1) also supports the concept that the molecules regulating the decision to differentiate are present in limiting quantities. The most direct evidence that I know of that factors regulating differentiated functions are present in limiting quantities comes from a recent study on the immunoglobulin heavy chain enhancer (30). This study demonstrated that cotransfection of myeloma cells with the enhancer linked to the CAT gene and increasing amounts of the enhancer linked to a null gene resulted in a progressive reduction in CAT expression. This implicated that the level of the enhancer-binding factors was limiting, so that competition from additional enhancer sequences resulted in a decreased binding to the target gene and a consequent reduction in transcription.

It is likely that the molecules regulating functions like the decision to differentiate are not subject to feedback regulation, since their purpose is not to maintain steady state levels of activity but rather to achieve a particular level of activity that causes a change in genetic expression. If the general pattern for such molecules is to be present in limiting quantities in the absence of feedback regulation, then the effects of BUdR on their functions would be large. Small reductions in the transcription of these molecules or their ability to bind to their regulatory targets would result in proportional changes of activity. If threshold mechanisms were involved, then even small reductions could result in a total elimination of the regulated function. The apparent specificity of BUdR for inhibiting cell differentiation would thus be a result of the different strategies used for regulating household and differentiated functions. Pathways in which most molecules were present in excess and in which feedback regulation controlled the few limiting steps would be insensitive to the effects of BUdR, whereas pathways involving limiting amounts of molecules not subject to feedback regulation and operating under threshold kinetics would be easily inhibited.

Regulating the Probability of Differentiation

Many differentiating systems appear to be controlled by stochastic mechanisms (31). This property has been difficult to rationalize because of the lack of known biochemical pathways exhibiting probabilistic kinetics. A variety of models to explain the stochastic aspects of the cell cycle have been advanced (32,33), but their applicability to cell differentiation has seemed remote. The experiments hybridizing differentiation defective and competent cells (Fig. 1) defined three key characteristics that the mechanism for regulating probability had to encompass: it had to be heritable, it had to be able to regulate probabilities ranging from 0.002% to 90%, and it had to result in a geometric average of the parental cells in cell hybrids. The model presented in Fig. 2 successfully explained these properties. One critical aspect of this model is that threshold effects must be able to regulate all-or-none phenomena. Biochemical mechanisms to produce such a result are known. For example, the cooperative effects of allostery change a linear response to a sigmoidal one. The sigmoidal curve becomes progressively steeper as the number of cooperative events increases. A mechanism involving multiple binding sites could thus exhibit very steep sigmoidal kinetics, which would effectively function as a threshold for an all-or-none response. Precedents for regulation involving multiple binding sites exist, such as the existence of multiple high affinity nuclear binding sites for the activated steroid receptor (34,35). Although the model presented in Fig. 2 is entirely speculative at present, it does provide a mechanism consistent with known molecular processes that effectively explains many of the stochastic aspects of cell differentiation. The proof of such a model is dependent upon the isolation of the relevant regulatory molecules, which may be accomplished as described below.

The Molecular Cloning of Regulatory Molecules

The availability of two related cell lines, one of which is overproducing a particular molecule and the other of which is not expressing it, provides a very powerful system for cloning the cDNA for the molecule using subtraction hybridization (36). In this approach, a cDNA made against mRNA from the overproducing cell is hybridized to an excess of mRNA from the null cell. The sequences common to both cell types will anneal and form double stranded structures, which can then be separated from the single stranded cDNA using hydroxyapetite. One is then left with a cDNA that is highly enriched for the sequence of interest, which can then be cloned and probed with a cDNA that has been similarly enriched. The BUdR-resistant cells and the differentiation defective cells thus provide the required starting material for this approach. Preliminary experiments indicate that we may have obtained about a hundredfold enrichment for the regulatory molecule. Even if its initial abundance were only 0.01% of total message, a hundred-fold enrichment to an abundance of 1% would make it sufficiently frequent to identify during screening procedures.

Clones isolated from such a library would be screened according to the following criteria: The level of their inserts should be low to absent in undifferentiated myoblasts, at maximal or near maximal levels during the period when the decision to differentiate is being made (following a stimulus to differentiate but prior to the onset of myotube formation), and should remain constant or decline in differentiated myotubes. Finally, and very importantly, the levels should be amplified during the decision phase in the BUdR-resistant cells. There are probably many molecules that could satisfy the first three criteria, given the changing physiology that must occur as a cell adapts from a dividing to a postmitotic state. However, since the only criteria used to select the BUdR-resistant cells was the ability to form morphologically identifiable myotubes in the presence of BUdR, only the expression of the regulatory molecules required prior to that step should have been amplified. Structural proteins with functions other than binding to DNA that are induced prior to myotube formation should have been unaffected by the selection protocol. The final screening criterion using the BUdR-resistant cells thus provides a very powerful method for identifying the most promising candidates for the regulatory molecules involved in the decision to differentiate.

We currently have no information that indicates whether the amplified expression presumably occurring in the BUdR-resistant cells is due to gene amplification or an increased transcription from a single gene. The isolation of cloned cDNAs will allow us to address that question.

Stages During Terminal Myogenesis

The number of regulatory steps or molecules controlling terminal differentiation is unknown. Because myotube formation was the sole selection criterion used during the isolation of the BUdR-resistant cells, upstream events should be resistant to BUdR whereas downstream events should remain inhibited by BUdR. If multiple stages are involved and if myotube formation is regulated by an early stage, then the expression of differentiated functions regulated by the later stages should be absent in the presence of BUdR. Preliminary experiments examining the levels of creatine kinase and the acetylcholine receptor in BU36 clone b cells that differentiated in the presence versus the absence of BUdR showed similar activities for both molecules. If these very preliminary results are sustained and similar results are obtained for other differentiated markers, it will imply either that myotube formation is regulated by a late event or that a common regulatory factor is responsible for the coordinate induction of the whole set of differentiated function.

Other Applications of Amplified Myoblasts

The existence of BUdR-resistant myoblasts should prove extremely useful for a variety of other approaches. Several investigators at this symposium are studying flanking sequences to differentiated messages or genes in order to identify regulatory regions. If the molecules that bind to such sequences are present in increased amounts in the BUdR-resistant cells, then the levels of transcription obtained from plasmids containing the flanking sequence should be increased during a transient expression transfection in BUdR-resistant cells as compared to normal cells. The demonstration of this phenomenon would confirm both our model of the nature of the BUdR-resistant phenotype and the function of the flanking sequence in question. If multiple regulatory factors are involved and the subtraction hybridization experiments described above do not succeed in isolating all of them, then the existence of an overproducing cell line would be valuable as the source of the initial cell extracts for alternate purification procedures.

The above discussion illustrates the many insights that have been obtained from somatic cell genetic investigations into the mechanisms controlling myogenesis. It is hoped that the experiments generated from these insights will ultimately result in a full molecular understanding of the process by which a cell changes the expression of its genetic program during terminal cell differentiation.

ACKNOWLEDGMENTS

I would like to thank the Rockefeller University Press for permission to reproduce Figs. 1-6.

REFERENCES

1. Yaffe D (1968). Retention of differentiation potentialities during prolonged cultivation of myogenic cells. Proc Nat Acad Sci USA 61:477.
2. Wright WE (1981). The synthesis of rat myosin light chains in heterokaryons formed between undifferentiated rat myoblasts and chick skeletal myocytes. J Cell Biol 91:11.
3. Wright WE (1978). The isolation of heterokaryons and hybrids by a selective system using irreversible biochemical inhibitors. Exp Cell Res 112:395.
4. Wright WE (1984). Toxin-antitoxin selection for isolating somatic cell fusion products between any cell types. Proc Natl Acad Sci USA 81:7822.
5. Wright WE (1984). The induction of muscle genes in neural cells. J Cell Biol 98:427.
6. Wright WE (1982). The BUdR content of DNA is decreased during the reversal of the inhibition of myogenesis by deoxycytidine. Somatic Cell Genetics 8:547.
7. Wright WE (1984). Control of differentiation in heterokaryons and hybrids involving differentiation-defective myoblast variants. J Cell Biol 98:436.
8. Coleman JR, Coleman AW, Hartline EJH (1969). A clonal study of the reversible inhibition of muscle differentiation by the halogenated thymidine analogue 5-bromodeoxyuridine. Develop Biol 19:527.
9. Stockdale F, Okazaki K, Nameroff M, Holtzer H (1964). 5-Bromodeoxyuridine: effect on myogenesis in vivo. Science 146:533.
10. Rutter WJ, Pictet RL, Morris PW (1973). Toward molecular mechanisms of developmental processes. Ann Rev Biochem 42:601.
11. Wright WE (1982). The regulation of myosin light chain synthesis in heterokaryons between differentiated and undifferentiated myogenic cells. In Dowben RM, Shay JW (eds): "Muscle and Cell Motility, 2" New York: Plenum Press, p 177.
12. Wright WE (1982). The induction of myosin light chain synthesis in heterokaryons between normal diploid cells. In Vitro 18:851.
13. Wright WE (1984). The expression of differentiated functions in heterokaryons between skeletal myocytes, adrenal cells, fibroblasts and glial cells. Expt'l Cell Res 151:55.

14. Blau HM, Chin C-P, Webster C (1983). Cytoplasmic activation of human nuclear genes in stable heterokaryons. Cell 32:1171.
15. Chiu CP, Blau H (1984). Reprogramming cell differentiation in the absence of DNA synthesis. Cell 37:879.
16. Wright WE, Aronoff J (1983). The regulation of rat myosin light chain synthesis in heterokaryons between BUdR-blocked rat myoblasts and differentiated chick myocytes. J Cell Biology 96:1571.
17. Carter SB (1967) Effects of cytochalasins on mammalian cells. Nature 213:261.
18. Wright WE (1985). The amplified expression of factors regulating myogenesis in L6 myoblasts. J Cell Biology 100:311.
19. Ringertz NR, Savage RE (1976). "Cell Hybrids." New York: Academic Press, p 180.
20. Davidson RL, Kaufman ER (1977). Deoxycytidine reverses the suppression of pigmentation caused by 5-BrdUrd without changing the amount of 5-BrdUrd in DNA. Cell 12:923.
21. Rogers J, Ng SK, Coulter MB, Sanwal BD (1975). Inhibition of myogenesis in a rat myoblast line by 5-bromodeoxyuridine. Nature 256:438.
22. Lin S-Y, Riggs AD (1971). Lac repressor binding to operator analogues: Comparison of poly[d(A-T)], poly[d(A-BrU)] and poly [d(A-U)]. Biochem Biophys Res Commun 45:1542.
23. Lin S-Y, Riggs AD (1976). The binding of lac repressor and the catabolite gene activator protein to halogen-substituted analogues of poly[d(AT)]. Biochem Biophys Acta 432:185.
24. Willing M, Nienhuis AW, Anderson WF (1979). Selective activation of human β but not δ-globin gene in human fibroblast x mouse erythroleukaemia cell hybrids. Nature 277:534.
25. Axelrod PE, Gopalakrishnan TV, Willing M, Anderson WF (1978). Maintenance of hemoglobin inducibility in somatic cell hybrids of tetraploid (2S) mouse erythroleukemia cells with mouse or human fibroblasts. Somatic Cell Genet 4:157.
26. Davidson RC (1972). Regulation of melanin synthesis in mammalian cells: effect of gene dosage on the expression of differentiation. Proc Natl Acad Sci USA 69:951.
27. Fougere C, Ruiz F, Ephrussi B (1972). Gene dosage dependence of pigment synthesis in melanoma x fibroblast hybrids. Proc Natl Acad Sci USA 69:330.

28. Brown JE, Weiss MC (1975). Activation of production of mouse liver enzymes in rat hepatoma-mouse lymphoid cell hybrids. Cell 6:481.
29. Darlington G, Bernhard HP, Ruddle FH (1974). Human serum albumin phenotype activation in mouse hepatoma-human leukocyte cell hybrids. Science 185:859.
30. Mercola M, Goverman J, Mirell C, Calame K (1985). Immunoglobulin heavy-chain enhancer requires one or more tissue-specific factors. Science 227:266.
31. Levenson R, Housman D (1981). Commitment: How do cells make the decision to differentiate? Cell 25:5.
32. Cooper, S (1979). A unifying model for the GI period in prokaryotes and eukaryotes. Nature 280:17.
33. Nurse B (1980). Cell cycle control - both deterministic and probabilistic? Nature 286:9.
34. Hayward MA, Mitchell RA, Shapiro DJ (1980). Induction of estrogen receptor and reversal of the nuclear-cytoplasmic receptor ratio during vitellogenin synthesis and withdrawal in Xenopus laevus. J Biol Chem 255:11308.
35. Shapiro D (1982). Steroid hormone regulation of vitellogenin gene expression. CRC Critical Rev in Biochem 12:187.
36. Hedrick SM, Cohen DI, Nielsen EA, Davis MM (1984). Isolation of cDNA clones encoding T cell-specific membrane-associated proteins. Nature 308:149.

Molecular Biology of Muscle Development, pages 105–116

ACTIVATION AND EXPRESSION OF MUSCLE GENES IN HETEROKARYONS[1]

Helen M. Blau, Grace K. Pavlath, Choy-Pik Chiu,[2] Edna Hardeman, and Cecelia Webster

Department of Pharmacology, Stanford University School of Medicine, Stanford, California 94305

ABSTRACT Highly specialized cells representative of three major lineages are reprogrammed to express muscle functions following fusion with muscle cells. Heterokaryons containing these cell types, differ in the rate of muscle gene expression. Activation of all eight muscle genes assayed is detected and the timecourse of expression and relative amounts of the gene products suggest that activation in heterokaryons parallels the myogenic program observed in pure muscle cultures. Activation occurs at a transcriptional level in the two cases examined. From studies of gene dosage, it appears that factor concentration plays an important role in the rate of muscle gene expression and that the muscle cell is frequently dominant in heterokaryons. In all of the primary diploid cell types examined, activation is independent of changes in chromatin structure requiring DNA synthesis. However, in a transformed cell type, Hela, changes induced by 5-azacytidine, presumably in the level of gene methylation, are necessary before the muscle genes can be activated. The activation of different muscle genes is not necessarily interdependent, since two functions which are normally expressed sequentially in myogenesis can be dissociated. Taken together, these results demonstrate that the nucleus of highly specialized cells is plastic and that the cytoplasm plays an important role in maintaining the differentiated state.

---

1 This work was supported by the National Institutes of Health and the Muscular Dystrophy Association.

2 Present address: Laboratory of Molecular Carcinogenesis, Dana-Farber Cancer Institute, Boston, MA.

INTRODUCTION

All cells contain the same chromosomes, or genetic material. Yet they only express a subset of the genes they contain. How genes are regulated to achieve the cell specialization characteristic of tissues is a question of major interest. Muscle is particularly attractive for studies of the mechanisms underlying the expression of tissue-specific functions, because the sequence of events accompanying these changes is well defined at a cellular and biochemical level and can be mimicked in tissue culture. A relatively simplistic scheme of the points at which muscle cell specialization can be regulated is depicted in Figure 1. A mesodermal stem cell gives rise to a determined myoblast destined for myogenesis: it is capable of spontaneously fusing with other myoblasts and gives rise to a differentiated myotube. The multinucleated myotube no longer divides or synthesizes DNA, but produces muscle proteins in large quantity. These include major constituents of the contractile apparatus and specialized cell surface components. Eventually the differentiated muscle cell exhibits characteristic striations and rhythmic contractions. A further step in this pathway is maturation: the myosins and actins of the contractile apparatus of muscle are known to differ at different stages of development.

POINTS OF REGULATION IN MUSCLE CELL SPECIALIZATION

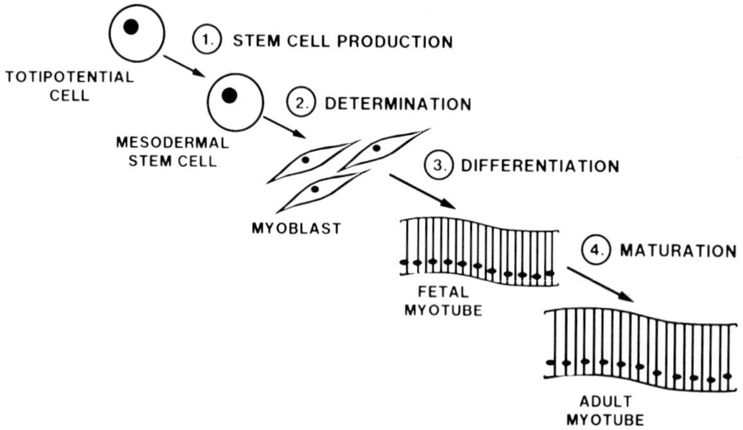

The regulatory mechanisms underlying these transitions are not well understood. Several theories have been proposed which could account for the development of specific phenotypes: changes in the genome which accompany quantal mitoses (1), sequential repression of genes (2), or modulation of gene expression by activators and repressors (3) with additional complex regulatory effects resulting from the relative concentration and affinity of these factors for specific genes (4). Some major questions are raised by these theories: How reversible is the differentiated state, or tissue-specific phenotype of a cell? What are the underlying mechanisms which regulate the transition from the determined to the differentiated to the mature cell phenotype? These are among the questions which we are attempting to address.

To study the mechanisms which regulate cell specialization, we combine muscle cells with cells of other phenotypes and examine whether the expression of previously silent muscle genes can be induced in response to tissue-specific factors present in the muscle. The nonmuscle cells are generally primary diploid human cells, with some noteworthy exceptions. These are fused with differentiated mouse muscle cells of the C_2C_{12} line using polyethylene glycol. The fusion product is a stable heterokaryon in which there is no cell division. As a result, the parental cell nuclei remain intact and retain a full complement of chromosomes. This is in contrast with typical interspecific hybrids (synkaryons), in which the parental nuclei fuse and chromosomes are progressively lost and rearranged during cell division. A further advantage of heterokaryons is that gene expression can be assayed immediately following fusion, since selection of the fused cells of interest is not necessary. Finally, since the heterokaryons can be maintained in culture for 10 to 12 days, changes in gene expression can be followed over a relatively long period of time.

Using this system, we have examined the reversibility of the differentiated state. We have taken human cells from the same lineage as muscle (mesodermal), or from different lineages (ectodermal and endodermal) and from human donors of different ages and fused them with muscle to see if they can be reprogrammed to express muscle functions (Figure 2). We find that each of these cell types can be induced to express muscle genes.

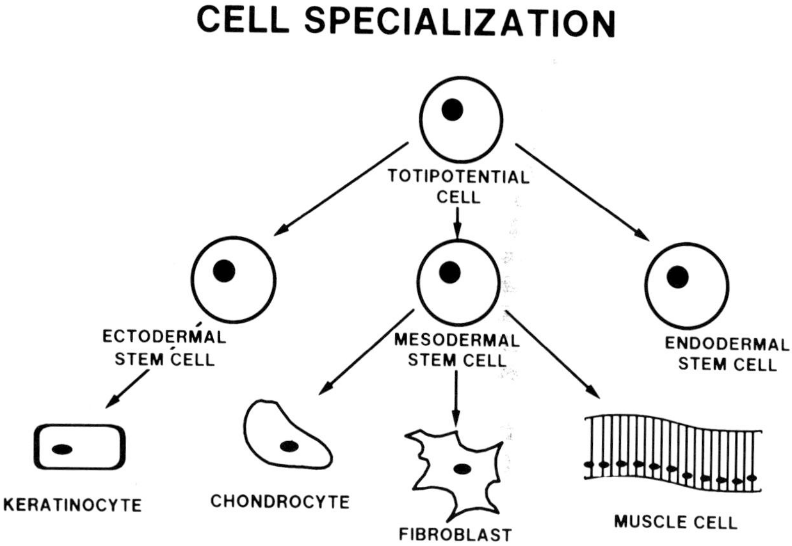

Fig. 2

Our results indicate that there is remarkable plasticity in the function of the nucleus of highly specialized and of aged cells. The potential of these nuclei is not fixed and irreversible, but susceptible to modulation by the cytoplasm surrounding them. If the feedback between the cytoplasm and the nucleus is altered, so is the expression of the genes responsible for the differentiated state. Since the nuclei of heterokaryons remain separate and intact, these changes in gene expression are mediated by factors that act in trans.

RESULTS

Histogenetic State and Gene Activation

We examined the influence of histogenetic state on muscle gene activation. In our initial experiments, only fibroblasts were used: from amniotic fluid, skin, and fetal lung (5,6). In the studies described below, heterokaryons were formed by fusing mouse muscle cells with well differentiated primary human nonmuscle cells (7). These included chondrocytes, which like muscle, are of the mesodermal lineage, keratinocytes of the ectodermal lineage, and hepatocytes of the endodermal lineage. Chondrocytes were isolated is our laboratory directly from neonatal cartilage and keratinocytes were from newborn foreskin (8). The hepatocytes (Hep G2) were not primary cells; however, they are not tumorigenic and, unlike most cell lines retain the ability to express at least 17 different liver-specific functions (9,10). The phenotype of each of these cells was confirmed in tissue culture by their reaction with specific antibodies to collagen type II, keratin, and albumin, respectively.

The expression of several human muscle genes was assayed and all were activated in the cells tested. Myosin light chains (1s, 2s, 2f, and fetal), and human isozymes of creatine kinase (MB, and MM and a functional mouse-human hybrid MM) were distinguished from their mouse muscle counterparts by electrophoresis (5,11). The human muscle-specific cell surface antigen, 5.1H11 (12,13), was induced and could be monitored on single heterokaryons using a monoclonal antibody (6). Finally, we assayed for the expression of two sarcomeric actin genes, α-cardiac and α-skeletal actin, which we previously showed to be expressed during myogenesis in vitro (14). The human muscle-specific transcripts of these genes were detected by Northern blot analysis using species and isotype-specific cDNA probes which were characterized in collaboration with the laboratory of L. Kedes (15). In heterokaryons, as in pure muscle cultures, the α-cardiac actin transcripts were detected earlier and reached a higher peak level than α-skeletal actin transcripts (7,16).

Altogether, the relative amounts and sequence of expression of the different muscle gene products paralleled normal myogenesis and was similar for heterokaryons containing the different cell types. There was one major difference: the kinetics of gene expression were markedly different (7). This difference in rate was first observed for the expression of the cell surface antigen. The t_{50}, or time required for half of the heterokaryons to express the antigen differed by several days. This could be a consequence of differences among cell types in post-transcriptional processing of membrane components. However, we observed the same kinetic differences at the transcriptional level: the time course for the accumulation of the human mRNAs for α-cardiac and α-skeletal actin paralleled that observed for 5.1H11. These results suggest that the differences may be intrinsic to the mechanism required for gene activation in the different cell types.

Developmental Stage and Muscle Gene Activation

Studies using synkaryons suggested that there are developmental differences which determine whether or not a genome is suceptible to regulation by exogenous activators or repressors. Accordingly, the developmental stage of a cell could affect the degree to which the epigenotype is stable or alterable. This could be due to a difference in the accessibility of the genes to regulatory factors or to a difference in the kinds of regulatory factors required at different stages of development.

To address the question of whether developmental stage alters the potential of a cell to be reprogrammed, we formed heterokaryons between mouse muscle and human skin fibroblasts of fetal and adult origin (17). These cells were obtained from a fetus (14 weeks) and from an adult (66 years). They differed significantly in cell density at confluence, total protein per cell, numbers of receptors for insulin and for insulin-like growth factor I per cell, labeling index and doubling time. We compared the expression of the genes coding for 5.1H11, creatine kinase, α-cardiac actin and α-skeletal actin in heterokaryons containing each of the two cell types. The activation of human muscle genes in fibroblasts from these two ages was indistinguishable: it occurred with equally high efficiency (>95%), and similar rapid kinetics. Thus, despite marked differences in cell size, rate of proliferation and numbers of mitogen receptors, the potential to activate previously silent muscle genes in response to trans-acting factors does not change in skin fibroblasts in the course of _in vivo_ aging.

Reprogramming Differentiation Without DNA Synthesis

We used the heterokaryon system to examine whether the expression of differentiated cell functions was heritable and stable due to changes in the genome. In these studies, we examined whether DNA replication was required to reprogram cells of different phenotypes to express muscle gene functions (6,7,17). For this purpose, human fibroblasts (MRC-5 from fetal lung, fetal skin, and adult skin), chondrocytes and keratinocytes were each fused with differentiated mouse muscle cells in the presence or absence of the DNA synthesis inhibitor, cytosine arabinoside. Individual heterokaryons were analyzed both for DNA synthesis by autoradiography following labeling with ^{3}H-thymidine and for the expression of the human muscle cell surface antigen, 5.1H11, by immunofluorescence. The frequency of activation was between 70 and 90% regardless of whether DNA synthesis occurred. Our results demonstrate that DNA replication is not a prerequisite for the activation of muscle genes and that alterations in chromatin configuration requiring DNA synthesis are not necessary for the muscle genes in primary human cells of different lineages or of different ages to be accessible to and respond to putative trans-acting regulatory factor(s) present in differentiated muscle cells.

Activation in Hela Cells: A Two-Step Mechanism

In contrast to our results with primary diploid cell types, muscle gene activation was never observed when we fused mouse muscle cells and human HeLa cells to form heterokaryons. To investigate whether 5-azacytidine (5AC) could alter the responsiveness of HeLa cells to muscle gene activators, HeLa cells were exposed to 5AC prior to fusion (18). Control HeLa cells were not exposed to 5AC at all. Heterokaryons were formed by fusing 5AC-treated or untreated HeLa cells with mouse muscle cells. Both an early (5.1H11 cell surface antigen) and a late (MM-creatine kinase) human muscle function were detected by immunofluorescence and electrophoresis, respectively. Both human muscle genes were activated in heterokaryons formed with 5AC-treated HeLa cells but never in control heterokaryons or in 5AC-treated HeLa cells fused to themselves. Furthermore, the two functions were frequently not coordinately expressed. 5.1H11 is typically expressed well before MM-creatine kinase in the course of muscle differentiation. The independent expression of these two gene products in heterokaryons containing 5AC-treated HeLa nuclei indicates that the expression of 5.1H11 is not a prerequisite for the expression of CK. We concluded from these results that prior treatment with 5AC is required to induce changes, presumably in the level of DNA methylation, which allow the muscle genes of HeLa cells to be activated in response to trans-acting muscle regulators. Thus, it appears that the muscle genes in Hela cells and in diploid primary cells are maintained in an inactive state by different mechanisms.

Concentration Dependence of Gene Activation

The regulatory factors in muscle cells act in a dominant, concentration-dependent manner. We examined the effect of gene dosage, or the nuclear ratio of the two cell types within a heterokaryon, on novel muscle gene expression. Heterokaryons were formed between mouse muscle cells and human fibroblasts (19). Each of three different types of fibroblasts from fetal lung (MRC-5), fetal skin and adult skin were used and gave essentially similar results. Nuclear composition was monitored using Hoechst 33258 and the activation of the human muscle cell surface antigen, 5.1H11, was assayed in the same heterokaryons by immunofluorescence. Our results showed that the ratio of the nuclei of the two cell types contained in the heterokaryon affected the kinetics of 5.1H11 expression. Prior to day 4, the percentage of total heterokaryons which expressed 5.1H11 on any given day was always higher when the ratio of muscle to nonmuscle nuclei was one, than when the relative numbers of nuclei of either parental cell type was increased. By day 4, an effect of nuclear ratio was no longer evident; 90% of the heterokaryons analyzed on that day expressed the human muscle antigen regardless of the ratio of muscle to non-muscle nuclei. In contrast with nuclear ratio, neither the total number of nuclei nor the relative position of the nuclei within a heterokaryon affected the kinetics of gene expression. We concluded from the timecourse of expression of 5.1H11 that the concentration of putative trans-acting factors significantly influences the rate of muscle gene expression: a threshold concentration is necessary, and an excess may be inhibitory.

DISCUSSION

Little is known regarding the nature of tissue-specific regulators involved in eukaryotic gene expression during differentiation. This is largely due to a lack of model experimental systems in which to probe this question. We have developed a muscle X nonmuscle heterokaryon system which is particularly advantageous for this kind of investigation. Using the heterokaryon system, we showed that following cell fusion, muscle genes in nonmuscle cells are responsive to trans-acting factors present in the muscle cytoplasm and can be activated.

The heterokaryon system is amenable to biochemical and molecular analyses of the mechanism of gene activation, since heterokaryons are efficiently produced and activate human muscle genes with high frequency. Between 80 to 90% of the myotubes in a fused culture contain nuclei of both cell types. This can be readily determined by the differential staining of the two types of nuclei observed with the dye Hoechst 33258. Furthermore, analyses of gene expression at the single cell level reveal that 95 $\pm$ 1% of all heterokaryons which contain human fibroblast nuclei activate human muscle genes. Thus, gene activation in heterokaryons is not a rare event and is quite efficient.

Muscle gene activation in heterokaryons appears to follow the myogenic program. We detect the human muscle products of eight distinct muscle genes for which we have species-specific assays, including enzymes essential to energy production, structural components of the contractile apparatus and membrane components. All are activated and the relative amounts and timecourse of production of these muscle gene products are typical of pure muscle cultures.

Using this system we are addressing questions regarding the plasticity of the nuclei of cell types of different lineages and different developmental stages. The choice of cells and culture conditions may be critical to the results obtained and may explain some of the ways in which our results differ from those of others (20,21). With the exception of the HeLa cells and hepatocytes, all of the human nonmuscle cells analyzed to date are derived directly from human tissues; consequently they are primary, diploid cells which have been subjected to tissue culture conditions for minimal periods of time. The muscle cells used are always mouse C_2C_{12} cells, a subclone of the line originally isolated by Yaffe (22).

Using these cells we determined that "terminally differentiated" chondrocytes, keratinocytes, and hepatocytes can be reprogrammed. Wright has obtained similar results using rat neural cell lines (23). Furthermore, fibroblast nuclei from fetal and adult stages of development are equally responsive to muscle regulatory factors. Yet, cells from different tissues differ in their kinetics, or rate of muscle gene expression following fusion, suggesting that there are tissue-specific differences in the mechanism of activation in these cell types. We are currently investigating the basis for these differences. In addition, experiments in our laboratory are underway to characterize the nature of the activators of muscle differentiation and to address other steps in the regulation of muscle cell specialization such as determination and maturation, (see for example 24,25).

ACKNOWLEDGEMENTS

I wish to thank the industrious and bright scientists in my laboratory for their contributions to the experiments described in this paper, my colleagues for stimulating discussions regarding the interpretation of results, and to Nancy Williams and Karen Benight for their secretarial help in the preparation of the manuscript. I also gratefully acknowledge the generous contribution of reagents: keratinocytes from Dr. Howard Green, skin fibroblasts from Drs. Peter Byer and Marvin Kerasek, C_2 myoblasts from Dr. David Yaffe, the 5.1H11 antibody from Dr. Frank Walsh, the collagen type II antibody from Dr. Robert Burgeson, and the keratin antibody from Dr. Tung-tien Sun.

REFERENCES

(1) Holtzer H. (1978) Cell lineages, stem cells and the 'quantal' cell cycle concept. In Lord, Potten, Cole (eds): Stem Cells and Tissue Homeostasis. Cambridge: Cambridge University Press, pp. 1-28.

(2) Caplan A.I., Ordahl C.P. (1978) Irreversible gene repression model for control of development. Science 201:120-130.

(3) Ptashne, M., Backman, K., Humayun, M.Z., Jeffrey, A., Maurer, R., Meyer, B., Sauer, R.T. (1976)

Autoregulation and function of a repressor in bacteriophage lambda. Science 194:156-161.
(4) Brown D.D. (1984) The role of stable complexes that repress and activate eucaryotic genes. Cell 37:359-365.
(5) Blau HM, Chiu C.-P., Webster C. (1983) Cytoplasmic activation of human nuclear genes in stable heterocaryons. Cell 32:1171-1180.
(6) Chiu C.-P., Blau H.M. (1984) Reprogramming cell differentiation in the absence of DNA synthesis. Cell 37:879-887.
(7) Pavlath G.K., Chiu C.-P., Blau H.M. (1985) Induction of muscle gene expression in differentiated cells of three lineages. In preparation.
(8) Rheinwald J.G., Green H. (1977) Epidermal growth factor and the multiplication of cultured human epidermal keratinocytes. Nature 265:421-424.
(9) Aden D.P., Fogel A., Plotkin S., Damjanov I., Knowles B.B. (1979) Controlled synthesis of HBsAg in a differentiated human liver carcinoma-derived cell line. Nature 282:615-616.
(10) Knowles B.B., Howe C.C., Aden D.P. (1980) Human hepatocellular carcinoma cell lines secrete the major plasma proteins and hepatitis B surface antigen. Science 209:497-499.
(11) Blau, H.M., Chiu, C.-P., Pavlath, G.K., Webster, C. (1985) Muscle gene expression in heterokaryons. In Strohman R.C., Wolf S. (eds): Gene Expression in Muscle. New York: Plenum Press, pp. 231-247.
(12) Walsh F.S. (1980) Identification and characterization of plasma membrane antigens of neurons and muscle cells using monoclonal antibodies. In Brzin M., Sket D., Bachelard H. (eds): Synaptic Constituents in Health and Disease. Oxford: Pergamon, pp. 285-320.
(13) Walsh F.S., Ritter M.A. (1981) Surface antigen differentiation during human myogenesis in culture. Nature 289:60-64.
(14) Bains W., Ponte P., Blau H., Kedes L. (1984) Cardiac actin is the major actin gene product in skeletal muscle cell differentiation in vitro. Mol. Cell. Biol. 4:1449-1453.
(15) Gunning P., Mohun T., Ng S.-Y., Ponte P., Kedes L. (1984) Evolution of human sarcomeric-actin genes: Evidence for units of selection within the 3'

untranslated regions of the mRNAs. J. Mol. Evol. 20:202-214.
(16) Hardeman E., Chiu C.-P., Blau H.M. (1985) Differential expression of mouse and human sarcomeric actins in heterokaryons. In preparation.
(17) Pavlath G.K., Chiu C.-P., Blau H.M. (1985) Plasticity of human skin fibroblasts does not change with developmental stage: Muscle gene expression is efficiently induced in heterokaryons. In preparation.
(18) Chiu C.-P., Blau H.M. (1985) 5-Azacytidine permits gene activation in a previously noninducible cell type. Cell 40:417-424.
(19) Pavlath G.K., Blau H.M. (1985) Expression of muscle genes in heterokayrons depends on gene dosage. In preparation.
(20 Wright W.E., Arnoff J. (1983) The suppression of myogenic functions in heterokaryons formed by fusing chick myocytes to diploid rat fibroblasts. Cell Differ. 12:299-306.
(21) Lawrence J.B., Coleman J.R. (1984) Extinction of muscle-specific properties in somatic cell heterokaryons. Dev. Biol. 101:463-476.
(22) Yaffe D., Saxel O. (1977) Serial passaging and differentiation of myogenic cells isolated from dystrophic mouse muscle. Nature 270:725-727.
(23) Wright W.E. (1984) Induction of muscle genes in neural cells. J. Cell. Biol. 98:427-435.
(24) Minty A.J., Blau H.M., Kedes L. (1985) Two-level regulation of cardiac actin expression: An upstream sequence interacting with muscle specific transcription regulatory factors. In preparation.
(25) Silberstein L., Blau H.M. (1985) Two fetal-specific fast myosin isozymes in human muscle. This volume.

Molecular Biology of Muscle Development, pages 117–131

PROTEOGLYCAN SYNTHESIS DURING SKELETAL MUSCLE DEVELOPMENT[1]

David A. Carrino and Arnold I. Caplan

Department of Biology
Case Western Reserve University
Cleveland, Ohio 44106

ABSTRACT Proteoglycans and their constituent glycosaminoglycans are important for many developmental processes. In order to elucidate the role of proteoglycans during skeletal muscle myogenesis, the proteoglycans synthesized by developing muscle were analyzed both in culture and *in vivo*. Early in myogenesis, unfused myoblasts and fused myotubes synthesize as the major proteoglycan species a very large chondroitin sulfate proteoglycan which has some unique structural features relative to other chondroitin sulfate proteoglycans. This skeletal muscle proteoglycan is proposed to structure large amounts of water in the extracellular matrix and, thereby, reserve space into which the myogenic cell population grows. Later in myogenesis, perhaps due to a regulatory interaction between the mature muscle cells and fibrogenic cells of the muscle tissue, synthesis of the chondroitin sulfate proteoglycan diminishes and synthesis of smaller proteoglycans, some of the heparan sulfate type, becomes predominant. These heparan sulfate proteoglycans, of which a proportion are produced by the fibroblasts, may take part in the splitting of the muscle masses into discrete muscle organs. Eventually, in the fully formed muscles, there are heparan sulfate proteoglycans as components of the basal lamina which surround each muscle fiber. This basal lamina seems to direct morphogenesis

[1]This work was supported by grants from the National Institutes of Health and the Muscular Dystrophy Association.

during regeneration of damaged mature muscle. Thus, there are stage-specific transitions in proteoglycan synthesis during myogenesis which are necessary in the shaping of the tissue.

INTRODUCTION

The development of skeletal muscle is a complex process in which several different cell types participate through cell-cell and cell-matrix interactions. Mature skeletal muscle is composed of multinucleated myotubes which result from fusion of mononucleated myoblasts. At a specific time during development, the mononucleated myoblasts adhere to each other in a specific way and their plasma membranes and cell surfaces become continuous; there is much evidence that cell surface components are involved in this process (1,2). The extracellular matrix also plays a role in myogenesis in that skeletal muscle myotubes are packaged in extracellular matrix wrappings, both individually and as small groups; in total, such wrappings confer upon muscles their discrete morphologies. There is evidence that extracellular matrix components are synthesized in a stage-specific fashion throughout myogenesis (3,4). The major changes in one component of the extracellular matrix, glycosaminoglycans, are manifested as a decrease in the synthesis of hyaluronic acid and an increase in the synthesis of heparan sulfate (4). A decrease in the tissue content of chondroitin sulfate, specifically chondroitin-6-sulfate, has also been reported to occur during myogenesis (5). Additionally, it has been shown that initiation of muscle cell cultures on a hyaluronic acid substrate results in inhibition of myogenesis (6); the cells continue to divide, remain at the myoblast stage, and show little or no evidence of either fusion or synthesis of muscle proteins characteristic of mature muscle (6,7).

Thus, glycosaminoglycans seem to be involved in the process of myogenesis. However, with the exception of hyaluronic acid, all glycosaminoglycans synthesized by cells have been shown to occur covalently attached to protein (8). Such proteins are referred to as core proteins and the entire molecule is called a proteoglycan (for review, see 8). Proteoglycans can contain as few as one or two or as many as one hundred or more individual glycosaminoglycan chains. The glycosaminoglycan chains range in molecular weight from two or three thousand to seventy thousand daltons. Other than cartilage proteoglycan, in which both chondroitin sulfate and keratan sulfate are covalently bound to the same core protein, proteoglycans contain only a single class of glycosaminoglycan: chondroitin sulfate, dermatan sulfate, heparan sulfate, or keratan sulfate. In addition to the glycosaminoglycans, many proteoglycans also contain saccharide in the form of small oligosaccharides of the type normally found in glycoproteins. These include both the mucin type or O-linked oligosaccharides which are covalently attached to the core protein through the hydroxyl groups of serine or threonine and the high mannose type of N-linked oligosaccharides that are covalently attached by N-glycosylamine bonds to asparagine residues. The entire molecule consisting of a central core protein and covalently linked saccharide is called a proteoglycan monomer and can be as large as two to four million daltons in molecular weight. In addition, some proteoglycan monomers, by virtue of a specialized region of the core protein, have the capacity to interact non-covalently with the glycosaminoglycan hyaluronic acid. By means of this interaction, which is stabilized by small proteins called link proteins, several to many monomers are anchored together to form a macromolecule of as much as 100 million daltons. Other proteoglycan monomers, most notably heparan sulfate proteoglycans, have been shown to be able to insert into the lipid bilayer of cell membranes, presumably through a hydrophobic region of the core protein (9,10). In this way, such proteoglycans can be localized at the cell surface where it is proposed they play a role in cell surface-mediated processes such as adhesion, motility, and growth control (11-13). Proteoglycans are also thought to participate in developmental processes such as morphogenesis (14,15).

STRUCTURAL ANALYSIS OF SKELETAL MUSCLE PROTEOGLYCANS

The observations described above suggest that proteoglycans are involved in many cell-cell and cell-matrix interactive processes which are important for myogenesis. Indeed, one component of proteoglycans, the glycosaminoglycans, have been shown to change during myogenesis (4), which implies a stage-specific function for these molecules. Because of these observations and because the overall structure of the parent proteoglycan molecules may be important for the function of muscle glycosaminoglycans, studies in our laboratory were directed at the elucidation of the structure of skeletal muscle proteoglycans (16-18). The protocol that was followed was to label embryonic chick leg muscle cells in culture with [^{35}S]sulfate, a highly specific precursor for proteoglycans, and then extract, isolate, and analyze the proteoglycans. With this scheme, the major proteoglycan found to be synthesized by both mononucleated myoblasts and multinucleated myotubes is a large chondroitin sulfate proteoglycan. Like most proteoglycans synthesized by non-cartilaginous tissues, the skeletal muscle chondroitin sulfate proteoglycan is of lower buoyant density than the very highly glycosylated and, therefore, higher buoyant density proteoglycans of cartilage. Nevertheless, the muscle proteoglycan is of higher buoyant density than proteins, as is typical of highly glycosylated macromolecules.

The high buoyant density form of the skeletal muscle proteoglycan has a structure which displays some unique features relative to other proteoglycans (Fig. 1). First, this proteoglycan is very large, even by proteoglycan standards; its molecular weight is estimated to be four million daltons. The large hydrodynamic size of these molecules is most likely due to the unusually large size of the chondroitin sulfate chains. Their molecular weight is estimated at seventy thousand daltons, which places them

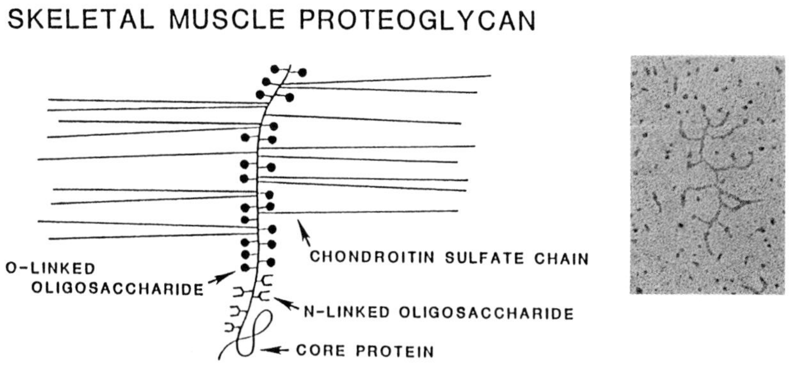

FIGURE 1. Diagrammatic representation of chick embryonic skeletal muscle chondroitin sulfate proteoglycan (left). On the right is an electron micrograph of skeletal muscle proteoglycan viaualized by the cytochrome c spreading procedure (x70,000, micrograph kindly provided by Dr. David G. Pechak of our laboratory).

among the largest glycosaminoglycans so far described. These chondroitin sulfate chains possess a unique chemical characteristic in that they consist of approximately 90% 6-sulfated, as opposed to 4-sulfated, chondroitin sulfate. Such a sulfation pattern is unusual among chondroitin sulfates. Similar to several other proteoglycans, the skeletal muscle proteoglycan is capable of forming link protein-stabilized aggregates with hyaluronic acid. Also similar to some other proteoglycans, the skeletal muscle proteoglycan possesses both O-linked and N-linked oligosaccharides whose distribution throughout the monomer is not random. The N-linked oligosaccharides are localized near or in that portion of the molecule which mediates the interaction with hyaluronic acid, the hyaluronic acid-binding region. The O-linked oligosaccharides are attached in other parts of the molecule with some localized in close proximity to the chondroitin sulfate chains.

As mentioned above, this skeletal muscle chondroitin sulfate proteoglycan is synthesized by muscle cells in culture at both the myoblast and myotube stages of myogenesis. However, as myogenesis proceeds in these cultures, an interesting phenomenon occurs. Early in the culture period, the muscle cells are the predominant cell type present, but after the end of fusion, as is true for all skeletal muscle, the muscle cells cease to proliferate. Because the small number of fibroblastic cells which contaminate the muscle cultures continue to undergo cell division, they eventually overgrow the cultures so that older cultures consist of cross-striated, mature myotubes enveloped by the fibroblasts. As the fibroblasts become the dominant phenotype, synthesis of the muscle chondroitin sulfate proteoglycan is no longer observed and instead the major proteoglycans synthesized are smaller molecules that are characteristic of fibroblasts. These proteoglycans comprise a mixture of dermatan sulfate proteoglycans and heparan sulfate proteoglycans. That the mature muscle is still capable of synthesizing the large chondroitin sulfate proteoglycan is evidenced by the observation of its synthesis in older cultures from which most of the fibroblasts are removed by treatment with an agent, cytosine arabinoside, that is toxic to proliferating cells.

The developmental transition in proteoglycan synthesis which is seen in the muscle cultures is even more interesting in light of recent studies concerning *in vivo* synthesis of proteoglycans in skeletal muscle (17). For this analysis, chick embryos were labeled *in ovo* with

[^{35}S]sulfate at various times of development, the leg muscle dissected free, and the proteoglycans isolated and characterized. Synthesis of both the large skeletal muscle proteoglycan and the small fibroblast proteoglycans is detected in the leg muscle *in ovo*. In addition, there is synthesis of a third proteoglycan which has been called mesenchymal proteoglycan, although its exact cell of origin has not yet been unequivocally identified. The mesenchymal proteoglycan, whose synthesis is not detected in the muscle cultures, is large, though not as large as the skeletal muscle proteoglycan. It is also a chondroitin sulfate proteoglycan, but in contrast to the skeletal muscle proteoglycan, its chondroitin sulfate contains only 60-65% 6-sulfated disaccharide. Moreover, the skeletal muscle proteoglycan and mesenchymal proteoglycan differ sufficiently in structure so that they display distinct morphologies in the electron microscope (19). In cytochrome c spreads, the standard format for visualizing proteoglycans in the electron microscope, the skeletal muscle proteoglycan appears as an image in which the chondroitin sulfate chains are highly extended (Fig. 1), while the mesenchymal proteoglycan has a condensed morphology. This condensed morphology is typical of other proteoglycans which have been visualized by this procedure and is further evidence of the uniqueness of the skeletal muscle proteoglycan. Of importance is the observation that as myogenesis proceeds *in ovo*, synthesis of the large proteoglycans, both muscle and mesenchymal, decreases and the small fibroblast proteoglycans become the major species produced in the muscle tissue so that in adult chicken leg muscle, no muscle or mesenchymal proteoglycan synthesis is detected and synthesis of only the fibroblast proteoglycans is seen (Fig. 2). This progression from synthesis of primarily large proteoglycans to production of predominantly small proteoglycans matches exactly that which occurs in the muscle cultures in spite of the fact that an additional large proteoglycan is produced *in ovo*.

In order to address the mechanism which regulates the developmental transition in proteoglycan synthesis from primarily large proteoglycans to predominantly small proteoglycans, an experiment was performed to assess the possible role that the fibroblasts might play in this process. In muscle cultures on day 3, the muscle cells have just completed fusion, and the fibroblast population is low. At this time, there is active synthesis of the skeletal muscle proteoglycan, which is the major proteogly-

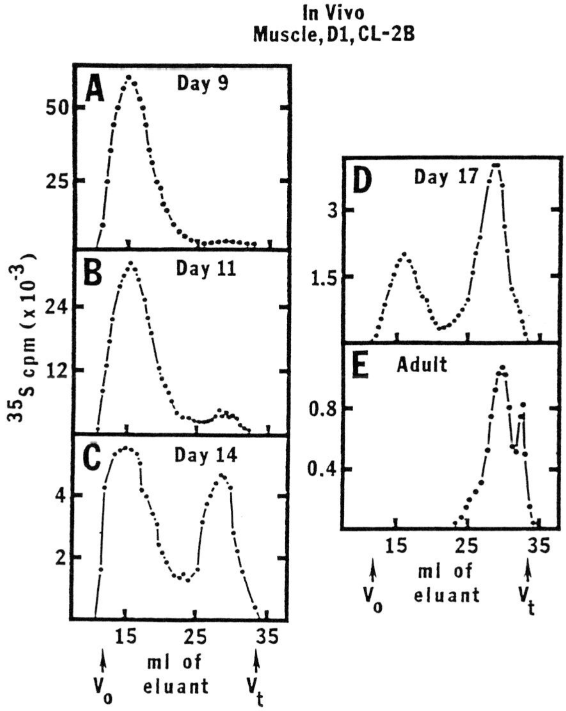

FIGURE 2. Sepharose CL-2B column chromatographic elution profiles of the high-buoyant density (i.e. D1) proteoglycans synthesized *in vivo* in chick leg muscle of various ages. The newly synthesized proteoglycans were analyzed in chick embryos at day 9 (A), day 11 (B), day 14 (C), and day 17 (D) and in adult chicken (E). The peak eluting in panel (E) at the column total volume was not retained in dialysis tubing and very likely represents unincorporated precursor.

can synthesized. Under normal conditions, the myotubes in day 3 cultures are not exposed to a large number of fibroblasts, which does not occur until a few days later. But in this experiment, fibroblasts were overplated into the day 3 muscle cultures, the cultures left overnight, and proteoglycan synthesis measured. In the control muscle cultures, synthesis of the muscle proteoglycan was abundant (Fig. 3), but in the overplated cultures, synthesis of the muscle proteoglycan was not seen (Fig. 3). To elicit this effect, the fibroblasts had to be overplated at a high density. Although the explanation for this effect may be a

FIBROBLASTS OVERPLATED ONTO MUSCLE		
	µg DNA/Culture	% Muscle PG
Day 3 Muscle	23.5	79
Fibroblasts (1x)	5.3	0
Fibroblasts (10x)	54.7	0
Day 3 Muscle + 1x Fibroblasts	28.7	68
Day 3 Muscle + 10x Fibroblasts	58.8	0

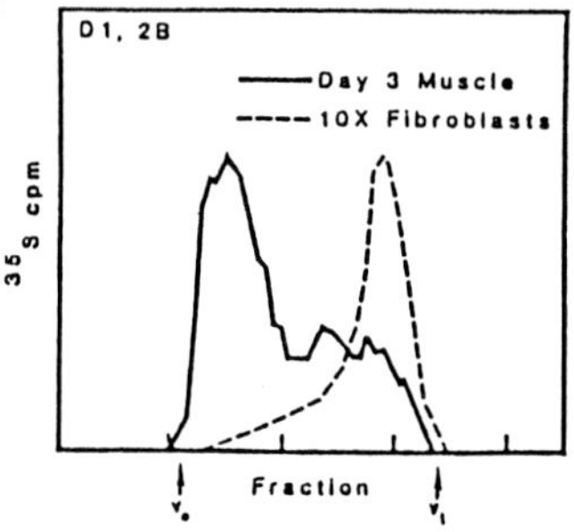

FIGURE 3. Effect of overplating fibroblasts into cultures of leg muscle myotubes. Fibroblasts were plated at two densities, designated 1X and 10X, either into fresh dishes or into muscle cultures. The cultures were labeled with [^{35}S]sulfate and the proteoglycans extracted and analyzed. The Sepharose CL-2B column elution profiles for 10X fibroblasts either alone or plated onto muscle cells were the same (% Muscle PG, percentage of incorporated [^{35}S]sulfate eluting from Sepharose CL-2B as muscle proteoglycan).

trivial one, such as metabolic competition between the muscle cells and the fibroblasts, the results of this experiment may indicate a regulatory interplay between muscle cells and fibroblasts that results in the observed transition of proteoglycan synthesis in developing skeletal muscle.

GENERALITY OF THE TRANSITION FROM LARGE TO SMALL PROTEOGLYCANS

Work in this laboratory has addressed proteoglycan synthesis in a number of other developing systems in the chick embryo. One such system comprises pre-chondrogenic mesenchyme, cartilage, hypertrophic cartilage, and bone (20). As with muscle development, there is a transition in proteoglycan synthesis from large to small proteoglycans. The pre-cartilage mesenchymal cells synthesize a proteoglycan that is somewhat different than that of muscle-mesenchymal cells (Fig. 4), although these may result from the fact that, in contrast to the muscle-mesenchymal proteoglycan whose synthesis was analyzed *in ovo* (17), the pre-cartilage mesenchymal proteoglycan was studied in chick embryo limb bud cultures

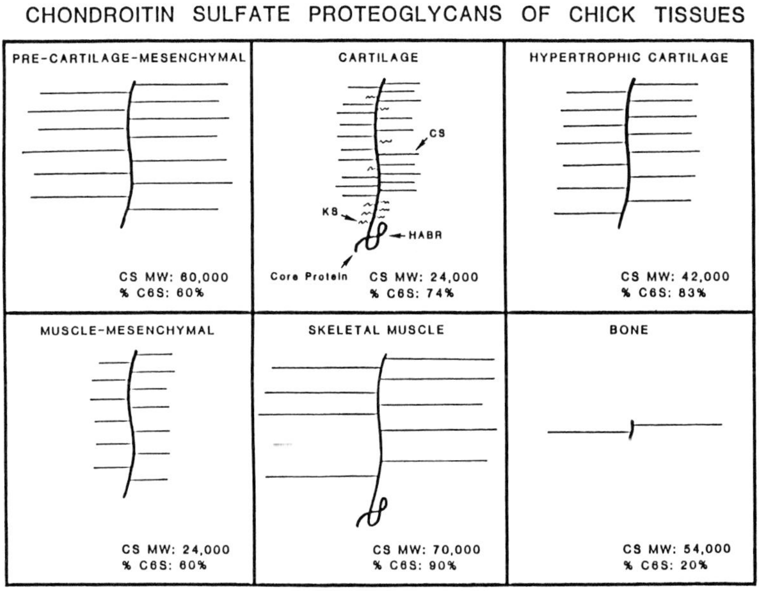

FIGURE 4. Diagrammatic representations of various chondroitin sulfate proteoglycans synthesized in chick tissues (CS, chondroitin sulfate; KS, keratan sulfate; HABR, hyaluronic acid-binding region of the core protein; MW, molecular weight).

---

in which the mesenchymal cells were plated under conditions which enhance chondrogenesis but were analyzed prior to the appearance of cartilage (21). However, it may be that these differences are inherent to the cells themselves since the limb myogenic cells and chondrocytes are thought to arise from distinct mesenchymal cell populations (21). After the differentiation to cartilage, the chondrocytes synthesize typical cartilage proteoglycans (Fig. 4) and proteoglycan synthesis by the chondrocytes progresses through a fixed developmental sequence in which the proteoglycans are synthesized with stage-specific structures (23,24). Eventually, the chondrocytes hypertrophy, at which time there is evidence (25) they synthesize a distinct proteoglycan species (Fig. 4), and then the cartilage is replaced by bone. Bone contains a tissue-specific proteoglycan (25) which is considerably smaller than those of the tissues which preceded this phenotype (Fig. 4). Thus, development of the hard tissues and the musculature of the limb are marked by a similar transition from synthesis of large to production of small proteoglycans. However, in contrast to the small proteoglycans of muscle tissue, the small bone proteoglycan is a chondroitin sulfate proteoglycan (25). Recent work in this laboratory has also revealed a transition from large to small proteoglycans in the development of embryonic chick brain (26).

FUNCTIONAL ASPECTS OF SKELETAL MUSCLE PROTEOGLYCANS

From the above proteoglycan analyses, it is clear that proteoglycan synthesis changes in a specific way during myogenesis from principally the large, chondroitin-6-sulfate muscle proteoglycan to mainly the small fibroblast proteoglycans of which heparan sulfate is a component. It should be noted that these changes correlate with previous studies that focused on merely the glycosaminoglycans produced during skeletal muscle development (4,5). Because the different proteoglycans are synthesized at distinct stages of myogenesis, it is reasonable to assume that they perform specific roles during muscle development. For example, the skeletal muscle chondroitin sulfate proteoglycan is synthesized principally in the earlier phases of myogenesis and it, therefore, probably is involved with some early aspect of muscle development. One possible function that it could fulfill derives from the unusually large size of its chondroitin sulfate chains. In cartilage tissue, which contains much more proteoglycans than other tissues and in which the proteoglycans are best characterized, the major role of the chondroitin sulfate proteoglycans seems to be to structure extracellular solvent (water). This process results in the filling of the extracellular space within the tissue's collagen network and confers upon the tissue its property of resiliency.

From work with chick embryo limb chondrocyte cultures which were treated with an agent, methylxyloside, that results in the production of proteoglycans with only one-third the normal number of chondroitin sulfate chains, it appears that it is the chondroitin sulfate moieties of the proteoglycans which are chiefly responsible for supplying the turgor that fills the space in the extracellular matrix (24). By analogy, then, it may be that the very large chondroitin sulfate proteoglycan of skeletal muscle is also required to fill extracellular space, not necessarily to provide resiliency to the tissue, but rather to reserve space for future expansion of the myogenic cells. At the time when cells begin to exhibit the characteristics of myoblasts, the mesenchymal tissue is quite loose with an abundance of extracellular space. This condition persists even during the early stages of myogenesis, and as further generations of myotubes come into being, they enlarge into the spaces that exist between the earlier myotubes. Hence, the skeletal muscle proteoglycan may be produced to reserve the extracellular space within the developing muscle for the incorporation of later generations of myotubes and/or the enlargement of existing myotubes. Although other functions may exist for muscle cell chondroitin sulfate (13,27), the role proposed above is consistent with both the structure of this proteoglycan and the timing of its synthesis.

It may, likewise, be argued that the large mesenchymal chondroitin sulfate proteoglycan is synthesized by the loose connective tissue around the forming muscle. These proteoglycans could similarly function to structure extracellular water and allow the developing muscle as a morphological unit to expand into that space. As the muscle continues to mature and become wrapped in its specialized connective tissue, this mesenchymal proteoglycan would also no longer be required and would be replaced by the smaller proteoglycans which may serve a highly specific morphological function. Since the developmental process is characterized by sequential replacement events (28), it is not unreasonable to suggest that the family of large chondroitin sulfate proteoglycans functions as space-filling or space-forming units which are replaced by smaller molecules that serve other highly specific roles.

At later stages of muscle development, when fusion is largely complete, the major process occurring in the tissue involves organizing the muscle masses into discrete muscle organs. The earlier myogenic phases of differentiation of myoblasts from mesenchymal cells and fusion of myoblasts into myotubes are, for the most part, autonomous processes. In contrast, the morphenetic phase of myogenesis is thought to involve the participation of fibroblasts which lay down the pattern for muscle formation during fusion and subsequently package the muscle tissue in fibrous connective tissue envelopes (22,29,30). It is at this phase of muscle development that the nature of the proteoglycans synthesized shifts from a majority of the muscle proteoglycan to primarily the fibroblast proteoglycans. Hence, it may be that these small fibroblast proteoglycans are somehow involved in encapsulation of the muscle tissue. Because fibroblast cell surface proteoglycans have been proposed to participate in cell adhesion and motility (11), perhaps the fibroblast proteoglycans synthesized at this stage of myogenesis take part in the organization of the muscle masses by the fibrogenic cells into distinct muscles.

After formation of the muscles themselves, one of the principal components of the extracellular matrix is the basement membrane or basal lamina which surrounds each muscle fiber (31). Interestingly, formation of this basal lamina seems to require the participation of both the muscle cells and the fibroblasts (32). Among the constituents of the basal lamina is a heparan sulfate proteoglycan as well as two structural proteins that are specific for the basal lamina, laminin and Type IV collagen (33). In mature muscle, the basal lamina serves a key function in regeneration of damaged tissue. It has been shown that when mature muscle is damaged and allowed to regenerate, the sites of innervation, the neuromuscular junctions, of the regenerated muscle are at the same location as those of the original muscle (31). Furthermore, the blueprint that allows this precise re-innervation to occur is preserved in the basal lamina which persists after degeneration of the nerve and muscle and directs the reformation of the neuromuscular junctions (31). Hence, heparan sulfate proteoglycans, by virtue of their integration into the basal lamina of muscle, assist in the reformation of muscle morphology during the regeneration of mature skeletal muscle. Indeed, a heparan sulfate proteoglycan localized within skeletal muscle basal lamina has been found to be concentrated at the neuromuscular junction (34). Thus, proteoglycans are synthesized during skeletal muscle development in a stage-specific fashion to fit the requirements of the tissue at that stage. Early in myogenesis, very large chondroitin sulfate proteoglycans are produced, presumably to reserve space for further increase of the muscle cells. Eventually heparan sulfate proteoglycans are synthesized, perhaps first as part of the ingrowth of fibrogenic cells during splitting of the muscle mass and then as a component of the basal lamina to retain the information necessary for normal muscle function and accurate morphogenesis during muscle regeneration.

ACKNOWLEDGEMENTS

This work was facilitated greatly by the valuable technical assistance of Mrs. Carol Ingle, and the creation of this manuscript was aided by the capable secretarial assistance of Mrs. Constance Fenters-Merritt.

REFERENCES

1. Pauw PG, David JD (1979). Alterations in surface proteins during myogenesis of a rat myoblast cell line. Dev Biol 70:27.
2. Cates GA, Brickenden AM, Sanwal BD (1984). Possible involvement of a cell surface glycoprotein in the differentiation of skeletal myoblasts. J Biol Chem 259:2646.
3. Bailey AJ, Shellswell GB, Duance VC (1979). Identification and change of collagen types in differentiating myoblasts and developing chick muscle. Nature (Lond) 278:67.
4. Pacifici M, Molinaro M (1980). Developmental changes in glycosaminoglycans during skeletal muscle differentiation in culture. Exp Cell Res 126:143.
5. Dietrich CP, Sampaio LO (1981). Changes of acidic mucopolysaccharides and mucopolysaccharidases during histogenesis and fetal development: possible involvement of chondroitin sulfate C and hyaluronidase with the processes of differentiation and cell division. In Yamakawa T, Osawa T, Handa S (eds): "Glycoconjugates: Proceedings of the Sixth International Symposium on Glycoconjugates", Tokyo: Japan Scientific Societies Press, p 407.
6. Kujawa MJ, Tepperman K (1983). Culturing chick muscle cells on glycosaminoglycan substrates: attachment and differentiation. Dev Biol 99:277.
7. Kujawa MJ, Pechak DG, Fiszman MY, Caplan AI (1985). Hyaluronic acid bonded to cell culture surfaces inhibits the program of myogenesis. Submitted to Dev Biol.
8. Hascall VC (1981). Proteoglycans: structure and function. In Ginsburg V (ed): "Biology of Carbohydrates, Vol. 1," New York: John Wiley and Sons, Inc., p 1.
9. Kjellen L, Pettersson I, Hook M (1981). Cell-surface heparan sulfate: an intercalated membrane proteoglycan. Proc Natl Acad Sci USA 78:5371.
10. Rapraeger AC, Bernfield M (1983). Heparan sulfate pro-

teoglycans from mouse mammary epithelial cells: a putative membrane proteoglycan associates quantitatively with lipid vesicles. J Biol Chem 258:3632.

11. Rollins BJ, Culp LA (1979). Preliminary characterization of the proteoglycans in the substrate adhesion sites of normal and virus-transformed murine cells. Biochem 18:5621.
12. Hook M, Robinson J, Kjellen L, Johansson S, Woods A (1982). Heparan sulfate: on the structure and function of cell associated proteoglycans. In Hawkes S, Wang JL (eds): "Extracellular Matrix," New York: Academic Press, p 15.
13. Markwald RR, Runyan RB, Kitten GT, Funderburg FM, Bernanke DH, Brauer PR (1984). Use of collagen gel cultures to study heart development: proteoglycan and glycoprotein interactions during the formation of endocardial cushion tissue. In Trelstad RL (ed): "The Role of Extracellular Matrix in Development." New York: Alan R Liss, p 323.
14. Bernfield M, Banerjee SB (1982). The turnover of basal lamina glycosaminoglycan correlates with epithelial morphogenesis. Dev Biol 90:291.
15. Thompson HA, Spooner BS (1982). Inhibition of branching morphogenesis and alteration of glycosaminoglycan biosynthesis in salivary glands treated with β-D-xyloside. Dev Biol 89:417.
16. Carrino DA, Caplan AI (1982). Isolation and preliminary characterization of proteoglycans synthesized by skeletal muscle. J Biol Chem 257:14145.
17. Carrino DA, Caplan AI (1984). Isolation and partial characterization of high-buoyant-density proteoglycans synthesized *in ovo* by embryonic chick skeletal muscle and heart. J Biol Chem 259:12419.
18. Carrino DA, Pechak DG, Caplan AI (1985). Structural characterization of skeletal muscle chondroitin sulfate proteoglycan. In preparation.
19. Pechak DG, Carrino DA, Caplan AI (1985). Electron microscopic characterization of chick embryonic skeletal muscle proteoglycans. J Cell Biol, in press.
20. Carrino DA, Weitzhandler M, Caplan AI (1985). Proteoglycans synthesized during the cartilage to bone transition. In: "Proceedings of the Second International Conference on the Chemistry and Biology of Mineralized Tissues," in press.
21. Hall DE, Caplan AI (1985). Manuscript in preparation.
22. Mauger A, Kieny M (1980). Migratory and organogenetic

capacities of muscle cells in bird embryos. Wilhelm Roux's Arch 189:123.

23. DeLuca S, Heinegard D, Hascall VC, Kimura JH, Caplan AI (1977). CHemical and physical changes in proteoglycans during development of chick limb bud chondrocytes grown *in vitro*. J Biol Chem 252:6600.
24. Lohmander LS, Hascall VC, Caplan AI (1979). Effects of 4-methyl umbelliferyl-β-D-xylopyranoside on chondrogenesis and proteoglycan synthesis in chick limb bud mesenchymal cell cultures. J Biol Chem 254:10551.
25. Carrino DA, Caplan AI (1985). Isolation and characterization of proteoglycans synthesized *in ovo* by embryonic chick cartilage and new bone. J Biol Chem 260:122.
26. Cooper C, Caplan AI (1985). Manuscript in preparation.
27. Schubert D, LaCorbiere M (1980). A role of secreted glycosaminoglycans in cell-substratum adhesion. J Biol Chem 255:11564.
28. Caplan AI, Fiszman MY, Eppenberger HM (1983). Molecular and cell isoforms during development. Science 221:921.
29. Shellswell GB, Wolpert L (1977). The pattern of muscle and tendon development in the chick wing. In Ede DA, Hinchliffe JR, Balls M (eds): "Vertebrate Limb and Somite Morphogenesis," Cambridge: Cambridge University Press, p 71.
30. Stopak D, Harris AK (1982). Connective tissue morphogenesis by fibroblast traction I. Tissue culture observations. Dev Biol 90:383.
31. Nitkin RM, Fallon JR, Godfrey EW, Wallace BG, McMahan UJ (1984). The role of extracellular matrix in regeneration of the neuromuscular junction. In Trelstad RL (ed): "The Role of Extracellular Matrix in Development," New York: Alan R. Liss, p 597.
32. Lipton BH (1977). A fine-structural analysis of normal and modulated cells in myogenic cultures. Dev Biol 60:26.
33. Lesot H, Kuhl U, von der Mark K (1983). Isolation of a laminin-binding protein from muscle cell membranes. EMBO Jour 2:861.
34. Anderson MJ, Fambrough DM (1983). Aggregates of acetylcholine receptors are associated with plaques of a basal lamina heparan sulfate proteoglycan on the surface of skeletal muscle fibers. J Cell Biol 97:1396.

Molecular Biology of Muscle Development, pages 133–142

HEPARIN-LIKE COMPONENTS MAY REGULATE SKELETAL MUSCLE GROWTH AND DIFFERENTIATION[1]

Elissavet Kardami, Sheila McClure and
Richard C. Strohman.

Department of Zoology, U.C. Berkeley, Ca 94720

ABSTRACT. We have found that selected chicken muscle extracts contain a growth factor,Chicken Muscle Growth Factor (CMGF) which stimulates myoblast proliferation and delays muscle fusion,and which binds strongly to heparin (1).When heparin or heparan sulfate proteoglycan (HSPG) are added to muscle primary cultures,a dramatic inhibition of growth ensues.This inhibition is partly lifted by Fibroblast Growth Factor (FGF) or CMGF,and completely reversed by selected muscle or nerve extracts.The effect of heparin is not mimicked by other extracellular matrix components and does not concern cell attachment.The interaction of CMGF with heparin-like components could modulate muscle growth and differentiation <u>in vivo</u>.

INTRODUCTION

Selected muscle or nerve extracts are highly myogenic in culture;this reflects in part the presence of transferrin in these extracts(2,3).We recently found another activity in the extracts which is distinct from transferrin (1,4): CMGF was purified 40,000-fold by heparin-affinity chromatography of the crude extracts (1) and can stimulate myoblast proliferation and delay the differentiation process in nanogram amounts.There are other growth factors with a similar affinity towards heparin,like FGF and Endothelial Cell Growth Factor,ECGF,(5,6);interestingly,the activity of the latter is potentiated by heparin (7).

We investigated the effects of heparin and other

[1]This work was supported by grants from the Muscular Dystrophy Association and the National Institute of Health.

proteoglycans on the growth and differentiation of muscle cells in culture,and the involvement of growth factors in these events.Our data,presented below,suggest a physiological role for the interaction of CMGF with heparin.

MATERIALS AND METHODS

Extracts from chicken Anterior Latissimus Dorsi-ALD-muscles were obtained as described previously (3).Skeletal muscle primary cultures from the pectoral muscle of 12-day old chick embryos were obtained as in ref.8.Total protein and myosin heavy chain (MHC) determination were performed as in ref.9.In most of the experiments described here, prolonged preplating of the trypsinized cells in the absence of serum (10),succeded in eliminating most of the contaminating fibroblasts.

Heparin-Sepharose CL-6B (Pharmacia Fine Chemicals) was processed according to the manufacturer's instructions. The sodium salts of heparin,grade I,chondroitin sulfate, grade III,and hyaluronic acid,grade III-s,were obtained from Sigma.HSPG was purchased from Miles.

FGF was a generous gift from Prof. D. Gospodarowicz, U.C.S.F. Purified chicken transferrin was purchased from Cappel laboratories.

Cells were routinely grown in complete medium(CM), consisting of 10% horse serum and 1.5% chick embryo extract,in Eagle's minimum essential medium (MEM),unless otherwise specified.

RESULTS AND DISCUSSION

Chicken muscle cells,grown in CM,go through a proliferative phase (24-48 h) before they cease dividing and start fusing,forming large multinucleated myotubes which predominate after 72 h in culture.The addition of ALD extract to these cultures,induces a profound stimulation of muscle growth (3) as well as a prolongation of the proliferative phase of muscle cells:

As can be seen in Figure 1,a, cells grown in the presence of ALD extract are still dividing after 4 days. When the extract has been passed through a heparin column, it is still myogenic but no longer prolongs the proliferation of myoblasts; large myotubes have formed after 4 days in the presence of heparin-absorbed extract (Fig.1,b).We have found that the component removed by heparin from the extract,CMGF,is a potent mitogen for muscle cells and is

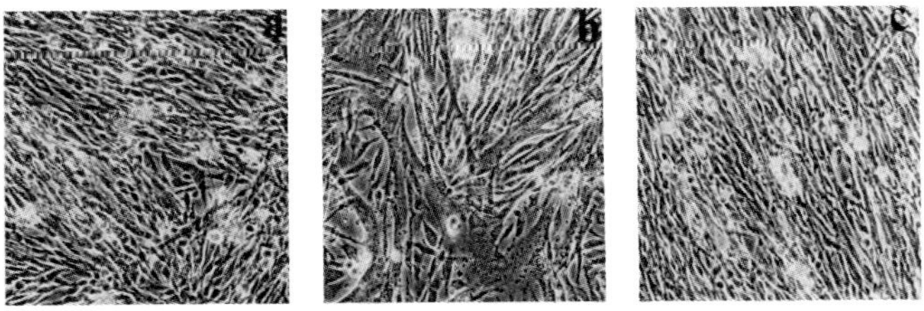

FIGURE 1. Skeletal muscle primary cultures after 4 days in the presence of transferrin-supplemented CM (10 ug/ /ml transferrin),and (a) 0.5 mg/ml ALD-extract before heparin absorption, (b) 0.5 mg/ml ALD extract after heparin absorption, and (c) 10 ng/ml CMGF.

responsible for the prolongation of proliferation,as can be seen in Figure 1, c.

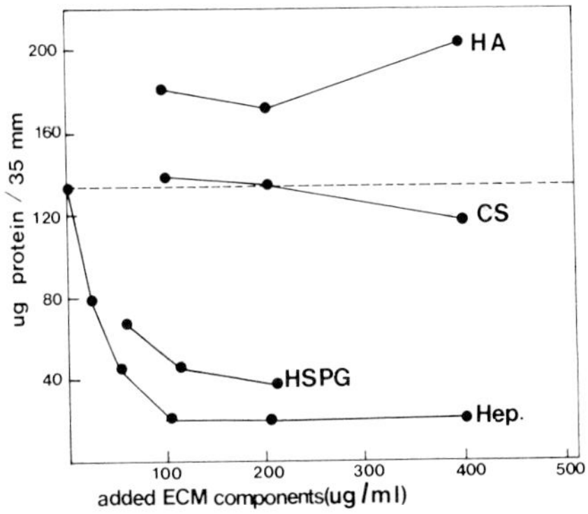

FIGURE 2. Effect of various proteoglycans on the growth of muscle cells,in CM,after 7 days in vitro. The ordinate denotes micrograms of accumulated protein (an adequate measure of cell growth in this case) and the abscissa,the micrograms of added proteoglycan.Each point is an average over 4 plates.The broken line indicates level of growth when no additions are made. HA,CS,HSPG,HEP. mark the levels of growth in the presence of CM and hyaluronic acid,chondroitin sulfate,heparan sulfate proteoglycan, and heparin, respectively.

The presence of such a potent retardant of differentiation in ALD muscles,and its affinity for heparin, pose intriguing questions about regulation of growth by components of the extracellular matrix. We therefore examined the effects of the various proteoglycans on the growth and differentiation of cultured muscle.

We found that chondroitin sulfate and hyaluronic acid did not have any significant effect on muscle growth,in the concentration range studied (Figure 2).Hyaluronic acid induced some stimulation of overall growth ,which varied greatly between batches.In contrast,the addition of heparin or HSPG,at concentrations above 20 ug/ml,induced a profound inhibition of muscle growth (Figures 2 and 3).

Heparin appears to affect specifically the myogenic cells in the culture.The fibroblasts,which contaminate the muscle cultures,continue to grow and predominate completely after 5 days in the presence of heparin (Figure 3).At this point,very few,thin myotubes can be detected in the cultures (Figure 3 ,b).

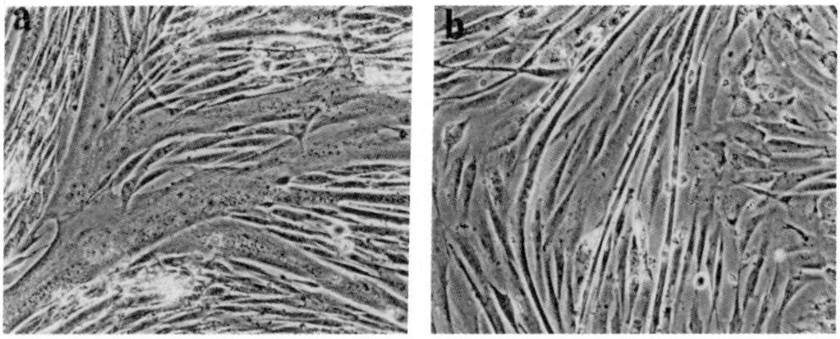

FIGURE 3. Skeletal muscle primary cultures,after 5 days in the presence of (a) CM , and (b) CM and heparin,to a final concentration of 90 ug/ml.

The effects of heparin can be counteracted by the addition of ALD extract to the cells,within 24 h from the time of plating,as shown in Table 1. Sciatic nerve and dystrophic pectoralis muscle extract exibit exactly the same ability for restoring growth,while extract from normal pectoralis muscle was far less effective (not shown here).

Preliminary observations in our laboratory indicate that heparin has no significant effect on muscle morphology or overall growth,if added to the cells after they have fused completely.It seems therefore that heparin

TABLE 1
HEPARIN EFFECT ON MUSCLE [a]

Additions to cells	ug MHC / 35 mm
-None-	35
90 ug/ml heparin	less than 5
0.5 mg/ml ALD-extract	65
0.5 mg/ml ALD-extract, + 90 ug/ml heparin.	52

[a] Chicken muscle primaries,after 7 days.

exerts its effect on pre- or post- mitotic myoblasts.

Heparin could inhibit growth either by depleting the medium of essential nutrients,or by acting directly on the cell surface,or both. Fibronectin is essential for the attachment of the muscle cells (11) , and has heparin-binding sites (12). As can be discerned from Table 2,there is no detectable difference in the number of attached cells in the presence of heparin.The same number of cells adhere to the collagen coated dish 6 h and 24 h after plating in the presence or absence of heparin.This implies that heparin may not bind enough fibronectin

TABLE 2
HEPARIN EFFECT ON ATTACHMENT [b]

Additions to cells	6 h	24 h
-None-	220,000	325,000
0.1 mg/ml heparin	210,000	330,000

[b] Cell density at the time of plating: 250,000/ /35 mm plate.The values shown are averaged over 3 plates.Cells were counted in a Coulter counter after brief trypsinization .

from the medium to affect attachment under our the conditions of our cultures.

When the complete medium is absorbed to heparin sepharose and fed to the cells within 6 h from the time of plating , it no longer supports muscle growth , as shown in Figure 4 (compare a and b).This shows that heparin inhibition involves the binding of essential components from

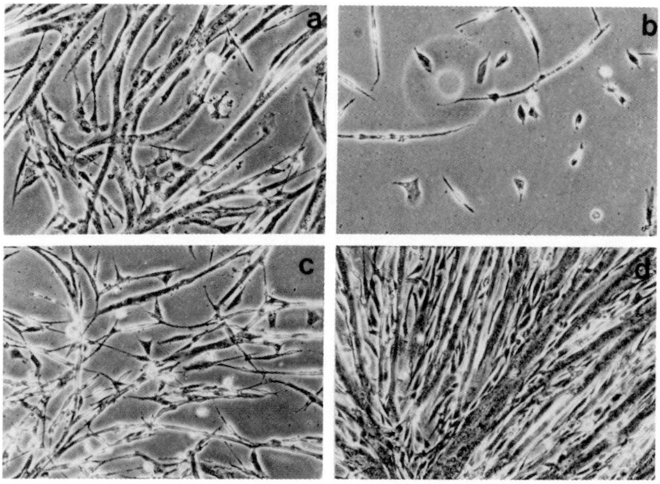

FIGURE 4. Effect of heparin-column absorption on the growth properties of CM. 4 day old cultures(preplated in the absence of serum),in the presence of,(a),non-absorbed CM,(b),heparin-absorbed CM,(c)heparin-absorbed CM and 10 ng/ml CMGF ,and,(d),heparin absorbed CM and 0.5 mg/ml ALD extract.

the medium,rendering them inaccessible or inactive for the cells.Heparin may also act directly on the cells but we do not explore this possibility further here.

Transferrin is one of the identified essential components of the growth medium (13). Neither transferrin,nor fibronectin can restore the growth promoting properties of the absorbed medium (not shown here).

When CMGF is added to the absorbed medium,at 10 ng/ml,a partial restoration of growth can be observed (Figure 4,c).The cells are viable and still dividing after 4 days, while some myotubes have formed.Intact ALD extract completely restores and stimulates even further muscle growth (Figure 4 , d).

Our results support strongly the notion that a CMGF-like mitogen is present in the growth medium ,which is essential in supporting myogenesis in vitro. Chicken embryo extract contains a differentiation - delaying activity which we feel may be CMGF-like (14).

The possibility exists that a heparin binding conta - minant of the CMGF preparation may be responsible for the partial restoration of growth. We have nevertheless found that pure FGF (CMGF shares many of the properties of FGF, (1,4),) also restores,to the same extent,the growth properties of CM (Table 3).It seems therefore likely that the growth factor activity and not a contaminant partially restores growth.

TABLE 3
PARTIAL RESTORATION OF GROWTH BY CMGF AND FGF[c]

Growth Medium	ug protein/35 mm
Non-absorbed CM	84
Absorbed CM	30
Absorbed CM + 10 ng/ml CMGF	60
Absorbed CM + 15 ng/ml FGF	65

[c] 6 day old cultures.

HSPG is one of the components of the basal lamina which surrounds muscle fibers and satellite cells (15). There is some evidence from <u>in vitro</u> experiments that HSPG synthesis may be developmentally regulated (16). Its effect on muscle is not as dramatic as that of heparin;at 10 ng/ml of HSPG,no significant inhibition is observed (Table 4). Nevertheless,if FGF,or CMGF , are added to the cells at the same time,the familiar stimulation of growth they induce on muscle is abolished,presumably by binding to HSPG(Table 4).Furthermore,the morphological differentiation of the untreated muscle cells appears identical to that of the cells that have been treated with the growth factor(s) in the presence of 10 ug/ml HSPG.

Since,as our results suggest,the CM contains CMGF-like growth factor(s),and the HSPG synthesis by muscle cells increases near the time of fusion (16), it seems possible that the medium mitogens become 'trapped' in a similar fashion,i.e. by complexing with heparin-like components, inducing,or accelerating the differentiation process.

TABLE 4
LOSS OF THE GROWTH FACTOR EFFECT [d]

Additions	ug protein / 35 mm
-None-	100
CMGF	129
FGF	135
HSPG,10 ug/ml	95
HSPG (10 ug/ml)+CMGF(10 ng/ml)	90
HSPG (10 ug/ml)+FGF(15 ng/ml)	99

[d] 6 day old cultures,preplated in the absence of serum.

One can speculate that a similar system is operant *in vivo*. ALD muscle, a small,oxidative,satellite cell rich muscle,accumulates two potent myogenic factors like transferrin (3) and CMGF (1),and yet retains its small size, although the satellite cells are stimulated by FGF(17).

The basal lamina,which contains HSPG, may trap the growth factor(s) ,releasing it when required,as in times of injury.The mechanism that would induce such a release is unknown. Interestingly,activated T-lympocytes synthesize a galactosidase which degrades HSPG side chains (18) ; the invasion therefore of the injured tissue by the cells of the immune system might also render the growth factor(s) accessible for the stimulation of the satellite cells.

In conclusion,we have found that heparin and HSPG inhibit the growth of muscle cells in culture,probably by binding essential components of the medium.CMGF-like factor(s) are among those components.The physiological consequences of the interaction between heparin and CMGF remain to be examined.

REFERENCES

1. Kardami E, Spector D, Strohman RC (1985). A myogenic growth factor present in skeletal muscle is purified by heparin affinity chromatography. Submitted for publication.
2. Markelonis GJ, Oh TH (1980). 'Sciatin':purification and characterization of a myotrophic protein from chicken sciatic nerves. J Biol Chem 255:8967.
3. Matsuda R, Spector D, Strohman RC (1984). There is selective accumulation of a growth factor in chicken skeletal muscle.I.Transefrrin accumulation in Anterior Latissimus Dorsi. Dev Biol 103:267.
4. Kardami E, Spector D, Strohman RC (1984). Muscle extracts contain a growth promoting factor which is different from transferrin. J Cell Biol 99:275a.
5. Gospodarowicz D, Cheng J, Lui G-M, Baird A, Bölent P (1984). Isolation of brain fibroblast growth factor by heparin-Sepharose affinity chromatography:Identity with pituitary fibroblast growth factor. Proc Natl Acad Sci USA 81:6963.
6. Maciag T, Mehlman T, Friesel R, Schreiber AB (1984). Heparin binds endothelial cell growth factor,the principal endothelial cell mitogen in bovine brain. Science 225:932.
7. Thornton SC, Mueller SN, Levine EM (1983). Human Endothelial cells: use of heparin in cloning and long serial cultivation. Science 222:623.
8. Bandman E, Strohman RC (1982). Increased K^+ inhibits spontaneous contractions and reduces myosin accumulation in cultured chick myotubes. J Cell Biol 93:698.
9. Strohman RC, Bandman E, Walker C (1981). Regulation of myosin accumulation by muscle activity in cell culture. J Muscle Res Motil 2:269.
10. Puri E, Turner D (1978). Serum-free medium allows chicken myogenic cells to be cultivated in suspension and separated from attached fibroblasts. Exp Cell Res 115: : 159.
11. Cliquet M , Puri E, Turner DC (1979). Fibronectin mediates attachment of chick myoblasts to a gelatin-coated substratum. J Biol Chem 254:5475.
12. Yamada KM, Kennedy DW, Kimata K, Pratt RM (1980). Characterization of fibronectin interactions with glycosaminoglycans and identification of active proteolytic fragments. J Biol Chem 255:6055.

13. Ii I, Kimura I, Osawa E (1982). A myotrophic protein chicken embryo extract:its purification,identity to transferrin and indispensability for avian myogenesis. Dev Biol 94:366.
14. Ii I, Osawa E (1984). Myoblast proliferation factor from embryo extract which delays myoblast fusion. International Cell Biology 1984.Tokyo: Satimaru Seno and Yoshio Okada, p 440.
15. Anderson MJ, Fambrough DM (1983). Aggregates of acetylcholine receptors are associated with plaques of a basal lamina heparan sulfate proteoglycan on the surface of skeletal muscle fibers. J Cell Biol 97:1396.
16. Noonan DM, Przybylski RJ (1984). Changes in heparan sulfate proteoglycans with skeletal muscle development. J Cell Biochem Suppl 8B:292.
17. Allen RE, Dodson MV, Luiten LS (1984). Regulation of skeletal muscle satellite cell proliferation by bovine Pituitary fibroblast growth factor. Exp Cell Res. 152: 154.
18. Naparstek Y, Cohen IR, Fuks Z, Vlodasky I (1984). Activated T-lymphocytes produce a matrix degrading heparan sulfate endoglycosidase. Nature (London) 310:241.

Molecular Biology of Muscle Development, pages 143–152

INORGANIC IRON ION SUPPORTS MUSCLE GROWTH AND DIFFERENTIATION AS IRON-BOUND TRANSFERRIN DOES[1]

Eijiro Ozawa*, Koji Saito*, Yasuko Hagiwara*, Akiko Shoji*, Kayoko Saito*, and Saoko Atsumi*+

Division of Cell Biology*, National Center for Nervous, Mental and Muscular Disorders, Kodaira, Tokyo 187 Japan and Department of Anatomy+, Yamanashi Medical College, Tamaho, Yamanashi, 409-38 Japan

ABSTRACT Myotubes were formed in vitro in a medium containing either Fe or FeTf and various effects of these substances on the growth and differenti- were compared. Examination of the differentiated state of both types of myotubes showed that the mechanism of their excitation-contraction and excitation-metabolism coupling was almost the same as that in adult muscle cells, and that Fe and FeTf served qualitatively in the same manner to support the differentiation. The amount of RNA in these cells was dependent on the presence of Fe or FeTf, and this may be the basis of the growth and differentiation of the myotube.

INTRODUCTION

We have shown that chick myogenic cells do not grow in a basal medium composed of Eagle's MEM and horse serum, but

[1]This work was partly supported by a grant from the National Center for Nervous, Mental and Muscular Disorders (NCMMD No.84-07) of the Ministry of Health and Welfare, and a Grants-in-Aid for Special Project Research from the Ministry of Education, Science and Culture (Project No. 59113008),Japan.

grow well when a small amount of chick transferrin, an Fe binding protein, is added to the basal medium (1). In the presence of Tf, myoblasts multiply to increase in number and fuse to form myotubes which then increase in size and eventually show cross striations and spontaneous twitching.

The growth promoting effect is proportional to the amount of Fe bound to Tf. In addition, inorganic Fe ion can replace FeTf in myotube formation; thus, Fe is essential for growth (2).

We have further shown that FeTf is required to prevent myotubes from degeneration. Fe can qualitatively replace FeTf in this process also (3).

In the present study, we compared myotubes grown in a basal medium in the presence of either Fe (Fe-myotubes) or FeTf (FeTf-myotubes), and examined to what extent Fe can replace FeTf in promoting growth and differentiation and in maintaining differentiated myotubes.

Materials and Methods

Chick myoblasts were obtained from 11-12 day embryo breast muscles and cultured and harvested as described (3). Chick Tf was prepared from chick serum as described (1). Creatine kinase (CK) activity was assayed by measuring the catalytic activity to transfer phosphate bond from creatine phosphate to ADP (4). CK isozymes were analyzed by the modified method of Rosalki (5). Activity of phosphorylase kinase (Ph K) of a simple extract of myotubes was assayed using rabbit phosphorylase b as a substrate (6). Myosin B was extracted from cultured cells by the method of Szent-Györgyi (7) and its ATPase activity was assayed by the method of Ohshima et al. (8). Membrane potential changes and contraction of the myotubes were recorded as previously described (9). The RNA content was assayed by the modified method of Schneider (10).

RESULTS

Myotubes accumulate proteins during growth. Since their amount has been proved to be proportional to the CK activity accumulated, this activity was employed to evaluate growth(11).

Microscopic photographs of myogenic cells cultured in the absence or presence of FeTf or Fe ion are shown in

Figure 1. Formation of myotubes is dependent on the presence of these substances.

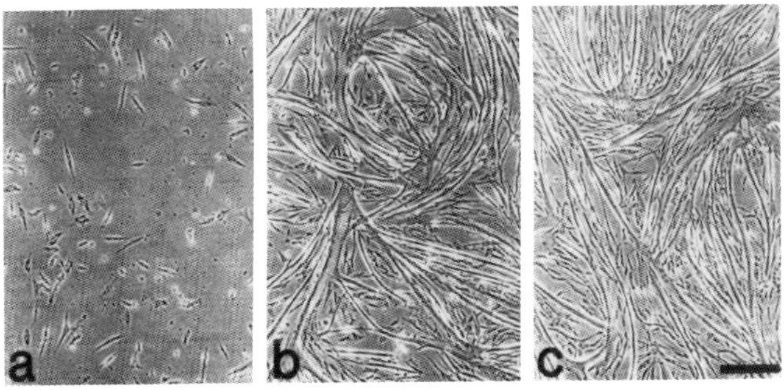

FIGURE 1. Myogenic cells cultured in the absence (a) or presence of 100 μM Fe^{3+} (b) or 1 μM FeTf (c). The cells were inoculated at $3X10^5$ cells/dish and incubated for 4 days. Scale bar:200 μm. (12)

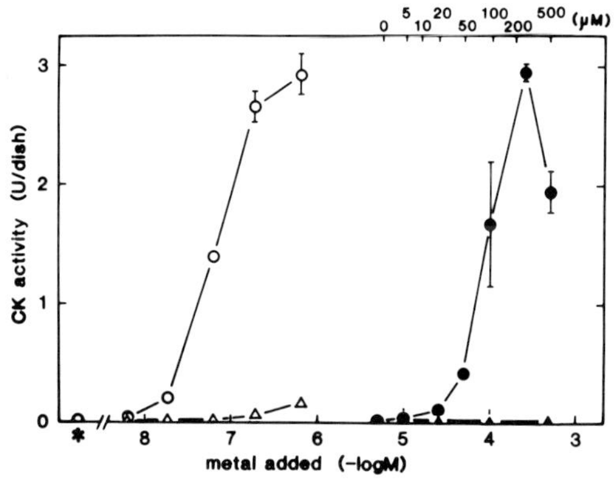

FIGURE 2. Effects of Fe^{3+} (●), FeTf (○), Cr^{3+} (▲) and CrTf (Δ) on myogenic cell growth. Each symbol and bar represents mean±SD of CK activity in 4 dishes. Circle above the asterisk shows the value of the controle cultured in the absence of the metals or the Tf complexes. The cells were inoculated at $5X10^5$ cells/dish and incubated for 4 days. (12)

As shown in Figure 2, Fe ion stimulated the growth of myogenic cells as did FeTf, although the concentration of Fe ion required was about 2 orders higher. Horse serum in the basal medium contained horse FeTf and apoTf, but horse FeTf does not exert a trophic effect on chick cells (11). Since the affinity of apoTf to Fe ion is very high, a part of the added Fe ion must be sequestered by the apoTf present. The actual effective concentration is, therefore, slightly lower than that of added Fe ion. The concentration calculated by subtracting the amount of sequestered Fe from that of added Fe ion is shown in the upper scale of Figure 2.

As any one of transition metals tested (Cr, Cu, Mn, Co, Cd, Zn, Ni) in either Tf-bound form or ion form does not stimulate growth, it is concluded that Fe is indispensable for its action (12). In order to confirm this indispensability, the growth promoting effect of heme-related substances was examined. Among δ-aminolevurinic acid, porphobilinogen, protopoporphyrin, hemin, bilirubin and biliverdine, only hemin promoted growth at concentrations between 10 and 100 μM. Hemoglobin and myoglobin promoted growth. Only Fe-containing substances were effective.

Muscle specific proteins accumulated in myotubes contribute not only enlargement but also structural and functional differentiation of the cells. In the following, various differentiation aspects of Fe- and FeTf-myotubes are compared.

1) Morphological differentiation

Myoblasts fused to form myotubes after multiplication. Following further incubation, Fe- and FeTf-myotubes grew large and showed cross striations. Electron microscopic observation revealed well developed myofibrils with A, I and Z bands arranged in order. The sarcoplasmic reticulum (SR) and T-system were well organized. Both the myotubes differentiated similarly, however, the myofibrils in Fe-myotubes looked slightly less developed.

2) Appearance of myofibrillar proteins

Myosin B which contained most of the myofibrillar proteins was prepared from both types of myotubes and analyzed by 2D-gel electrophoresis. Spots corresponding to actin(A), tropomyosin (TM), and 3 kinds of myosin light chains (fast 1 (Lf1), fast 2 (Lf2) and slow 2 (Ls2)) and other undefined spots were found in both preparations.

There were also a few additional spots in the basic domain in myosin B from Fe-myotubes, although their nature was not known.

These myosin B preparations showed ATPase activity, and this activity increased with the increase of Ca ion concentration from 10^{-6} to 10^{-5} M. The activity-Ca ion concentration curves fitted with the curve of myosin B preparation from 15 day chick embryo breast muscle. Even in a Ca-free environment, myosin B from cultured myotubes showed fairly high activity, whereas the preparation from 15 day chick embryo had virtually no activity (data not shown).

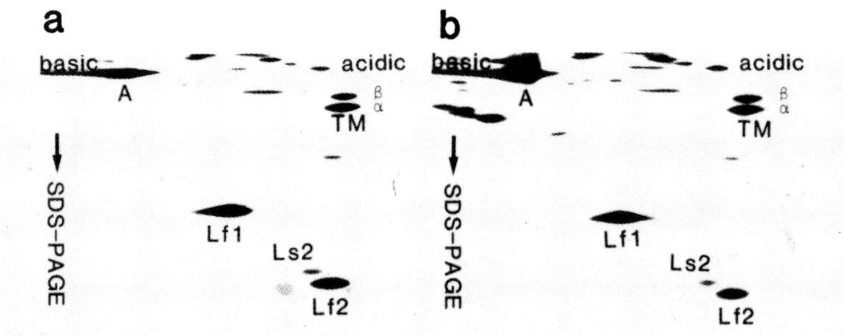

FIGURE 3. Two dimensional gel electrophoretic patterns of myosin B from myotubes cultured for 144 h in the presence of 0.2 μM FeTf (a) or 100 μM Fe^{3+} (b).

3) Differentiation and appearance of enzymes related to ATP production

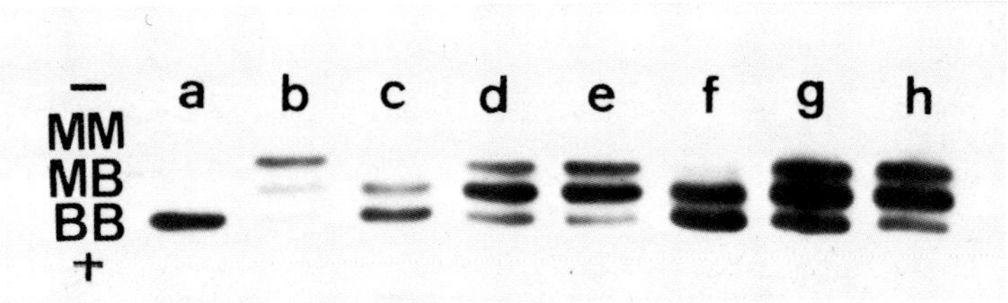

FIGURE 4. Electrophoretic patterns of CK isoenzymes on cellulose acetate membrane. CK preparations were obtained from the brain (a) and breast muscle (b) of 13th day chick embryo and myotubes cultured in the presence of 0.2 μM FeTf for 72 h (c), 120 h (d) and 157 h (e) or 100 μM Fe^{3+} for the same periods (f, g, h, respectively).

ADP, which is degraded from ATP on contraction, is furnished with "high energy phosphate bond" from creatine phosphate (CP) to produce ATP. This reaction is catalyzed by CK. After 72 h of incubation, both myotubes contained brain type (BB) and brain-muscle hybrid type (BM) of CK; muscle type (MM) activity was not yet elicited. With the advance of incubation period, MM type appeared and increased and the relative amount of BB type decreased.

Since the amount of CP in the cells is limited, muscle cells degrade glycogen to glucose 1-phosphate on contraction, which is further broken down to produce ATP. The key enzyme of the process is Ph K which catalyzes conversion of phosphosylase b (an inactive form) to phosphorylase a (an active form) in the presence of a small amount of Ca ion.

Ph K activities were detected in both types of myotubes. Their activity-Ca ion concentration curves fitted with the curve of the kinase prepared from 15 day chick embryo breast muscle (data not shown).

4) Development of excitation of cell membrane and muscle contraction

In the chick myotubes grown in vivo and in the presence of chick embryo extract in vitro, various kinds of myotubes have been found which show only passive depolarization, a prolonged depolarization, spike discharges followed by prolonged depolarization and only the spike discharges (13,14). Excitability of the membrane is considered to develop in this order in the chick embryo (14). Figure 5 shows the responses observed in the FeTf- and

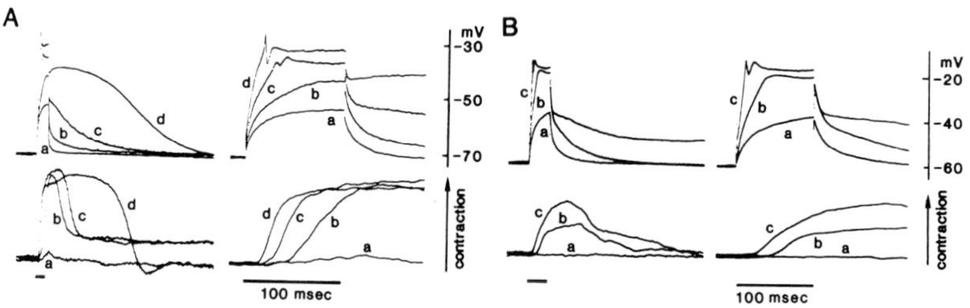

FIGURE 5. Electrical and mechanical responses to electrical stimulation of myotubes cultured in the presence of 0.2 μM FeTf for 12 days (A) or 100 μM Fe^{3+} for 9 days (B). Intensities of current passed were 5 (a), 7 (b), 10 (c) and 15 nA (d) in A, or 5 (a), 10 (b) and 20 nA (c) in B.

Fe-myotubes. Only a passive response was elicited with a small current. A prolonged depolarization and spike discharge appeared with an increase in current intensity. Clear contraction followed the prolonged depolarization, and its latency and rising phase were shortened as the rate of rise of electrical response increased. The electrical and mechanical responses were thus similar between FeTf- and Fe-myotubes. Resting membrane potential did not differ between the two myotube types cultured for 3-10 days: FeTf-myotube, 69.1±12.3 mV (mean±SD, n=49); Fe-myotube, 69.4 ±12.2 mV (n=19). There was no relationship between the culture period and the resting membrane potential.

5) Reduction in RNA synthesis due to FeTf removal

Myotubes degenerate when cultured for a few days without Fe ion or FeTf (3), showing fatty droplets and vacuoles in the cytoplasm, leakage of blebs through cell membrane and atrophy of cell body. We further found that the total amount of RNA was greatly reduced without reduction in the DNA content in parallel with the degenerative process. Reduction and degeneration were both prevented by providing Fe ion or FeTf (data not shown).

Discussion

Fe or FeTf at appropriate concentrations stimulated sufficient growth and differentiation of myotubes and prevented the cells from degenerating. The concentration of Fe ion required to achive this was about 2 orders higher than that of FeTf. This difference may be due to the role of Tf protein moiety which serves as a vehicle of Fe ion to facilitate its entrance into the cells. Apart from this, the effects of Fe and FeTf were almost the same.

This paper provides additional evidence to show that the mechanisms of coupling between excitation of cell membrane and activation of contractile and metabolic systems are almost the same in myotubes as those in adult muscle cells. It has been apriorily believed that cultured myotubes contract with the same molecular mechanism.

As for the ATP-consuming system, proteins composed of myofibrils were detected. Myosin B, the assembly of these proteins, showed Ca-dependent ATPase activity; myosin B has troponin C as the receptor of Ca ion. For the ATP-producing system, CK and Ca-dependent Ph K activity was elicited. Ph K is known to contain calmodulin as one of its subunits (16).

SR, which releases Ca ion following membrane excitation in adult muscle, was morphologically found in myotubes. Further, we found calsequestrin in homogenate of myotubes which sequesters Ca ion during relaxation, and have fairly reliable evidence showing that the contraction of myotubes is regulated by Ca ion stored inside the cells (unpublished data). Therefore, it is plausible that Ca ions, released from SR when myotubes contract, binds simultaneously to troponin C and calmodulin to activate the ATP-consuming and ATP-producing systems, respectively, as in the case of the adult muscle (17,18).

Excitation of myotube cell membrane which gives rise to Ca release from SR is not necessarily induced by the same ionic mechanism as that in adult muscle cells. In the latter, the inward current is conveyed almost solely by Na ions. In myotubes, while spike discharge is due to Na inward current, prolonged depolarization is reported to result mainly from an increase in calcium (13,14) or chloride (15) conductance. Although the prolonged depolarization may, at least in part, be due to Ca current, we could not obtain the evidence which suggested that Ca ion carried into the cell elicited contraction.

These excitatory, contractile and metabolic systems develop after myotube formation. The development is clearly dependent on the synthesis of a certain amount of various kinds of proteins. We proved that synthesis of total RNA and also myosin heavy and light chain mRNAs is dependent on the presence of FeTf. Following deprivation of FeTf from the medium, the rate of synthesis was reduced (in preparation). Presumably, the effects of both FeTf and Fe can be explained on the basis of their influence on RNA synthesis.

In conclusion, Fe replaced FeTf in promoting growth and differentiation and in preventing the degeneration of myogenic cells. Protein moiety of FeTf plays a role in facilitating the transport of Fe into the cells, though the possibility cannot be excluded that the protein modifies the effect of Fe.

REFERENCES

1. Kimura I, Hasegawa T, Miura T and Ozawa E (1981). Muscle trophic factor is identical to transferrin. Proc Japan Acad 57B:200.

2. Hasegawa T, Saito K, Kimura I and Ozawa E (1981). Fe promotes in vitro growth of myoblasts and other cells from chick embryo. Proc Japan Acad 57B:206.
3. Ozawa E and Hagiwara Y (1982). Degeneration of large myotubes following removal of transferrin from culture medium. Biomed Res 3:16.
4. Nielsen L and Ludvigsen B (1963). Improved method for determination of creatine kinase. J Lab Clin Med 62:159.
5. Rosarki S (1965). Creatine phosphokinase isozyme. Nature 207:414.
6. Ozawa E (1972). Activation of muscular phosphorylase b kinase by a minute amount of Ca ion. J Biochem 71:321.
7. Szent-Győrgyi (1951). "Chemistry of Muscular Contraction." New York: Academic Press.
8. Ohshima Y, Maruyama K and Noda H (1965). "Developmental changes in chick muscle contractile proteins." in Ebashi S, Oosawa F, Sekine T and Tonomura Y (eds): "Molecular Biology of Muscular Contraction." Tokyo: Igaku Shoin, p 132.
9. Saito K and Ozawa E (1983). Contraction of cultured chick myotube before acquisition of cell membrane excitability. Proc Japan Acad 59B:367.
10. Schneider WC (1957). "Determination of nucleic acids in tissues by pentose analysis." in Colowick SP and Kaplan NO (eds): "Methods in Enzymology III." New York: Academic Press, p 680.
11. Shimo-Oka T, Hagiwara Y and Ozawa E. Class-specificity of transferrin as a muscle trophic factor. (submitted)
12. Saito K, Hagiwara Y, Hasegawa T and Ozawa E (1982). Indispensability of iron for the growth of cultured chick cells. Devel growth Differ 24:571.
13. Kano M, Shimada Y and Ishikawa K (1972). Electrogenesis of embryonic chick skeletal muscle cells differentiated in vitro. J Cell Physiol 79:363.
14. Kano M (1975). Development of excitability in embryonic chick skeletal muscle cells. J Cell Physiol 86:503.
15. Fukuda J (1974). Chloride spike: A third type of action potential in tissue-cultured skeletal muscle cells from the chick. Science 185:76.
16. Cohen P, Burchell J and Foulkes G, Cohen PTW, Vanaman TC and Nairn AC (1978). Identification of the Ca^{2+}-dependent modulator protein as the fourth subunit of rabbit skeletal muscle phosphorylase kinase. FEBS Letters 92:287.
17. Ebashi S and Endo M (1968). "Calcium ion and muscle

contraction." in Buther JAV and Noble D (eds): "Progress in Biophysics and Molecular Biology." Oxford: Pergamon Press, p 123.

18. Ozawa E, Hosoi K and Ebashi S (1967). Reversible stimulation of muscle phosphorylase b kinase by low concentration of calcium ions. J Biochem 62:285.

Molecular Biology of Muscle Development, pages 153-165

FACTORS AND CONDITIONS REGULATING MUSCLE MATURATION *IN VITRO*[1]

Richard C. Strohman, Elissavet Kardami, Paul Moss Julie Micou-Eastwood and Dennis Spector

Department of Zoology, University of California, Berkeley CA 94720

ABSTRACT We have found that myogenesis *in vitro* is profoundly influenced by (A) a growth factor isolated from adult muscle and (B) conditions allowing extensive fibroblast matrix-muscle interaction.The latter conditions permit synthesis of adult patterns of myosin light chains. The growth factor binds to heparin and the muscle extracellular matrix may therefore serve as a powerful regulator of both muscle cell growth and of fiber maturation.

INTRODUCTION

Cell culture conditions have been found which are permissive for complete terminal differentiation of skeletal muscle (1). These cultures require a fibronectin-mediated attachment to the substratum (2) and while most conditions employ a complex medium consisting of serum and embryo extract it is clear that terminal differentiation can occur in a defined medium (3).

---

[1]This work was supported by grants from the Muscular Dystrophy Association and from the National Institutes of Health.

It is also apparent however that most culture conditions currently in use are inadequate in supporting regulation of muscle fiber maturation. For example, it has been our experience that while defined media may support cell fusion and differentiation of myotubes, such media do not support long term fiber survival. Fibers in defined medium remain thin and ultimately detach from the substrate. At the same time, antibiotics normally used in muscle cultures actually inhibit the time course and extent of muscle protein expression (4). We have not been able to maintain active, spontaneously contracting muscle fiber cultures for more than several days post fusion in defined medium. For long-term cultures in which one can adequately study fiber maturation and effects of heterologous cell types and environmental shifts on maturation we need a complex media. Finally, even in complex media it has not been possible to produce in culture the conditions permissive for full muscle fiber maturation and expression of a complete adult phenotype (see below). This is an important message to those involved in defined media work and in screening serum and embryo extract for use in culture. Success is not to be defined in terms of short-term growth or even in terms of high levels of cell fusion and terminal differentiation; success must be measured as well in terms of long-term survival and in terms of fiber maturation.

MUSCLE FIBER MATURATION

The early skeletal muscle myotube after several days in culture may become cross striated and contractile. However, the major pattern of gene expression in these fibers is an embryonic pattern. The myosin heavy chain is the embryonic isoform (5,6) and a cardiac myosin isoform may also be present (7); cardiac actin is the first myofibrillar actin expressed (8). Fast light chain 3 remains at quite low levels (less than 5%) (5), slow light chains abound in otherwise fast fibers and troponin expression shows lack of fiber type specificity (10).

The differentiated chicken muscle fiber in culture appears therefore to be highly differentiated from a morphological point of view. But on close analysis it is shown to be amorphous and relatively undifferentiated at the molecular level. Indeed this patchwork of functional differentiation may reflect things as they occur *in vivo*. But in the animal things are quickly sorted out and this sorting out process is what we call maturation. Slow light chains are quickly repressed in very young fast fibers, FLC3 accumulation achieves parity with FLC1 (11), cardiac actin disappears to be replaced with the skeletal isoform (8), beta Tm disappears from fast fibers and only alpha-fast Tm is seen (12), and there is a precise sequential replacement of embryonic and neonatal isoforms of myosin heavy chain as the animal achieves postnatal stages of growth and maturation (5,6).

MUSCLE FIBER REGENERATION *IN VITRO*

If normal muscle quickly sorts out the maturation process then regenerating muscle appears to do so even more efficiently. Regenerating skeletal muscle will complete myosin isoform shifts in a matter of a few weeks where the same shift to completely adult patterns may require a month or more of normal development and maturation (9).

We now know that fiber regeneration proceeds biochemically along Pathway 1 in Figure 1. It appears obligatory for satellite cell nuclei to read through a program of sequential expression of embryonic genes prior to expression of adult genes. While this sequential gene expression is apparent when one begins with regeneration *de novo*, it is not at all clear what happens when a satellite cell fuses into an adult fiber cytoplasm in the process of normal growth and hypertrophy. Is that nucleus deflected by an adult cytoplasm to go through Pathway 1, or is reading through an obligatory program of sequential gene expression an undeviating rule?

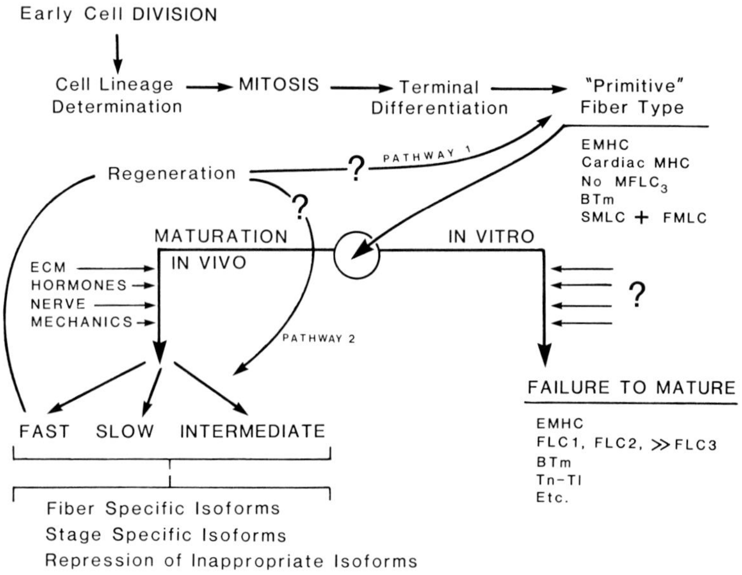

Figure 1. Muscle fiber maturation and regeneration

We may define a form of regeneration <u>in vitro</u> as that process of muscle fiber formation that begins not with embryonic myoblasts but with adult muscle fiber satellite cells. If we culture satellite cells from adult chicken muscle then myogenesis occurs normally, striated fibers are formed in culture, and these will develop into fibers that may be maintained for one month. But once again, these fibers in culture do not progress beyond the embryonic pattern of myosin gene expression (9). The questionmark in Figure 1 refers to the fact that conditions have not been defined *in vitro* which permit fiber maturation. One other interesting point emerging from this work is that muscle fibers developing in culture from slow muscle satellite cells appear to be different from similar fibers that develop from fast muscle fiber satellite cells. Slow muscle fibers synthesize almost equal levels of both fast and slow myosin light chains while the fast muscle fibers in culture synthesize extremely low levels of slow light chains. In addition, the slow fibers in culture synthesize a slow myosin heavy chain in addition to an embryonic myosin heavy chain that is found in both fast and slow muscle fibers in culture(9). It appears therefore that satellite cells from adult fast and slow muscle have diverged from one another along a commitment pathway for fast and slow muscle respectively.

NEW DIRECTIONS IN MUSCLE MATURATION IN CULTURE

I.Muscle Growth Factors And The Extracellular Matrix

Muscle cell cultures also show very different growth characteristics compared with those seen *in vivo*. For example, presumptive myoblasts and satellite cells in culture withdraw very early from the cell cycle and fuse. In the animal however these cells may undergo large numbers of cell divisions before fusing. On the other hand, muscle fibroblasts *in vivo* are growth regulated so that they do not overgrow myogenic cells within the muscle perimycium. But *in vitro* fibroblast cell growth is relatively unregulated and these cells will quickly overgrow myogenic cells and differentiated muscle fibers.

TABLE 1

STIMULATION OF MYOGENESIS *IN VITRO* BY MUSCLE AND NERVE EXTRACTS DEPLETED OF TRANSFERRIN[a]

Additions	ug Protein/Dish
Control	60
ALD muscle extract	165
Tf-depleted ALD extract	120
	Myosin HC/Total Protein
Control	35
Sciatic nerve extract	52
Tf-depleted sciatic extract	48

[a]All extracts were added to cultures at a final concentration of 0.3 mg/ml. Values are averages of 4 culture dishes with a mean variation of 10%. Extracts were prepared as previously described (13). Tf was removed by passage over an anti-Tf column. Measurements were made on well fused cultures.

We have recently been interested in the role that growth factors play in regulating fibroblast, myoblast and satellite cell replication in maturing and in regenerating muscle. This work began with the finding in our laboratory that transferrin (Tf) was preferentially localized in slow, tonic muscles of the chicken. But, more than that, Tf appears to be concentrated in the extracellular space of the muscle and the method of extraction suggested that Tf may indeed be bound in the muscle extracellular matrix (13). Tf is essential for growth and differentiation of muscle cells in culture (14).

TABLE 2

STIMULATION OF MYOGENESIS IN CULTURE BY HEPARIN-PURIFIED MUSCLE GROWTH FACTOR[b]

Additions to culture	Myosin heavy chain accumulation in ug per culture	
	Day 5	Day 8
Heparin-chromatographed muscle growth factor at final concentration of 5 ng/ml + Tf at 20 ug/ml.	28	80
Tf only at 20 ug/ml	10	48

[b]ALD muscle extracts were chromatographed on heparin-sepharose as described(15). Additions were made at cell plating and MHC accumulation was measured at times indicated.

This work has been continued in our laboratory. We are now able to demonstrate that muscle extracts contain growth factor(s) other than Tf. Using heparin-sepharose chromatography, an activity (CMGF or chicken muscle growth factor) is purified some 40,000 fold from ALD muscle; CMGF increased myogenic cell growth in culture and produced sizable increases in myosin heavy chain accumulation. Heparin columns are eluted with 2M NaCl and active fractions yield bands on SDS/PAGE in the 14-18,000 MW range. When this material is added to myogenic cell cultures in complete medium supplemented with Tf there is a stimulation of DNA synthesis and total protein and myosin heavy chain (MHC) accumulation. (Table 1 & 2).

The finding of a growth factor accumulated in adult muscles at concentrations high enough to stimulate myogenesis *in vitro* raises more questions than it answers. How is the effect of this factor modulated *in vivo*? That is, what is the physiological barrier between the factor and the fibroblasts and satellite cells of the muscle which in fact are more growth-arrested than in growth. Our extraction procedure is very gentle and its success suggests the factor may be, as Tf was shown to be, associated with the muscle extracellular space. The factor binds strongly to heparin, a major component of the extracellular matrix. What is the role of extracellular matrix in reversibly binding growth factors and thereby regulating cell growth at the tissue level? For muscle we have preliminary evidence that heparin will indeed inhibit myogenesis *in vitro* (see accompanying article by Kardami et al. in this volume). In endothelial tissues the role of heparin is clearly regulatory since this extracellular matrix component will accelerate growth of cloned endothelial cells (16) but will also inhibit growth of arterial smooth muscle cells (17).

II. Mechanical Factors In Culture Determine Extent Of Contractility And Patterns Of Gene Expression

Vandenburgh and Kaufman showed several years ago that muscle cultures exposed to repetitive stretch and relaxation would display hypertrophy (18). Work in our own laboratory has developed a clear relationship between muscle fiber activity in culture and rate of turnover of myosin heavy chain. If cultured fibers are relaxed either with tetrodotoxin or with depolarizing levels of external K^+ then MHC synthesis continues but turnover increases 3-5 fold (19). This effect appears to be specific for myofibrillar entities since turnover of intermediate filament proteins was not affected by prolonged relaxation(19).

In recent years we have attempted to define conditions in long term cultures for enhanced spontaneous muscle contractions. We wanted to know if such conditions would partly or wholly support expression of adult myosin genes and, generally speaking, an adult muscle phenotype. One of the major problems encountered was that the muscle in culture developed as a continuous tissue sheet in which the muscle fibers were embedded in a fibroblast matrix. The matrix-sheet turns out to be a great advantage over pure muscle cultures since the fibroblasts evidently secrete extracellular matrix components and contribute to muscle basal lamina formation both of which contribute to long term adhesion of the fibers to the matrix and to the culture dish. But what also tends to happen is that the sheet contracts and detaches itself from the culture dish becoming a floating, tenotomized preparation which quickly deteriorates.

TABLE 3

EFFECT OF MUSCLE FIBRILLATION ON MYOSIN FAST LIGHT CHAIN 3 SYNTHESIS[c]

Myosin Light Chain Synthesis	Days In Culture 10	14	21	24
SLC1	0	0	2	0
FLC1	23	42	30	34
SLC2	5	2	2	2
FLC2	68	49	49	43
FLC3	5	7	18	21

[c] In the above cultures intense contractile activity began on day 16-17. Myosin light chain synthesis was measured as described previously (19). Numbers indicate percent CPM of each light chain in the total myosin light chain population. SLC1=slow myosin light chain,etc.

This problem was solved somewhat by one of us (PM) by detaching the sheet at certain points at the periphery so that there was some lessening of tension throughout the network. We call this a "partially detached" network. Such a network remains attached over 50-70% of its periphery and contractions continue for many days. But what also happens is that the lessening of tension in the network somehow provides conditions for increased spontaneous frequency of contractions. What we could actually observe for many days was a fibrillation over large areas of the network. Under these conditions we were able to measure a reproducible, large increase in at least one aspect of muscle maturation; FLC3 synthesis went from 5% in attached networks which were contractile but at low frequency to over 20% in partially detached networks in which fibrillation was occurring (Table 3).

Recently we have developed techniques for growing long-term cultures on elastic membranes treated with colagen and/or attachment factors such as fibronectin. These elastic substrates are superior to collagen-coated plastic in many ways and provide conditions for remarkable tissue level organization of cultured muscle. For example, we are able to observe bundles of muscle fibers being surrounded by a thin layer of fibroblasts so that an actual perimysium forms in culture. This muscle "tissue" forms *de novo* after several weeks in cultures seeded with a population of single embryonic myogenic cells. This system will be the subject of a forthcoming publication.

SUMMARY Two major conditions emerge as important determinants of muscle growth and maturation in cell culture. First, muscle cell growth *in vitro* may be accelerated by a muscle growth factor that is apparently localized in tonic muscles of adult chickens. The same factor is apparently also localized in chicken sciatic nerve. The factor is quite distinct from transferrin but works synergistically with transferrin in regulating myogenic cell growth *in vitro*. We suggest that localization of this material takes place by binding to muscle extracellular matrix components since, for example, the factor is shown to bind strongly to heparin. Second, muscle cultures allowed to develop within an extensive fibroblast matrix will form a three-dimensional network in which we can create conditions for high-frequency spontaneous muscle contractions. Under these conditons cultured fibers begin to express an adult pattern of myosin light chain synthesis. The encouragement of a native muscle extracellular matrix in vitro appears therefore to be of prime importance in future studies of muscle maturation.

REFERENCES

1. Holtzer H, Sanger J, Ishikawa H & Strauss K (1973). Selected topics in skeletal myogenesis. Cold Spring Harbor Symposia on quantitative Biology 23:549.

2. Chiquet M, Puri EC, & Turner DC (1979). Fibronectin mediates attachment of chicken myoblasts to a gelatin-coated substratum. J Biol Chem 254:5475.

3. Dollenmeier P, Turner DC & Eppenberger HM (1981). Proliferation and differentiation of chick skeletal muscle cells cultured in a chemically defined medium. Exp Cell Res 135:47.

4. Moss PS, Spector DH, Glass CA & Strohman RC (1984). Streptomycin retards the phenotypic maturation of chick myogenic cells. In Vitro 20:473.

5. Bandman E, Matsuda R & Strohman RC (1982). Developmental appearance of myosin heavy and light chain isoforms in vivo and in vitro in chicken skeletal muscle. Dev Biol 93:508.

6. Bader D, Masaki T & Fischman DA (1982). Immunological analysis of myosin heavy chain during avian myogenesis in vivo and in vitro. J Cell Biol 95:763.

7. Sweeney LJ, Clark WA, Umeda PK, Zak R & Manasek FJ (1984).Immunofluorescence analysis of the primordial myosin detectable in embryonic skeletal muscle. Proc Natl Acad Sci USA 81:797.

8. Paterson BM & Eldridge JD (1984). Alpha cardiac actin is the major sarcomeric isoform expressed in embryonic avian skeletal muscle. Science 224:1436.

9. Matsuda R, Spector D & Strohman RC (1983). Regenerating adult chicken skeletal muscle and satellite cell cultures express embryonic patterns of myosin and tropomyosin isoforms. Dev Biol 100:478.

10. Toyota,N & Shimada,Y (1983).Isoform varients of troponin in skeletal and cardiac muscle cells cultured with and without nerves. Cell 33:297.

11. Crow,MT, Olson PS, Conlon SB & Stockdale FE (1983). Myosin light chain expression in the developing avian hindlimb. In Limb Development and Regeneration. New York: Alan R. Liss p417.

12. Matsuda R, Bandman E & Strohman RC (1983). Regional differences in expression of myosin light chains and tropomyosin subunits during development of chicken breast muscle. Dev Biol 95:484.

13. Matsuda R, Spector D, Micou-Eastwood J & Strohman RC (1984).There is selective accumulation of a growth factor in chicken skeletal muscle. Dev Biol 103:276.

14. Ii I, Kamura I & Ozawa E (1982). A myotrophic protein from chick embryo extract: Its purification, identity to transferrin,and indispensibility for avian myognenesis. Dev Biol 94:366.

15. Gospodarowicz D, Cheng J, Lui G-M, Baird A & Bohlent P (1984).Isolation of brain FGF by heparin-sepharose affinity chromatography: Identity with pituitary FGF. Proc Natl Acad Sci USA 81:6963.

16. Thornton SC, Mueller SN & Levine EM (1983). Human endothelial cells: Use of heparin in cloning and long-term serial cultivation. Science 222:623.

17. Hoover RL, Rosenberg R, Haering W & Karnovsky MJ (1983).Inhibition of rat arterial smooth muscle cell proliferation by heparin. Circ Res 47:578.

18. Vandenburgh HH & Kaufman S (1979). In vitro model for stretch-induced hypertrophy of skeletal muscle. Science 203:265.

19. Crisona NJ & Strohman RC (1983). Inhibition of contraction of cultured muscle fibers results in increased turnover of myofibrillar proteins but not of intermediate-filament proteins. J Cell Biol 96:684.

Molecular Biology of Muscle Development, pages 167–178

MYOBLAST METALLOENDOPROTEASE: ROLE IN MYOBLAST FUSION

Warren J. Strittmatter, Christine B. Couch

Department of Neurology and Program in Neuroscience, Baylor College of Medicine, Houston, Texas 77030

ABSTRACT During muscle development, mononucleated myoblasts fuse to form multinucleated myotubes. As myoblasts fuse, muscle cell surface proteins undergo many changes in location and molecular weight. Some of these changes could be caused by proteolysis. Studies presented here explore the hypothesis that proteolytic activity is required for myoblast fusion.

Synchronous myoblast fusion may be initiated in primary tissue culture by addition of calcium to fusion-competent myoblasts grown in the absence of calcium. Metalloendoprotease inhibitors and substrates prevent myoblast fusion when added to fusion-competent myoblasts at the time of calcium addition. Myoblast metalloendoprotease activity was assayed, both in primary cultures and in the L_6 myoblast cell line, using the fluorogenic protease substrate, 2-aminobenzoyl-alanine-glycine-leucine-alanine-4-nitrobenzylamide. Myoblasts contain both soluble and particulate metalloendoproteases, and these proteases have different inhibitor specificities. Several inhibitors which block myoblast fusion inhibit only the soluble, and not the particulate metalloendoprotease activity in myoblasts. Another metalloendoprotease inhibitor, phosphoramidon, which has no effect on fusion, inhibits only the membrane-associated metalloendoprotease. These observations implicate a soluble metalloendoprotease in myoblast fusion.

Two soluble metalloendoproteases can be demonstrated by column chromatofocusing at pI 4.8 and at pI 5.9. The soluble metalloendoprotease eluted at pH 5.9 is not inhibited by an inhibitor which blocks fusion, while the soluble metalloendoprotease eluted at pH 4.8 is inhibited. Therefore, of the two soluble metalloendoproteases, only the one with a pI of 4.8 seems involved in fusion. Gel filtration chromatography indicates that the soluble metalloendoprotease is about 70,000 daltons. The above observations are consistent with the hypothesis that a soluble metalloendoprotease of 70,000 daltons with a pI of 4.8 is required for myoblast fusion.

INTRODUCTION

The fusion of mononucleate myoblasts to multinucleate myotubes represents an important step in muscle development. In culture, myoblasts proliferate, aggregate, align, and form gap junctions prior to fusion (Bischoff, 1978; Rash and Fambrough, 1973.) Calcium-deficient medium permits proliferation and alignment of myoblasts, but does not permit fusion. Reintroducing calcium causes extensive, synchronized fusion. The requirement for calcium thus provides a simple means of initiating and synchronizing cell fusion (Shainberg et al., 1969; Turner et al., 1976).

Inhibitor studies indicate that glycoproteins are involved in myoblast fusion. Also, during fusion, myoblast proteins change in molecular weight and distribution (Strittmatter et al., 1984). Proteolysis is required for membrane fusion in other biological systems, such as virus-host cell fusion, either to allow membrane glycoproteins freedom of movement or to generate a hydrophobic peptide which mediates fusion (White et al., 1983; Lucy, 1984.) In light of these findings, we have examined the possibility that myoblast fusion requires protease activity.

Metalloproteases are proteases which are inhibited by metal chelators such as EDTA and 1,10-phenanthroline and thiol compounds (e.g. dithiothreitol) (Barrett, 1977). Some aminopeptidases and carboxypeptidases (exopeptidases which cleave one amino acid off an unblocked end of a protein) are metal-dependent. Metal-dependent dipeptidases, peptidyl dipeptidases, and aminotripeptidases cleave amino acids off of unblocked substrates. Metalloendoproteases, which cleave central portions of peptides and proteins, include neutral proteases, collagenases, gelatinases, and elastases. These proteases hydrolyze peptide bonds at the amino group of uncharged aromatic and aliphatic amino acids (Morihara, 1974; Chlebowski and Coleman; 1976). High affinity inhibitors have been specially designed which interact with the active site of metalloendoprotease. These include hydroxamic derivatives such as 2-(N-hydroxy-carboxamido)-3-phenylpropanoyl Ala-Gly-NH_2 (Nishino and Powers, 1979), and N-carboxymethyl derivatives such as Homo-Phe-[N]-L-Phe-B-Ala (Mumford et al., 1982). The naturally occurring inhibitor phosphoramidon (rhamnosyl-p-Leu-L-Trp-OH) also inhibits thermolysin-like metallo-endoproteases with high specificity (Komiyama et al., 1975). The hydroxamic acid derivative, $ClCH_2CO$ (N-OH)-Phe-Ala-Ala-NH_2 irreversibly inhibits thermolysin by alkylation with a Ki of 7.5 mM (Rasnick and Powers, 1978).

Inhibitors of Metalloendoproteases Block Myoblast Fusion.

To determine whether metalloprotease activity is required for the fusion of myoblasts, metalloprotease inhibitors were added to the medium of calcium-deprived rat myoblasts in primary culture just before fusion was triggered by addition of calcium. The metal chelator 1,10-phenanthroline prevented myoblast fusion with half maximal inhibition requiring approximately 20 µg/ml. Normal fusion occurred if 1,10-phenanthroline was chelated with an equimolar concentration of $ZnCl_2$ before being added to the cultures. Increasing the calcium concentration did not alter inhibition by 1,10-phenanthroline. To further examine whether a metalloprotease is involved in fusion, other types of metalloprotease inhibitors were used. CBZ-L-phenyl-alanine, a competitive inhibitor of metalloendoproteases (Burstein et al., 1974) prevented myoblast fusion with half-maximal inhibition of fusion at approximately 1.6 mM, while CBZ-L-leucine, CBZ-L-arginine, CBZ-L-proline, CBZ-L-tyrosine, and CBZ-L-tryptophan had little effect at this concentration. The inhibition by CBZ-L-phenylanine was stereospecific since CBZ-D-phenylalanine had no effect. Inhibitors of metallo-exoproteases, hippuryl-L-arginine and hippuryl-L-phenyl-alanine (Chlebowski and Coleman, 1977) had no effect on myoblast fusion at 3 mM. Captopril, an inhibitor of the metallodipeptidase angiotensin-1-converting enzyme (Ondetti et al., 1977), had no effect on myoblast fusion even at 5 mM (Couch and Strittmatter, 1983).

If a metalloendoprotease is required for myoblast fusion, substrates of known metalloendoproteases should compete with the natural protein substrate and thus prevent fusion. Neutral metalloendoproteases, such as thermolysin, hydrolyze peptide bonds at the amino group of uncharged aromatic and aliphatic amino acids (Morihara, 1974; Chlebowski and Coleman, 1976). These metalloendoproteases will hydrolyze synthetic carbobenzoxy (CBZ)-dipeptide-amides only with large aliphatic amino acids, such as leucine, or noncharged aromatic amino acids, such as phenylalanine, on the amino terminus of the hydrolyzed peptide bond. The interaction of dipeptide substrates with this protease class is also altered by the amino acid on the carboxy terminus of the hydrolyzed peptide bond, with these proteases having a higher affinity for serine and tyrosine than for glycine. Therefore, CBZ-Ser-Leu-amide and CBZ-Tyr-Phe-amide have high affinities for metalloendoproteases, and CBZ-Gly-Leu-amide and CBZ-Gly-Phe-amide have intermediate affinities, while CBZ-Gly-Gly-amide is not a substrate. The affinity of a dipeptide substrate varies according to the particular metalloendoprotease, and is in the millimolar range. Carbobenzoxy-dipeptide-amide derivatives that interact with metalloendoproteases inhibited myoblast fusion. Cultures incubated in 0.75 mM CBZ-Gly-Phe-amide and calcium demonstrated virtually no fusion after 48 hr, compared with cultures incubated with calcium alone. Inhibition of fusion by these dipeptides was completely reversible by the removal of the dipeptide from the medium. The ability of CBZ-dipeptide-amides to inhibit fusion matched their specificity for the metalloendoproteases. CBZ-Tyr-Leu-amide and CBZ-Ser-Leu-amide are high-affinity metalloendoprotease substrates, and were more potent than both CBZ-Gly-Leu-amide and CBZ-Gly-Phe-amide in inhibiting fusion. CBZ-Gly-Gly-amide is not a metalloendoprotease substrate and had no effect on fusion. The ability of these dipeptides to inhibit fusion required both the amide and the carbobenzoxy groups, which are also required for recognition by the metalloendoproteases. CBZ-Gly-Phe-amide inhibited myoblast fusion, but neither CBZ-Gly-Phe nor Gly-Phe-amide had an effect on fusion. The inhibition by CBZ-Ser-Leu-amide was stereospecific. Only CBZ-Ser-L-Leu-amide blocked fusion. CBZ-Ser-D-Leu-amide had no effect even at 0.5 mM (Couch and Strittmatter, 1983).

The metalloendoprotease inhibitor $ClCH_2CO$-(N-OH)-Phe-Ala-Ala-NH_2 irreversibly inhibits thermolysin-like metalloendoproteases (Rasnick and Powers, 1978). The phenylalanine of the inhibitor occupies the active site of the protease, and the hydroxamic acid functional group complexes the endoprotease metal atom. The protease is then irreversibly inactivated by alkylation. This metalloprotease inhibitor prevented myoblast fusion (Couch and Strittmatter, 1983).

The fluorogenic substrate, 2-aminobenzoyl-Ala-Gly-Leu-Ala-4-nitrobenzylamide (AAGLAN), is used to assay metalloendoproteases (Kam et al., 1979). If the fluorescent protease substrate, AAGLAN, is hydrolyzed by the metalloendoprotease involved in myoblast fusion, it should compete with the natural protein substrate for the endoprotease and prevent fusion. When added to myoblast culture medium 30 min before the addition of calcium, this protease substrate inhibited myoblast fusion, with half-maximal inhibition at 500 µM, similar to the concentration of CBZ-Gly-Leu-amide that inhibited fusion. Incubation of AAGLAN with the bacterial metalloendoprotease, thermolysin, hydrolyzes the Gly-Leu peptide bond of this substrate to yield a 2-aminobenzoyl-Ala-Gly fragment and a Leu-Ala-4-nitrobenzylamide fragment. This prehydrolyzed substrate no longer inhibited myoblast fusion (Couch and Strittmatter, 1983).

The metalloendoprotease inhibitors and substrates appear to block fusion by interacting with a protease, rather than by nonspecific effects, for several reasons. Myoblasts preincubated with 1,10-phenanthroline remained viable, and unfused for 3 days. The inhibition by 1,10-phenanthroline was not due to decreased protein synthesis. Phenanthroline inhibition was due to its metal chelation properties since normal fusion occurred if 1,10-phenanthroline was chelated with an equimolar concentration of $ZnCl_2$. 1,10-phenanthroline did not prevent fusion by chelating calcium since 1,10-phenanthroline was removed before calcium was added and increasing the calcium concentration did not alter inhibition by 1,10-phenanthroline. 1,10-phenanthroline is blocking fusion by metal chelation, and that the inhibition is not due to calcium chelation, decreased protein synthesis, decreased cell density, or decreased viability.

Inhibition of myoblast fusion by CBZ-L-phenylalanine is unlikely to be nonspecific since it was stereospecific and since many other CBZ amino acids and hippuryl-L-phenylalanine did not inhibit fusion. The ability of dipeptide derivatives to inhibit myoblast fusion matched their ability to bind to characterized neutral metalloendoproteases, with the potency series CBZ-Tyr-Leu-amide $\approx$ CBZ-Ser-Leu-amide > CBZ-Gly-Phe-amide $\approx$ CBZ-Gly-Leu-amide >> CBZ-Gly-Gly-amide. The dipeptide substrate CBZ-Gly-Phe-amide inhibits fusion and interacts with the metalloendoproteases only if both the carbobenzoxy or amide groups are present. The close correlation between the potency of these dipeptides and their ability to interact with metalloendoproteases makes it likely that they are inhibiting fusion by interacting with the metalloendoprotease. Nonspecific effects are unlikely since inhibition was reversible and stereospecific in the cases tested.

Biochemical Identification of Myoblast Metalloendoprotease.

Metalloendoprotease activity was assayed using the fluorogenic protease substrate, 2-aminobenzoyl-alanine-glycine-leucine-alanine-4-nitrobenzylamide (AAGLAN) in both primary cultures of rat myoblasts, as well as in L_6 myoblast cultures (Couch and Strittmatter, 1983, 1984). The fluorescence of the 2-aminobenzoyl detector group is quenched in the intact substrate by intramolecular energy transfer (Kam, 1979). Proteolytic hydrolysis of a peptide bond generates an unquenched detector. The fluorescence of the unquenched detector group therefore quantitatively indicates hydrolysis of this substrate. The hydrolyzed product was assayed spectrofluorometrically.

Both the soluble and the particulate fractions of L_6 myoblasts hydrolyzed the fluorogenic protease substrate AAGLAN at a linear rate for 6½ hr. Hydrolysis of the protease substrate by both the soluble and the particulate fractions was partially inhibited by the metal chelator 1,10-phenanthroline, with maximal inhibition at 100 μg/ml. This 1,10-phenanthroline-sensitive protease activity was 24.1 ± 1.0 nmol of AAGLAN hydrolyzed/hr/mg of protein (mean ± S.E.) in the soluble fraction and 17.1 ± 0.7 nmol of AAGLAN hydrolyzed/hr/mg of protein in the particulate fraction (Couch and Strittmatter, 1984). Both the soluble and particulate fractions of metalloendoprotease have pH optima about pH 7.5. To test further whether or not the protease activity in both the soluble and particulate fractions inhibited by 1,10-phenanthroline is a metalloendoprotease, other metalloendoprotease inhibitors were also tested. The metal chelator EDTA inhibited both activities as did the higher affinity reversible inhibitors 2-(N-hydroxycarboxamido)-3-phenyl-propanoyl-Ala-Gly-NH_2 and Homo-Phe-[N]-Phe-β-Ala. The thiol compound dithiothreitol also blocked both activities. These inhibitors blocked only protease activity in both preparations which was inhibited by 1,10-phenanthroline, and had no effect on protease activity not inhibited by 1,10-phenanthroline. These observations demonstrate that L_6 myoblasts contain metalloendoprotease activity in both the supernatant and particulate fractions (Couch and Strittmatter, 1984). One inhibitor, phosphoramidon, inhibited the particulate metalloendoprotease, and not the soluble metalloendoprotease of myoblasts. Several compounds inhibited the soluble, but not the particulate metalloendoprotease activity. The synthetic dipeptide CBZ-Gly-Leu-Amide is a substrate for thermolysin and thermolysin-like metalloendoproteases, whereas CBZ-Gly-Gly-amide is not a substrate for these enzymes (Chlebowski and Coleman, 1976; Orlowski and Wilk, 1981; Kerr and Kenny, 1974). CBZ-Gly-Leu-amide inhibited the myoblast soluble, but not the particulate, metalloendoprotease activity. CBZ-Gly-Gly-amide had no effect on either soluble or particulate metalloendoproteases. Additionally, the irreversible active site metalloendoprotease inhibitor $ClCH_2CO$(N-OH)-Phe-Ala-Ala-NH_2 (Rasnick and Powers, 1978) inhibited only the soluble, and not the particulate metalloendoprotease activity (Couch and Strittmatter, 1984).

Chromatofocusing of Soluble Metalloendoprotease.

To test whether or not the soluble metalloendoprotease activity consists of more than one enzyme, two methods were used which separate proteins on the basis of isoelectric point. Isoelectric focusing was not useful since the ampholines irreversibly inhibited the enzyme activity. Column chromatofocusing of the soluble metalloendoprotease activity revealed two peaks of metalloendoprotease activity, with pI values of 5.9 and 4.8. Both of these soluble metalloendoproteases were inhibited by 1,10-phenanthroline and the active site metalloendoprotease inhibitor 2-(N-hydroxycarboxyamido)-3-phenylpropanoyl-Ala-Gly-NH_2. The soluble metalloendoprotease with a pI of 5.9, however, was not inhibited by CBZ-Gly-Leu-amide, while the metalloendoprotease with pI 4.8 was partially inhibited (Couch and Strittmatter, 1984).

In summary, metalloendoprotease activity was assayed using the fluorogenic protease substrate 2-aminobenzoyl-Ala-Gly-Leu-Ala-4-nitrobenzylamide. This compound inhibits myoblast fusion, indicating that it is recognized by the myoblast metalloendoprotease involved in myoblast fusion. This substrate is quite specific for metalloendoproteases since both ends are blocked, therefore, exoproteases cannot cleave it which is important since some carboxyproteases and aminoproteases are metalloproteases. Also, the four amino acid peptide is not easily cleaved by most other classes of proteases. To further assure that only metalloendoproteases are studied, most experiments have been done with and without saturating concentrations of the metal chelator 1,10-phenanthroline. Only the protease activity inhibited by phenanthroline is referred to as metalloendoprotease activity. In addition, the irreversible inhibitor $ClCH_2CO$-(N-OH)-Phe-Ala-Ala-NH_2 was shown to inhibit the primary culture metalloendoprotease. The active site inhibitors 2-(N-hydroxycarboxamido)-3- phenyl-propanoyl-Ala-Gly-NH_2 and Homo-Phe-[N]-Phe-β-Ala were shown to block the metalloendoprotease activity in L_6 cultures. This is further evidence that the phenanthroline-inhibitable hydrolysis of AAGLAN is due to a true metalloendoprotease.

Crude fractionation of L_6 myoblasts reveals two populations of metalloendoprotease activity, soluble and particulate. These two populations seem to be truly separate enzymes based on the following information:

1. Modulation of membrane preparation by changes in ionic strength, pH, buffer types, 37°C preincubation, and divalent cations has little effect on the proportion of enzyme activity found in the two populations. If the particulate protease was actually a peripheral membrane protein, or the soluble protease was a cleavage product of integral membrane protease, changes

in the relative activities should have been seen.

2. While many metalloendoprotease inhibitors block both soluble and particulate metalloendoprotease activity, some inhibitors block only the soluble (chymostatin, CBZ-Gly-Leu-amide, and $ClCH_2CO$-(N-OH)-Phe-Ala-Ala-NH_2) or only the particulate (phosphoramidon) metalloendoproteases.

Thus, there seem to be two separate metalloendoproteases in L_6 myoblasts which differ in localization and specificity.

The soluble metalloendoprotease does not seem to be a single protease since chromatofocusing reveals two peaks of activity. The pI 5.9 peak is variable and very unstable. Both seem to be metalloendoproteases (phenanthroline and 2-(N-hydroxycarboxamido)-3-phenyl-propanoyl-Ala-Gly-NH_2 inhibitable), but they may have different specificities since CBZ-Gly-Leu-amide inhibited the pI 4.8 activity and not the pI 5.9 activity. The pI 4.8 metalloendoprotease is probably the major protease seen in soluble protease assays due to the low levels and instability of the pI 5.9 activity.

Myoblasts therefore contain both soluble and particulate metalloendoproteases. The relationship of these proteases to known classes of metalloendoproteases is of interest. The particulate metalloendoprotease resembles the class EC 3.4.24.11 proteases in that it is particulate (i.e., membrane-bound probably) and is inhibited by phosphoramidon, a hydroxamic acid derivative, and Homo-Phe-[N]-L-Phe-β-Ala.

Several metalloendoproteases present in myoblasts at the time fusion have been partially characterized. The soluble metalloendoprotease was blocked by 1,10-phenanthroline, CBZ-Gly-Leu-amide, and the irreversible inhibitor $ClCH_2CO$(N-OH)-Phe-Ala-Ala-NH_2, all of which also blocked myoblast fusion. However, the particulate metalloendoprotease, while blocked by 1,10-phenanthroline, was not blocked by CBZ-Gly-Leu-amide or the irreversible inhibitor $ClCH_2CO$(N-OH)-Phe-Ala-Ala-NH_2. Furthermore, the particulate protease was blocked by phosphoramidon, which does not block fusion. Therefore, the soluble metalloendoprotease could be involved in myoblast fusion, whereas the particulate protease is unlikely to be involved in fusion.

The myoblast soluble metalloendoprotease preparation is stable at -80°C for several weeks. It is also stable at 5°C for several days, and does not lose activity upon lyophilization and dialysis. The soluble metalloendoprotease is precipitated by ammonium sulfate between 45% and 70%.

Molecular exclusion HPLC of the soluble metalloendoprotease gives one peak of phenanthroline-inhibitable metalloendoprotease activity with a molecular weight of about 60-100K daltons.

Molecular exclusion chromatography of the soluble myoblast metalloendoprotease preparation on a G-200 column gives one peak of phenanthroline-inhibitable activity with a molecular weight of about 70-80K. Non-phenanthroline-inhibitable activity is seen both at the void volume and within the phenanthroline-inhibitable peak (at slightly lower molecular weight). When G-200 chromatography is performed after ammonium sulfate fractionation, the G-200 step alone gives about 71-77% recovery and about 1.7-1.9-fold purification.

Ion exchange chromatography of the soluble myoblast metalloendoprotease on a DE52 column gives one major peak of activity at about 150 mM NaCl. The non-phenanthroline-inhibitable activity comes off in the same fractions. When myoblast soluble preparation was run directly on a DE52 column, a 30% recovery and 3.3-fold purification was attained. When the DE52 column was run after the G-200 column, a 44% recovery and 4.9-fold purification resulted for this step alone (Couch, 1984).

SUMMARY

The purpose of this study is to explore the hypothesis that proteolytic activity is required for myoblast fusion. The following three predictions which follow from this hypothesis have been investigated:

1. Inhibition of the protease should inhibit myoblast fusion. Metalloendoprotease inhibitors and substrates prevented myoblast fusion when added to fusion-competent myoblasts with the addition of calcium.
2. Protease activity should be present during myoblast fusion. Metalloendoprotease activity was demonstrated in both primary cultures and the L_6 myoblast cell line at the time of fusion. Metalloendoprotease levels do not

change over the course of L_6 myoblast fusion. Several different metalloendoproteases were characterized in L_6 myoblasts, and inhibitor studies implicated the soluble metalloendoprotease with a pI of 4.8 as the protease most likely to be involved in fusion.

3. Isolated protease should cause myoblast fusion when properly added to an otherwise fusion-competent system. The soluble myoblast metalloendoprotease was partially isolated by classical methods on the basis of solubility, size and charge. Further work in progress is attempting to generate antibodies to the myoblast metalloendoprotease.

REFERENCES

1. Barrett AJ (1977). Introduction to proteinases. In Barrett AJ (ed): "Proteinases in Mammalian Cells and Tissues," Amsterdam: Elsevier, p 1-55.
2. Burstein Y, Walsh KA, Neurath H (1974). Evidence of an essential histidine residue in thermolysin. Biochemistry 13:205-210.
3. Chlebowski J, Coleman JE (1976). Zinc and its role in enzymes. In Siegel H (ed): "Metal Ions in Biological Sciences," 6, p 1-7.
4. Couch CB (1984). Characterization of Myoblast Metalloendoprotease and Their Involvement in Myoblast Fusion. Ph.D. Thesis.
5. Couch CB, Strittmatter WJ (1983). Rat myoblast fusion requires metalloendoprotease activity. Cell 32:257-265.
6. Couch CB, Strittmatter WJ (1984). Specific blockers of myoblast fusion inhibit a soluble and not the membrane-associated metalloendoprotease in myoblasts. J Biol Chem 259:5396-5399.
7. Kam C-M, Nishino N, Powers JC (1979). Inhibition of thermolysin and carboxypeptidase A by phosphoramidates. Biochemistry 18:3032-3038.
8. Komiyama T, Suda H, Aoyagi T, Takeuchi T, Umezawa H (1975). Studies on inhibitory effect of phosphoramidon and its analogs on thermolysin. Arch Biochem Biophys 171:727-731.
9. Lucy JA (1984). Do hydrophobic sequences cleaved from cellular polypeptides induce membrane fusion reactions <u>in vivo</u>? Febs Letters 166:223-231.

10. Morihara K (1974). Comparative specificity of microbial proteinases. Adv Enzymol 41:179-243.
11. Mumford RA, Zimmerman M, Broeke J, Taub D, Joshua H, Rothrock JW, Hershfield JM, Springer JP, Patchett AA (1982). Inhibition of porcine kidney "enkephalinase" by substituted N-carboxymethyl dipeptides. Biochem Biophys Res Comm 109:1303-1309.
12. Nishino N, Powers JC (1979). Design of potent reversible inhibitors for thermolysin. Peptides containing zinc coordinating ligands and their use in affinity chromatography. Biochemistry 18:4340-4347.
13. Ondetti MA, Rubin B, Cushman DW (1977). Design of specific inhibitors of angiotensin-converting enzyme: New class of orally active antihypertensive agents. Science 196:441-444.
14. Rasnick D, Powers JC (1978). Active site directed irreversible inhibition of thermolysin. Biochemistry 17:4363-4369.
15. Shainberg A, Yagil G, Yaffe D (1969). Control of myogenesis _in vitro_ by Ca^{2+} concentration in nutritional medium. Exp Cell Res 58:163-167.
16. Strittmatter WJ, Couch CB, Elias SB (1982). Role of specific proteases in rat myoblast fusion. In Pearson ML, Epstein HF (eds): "Muscle Development: Molecular and Cellular Control," New York: Cold Spring Harbor Laboratory, p 311-318.
17. Strittmatter WJ, Couch CB, Mundy DI (1984). Role of proteins in the fusion of biological membranes. In Aloia RC, Boggs J (eds): "Membrane Fluidity in Biology, VIII, Academic Press, in press.
18. Turner DC, Gmür R, Sieguist M, Burckhardt E, Eppenberger HM (1976). Differentiation in cultures derived from embryonic chicken muscle. Dev Biol 48:258-283.
19. White J, Kielian M, Helenius A (1983). Membrane fusion proteins of enveloped animal viruses. Quart Rev of Biophys 16:151-195.

ACKNOWLEDGMENT

This work was approved by the Jerry Lewis Neuromuscular Disease Center Grant from the Muscular Dystrophy Association and by the National Institute of Neurologic and Communicative Disorders and Stroke.

Molecular Biology of Muscle Development, pages 179–187

EXPRESSION OF NEURAL CELL ADHESION MOLECULE N-CAM IN SKELETAL MUSCLE[1]

Frank S. Walsh and Stephen E. Moore

Institute of Neurology, Queen Square
London WC1N 3BG, UK

ABSTRACT The expression of the N-CAM cell adhesion molecule was studied in skeletal muscle. Developing myofibres from rat and human muscle reacted with anti-N-CAM at the cell surface. Levels of N-CAM decreased during development such that by day 9 in the neonatal rat it had disappeared and adult fibres were uniformly negative. N-CAM was however re-expressed when muscle was damaged and regenerating myofibres expressed high levels of the antigen. In dissociated muscle cell cultures N-CAM was found at the cell surface of myoblasts and myotubes but not fibroblasts. N-CAM levels can however be down regulated in myoblasts by altering culture conditions.

INTRODUCTION

Glycoconjugates expressed at the cell surface play vital roles in the development and functioning of cells and tissues[1]. A number of classes of adhesion molecules are likely to be involved in such processes and include molecules involved in morphologically definable cell-cell interactive structures, components of the extracellular matrix and a family of specific cell adhesion molecules (CAMs).

To date, three specific classes of CAMs have been identified, N-CAM (neural), L-CAM (liver) and Ng-CAM (neuron-glial)[2]. Of these, the properties of N-CAM have been best defined[3]. A number of laboratories have studied N-CAM biochemistry and expression[2] and recently cDNA clones en-

[1]This work was supported by the Medical Research Council and the Muscular Dystrophy Group of Great Britain.

coding this molecule have been described[4,5], suggesting that a full molecular description of the N-CAM gene family and its relationship to other CAMs will soon be at hand. N-CAM is in many ways ideally suited to play a key role in morphogenesis and development since it has a specific pattern of expression during embryogenesis, binds in a homophilic manner and has a variable carbohydrate composition which is thought to play a role in modifying binding affinity[2]. There have however only been a few studies on the expression of N-CAM in muscle. In the chick embryo it has been shown that N-CAM is present at sites determined to differentiate into muscle[6], is expressed by chick muscle cells in culture[7] and has been implicated in the initial interaction of nerve with muscle during in vitro synaptogenesis[8]. We therefore wished to study in some detail the expression of N-CAM in developing and mature muscle as a prelude to functional studies.

METHODS

Cell Culture.

Cell culture of foetal and adult human muscle cells were routinely produced as described previously[9,10].

Indirect Immunofluorescence of Cells and Tissues With Antibodies.

Indirect immunofluorescence analysis of fibroblasts and muscle cells was carried out as described previously[9,10]. Cryostat sections of normal, diseased human muscle, rat muscle and baboon intercostal muscle were prepared according to Hurko and Walsh[11] and Moore et al[12]. All polyclonal and monoclonal antibody (McAb) reagents have been described by us previously[13].

Immunoblotting.

Immunoblotting was performed using the procedure of Towbin et al[14]. The detecting antibody used was ^{125}I-labelled donkey anti-rabbit $F(ab')_2$ (Amersham International plc).

RESULTS

N-CAM is Expressed by Developing and Regenerating but Not Adult Muscle Fibres.

Thiery et al[6] have shown that N-CAM is expressed at sites in the early chick embryo that develop subsequently into skeletal muscle. We wished to extend these observations by studying later stages of development. Neonatal rat skeletal muscle at days 1, 3, 9 and 15 were tested for reactivity with anti-N-CAM. Figure 1 (a-d) shows that on days 1 and 3 the sarcolemma of all muscle fibres were found to be positive for N-CAM. By day 9 the staining had decreased to background levels while by day 15 this down regulation was complete. Adult rat muscle did not express N-CAM at the sarcolemma. Similar profiles of immunoreactivity were found in human muscle where foetal muscle of up to 20 weeks expressed N-CAM but mature adult muscle did not. We have found this type of reactivity previously with McAb 5.1H11[11] and wished to compare their reactivities. Samples of muscle from patients with muscle disease were therefore assessed for reactivity as McAb 5.1H11 reacted with regenerating fibres in certain myopathies. Regenerating muscle fibres were identified by hematoxylin and eosin staining and these same fibres reacted with McAb 5.1H11 and anti-N-CAM (Fig 1e,f). Interestingly, N-CAM does not appear to be expressed by muscle satellite cells. McAb 24.1D5 reacts specifically with muscle satellite cells[13] but these same cells did not react with anti-N-CAM when double stains were carried out. These data show that although N-CAM is not expressed by adult myofibres, when muscle is damaged by injury or disease and new myofibres form, these regenerating fibres are positive for N-CAM although the precursor satellite cell pool is negative for N-CAM.

Expression of N-CAM at the Neuromuscular Junction (NMJ).

N-CAM has been implicated in the initial recognition of muscle by nerve[8]. Also its profile of expression during skeletal muscle development and in muscle regeneration coincides with times of synaptogenesis. We therefore wished to determine whether N-CAM was expressed at the NMJ. Baboon intercostal muscle was used as a rich source of NMJs and these were positively identified using rhodamine-labelled α-bungarotoxin (Fig 1g,h). Although N-CAM is not found in the sarcolemma of adult muscle, N-CAM can be shown to be

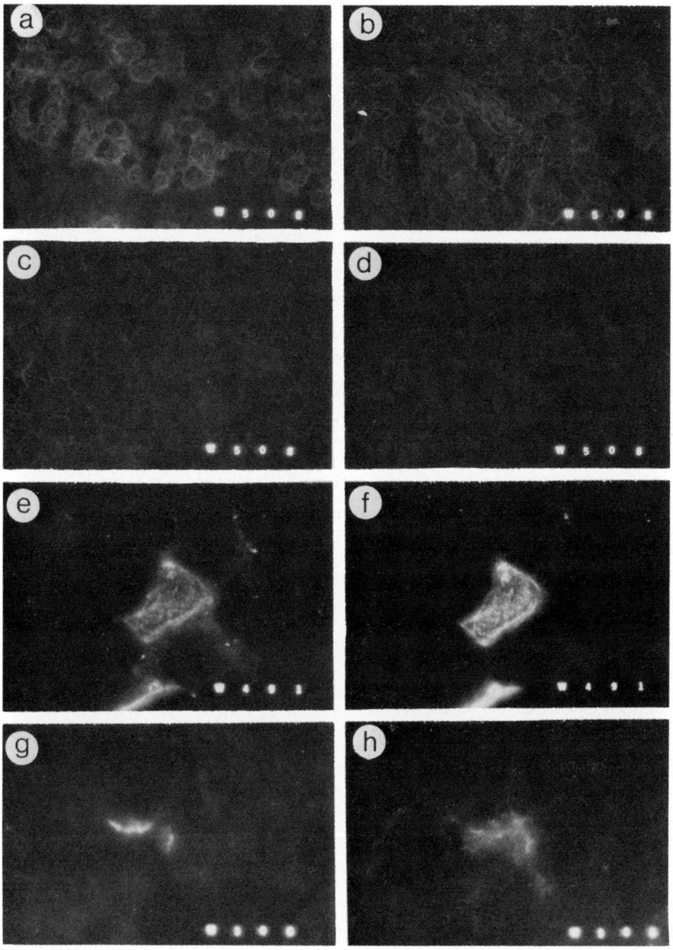

FIGURE 1. Expression of N-CAM on cryostat sections of normal and diseased skeletal muscle. Indirect immunofluorescent staining of neonatal rat skeletal muscle aged 1 day (a), 3 days (b), 9 days (c), 15 days (d). (e) (f) shows a double stain of regenerating muscle fibres from a patient with polymyositis, stained with (e) McAb 5.1H11 and (f) anti-N-CAM. (g) (h) shows a double stain of intercostal muscle with α-bungarotoxin (g) and anti-N-CAM (h).

present at the NMJ but there is not an exact codistribution with α-bungarotoxin binding. The N-CAM reactivity is more diffuse and extends beyond the NMJ. By comparison McAbs to acetylcholinesterase closely parallel the pattern of reactivity found with α-bungarotoxin. In low power micrographs the diffuse nature of the N-CAM pattern of reactivity is highlighted (not shown). Here, all the α-bungarotoxin reactivity is associated with N-CAM reactivity. In contrast, the N-CAM reactivity extends well beyond the NMJ and onto other areas of the muscle section, and it is likely that this reactivity coincides with that of nerve.

Molecular Nature of N-CAM in Muscle Cells.

Immunoblotting experiments have been carried out to determine the nature of the molecule being detected in the experiments outlined above. A broad band of 150000-200000 daltons was found for brain in immunoblots. In contrast, human muscle gave one tight banding of 180000 daltons. Although the embryonic brain N-CAM molecule identified here appears identical to that reported by Rutishauser et al[8], our results in muscle cells vary. This may be due to the molecules originating from muscle cells derived from different tissue ages or due to species variation.

N-CAM is Specifically Expressed by Human Muscle Cells but Not Fibroblasts in Cell Culture.

Previous studies by Grumet et al [7] showed that N-CAM was expressed by muscle cells in culture, but what cells were reactive was not rigorously determined. We have determined the profile of reactivity of N-CAM antibody with human skeletal muscle and fibroblast cultures using double labelling with antibodies which distinguish fibroblasts from myoblasts in cultures of human muscle[9]. Indirect immunofluorescence double staining of human skin fibroblasts with anti-N-CAM and anti-Thy-1 antibodies show that all fibroblasts were positive for Thy-1 antigen but negative for N-CAM. In cultures of adult and foetal human muscle only the mononucleate cells were positive with Thy-1 antibody while myoblasts and myotubes, but not fibroblasts, react with monoclonal McAb 5.1H11[9]. The large multinucleate myotubes were clearly negative for Thy-1 reactivity but positive for N-CAM. In addition only some of the Thy-1 positive mononucleate cells were positive for N-CAM. To determine whether those N-CAM positive mononucleate cells were myoblasts or fibroblasts, a double stain was carried out with McAb 5.1H11. McAb 5.1H11 and anti-N-CAM antibody reacted with the same cells in human muscle cultures. These results indicate that N-CAM expression by mononucleate cells in these cultures is restricted to myoblasts. These experiments indicate that in skeletal muscle cell cultures, N-CAM is specifically expressed by muscle cells and not by fibroblasts.

Temporal Regulation of N-CAM Expression in Cell Culture.

As N-CAM was expressed by skeletal muscle cells we wished to determine in some detail when it is first expressed by myoblasts and how its detailed profile of reactivity compared with other antibodies we have found that react with myoblasts such as anti-Thy-1 and McAb 24.1D5[13].

After dissociation of human foetal muscle, N-CAM can be found on some myoblasts in suspension as early as 3 hours. At this time only a subpopulation of the N-CAM$^+$ cells were Thy-1$^+$ and none were 24.1D5$^+$. Over the next 24 hours the number of N-CAM$^+$/Thy-1$^+$ cells increases, but it is not until 40 hours that N-CAM$^+$/24.1D5$^+$ cells appear. These data suggest that N-CAM is one of the first muscle specific cell surface antigens to be expressed after dissociation, and shows that there is not co-ordinate activation of the expression of muscle surface antigens.

We have studied these antigenic phenotypes further in cell cultures of foetal muscle grown on gelatin coated coverslips, and some representative micrographs are shown in Figure 2. After 2 days of culture in media containing 20% foetal calf serum and 2% chick embryo extract, 98% of the cells present were N-CAM$^+$/Thy-1$^+$, while only 16% were N-CAM$^+$/24.1D5$^+$. After a further 3 days in culture, when 98% of the cells were N-CAM$^+$/Thy-1$^+$, the N-CAM$^+$/24.1D5$^+$ population had increased to 59%. If instead of growing the myoblasts in foetal calf serum they were grown in serum free Sato media for 3 days, there was a dramatic reduction in the percentage of N-CAM$^+$ cells. While 96% of the myoblasts expressed N-CAM in foetal calf serum, only 59% expressed N-CAM in Sato media. In addition, there was the appearance of a large (41%) of N-CAM$^-$/24.1D5$^+$ population. It would appear that the N-CAM and 24.1D5 reactivities can be modulated by cell density and/or factors in serum. Whether there is an obligatory sequence of antigenic expression during the development of myoblast clones in cell culture will be the subject of further study.

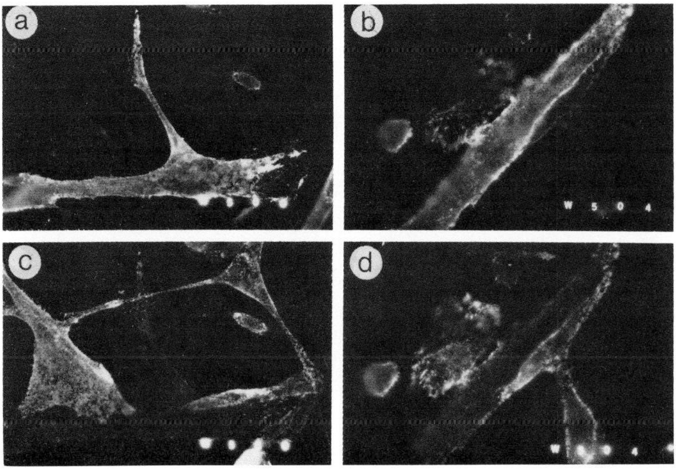

FIGURE 2. Expression of N-CAM in cell cultures of human skeletal muscle. Indirect immunofluorescent double stains showing a & b) Anti-N-CAM with c) Anti-Thy-1 or d) McAb 24.1D5.

DISCUSSION

A variety of roles have been ascribed to N-CAM in the nervous system. Thus it has been implicated in the formation of cell layers in the retina, neurite fasiculation and the tracking of growth cones along nerves[2,3]. Little attention has however been paid to the role of N-CAM in skeletal muscle although Rutishauser et al[8] have suggested that it may be implicated in the initial interaction of nerve with muscle. We show here that N-CAM expression correlates well with times of ongoing synaptogenesis. N-CAM levels in muscle are high during synaptogenesis and only after functional contacts have been made is its synthesis down regulated. The only area of a mature muscle fibre where N-CAM remains in adult muscle is the NMJ and it is not possible to state whether the N-CAM found here has its origin in nerve or muscle. Skeletal muscle can regenerate itself after injury and many of the features of regenerating muscle are similar to developing muscle. As N-CAM was expressed by histochemically identifiable regenerating muscle fibres it provides an additional phenotypic marker for these muscle fibres. It may also provide further support for the notion that N-CAM is involved in nerve-muscle interactions although more biopsies will have to be analysed to determine the spectrum of disease states that are permissive for activation of expression.

In cell cultures of skeletal muscle N-CAM is expressed by cells of only the myogenic lineage and not by fibroblasts. Although populations of cycling myoblasts can be obtained that react with a number of antibodies that define muscle, differentiation antigens such as McAb 5.1H11, Thy-1, 24.1D5 and anti-N-CAM, the regulation of synthesis of these gene products is not co-ordinate. In preliminary experiments we have found in freshly dissociated myoblast populations that N-CAM expression is activated first followed by Thy-1 and only much later is 24.1D5 reactivity found. In addition, it appears that populations of myoblasts that all express N-CAM can lose it but continue to express 24.1D5. As these changes occur when cultures are dense and fusion is occurring, it will be of some interest to find whether N-CAM plays any role in the fusion of myoblasts to form myotubes. It also seems likely that the above antibodies should allow us to analyse the lineage relationship of myoblasts in human muscle.

REFERENCES

1. Schnaar RL (1984). Immobilized glycoconjugates for cell recognition studies. Analyt Biochem 143:1.
2. Edelman GM (1984). Modulation of cell adhesion during induction, histogenesis, and perinatal development of the nervous system. Ann Rev Neurosci 7:339.
3. Rutishauser U (1984). Developmental biology of a neural cell adhesion molecule. Nature 310:549.
4. Murray BA, Hemperly JJ, Galin WJ, MacGregor JS, Edelman GM, Cunningham BA (1984). Isolation of cDNA clones for the chicken neural cell adhesion molecule (N-CAM). Proc Natl Acad Sci (USA) 81:5584.
5. Goridis C, Hirn M, Santoni M-J, Gennarini G, Deagostini-Bazin H, Rjordan B, Kiefer M, Steinmetz M (1984). Isolation of mouse N-CAM-related cDNA: detection and cloning using monoclonal antibodies. EMBO J (in press).

6. Thiery J-P, Duband J-L, Rutishauser U, Edelman GM (1982). Cell adhesion molecules in early chick embryogenesis. Proc Natl Acad Sci (USA) 79:6737.
7. Grumet M, Rutishauser U and Edelman GM (1982). Neural cell adhesion molecule is on embryonic muscle cells and mediates adhesion to nerve cells in vitro. Nature 295:693.
8. Rutishauser U, Grumet M and Edelman GM (1983). Neural cell adhesion molecule mediates initial interactions between spinal cord neurons and muscle cells in vitro. J Cell Biol 97:145.
9. Walsh FS, Ritter MA (1981). Surface antigen differentiation during human myogenesis in culture. Nature 289:60.
10. Walsh FS, Phillips E, Dhut S, Moore SE (1983). Extracellular matrix antigen of human muscle defined by a monoclonal antibody. J Neuroimmunol 5:11.
11. Hurko O, Walsh FS (1983). Human fetal muscle-specific antigen is restricted to regenerating myofibres in diseased adult muscle. Neurology 33:737.
12. Moore SE, Hurko O, Walsh FS (1984). Immunocytochemical analysis of fibre type differentiation in developing skeletal muscle. J Neuroimmunol 7:137.
13. Moore SE, Walsh FS (1985). Specific regulation of N-CAM/D2-CAM cell adhesion molecule during skeletal muscle development. EMBO J (in press).
14. Towbin H, Staehelin T and Gordon J (1979). Electrophoretic transfer of proteins from polyacrylamide gels to nitrocellulose sheets: procedure and some applications. Proc Natl Acad Sci (USA) 76:4350.

Molecular Biology of Muscle Development, pages 189–197

TORPEDO ACETYLCHOLINESTERASE: ISOLATION OF cDNA CLONES WITH ANTIBODIES RAISED AGAINST AN ACHE ACTIVE SITE SYNTHETIC DECAPEPTIDE*

Sandra M. Gaston, John Tamkun, Steve Burden, Richard Lappin,** and Phillips W. Robbins

The Massachusetts Institute of Technology, Cambridge, MA 02139, and **The Rockefeller University, New York, NY 10021

ABSTRACT A lambda gt11 library generated from Torpedo electric organ poly A+ RNA has been screened with a polyclonal antisera directed against the active site of Electrophorus AChE. The antisera was prepared by immunization with a synthetic decapeptide whose sequence corresponds to the active site serine region of Electrophorus AChE (Schaffer et al., Biochemistry 12: 2946, 1973). The synthetic polypeptide preparation was crosslinked to thyroglobulin carrier prior to injection. The resulting anti-decapeptide antisera reacts with a 68K MW polypeptide in Western blots of Torpedo electric organ total protein. Three independant recombinant lambda gt11 clones have been isolated with this antisera, with cDNA inserts of approximately 170, 290, and 350 base pairs. These cDNA inserts are being further characterized prior to their use as molecular probes.

INTRODUCTION

The muscle and neuronal acetylcholinesterases (EC 3.1.1.7, AChE) are highly polymorphic enzymes containing nearly identical catalytic subunits. The molecular basis of this polymorphism is of keen interest,

*This work was supported by grants from the National Cancer Institute, USPHS. S. Gaston was the recipient of an American Cancer Society postdoctoral fellowship.

both because the pattern of AChE expression is specific to tissue and to subcellular compartment and because it is subject to developmental regulation (reviewed in Massoulie and Bon, 1980; Brimijoin, 1983). As part of an ongoing study of the expression and glycosylation of the different neuronal AChE molecular forms, we are preparing a series of AChE cDNA clones from different tissue sources. These clones will be used to attempt to determine how much of the AChE diversity arises from multiple genes and/or mRNA's, and how much from post-translational mechanisms.

We relied upon AChE specific antibodies to identify our first set of AChE cDNA clones. Successful screening with antibodies is facilitated by selection of a tissue source that is highly enriched in the protein of interest. For that reason we began our cloning project with a lambda expression vector cDNA library prepared from the electric organ of the eel Torpedo. To increase the probability of selecting clones that would be useful as probes in neuronal preparations, the Torpedo cDNA library was screened with the aid of an anti-active site peptide antibody. This anti-peptide antibody is specific for a synthetic decapeptide whose sequence corresponds to the active site sequence of AChE from a second electric eel species, Electrophorus electricus (Schaffer et al., 1973). This antibody cross reacts with both Torpedo and mouse neuronal AChE, and has been very useful as a screening tool.

MATERIALS AND METHODS

The decapeptide used to generate the antibody used in the screening of the Torpedo electric organ cDNA library was synthesized at the Children's Hospital Medical Center Peptide Synthesis Facility in Boston by automated stepwise solid-phase peptide synthesis (Young et al., 1983). Because thin layer chromatography performed on 0.1 mm cellulose (Polygram Cel 300, Macherey-Nagel and Co.) with butanol: acetic acid: water (3:1:1) revealed that the peptide product was heterogeneous, the product was subjected to additional analysis.

Samples of the peptide product were fractionated first on a 1 x 100 cm Biogel P2 column (Biorad Laboratories) in 0.1M pyridine acetate, and then on HPLC using a C_{18} microBondapak column (Waters Associates) with a linear gradient from 0.05% TFA in water to 0.035% TFA in acetonitrile. Mass spectrophotometry was used to analyse the idividual components. Fast atom bombardment spectra were obtained and interpreted at the MIT Mass Spectrophotometry Facility. The peptide fraction with the protonated molecular weight (821 daltons) and amino acid composition of the desired decapeptide was partially sequenced from the mass spectra to confirm its identity. Ultimately, the peptide preparation selected for rabbit immunizations was a fraction from the Biogel P2 column containing the 821 MW peptide and two other peptides that have the expected amino acid composition but slightly higher molecular weights (835 and 849 daltons). The nature of the peptide modifications that produced these deviations in molecular weight have not been determined.

For immunization of rabbits, the synthetic peptide preparation was coupled to the carrier protein thyroglobulin with gluteraldehyde, using equal parts w/w polypeptide:carrier (Sutcliffe et al., 1983). Rabbits were immunized by standard procedures, using Freund's adjuvants.

Details of the protocol used in the preparation and screening of the Torpedo electric organ cDNA library followed closely those described by Schwarzbauer et al. (1983). The charateristics of the lambda gtll expression vector have been described by Young and Davis (1983a, 1983b). The Torpedo electric organ cDNA library had been amplified once prior to the screening described here. The anti-peptide antibody was used as a 1:100 dilution of immune sera. Control experiments in which the antisera was exhaustively absorbed with thyroglobulin prior to use showed that the presence of anti-carrier protein activity in the sera had no effect on its detection of AChE or of fusion proteins obtained from the positive clones.

RESULTS AND DISCUSSION

The vector chosen for the construction of the Torpedo electric organ cDNA library, lambda gtll, was designed for selection of low frequency clones from a large pool of recombinants. Because the cDNA's are cloned into a unique EcoRI site within the phage *lacZ* gene, those inserts which are in correct reading frame and orientation give rise to fusion proteins. These fusion proteins, composed of beta galactosidase and the protein sequence encoded by the cDNA insert, make it possible to screen large numbers of recombinant phage immunochemically (figure 1).

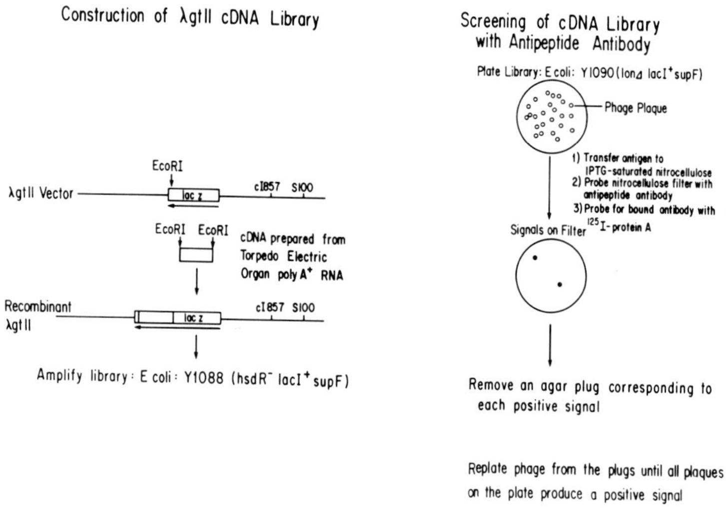

FIGURE 1. Strategy for preparation and immunological screening of lambda gt11 libraries.

Three independant cDNA clones were detected with the antibody raised against the AChE active site peptide "gly-gly-glu-ser-ser-glu-gly-ala-ala-gly" (figure 2). To further characterize the fusion proteins made by these positive clones, lysogens of 10aa, 10cd and 9xc were prepared and induced with isopropyl-beta-D-thiogalactopyranoside (IPTG). Because the phage beta galactosidase gene is under the control of the lac operon, IPTG results in the synthesis of large amounts of the beta-galactosidase fusion protein (figure 3A). The three fusion proteins are only slightly larger than the phage vector beta galactosidase. The presence of the cDNA encoded polypeptide within each of these fusion proteins was confirmed by their lack of beta galactosidase activity and their positive reaction with the anti-peptide antibody (figure 3B). As expected, the phage vector beta galactosidase shows no reaction with the antibody.

It was anticipated from the apparent molecular weights of the fusion proteins that the cDNA inserts of 10aa, 10cd and 9xc were likely to be less than 500 base pairs (bp) in

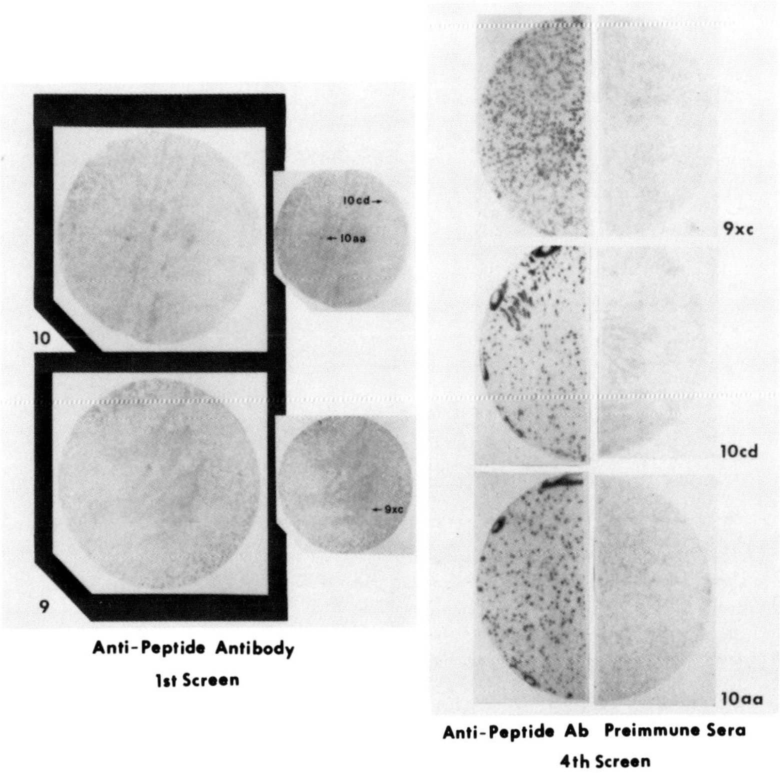

FIGURE 2. Immunochemical screening of recombinant lambda gt11 plaques with Torpedo electric organ cDNA inserts, using antibody raised against the AChE active site decapeptide "gly-gly-glu-ser-ser-glu-gly-ala-ala-gly". Ten 150 mm plates with approximately 2 x 10^4 plaques/plate were evaluated in the initial screen. Autoradiographs of plates 9 and 10, first screen, show the initial detection of the three immunopositive clones: 10aa, 10cd and 9xc. Each clone was subjected to successive rounds of plaque purification, and autoradiographs of the fourth round of screening show that all of the plaques react positively with the anti-peptide antisera. This positive reaction is easily distinguishable from the background seen with the pre-immune sera.

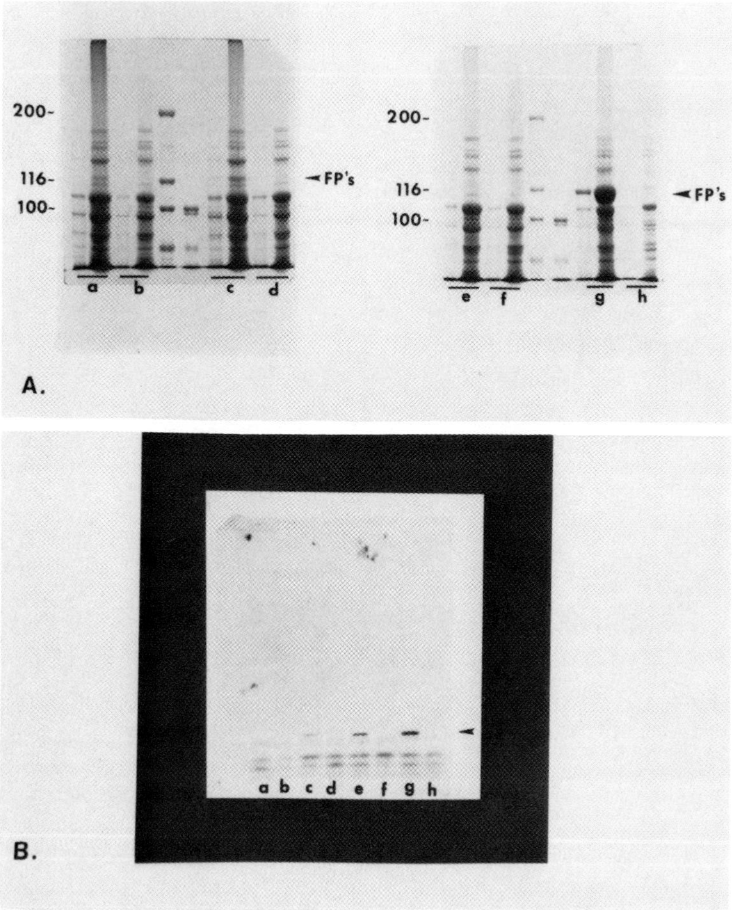

Figure 3. Characterization of beta galactosidase fusion proteins generated by recombinant lambda gtll clones. Y1089 lysogens of lambda gtll and the 10aa, 10cd and 9xc recombinants were prepared to facilitate production of large quantities of fusion proteins. The temperature sensitive repressor was inactivated at 42°, and the lysogens grown either in the presence or the absence of IPTG. It is expected that synthesis of both the lambda gtll beta galactosidase and any fusion proteins derived from that beta galactosidase will be induced by IPTG.

Figure 3 (continued)
3A. Coomassie blue stained 5% SDS polyacrylamide gel of the lysogen bacterial lysates. Y1089/9xc is in lanes A (IPTG) and B (no IPTG), Y1089/10cd is in lanes C (IPTG) and D (no IPTG), Y1089/10aa is in lanes E (IPTG) and F (no IPTG), and Y1089/lambda gt11 vector is in lanes G (IPTG) and H (no IPTG). The position of the beta galactosidase and fusion proteins is indicated (FP).
3B. Western blot autoradiograph of a separate 5% SDS polyacrylamide gel of lysogen cell lysates. The blot was probed with the anti-AChE active site antibody and I^{125} protein A. Fusion proteins (arrow) were detected by the anti-peptide antibody, this detection is dependant on IPTG induction (lanes C, E, and G). Background staining is present in all lysate samples and this is independant of IPTG. Y1089/lambda gt11 is in lanes A (IPTG) and B (no IPTG), Y1089/10aa is in lanes C (IPTG) and D (no IPTG), Y1089/10cd is in lands E (IPTG) and F (no IPTG), Y1089/9xc is in lanes G (IPTG) and H (no IPTG).

---

length. From a series restriction mapping experiments, it was determined that a digestion of the recombinant phage DNA preparations with Kpn I and Sac I would produce fragments that could be used to estimate the complete length of each insert (figure 4). Clone 10aa is estimated at 170 bp, clone 10cd is estimated at 350 bp, and clone 9xc is estimated at 290 bp.

Currently, these three positive clones are being subjected to DNA sequencing in order to permit comparison of the insert encoded amino acid sequences with the available partial amino acid sequence of Torpedo AChE. Two of our immediate objectives in this study are the identification of the active site region of the Torpedo AChE amino acid sequence, and the derivation of a complete Torpedo AChE amino sequence from additional clones selected with the aid of the 10aa, 10cd and 9xc. We are also developing probes derived from our AChE cDNA clones inorder to begin to determine the nature of AChE genes and mRNA's.

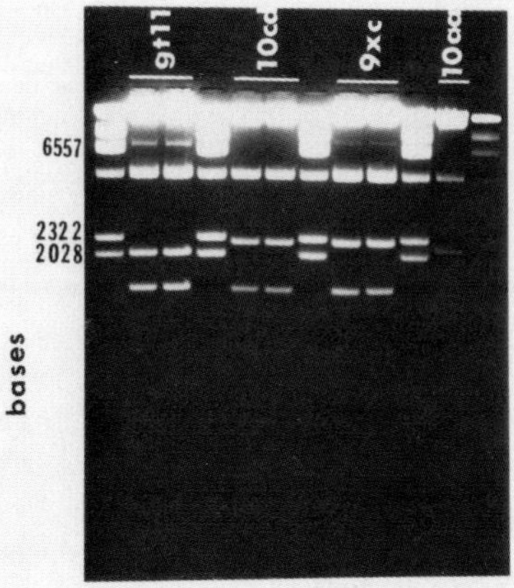

FIGURE 4. Restriction digestions to estimate the size of the cDNA inserts in the 10aa, 10cd and 9xc recombinant clones. Purified phage DNA preparations were digested with Kpn I and Sac I. The 2,030 Kpn I/Sac I fragment of the vector lambda gt11 DNA is the fragment that overlaps the Eco RI insertion site. The position of this overlap fragment in each of the recombinant clones, none of which contains a Kpn I or Sac I site within the insert, was used to estimate the size of the cloned cDNA's. Estimates are 170 base pairs for 10aa, 290 base pairs for 9xc, and 350 base pairs for 10cd. 1.4% Agarose gel, stained with ethidium bromide.

REFERENCES

1. Brimijoin S (1983). Molecular Forms of Acetylcholinesterase in Brain, Nerve and Muscle: Nature, Localization and Dynamics. Prog. in Neurobiol. 21:291.

2. Massoulie J, Bon S (1982). The Molecular Forms of Cholinesterase and Acetylcholinesterase in Vertebrates. Ann. Rev. Neurosci. 5:57.

3. Schwarzbauer JE, Tamkun JW, Lemischka IR, and Hynes RO (1983). Three different fibronectin mRNA's Arise by Alternate Splicing Within the Coding Region. Cell 35:421.

4. Sutcliffe JG, Milner RJ, Shinnick TM, and Bloom FE (1983). Identifying the Protein Products of Brain-Specific Genes with Antibodies to Chemically Synthesized Peptides. Cell 33: 671.

5. Young JD, Costello CE, Van Langenhove A, Haber E, and Matsueda GR (1983). Synthesis of a cyclic fibrin-like peptide and its analysis by fast atom bombardment mass spectrometry. Int. J. Peptide Protein Res. 22: 374.

6. Young, RA, Davis, RW (1983a). Efficient isolation of genes by using antibody probes. Proc. Natl. Acad. Sci. 80:1194.

7. Young, RA, Davis, RW (1983b). Yeast polymerase II genes: Isolation with antibody probes. Science 222:778.

Molecular Biology of Muscle Development, pages 199–212

CLONING OF DNA CODING FOR Ca^{2+} + Mg^{2+} DEPENDENT ATPases OF SARCOPLASMIC RETICULUM

Christopher J. Brandl*, Bozena Korczak*, N. Michael Green# and David H. MacLennan*

*Banting and Best Department of Medical Research, University of Toronto, C.H. Best Institute, 112 College Street, Toronto, Ontario M5G 1L6, Canada.
and
#National Institute for Medical Research, Mill Hill, London, U.K. NW7 1AA

ABSTRACT Cloning and sequence analysis of cDNA encoding the Ca^{2+}+Mg^{2+}-dependent ATPase of sarcoplasmic reticulum shows that transcripts for an alternatively spliced skeletal form and a cardiac form of the enzyme exist in neonatal rabbit skeletal muscle and are developmentally regulated. The deduced amino acid sequence of one of these, believed to be the cardiac form, shows that the protein is 997 amino acids in length. Analysis of secondary structure indicates that the protein contains two transmembrane loops in the amino-terminal quarter which are separated on the cytoplasmic side by an antiparallel beta domain. The carboxyl-terminal quarter contains six potential transmembrane segments. The central half is entirely cytoplasmic and is composed of two parallel beta domains which form a hinged pair such that ATP bound to one phosphorylates an Asp on the other. The three cytoplasmic, globular domains are linked to the transmembrane helices by a multi-helical stalk which is highly negatively charged. The stalk probably contains the two high affinity Ca^{2+} binding sites from which translocation is initiated.

INTRODUCTION

The contraction and relaxation of muscle are controlled through variations in cytoplasmic Ca^{2+} concentrations (1). These Ca^{2+} concentrations are, in turn, regulated by an extensive membrane system within muscle cells, the sarcoplasmic reticulum. Ca^{2+} is released from these organelles in response to an electrical depolarization of the muscle cell membrane and is taken up again by the action of a Ca^{2+} transporting ATPase which makes up more than 60% of the total protein in the sarcoplasmic reticulum membrane.

The Ca^{2+} ATPase was isolated in 1970 (2) and has been the subject of extensive studies on structure (3), function (4) and biosynthesis (5,6) since that time. Many studies on development of the sarcoplasmic reticulum have been concerned with the temporal appearance of either Ca^{2+}-dependent ATPase activity or with the immunological identification of the newly synthesized protein. These studies, when carried out with differentiating muscle cells in culture (7,8), have revealed that the synthesis of the Ca^{2+}-ATPase is initiated approximately coincidentally with cell fusion and with the coordinate synthesis of a series of other muscle cell-specific proteins (9).

In order to learn more about the primary stucture of the Ca^{2+} ATPase, about the number of ATPase genes and about transcription of these genes during development, it became essential to clone DNA coding for Ca^{2+} ATPases. In this paper we describe the cloning of DNA coding for two forms of the Ca^{2+} ATPase and we describe new observations on the structures of the enzymes and on their expression in neonatal life.

RESULTS AND DISCUSSION

Isolation of cDNA Clones.

In earlier studies of the biosynthesis of the ATPase and calsequestrin, we carried out cell free synthesis of these proteins using mRNA isolated from neonatal rat muscle (10). While this proved to be a good source of ATPase mRNA, we recognized that cloning of cDNA from this source would provide us with information on a protein for which relatively little amino acid sequence was available. By contrast, cloning of cDNA from neonatal rabbit muscle should yield sequence information on a well characterized protein (11). Accordingly, we investigated whether neonatal rabbit muscle would be a suitable source of ATPase mRNA. Our finding (12) that excellent preparations of ATPase could be translated from neonatal rabbit muscle mRNA led us to use this source for cDNA cloning.

Several cDNA libraries have been made, using standard techniques (13), as follows: a library of 2200 clones in the vector pBR322 constructed in collaboration with Dr. John Nilson (Case Western Reserve University); a library of 3,500 clones constructed in pBR322 by Drs. M. Powers, P. Hwang and R. Fletterick (UCSF) using adult rabbit muscle; a library of 46,000 clones in pBR322; and a library of 1,600,000 clones, constructed in collaboration with Dr. H. Okayama (NIH), in the expression vector pCD-X (14).

These libraries were initially screened with oligonucleotides (14 mers) generated from ATPase sequences. The first library did not yield any ATPase clones while the second library contained a single 75 bp ATPase clone coding for 25 amino acids recently identified in the carboxyl-terminal region of the ATPase sequence (15). This clone, when used as a probe in the third library, hybridized with several clones, six of which coded for the skeletal muscle ATPase and two of which coded for an homologous clone. Two of the six skeletal ATPase clones, when overlapped, coded for 1117 bp of ATPase sequence of which 918 bp coded for 306 carboxyl-terminal amino acids. One of the homologous clone of 1900 bp encoded 498 carboxyl-terminal amino acids.

Several long ATPase clones were isolated upon screening of the library in the vector pCD-X. The longest skeletal ATPase clone was 3000 bp and coded for the 3' untranslated region plus 905 of the approximately 1000 amino acids which make up the entire ATPase sequence (fig. 1). The longest clone for the homologous ATPase was 3,722 bp and coded for 997 amino acids plus the 5' and 3' untranslated regions (fig. 1). In the cDNA clones in pBR322, the skeletal ATPase was present in the ratio of 3:1 over the homolog; in the pCD-X-clones the ratio was 1:8.

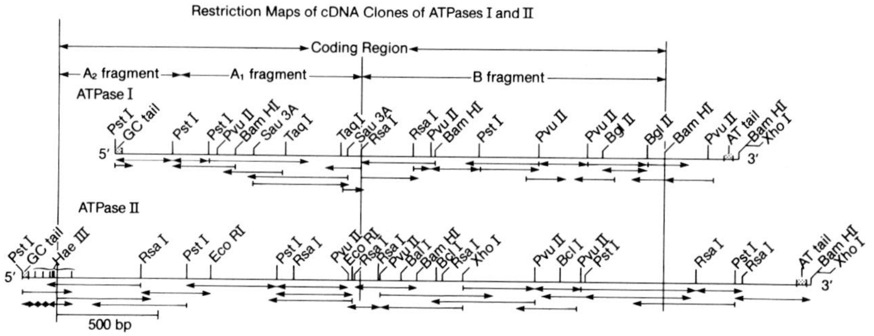

Figure 1. Restriction maps and sequencing strategies for cDNA clones encoding two forms of the Ca^{2+} ATPase of sarcoplasmic reticulum. ATPase I is an alternatively transcribed form of the skeletal enzyme: ATPAse II is probably the cardiac form of the enzyme. The coding region is divided into three segments A2, A1 and B which correspond to the fragments generated by brief tryptic digestion of the skeletal Ca^{2+} ATPase in intact sarcoplasmic reticulum (20).

Sequence Analysis of the two ATPase Clones.

Sequencing of the skeletal ATPase clone has shown virtually an exact concordance between the deduced amino acid sequence and the amino acid sequence published for the Ca^{2+} ATPase by Allen et al (11). In addition, several fragmentary amino acid sequences that were not overlapped, (16-18) have been aligned in the amino acid sequence deduced from cDNA analysis. The cDNA sequences enabled the acidic Asp and Glu residues to be distinguished from their amides, corrected two "secretarial" errors in the sequence and demonstrated two real differences in amino acid sequence, one near the amino-terminal end and one right at the carboxyl-terminus. These alignments provided unambiguous evidence that our clones contained cDNA coding for the Ca^{2+} ATPase of skeletal muscle sarcoplasmic reticulum.

The homologous clone coded for a protein with 84% overall sequence identity with the skeletal ATPase but two stretches of 90 and 87 residues, which included parts of Allen's sequences 2 and 4, respectively, were identical (19). The nonhomologous residues were frequently clustered.

A major sequence difference was observed in the center of the molecule, in a region sensitive to tryptic digestion (20,21). A few seconds exposure of the sarcoplasmic reticulum to trypsin results in quantitative cleavage of an Arg-Ala bond at residues 505-506 in the skeletal sequence (22) and of an Arg-Thr bond in the cardiac sequence (23). The Lys residue which binds fluorescein isothiocyanate (FITC) (24) is found 10 residues downstream of this tryptic cleavage site and the FITC binding residue can be isolated as a component of a relatively short tryptic fragment. Briggs *et al.* (23) have recently shown that the sequence of an FITC-containing tryptic peptide from the dog cardiac sarcoplasmic reticulum Ca^{2+} ATPase contains the sequence (R)TSMSKMFVK*GAPEGVIDR(*=FITC binding site). This precise sequence was found in our homologous clones while the comparable sequence RAAVGNKMFVKGAPEGVIDR was found in the skeletal Ca^{2+}-ATPase protein (11) and in our skeletal ATPase clones (C.J. Brandl *et al.*, in preparation). We suggest that the sequence TSMS, in the center of the ATPase molecule, may be diagnostic of the cardiac form of the Ca^{2+} ATPase, and, therefore, that our homologous clones encode the cardiac form of the enzyme.

In previous studies of the *in vitro* synthesis of the ATPase (10) we found no evidence for an amino-terminal signal sequence. We were very interested in confirming this observation by analysis of the 5' end of the cDNA clone. We found that the 5' untranslated region was 128 bp in length and was very GC rich. Thus the first ATG in the sequence encoded the Met found at the amino-terminus of the ATPase, confirming our hypothesis that the ATPase is made without an amino terminal signal sequence.

Deduction of Secondary Structure in the Proteins.

The complete cDNA sequence of the cardiac clone and the nearly complete sequence of the skeletal clone have provided us with the deduced amino acid sequences for these proteins. We have used these sequences to predict the secondary structures and folding patterns of these transport enzymes (19). In our analysis of secondary structure in the ATPases, we have used the method of Taylor and Thornton (25) which begins with the prediction of alpha helices, beta strands and beta turns. Some of the ambiguities in priority of beta strand and alpha helix prediction are resolved by comparing patterns of hydrophobicity and of supersecondary structure derived from a database of known structures. Unfortunately hydrophobic transmembrane sequences frequently appear as beta strands in these predictions. Our approach to this problem has been to identify the transmembrane sequences by using both polarity plots and hydropathy plots (26) to detect nonpolar sequences of sufficient length to traverse the membrane. We have predicted that these transmembrane sequences, like those of other membrane proteins, form alpha helices.

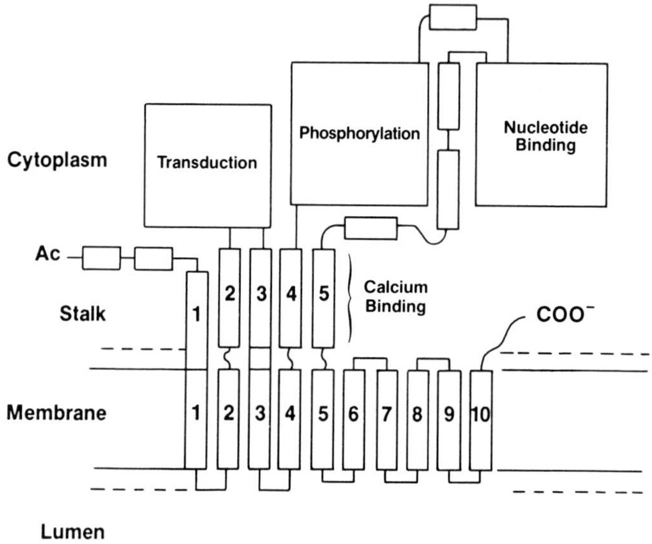

Figure 2. Planar diagram for the folding pattern of the Ca^{2+} ATPase. The arrays of helices are indicated as rectangles while the large globular domains appear as squares. In a three dimensional structure, the transmembrane helices would form tight clusters while the extramembranous domains would be folded beside each other permitting interaction among all three. ATP is bound in the nucleotide binding domain and phosphorylates an Asp in the phosphorylation domain. The transduction domain is involved with the other two domains in the transduction of the energy of ATP hydrolysis to Ca^{2+} transport. Ca^{2+} is bound in the five-membered stalk region and is translocated through channels created by ten transmembrane helices.

We have found four transmembrane sequences in the amino-terminal quarter of the molecule. These sequences form two hairpin loops in which the four alpha helices in the membrane extend 3 nm into the cytoplasm forming a stalk on top of the transmembrane sequences. The first three alpha helices that extend into and out of the various transmembrane segments contain 18 Glu and 1 Asp residues in close proximity to the membrane. These acidic residues are not aligned in EF hand conformations typical of high affinity Ca^{2+} binding sites (27) but they are the best candidates for the Ca^{2+} binding sites in the enzyme. We refer to this stalk region as the Ca^{2+} binding domain (fig. 2). The first hairpin loop contains an additional 5 Glu residues on the luminal surface. The high density of negative charges in the stalk section on the cytoplasmic surface and on the first luminal loop strongly indicates that these structures are involved in the movement of Ca^{2+} across the membrane. Indeed, tryptic and CNBr fragments from this region of the enzyme have been shown to increase Ca^{2+} conductance in a black lipid membrane (28).

The central portion of the molecule, 435 amino acids long, consists predominantly of alternating beta strands and alpha helices generating parallel beta sheets but it extends at its carboxyl-terminal end into a series of alpha helices and eventually joins the alpha helix of transmembrane sequence 5. This extramembranous region contains the site of phosphorylation at residue 351 (11), the highly sensitive tryptic cleavage site between residues 505 and 506 (22) and the site where FITC binds at residue 514 and inhibits nucleotide binding (and, thereby, ATP hydrolysis) without inhibiting acetyl phosphate hydrolysis (24). The fact that the phosphorylation site and the nucleotide binding site are so far apart suggests that this region of the protein is folded into two globular domains separated by an exposed, protease-sensitive sequence. In this model (fig. 2), ATP bound to one domain (nucleotide binding domain) would phosphorylate an Asp residue on the other domain (phosphorylation domain).

There is a fourth extramembranous domain in the amino-terminal quarter of the molecule (fig. 2). This domain lies between the two extended helical hairpin loops that traverse the membrane and is composed of seven beta strands separated by short loops and likely to fold into an antiparallel beta sandwich. This globular domain must participate in the transduction of the energy of ATP hydrolysis into Ca^{2+} translocation since Scott and Shamoo

(29) noted a close correlation between the tryptic cleavage which occurs in this domain and the loss of the ability of the enzyme to transport Ca^{2+} even though it still hydrolysed ATP. On the basis of these observations, we have referred to this extramembranous domain as the transduction domain.

The carboxyl-terminal quarter of the protein contains at least six potential transmembrane sequences. The first two and the last two could form helical hairpin loops but the central two would be separated by about 30 residues which might constitute the only significant luminal sequence.

The structure that emerges is that of a headpiece constituted of the three globular domains, a stalk composed of a cluster of five alpha helices and a basepiece composed of 10 transmembrane sequences (fig. 2). ATP is bound and hydrolysed in the headpiece. Ca^{2+} is bound in the stalk sector and translocated through the transmembrane sequences in the basepiece.

These structural predictions allow us to begin to formulate theories on Ca^{2+} translocation across the membrane. We suggest that Ca^{2+} bound to hydrophilic faces of the stalk sector could be occluded (30) in the interior of the stalk by conformational changes induced by ATP hydrolysis. This occlusion step might lead to exposure of hydrophobic surfaces in the stalk thus favoring the movement of the stalk into the lipid bilayer as has been suggested by morphological (31) as well as X ray and neutron diffraction analyses (32). This movement of the bound Ca^{2+} ion into the bilayer region would be followed by a change in binding affinity and by the release of Ca^{2+} into an environment continuous with the lumen of the membrane. These ideas will require careful future evaluation.

The Ca^{2+} ATPase during Development.

An exciting finding in our studies was the discovery that a putative cardiac form of the ATPase was expressed in neonatal rabbit muscle. In other chapters in this book, it is evident that, as development proceeds, switching occurs among genes expressing different forms of various muscle proteins. Cardiac forms of actin (33) and of myosin (34) are expressed in skeletal muscle during fetal or neonatal development. We suspect that the same pattern is true for the Ca^{2+} ATPase. Our probes of genomic DNA indicate that there may be seven different cross hybridizing ATPase genes (B. Korczak et al. in preparation). Thus switching among these genes may eventually be shown to be even more complex than we now visualize it to be. Although we have not yet analysed these developmental patterns, it is probable that the widely differing ratios between cardiac and skeletal ATPase clones in our different cDNA libraries results from the fact that there is a rapid switch from cardiac to skeletal gene expression early in neonatal life.

An interesting aspect of the deduced skeletal ATPase sequence was the presence in the neonatal skeletal transcript of the carboxyl-terminal sequence -DPEDERRK. Allen et al.(11) found that the protein isolated from mature English rabbit muscle ended with -G and we have confirmed this observation with Canadian rabbits (21). Thus the terminal Gly of the adult rabbit muscle enzyme was replaced in the neonatal rabbit enzyme by a new sequence of 8 residues of which 4 were negatively charged and 3 postively charged. These observations suggested that the neonatal muscle sequence may have arisen from alternate splicing of a single gene transcript coding for both neonatal and mature forms of the protein. Nabeshima et al. (35) have described this mechanism for the generation of myosin light chains 1 and 3 in chick muscle. We sought proof of this possibility through isolation and analysis of the rabbit skeletal muscle Ca^{2+} ATPase gene (B. Korczak et al., in preparation).

We used a recombination assay (36) for the isolation of the skeletal Ca^{2+} ATPase gene from a rabbit liver genomic library prepared (37) and generously provided to us by Dr. T. Maniatis. We used both π AN7 and π VX as vectors for the insertion of a 71 bp Sau 3a - Hae III fragment from our first cDNA clone, pSAI, into the lambda phage library. Repeated screening of this library yielded a single clone λ Al of 17 kb.

The 3' end of the genomic clone λ AI has been extensively analysed. We found a penultimate exon that coded for the sequence lARNYLE (fig. 3), the carboxyl-terminal sequence common to both our sequence and to Allen's (11). The penultimate exon did not end with a Gly followed by a termination codon, however: This sequence was found 230 bp downstream. The sequence coding for a Gly-stop was preceeded by a polypyrimidine tract and by an AG which is mandatory for the 3' end of an intron. Some 600 bp downstream from the penultimate exon, we found another terminal exon coding for DPEDERRK-stop. Again, these codons were preceded by a polypyrimidine tract and by the mandatory AG splice residues. The splicing, in this case, created the nucleotide sequence GGATCC, a BamH1 restriction site, which was found in the 3' end of the coding sequence in our cDNA clones but was absent from the genomic clone. These results strongly support our hypothesis that DPEDERRK and G are coded in different exons that are alternately spliced into the mature transcript as muscle development proceeds.

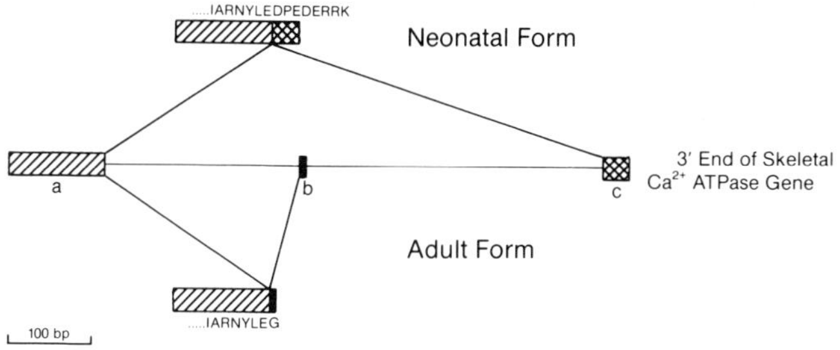

Figure 3. Model for the generation of neonatal and mature skeletal Ca^{2+} ATPase mRNAs by alternative splicing. a, penultimate exon (118 bp); b, terminal exon (3 bp) coding for the mature form of the protein (terminal G); c, terminal exon (24 bp) coding for the neonatal form of the protein (terminal DPEDERRK).

These data offer provocative evidence that muscle membrane proteins are subjected to the same type of developmental regulation that has previously been found for muscle contractile proteins. The further investigation of these early observations will undoubtedly uncover new and interesting insights into membrane assembly during development.

ACKNOWLEDGEMENTS

This work was supported by Grants to D.H.M. from the Ontario Heart Foundation, the Medical Research Council of Canada (MRCC) and the Muscular Dystrophy Association of Canada (MDAC). C.J.B. is a predoctoral fellow of the MRCC; B.K. is a postdoctoral fellow of the MDAC on leave from the

Institute of Experimental Biology, Polish Academy of Science, Warsaw, Poland. The work of N.M.G. was supported by the Medical Research Council (U.K.).

REFERENCES

1. Ebashi, S., Endo, M., and Ohtsuki, I., (1969). Control of Muscle Contraction. Quart. Rev. Biophys. 2, 351-384.
2. MacLennan, D.H., (1970). Purification and Properties of an Adenosine Triphosphatase from Sarcoplasmic Reticulum. J. Biol. Chem. 245, 4508-4518.
3. MacLennan, D.H., and Reithmeier, R.A.F., (1982). The Structure of the Ca^{2+}+Mg^{2+}-ATPase of Sarcoplasmic Reticulum. in "Membranes and Transport". (Ed. A.N. Martonosi) Vol. 1, pp. 567-571, Plenum Press, New York.
4. de Meis, L. and Vianna, A.L., (1979). Energy Interconversion by the Ca^{2+}-Dependent ATPase of the Sarcoplasmic Reticulum. Annu. Rev. Biochem. 48, 275-292.
5. Martonosi, A., (1982). The Development of Sarcoplasmic Reticulum Membranes. Annu. Rev. Physiol. 44, 337-355.
6. MacLennan, D.H., Zubrzycka-Gaarn, E., and Jorgensen, A.O. (1985). Assembly of the Sarcoplasmic Reticulum during Muscle Development. Curr. Top. Membr. Transport 24, 337-368.
7. Holland, P.C. and MacLennan, D.H. (1976). Assembly of the Sarcoplasmic Reticulum. Biosynthesis of the Adenosine Triphosphatase in Rat Skeletal Muscle Cell Culture. J. Biol. Chem. 251, 2030-2036.
8. Martonosi, A., Roufa, D., Boland, R., Reyes, E., and Tillack, T.W., (1977) Development of Sarcoplasmic Reticulum in Cultured Chicken Muscle. J. Biol. Chem. 252, 318-332.
9. Shainburg, A., Yagil, G., and Yaffe, D., (1971). Alterations of Enzymatic Activities During Muscle Differentiation *in vitro*. Dev. Biol. 25 1-29.
10. Reithmeier, R.A.F., de Leon, S., and MacLennan, D.H. (1980). Assembly of the Sarcoplasmic Reticulum. Cell free Synthesis of the Ca^{2+}+Mg^{2+} Adenosine Triphosphatase and Calsequestrin. J. Biol. Chem. 255, 11839-11846.

11. Allen, G., Trinnaman, B.J. and Green, N.M. (1980). Primary Structure of the Calcium-Transporting Adenosine Triphosphatase of Rabbit Skeletal Sarcoplasmic Reticulum. Peptides Derived from Digestion with Cyanogen Bromide and the Sequence of Three Long Extramembranous Segments. Biochem. J. 187, 591-616.
12. MacLennan, D.H., and de Leon, S., (1983). Biosynthesis of Sarcoplasmic Reticulum Proteins. Methods Enzymol. 96, 570-579.
13. Maniatis, T., Fritsch, E.F., and Sambrook, J. (1982) "Molecular Cloning, a Laboratory Manual;". Cold Spring Harbor laboratory.
14. Okayama, H., and Berg, P., (1983). A cDNA Vector that permits Expression of cDNA Inserts in Mammalian Cells. Mol. Cell. Biol. 3, 280-289.
15. Green, N.M. and Toms, E.J. (1985). The Sequence of two Peptides from the Ca^{2+}-Transporting ATPase of Sarcoplasmic Reticulum after Cleavage at Tryptophan. Biochem. J. in press.
16. Allen, G., (1980). Primary Structure of the Calcium-Transporting Adenosine Triphosphatase of Rabbit Skeletal Sarcoplasmic Reticulum. Soluble Tryptic Peptides from the Succinylated Carboxymethylated Protein. Biochem. J. 187, 545-563.
17. Allen, G., (1980). Primary Structure of the Calcium-Transporting Adenosine Triphosphatase of Rabbit Skeletal Sarcoplasmic Reticulum. Soluble Peptides from the Chymotryptic Digest of the Carboxymethylated Protein. Biochem. J. 187, 565-575.
18. Allen, G., Bottomley, R.C. and Trinnaman, B.J. (1980). Primary Structure of the Calcium-Transporting Adenosine Triphosphatase of Rabbit Skeletal Sarcoplasmic Reticulum. Some Peptic, Thermolytic, Tryptic and Staphylococcal-Proteinase Peptides. Biochem. J. 187, 577-589.
19. MacLennan, D.H., Brandl, C.J., Korczak, B., and Green, N.M. (1985). The Structure of a Calcium-Transporting ATPase from Rabbit Muscle Sarcoplasmic Reticulum Deduced from cDNA Sequence. Submitted to Nature.
20. Thorley-Lawson, D.A., and Green, N.M., (1973). Studies on the Location and Orientation of Proteins of the Sarcoplasmic Reticulum. Eur. J. Biochem. 40, 403-413.
21. Stewart, P.S., and MacLennan, D.H., (1974). Surface Particles of Sarcoplasmic Reticulum Membranes.

Structural Features of the Adenosine Triphosphatase. J. Biol. Chem. 249, 985-993.

22. Klip, A., Reithmeier, R.A.F. and MacLennan, D.H. (1980). Aligment of the Major Tryptic Fragments of the Adenosine Triphosphatase from Sarcoplasmic Reticulum. J.Biol. Chem. 255, 6562-6568.

23. Briggs, F.N., Cable, M.B., Geisow, M.J. and Green, N.M. (1985). Sequence of a Fluorescein-Labelled Tryptic Peptide Isolated from Canine Cardiac Sarcoplasmic Reticulum. submitted to Eur. J. Biochem.

24. Mitchinson, C., Wilderspin, A.F., Trinnaman, B.J. and Green, N.M., (1982) Identification of a Labelled Peptide after Stoicheiometric Reaction of Fluorescein Isothiocyanate with the Ca^{2+}-Dependent Adenosine Triphosphatase of Sarcoplasmic Reticulum. FEBS. Lett. 146, 87-92.

25. Taylor, W.R., and Thornton, J.M. (1984). Recognition of Super-secondary Structure in Proteins. J. Mol. Biol. 173, 487-514.

26. Kyte, J., and Doolittle, R.F. (1983). A Simple Method for Displaying the Hydropathic Character of a Protein. J. Mol. Biol. 157, 105-132.

27. Kretsinger, R.H., (1976). Calcium-Binding Proteins. Annu. Rev. Biochem. 45, 239-266.

28. MacLennan, D.H., Reithmeier, R.A.F., Shoshan, V., Campbell, K.P., Le Bel, D., Herrmann, T.R., and Shamoo, A.E., (1980). Ion Pathways in Proteins of the Sarcoplasmic Reticulum. Ann. N.Y. Acad. Sci. 358, 138-148.

29. Scott, T.L., and Shamoo, A.E., (1982). Disruption of Energy Transduction in Sarcoplasmic Reticulum by Trypsin Cleavage of (Ca^{2+}+Mg^{2+})-ATPase. J. Mem. Biol. 64, 137-144.

30. Dupont, Y., (1980). Occlusion of Divalent Cations in the Phosphorylated Calcium Pump of Sarcoplasmic Reticulum. Eur. J. Biochem. 109, 231-238.

31. Perrachia, C., Dux, L., and Martonosi, A.N. (1984). Crystallization of Intramembrane Particles in Rabbit Sarcoplasmic Reticulum Vesicles by Vanadate. J. Mus. Res. Cell Motil. 5, 431-442.

32. Blasie, J.K., Herbette, L., Pierce, D., Pascolini, D., Scarpa, A., and Fleischer, S., (1982). Static and Time-Resolved Structural Studies of the Ca^{2+} ATPase of Isolated Sarcoplasmic Reticulum. Ann. N.Y. Acad. Sci. 402, 478-484.

33. Minty, A.J., Alonso, S., Carvatti, M., and Buckingham, M.E., (1982) A Fetal Skeletal Muscle Actin mRNA in the Mouse and Its Identity with Cardiac Actin mRNA. Cell 30, 185-192.
34. Whalen, R.G., Sell, S.M., Eriksson, A., and Thornell, L.E., (1982). Myosin Subunit Types in Skeletal and Cardiac Tissues and Their Developmental Distribution. Dev. Biol. 91, 478-484.
35. Nabeshima, Y-I., Fujii-Kurayama, Y., Muramatsu, M., and Ogata, K., (1984) Alternative Transcription and Two Modes of Splicing Result in Two Myosin Light Chains from One Gene. Nature 308, 333-338.
36. Seed., B., (1983). Purification of Genomic Sequences from Bacteriophage Libraries by Recombination and Selection *in vitro*. Nucleic Acids Res. 11, 2427-2445.
37. Maniatis, T., Hardison, R.C., Lacy,E., Lauer, J., O'Connell C., Quon, D., Sim, G.K., and Efstradiatis, A. (1978). The Isolation of Structural Genes from Libraries of Eucaryotic DNA. Cell 15, 687-701.

Molecular Biology of Muscle Development, pages 213–223

MYOSINS, MYOTUBES, AND MYOBLASTS ORIGINS OF FAST AND SLOW MUSCLE FIBERS[1]

F.E. Stockdale, J.B. Miller, D.A. Schafer, and M.T. Crow[2]

Department of Medicine, Stanford University
Stanford, CA 94305

ABSTRACT Most early developing muscles of the avian thigh contain both a slow and a fast class of myosin heavy chains and two primary fiber types. The predominate primary fibers are the F fibers which contain only the fast class of myosin heavy chain and the minority primary fibers are F/S fibers which contain both a fast and slow class of myosin heavy chain. Each muscle has a specific localization of the F/S fibers which gives it a distinctive "blueprint" that is established at the time individual muscles are first demarcated from the dorsal and ventral muscle masses. The formation of primary F and F/S fibers and their distribution within each muscle is independent of functional innervation.

INTRODUCTION

Adult muscles of all organisms are now known to contain multiple isoforms of the contractile proteins (1-7). This diversity in the differentiated function of the muscle is increasingly recognized to be rooted in a diversity of muscle fiber types (8). However, there have remained differences in reports on the developmental basis of the diversity based on the expression of myosin in extracts from embryonic muscles and on the identification of fiber types in early developing muscles (9-19). We have, therefore, investigated the expression of myosin both in extracts of muscles and in muscle cells (fibers) as development of the limb proceeded from the time individual muscles are first recognized in the avian thigh.

[1]This work was supported by the National Institutes of Health and the Muscular Dystrophy Association of America.
[2]Present address: Department of Biology, University of Houston, Houston, Texas 77004

We wished to determine when in development different myosin heavy chains were first expressed, when fibers of different types (those containing different myosin heavy chains) were first seen, and what myosin heavy chain(s) were expressed in myotubes (muscle fibers) formed in cell culture. These studies show that diversity of myosin heavy chain expression occurs as one of the earliest events in muscle formation and that this diversity comes from differences in the proportions of two primary fiber types in a muscle, which in turn is rooted in differences in the myoblast populations that produce the primary muscle fibers.

METHODS

White leghorn chickens and embryos were used for these studies. The techniques of hybridoma formation, immunohistochemistry and immunoblotting are those reported previously (19). The methods of cell culture are those of O'Neill and Stockdale (20).

RESULTS

Myosin Heavy Chain Isoforms in Extracts of Embryonic Muscles

To determine the content of myosin heavy chains in the developing hindlimb muscles of the avian embryo, the thighs of progressively older embryos were removed and extracted. Extracts for myosin were made from embryos beginning at the age where muscles in the thigh are first demarcated (day 6 of development) and at 8, 10, and 16 days of embryonic development, and also from several muscles in the adult thigh. Muscles of the thigh at all ages including the adult contained two classes of myosin heavy chain of different molecular weight. One was of the high molecular weight "slow" class and one was of the lower molecular weight "fast" class. These two classes of heavy chain were identified on nitrocellulose transfers of 5% gels (2) after reaction with antibodies to myosin heavy chains (Figure 1). Two monoclonal antibodies were used for this analysis. Monoclonal antibody F59 reacted with the lower molecular weight heavy chain found at all developmental ages in all fast muscles, including the pectoral muscle and the superficial biceps of the thigh, as well as in all mixed muscles, such as the sartorius of the thigh. Monoclonal antibody S58 reacted with the higher molecular weight myosin heavy chain of slow muscles, such as the anterior latissimus dorsi, and the mixed sartorius. Therefore, from the earliest to the latest stages of avian muscle development at least two classes of myosin heavy chain are expressed.

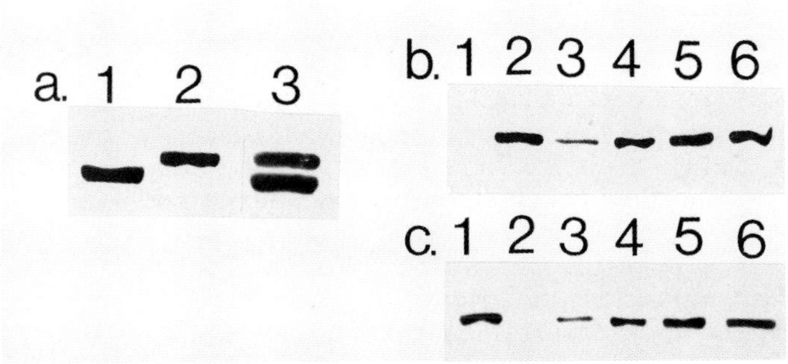

FIGURE 1. Myosin heavy chains of the fast and slow class are found at all times in development. Myosins were extracted, separated by electrophoresis, and visualized by immunoblotting. Myosin heavy chains from the pectoralis major (a1), a fast muscle, and the anterior latissimus dorsi (a2), a slow muscle, were reacted with McAb F59 and S58 respectively. Binding in (a1) and (a2) was visualized with an HRP-conjugated secondary antibody. When the myosins from these two muscles were mixed and first reacted with S58 and an HRP-conjugated secondary antibody and then with F59 and a glucose oxidase-conjugated secondary antibody, it was evident that two molecular size classes of MHC exist. Immunoblots of myosin heavy chain extracted from thigh muscles are shown in (b) and (c). Extracts of thigh muscles were analyzed from ED 6 (Lane 3), ED 8 (Lane 4), ED 12 (Lane 5), and ED 16 (Lane 6) in addition to control adult myosin from the PM (Lane 1) and ALD (Lane 2). All the lanes in (b) were reacted with McAb S58 and all the lanes in (c) were reacted with McAb F59. Visualization of (b) and (c) are by a secondary HRP-conjugated antibody.

Myosin Heavy Chain Isoforms in Individual Fibers of Embryonic Muscles

To determine whether all muscles of the developing limb contain both a fast and a slow class of myosin heavy chain because they are composed of different types of muscle cells (fibers) or because all muscle cells (fibers) contain more than a single class of myosin heavy chain, serial cryostat sections of the thigh were reacted with monoclonal antibodies F59 and S58 (21). Figure 2 shows cross sections of the 7 day embryonic thigh stained with both the fast and slow myosin heavy chain antibodies. These fibers have been designated the primary generation of muscle fibers, or primary fibers (22). These data show that the presence of two myosin heavy chain classes in extracts of developing muscles is due to the presence of at least two primary fiber types. Beyond 5 days of development two types of muscle fibers were distinguished by the antibodies; those that contained only fast myosin heavy chain (F primary fibers) and those that contained both fast and slow heavy chains (F/S primary fibers). Thus, there were two types of primary generation fibers which contained different isoforms of myosin heavy chain. Beginning at about 10 days of development, additional fibers appeared which have been designated as secondary generation fibers (22, 23). Second generation fibers stained predominantly with F59 and were most abundant in fast and mixed muscles (data not shown).

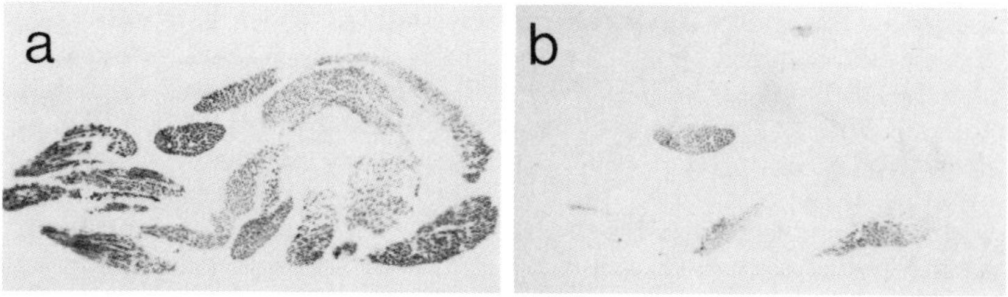

FIGURE 2. Serial cross section of the thigh of a seven day chick embryo. (left) This section was stained with McAb F59 which recognizes fast MHC. (right) This section was stained with McAb S58 which recognizes slow MHC. The binding of each antibody was visualized with a secondary antibody conjugated to horseradish peroxidase. All fibers in the early thigh react with antibodies to the fast class of MHC, and select fibers in a specific pattern also react with the slow class of MHC. In addition to predominately fast and fast/slow fibers, it is possible that a very small population of slow fibers may exist.

The F and F/S primary generation fibers were distributed in a characteristic pattern in each muscle such that all muscles could easily be identified when they first formed. This "blueprint" was based upon the location of F/S primary fibers within the total primary fiber population. The location of these fibers remained nearly constant throughout embryonic and fetal development (21). The addition of the secondary generation of fibers served to increase the size of the embryonic muscles, but did not alter these early muscle fiber patterns.

The finding that a slow myosin heavy chain is expressed early in muscle development and that the fibers containing slow heavy chain isoforms are distributed in a specific topographic pattern suggests that innervation could control the emergence of primary F/S fibers. We find this unlikely since in embryos treated with d-tubocurare from day 4 of development, a time before fibers first appear in the limb, and daily thereafter until day 10, all muscles formed on schedule, and contained F and F/S primary fibers distributed in a pattern indistinguishable from untreated animals. Therefore, neither the formation of the two primary fiber types nor their specific localization within a muscle is dependent upon functional innervation. However, when d-tubocurare treatment continued beyond the 10th day of development, changes were noted. The secondary population of fibers failed to form, muscles that contained the most primary F/S fibers atrophied, and the staining with monoclonal antibody to slow myosin heavy chain began to disappear. Thus, the formation of secondary fibers and the maintenance of slow isoform expression in fetal development appeared dependent upon functional innervation.

Myosin Heavy Chain Isoforms in Individual Muscle Fibers in Cell Culture

The fact that functional innervation was not required for the formation of embryonic muscle fibers that express either exclusively fast or fast and slow myosin heavy chain isoforms, suggested that either intrinsic factors within muscle cells themselves or environmental factors other than innervation control the expression of these isoforms. To investigate the possibility that the presence of F and F/S primary fibers in a muscle are the result of intrinsic differences within fibers, myoblasts were isolated from the limb buds of 4 day embryos before the first fibers were formed. These myoblasts were plated on collagen-coated tissue culture dishes in standard culture medium (2.5% embryo extract, 10% horse serum and 87.5% MEM) at a density of 10^4 cells/cm^2 and incubated for 5 days. These mass cultures were then reacted with F59 or S58 antibodies. Two types of myotubes (fibers) formed in all cultures (Figure 3). About 10% of the fibers stained with S58 and more than 90% stained with F59. It is not known if those myotubes containing slow class of myosin heavy chain contained fast as well. Analysis

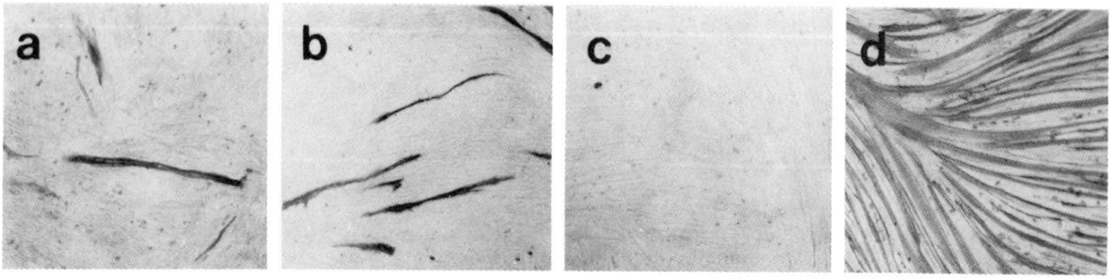

FIGURE 3. Myosin heavy chains in myotubes formed in cell culture from myoblasts isolated from 4 day (a and b) or 12 day (c and d) embryonic chick thigh muscles. On the sixth day of incubation parallel cultures were stained with McAb S58 (a and c), or McAb F59 (b and d) and antibody binding was visualized with an HRP-conjugated secondary antibody. These are phase contrast micrographs which illustrate that most but not all myotubes formed in ED 4 cultures contain a fast clas of MHC, whereas, all do in ED 12 cultures and that there are fibers that contain a slow class of MHC in ED 4 cultures but none in ED 12 cultures.

of extracts of such cultures by immunoblotting with antibodies F59 and S58 showed that the two molecular size classes of myosin heavy chain (a fast and slow heavy chain) were present in all cultures (data not shown). Cultures of myoblasts isolated from progressively older limbs revealed that by day 7 of limb development no myoblasts remained that formed myotubes which contained a slow class of myosin heavy chain. Those myotubes which formed in cultures of myoblasts of older limbs (beyond day 7) were much longer than those from limb bud cultures of day 4 through 6 of development and contained only a fast class of myosin heavy chain.

DISCUSSION

These results show that the developing limb contains two classes of myosin heavy chain--a fast and a slow class that differ in molecular weight and antibody reactivity. These two myosin heavy chains are present in at least two different types of primary fibers: one which expresses only the lower molecular weight fast form and a second which expresses both the fast and the higher molecular weight slow forms. The origin of the two primary fiber types is dependent on events unrelated to functional innervation. The fact that at least two fiber types also formed in tissue cultures of early limb myoblasts, one of which expresses the low and the other the high molecular weight classes of myosin heavy chain, suggests that the primary fiber population has its origins in at least two separate muscle cell lineages.

Our observations impose an important framework for conceptualization of muscle differentiation. They emphasize the importance of the muscle cell itself in defining the early development and function of muscle and demonstrate that fiber diversity has its origin not in events that transpire after a muscle forms, but in antecedent events that occur within the myoblast population. In addition, these results indicate that muscles of different function are composed of different proportions of cells of different lineages. Thus, although function is undoubtedly controlled by regulating the expression of different isoforms of myosin, such regulatory processes are imposed on cells of differing developmental origin. These concepts are not meant to imply that the nervous system plays no role in these events, but to emphasize that diversity in function does not emerge by the nervous system imposing diversity on a uniform population of embryonic muscle fibers.

Studies are needed to define the myoblast populations from which primary and secondary myotube populations arise. Undoubtedly the myoblasts that produced the two myotube types in our cultures are in the "early" (CMR I) myoblast population that Bonner and Hauschka (24) and their colleagues have described. We find that early myoblasts must be a heterogeneous population because they produce at least two types of fibers with differing myosin heavy chain content. It is likely that the different types of early myoblasts are exhausted before day 8 of limb development. An emerging myoblast population, perhaps the "late" type (24), which begins to appear at this time is responsible for the exclusively F type myotubes that form in cultures of myoblasts from 7 to 8 days onward. Clonal analysis of the myoblasts that produce myotubes in cell culture that contain a fast or slow class of myosin heavy chain will help to clarify these relationships.

The formation of muscles should be divided into two phases, an embryonic and a fetal phase. The embryonic phase is that period during which the limb is undergoing morphogenesis and the basic anatomy of an individual muscle is forming by the development of at least two types of primary fibers which in turn probably originate from different myoblast populations. Those factors which regulate the particular topographic location of the two primary populations of primary fibers within each muscle must be dependent upon some environmental cues other than those provided by innervation. The fetal phase of muscle development which has been most commonly studied, begins as morphogenesis of the limb is completed and as the secondary generation of fibers begins to appear. This phase of development is responsible for the pronounced growth of muscles through formation of a new generation of fibers and the hypertrophy of descendents of the primary and secondary fiber populations. The fetal phase differs from the embryonic phase because it is initiated and sustained by functional innervation, and emerges from a different population of myoblasts.

Since no muscle in the thigh (and for that matter no other muscle of the trunk, wing, or pectoral girdle, unpublished results) contains a single form of heavy chain in its embryonic phase of development, there is a need to more clearly define what is meant by "embryonic" myosin heavy chains. Most of the analyses on early myosin heavy chain isoforms were done on the fetal phase of muscle development (25-27). It is likely that the isoforms that are present in the fibers in the embryonic phase are not those of the fetal phase, because the myoblasts that give rise to the fibers forming at this time are different. It is clear that a high molecular weight heavy chain (a slow class) is as "embryonic" as the low molecular weight fast isoform. Thus, if the definition of embryonic, neonatal and adult myosin heavy chain isoforms is based on chronology, then at least two embryonic myosin heavy chain isoforms must be present when muscles are first formed.

Recently, the work of several laboratories has served to clarify the embryonic development of skeletal msucle. Mc Lennan and Phillips and Bennett (17, 28), using ATPase staining have shown that at very early stages of muscle development there are fibers that contain ATPase with differing pH stability; these fibers have different locations with particular muscles of the limbs, and they form independently of innervation. It is very likely that the fibers they call type I_{emb} are fibers that contain two heavy chains (F/S primary fibers) and those they call type II_{emb} contain only the fast class of isoforms (F primary fibers). They also show that later muscle development and growth is dependent upon innervation. These observations and those reported here may reconcile the observation of others on avian and mammalian muscle development (15, 16, 29) by suggesting that the same events would be seen in mammalian muscle development if studies were to be conducted at the embryonic phase rather than the fetal phase of muscle development.

Our studies emphasize the continuity of fast and slow myosin heavy chain expression throughout development. We have attempted to redefine myosin heavy chain isoform expression into two classes, a fast and a slow class. Although we recognize that there is more than a single isoform of fast and slow myosin heavy chain, it is also true that at every stage of development, many if not most muscles express at least two heavy chain classes which have distinctive molecular weights and antibody reactivity. Thus, even though myosin isoform expression within the fast and slow classes may go through developmentally regulated patterns, we think that the particular class of heavy chain expressed in a muscle may be restricted or specified by the developmental origins of the fibers in that muscle.

ACKNOWLEDGMENTS

We thank Sandra Conlon and Barbara Hill for excellent assistance in this work and Gloria Garcia for her preparation of this manuscript.

REFERENCES

1. Hoh JYS, McGrath PA, White RI (1976). Electrophorectic analysis of multiple forms of myosin in fast-twitch and slow-twitch muscles of the chick. Biochem J 157:87.
2. Rushbrook JI, Stracher A (1979). Comparison of adult, embryonic and dystrophic myosin heavy chains from chicken muscle by sodium dodecyl sulfate/polyacrylamide gel electrophoresis and peptide mapping. Proc Natl Acad Sci USA 76:4331.
3. Whalen RG, Sell SM, Butler-Browne GS, Schwartz K, Bouveret P, Pinset-Harstrom I (1981). Three myosin heavy-chain isozymes appear sequentially in rat muscle development. Nature (London) 292:805.
4. Bandman E, Matsuda R, Strohman RC (1982). Developmental appearance of myosin heavy and light chain isoforms in vivo and in vitro in chicken skeletal muscle. Dev Biol 93:508.
5. Nadal-Ginard B, Medford RM, Nguyen HT, Periasamy M, et al (1982). Structure and regulation of a mammalian sarcomeric myosin heavy-chain gene. In Pearson ML, Epstein, HJ (eds): "Muscle Development: Molecular and Cellular Control", Cold Spring Harbor Laboratory, p 143.
6. Benfield PA, Lowey S, Le Blanc DD, Waller GS (1983). Myosin isozymes in avian sekeletal muscles II. Fractionation of myosin isozymes from adult and embryonic chicken pectoralis muscle by immuno-affinity chromatography. J Musc Res Cell Motil 4:717.

7. Bader D, Masaki T, Fischman DA (1982). Immunochemical analysis of myosin heavy chain during avian myogenesis _in vivo_ and _in vitro_. J Cell Biol 95:763.
8. Stockdale FE (1982). Introduction: Myoblast commitment and the embryogenesis of skeletal muscle. In Pearson, ML, Epstein HJ (eds): "Muscle Development: Molecular and Cellular Control", Cold Spring Harbor Laboratory, p. 339.
9. Gauthier GF, Lowey S (1977). Polymorphism of myosin among skeletal muscle fiber types. J Cell Biol 74:760.
10. Gauthier GF, Lowey S, Hobbs AW (1978). Fast and slow myosin in developing muscle fiber types. Nature (London) 274:25.
11. Gauthier GF, Lowey S (1979). Distribution of myosin isoenzymes among skeletal muscle fiber types. J Cell Biol 81:10.
12. Hoh, JFY, Yeoh GPS (1979). Rabbit skeletal myosin isoenzymes from fetal fast-twitch and slow-twitch muscles. Nature (London) 280:321.
13. Whalen R, Butler-Brown G, Sell S, Gros F (1979). Transitions in contractile protein isozymes during muscle cell differentiation. Biochimie 61:5197.
14. Stockdale FE, Raman N, Baden H (1981). Myosin light chains and the developmental origin of fast muscle. Proc Natl Acad Sci USA 78:931.
15. Rubinstein NA, Kelly AM (1981). Development of muscle fiber specialization in the rat hindlimb. J Cell Biol 90:128.
16. Gauthier GF, Lowey S, Benfield PA, Hobbs AW (1982). Distribution and properties of myosin isozymes in developing avian and mammalian skeletal muscle fibers. J Cell Biol 92:471.
17. McLennan IS (1983a). Differentiation of muscle fiber types in the chicken hindlimb. Dev Biol 97:222.
18. Winkelmann DA, Lowey S, Press JL (1983). Monoclonal antibodies localize changes on myosin heavy chain isozymes during avian myogenesis. Cell 34:295.
19. Crow MT, Olson PS, Stockdale FE (1983). Myosin light-chain expression during avian muscle development. J Cell Biol 96:736.
20. O'Neill MC, Stockdale FE (1974). 5-Bromodeoxyuridine inhibition of differentiation. Kinetics of inhibition and reversal in myoblasts. Dev Biol 37:117.
21. Crow MT, Stockdale FE (1985). Specialization of muscle fiber types in the developing chicken hindlimb. Dev Biol (in press).

22. Ashmore CR, Robinson DW, Rattray P, Doerr L (1972). Biphasic development of muscle fibers in the fetal lamb. Exp Neurol 37:241.
23. Kelly AM, Zachs SI (1969). The histogenesis of rat intercostal muscle. J Cell Biol 42:135.
24. Bonner PH, Hauschka SD (1974). Clonal analysis of vertebrate myogenesis I. Early developmental events in the chick limb. Dev Biol 37:317.
25. Rubinstein NA, Pepe FA, Holtzer H (1977). Myosin types during the development of embryonic chicken fast and slow muscles. Proc Natl Acad Sci USA 74:4524.
26. Whalen R, Schwartz K, Bouveret P, Sell S, Gros F (1979). Contractile protein isozymes in muscle development: identification of an embryonic form of myosin heavy chain. Proc Natl Acad Sci USA 76:5197.
27. Silberstein L, Lowey S (1981). Isolation and distribution of myosin isoenzymes in chicken pectoralis muscle. J Mol Biol 148:153.
28. Phillips WD, Bennet MR (1984). Differentiation of fiber types in wing muscles during embryonic development: effect of neural tube removal. Dev Biol 106:457.
29. Gauthier GF, Ono RD, Hobbs AW (1984). Curare-induced transformation of myosin pattern in developing skeletal muscle fibers. Dev Biol 105:144.

Molecular Biology of Muscle Development, pages 225–236

MYOSIN ISOZYME TRANSITIONS IN EMBRYONIC CHICKEN PECTORALIS MUSCLE[1]

Susan Lowey, Saverio Sartore,[2] Geraldine F. Gauthier,* Guillermina S. Waller, and Ann W. Hobbs*

Rosenstiel Basic Medical Sciences Research Center, Brandeis University, Waltham, Massachusetts 02254
and
*Department of Anatomy, University of Massachusetts Medical School, Worcester, Massachusetts 01605

ABSTRACT Using monoclonal antibodies specific for chicken pectoralis myosin, we have shown by immunofluorescence and by radioimmunoassay that several isozymes of myosin appear sequentially during the embryonic stages of avian muscle development. Although pectoralis myosin at 10 days' incubation is immunologically distinct from that present at 18 days' incubation, both have a high level of ATPase activity, which is equivalent to that found for adult pectoralis myosin. The myosin heterogeneity seen in ovo is also evident in cells grown in culture. Whereas all myotubes reacted with an antibody specific for the light chain, differential staining was observed with an antibody specific for the heavy chain. These results suggest that as many as four or more myosin isozymes may be expressed in the chicken pectoralis during the course of its development into a mature muscle.

INTRODUCTION AND BACKGROUND

A major problem in muscle differentiation has been to understand the functional significance of the extensive polymorphism found among contractile proteins. Biochemical

[1]This work was supported by grants from the National Institutes of Health, the National Science Foundation and the Muscular Dystrophy Association.

[2]Present address: Institute of General Pathology, University of Padua, Padua, Italy 35100

and immunological techniques have provided evidence for at least three classes of myosin heavy chain isozymes: an embryonic form followed by neonatal myosin that persists until the synthesis of adult myosin, Table 1 (1,2,3,4).

TABLE 1
SUBUNITS OF MYOSIN DURING CHICKEN PECTORALIS DEVELOPMENT

Embryonic	Posthatch	Adult
Heavy chains	Heavy chains	Heavy chains
MHC_{emb}	MHC_p	MHC_f
Light chains	Light chains	Light chains
$LC1_f$	$LC1_f$	$LC1_f$
$LC2_f$	$LC2_f$	$LC2_f$
~10% {$LC1_s$, $LC2_s$}	$LC3_f$	$LC3_f$

Although an embryonic light chain has been demonstrated in rat muscle (5), the predominant light chain in chicken muscles is of the fast, adult type. About 10% adult slow light chains are transiently observed in the pectoralis muscle of 10-day chick embryos, but these disappear by 18 days' incubation (6,7). The reactivity of 10-day embryonic pectoralis myosin with polyclonal antibodies against the slow heavy chain of ALD (anterior latissimus dorsi) myosin has been ascribed, in part, to the presence of a minor slow component (7,8). This "slow" isozyme is distinct from adult ALD myosin and from the major embryonic component (8). Its presence did suggest, however, that embryonic myosin might not be a homogeneous species.

The advent of monoclonal antibodies made it possible to localize some of the chemical differences among the developmental isozymes, and to provide for a more accurate time course for the appearance of the three classes of myosin isozymes (1,9). An antibody (12C5) specific for the amino-terminal 25 kD region of the heavy chain of chicken myosin was shown to bind to the myosin head (S1) at a distance of about 140 Å from the head-rod junction, Figure 1a. This anti-S1 antibody reacted strongly with adult and 18-day embryonic myosin, but not with 13-day posthatch myosin (9). This property confirmed the existence of a posthatch

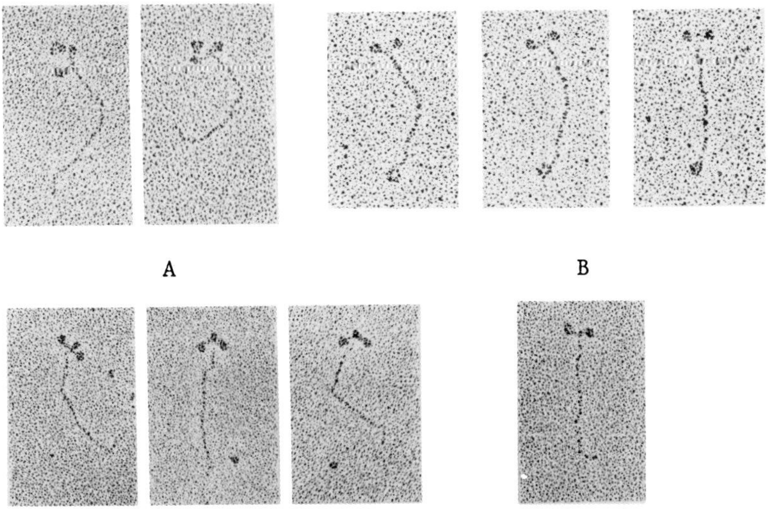

A B

C D

Figure 1. Gallery of rotary-shadowed myosin molecules labeled with anti-S1 (A), anti-LMM (B), anti-LC2 (C) and control myosin (D). (Reprinted from ref. 9)

myosin, and provided a means to follow the disappearance of embryonic myosin and the emergence of adult myosin (9), Figure 2. Another antibody (5C3), specific for the carboxyl terminus of the myosin rod, reacted only with adult myosin and not with any of the developmental isozymes, Figure 1b. This anti-LMM antibody could therefore also be used to detect the appearance of adult myosin, Figure 2. An antibody (7C10) specific for $LC2_f$ reacted with all isozymes throughout myogenesis (9), Figure 1c, consistent with electrophoretic analyses which show no change in size or charge of this light chain with age. In the discussion that follows, we will show how these and other antibodies have been used to analyze the composition of embryonic chicken myosin at several stages of development, thereby leading to the conclusion that embryonic myosin is far more heterogeneous than previously suspected.

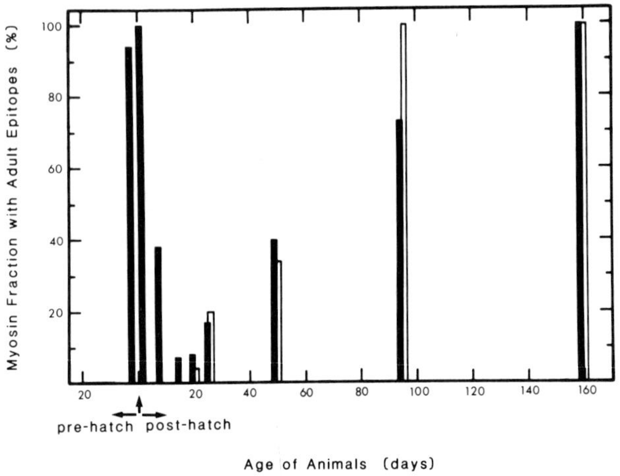

Figure 2. Histogram relating the myosin fraction having the LMM determinant (open bars) and the S1 determinant (shaded bars) to the age of the animals from which the myosin samples were isolated. (Reprinted from ref. 9).

RESULTS

When monoclonal antibodies specific for chicken pectoralis myosin (anti-S1, anti-LMM and anti-LC2) were reacted with cryostat sections of pectoralis muscle, the staining pattern observed at 18-days' incubation was consistent with the reactivity found for isolated myosin (9), Figure 3. All the fibers reacted with anti-S1 and anti-LC2, but no staining was observed with anti-LMM (data not shown). The polyclonal anti-ALD antibody also failed to react with 18-day embryonic muscle, as previously reported (7). At day 14 posthatch, all the fibers reacted strongly with anti-LC2, but little or no staining occurred with anti-S1, anti-ALD and anti-LMM (data not shown). These results are in good agreement with the data summarized in the histogram in Figure 2.

The earlier studies had assumed that the major component of 10-day embryonic myosin would have the same immunological cross-reactivity as 18-day embryonic myosin (3,8, 9). This turned out not to be the case. Frozen sections of pectoralis muscle at day 10 in ovo stained weakly with anti-S1 in contrast to the strong reactivity observed with anti-LC2 and anti-ALD, Figure 4. In order to rule out the possibility that the weak reactivity was due to a masking

DAY 18 IN OVO

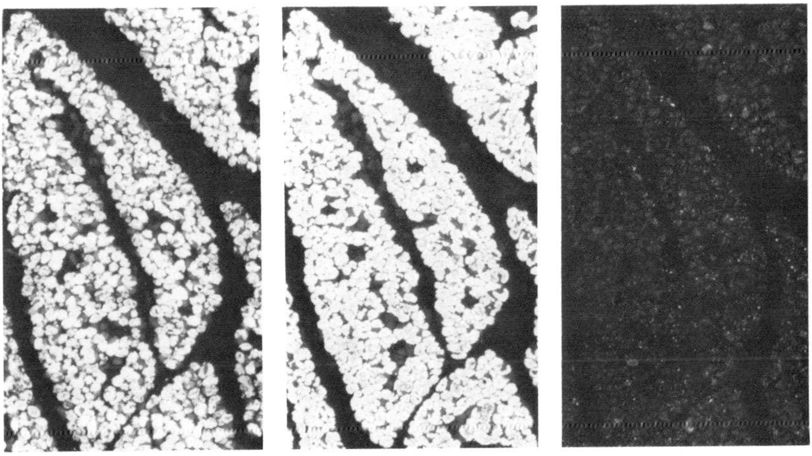

anti-S1 anti-LC2 anti-ALD

DAY 14 POSTHATCH

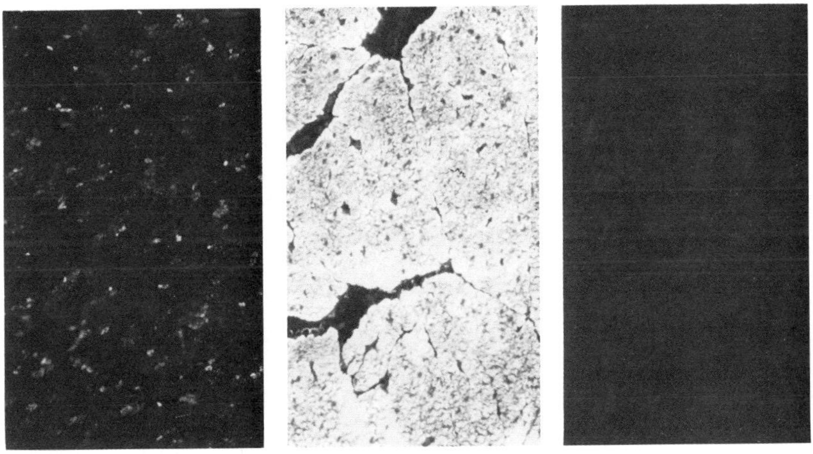

anti-S1 anti-LC2 anti-ALD

Figure 3. Frozen sections of chicken pectoralis muscle fibers reacted with the designated antibodies. Day 18 *in ovo* (top) and day 14 posthatch (bottom). Indirect immunofluorescence.

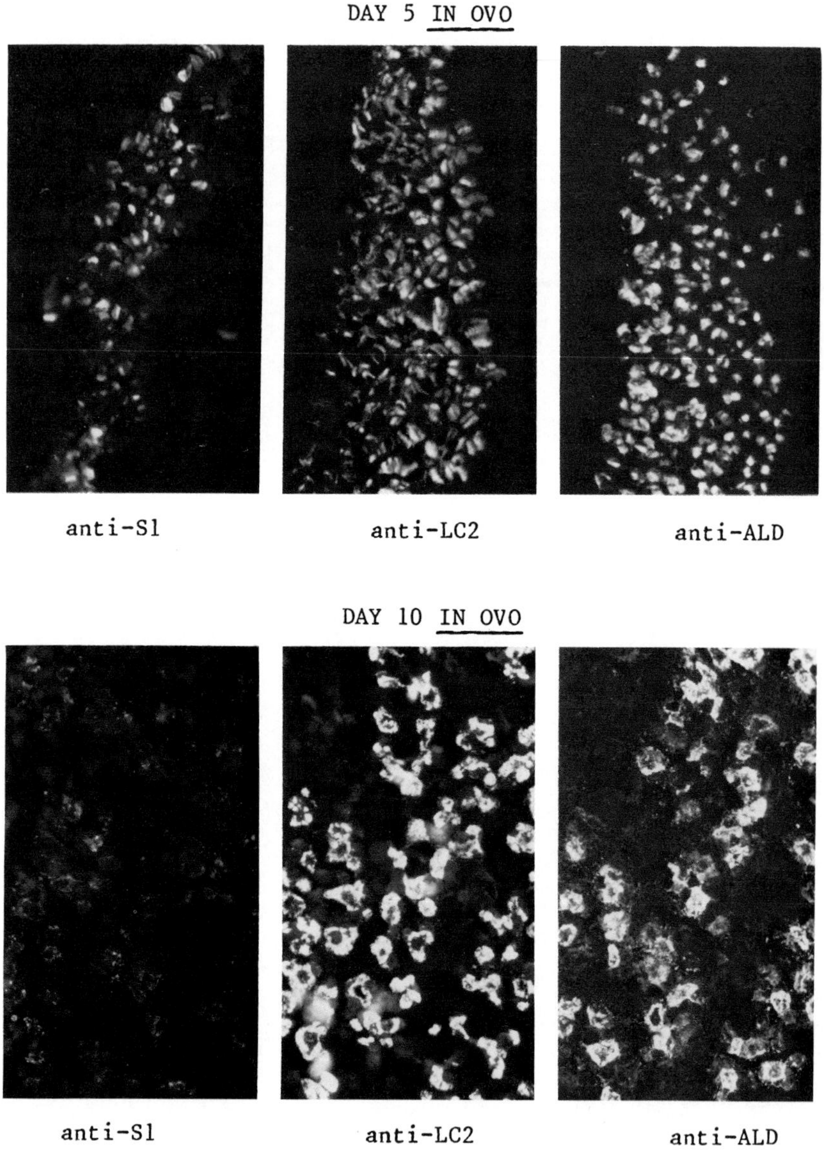

Figure 4. Frozen sections of chick embryos reacted with the designated antibodies. Day 5 *in ovo* (top) and day 10 *in ovo* (bottom). Indirect immunofluorescence.

of antigenic determinants in the tissue, we isolated the myosin from 10-day embryos and compared its reactivity to myosins from later developmental stages, Figure 5.

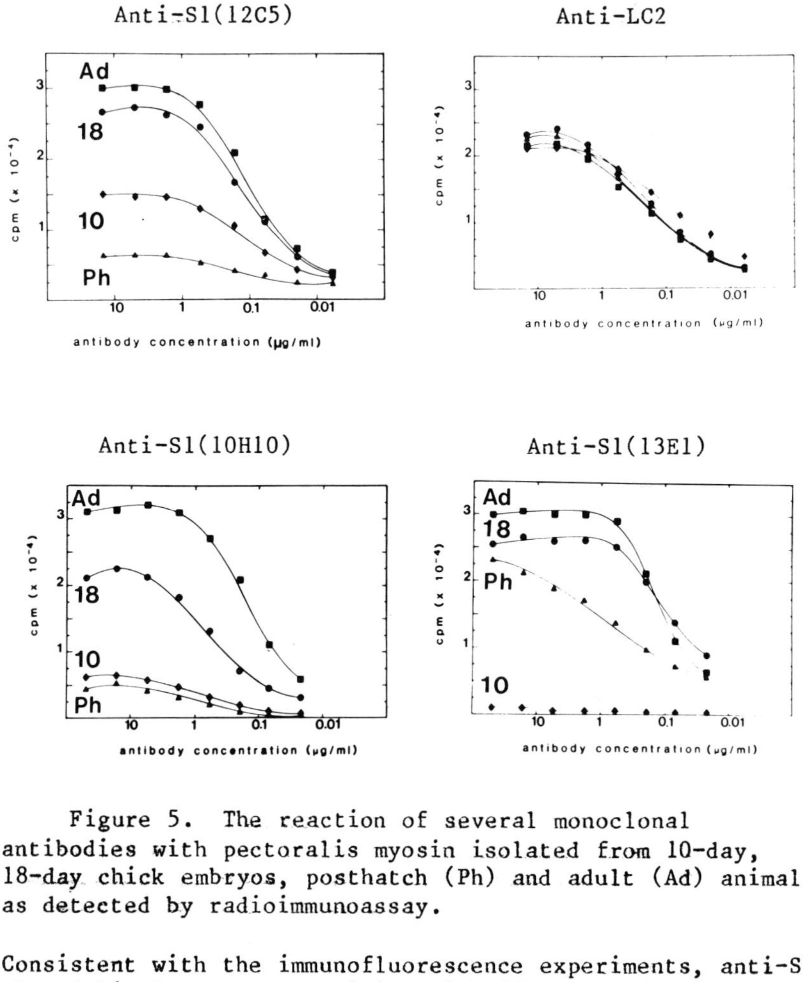

Figure 5. The reaction of several monoclonal antibodies with pectoralis myosin isolated from 10-day, 18-day chick embryos, posthatch (Ph) and adult (Ad) animals as detected by radioimmunoassay.

Consistent with the immunofluorescence experiments, anti-S1 showed little or no reactivity with 10-day embryonic and posthatch myosin, but strong reactivity with 18-day and adult myosin. Anti-LC2 reacted uniformly with the myosin from all development stages.

It has been suggested in the literature that the earliest myosin expressed in ovo might be a cardiac-type myosin, and that the skeletal-type embryonic myosins followed at a later stage (10). To investigate this hypothesis further, myotomes from day 5 in ovo were reacted with the fluorescein-labeled antibodies, Figure 4. Unexpectedly, the cells reacted strongly with anti-S1 as well as with anti-LC2 and anti-ALD. When these antibodies were reacted with cardiac tissue from the same embryo, only anti-ALD cross-reacted. These results demonstrate that skeletal-type embryonic myosins are already present by day 5 in ovo.

The data described above are summarized in Table 2.

TABLE 2
MYOSIN ISOZYME EXPRESSION IN AVIAN MUSCLE[a]

Time	Anti-LC2 (7C10)	Anti-LMM (5C3)	Anti-ALD (MHC)	Anti-S1 (12C5	Anti-S1 10H10	Anti-S1 13E1)
5d	+	-	+	+	ND	ND
10d	+	-	+	-	-	-
18d	+	-	-	+	+	+
posthatch	+	-	-	-	-	-
adult	+	+	-	+	+	+

[a]All antibodies are monoclonal, with the clone designation given in parentheses, except for anti-ALD which is a polyclonal antibody specific for the myosin heavy chain. Reactivity is indicated by positive (good to strong) or negative (none or weak). ND = not determined.

It can be seen that as many as five fast-type myosins may be expressed in chicken muscle. This heterogeneity is, moreover, not confined to just in vivo expression of myosin. Cells grown in culture also show differential staining with monoclonal antibodies. Whereas all myotubes reacted with anti-LC2 and anti-ALD antibodies, only certain myotubes were stained with antibody against S1, Figure 6. Thus the presence of populations of myoblasts capable of synthesizing different myosins in vivo is reflected in the expression of different isoforms in culture.

The existence of multiple forms of myosin raises the question of their functional significance. It has generally been tacitly assumed that embryonic muscles are slow and therefore the enzymatic activity of their myosins should be correspondingly low. In an earlier comparison of the ATPase activity of 10-day embryonic and adult myosin, we showed that this was not the case, and that both myosins

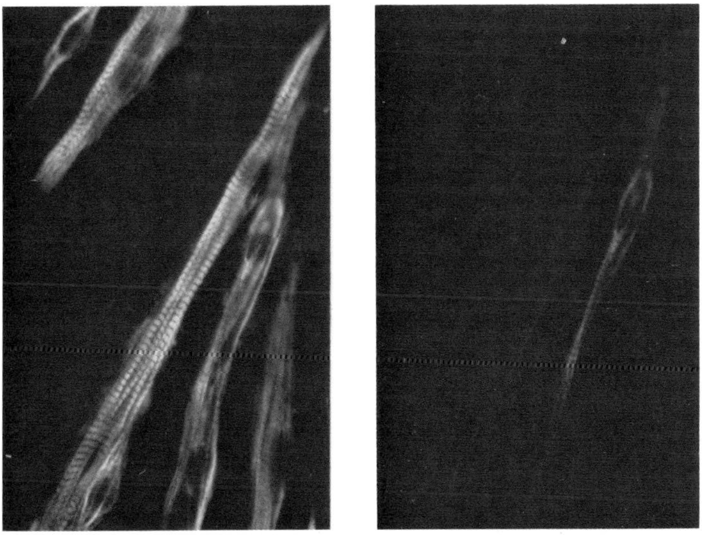

Figure 6. Muscle culture derived from 9-day embryonic pectoralis muscle after 8 days _in vitro_, reacted simultaneously with anti-ALD (left) and anti-S1 (12C5) (right). Indirect immunofluorescence.

had equally high activities (3). Here we have extended these studies to include 18-day embryonic and posthatch myosin, as well as PLD and ALD myosin. Among all these isoforms, ALD myosin is the only example of a low actin-activated ATPase; the other myosins all showed uniformly high ATPase activities, Table 3.

DISCUSSION

The experiments described here lead to the conclusion that a minimum of two fast-type isozymes are expressed sequentially in embryonic chicken pectoralis muscle. Myosin from 10-day chick embryos reacted weakly or not at all with three different monoclonal antibodies specific for S1, whereas myosin from 18-day embryos reacted almost as strongly as adult myosin. Posthatch myosin also reacted weakly with these same antibodies, but peptide mapping has

TABLE 3
KINETIC CONSTANTS OF CHICKEN MYOSIN ISOZYMES[a]

Myosin		K^+(EDTA)ATPase (s^{-1})	Actin-activated Mg^{2+} ATPase V_{max}(s^{-1})	K_{app}(μM)
Pectoralis	adult	11	9	<23
PLD	adult	12	9	< 4
ALD	adult	7	1	<25
Pectoralis (10-d embryo)		10	11	--
Pectoralis (18-d embryo)		12	10	< 3
Pectoralis (12-d posthatch)		13	10	<14

[a]The data given here represent a range of values from several experiments; a detailed analysis will be presented elsewhere (Lowey & Waller, in preparation); the activity is expressed in (mole P_i) (mole protein active sites)$^{-1}$ (s)$^{-1}$. Conditions for the K^+ATPase were 0.6 M KCl, 2 mM EDTA, 50 mM Tris (pH 8.2), 2.5 mM ATP, 25°C. Assays in the presence of actin and minifilaments were in 5 mM KCl, 5 mM citric acid, 28 mM Tris (pH 8.2, 25°C), 3.75 mM $MgCl_2$, 2.5 mM ATP.

shown that embryonic myosin and posthatch myosin are distinct species (2,3). Myotomes at 5 days in ovo appeared to react more strongly with anti-S1 (12C5) than did pectoralis fibers at 10 days in ovo, suggesting that yet another myosin isozyme may be present at this stage of development. However, no biochemical data are available for the earliest stages, and more immunocytochemical experiments are needed before a convincing case can be made for a unique myosin at day 5.

The possibility has been raised that the initial developmental program for myosin expression in skeletal muscle may involve a cardiac-like myosin (10). When myotomes at day 2.5 were reacted with monoclonal antibodies specific for ventricular cardiac myosin, strong fluorescence was observed. In contrast, no reaction occurred with polyclonal antibodies against fast or slow skeletal myosin (10). Although the presence of a cardiac-like myosin at the earliest stages of development is not addressed by our experiments, we can state that determinants for skeletal myosin are present at least by day 5 in ovo. Moreover, cardiac myosin cannot be present in significant amounts by day 10 in ovo since the ATPase activity measured for myosin isolated from that stage is so high, Table 3. Chicken cardiac myosin has been reported to have an actin-activated MgATPase activity similar to that of ALD myosin (11), which is clearly far lower than the values found for the embryonic myosins. It is possible, however, that the minor "slow component" in 10-day embryonic myosin, detected by its cross-reactivity with anti-ALD, may be related to the putative cardiac-like primordial myosin.

If we assume that as many as four distinct myosins could exist in the embryonic period of development, then the number of isozymes including posthatch and adult myosin rises to six or more. This may seem at first glance as an unrealistically high degree of polymorphism, but recent genetic experiments indicate that the size of the chicken myosin heavy chain gene family may be far larger than previously thought (Robbins et al., this volume). Although the functional significance of this extensive polymorphism remains unknown, it seems plausible that a variety of myosins could be required in the development of a muscle cell from a myotube to a fully differentiated fiber.

ACKNOWLEDGEMENTS

We thank Louise Seidel for expert secretarial assistance in preparing this camera-ready manuscript.

REFERENCES

1. Bader D, Masaki T, Fischman DA (1982). Immunochemical analysis of myosin heavy chain during avian myogenesis in vivo and in vitro. J Cell Biol 95:763.
2. Bandman E, Matsuda R, Strohman RC (1982). Developmental appearance of myosin heavy and light chain isoforms in vivo and in vitro in chicken skeletal muscle. Dev Biol 93:508.
3. Lowey S, Benfield PA, LeBlanc DD, Waller GS (1983). Myosin isozymes in avian skeletal muscles. I. Sequential expression of myosin isozymes in developing chicken pectoralis muscles. J Musc Res Cell Motility 4:695.
4. Whalen RG, Sell SM, Butler-Browne GS, Schwartz K, Bouveret P, Pinset-Harstrom I (1981). Three myosin heavy chain isozymes appear sequentially in rat muscle development. Nature (Lond.) 292:805.

5. Whalen RG, Butler-Browne GS, Gros F (1978). Identification of a novel form of myosin light chain present in embryonic muscle tissue and cultured muscle cells. J Mol Biol 126:415.
6. Stockdale FE, Raman N, Baden H (1981). Myosin light chains and the developmental origin of fast muscle. Proc Natl Acad Sci USA 78:931.
7. Gauthier GF, Lowey S, Benfield PA, Hobbs AW (1982). Distribution and properties of myosin isozymes in developing avian and mammalian skeletal muscle fibers. J Cell Biol 92:471.
8. Benfield PA, Lowey S, LeBlanc DD, Waller GS (1983). Myosin isozymes in avian skeletal muscles. II. Fractionation of myosin isozymes from adult and embryonic chicken pectoralis muscle by immuno-affinity chromatography. J Musc Res Cell Motility 4:717.
9. Winkelmann DA, Lowey S, Press JL (1983). Monoclonal antibodies localize changes on myosin heavy chain isozymes during avian myogenesis. Cell 34:295.
10. Sweeney LJ, Clark WA Jr, Umeda PK, Zak R, Manasek FJ (1984). Immunofluorescence analysis of the primordial myosin detectable in embryonic striated muscle. Proc Natl Acad Sci USA 81:797.
11. Marston SB, Taylor EW (1980). Comparison of the myosin and actomyosin ATPase mechanisms of the four types of vertebrate muscles. J Mol Biol 139:573.

Molecular Biology of Muscle Development, pages 237–251

CONTROL OF MYOSIN ISOFORM EXPRESSION DURING SKELETAL MUSCLE DEVELOPMENT IN RODENTS [1]

Robert G. Whalen, Gillian S. Butler-Browne, Christian Pinset, Madeleine Toutant, Simon C. Watkins, Janet Ajioka, Christine Laurent, Danny McCormick, and Graham P. Riley

Department of Molecular Biology
Pasteur Institute
75724 Paris, France

ABSTRACT Muscle development involves an initial process by which a precursor myoblast cell is induced to fuse and to begin expressing a number of muscle-specific characteristics. The process of induction of differentiation has been studied in cell culture using the rodent myoblast cell lines L6 and C2. Differentiation can be induced in populations of quiescent myoblasts using serum-free medium, and we can show that the process of induction occurs uniquely in the G1 phase of the cell cycle. This induction of differentiation is followed by a maturational process, most easily studied in the animal, during which the muscle fiber acquires adult properties including those myosin isozymes appropriate for the adult phenotype of different muscles. Transitions between isozymes are subject to two general types of control: some transitions are nerve-dependent while others are more influenced by endocrinological factors, in particular thyroid hormone levels. We have used the mouse mutant dwarf to demonstrate that thyroid hormone control of the neonatal to adult fast myosin transition does not operate via stimulation of growth hormone production.

[1]This work was supported by grants from the French Ministry of Research, the CNRS, INSERM, the American Heart Association and the Muscular Dystrophy Association of America.

INTRODUCTION

Two general phases of skeletal muscle differentiation can be distinguished. In the first, myoblast cells fuse to form multinucleated myotubes, undergoing changes in gene expression leading to the induction of many muscle-specific proteins. This stage of differentiation can be best studied in cell culture, where the behavior of the myogenic cell population is relatively more synchronous than in the embryo. The second phase of development concerns the maturation of myotubes as they acquire the properties of adult muscle fibers. Clearly, this process must be studied in animal models where those neuronal and hormonal factors which influence maturation are most effective.

In our studies of cell culture models for the induction of muscle differentiation, we have placed emphasis on manipulating culture conditions so that the separate stages of cell growth and myogenic expression might be controlled. We reasoned that we would be better able to evaluate the factors or events required for the induction of muscle-specific gene expression. Most of this work has been carried out with established myogenic cell lines, particularly the rat L6 line (1) and mouse C2 cells (2), because of the greater ease with which experiments can be done. Experimental protocols developed with these lines could then be applied to primary muscle cell cultures. We summarize here the results of our attempts to control the expression of differentiation using this approach.

We study the maturation of muscle fibers by using the several isoforms of myosin heavy and light chains as markers (3). Four principal isoforms of myosin are found in developing and adult rat hindlimb muscles, and investigation of the transitions between these various forms has led to working hypotheses on how the adult muscle fiber types arise (3,4). Some of these transitions seem to depend on the type of innervation that the muscle fiber receives, or more specifically on the activity imposed on the muscle. Other transitions can occur in the absence of any innervation, although some of these transitions are affected by hormonal influences, such as hypothyroidism. In this paper, we present a new animal model for studying endocrinological effects on myosin isozyme transitions, that of hereditary pituitary dwarfism in mice. This model clarifies some ambiguities about hormonal control of myosin transitions and may allow further study of these phenomena.

INDUCTION OF MYOGENIC DIFFERENTIATION

Production of Quiescent, Undifferentiated Rodent Myoblasts

The standard conditions for growing the rat myogenic cell line L6 consist of a nutrient medium, often Dulbecco's modified Eagle's (DME) medium, to which is added fetal calf serum. The serum provides mitogenic factors which allow the cells to proliferate; when cells become confluent, and have presumably exhausted the mitogens, they will fuse and express a differentiated phenotype. Thus, in these conditions the phenomena of growth and differentiation for a given myoblast appear to be mutually exclusive. Studies with mouse myogenic cells growing in the presence of purified growth factor, FGF, have supported this general idea.(5)

However, it is possible to find medium conditions which are non-permissive for myogenic differentiation. We have shown that changing the nutrient medium to Ham's F12, supplemented as usual with fetal calf serum, will allow L6 cells to grow to confluence but will not permit fusion to occur (6). These confluent cells have not irreversibly lost the ability to differentiate, since changing the medium to DME plus serum will induce fusion. In addition to remaining mononucleate, these confluent cells do not synthesize detectable amounts of muscle-specific contractile proteins. Finally, the majority of cells in these growth-arrested cultures are in the G1 phase of the cell cycle (6).

Taken together, these observations mean that simply prolonging the G1 phase of the myoblast cell cycle is not itself sufficient to induce myogenic differentiation, at least in the non-permissive conditions represented by F12 + serum. This is different therefore from the use of permissive conditions in which differentiation of cells in the G1 phase has been demonstrated to be an apparently stochastic event, the probability of which is dependent on the length of time spent in G1 (7).

We have identified a family of nutrient media which all produce quiescent, mononucleated, undifferentiated L6 myoblasts. These include Coon's modification of F12, as well as several of the MCDB series of media (8), including 202, 301, 402 and 411. In some cases, minor modifications had to be made to achieve efficient myoblast growth; for example, the cysteine concentration of MCDB 202 must be increased 10-fold. However, in all cases the results were analogous to those obtained with F12. In MCDB 202, the number of cells traversing the cell cycle in confluent cultures appears to be even lower than in F12, and no thymidine-labeled nuclei have been detected out of several thousand cells examined.

Although all of the above results were obtained with rat L6 cells, it is possible to demonstrate the same phenomena using cells derived from the C2 line, prepared from mouse satellite cells (2). The parent C2 cell line has a high spontaneous tendency to fuse and differentiate, rendering the problems of asynchrony of the myogenic process in culture even greater than with L6 under standard conditions. The use of F12 or MCDB 202 nutrient media will slow down but not entirely block differentiation in these cells. However, it has been possible to isolate a subclone of C2 cells which can be obtained as quiescent, undifferentiated myoblasts with the same features as the L6 system. When fusion is induced in the C2 subclone, the qualitative parameters of myogenic expression such as accumulation of acetylcholine receptor and contractile protein synthesis are all similar to the parent line. Thus the phenomena described here are not unique to rat L6 cells.

In comparing the composition of the permissive and non-permissive media, it is apparent the latter contain a broader spectrum of nutrient components. For example, there are more different non-essential amino acids, lipid compounds and trace mineral elements in F12 and MCDB media than in DME. Preliminary results suggest that the trace mineral elements, when added to DME, will inhibit fusion; the presence of these metal ions may therefore explain, at least partly, the non-permissive nature of certain media.

Induction of Differentiation in the Absence of DNA Synthesis

Whatever the explanation for the phenomenon, the availability of the quiescent myoblast system allows certain types of experiments to be carried out which would otherwise be difficult in cultures where myoblasts are proliferating and beginning to fuse in an asynchronous fashion with respect to the cell cycle. For example, since the quiescent myoblasts are arrested in the G1 phase of the cell cycle, we were able to test directly whether another round of DNA synthesis was required for the induction of myogenic differentiation.

For this experiment, it was necessary to use conditions to induce differentiation which would not adventitiously stimulate cell division in the quiescent population. Therefore, we used a serum-free medium composed of DME plus insulin at 10 μg/ml as the only protein component (9). Although insulin is a mitogen in other cell culture systems, it stimulates fusion and differentiation when used alone in the quiescent L6 system (10).

Table 1 shows results of two experiments in which fusion was induced in a population of quiescent L6 myoblasts by DME + insulin.

TABLE 1
INDUCTION OF FUSION BY DME + INSULIN IN THE PRESENCE OF ^{3}H-THYMIDINE

	Day 0	Day 4		
	Total Nuclei per cm^2	Total nuclei per cm^2	% Fused	% Myotube Unlabeled
Without AraC	9.5×10^4	9.7×10^4	84.5%	85%
With AraC	9.5×10^4	7.6×10^4	89.5%	100%

Labeling with 3H-thymidine during the induction shows that more than two-thirds of the cells in the initial quiescent population are capable of fusing without entering another S phase. As expected, the presence of an inhibitor of DNA synthesis, cytosine arabinoside (AraC), during the induction does not block fusion. Using quiescent mouse C2 cells (see above), equivalent results have also been obtained. We conclude therefore that for rat L6 or mouse C2 cells the expression of the myogenic program does not require any event occurring uniquely in S, G2 or M phase.

It could be argued that the growth-arrested L6 cells in non-permissive media are in some way primed for differentiation to occur. Although difficult to exclude directly, several arguments suggest that this is not the case. The cells can be reinduced to divide by changing to fresh medium, showing that they are not irreversibly committed to differentiate. No detectable synthesis of several contractile protein markers is found in the quiescent cells, even when labeling is carried out for

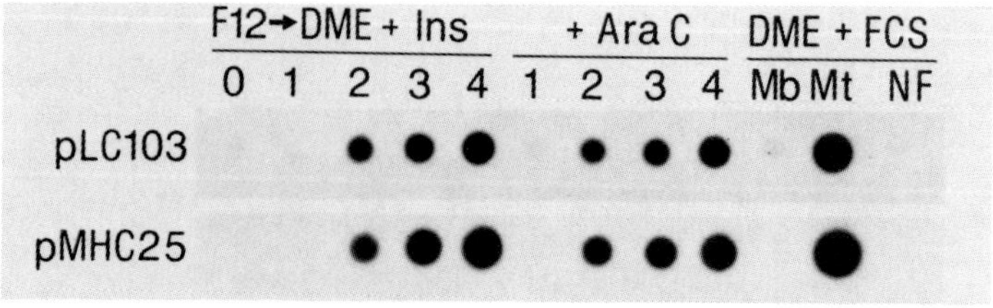

Figure 1. Accumulation of myosin mRNAs during induction of fusion. RNA was prepared from quiescent L6 myoblasts in F12 + serum, and then each day for 4 days after a switch to DME + insulin (Ins) in the absence or presence of AraC. For comparison, RNA was also made from L6 cells in DME + fetal calf serum (FCS), either at the proliferating myoblast stage (Mb) or at the myotube stage (Mt). A non-fusing variant (NF) derived from L6 was also used. The RNA samples were spotted onto nitrocellulose paper and hybridized with nick-translated cloned cDNA probes to myosin light (pLC103) or heavy chain (pMHC25) subunits (11,12).

short (5-30 minute) periods (6). The accumulation of messenger RNAs for myosin heavy and light chains cannot be detected, upon hybridization of specific cDNA probes (11,12) to RNA from quiescent L6 cells. As expected however, a clear accumulation of these mRNA species is observed upon induction of fusion by DME + insulin, with or without AraC, as figure 1 shows.

The time course for the induction of the messenger RNAs for the heavy and light chains is not modified by the presence of AraC, suggesting that not only fusion and muscle protein synthesis are independent of DNA synthesis but also the events which lead to the changes in gene expression at the level of mRNA accumulation and possibly transcription. In particular, if induction of differentiation in L6 cells requires changes in the state of DNA methylation or in the conformation of chromatin, then these changes must be able to occur in the G1 phase of the cell cycle and do not require DNA synthesis in order for the reprogramming to occur (9).

The quiescent L6 and C2 myoblast systems in which fusion can be induced with a serum-free medium allow the processes of growth and differentiation to be separated and may be well suited to detailed study of the mechanisms leading to gene activation. Other types of investigations could also profit from the availability of serum-free conditions for induction of differentiation, such as evaluation of the effects of growth factors, hormones or other pharmacological agents on these processes.

MYOSIN ISOFORMS AS MARKERS FOR MUSCLE MATURATION

The combined results of protein chemical, immunochemical and molecular cloning approaches have demonstrated the existence of at least 10 myosin heavy chains in vertebrate animals, of which 7 are found in skeletal and cardiac muscles (see 3). These striated muscle forms are the predominant myosin types either in different adult muscles, or in developing muscles. For example, in rat hindlimb muscles the major myosin type found at late gestation is the embryonic form. This isoform is replaced by neonatal myosin which is the predominant type during the first 10 days after birth (13,14). Eventually, adult fast myosin replaces neonatal myosin in those muscles which are fast-contracting in adult rats. Certain slow-contracting muscles have another adult myosin type, and this slow isoform can also be found as a minor component in other muscles.

The descriptive names that are given to the various isoforms are probably necessary, but they should not be interpreted in a strict sense. For example, slow myosin is an adult isoform which is apparently also found in fetal muscle (15). Likewise, the form(s) which constitute(s) the earliest muscles arising directly from somites may not necessarily be the embryonic isoform.

Another aspect of this problem is the possible presence of developmental forms of contractile proteins in certain adult muscles. In the case of actin and the myosin light chains, examples of this possibility have already been found. The LC1 light chain which is the embryonic form in skeletal muscle (16) has turned out to the be identical to the adult cardiac atrial light chain (17, Robert et al., this volume). Cardiac actin is expressed during development of skeletal muscle, and therefore can be considered a fetal form in this tissue (18).

Indications that this type of situation may also be true for the myosin heavy chain gene family have now been found. Weiczorek et al. (19) have studied the extraocular

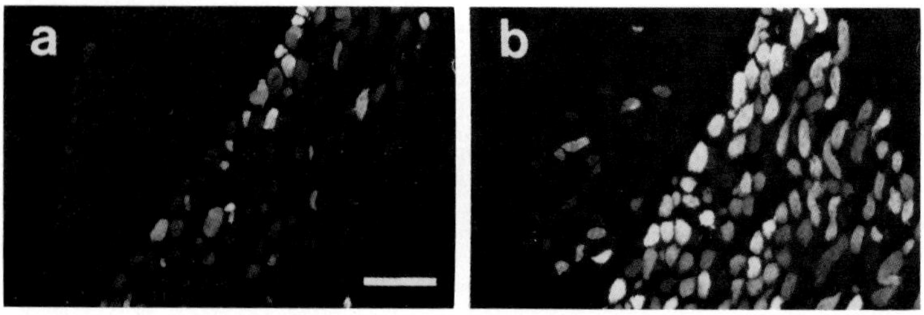

Figure 2. Immunocytochemistry of adult rat extra-ocular muscles. Serial sections of the muscle were reacted either with antibody to embryonic (a) or neonatal (b) myosins. The majority of fibers not reacting in either section are stained by a fast antibody, while a few scattered fibers react with anti-slow (results not shown). The scale bar in (a) represents 100 μm.

muscles of the rat and found that not only is a new heavy chain gene expressed, but all other skeletal muscle forms are also expressed in the adult muscle at least at the messenger RNA level (Mahdavi et al, this volume). We have confirmed the presence of at least four of these isoforms at the protein level: not only were the adult fast and slow types detected, but also the embryonic and neonatal forms were present (19). Using immunocytochemistry, we can also show that the various forms are generally segregated into different fiber populations; this is shown in figure 2 for embryonic and neonatal myosins.

Thus, for several types of contractile proteins the existence of multiple isoform genes is utilized by the organism to provide developmentally-regulated forms for some muscles while using the same forms as adult types in other muscles.

Developmental Transitions Between Different Myosin Isoforms: Formation of Adult Fiber Types

A simple model to explain the developmental transitions of myosin isoforms would be a sequential one, with each isozyme appearing and disappearing in a homogeneous fashion until the appropriate adult form were present. However, this simple model is complicated by a heterogeneity in the population of developing fibers in several types of muscles, as revealed by immunocytochemical techniques (4,15,20,21). At its simplest, this heterogeneity is due to the presence of fibers containing slow myosin intermixed with fibers containing embryonic and/or neonatal myosin (e.g. 21). Describing the pathways by which these fibers attain the adult phenotype requires information on the myosin types in individual fibers as a function of time of development. For example, the developing rat soleus muscle is a predominantly slow muscle in adults which manifests a complex immunocytochemical staining pattern with myosin antibodies throughout the first month after birth (4). Our interpretation of the myosin transitions occurring in this muscle can be illustrated schematically:

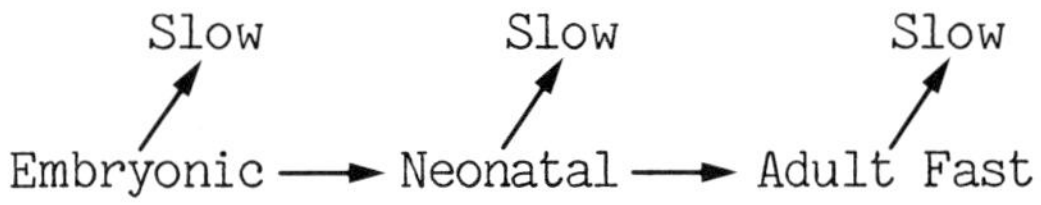

We have proposed (3,4) that a basic transition pathway is the embryonic to neonatal to adult fast one, and that the induction of slow myosin can result from the arrival of signals to the muscle fiber at any time during its development. Although derived from studies of the soleus muscle, this transition scheme can be used to explain the pathways leading to the formation of adult fiber types in any muscle, simply by fixing the number of fibers that follow one or another pathway. This model accounts for the two principal fiber types seen in fetal and neonatal muscles since one type would have already been induced to accumulate slow myosin. It should be emphasized that this scheme describes events seen after the large burst in fiber production which occurs in late gestation and early postnatal development (22). It is therefore possible that in the earliest stages of muscle development the myosin transitions will not follow this scheme.

The factors which control these transitions, or provide the signals for induction of each isoform type, are known in general outline only (reviewed in 3). The induction of slow myosin is established to be nerve-dependent, at least in the post-natal period (23). The role of the nerve in this case apparently is to provide an activity pattern which forces the muscle to induce the accumulation of slow myosin forms (24). It is not at all clear whether a similar mechanism accounts for the induction of slow myosin in fetal muscles, and a nerve-dependent mechanism should probably not be assumed.

The transitions leading to the appearance of adult fast myosin are, on the other hand, independent of nerve in muscle tissue. Denervation of muscles of newborn rats at a time when neonatal myosin is the predominant species does not block the subsequent appearance of adult fast myosin (25,26). Either the induction of fast myosin is truly an endogenously programmed event, or the factors required can be furnished by a source other than the nerve. The latter explanation may be nearer the truth based on some recent observations with an organotypic nerve-muscle culture system, in which adult fast myosin is accumulated (27). Experiments using inhibitors of nerve-muscle interaction and using denervated cultures have shown that fast myosin appearance in this system is nerve-dependent (Ecob and Whalen, this volume). This apparent contradiction with the in vivo data can be reconciled by suggesting that the stimuli needed for adult fast myosin accumulation can be provided in the animal by sources other than the nerve whereas in the organotypic system, the nerve might be the major source. Clearly however, these phenomena need further detailed study, and the organotypic system might prove a more suitable one for varying the different parameters.

An important factor controlling the appearance of adult fast myosin seems to be the levels of thyroid hormone: chemical induction of hypothyroidism in newborn rats blocks the accumulation of fast myosin whereas injections of thyroid hormone can speed up the transition (26,28). The presence of slow myosin is not inhibited by hypothyroidism however. On the contrary, the slow isoform is induced in some fibers in adult hypothyroid rats (29), and the contingent of slow myosin-containing fibers in the developing rat is still found in neonatal hypothyroidism. However, the remaining fibers of the developing soleus are apparently blocked at the neonatal myosin stage (28).

Taken together, these observations on neonatal and adult hypothyroidism highlight an interesting paradox concerning developing fast fibers: depending on the stage of development, or on the myosin isoform present, these fibers are either blocked (neonatal stage) or induced to accumulate slow myosin (adult stage). Why the neonatal fibers do not simply accumulate slow type myosin is not clear from current data.

The effects of hypothyroidism on developing animals are undoubtedly complex, and the use of models in which hypothryoidism is chemically induced by anti-thyroid drugs does not make the situation any less complex and may even introduce further complications. We have therefore sought new models for the study of the endocrinological control of myosin transitions during development, and in which some of the elements of complexity might be dealt with. The mouse mutant dwarf offers some advantages in this regard.

Dwarf Mice: A New Model for the Study of Endocrinological Control of Myosin Transitions

A mouse mutant called dwarf was described over 50 years ago by Snell (30). These animals are only about one-quarter the size of their control littermates, and they present immature body features. The effect of this mutation is to impair development of several types of anterior pituitary cells, including those responsible for the production of growth hormone (GH) and thyroid-stimulating hormone (31). Thus the external phenotype can be explained by the resulting low levels of both GH and thyroid hormone.

In the course of normal postnatal development, increases in thyroid hormone levels are required to stimulate GH production. Thus in chemically-induced hypothyroidism, animals are deficient not only in thyroid hormone but also GH, and injection of thyroid hormone stimulate GH synthesis (32). In the dwarf mouse model however, pituitary GH-producing cells are absent and therefore cannot respond even to exogenous thyroid hormone (33). Dwarf mice are useful therefore because they are a model of the hypothyroid state in which the stimulatory effects of thyroid hormone on pituitary GH production no longer exist.

We have analyzed the myosin types in dwarf mice using electrophoretic and immunochemical techniques (34). In hindlimb muscle, neonatal myosin persists for an abnormally long period. Using immunopolypeptide mapping, this isoform is detectable in normal mice only until 16 days after birth whereas in dwarf mice it is detected up to 75 days. Nonetheless, adult fast myosin appears in the dwarfs, although the neonatal - fast transition has a slower time course. Immunocytochemistry using antibodies to neonatal and fast myosin demonstrates that the abnormal persistence of neonatal myosin occurs in the majority of fibers and is not just limited to a minor population. Electron microscopic observation of dwarf skeletal muscle reveals signs of general immaturity but no obvious pathological features.

As an internal control for the effects of hypothy-

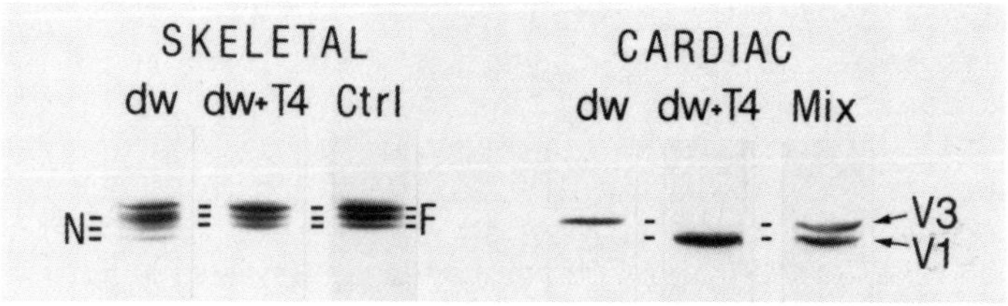

Figure 3. Analysis of the skeletal neonatal (N) and adult fast (F) myosins, and the cardiac V1 and V3 isozymes, in dwarf and control mice. Electrophoresis was performed by under non-denaturing conditions on extracts from 38 day old mice. The samples were from control (Ctrl) or dwarf (dw) mice, or from dwarf mice injected with thryoxine (dw+T4) every 2-3 days for 11 days prior to analysis.

roidism in these animals, we also examined the cardiac myosin isozymes since they are known to be regulated by thyroid hormone levels in rats (35). In dwarf mice the fetal myosin form (V3) does not disappear and the adult form (V1) cannot be detected up to at least 70 days after birth. Thus the effects of hypothyroidism are different for these two muscle tissues.

By injecting dwarf mice with thyroxine, we determined whether thyroid hormone alone would restore a normal myosin phenotype among all the hormones lacking in this mutant. Indeed, after 11-15 days of hormone treatment only the adult myosin isozymes were detected in both the heart and skeletal muscle tissue (figure 3). Thus the effects of the mutation can be explained by the lack of thyroid hormone in dwarf mice. However, this observation does not indicate whether thyroid hormone acts directly on skeletal muscle tissue, and one obvious possibility, as yet untested, is that the underdeveloped nervous system could have a negative effect on skeletal muscle development.

Nonetheless, these results with dwarf mice would seem to indicate that the stimulation of myosin transitions in both the heart and skeletal muscle by thyroid hormone is independent of GH production. Although GH influences body growth possibly via direct action as well as via somatomedins (36), this complex system might not play a major role in promoting myosin transitions or, more generally, maturation of muscle tissue.

Skeletal and cardiac muscle tissue respond differently to the hypothyroidism in dwarf mice, and this suggests that the action of thyroid hormone in influencing these transitions is different. Comparative studies of these two tissues in the dwarf mouse model might give some insight into the mechanisms of these processes.

REFERENCES

1. Yaffé D (1968) Retention of differentiation potentialities during prolonged cultivation of myogenic cells. Proc Natl Acad Sci USA. 61:477-483.
2. Yaffé D, Saxel O (1977) Serial passaging and differentiation of myogenic cells isolated from dystrophic mouse muscle. Nature 270:725-727.
3. Whalen RG (1985) Myosin isoenzymes as molecular markers for muscle physiology. J Exp Biol, in press.
4. Butler-Browne GS, Whalen RG (1984) Myosin isozyme transitions occurring during the postnatal development of the rat soleus muscle. Develop Biol 102:324-334.
5. Linkhart TA, Clegg CH, Lim RW, Merrill GF, Chamberlain JS, Hauschka SD (1982) Control of mouse myoblast commitment to terminal differentiation by mitogens. In Pearson ML, Epstein HF (eds): "Muscle Development: Molecular and Cellular Control," Cold Spring Harbor Laboratory, New York, pp 377-382.
6. Pinset C, Whalen RG (1984) Manipulation of medium conditions and differentiation in the rat myogenic cell line L6. Develop Biol 102:269-277.
7. Devlin BH, Konigsberg IR (1983) Reentry into the cell cycle of differentiated skeletal myocytes. Develop Biol 95:175-192.
8. Ham RG, McKeehan WL (1979) Media and growth requirements. Methods Enzymol 58:44-93.
9. Pinset C, Whalen RG (1985) Induction of myogenic differentiation in serum-free medium does not require DNA synthesis. Develop Biol 108:284-289.
10. Ewton DZ, Florini JR (1981) Effects of the somatomedins and insulin on myoblast differentiation in vitro. Develop Biol 86:31-39.
11. Katcoff D, Nudel U, Zeven-Sonkin D, Carmon Y, Shani M, Lehrach H, Frischauf AM, Yaffé D (1980) Construction of recombinant plasmids containing rat muscle actin and myosin light chain DNA sequences. Proc Natl Acad Sci USA 77: 960-964.

12. Medford RM, Wydro RM, Nguyen HT, Nadal-Ginard B (1980) Cytoplasmic processing of myosin heavy chain messenger RNA: evidence provided by using a recombinant DNA plasmid. Proc Natl Acad Sci USA 77:5749-5853.
13. Whalen RG, Schwartz K, Bouveret P, Sell SM, Gros F (1979) Contractile protein isozymes in muscle development: Identification of an embryonic form of myosin heavy chain. Proc Natl Acad Sci USA 76:5197-5201.
14. Whalen RG, Sell SM, Butler-Browne GS, Schwartz K, Bouveret P, Pinset-Härström I (1981) Three myosin heavy-chain isozymes appear sequentially in rat muscle development. Nature (Lond) 292:805-809.
15. Whalen RG, Johnstone D, Bryers PS, Butler-Browne GS, Ecob MS, Jaros E (1984) A developmentally regulated disappearance of slow myosin in fast-type muscles of the mouse. FEBS Letters 177:51-56.
16. Whalen RG, Butler-Browne GS, Gros F (1978) Identification of a novel form of myosin light chain present in embryonic muscle tissue and cultured muscle cells. J Mol Biol 126:415-431.
17. Whalen RG, Sell SM, Eriksson A, Thornell L-E (1982) Myosin subunit types in skeletal and cardiac tissues and their developmental distribution. Develop Biol 91:478-484.
18. Minty A, Alonso S, Caravatti M, Buckingham, ME (1982) A fetal skeletal muscle actin mRNA in the mouse and its identity with cardiac actin mRNA. Cell 30:185-192.
19. Wieczorek DF, Periasamy M, Butler-Browne GS, Whalen RG, Nadal-Ginard, B (1985) Co-expression of multiple myosin heavy chain genes, in addition to a tissue-specific one, in extraocular muscle. J Cell Biol, in press.
20. Rubinstein NA, Kelly AM (1981) Development of muscle fiber specialization in the rat hindlimb. J Cell Biol 90:128-144.
21. Thornell L-E, Billeter R, Butler-Browne GS, Erikson PO, Ringqvist M, Whalen, RG (1984) Development of fiber types in human fetal muscle: an immunocytochemical study. J Neurol Sci 66:107-115.
22. Ontell M, Kozeka K (1984) The organogenesis of murine striated muscle: a cytoarchitectural study. Am J Anat 171:133-148.
23. Jolesz F, Sreter, FA (1981) Development, innervation and activity-induced changes in skeletal muscle. Annu Rev Physiol 43:531-552.
24. Brown WE, Salmons S, Whalen RG (1983) The sequential replacement of myosin subunit isoforms during muscle

type transformation induced by long term stimulation. J Biol Chem 258:14686-14692.
25. Butler-Browne GS, Bugaisky LB, Cuénoud S, Schwartz K, Whalen RG (1982) Denervation of newborn rat muscle does not block the appearance of adult fast myosin. Nature (Lond) 299:830-833.
26. Gambke B, Lyons GE, Haselgrove J, Kelly AM, Rubinstein NA (1983) Thyroidal and neural control of myosin transitions during development of rat fast and slow muscles. FEBS Letters 156:335-339.
27. Ecob MS, Butler-Browne GS, Whalen RG (1983) The adult fast isozyme of myosin is present in a nerve-muscle tissue culture system. Differentiation 25:84-87.
28. Butler-Browne GS, Herlicoviez D, Whalen RG (1984) Effects of hypothyroidism on myosin isozyme transitions in developing rat muscle. FEBS Letters 166:71-75.
29. Johnson MA, Mastaglia FL, Montgomery A, Pope B, Weeds AG (1980) A neurally mediated effect of thyroid hormone deficiency on slow-twitch skeletal muscle? In Pette D (ed): "Plasticity of Muscle," Berlin: Walter de Gruyter, pp 607-615.
30. Snell GD (1929) Dwarf, a new mendelian recessive character of the house mouse. Proc Natl Acad Sci USA 15:733-734.
31. Bartke A (1979) Genetic models in the study of anterior pituitary hormones. In Shire JGM (ed): "Genetic Variation in Hormone Systems," Vol. I, Cleveland: CRC Press, pp 113-126.
32. Seo H, Wunderlich C, Vassart G, Refetoff S (1981) Growth hormone responses to thyroid hormone in the neonatal rat. Resistance and anamnestic response. J Clin Invest 67:569-574.
33. Slabaugh MB, Lieberman ME, Rutledge JJ, Gorski, J. (1981) Growth hormone and prolactin synthesis in normal and homozygous Snell and Ames dwarf mice. Endocrinol 109:1040-1046.
34. Whalen RG, Toutant M, Butler-Browne GS, Watkins SC (1985) Hereditary pituitary dwarfism in mice affects skeletal and cardiac myosin isozyme transitions differently. J Cell Biol, in press.
35. Chizzonite RA, Zak R (1984) Regulation of myosin isoenzyme composition in fetal and neonatal rat ventricle by endogeneous thyroid hormones. J Biol Chem 259:12628-12632.
36. Mittra I (1984) Somatomedins and proteolytic bioactivation of prolactin and growth hormone. Cell 38:347-348.

Molecular Biology of Muscle Development, pages 253–262

TWO FETAL-SPECIFIC FAST MYOSIN ISOZYMES IN HUMAN MUSCLE

Laura Silberstein and Helen M. Blau[1]

Department of Pharmacology, Stanford University School of Medicine, Stanford, California 94305

ABSTRACT We demonstrate the existence of two temporally distinct fetal myosin isozymes of different heavy chain composition in human striated muscle using monoclonal antibodies. The fetal specificity of these monoclonal antibodies was determined by solid phase titration assays, immunoblots, and reactions in cryostat sections of muscles at different stages of human development. The two myosins are sequentially down-regulated in slow fibers early in development, but persist in other fibers until birth suggesting that these isozymes are fetal-specific "fast" myosins.

INTRODUCTION

Skeletal muscle fiber development progresses through a series of phases distinguished by the expression of particular myosin isoforms. At least three sequential myosin heavy chain isoforms have been demonstrated in rat (1,2) and chicken (3,4) muscles by peptide mapping and immunohistochemical reactions. We have isolated a collection of monoclonal antibodies (mAbs) to mammalian myosin heavy chain in order to define stages in muscle differentiation. Our immunochemical analysis demonstrates that the broad categories of "embryonic", "neonatal" and "adult" myosins include multiple isoforms characteristic of each of these developmental phases. In this paper, we present evidence

---

[1] This work was supported by the National Institute of Health and the Muscular Dystrophy Association of America.

using monoclonal antibodies for two distinct fetal myosins which are specifically and sequentially down-regulated in developing slow fibers.

METHODS

Monoclonal antibodies to human myosin were isolated by the procedure of Kohler and Milstein (5) as described elsewhere (6). Myosin was purified from myofibrils of normal human quadriceps muscle.

ELISA antigens (fetal: 15 wks., neonatal: five days, adult: 21 yrs.) were diluted to 7 ug/ml in 0.1 M sodium carbonate, pH 8.6 and applied to plastic microtiter plates. The bound myosin fraction retained after rinsing the plates with wash buffer (0.5 M NaCl, 20 mM sodium phosphate, pH 7.4 0.05% Tween-80) was then used to adsorb monoclonal antibody. Hybridoma supernatants were used in the range of 0.1 to 50 ug/ml; concentrated immunoglobulin fractions prepared from either cell culture (mAb f1) or from ascites supernatants (mAb f2) were used in the range of 2 ug/ml to 1 mg/ml. Bound antibody was detected following incubation with a peroxidase-conjugated goat anti-mouse immunoglobulin. The concentration of antibody in each preparation was independently determined using as a standard a purified mouse monoclonal antibody of the same subclass, IgG(1) (gift of R Levy), by methods described elsewhere (6).

Immunoblots were prepared by separation of myosin by SDS gel electrophoresis (5% stacking gel, 7.5% separating gel, Laemmli discontinuous system) and electrophoretic transfer to nitrocellulose paper. Bound antibody was detected following sequential incubations with biotinylated goat anti-mouse IgG, peroxidase Avidin D (Vector ABC), and 4-chloro-1-napthol (BioRad).

Blocks of normal human quadriceps muscle were frozen in melting isopentane and cut as 6 micron sections on a cryostat. Adult muscle biopsies were histochemically analyzed by the Stanford Neuromuscular Laboratory for pathological abnormalities, and found negative. Sections were incubated with monoclonal antibodies, washed in phosphate buffered saline, pH 7.2, and then reacted with fluorescein-labeled goat anti-mouse IgG, together with rhodamine-labeled goat anti-mouse IgM. MAbs f1, f2, 25 and 74 are IgGs, whereas the anti-slow monoclonal is an

IgM. All photographic images were obtained under parallel conditions of exposure, development, and printing.

RESULTS

Solid phase immunoassays were used to compare the relative reactivity of mAbs f1 and f2 in myosin preparations isolated from quadriceps muscle at different developmental stages. The results obtained using an enzyme-linked immunoadsorbent assay ("ELISA") revealed the presence of the f1 epitope in fetal, but not neonatal or adult myosin preparations (Fig. 1). The absolute specificity of mAb f1 for fetal myosin was observed over four orders of magnitude in antibody concentration. Qualitatively similar results were obtained with mAb f2 over a 100-fold range in antibody concentration. Binding to neonatal myosin is observed only after increasing the mAb concentration 500-fold (Fig. 2b). The decreased reaction of mAb f2 with neonatal, as compared to fetal, myosin suggests the presence of relatively low concentrations of the f2 epitope in the neonatal antigen preparation.

The specificity of mAbs f1 and f2 for fetal myosin was further demonstrated in immunoblot reactions (Fig. 2). Both antibodies react strongly with antigen(s) present in fetal, but not neonatal or adult myosin preparations. The molecular weight of the antigens corresponds to that of the myosin heavy chain. Thus, both mAbs f1 and f2 recognize epitopes associated with myosin molecules primarily expressed in early stages of human skeletal muscle development. We refer to these fetal isozymes as F1 and F2 myosin, indicating reaction with mAbs f1 and f2, respectively.

The fetal-specific expression of the F1 and F2 myosin epitopes was also determined by immunohistochemical reactions *in situ*. The early appearance of these myosins in all myofibers is evident in cryostat sections of developing limb muscle at 12 weeks gestation (Fig. 3). At this stage, the developing muscles consist of few fibers, well separated by connective tissue from each other and from neighboring clusters of myofibers. Rapid growth of the muscle occurs during the next several weeks, largely by the formation of new fibers. By 17 weeks gestation, all fibers still have F2 myosin (Fig. 4C). In contrast,

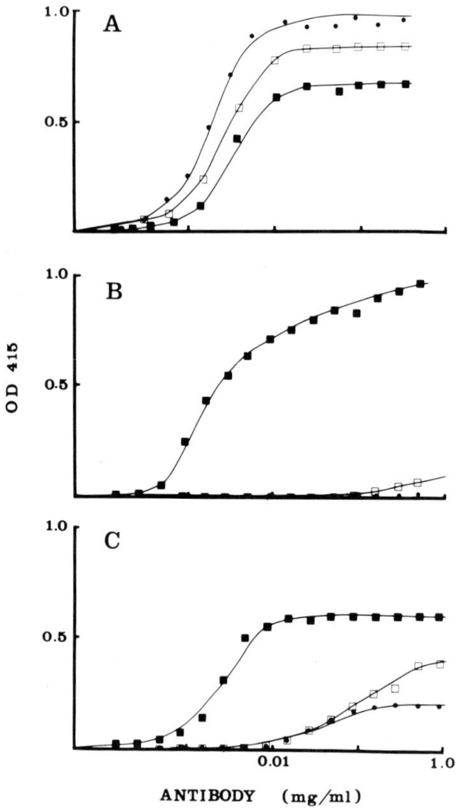

FIGURE 1. ELISAs showing reactions of monoclonal antibodies with fetal, neonatal, and adult myosins. (●) Adult myosin, (□) neonatal myosin, (■) fetal myosin. A) Positive control: mAb 25; B) mAb f1; C) mAb f2.

F1 myosin is no longer present in a minor sub-population of fibers. This minor population of fibers reacts intensely with antibody to slow myosin (Fig. 4A and B).

At birth, the F1 myosin epitope is undetectable in tissue sections. In contrast, the F2 epitope is still present at this stage, although clearly restricted to a subpopulation of fibers (Fig. 5). A comparison of the same sections, double-labeled with antibody to slow myosin and either f1 or f2, reveals that fibers containing slow myosin no longer express F2 myosin.

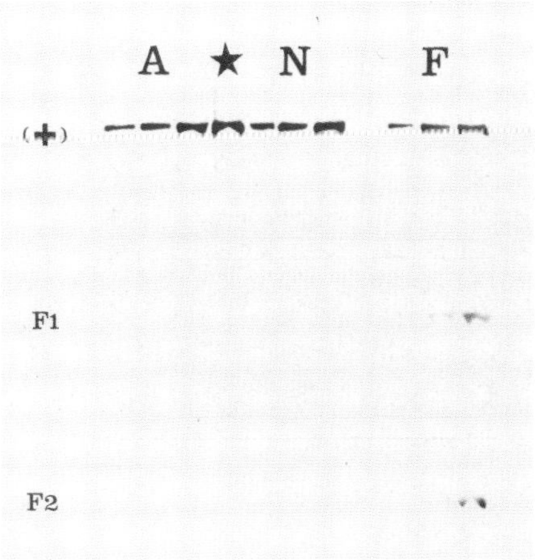

FIGURE 2. The specificity of mAbs f1 and f2 for fetal myosin is confirmed in immunoblot reactions. The top third of each transferred gel pattern is shown here. One, two, and four micrograms of each myosin preparation were applied: adult, A, neonatal, N, and fetal, F. Lane 4 (star) contains 2 micrograms of purified chicken myosin.

In adult muscle, the F1 and F2 myosins are essentially not detectable by immunohistochemistry (Fig. 6). A comparison of the same sections under rhodamine and fluorescein filters demonstrates that the lack of reaction of adult fibers with mAbs f1 and f2 is apparent in both fast and slow fibers.

The differential expression of F1 and F2 epitopes at 17 weeks gestation and in neonatal muscle demonstrates the existence of two distinct fetal myosin isozymes, recognized by mAbs f1 and f2, respectively. Both of these epitopes are initially expressed in all muscle fibers, but are later restricted to fast fibers. It appears that slow muscle maturation involves the down-regulation of first, F1 myosin expression (Fig. 4) and later, of F2 myosin expression (Fig. 5).

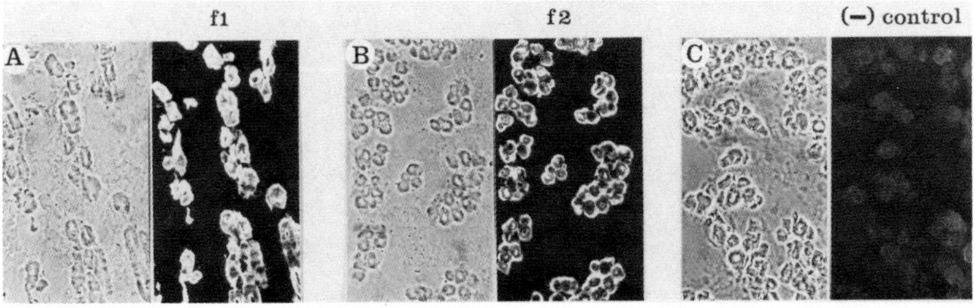

FIGURE 3. Myosin isozymes in fetal myofibers at 12 weeks. Each pair shows phase (left) and fluorescence (right) images of the same field. A) mAb f1; B) mAb f2; C) negative control mAb. Note that all myofibers at this stage of fetal muscle development react with both mAbs f1 and f2.

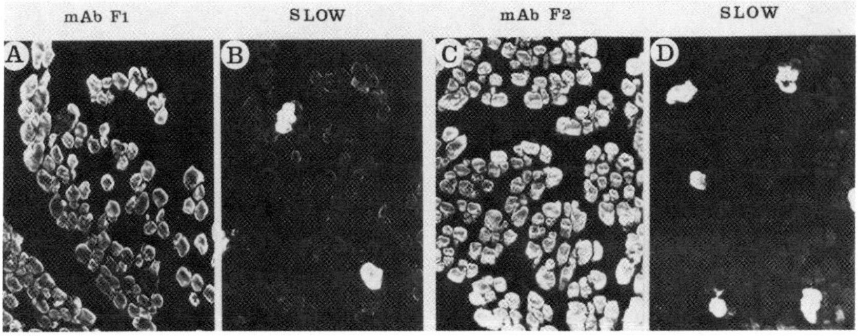

FIGURE 4. Myosin isozymes in fetal myofibers at 17 weeks. (A, B) and (C, D) are paired micrographs of the same fields in fluorescein optics (A, C) to show mAbs f1 and f2, respectively, or rhodamine optics (B, D) to show fibers reacting with antibody to slow myosin. Note that "slow" fibers contain F2 myosin (C, D) but not F1 myosin (A, B).

DISCUSSION

The broad categories of "embryonic", "neonatal" and "adult" myosin isoforms have been identified by peptide mapping (1,3,4). Fast and slow isozymes are present in the adult; these may be readily detected in fast or slow fibers using antisera prepared against the relatively pure fast and slow chicken muscles (7). Even immature muscles, such as the chicken pectoralis, may contain distinct fetal myosin isozymes with fast and slow epitopes (4,8). Here we describe a further level of complexity: there appear to be multiple developmentally distinct isoforms which are fiber-type specific. We have shown that there are two distinct myosins present in the fibers of developing human muscle which are not recognized by antibodies to slow myosin and are therefore presumably specific to maturing fast fibers. The time course of expression of these two myosins in fetal development differ from one another. These data suggest that human fetal muscle contains at least three distinct myosin isozymes: one "slow" and two "fast".

The sequential down-regulation of F1 and F2 myosins might suggest that mAbs f1 and f2 recognize "embryonic" and "neonatal" isozymes, respectively, analogous to the myosin isozymes described in rat muscle (1). However, significant amounts of "embryonic" myosin remain in the early post-natal weeks in rat limb muscle (2), whereas F1 myosin is not detectable by birth (Figs. 1, 2 and 5). In addition, the rat neonatal isozymes(s) are not detectable until just before birth (1), whereas both F1 and F2 myosins are present in all developing myofibers at 12 weeks gestation. Thus, it is unlikely that F1 and F2 myosins are homologous to the rat embryonic and neonatal isozymes.

The monoclonal antibodies described here recognize myosin isozymes which differ in their temporal expression and spatial distribution in muscle fibers. As shown here, these probes permit us to examine the expression of myosin isozymes in the course of normal <u>in vivo</u> muscle development. In addition, they may have useful application in the diagnosis and study of the etiology of human neuromuscular diseases. They have also permitted us to investigate questions of myoblast plasticity or commitment <u>in vitro</u>. For example, using a clonal analysis of myogenesis, we have determined that individual myoblast

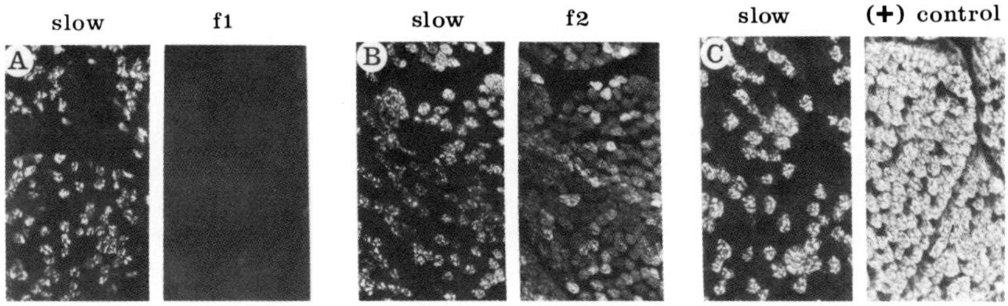

FIGURE 5. Myosin isozymes in neonatal muscle (1 day post-natal). Each pair shows the reaction of the same cells with the IgM antibody specific for slow fibers (left) and with the test IgG (right). A) mAb f1; B) mAb f2; C) positive control mAb 25. The muscle shown here is psoas.

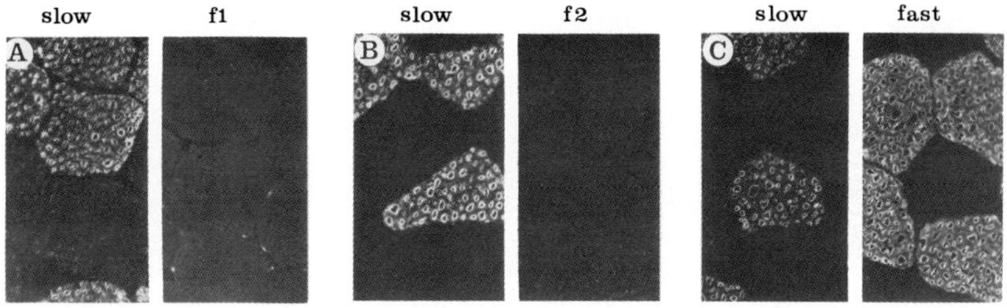

FIGURE 6. Myosin isozymes in adult muscle. Each pair shows the reaction of the same cells with the IgM (left) and IgG (right) mAbs. A) mAb f1; B) mAb f2; C) mAb 74, an IgG specific for adult fast myosin.

clones are capable of expressing multiple myosin heavy chain isozymes. These include the F1 and F2 myosins described here, as well as other fast muscle myosins which appear at more mature stages *in vivo* (9). Developmental transitions in the degree of plasticity within myoblast populations, as well as the role of innervation in the modulation of this plasticity, are the subject of current investigations.

ACKNOWLEDGEMENTS

We thank Evelyn Nunez and Dennis Arvanitis for technical assistance, John Oehlert and Dr. B. Adornato for histochemical analysis of human adult muscle sections, Dr. R. Levy for generous gifts of control antibodies, and Nancy Williams and Karen Benight for preparation of the manuscript. We gratefully acknowledge the co-operation of Drs. Adriano, Carrington, Dworsky, and Shumway in providing the human muscle samples, under the guidelines of the Stanford Medical School Human Experimentation Committee.

REFERENCES

1. Whalen RG, Sell SM, Buler-Browne GS, Schwartz K, Bouveret P, Pinset-Harstrom I (1981). Three myosin heavy-chain isozymes appear sequentially in rat muscle development. Nature 292:805.
2. Gambke B, Rubinstein NA (1984). A monoclonal antibody to the embryonic myosin heavy chain of rat skeletal muscle. J Biol Chem 259:12092.
3. Bandman E, Matsuda R, Strohman RC (1982). Developmental appearance of myosin heavy and light chain isoforms in vivo and in vitro in chicken skeletal muscle. Dev Biol 93:508.
4. Lowey S, Benfield PA, Gauthier PA, LeBlanc DD, Waller G, Winkelman DA (1982). Characterization of myosins from embryonic and developing chicken pectoralis muscle. In Pearson ML, Epstein HJ (eds): Muscle Development: Molecular and Cellular Control, Cold Spring Harbor Laboratory, p. 15.
5. Kohler G, Milstein C (1975). Continuous cultures of fused cells secreting antibody of predefined specificity. Nature 256:495.
6. Silberstein L, Blau HB (in preparation).
7. Gauthier GF, Lowey S (1979). Distribution of myosin isozymes among skeletal muscle fiber types. J Cell Biol 74:81.
8. Benfield P, Lowey S, LeBlanc DD, Waller G (1983). Myosin isozymes in avian skeletal muscles. II. Fractionation of myosin isozymes from adult and embryonic chicken pectoralis muscle by immuno-affinity chromatography. J Mus Res Cell Motil 4:717.

9. Silberstein L, Webster SG, Blau HB (1985). Expression of fetal and neonatal myosin heavy chain isozymes in myotube cultures derived from a single myoblast clone. Abstract #316, AAAS Publication #85-7 (Abstracts Volume), Annual Meeting, Los Angeles, CA.

Molecular Biology of Muscle Development, pages 263–272

CLONING AND CHARACTERIZATION OF MYOSIN cDNAS IN ADULT HUMAN SKELETAL MUSCLE[1]

Lino J. Saez and Leslie A. Leinwand

Department of Microbiology and Immunology and Genetics
Albert Einstein College of Medicine
New York, New York 10461

ABSTRACT To define myosin heavy chain (MHC) gene organization and expression in adult human skeletal muscle, we have isolated and characterized MHC genomic and cDNA sequences. We present here the analysis of 2 MHC cDNA clones whose mRNAs are the products of 2 different genes. The cDNA sequences encode MHC proteins that are closely related to those previously described from other vertebrate organisms. Sequence comparisons show the human cDNA clones to be less homologous to each other than to MHC from other organisms. The 3' untranslated regions of the 2 human clones are completely nonhomologous, but one of them shows considerable homology to other MHC 3' untranslated regions. These data indicate that myosin isotypes have been preserved through evolution and have been maintained as distinct genes. A synthetic oligonucleotide from the 3' untranslated sequence of one human cDNA clone has been used to identify the gene encoding this MHC mRNA.

INTRODUCTION

Muscle development is accompanied by many cellular events including the induction of a number of contractile protein genes, many of which are tissue specific and

[1]This work was supported by NIH GM 29090. L.S. is supported by a Dupont Science and Engineering Grant, L.L. is an American Heart Association Established Investigator.

developmentally regulated. The contractile apparatus in muscle consists of myosin heavy chain, myosin light chain, α tropomyosin and troponins I, C and T. Muscle is highly specialized in its use of these proteins to transform chemical energy into mechanical force through the hydrolysis of ATP. Myosin consists of 2 heavy chains (200,000 daltons each) and 2 pairs of light chains. The carboxy terminus of the molecule forms an α helical tail in dimeric form which is responsible for the assembly of myosin into thick filaments. The globular amino terminal portion binds two pairs of light chains, contains ATPase activity and the actin-binding site.

Protein studies have established the existence of multiple forms of MHC within different types of muscle and in different developmental states (e.g.,1). It is presumed that the diversity of MHC proteins has functional significance. Different ATPase activities are correlated with the contractile properties of individual muscles (2,3). Also, results of a recently developed motility assay show that myosins purified from different sources can be functionally distinguished by their velocity on actin cables (4). Molecular clones corresponding to some of these forms have been obtained and have been invaluable in elucidating features of MHC regulation and expression (5-9). α and β ventricular cardiac MHC cDNA clones have been isolated from rat and rabbit (6,7). Rat embryonic and perinatal skeletal muscle MHC and chicken embryonic skeletal MHC clones have been isolated (5,8). These and other studies have demonstrated that MHC is encoded by a highly conserved multigene family (10-13). The exception to this is Drosophila which has one MHC gene that encodes 3 mRNAs (14). In mouse and human the genes are localized to a single chromosome (15).

Protein data indicate that during skeletal muscle development at least 5 MHC forms are expressed. These include fetal, neonatal, 2 adult fast and 1 slow MHC forms (1,16,17). The full extent of MHC diversity within each muscle type and developmental state has yet to be completely elucidated. We report here the isolation and characterization of 2 MHC cDNA clones expressed in adult human skeletal muscle. These clones are the first reported human MHC cDNA clones. They have been used to analyze the degree of sequence homology among the different sarcomeric MHC mRNAs.

RESULTS

Diversity of Adult Human Skeletal Muscle MHC Expression

An adult human skeletal muscle cDNA library was screened for MHC clones using a previously described human genomic MHC DNA fragment as a probe (13). This library was kindly provided by L. Kedes, Stanford University, and was constructed in the Okayama-Berg vector system (18). 9 clones were isolated from 2000 screened. Several distinct restriction patterns result from digestion of these cDNA PstI and PvuII (Figure 1). PstI is known to often cleave sequences encoding the carboxy terminal portion of MHC (9). Its recognition sequence encodes the dipeptide, leucine-glutamine, which is a frequently occurring sequence in the MHC rod (9,19). The 2 bands at 1600 and 1100 bp that are present in all clones correspond to vector bands. There are multiple PstI sites in the 9 clones analyzed (Figure 1). The insert sizes of the MHC clones range from .6Kb to 2.6Kb. The clones in lanes 3, 7, and 9 show one pattern of fragments while clones in lanes 1 and 6 show a second pattern. Clones in lanes 2, 4, 5 and 8 each show distinct patterns. These data suggest that multiple MHCs (possibly 4-6) are expressed in adult human skeletal muscle. DNA sequence analysis on portions of clones representing 3 of the 6 patterns has shown that the 3 clones analyzed all contain the 3' end of different MHC sequences and do not merely represent different portions of one sequence (in preparation). Two clones were chosen for further analysis (lanes 3 and 6) because they have long inserts and because their restriction digestion patterns appear to be distinct. These clones are designated pSMHCA and pSMHCZ and have insert sizes of 1.5 and 2.0 kb respectively. DNA from both clones hybridizes to a single 7.1 Kb mRNA species from adult skeletal muscle (Figure 2A). pSMHCZ hybridizes very strongly to both cardiac and skeletal muscle RNA while pSMHCA shows strong hybridization to skeletal muscle RNA and very faint hybridization to cardiac muscle RNA. The patterns of the two cDNA clones when hybridized to human genomic DNA are shown in Figure 2B. Each cloned DNA hybridizes to multiple EcoRI and BamHI genomic fragments. The number of bands is 10-15. As expected, many of the same fragments are recognized by both clones. In addition, there are bands that are specific to each clone. Some of these fragments presumably represent the genes encoding these mRNAs.

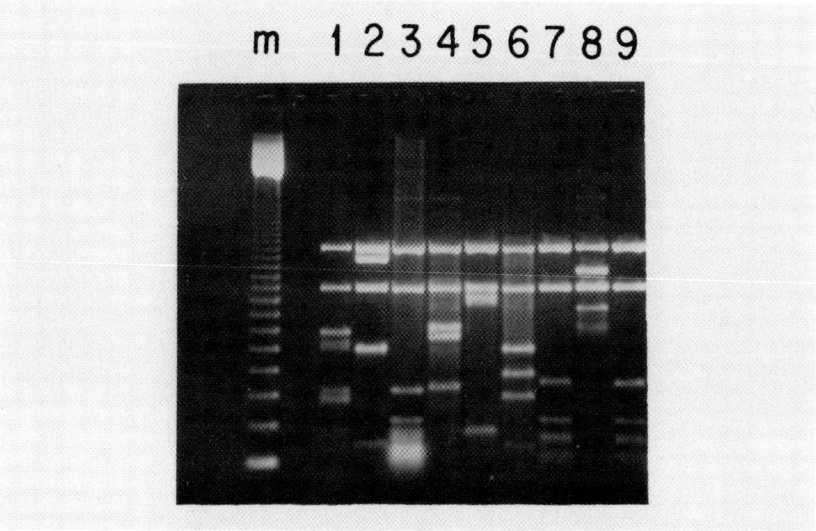

FIGURE 1. Ethidium bromide-stained gel of restriction endonuclease digestion of human skeletal muscle MHC cDNA clones with PvuII + PstI (lanes 1-9). M indicates the marker lane with a 123 bp ladder.

Graphic Matrix Analysis of MHC DNA Sequences

DNA sequence analysis was carried out on clones pSMHCA and pSMHCZ. Complete sequence data will be presented elsewhere. Amino acid sequences derived from these clones are closely related to previously described MHCs from chicken skeletal muscle, rat cardiac and skeletal muscle (data not shown). Graphic matrix analysis for DNA sequence homology was carried out (Figure 3; Panels A-E). The 2 human skeletal muscle MHC clones were compared with themselves (Panels A and B), with each other (Panel E) and with a chicken skeletal muscle MHC cDNA clone (Panels C and D). This analysis was carried out with a window of 10 at a 75% stringency level. A match of 8 out of 10 nucleotides in each window is required for placement of a dot. The results

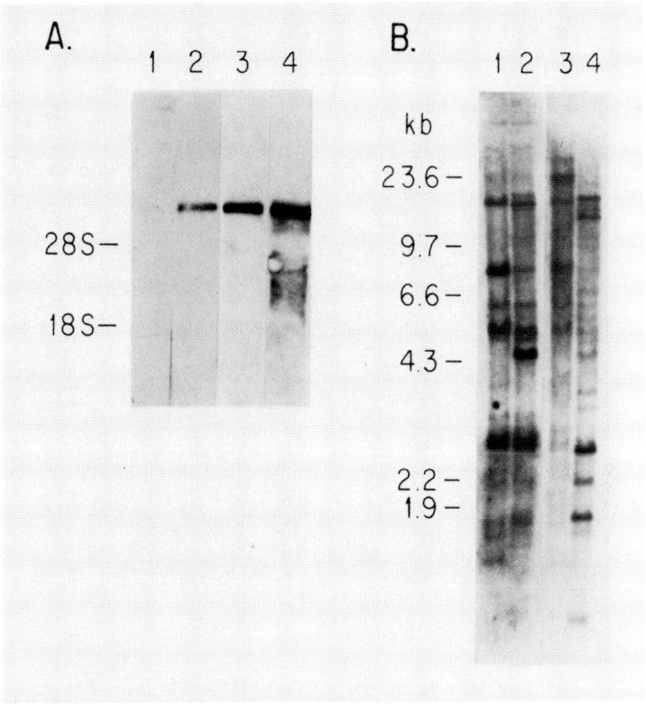

FIGURE 2. Panel A. Hybridization of human MHC cDNA clones to cardiac (lanes 1 and 3) and skeletal (lanes 2 and 4) muscle RNA. pSMHCA was hybridized to lanes 1 and 2. pSMHCZ was hybridized to lanes 3 and 4. 28S and 18S refer to migration of ribosomal RNAs.
Panel B. Hybridization of human MHC cDNA clones to human genomic DNA digested wth EcoRI (lanes 1 and 3) and BamHI (lanes 2 and 4). pSMHCA was hybridized to lanes 1 and 2 and pSMHCZ was hybridized to lanes 3 and 4. Size markers are indicated by horizontal lines.

obtained by such analysis are somewhat surprising in that the 2 human clones are less homologous to each other (70%) (Panel E) than for example, pSMHCA is to the embryonic chicken skeletal muscle clone (80%) (Panel D). Northern blot analysis presented in Figure 2 is consistent with this observation since pSMHCA appears to be skeletal muscle-specific. The sequence differences between the 2 human MHC clones are more pronounced toward the 5' end. (Panel E).

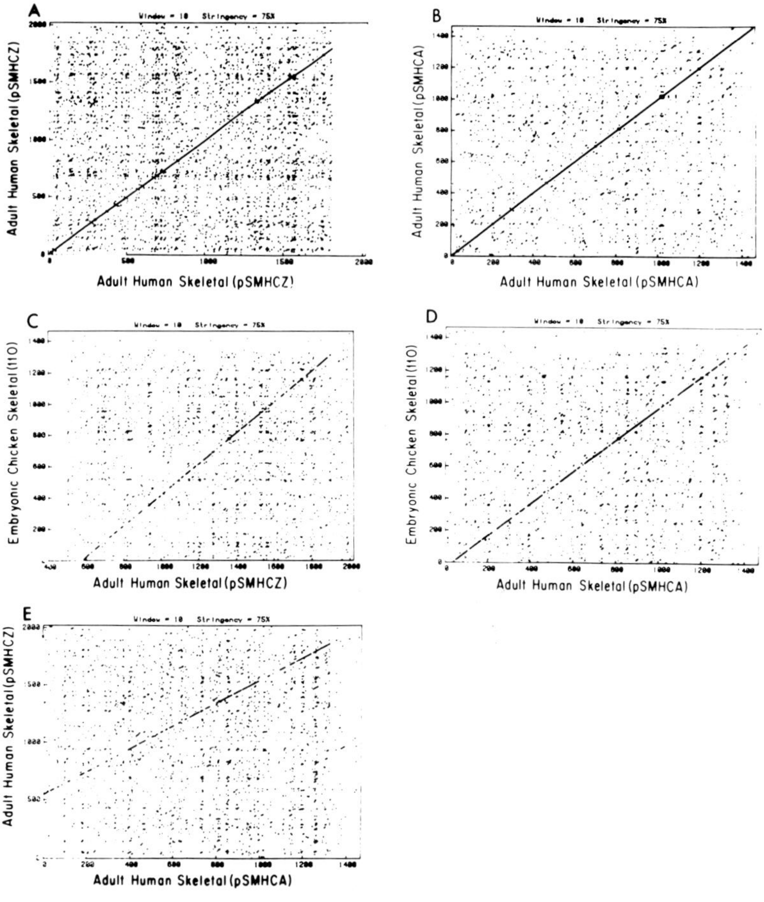

FIGURE 3. DNA sequence comparisons of MHC cDNA clones were carried out by graphic matrix analysis. A window of 10 with a 75% stringency level was used. A match of 8 out of 10 nucleotides in each window is required for placement of a dot. Sequences used in the comparisons are named on the X and Y axes. The chicken sequence was obtained from (9).

The entire 3' untranslated regions of both mRNAs are included in the clones including a portion of the poly(A) tail. Despite considerable homology in the protein-coding portions of pSMHCA and Z, their 3' untranslated regions are completely nonhomologous (Figure 4A). When compared with 3' untranslated MHC regions from other organisms, pSMHCA shows extensive blocks of homology to chicken and rat skeletal muscle MHC clones while pSMHCZ shows no homology to previously described MHC sequences (data not shown).

Isolation of the Gene Encoding pSMHCZ by Oligonucleotide Hybridization

Our results indicate that the mRNAs corresponding to pSMHCA and Z are the products of different genes. Given the highly conserved nature of the sarcomeric MHC gene family, it is difficult to identify the genes encoding specific MHC mRNAs and to differentiate expressed genes and pseudogenes. We have previously described 4 nonoverlapping human genomic MHC clones (13). Subsequent screenings have resulted in at least 12 nonoverlapping genomic clones. Based on our sequence data, it seemed likely that oligonucleotides from the 3' untranslated regions of MHC cDNA clones would be gene-specific. Most probes from the protein coding regions of MHC clone would cross-hybridize with other MHC genes. Oligonucleotides were synthesized from the 3' untranslated regions of pSMHCA and Z indicated by overlining (Figure 4A). In order to determine whether any of the genomic clones we had isolated corresponded to the genes encoding pSMHCA and Z, oligonucleotide hybridizations were carried out. 12 different human MHC genomic clones were digested with EcoRI (Figure 4B) and hybridized to the entire pSMHCA clone (Figure 4C) and the entire pSMHCZ clone (Figure 4D). The cDNA clones hybridize to multiple λ genomic clones, as expected. When the oligonucleotide from pSMHCA was hybridized to this filter, no signal was seen (data not shown). When the oligonucleotide from pSMHCZ was used, one clone showed positive hybridization (Figure 4E). This clone, designated λHMHC8, is likely to be the gene encoding the mRNA for cDNA clone pSMHCZ.

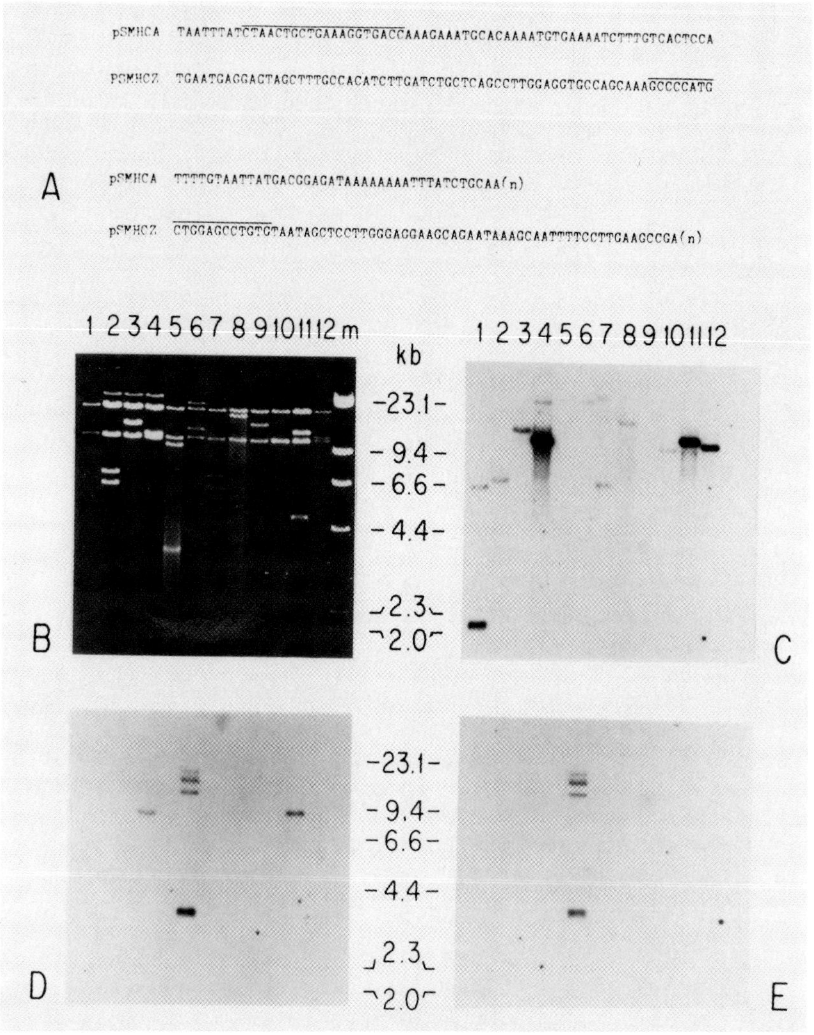

FIGURE 4. Panel A. 3' untranslated regions of clones pSMHCA and Z, beginning with the termination codons. Sequences used for construction of gene-specific oligonucleotides are indicated by overlining. EcoRI digests of 12 human genomic MHC clones were run on an ethidium-bromide stained gel (Panel B), blotted and hybridized to pSMHCA (Panel C), pSMHCZ (Panel D) and the oligonucleotide from the 3' untranslated region of pSMHCZ (Panel E).

DISCUSSION

We have demonstrated that there is considerable diversity of MHC expression in adult human skeletal muscle. There may be as many as six different forms of MHC expressed simultaneously. DNA sequence analysis of all MHC clones with different restriction patterns is required to definitively settle the issue of MHC diversity within one muscle type. The first two human clones chosen for our analysis clearly represent distinct MHC forms. In fact, their homology to each other is only 70% at the DNA sequence level while one of the human clones, pSMHCA, is 80% homologous to a chicken skeletal muscle MHC sequence. This is in contrast to two rat cardiac muscle MHC cDNA clones which are 95% homologous in their coding regions (6) and two chicken skeletal muscle MHC cDNA clones which are 89% homologous in their coding regions (5). In view of the substantial differences observed between human clones pSMHCA and Z, it is likely that these sequences are products of two separate and distinct genes.The chicken MHC sequence to which pSMHCA is homologous represents a "fast" twitch skeletal muscle MHC form (5). pSMHCA most likely represents a "fast" skeletal MHC form. Since pSMHCZ hybridizes equally to cardiac and skeletal muscle RNA, it may represent an MHC form that shows overlapping patterns of expression. This has been observed in the rat in which a β cardiac MHC form was also expressed in soleus muscle (20).

We have used oligonucleotide hybridization to identify the gene encoding the pSMHCZ MHC form. This approach may prove to be generally useful in the study of highly conserved multigene families. It remains to be seen whether multiple MHC genomic sequences will have exact homology in their 3'-untranslated regions, or whether these probes will be truly gene-specific. The analysis of MHC expression should lead to an increased understanding of structure-function relationships in motility. We hope to identify those regions of the MHC molecule that confer functional diversity on myosin isotypes.

ACKNOWLEDGEMENTS

We acknowledge the excellent secretarial assistance of D. Jackson and L. Vitagliano.

REFERENCES

1. Whalen RG, Schwartz K, Bouveret P, and Gros F (1979). Proc. Nat'l. Acad. Sci. U.S.A. 76:5197.
2. Wagner PD, and Giniger E (1981). Nature 292:560.
3. Barany M (1967). J. Gen. Physiol. 50:197.
4. Sheetz MP and Spudich JA (1983). Nature 303:31.
5. Kavinsky CJ, Umeda PK, Sinha AM, Elzinga M, Wing Fong S, Zak R, Jakovcic S and Rabinowitz M (1983). J. Biol. Chem. 258:5796.
6. Mahdavi V, Periasamy M and Nadal-Ginard B (1982). Nature 297:657.
7. Sinha AM, Umeda PK, Kavinsky CJ, Rajamanickam C, Hsu HJ, Jakovcic S and Rabinowitz M (1982). Proc. Nat'l. Acad. Sci. 79:5847.
8. Periasamy M, Wieczorck DF and Nadal-Ginard B (1984). J. of Biol. Chem. 259:13573.
9. Kavinsky CJ, Umeda PK, Lenin JE, Sinha AM, Nigro JM, Jakovcic S and Rabinowitz M (1984). J. of Biol. Chem. 259:2775.
10. Wydro RM, Nguyen HT, Gubits R and Nadal-Ginard B (1983). J. Biol. Chem. 258:670.
11. Nguyen HT, Gubits R, Wydro RM, and Nadal-Ginard B (1982). Proc. Nat'l. Acad. Sci. 79:5230.
12. Friedman DJ, Umeda PK, Sinha AM, Hsu HJ, Jakovcic S and Rabinowitz M (1984). Proc. Nat'l. Acad. Sci USA 81:3044.
13. Leinwand L, Saez L, McNally E and Nadal-Ginard B (1983). Proc. Nat'l. Acad. Sci. USA 80:3716.
14. Rozek CE and Davidson N (1983). Cell 32:23.
15. Leinwand LA, Fournier REK, Nadal-Ginard B and Shows TB (1983). Science 221:766.
16. Whalen RG, Sell SM, Butler-Browne GS, Schwartz K, Bouveret P, and Pinset-Harstrom I (1981). Nature 292:805.
17. Gauthier GF, Lowey S, Benfield PA and Hobbs AW (1982). J. Cell Biol. 92:471.
18. Okayama H and Berg P (1983). Mol. Cell Biol. 3:280.
19. McLachlan AD and Karn J (1982). Nature 229:226.
20. Lompre A, Nadal-Ginard B and Mahdavi V (1984). J. Biol. Chem 259:6437.

Molecular Biology of Muscle Development, pages 273–282

THE EXPRESSION OF MYOSIN HEAVY CHAIN ISOFORMS IN A NERVE-MUSCLE CULTURE SYSTEM[1]

Marion Ecob and Robert Whalen

Muscular Dystrophy Research Laboratories, Newcastle, England
Departement de Biologie Moleculaire, Institut Pasteur
Paris

ABSTRACT When adult mouse muscle fibers are cocultured with embryonic mouse spinal cord, the muscle regenerates to form myotubes which develop cross-striations and contractions. The adult isoform of fast myosin is present in cultures which have been contracting for at least 10 days (Ecob et al, Differentiation 25, 84-87, 1983). We have now shown that the adult fast isoform was not present in young myotubes, but that its appearance was coincident with the development of cross-striations and contractions, both of which are cord-dependent phenomena. The adult isoform could be found as early as 7 days in culture. This was not due to persistence of the explanted adult muscle fibers. The combinations of myosin isoforms in individual regenerated fibers suggested that there was a neonatal-to-adult-fast transition in this system. Moreover, the adult fast isoform was never found in cultures without spinal cord, and so its expression appeared to be spinal cord-dependent.

When cultures were established and maintained in the continuous presence of d-tubocurarine or α-bungarotoxin, adult fast myosin heavy chains were still expressed, although there were virtually no contradictions and only a transient appearance of cross-striations (curare only). Nerve-induced muscle contraction was therefore not a prerequisite for the expression of the adult fast isoform.

This work was supported by grants from the Muscular Dystrophy Group of Great Britain, the Institut National de la Sante et de la Recherche Medicale, and MDA of Amercia.

INTRODUCTION

The expression of the different myosin heavy chain isoforms in skeletal muscle of rats, rabbits, mice, humans and chickens is developmentally regulated. For example, in rat fast contracting muscle, the initial embryonic heavy chain is replaced by neonatal isoform (predominant by 7-11 days after birth) which is subsequently replaced by the adult fast heavy chain isoform (1,2,3,4). Development of slow muscle fibers similarly involves myosin isoenzyme transitions (2,5).

Such transitions had not, however, been shown in cultured muscle cells which appeared to express only embryonic (3,6,7,8,9) or embryonic and neonatal myosin even after long periods in culture (10,11). Nevertheless, in organotypic nerve-muscle cocultures, we found that adult fast myosin heavy chains were present in myotubes which had been cross-striated and contracting for at least 10 days, some 3-5 weeks in culture (12).

We have now investigated whether the expression of adult fast myosin was associated with time in culture, the diameter of the myotubes or the presence of the spinal cord. We also wanted to know if the neonatal-to-fast isoenzyme transition occurred in this system. The results indicate that such a transition does occur and that the expression of the adult fast isoform does require the presence of the spinal cord, but is not dependent upon nerve-induced activity in the muscle fibers.

MATERIALS AND METHODS

In most experiments, 3-5 adult mouse muscle fibers, about 7 mm long, were cultured with a section of embryonic mouse spinal cord (13,14). In some experiments, d-tubocurarine (1, 50 or 100 μgm/ml) or α-bungarotoxin (10^{-7} or 10^{-8} M) was incorporated into the growth medium throughout the culture period. When muscle was cultured without spinal cord, myotube formation was stimulated by either the addition of a piece of embryonic lung, or by denervating the muscle in vivo prior to explantation. (Cut-fiber preparations of adult mouse muscle in this system do not regenerate spontaneously on their own.)

The polyclonal antibodies used were prepared against rat myosins. The preparation and characterization of the antibodies to neonatal and adult fast myosin, or to slow myosin have been described (5 and 15 respectively). Their cross-specificity to the analogous mouse isoforms was shown using immunoblots on mouse muscle extracts as described elsewhere (12,15), and control mouse muscle samples were incorporated into all the frozen blocks of cultures.

Nerve-muscle cultures were observed by light microscopy every 3-4 days, and at appropriate times the bundle of regenerated myotubes was excised. The bundle was sandwiched between 2 blocks of liver together with part of a soleus from a 3 week old mouse and a piece of adult mouse EDL, and frozen. A 3 week soleus contains neonatal, slow and fast myosin, but the EDL contains predominantly fast myosin and is negative for neonatal; they therefore act as internal controls for the antibody staining.

Frozen serial sections were reacted with each of the rabbit antibodies, followed by a rhodamine-conjugated goat anti-rabbit second antibody. After microscope observation and photographic recording, each section was stained by hematoxylin and eosin to determine the number of muscle fibers present. The minimum fiber diameters were measured from the photographs of the fluorescence using a Videoplan semi-automatic image analyser.

RESULTS

In these nerve-muscle cultures, the explanted fibers lose their plasma membrane and myonuclei within the first 24 hours (16) and are subsequently cleared away by phagocytic cells. Satellite cells become activated and form myotubes beneath the basal lamina which is all that remains of the central part of the explanted fibers. The living cultures are viewed by light microscopy every 3-4 days to monitor the loss of the old and the regeneration of the new muscle fibers. Thus, we can be sure that <u>only</u> regenerated myotubes are being excised and that these bundles do not contain even atrophied explanted adult muscle fibers. Cultures in which explanted fragments remained were not used in this study.

Frozen serial transverse sections of the bundle of myotubes from individual cultures were cut, and were each reacted with one of the 3 antibodies to myosin heavy chains (neonatal, adult fast, adult slow). It was then possible to determine which isoforms were present in a particular fiber. It was not always possible to follow the small diameter (<3 μm) fibers in serial section because they changed their position or were not present in all sections. However, it was possible to determine the combinations of myosin present in the larger fibers. An example of serial sections stained with each of the 3 antibodies is shown in Figure 1.

Myosin Isoforms in Developing Nerve-Muscle Cultures

Fifteen nerve-muscle cultures were frozen when the muscle bundle was at one of three different stages of development (7-23 days *in vitro*), and their myosin content was determined immunocytochemically. Cultures were judged by light microscopy to contain either 1) immature myotubes without striations, 2) myotubes with some striations, or 3) myotubes that were cross-striated and had *just begun* contracting, before they were frozen. When the cultures were ranked on these morphological groups and not on time *in vitro*, it became obvious that adult fast myosin was absent from immature myotubes (1/5 had "pale" staining), began to appear with the appearance of striations (2/5 cultures had adult fast), and was present in all (5/5) cultures with clearly cross-striated fibers which were beginning to contract irrespective of their age in culture. Thus the

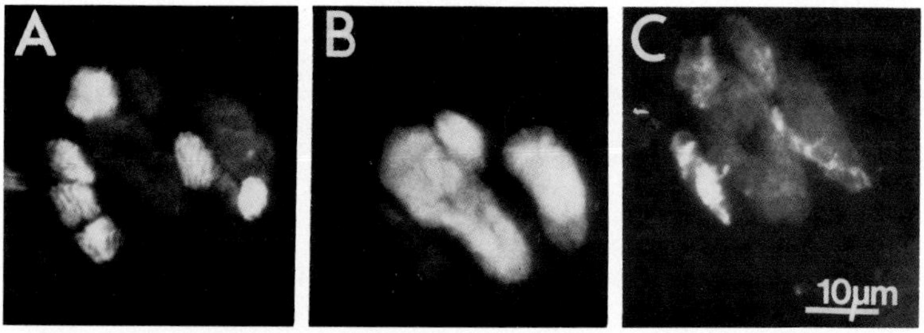

FIGURE 1. Frozen serial sections of culture 248-24 which have been stained with antibodies to neonatal (A), adult fast (B) or adult slow myosin (C). Three fibers contain fast alone. The culture was 21 days old and contained 2 large and 2 small striated fibers which had been contracting 7 days, as well as several undifferentiated small myotubes.

appearance of adult fast myosin coincided with the appearance of cross-striations and contractions, both of which are nerve dependent phenomena (13,14). It also showed that adult fast myosin could be expressed as early as 7 days *in vitro*, and that it was not therefore simply a product of long periods in culture.

Myosin Isoform Transitions in Nerve-Muscle Cultures

Immature myotubes contained neonatal with or without adult slow myosin. Striated myotubes which had just begun to contract contained different combinations of myosin: neonatal + slow + fast, neonatal + slow, neonatal + fast, or neonatal alone. However, in each of 12 cultures which had been contracting for 11-17 days, their were fibers which contained only fast myosin, in the absence of neonatal or slow isoforms, as we had found before (12). This is compatible with the concept that there is a neonatal to neonatal + fast to fast alone transition of myosin expression in these cultures.

Proportions and Diameters of Fibers which Contain Adult Fast Myosin

In the 12 cultures which had been contracting for 11-17 days, there was a total of 303 regenerated fibers, 93% of which contained neonatal myosin, 38% contained adult slow myosin and 24% contained adult fast myosin. Of the 73 fibers containing adult fast myosin, 21 (7% of total number) contained fast in the absence of neonatal or slow myosin.

The diameter of the fibers containing either fast, slow or neonatal myosin all had a range of 0.5 μm to 9.0 μm. However, the proportion of fibers with a diameter >5μm was greater in the population of fibers containing adult fast myosin (26% of the total) than in the fibers containing either slow or neonatal myosin (5% or 10% respectively). This association of large diameter and the presence of fast myosin was even more striking in those 21 fibers which contained fast without neonatal or slow myosin, where 67% were >5μm in diameter. Nevertheless, despite this association, there was no evidence that fast myosin was only produced in large-diameter fibers.

Myosin Isoforms in Cultures without Spinal Cord Cells

In the 7 cultures used in this study, there were 60 regenerated myotubes of a similar age in culture (20-35 days) as the "mature" nerve-muscle cultures (25-39 days). They all contained neonatal myosin, 77% contained adult slow, and none contained adult fast myosin. Similarly, none of the myotubes developed cross-striations or contractions as judged by light microscopic observation of the living cultures.

Myosin Isoforms in Nerve-Muscle Cultures Containing d-Tubocurarine or α-Bungarotoxin

Twelve cultures, containing 240 myotubes, were cultured in the continuous presence of d-tubocurarine for 20-31 days. Neonatal myosin was present in 95% of the fibers, slow myosin was in 18%, and 5% contained adult fast myosin. Patches of transient cross-striations were seen in 5 of the living cultures, and contractions were observed in only one of these cultures at one time. Adult fast myosin was found in only 2 of these cultures, and was present in 2 which had no morphological differentiation.

Five cultures, containing a total of 70 myotubes, were cultured in the continuous presence of α-bungarotoxin for 21-42 days. Neonatal myosin was present in 96% of the fibers, slow myosin was in 53%, and 27% contained adult fast myosin. Cross-striations or contractions were not seen in the living cultures at any time.

Thus, in the presence of drugs which block nerve-induced contractile activity in these cultures, adult fast myosin was still produced. Spontaneous contractions were rarely observed by light microscopy during the culture period, but the possibility that they were more frequent cannot be excluded.

DISCUSSION

Neonatal and slow myosin were present in myotubes which had regenerated in the absence of spinal cord tissue, and so their expression was clearly not nerve-dependent. However, neonatal myosin has only been described in long-term aneural cultures of primary rat myotubes (3,10) and not in short-term primary cultures or cell lines (3,11) and so the finding of neonatal myosin as early as 7 days in culture is a little surprising, but may reflect the different conditions used in this study. Experiments *in vivo* have suggested that the expression of adult slow myosin is controlled by a specific activity pattern of the nerve (17,18). However, muscles of fetal and newborn mammals have slow myosin containing fibers (2,15,19) and it has been suggested that this is a nerve-independent process (15). The expression of slow myosin in these cultures is probably similar to that in developing muscles rather than a reflection of specific nerve activity.

Adult fast myosin was not found in myotubes which had regenerated in the absence of spinal cord, and it was not present in young immature (lacking striations and/or contractions) myotubes. It was, however, found in fully cross-striated myotubes which were beginning to contract, usually in association with neonatal and occasionally with adult slow myosin. In cultures which had been contracting at least 10 days, fibers were present which contained adult fast myosin without either neonatal or slow myosin (13 and this study). The development of striations and contractions in this system is known to be a spinal cord-dependent phenomenon (14,15). It therefore appears that there is a neonatal-to-fast transition, and that the expression of the adult fast isoform is spinal cord-dependent in these cultures.

However, studies in animals have indicated that the expression of adult fast myosin in developing muscle (1,20, 21) or regenerating adult (22) is not nerve-dependent. We have no simple explanation for this enigma, but it may be that trophic influences necessary for adult fast myosin expression can be supplied in the animal by some source other than the nerve, or that all tissue culture media currently in use lack some factor which is supplied in this system by the spinal cord. The passive stretch of developing or regenerating muscles that must occur *in vivo* (and which does not occur in culture) is probably not responsible for the expression of the adult fast isoform in denervated limbs of animals as we found that inhibition, at least of nerve-induced activity, did not prevent the expression of fast myosin.

There were, however, much lower levels of adult fast myosin in cultures maintained in the continuous presence of d-tubocurarine. Neuromuscular junctions can form in the presence of the drug (eg. 23-25), but d-tubocurarine is also known to cause a depolarization of embryonic rat muscle (26) and to selectively interfer with the accumulation of myosin during development of the chick embryo (27). These secondary effects probably account for the reduced proportions of fibers containing either fast or slow myosin which we found. Levels of adult fast myosin in cultures maintained in α-bungarotoxin were similar to those of controls.

We would therefore conclude that the expression of neonatal or slow myosin in these cultures is nerve-independent. The expression of adult fast myosin, on the other hand, requires the presence of the spinal cord but does not require nerve-induced activity, and there appears to be a neonatal-fast myosin isoenzyme transition in the culture system.

ACKNOWLEDGEMENTS

We wish to thank Aileen Brown for culture work, Carol Young for sectioning, and Gill Butler-Browne and Graham Riley for producing and characterizing the antibodies.

REFERENCES

1. Butler-Browne GS, Bugaisky LB, Cuénod S, Schwartz K, Whalen RG (1982). Denervation of newborn rat muscles does not block the appearance of adult fast myosin. Nat (Lond) 299:830.
2. Lyons GE, Haselgrove J, Kelly AM, Rubinstein NA (1983). Myosin transitions in developing fast and slow muscles of the rat hindlimb. Differentiation 25:168.
3. Whalen RG, Schwartz K, Bouveret P, Sell SM, Gross F (1979). Contractile protein isoenzymes in muscle development: Identification of an embryonic form of myosin heavy chain. Proc Nat Acad Sci 76:5197.
4. Whalen RG, Sell SM, Butler-Browne GS, Schwartz K, Bouveret P, Pinset-Harström I (1981). Three myosin heavy-chain isoenzymes appear sequentially in rat muscle development. Nat (Lond) 292:805.
5. Butler-Browne GS and Whalen RG (1984). Myosin isozyme transitions occurring during postnatal development of the rat soleus muscle. Develop Biol 102:324.
6. Bader DT, Masaki T, Fischman DA (1982). Immunochemical analysis of myosin heavy chain during avian myogenesis in vivo and in vitro. J Cell Biol 95:763-770.

7. Bandman E (1984) Myosin components of the latissimus dorsi and the pectoralis major muscles in the dystrophic chicken. Muscle & Nerve 7:312-326.
8. Bandman ER, Matsuda R, Strohman RC (1982) Developmental appearance of myosin heavy and light chain isoforms in vivo and in vitro in chicken skeletal muscle.
9. Matsuda R, Spector DH, Strohman RC (1983) Regenerating adult chicken skeletal muscle and satellite cell cultures express embryonic patterns of myosin and tropomyosin isoforms. Develop Biol 100:478-488.
10. Bugaisky LB, Butler-Browne GS, Laurent C, Whalen RG (1983) Myofibrillar protein synthesis in satellite cell cultures from rat gastrocnemius and soleus muscles. J Cell Biol 97:45a.
11. Whalen RG, Bugaisky LB, Butler-Browne GS, Pinset C (1984) Control of myogenesis and myosin isozyme transitions in rat muscle cells. In Eppenberger HM, Perriard JC (eds): Exptl Biol and Medicine, Vol 9; Basel p 155.
12. Ecob MS, Butler-Browne GS, Whalen RG (1983) The adult fast isozyme of myosin is present in a nerve-muscle tissue culture system. Differentiation 25:84-87.
13. Ecob MS (1983) The application of organotypic nerve cultures to problems in neurology with special reference to their potential use in research into neuromuscular diseases. J Neurol Sci 58:1-15.
14. Peterson ER, Crain SM (1972) Regeneration and innervation of cultures of adult mammalian skeletal muscle coupled with fetal rodent spinal cord. Exp Neurol 36: 136-159.
15. Whalen RG, Johnstone D, Bryers P, Butler-Browne GS, Ecob MS, Jaros E (1984) A developmentally-regulated disappearance of slow myosin in fast-type muscles of the mouse. FEBS Letters 177:51-56.
16. Jenkison M, Ecob M (1982) Electron microscopic observation of satellite cell activation in regenerating adult mouse muscle fibres cocultured with foetal neurones. Fifth Int Cong Neuromusc Dis, Marseilles, France. Abst No MO11.
17. Jolesz F, Sreter FA (1981) Development, innervation and activity-induced changes in skeletal muscle. Annu Rev Physiol 43:531-552.
18. Salmons S, Henriksson J (1981) The adaptive response of skeletal muscle to increased use. Muscle & Nerve 4: 94-105.
19. Thornell LE, Billeter R, Butler-Browne GS, Erikson PO, Ringqvist M, Whalen RG (1984) Development of fiber

types in human fetal muscle: an immunocytochemical study. J Neurol Sci, in press.

20. Gambke B, Lyons GE, Haselgrove J, Kelly AM, Rubinstein NA (1983) Thyroidal and neural control of myosin transitions during development of rat fast and slow muscles. FEBS Letters 156:335-339.
21. Matsuda R, Spector D, Strohman RC (1984) Denervated skeletal muscle displays discoordinate regulation for the synthesis of several myofibrillar proteins. Proc Natl Acad Sci, USA 81:1122-1125.
22. Carraro U, Dalla Libera L, Catani C (1983) Myosin light and heavy chains in muscle regenerating in absence of the nerve: transient appearance of the embryonic light chain. Exp Neurol 79:106-117.
23. Cohen MW (1972) The development of neuromuscular connections in the presence of D-tubocurarine. Brain Research 41:457-463.
24. Obata K (1977) Development of neuromuscular transmission in culture with a variety of neurons and in the presence of cholinergic substances and tetrodotoxin. Brain Research 119:141-153.
25. Pappas GD, Peterson E, Masurovsky E, Crain S (1971) The fine structure of developing neuromuscular synapses *in vitro*. Anat Rec 169:395.
26. Ziskind L, Dennis MJ (1978) Depolarizing effect of curare on embryonic rat muscles. Nature 276:622-623.
27. Roufa D, Martonosi AN (1981) Effect of curare on the development of avian chicken skeletal muscle *in ovo*. Biochem Pharmac 30:1501-1505.

Molecular Biology of Muscle Development, pages 283–292

ACCUMULATION OF MLC_{3f} IN EMBRYONIC QUAIL MUSCULATURE AND IN LIMB BUDS GRAFTED TO THE CHORIOALLANTOIC MEMBRANE: EVIDENCE FOR THE NERVE-DEPENDENT MATURATION OF MUSCLE[1]

Peter A. Merrifield[2] and Irwin R. Konigsberg

Department of Biology, University of Virginia, Charlottesville, VA 22901

ABSTRACT In the quail, as in the chick, the accumulation of fast myosin light chain 3 (MLC_{3f}) in limb musculature is developmentally regulated. Although MLC_{1f} accumulates in myotubes cultured *in vitro* and in the limb musculature of early embryos, MLC_{3f} is only detected in quail embryos older than 11 days. We have grafted aneural and innervated limb buds from quail embryos to the chorioallantoic membrane of chick hosts in order to examine the role of hormones and nervous influences on myosin light chain isoform switching. While aneural grafts accumulate only MLC_{1f}, innervated grafts accumulate both the MLC_{1f} and MLC_{3f} subunits characteristic of mature muscle. This is the first positive demonstration that the isoform switches which normally occur during the maturation of embryonic muscle *in vivo* are nerve-dependent.

INTRODUCTION

Multiple isoforms have been described for each of the contractile proteins and shown to be derived from multiple genes, from differential splicing of single genes or from a combination of these processes. In many instances the expression of these isoforms has been shown to be, not only tissue specific, but also developmentally regulated. For

[1]This work was supported by grants from NIH (HD 07083) and the Muscular Dystrophy Association to I.R. Konigsberg.
[2]Author to whom correspondence should be addressed.

example, while both myosin alkali light chains 1 and 3 (MLC_{1f}, MLC_{3f}) are transcribed from the same gene in the chick (1), only MLC_1 can be detected in myotubes cultured in vitro and in embryonic muscle prior to day 16 in ovo (2). In those muscles of the chick which will become physiologically fast in the adult (i.e. pectoralis or medial adductor), MLC_{3f} accumulation begins just prior to hatching. At approximately the same time, the synthesis of slow myosin light chains ceases (3) and changes in other contractile protein isoforms reportedly occur (4,5,6,7). These coordinated changes in isoform switching during late embryonic development suggest that many of these isoform switches may be orchestrated by some common influence in the embryo. While early studies of nerve (8) and thyroid hormone (9) dependent regulation of isoform switching in skeletal muscle have demonstrated that these influences are important in determining the expression of contractile protein isoforms in adult muscle, their role in the maturation of embryonic muscle is unclear.

We have used the technique of chorioallantoic grafting popularized by experimental embryologists during the 1920's (10), in conjunction with modern immunological techniques, to examine the role of hormonal and nervous influences in the maturation of embryonic muscle in vivo. Using a monoclonal antibody specific for MLC_{1f} and MLC_{3f}, we have characterized the accumulation of both myosin light chains in the limb musculature of developing quail embryos and in aneural and innervated limb buds from 3 day quail embryos cultured on the chorioallantoic membrane (CAM) of chick hosts. Previous studies have shown that limb buds grafted onto the CAM become vascularized by the blood vessels of the host and develop differentiated muscle, cartilage, bone and feathers (11,12). One would anticipate, therefore, that if humoral influences are sufficient to induce the accumulation of MLC_{3f} in developing muscle, then aneural grafts should develop mature muscle containing MLC_{3f}. Alternatively, if nervous influences are crucial, one would predict that only innervated limb bud grafts would accumulate MLC_{3f}. Our results indicate that the accumulation of MLC_{3f} in developing limb musculature is, indeed, nerve-dependent.

MATERIALS AND METHODS

Myoblasts from the breast muscle of 9 day quail embryos were cultured in 1-10 MEM medium for 6 days, at which point cultures were fully differentiated (13). Grafting of limb buds from 3-6 day quail embryos was carried out as described by Hamburger (14). Aneural limb buds were dissected to contain the somatopleure lateral to the somites at the level of the limb. Innervated limb buds were excised to include both somite and neural tube (see Fig. 2). Prior to analysis, cultured myotubes, complete limbs and grafts were homogenized in 10 vol (w/v) of high salt extraction buffer [0.6 M NaCl, 15 mM Tris HCL (pH 7.4), 1 mM PMSF] in a ground glass homogenizer and extracted for 15 min on ice. Insoluble material was sedimented by centrifugation (12,800g for 15 min) and the supernatant was adjusted to contain 50% glycerol and stored at -20°C. Actomyosin enriched in myosin was subsequently precipitated from these glycerinated samples by dialysis against 50 volumes of water for 6 hours at 4°C. Precipitates were collected at 8000g for 10 min, solubilized in 0.1% SDS and assayed for protein content (15). Electrophoresis of actomyosin samples on 12.5% gels and immunoblot analysis with antibody QBM-2 were carried out as previously described (16). We have previously demonstrated that QBM-2 specifically recognizes the alkali light chains of only fast muscle myosin (16).

RESULTS

Accumulation of Myosin Light Chains *In Vitro* and *In Vivo*.

Myotubes cultured *in vitro* are similar with respect to myosin light chain accumulation to myotubes which develop in the limbs of early embryos. This is apparent from immunoblots loaded with a constant amount of actomyosin (25 ug) extracted from cultured myotubes and from whole limbs (Fig. 1). Although both cultured myotubes and limbs from 9 day old embryos accumulate large amounts of MLC_{1f}, in neither muscle source do detectable levels of MLC_{3f} accumulate. MLC_{3f} is first detected in limb extracts from day 11 quail embryos and accumulates to increased levels at later stages. The developmentally regulated accumulation of MLC_{3f} in our assay cannot be attributed to changes in the incorporation of MLC_{3f} into actomyosin aggregates, since the same qualitative result is obtained if one analyzes the crude limb extracts (data not shown).

The immunoblot assay is sensitive enough to detect 15 ng of either light chain, so while we cannot rule out the synthesis of MLC_{3f} completely, we can say with assurance that,

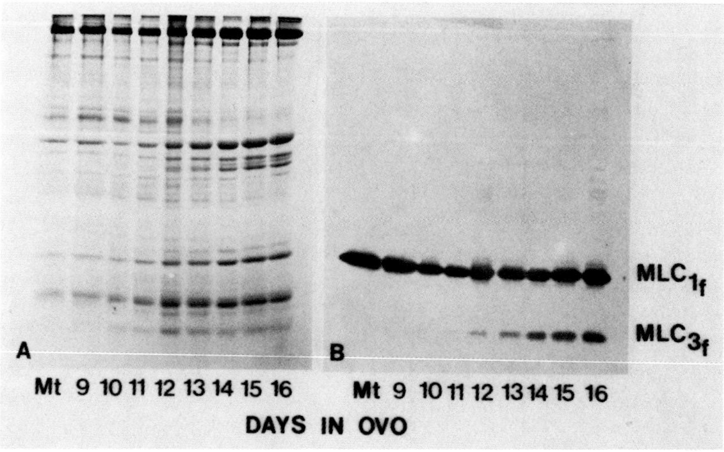

Figure 1. Myosin alkali light chain content of actomyosin samples prepared from cultured myotubes (Mt) and limb muscle of day 9 to 16 quail embryos. Actomyosin samples (25 ug) were run on 12.5% gels and stained with Coomassie blue (A) or transferred to nitrocellulose for immunoblot analysis with monoclonal antibody QBM-2 (B). Although MLC_{1f} is detected at all embryonic stages, MLC_{3f} is only detected after 11 days of development *in ovo*.

if MLC_{3f} is present in cultured and early embryonic myotubes, it is present at less than 4% of the level normally found in adult muscle. The absence of MLC_{3f} has also been described for chick myotubes cultured *in vitro* (17) and in the most exhaustive analysis of avian light chain synthesis done to date, Keller and Emerson (18) reported that although MLC_{3f} synthesis can be detected in cultured myotubes, this level is so low as to be physiologically insignificant.

Characteristics of Aneural and Innervated Limb Buds Grafted onto the CAM of Chick Hosts.

In order to examine the developmental potential of limb musculature grown in the total absence of nerve, limb buds were initially obtained from 3 day quail embryos. The best evidence suggests that nerves do not penetrate into the limb bud of the quail until at least day 4 *in ovo* (19). The grafts obtained from 3 day embryonic quail hind limb, grown on the

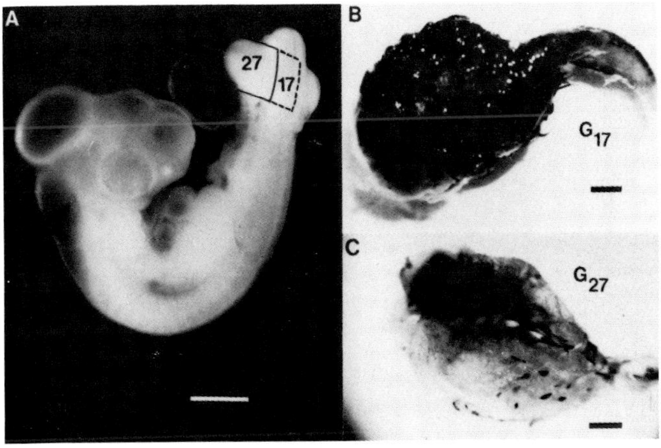

Figure 2. Aneural and innervated limb buds of a 3 day quail embryo before (A) and after (B and C) incubation on the chorioallantoic membrane. The area contained by the solid line in panel A represents a typical dissection of a limb bud which excluded somites and neural tube (i.e. an aneural graft, see G_{27}) while the broken line denotes all of the tissue which was typically included in an innervated graft such as G_{17}. Two representative grafts retrieved after 12 days of growth on the CAM are shown in panels B (G_{17}) and C(G_{27}). G_{17} was subsequently shown to contain both MLC_{1f} and MLC_{3f} while G_{27} contained only MLC_{1f}. The bar represents 1 mm in each respective field.

chick CAM for 12 days, ranged in weight from 22 to 266 mg and from feathered tissue masses of approximately 2 mm in diameter, to normal but stunted limbs of 10 mm in length with normal scale, feather and digit formation. Although grafts which included neural tube seemed to survive better (Table 1), their average size and morphological development were very similar to aneural grafts (Fig. 2).

The most striking difference between aneural and innervated grafts was in their myosin light chain content. When a constant amount of actomyosin from each graft was analysed with QBM-2 following immunoblotting, MLC_1 was detected in all grafts. MLC_3 however, was never detected in aneural grafts (Table 1). In this respect, actomyosin preparations from aneural grafts had the same isoform content as those prepared from cultured myotubes or from the limbs of normal

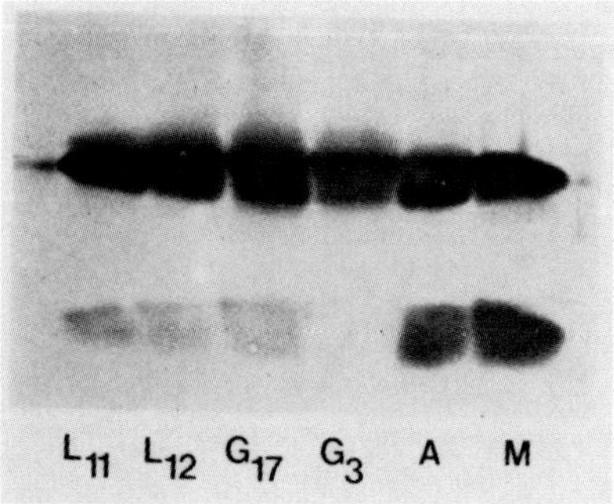

Figure 3. Immunoblot analysis of actomyosin prepared from normal day 11 (L_{11}) and day 12 (L_{12}) quail limb buds, aneural (G_3) and innervated (G_{17}) grafts and standard actomyosin (adult) and myosin preparations of adult muscle using antibody QBM-2. 50 ug of each actomyosin and 10 ug of myosin were loaded. Only the light chain portion of the blot is shown.

quail embryos sampled prior to 11 days in ovo. In contrast, all of the innervated grafts contained detectable levels of both MLC_{1f} and MLC_{3f} when analysed under the same conditions. This is most apparent when similar amounts of actomyosin from an innervated (G_{17}) and an aneural (G_3) graft are analyzed on the same immunoblot along with actomyosin from normal limbs (Fig. 3). While MLC_3 cannot be detected in the aneural graft extract, the amount of MLC_3 in the innervated graft is comparable to that seen in normal limbs from day 12 embryos.

Role of Neural Tube and Somites in Isoform Switching.

These results are consistent with the idea that the inclusion of neural tube (i.e. nervous influence) is responsible for the maturation of limb musculature from an embryonic to an adult, fast muscle phenotype. The only alternative explanation is that the inclusion of somite material in these innervated grafts, necessitated by the dissection, somehow affects the MLC_{3f} content of innervated grafts. The inclusion of somite material might affect our results if, in three day old embryos, muscle precursor cells are still migrating from the somite to populate the

TABLE 1
SUMMARY OF CAM GRAFT EXPERIMENTS

Age of Quail Donor/Chick Host (Days)	Presence of Neural Tube (+/-)	Age (Days) At Sacrifice (Host/Graft)	No. Grafts Survived/ Attempted	Presence* of MLC_1/ MLC_3
3/7	-	12/15	12/45	8/0
3/7	+	12/15	13/41	8/8
6/8	-	10/16	7/43	4/0

*Only successful grafts >50 MG were processed for MLC analysis.

limb, as has been described in younger embryos (20). To test this possibility, we obtained limb buds from 6 day embryos and allowed these to develop as CAM grafts as described above. In every case, these grafts had the same characteristics as aneural grafts obtained from younger birds (Table 1). They contained muscle (as indicated by the presence of MLC_1) but no MLC_3. Since all cell migration from the somite is completed by day 6, this strongly suggests that the presence of neural tube, not somite, is the factor responsible for MLC_3 activation. In addition, these experiments indicate that the nerve muscle interactions which are responsible for the activation of MLC_{3f} accumulation occur after 6 days of development in ovo.

DISCUSSION

These studies indicate that the accumulation of myosin fast alkali light chains can be easily assayed using monoclonal antibody QBM-2 and that the presence of MLC_{3f} in tissue extracts is a good indication of muscle maturation. MLC_{3f} cannot be detected in embryonic myotubes cultured in vitro or in the muscle of early embryos, but can be detected first in the muscle of day 11 embryos. This is approximately the same developmental stage at which neonatal myosin heavy chain is first detected in the muscle of the developing chick (4). Quantitative changes in α actin (5), tropomyosin (6), and troponin (7) isoforms have also been reported at this stage of development in the chick.

Chorioallantoic grafting of limb buds has long been used to study the capacity of limb bud cells to differentiate normally in a neutral environment (10). It has previously been shown that limb buds incubated on the CAM will form differentiated muscle similar to the muscle in normal limbs of late stage embryos (11,12). We wanted to determine, however, whether this muscle is arrested at an embryonic stage (like cells cultured *in vitro*) or matures and expresses adult patterns of contractile protein isoforms. Our observation that aneural limb buds cultured on the chorioallantoic membrane fail to express MLC_{3f} suggests that, if hormonal influences play a role in the accumulation of MLC_{3f}, this role is entirely permissive and not sufficient, in itself, to induce MLC_{3f} synthesis. This observation is in contrast to that previously reported in the developing rat, where the level of thyroid hormone has been shown to regulate the switch from embryonic to neonatal forms of myosin heavy chain (21,22). These differences suggest that either the switching of myosin heavy and light chains are regulated by different mechanisms, or that the mechanisms responsible for the maturation of muscle in birds and mammals are not the same.

While the aneural CAM graft experiments use negative evidence - the absence of MLC_{3f} accumulation - to argue for the importance of innervation, such arguments are not compelling. However, our demonstration that the inclusion of neural tube with the limb bud results in the accumulation of MLC_{3f} in such grafts is the first positive demonstration of the importance of nerve in isoform switching *in vivo*. This observation cannot be attributed to differences in the amount of muscle between aneural and innervated grafts, since identical amounts of actomyosin were routinely analysed in each lane of our immunoblots. We have also shown that the accumulation of MLC_{3f} in innervated grafts cannot be attributed to the differentiation of a distinct myoblast lineage which migrates from the somite after day 3 of development, since aneural grafts from older embryos are also deficient in MLC_{3f}. Evidence from other laboratories suggests that isoform switching of contractile proteins other than myosin light chains may also be nerve-dependent. For example, it has recently been shown that cultured myotubes will express adult-specific patterns of troponin (23) and myosin heavy chain (24) accumulation if co-cultured with neurons. In fact, Toyota and Shimada have shown that the selective repression of cardiac isoforms of troponin T and C can be accomplished in cultures of skeletal myotubes by the simple addition of nerve extract to the culture medium (23). We are currently analysing the myosin light chain content of quail myotubes cultured in the presence of motoneurons or nerve extract in order to establish whether contractile activity, trophic factors from nerve or a combination of these factors are responsible for the nerve induced accumulation of myosin light chain 3 in embryonic muscle.

REFERENCES

1. Nabeshima Y, Fujii-Kuriyama Y, Muramatsu M and Ogata K (1984). Alternative transcription and two modes of splicing result in two myosin light chains from one gene. Nature 308: 333.
2. Roy RK, Sreter FA and Sarkar S (1979). Changes in tropomyosin subunits and myosin light chains during development of chicken and rabbit striated muscles. Devel Biol 69: 15.
3. Crow MT, Olson PS and Stockdale FE (1983). Myosin light chain expression during avian muscle development. J Cell Biol 96: 736.
4. Bader D, Masaki T and Fischman D (1982). Immunochemical analysis of myosin heavy chain during avian myogenesis in vivo and in vitro. J Cell Biol 95: 763.
5. Paterson BM and Eldridge JD (1984). α Cardiac actin is the major sarcomeric isoform expressed in embryonic avian skeletal muscle. Science 224: 1436.
6. Montarras D, Fiszman MY and Gros F (1982). Changes in tropomyosin during development of chick embryonic skeletal muscles in vivo and during differentiation of chick muscle cells in vitro. J Biol Chem 257: 545.
7. Matsuda R, Obinata T and Shimada Y (1981). Types of troponin components during development of chicken skeletal muscle. Devel Biol 82: 11.
8. Sreter FA, Gergely J, Salmons S and Romanul F (1973). Synthesis of fast muscle of myosin light chains characteristic of slow muscle in response to long term stimulations. Nature New Biol 241: 17.
9. Johnson MA, Mastaglia FL, Montgomery AG, Pope B and Weeds AG (1980). Changes in myosin light chains in the rat soleus after thyroidectomy. FEBS Letters 110: 230.
10. Murray PDF and Huxley JS (1925). Self differentiation in the grafted limb bud of the chick. J Anat 59: 379.
11. Hunt E (1932). The differentiation of chick limb buds in chorioallantoic grafts, with special reference to the

muscles. J Exp Zool 62, 57.
12. Eastlick HL (1943). The development of muscle in nerveless and innervated grafts. J Exp Zool 93: 27.
13. Konigsberg IR (1979). Skeletal myoblasts in culture. In "Methods in Embryology" (W.B. Jakoby and IR Pastan, eds. Vol 58: p 511. Academic Press, NY
14. Hamburger V (1942). Chorioallantoic Grafts. In " A Manual of Experimental Embryology". (by Viktor Hamburger) p 143. University of Chicago Press.
15. Lowry OH, Rosebrough NJ, Farr AL and Randall RJ (1951). Protein measurement with the Folin phenol reagent. J Biol Chem 193: 265.
16. Merrifield PA, Payne MR and Konigsberg IR (1983). Isoform specificity of monoclonal hybridoma antibodies to quail skeletal muscle myosin subunits. Biochem Biophys Res Commun 113: 407.
17. Bandman E, Matsuda R and Strohman RC (1982). Developmental appearance of myosin heavy and light chain isoforms in vivo and in vitro in chicken skeletal muscle. Devel Biol 93: 508.
18. Keller LR and Emerson CP Jr (1980). Synthesis of adult myosin light chains by embryonic muscle cultures. Proc Nat Acad Sci USA 77: 1020.
19. Fouvet B (1973). Innervation et morphogenese de la patte chez l'embryon de poulet. Arch Anat Microscop 62: 281.
20. Christ B, Jacob HJ and Jacob M (1977). Experimental analysis of the origin of the wing musculature in avian embryos. Anat Embryol 150: 171.
21. Gambke B, Lyons GE, Haselgrove J, Kelly AM and Rubenstein NA (1983). Thyroidal and neural control of myosin transitions during development of rat fast and slow muscles. FEBS Letters 156: 335.
22. Gambke B and Rubenstein N (1984). A monoclonal antibody to the embryonic myosin heavy chain of rat skeletal muscle. J Biol Chem 259: 12092.
23. Toyota N and Shimada Y (1983). Isoform variants of troponin in skeletal and cardiac muscle cells cultured with and without nerves. Cell 33: 297.
24. Ecob MS, Butler-Browne G and Whalen RG (1983). The adult fast isozyme of myosin is present in a nerve muscle tissue culture system. Differentiation 25: 84.

Molecular Biology of Muscle Development, pages 293–307

EXPRESSION OF C-PROTEIN ISOFORMS DURING CHICKEN STRIATED MUSCLE DEVELOPMENT AND ITS DEPENDENCE ON INNERVATION[1]

Takashi Obinata*, Maho Kawashima*, S. Kitani*, O. Saitoh*, T. Masaki+ D. M. Bader** and D. A. Fischman**

*Department of Biology, Chiba University, Yayoi-cho Chiba 260, Japan, +Institute of Basic Medical Sciences, University of Tsukuba, Ibaraki 305, Japan, and **Department of Cell Biology and Anatomy Cornell University Medical College, N.Y. 10021.

ABSTRACT The sequential changes in C-protein isoform expression during chicken skeletal muscle development in vivo and in vitro were examined using monoclonal and polyclonal antibodies to cardiac, slow skeletal and fast skeletal C-proteins. During embryonic development of breast (pectoralis) muscle, a C-protein recognized by anti-cardiac C-protein appears first and transiently, and subsequently, two other isoforms are expressed, recognized by the antibodies to slow or fast skeletal muscle C-proteins. Three C-protein variants also appears sequentially in myotubes in culture and continue to be expressed. During postnatal development of the pectoralis muuscle, synthesis of slow C-protein is down regulated, and we suggest that this process is selectively regulated by the nervous system since the denervation of the pectoralis muscle causes the expression of slow C-protein but not the expression of cardiac C-protein.

[1]This work was supported in part by grants from the Ministry of Education, Science and Culture, Japan, the NCNMMD of the Ministry of Health and Welfare of Japan, Muscular Dystrophy Association of America (032771) and the NIH (AM31305 and AM32147).

INTRODUCTION

C-protein is a regulatory protein located in A-band region of myofibrils in association with myosin filaments (1-3). Since it has been shown that C-protein modifies myosin polymerization in vitro (4, 5), expression of C-protein isoforms during myogenesis is of interest in relation to thick filament assembly during myofibrillogenesis.

Recently, it has been shown that different isoforms of C-protein exist in striated muscles of various fiber type composition. In the rabbit, C-protein from soleus muscle differs in molecular weight and amino acid composition from that in the psoas (6) and the slow and fast muscle C-proteins can be distinguished with polyclonal antibodies to the respective proteins (6). Rabbit cardiac C-protein is also different from that in both red and white muscles (7). C-protein isoforms in chicken fast (pectoralis) and slow (ALD) skeletal muscles were distinguished by monoclonal antibodies (McAbs) specific for the respective fast (MF-1 or MF-21) and slow (ALD-66) protein variants (8, 9). Using these McAbs, the distribution of two C-proteins among chicken striated muscles has been examined (8, 9) and we have observed sequential changes in the C-protein isoform expression during chicken skeletal muscle development (10, 11). In this investigation, we report the preparation of a monoclonal antibody specific to cardiac type C-protein and, in combination with the other anti-C-protein antibodies, have studied the expression of C-protein isoforms in developing muscle cells both in vivo and in vitro.

METHODS

Antibodies. The monoclonal antibodies (McAbs) against fast (MF-1) and slow (ALD-66) skeletal muscle C-protein have been described (8). The McAb to cardiac C-protein (C-315) was prepared as described (8) using crude chicken cardiac C-protein, separated from myosin by DEAE-Sephadex column chromatography (2), as an immunogen. The specificity of C-315 was established by immunocytochemistry, immunobloting and radioimmunoassay as previously shown for MF-1 and ALD-66 (8): C-315 stained cryosections of only cardiac muscle in the central and narrow region of A-band of cardiac myofibrils and reacted with a 140 Kd peptide in the whole protein extract from cardiac muscle. This McAb did not exhibit any reaction with the cryosections, myofibrils or protein extracts of either slow or fast skeletal muscles. Polyclonal antibody to fast-skeletal C-protein (anti-FC) was prepared by immunizing rabbits with C-protein purified from pectoralis muscle by DEAE-Sephadex, hyroxylapatite (12), and finally, by preparative SDS-polyacrylamide gel electrophoresis. The specific immunoglobulins were isolated by affinity chromatography using Sepharose 4B coupled to fast-skeletal C-protein. Anti-FC was highly specific to fast-skeletal C-protein.

Muscle denervation. The pectoralis major muscle of White Leghorn chicken at various ages were denervated by cutting the brachial plexus on one side (13). The contralateral side was used as a control. To avoid reinnervation, the proximal cut end was sewn to skin at the back of the chicken. After denervation, flapping movements of the wing on the side of denervation were not observed. To avoid regional differences in fiber types (14, 15), the central and superficial regions of the muscle were sampled at specific times after operation and used for analyses of the muscle protein isoforms.

Muscle cell cultures. Myogenic cells were obtained from 11-day old chicken embryo breast muscle or 18-old embryo ALD muscle and cultured by standard procedures (16). For immunofluorescent staining, the cells were fixed with 4 % paraformaldehyde in PBS and treated with 0.2 % Triton X-100.

The other procedures were described elsewhere (8, 10).

RESULTS AND DISCUSSIONS

1. C-protein Isoforms Expressed in Developing Chicken Striated Muscles.

To identify the types of C-protein present in developing chicken striated muscles, indirect immunofuorescent staining of serial frozen sections was performed at different developmental stages.

In frozen sections of 10-day old embryonic breast muscle, the myofibers were stained brightly with anti-cardiac C-protein (C-315), but did not react with either ALD-66 or anti-FC (Fig. 1). At 15 days of incubation all myofibers reacted strongly with ALD-66 and faintly with anti-FC, but no staining with C-315 was

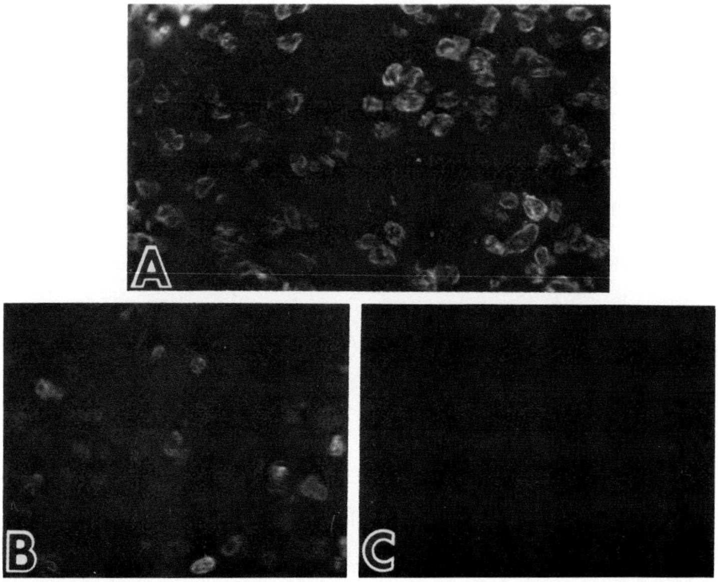

FIGURE 1. Immunofluorescent staining of the cryosections of 12-day old embryo breast muscle with C-315 (A), ALD-66 (B), and anti-FC (C). Magnification, x200.

observed. From 17 days in ovo to a few days after hatching, all myofibers of the breast muscle reacted positively with both ALD-66 and anti-FC (or MF-1). Occasionaly variations of staining intensity was observed between different myofibers and thereby producing a "checker-board"-like pattern of fluorescence (11). From the time of hatching until 10 days afterwards, the binding of ALD-66 to breast muscle decreases progressively; by two weeks after hatching the fibers of the breast muscle are completely negative with ALD-66 (Fig. 2). All myofibers continued to react strongly with anti-FC (and MF-1) thru adulthood. These changes in reactivity of the developing muscle with the three anti-C-protein antibodies were confirmed with parallel immunoblots (11) and by radioimmunoassay. The peptides recognized by the three antibodies differed slightly in size both in embryonic and

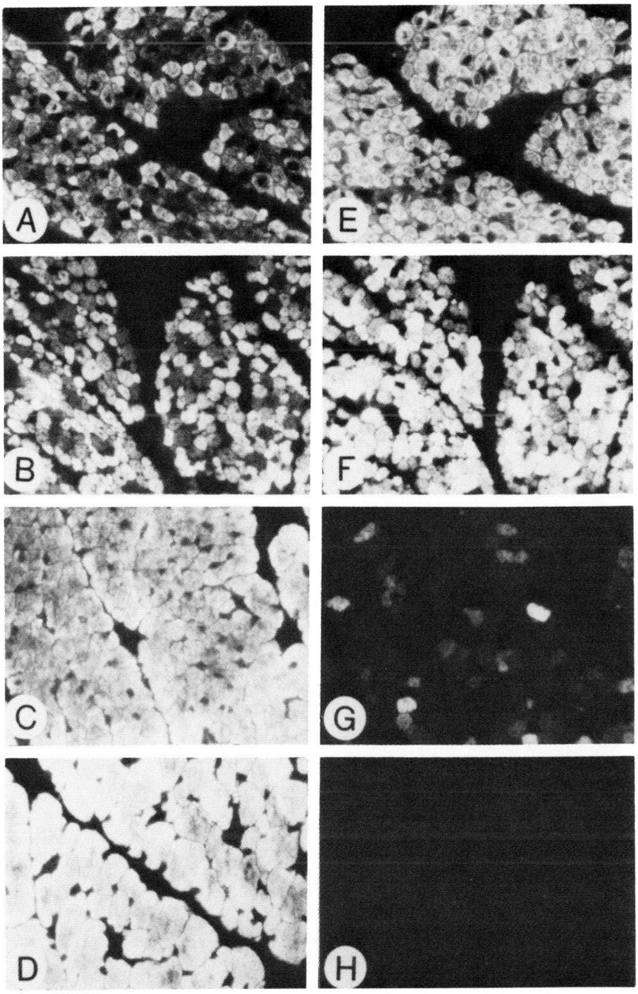

FIGURE. 2. The reactivity of developing chicken pectoralis muscle to McAbs against fast C-protein (MF-1) and anti-slow C-protein (ALD-66). Frozen serial sections of the pectoralis muscle from 17-day-old embryo (A and E), 1-d posthatched (B and F), 6-d posthatched (C and G) and adult (D and H) chicken were reacted with MF-1 (A-D) or ALD-66 (E-H). Photograghs of the sections were taken with indirect immunofluorescence. Magnification, x 160.

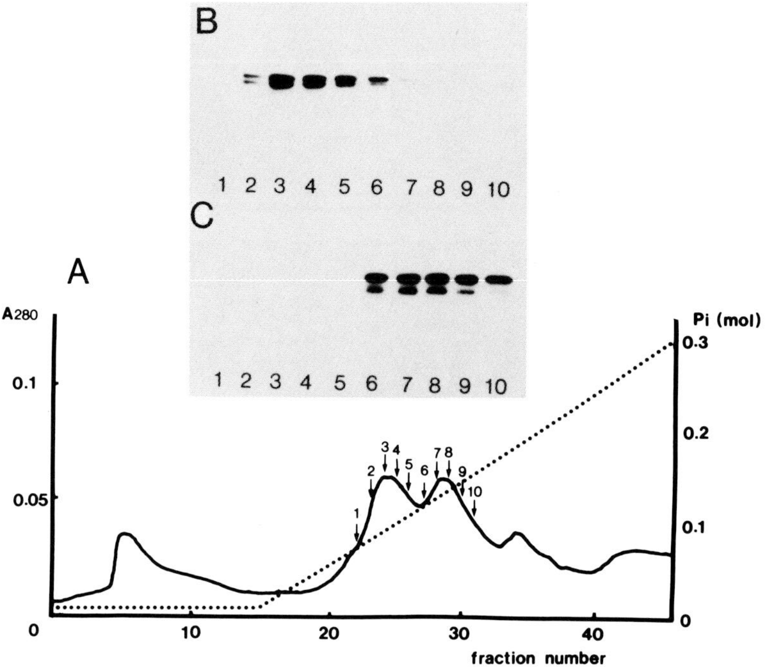

FIGURE 3. Isolation of C-protein from 2-d post-hatched chicken pectoralis muscle. (A), Hydroxylapatite chromatography of crude C-protein obtained by DEAE-Sephadex fractionation of myosin. Solid line, absorbance at 280 nm; dotted line , phosphate concentration. (B) & (C), Immunoblot analysis. The proteins in the fractions 1 to 10 in (A) were displayed by SDS-PAGE, transfered to nitrocellulose paper and reacted with MF-1 (B) and ALD-66 (C) followed by the treatment with ^{125}I-labeled anti-mouse IgG. (From ref. 11 with permission).

adult tissues.

We have obtained evidence that the C-protein isoforms recognized by ALD-66 and by MF-1 in the neonatal breast muscle are probably the same isoforms as the slow-type and fast-type C-proteins of adult muscle. As shown in Fig. 3, two types of C-protein isoforms can be separated from neonatal chicken muscle by hydroxylapatite column chromatography. The major part of the first peak reacted with MF-1, but failed to react with ALD-66, whereas the latter half of the second peak reacted only with ALD-66 (Fig. 3-B, C). The middle fractions between the first and the second peaks reacted with both McAbs. The molecular weights of the major polypeptide chains recognized by each McAb was roughly the same as that of C-protein, as judged by SDS-PAGE. All of myofibrils isolated from the neonatal chicken muscle stained positively stained with both McAbs using indirect immofluorescence. In contrast, myofibrils from the adult chicken breast muscle reacted only with MF-1 as shown previously (8). Staining of neonatal myofibrils with either McAb was restricted to a specific region of A-band in a pattern characteristic of C-protein (1-3). We suggest that in the neonatal muscle both fast-type and slow-type C-proteins are intermingled and assembled in the same myofibrils (11). It was previously demonstrated that two C-protein isoforms, a fast-type and a slow-type, are contained in all myofibers of the adult PLD muscle and all myofibrils (9). Therefore, the C-protein isoforms present in the neonatal pectoralis muscle and their distribution within myofibrils are very similar to those in the adult PLD muscle.

The earliest form of C-protein which is expressed in breast muscle may either be a cardiac-type isoform or a unique embryonic protein containing epitopes in common with the adult cardiac-type protein. If we assume that the young embryonic breast muscle expresses an adult cardiac-type C-protein, then this would be another example of a contractile protein encoded by a cardiac gene which is expressed at an early stage of skeletal muscle differentiation (17-22). Tentatively, we have made this assumption and suggest that during breast muscle development, these occur two transition steps in C-protein expression: first, a single cardiac-type isoform is replaced by two skeletal muscle isoforms (one a slow-type, the other a fast-type), then a second transition in which the slow-type disappers leaving continued expression of only the fast-type C-protein (10, 11, 23). The first transition occurs in late embryonic development and the second during 1 - 2nd weeks posthatch.

C-protein isoform transitions were also examined during the development of a slow muscle (ALD). The epitopes recognized by C-315 first appeared in this muscle

TABLE 1
IMMUNOFLUORESCENT STAINING OF DEVELOPING CHICKEN STRIATED MUSCLES WITH THREE DIFFERENT ANTI-C-PROTEIN ANTIBODIES.

Muscle	antibody	12dE	17dE	1d	14d
	C-315	+	-	-	-
Breast	ALD-66	+	+	+	-
	anti-FC	-	+	+	+
	C-315	+	-	-	-
ALD	ALD-66	+	+	+	+
	anti-FC	-	-	-	-
	C-315	+	+	+	+
Cardiac	ALD-66	-	-	-	-
	anti-FC	-	-	-	-

Note. Column heads 12dE, 17dE represent days in ovo, remaining column heads are stages after hatching. Antibody binding data to chicken striated muscles were obtained with McAbs specific for slow-type (ALD-66) and cardiac-type C-protein (C-315) and polyclonal antibody.

at the same embryonic stages as in breast muscle. We observed only one transition from a cardiac isoform (or a unique embryonic isoform with shared epitopes common to the cardiac-type protein) to a slow muscle isoform in this muscle (Table 1). In contrast to skeletal muscle, in cardiac muscle only a cardiac-type C-protein was expressed throughout its development.

It is of interest that the patterns of C-protein isoform expression during pectoralis muscle development seemed to mimic the sequential expression of those C-protein isoforms seen in cardiac, ALD and PLD muscles.

2. C-protein Isoform Expression in Myogenic Cells in Culture.

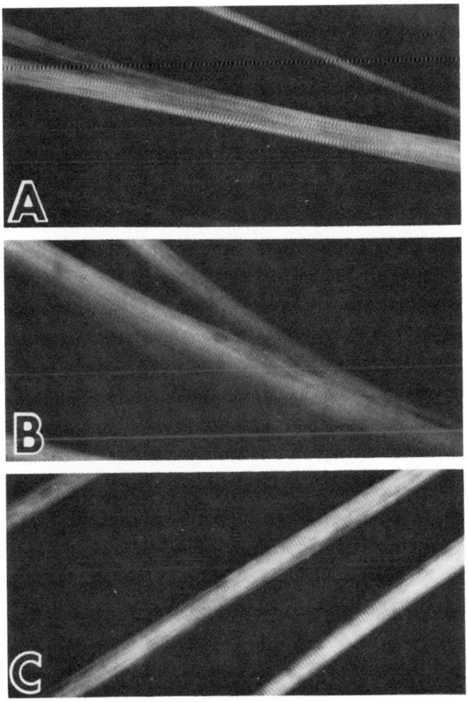

FIGURE 4. Immunofluorescent staining of breast muscle myotubes cultured for 1 weeks in vitro with C-315 (A), ALD-66 (B),and anti-FC (C). Magnification, x200.

Immunochemical studies have been performed on the expression of C-protein in monolayer cultures of breast, ALD and cardiac muscles. Mononucleated myoblasts of skeletal muscles were completely unreactive with any of anti-C-protein antibodies against cardiac-type (C-315), slow-type (ALD-66) or fast-type (anti-FC) C-proteins. When multinucleated myotubes were formed, typically in the 3rd day of culture with breast muscle, C-315 reactivity appeared first even in myotubes which were unstained with either ALD-66 or anti-FC. When muscle cultures were maintained for about one week, most of myotubes which were cross-striated stained positively with antibodies of all three C-protein isoforms. Each antibody stained A-bands of myofibrils in a typical C-protein pattern (Fig. 4). The C-protein isoforms recognized by anti-FC and by C-315 or by ALD-66 were co-localized in a single myotube by performing dual treatment with rhodamine-labelled anti-rabbit IgG for polyclonal antibody (anti-FC) and FITC-anti-mouse IgG for McAbs (C-315 and ALD-66) as the second antibodies. The C-protein recognized by C-315 appeared earlier than the proteins recognized by the other anti-C-protein antibodies in ALD muscle cultures as well. Thus, as in vivo, the first C-protein detectable in muscle cultures of 11-day-old pectoralis muscle or 18-day-old ALD muscle appeared to be the cardiac isoform, followed by the expression of skeletal muscle C-proteins. However, in the cell cultures, the reactivity of C-315 with myotubes remained unchanged even in long-term cultures maintained for up to two weeks, suggesting that the transitions of C-protein isoforms specific to each fiber-type does not occur in culture.

3. Effect of Denervation on C-protein Isoform Expression *in vivo*.

It has been reported that the expression of several contractile proteins in mammalian skeletal muscles is controled by neural factors, for example the isoforms of myosin and tropomyosin in mammalian skeletal muscle are altered by chronic nerve stimulation (24-26), cross-reinnervation (27), or denervation of the mature muscles (28). Thyroid hormone is as another factor which affects myosin isoform expression in mammalian skeletal muscle (29). Little informations is now available on the regulation of the isoform transitions which occur during avian muscle development. Recently, neural effects on tropomyosin and troponin T isoform expression in chicken breast muscle have been reported (13, 30). Here we demonstrate that C-protein isoform expression during postnatal development of chicken breast muscle is markedly affected by the denervation. Nerves to the pectoralis muscle of neonatal or mature chicken were transected on one side and C-protein expression was analyzed at selected times after surgery using type-specific anti-C-protein antibodies.

As already described, neonatal chicken breast muscle contains both fast-type and slow-type C-proteins (Fig. 2-B, F & Fig. 5-A) but as the muscle develops, the slow-type C-protein disappears (Fig. 2-G & Fig. 5-B). When the

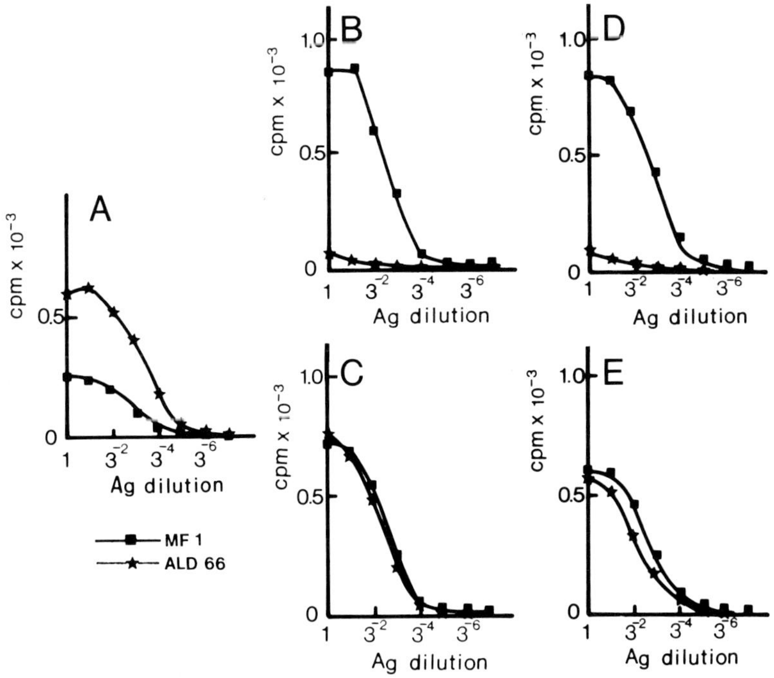

FIGURE 5. Solid-phase indirect RIA analysis of C-protein with MF-1 and ALD-66. Pectoralis muscle was denervated 2-days (B & C) or 2-months (D & E) after hatching and crude myosin was extracted 3 weeks after the surgery. A fixed amount of McAbs was reacted with the protein serially diluted from 10 µg. A, 2-day old chicken without denervation. B & C, contralateral muscles in each experiment; D & E, denervated muscles.

pectoralis muscle was denervated at the neonatal stage, however, this transition of C-protein expression prevented (Fig. 5-D). For example, slow-type C-protein persisted in all myofibers of the denervated muscle (Fig. 6-D), even after 8 weeks post-hatching, whereas slow-type C-protein had been lost completely from the contralateral muscle by 2 weeks post-hatching (Fig. 5-B & Fig. 6-B). When 2-month old chicken breast muscle, which contains no slow-type

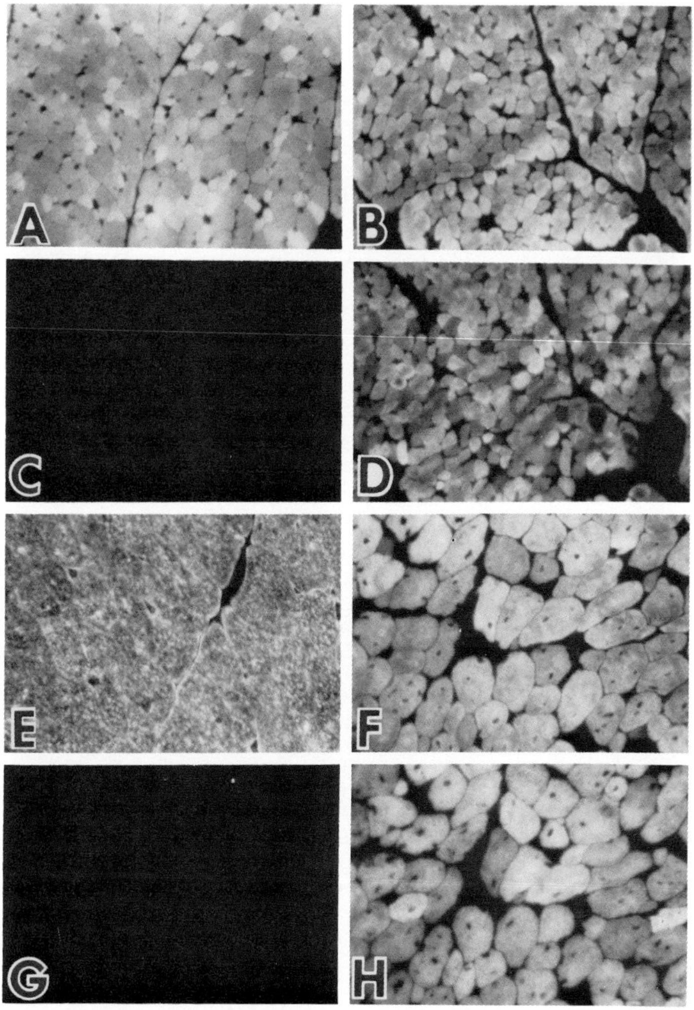

FIGURE 6. Chicken pectoralis muscle was denervated on one side 2-days (A-D) or 2-months (E-H) after hatching and serial cryosections were prepared 3-weeks after the surgery. The sections were reacted with McAbs to fast C-protein, MF-1 (A, B, E, and F), or to slow C-protein, ALD-66 (C, D, G, and H). Photographs of sections were taken with indirect immunofluorescence. A, C, E, & G, contralateral control muscles; B, D, F, & H, denervated muscles. Magnification: x240.

C-protein, was denervated, a reappearance of the slow-type isoform was observed within 2 weeks after the operation. Furthermore, the relative proportion of slow-type C-protein in the breast muscle increased with age after the operation (Fig. 6-E). Slow-type C-protein appeared in a majority of the myofibers of large diameter in the denervated muscle. C-315 scarcely reacted with the denevated muscle, even though this McAb clearly stained regenerating muscle cells after focal cold injury of the muscle. These results indicate that the reappearance of slow-type C-protein in the denervated muscle is not due to muscle regeneration, but to an alteration of preexisting myofibers induced to reexpress slow-type C-protein.

We have also examined the changes in other muscle proteins after breast muscle denervation. Similar changes in tropomyosin and troponin T, namely the reexpression of β-tropomyosin and leg-type troponin T, after denervation (13). However, myosin isoform expression was not markedly affected. Three McAbs (MF-14, MF-30 and MF-20) to myosin heavy chains have been shown to exhibit differential reactivity with the myosin isoforms in the developing breast muscle (31). Our results indicate that changes in myosin expression did not occur in parallel with the C-protein isoform transitions. Thus, we suggest that myosin and C-protein, both of which are the constituents of the thick filament, are regulated independently by the nervous system.

REFERENCES

1. Offer G (1972). C-protein and periodicity in the thick filaments of vertebrate skeletal muscle. Cold Spring Harbor Symp Quant Biol 37: 87.
2. Offer G, Moos C, Starr R (1973). A new protein of the thick filaments of vertebrate skeletal myofibrils. J Mol Biol 74: 653.
3. Pepe FA, Drucker B (1975). The myosin filament. III. C-protein. J Mol Biol 99: 609.
4. Koretz JF (1979) Structural studies of synthetic filaments prepared from column-purified myosin. Biophys J 27: 423.
5. Koretz JF (1979). Effects of C-protein on synthetic myosin filament structure. Biophys J 27: 433.
6. Callaway JE, Bechtel PJ (1981). C-protein from rabbit soleus (red) muscle. Biochem J 195: 463.

7. Yamamoto K, Moos C (1983). The C-protein of rabbit red, white, and cardiac muscles. J Biol Chem 258: 8359.
8. Reinach FC, Masaki T, Shafiq S, Obinata T, Fischman DA (1982). Isoforms of C-protein in adult chicken skeletal muscle: Detection with monoclonal antibodies. J Cell Biol 95: 78.
9. Reinach FC, Masaki T, Fischman DA (1983). Characterization of the C-protein from posterior latissimus dosi muscle of the adult chicken: Heterogeneity within a single myofibril. J Cell Biol 96: 297.
10. Obinata T, Reinach FC, Bader FC, Masaki T, Kitani S, Fischman DA (1984). Immunochemical analysis of C-protein isoform transitions during the development of chicken skeletal muscle. Develop Biol 101: 116.
11. Obinata T, Kitani S, Masaki T, Fischman DA (1984). Coexistence of fast-type and slow-type C-proteins in neonatal chicken breast muscle. Develop Biol 105: 253.
12. Starr R, Offer G (1982). Preparation of C-protein, H-protein, X-protein, and phosphofructokinase. Methods in Enzymol 85: 130.
13. Obinata T, Saitoh O, Takano-Ohmuro H (1984) Effect of denervation of tropomyosin, troponin T, and myosin isozymes in chicken breast muscle. J Biochem 95: 585.
14. Gauthier GF, Lowey S (1977). Polymorphism of myosin among skeletal types. J Cell Biol 74: 760.
15. Matsuda R, Bandman E, Strohman RC (1983) Regional differences in the expression of myosin light chains and tropomyosin subunits during development of chicken breast muscle. Develop Biol 95: 484.
16. Shimada Y, Fischman DA, Moscona AA (1967). The fine strcture of embryonic chick skeletal muscle cells differentiated in vitro. J Cell Biol 35: 445.
17. Masaki T, Yoshizaki C (1974) Differentiation of myosin in chick embryos. J Biochem 76: 123.
18. Toyota N, Shimada Y (1983) Isoforms variants of troponin in skeletal and cardiac muscle cells cultured with and without nerve. Cell 33: 297.
19. Sweeney LJ, Clark WA, Umeda PK, Zak R, Manasek FJ (1984). Immunofluorescence analysis of the primordial myosin detectable in embryonic striated muscle. Proc Natl Acad Sci USA 81: 797.
20. Minty AJ, Alonso S, Caravatti M, Buckingham ME (1982) A fetal skeletal muscle actin mRNA in the mouse and its identity with cardiac actin mRNA. Cell 30: 185.
21. Cooper TA, Ordahl CP (1984) A single troponin T gene

reglated by different program in cardiac and skeletal muscle development. Science 226: 979.
22. Paterson BM, Eldridge JD (1984) α-cardiac actin is the major sarcomeric isoform expressed in embryonic avian skeletal muscle. Science 224: 1436.
23. Obinata T, Kawashima M, Saitoh O, Bader DM, Fischman DA (1984) C-protein isofrom expression is regulated by muscle innervation. J Cell Biol 99: 347a.
24. Sreter FA, Gergely J, Salmons S, Romanul F (1973) Synthesis by fast muscle of myosin light chains characteristic of slow muscle in response to long-term stimulation. Nature New Biol 241: 17.
25. Rubinstein N, Mabuchi K, Pepe F, Salmons S, Gergely J, Sreter F (1978) Use of type-specific anti-myosins to demonstrate the transformation of individual fibers in chronically stimulted rabbit fast muscles. J Cell Biol 79: 252.
26. Roy RK, Mabuchi K, Sarkar S, Mis C, Sreter FA (1979) Changes in tropomyosin subunit pattern in chronic electrically stimulated rabbit fast muscles. Biochem Biophys Res Commun 89: 181.
27. Weeds AG, Trentham DR, Kean CJ, Buller AJ (1974) Myosin from cross-reinnervated cat muscles. Nature 247: 135.
28. Carraro U, Catani C, Libera LS, Vascon M, Zanella G (1981) Differential distribution of tropomyosin subunits in fast and slow rat muscle and its changes in long-term denervated hemidiaphragm. FEBS Lett 128: 233.
29. Gambke B, Rubinstein, NA (1984) A monoclonal antibody to the embryonic myosin heavy chain of rat skeletal muscle. J Biol Chem 259: 12092.
30. Matsuda R, Spector D, Strohman RC (1984) Denervated skeletal muscle displays discoordinate regulation for the synthesis of several myofibrillar proteins. Proc Natl Acad Sci USA 81: 1122.
31. Bader D, Masaki T, Fischman DA (1982). Immunochemical analysis of myosin heavy chain during avian myogenesis in vivo and in vitro. J Cell Biol 95: 763.

Molecular Biology of Muscle Development, pages 309–322

EXPRESSION OF MYOSIN AND C-PROTEIN ISOFORMS IN REGENERATING MUSCLE

A. D. Saad[1], T. Obinata[2] and D. A. Fischman[1]

[1]Department of Cell Biology and Anatomy, Cornell University Medical College New York, NY 10021 and [2]Department of Biology, Chiba University, Yayoi-cho Chiba 260, Japan

ABSTRACT The expression of MHC[3] and C-protein isoforms has been examined immunocytochemically in regenerating skeletal muscles of adult chickens. Two, five and eight days after focal freeze injury to the ALD, PLD and pectoralis muscles, cryostat sections of injured and control tissues were reacted with a series of McAbs previously shown to specifically bind MHC or C-protein isoforms in adult or embryonic muscles (1-6). We observed that during the course of regeneration in each of these muscles there was a reproduceable sequence of antigenic changes consistent with differential isoform expression for these two proteins. These isoform switches appear to be tissue specific; i.e., the isoforms of MHC and C-protein which are

This work was supported by grants from NIH (AM 32147) and the Muscular Dystrophy Association. A.D.S. was a Revson Fellow during the course of this work.

[3]Abbreviations: MHC, myosin heavy chain; ALD, anterior latissumus dorsi; PLD, posterior latissumus dorsi; McAbs, monoclonal antibodies; ABC, Avidin-Biotin complex (Vector Labs., California).

expressed during the regeneration of a "slow" muscle (ALD) differ from those which are synthesized in a regenerating "fast" muscle (PLD). Evidence has been obtained for the transient expression of a "fast-type" MHC and C-protein during ALD regeneration. Furthermore, during early stages of PLD regeneration this muscle contains MHCs which antigenically resemble those found in the pectoralis muscle at embryonic and early posthatch stages of development (1,7,8). All three of these regenerating muscles express an isoform of C-protein which appears identical immunochemically to that normally expressed in embryonic and adult cardiac muscle (9). These results support the concept that isoform transitions in regenerating skeletal muscles qualitatively resemble those found in developing muscles but differences may exist in temporal and tissue specific patterns of gene expression.

INTRODUCTION

Variant isoforms of the MHC and C-protein have been identified in specific adult striated muscles and at selected stages of embryonic and postnatal development (1-16). Several recent reports have also detected changes in the expression of contractile protein isoforms during skeletal muscle regeneration (17-19). However, relatively little is known about whether these isoform transitions during regeneration are either fiber-type specific or a faithfull recapitulation of similar transitions during ontogenesis. To explore these questions, we have examined the expression of MHC and C-protein in ALD, PLD and pectoralis muscles of adult chickens at selected time points after focal freeze injury to these tissues. Expression of these two proteins was measured immunocytochemically using a panel of McAbs prepared in our laboratory and previously shown to exhibit specificity to one or another isoform of the MHC or C-protein (1-6,9). A preliminary report of this work has been presented in abstract form (20).

MATERIALS AND METHODS

Adult White Leghorn chickens were narcotized with sodium pentobarbital (60 mg/kg body weight) and the ALD, PLD or pectoralis muscles exposed surgically. Muscle was frozen focally by holding a steel rod (3x4 mm) against the tissue surface for 20 sec and then the skin was resutured. At selected times after injury the muscles were excised, frozen in liquid N_2-cooled isopentane and stored at -80^oC.

Protein localization in 10um frozen sections was performed with either indirect immunoperoxidase or immunofluorescence cytochemical procedures. The immunoperoxidase reaction utilized the ABC (Avidin-Biotin-Complex) method (Vector Labs., California). Sections were blocked with 10% horse serum, then incubated sequentially with a McAb (30 min), biotinylated horse-anti-mouse second antibody (30 min), followed with the ABC reagent (30 min). Peroxidase activity was visualized with diaminobenzidine and H_2O_2. After fixation in OsO_4 the sections were dehydrated and mounted in Permount. Immunofluorescent staining was performed as described previously (2,4). Light microscopy was conducted with a Zeiss Standard Microscope equipped with epifluorescent optics.

The McAbs used for this study were specific for either the MHC (MF20, MF30, MF14 and ALD58) or C-protein (MF1, ALD66 and C-315). Of these, MF20 and MF30 crossreact with myosin isoforms in both pectoralis (fast) and ALD (slow) muscles, MF14 is specific for the MHC of fast muscle, whereas ALD58 binds only to the MHC from slow muscles. Antibodies MF20, MF30 and MF14 were of particular interest since these have been shown to react with MHC epitopes which appear sequentially during pectoralis muscle development (1). MF20 reacted with myosin at all stages of development; binding of MF30 was observed only after 18 days in ovo while reactivity of muscle with MF14 only occurred after 7-14 days posthatching. Of the C-protein antibodies, C-315 is specific for that isoform found in cardiac muscle of adult chickens, whereas ALD66 and MF1 bind to C-proteins found in slow and fast muscles respectively. Interestingly, some fast muscles (e.g., the PLD muscle) coexpress both the fast and slow-type C-protein isoforms and these have been colocalized in the same sarcomeres (2, 21). Myofibers in embryonic pectoralis muscle have been shown to express a cardiac-type C-protein reactive with McAb C-315 (9).

RESULTS

MHC EXPRESSION

Five days after freeze injury myofibers of the PLD muscle were clearly reactive with McAbs MF20 and MF30 (Fig. 1a,b); no staining with MF14 was observed in the zone of injury even though this antibody stained the adjacent, uninjured PLD fibers Fig. 1c). Three days later a change in MHC immunoreactivity was apparent for the regenerating myofibers which were now reactive with MF14 (Table 1). Additional experiments using a McAb (2E9A), specific for the neonatal MHC of pectoralis muscle, kindly provided by Dr. E. Bandman (22), have demonstrated that an antigen reactive with McAb 2E9A is present in 5 day PLD regenerates (results not shown). These results are compatible with the hypothesis that sometime between days 5 and 8, postinjury, a new MHC isoform is expressed in the PLD muscle fibers, presumably identical to that expressed in uninjured myofibers of this muscle (1). Since no reactivity of uninjured or regenerating PLD fibers was observed with McAb ALD58 (Fig.1d and Table 1), it is unlikely that a slow-type MHC isoform is expressed in this muscle.

In uninjured ALD muscle all of the fibers were reactive with MF20, MF30 and ALD58 (Fig. 2a,b,d); no staining of these cells was observed

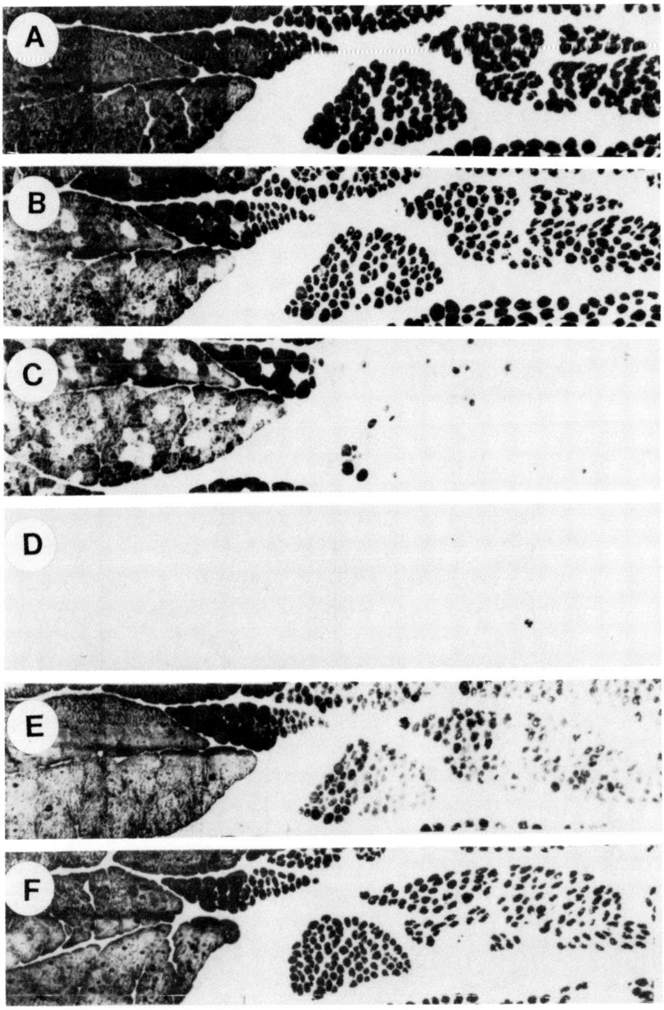

FIGURE 1. Frozen serial sections of 5-day regenerating PLD muscles reacted with MHC McAbs MF20 (a), MF30 (b), MF14 (c) ALD58 (d); or C-protein McAbs ALD66 (e) MF1 (f) and visualized by immunoperoxidase staining. Photomicrographs were taken with bright field optics at 160x magnification.

with MF14 (Fig. 2c). Five days postinjury, many of the regenerating fibers in the center of the wound did not stain with ALD 58 (Fig. 2d), however, the newly formed myofibers in these areas were reactive with MF14 (Fig. 2c). By 8 days after injury, many of the fibers at the periphery of the wound had acquired reactivity with ALD58 and exhibited a corresponding loss of staining with MF14 (Table 1); cells in the central zone of injury were similar in immunoreactivity to those present in 5-day-old wounds. No change in reactivity was seen with MF20 or MF30 throughout the course of regeneration. Fifteen days postinjury, the antigenic properties of myosin in control and regenerated fibers appeared identical; all of the fibers were stained with ALD58 but none were reactive with MF14 (Table 1).

C-PROTEIN EXPRESSION

During the course of regeneration, PLD myofibers exhibited no change in reactivity with McAbs MF1 or ALD66 (Fig. 1e,f, Table 1). However, there was a significant change in staining with McAb C-315. Uninjured fibers of the PLD, ALD and pectoralis muscles were unreactive with C-315 but within 2-5 days after freeze injury all 3 of these muscles stained strongly with this antibody. Results for the pectoralis are presented (Fig. 3); similar experiments for the PLD and ALD muscles have been completed but not illustrated (Table 1). This reactivity was confined to the C-region (23) in A-bands of the myofibrils (results not shown). Interestingly, all of these sarcomeres also contained at least one other isoform of C-protein. In fact, regenerating fibers of the PLD muscle, 5-8 days after injury, were reactive with McAbs MF1 and ALD66, in addition to C-315 (Table 1). We consider it likely that these sarcomeres each contain 3 isoforms of C-protein.

Uninjured myofibers of the ALD muscle were unreactive with MF1 and have been shown to contain one isoform of C-protein which binds to

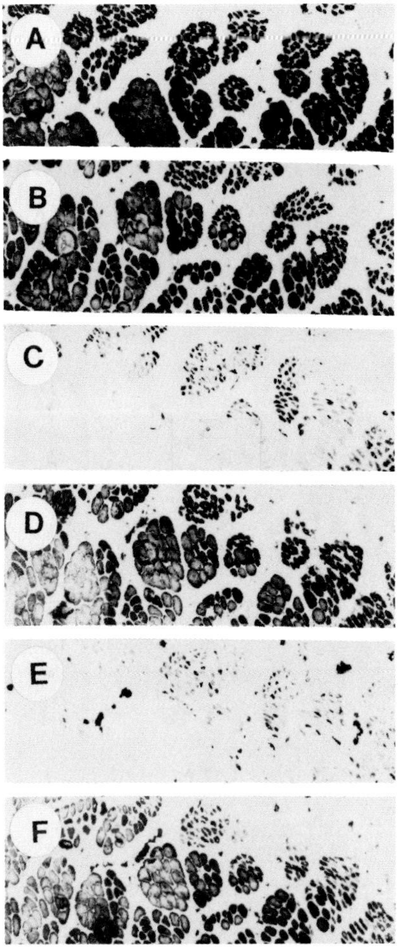

FIGURE 2. Frozen serial sections of 5-day regenerating ALD muscles reacted with MHC McAbs MF20 (a), MF30 (b), MF14 (c), ALD58 (d); or C-protein McAbs ALD66 (e), MF1 (f) and visualized by immunoperoxidase staining. Photomicrographs were taken with bright field optics at 160x magnification.

McAb ALD66 (3). We have termed this isoform "slow-type" C-protein (4,5). Day 5 postinjury, regenerating ALD fibers were definitely reactive with MF1 (Fig. 2f) and we assume that these cells were transiently expressing a "fast-type" isoform of C-protein. The isoform transitions of the MHC and C-protein were not temporally coupled. Many of the regenerating fibers of the ALD muscle, 8 days postinjury, were unreactive with ALD58 yet most of these same cells were stained with ALD66. Thus, the expression of a mature slow-type isoform of C-protein preceded the appearance of the mature slow-type isoform of myosin.

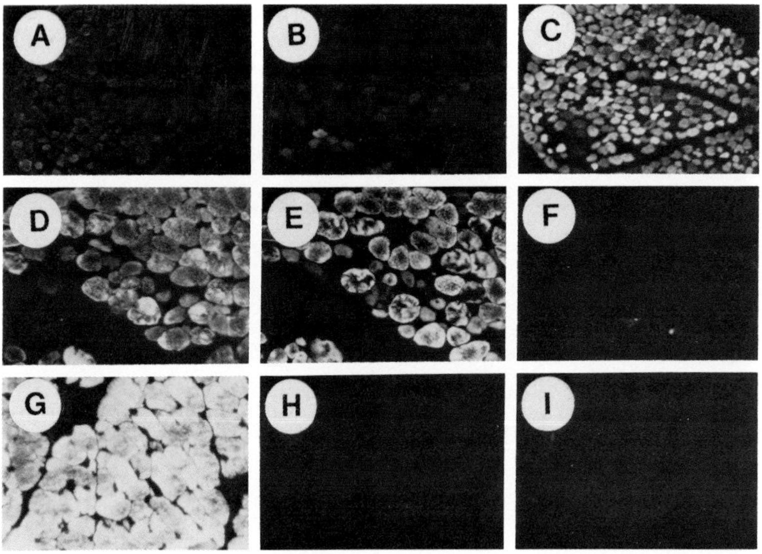

FIGURE 3. Frozen serial sections of regenerating pectoralis muscle at 3(a-c), 7(d-f) and 14(g-i) days after injury reacted with the C-protein antibodies anti-FC(similar to MF1;a,d,g), ALD66(b,e,h) and C-315(c,f,i) and visualized by immunofluorescence microscopy. Photomicrographs were taken with fluorescence optics at 160x magnification.

TABLE I

(a) PLD

Antibody	Antigen	Uninjured Muscle	Regenerating Muscle			
			2 day*	5 day	8 day	15 day
MF20	MHC	+	+	+	+	+
MF30	MHC	+	+	+	+	+
MF14	MHC	+	+	–	+	+
ALD58	MHC	–	–	–	–	–
ALD66	C-Pr	+	+	+	|	+
MF1	C-Pr	+	+	+	+	+
C315	C-Pr	–	ND	+	–	–

(b) ALD

Antibody	Antigen	Uninjured Muscle	Regenerating Muscle			
			2 day*	5 day	8 day	15 day
MF20	MHC	+	+	+	+	+
MF30	MHC	+	+	+	+	+
MF14	MHC	–	–	+	a	–
ALD58	MHC	+	+	a	a	+
ALD66	C-Pr	+	+	a	a	+
MF1	C-Pr	–	–	+	+	–
C315	C-Pr	–	ND	+	–	–

*Muscle at this stage is predominantly degenerating; signs of regeneration are only beginning

[a]Reactivity in the regenerating fibers is decreasing

DISCUSSION

The expression of MHC and C-protein has been analyzed by immunocytochemistry in the regenerating myofibers of ALD, PLD and pectoralis muscles. Results have been obtained that are compatible with isoform conversions for both of these proteins during regeneration, and these conversions qualitatively resemble those previously demonstrated in developing muscle. In ovo, the MHC epitope specified by MF30 was only detectable in the pectoralis muscle 3 days before hatching. In the PLD muscle this epitope is clearly detectable within 5 days after freeze injury, the earliest time point of regeneration examined in the present study. The binding of MF14 was detected 8 days after injury. Assuming that the PLD muscle undergoes a similar pattern of MHC isoform conversions to that of the pectoralis muscle, our results would suggest that expression of "neonatal" and "adult" MHC isoforms occurs precociously during regeneration when compared with timing of these conversions in normally developing muscle. Alternatively, the types of MHC, and their sequence of expression, may differ fundamentally in the PLD muscle from that in the pectoralis. Evidence for different MHCs in these two "fast" muscles has been obtained (24). An acceleration of isoform transitions in regenerating breast muscle has also been suggested by Matsuda et al. (19), but these same authors have concluded that little or no "neonatal" MHC is expressed during the regenerative process. Our results suggest that this is not the case for the PLD muscle.

During regeneration of the "slow" ALD muscle transient expression of a fast-type MHC and a fast-type C-protein was observed. The appearance of the "fast" isoforms preceded that of the "slow" isoforms and, at intermediate stages of regeneration, fibers were present which contained both the fast-type and slow-type isoforms of myosin and C-protein. As regeneration progressed, the fast-type isoforms were no longer detectable and the mature pattern of MHC and C-protein expression emerged. Since the disappearance of fast-type and appearance of slow-type isoforms of MHC and C-protein were out of phase, it is unlikely that their transcriptional regulation is tightly coupled. A similar conclusion has been reached for regulation of these proteins during embryonic development (4).

For all three muscles examined in this study, unambiguous evidence has been obtained for the expression of a cardiac-type C-protein during regeneration. In the regenerating pectoralis three distinct stages of C-protein expression were defined. Initially, only the cardiac-type isoform was detectable; this was then followed by an intermediate stage in which the cardiac-type isoform was not observed, but both fast- and slow-type skeletal isoforms were expressed. The final stage of regeneration was characterized by the sole presence of fast-type C-protein. The mechanisms underlying both C-protein and MHC insertion or replacement in preexisting myofibrils are completely unknown. Can new isoforms be inserted without myofilament or A-band disassembly? Recent work in our laboratory would suggest that filamentous myosin is in equilibrium with a monomeric pool (25); it remains to be established if a similar equilibrium exists for C-protein.

Finally, suggestive evidence has been obtained for the existence of differing "embryonic" isoforms in regenerating muscle of PLD and ALD fibers. Five days postinjury, regenerating fibers of the ALD muscle contained an epitope reactive with MF14, whereas regenerating fibers of the PLD muscle were unreactive with this antibody. Presumably, a fast-type MHC isoform is synthesized in both of these muscles during regeneration, but these isoforms appear to be immunologically distinct.

REFERENCES

1. Bader D, Masaki T, Fischman DA (1982) Immunochemical analysis of myosin heavy chain during avian myogenesis in vivo and in vitro. J Cell Biol 95:763.

2. Reinach FC, Masaki T, Shafiq SA, Obinata T, Fischman DA (1982) Isoforms of C-protein in adult chicken skeletal muscle: Detection with monoclonal antibodies. J Cell Biol 95:78.

3. Reinach FC, Masaki T, Fischman DA (1983) Characterization of the C-protein from PLD muscle of the adult chicken: Heterogeneity within a single sarcomere. J Cell Biol 96:297.

4. Obinata T, Reinach FC, Bader DM, Masaki T, Kitani S, Fischman DA (1984) Immunochemical analysis of C-protein isoform transitions during the development of chicken skeletal muscle. Dev Biol 101:116.

5. Obinata T, Kitani S, Masaki T, Fischman DA (1984) Coexistence of fast-type and slow-type C-proteins in neonatal chicken breast muscle. Dev Biol 105:253.

6. Shafiq SA, Shimizu T, Fischman DA (1984) Heterogeneity of type 1 skeletal muscle fibers revealed by monoclonal antibody to slow myosin. Muscle and Nerve 7:380.

7. Bandman E, Matsuda R, Strohman RC (1982) Developmental appearance of myosin heavy and light chain isoforms in vivo and in vitro in chicken skeletal muscle. Dev Biol 93:508.

8. Lowey S, Benfield PA, Leblanc DD, Waller G, Winkelmann D, Gauthier GF (1982) Characterization of myosin from embryonic and developing chicken pectorialis muscle. In: Muscle Development (ML Pearson and

Epstein, eds) pp. 15-25, Cold Spring Laboratory, Cold Spring Harbor, NY.

9. Kawashima M, Kitani B, Tanaka T, Obinata T (1984) Expression of C-protein isoforms during chicken skeletal muscle development in vitro. Proc Third Int Cong on Cell Biol p. 511.

10. Whalen RG, Sell SM, Butler-Brown GS, Schwartz K, Bouvert P, Pinset-Harstrom I (1981) Three myosin heavy chain isozymes appear sequentially in rat muscle development. Nature (London) 292:805.

11. Rushbrook JL and Stracher A (1979) Comparison of adult, embryonic and dystrophic myosin heavy chains from chicken muscle by sodium dodecyl sulfate/polyacrylamide gel electrophoresis and peptide mapping. Proc Natl Acad Sci (USA) 76:4331.

12. Wydro RM, Nguyen HT, Gubits R, Nadal-Ginard B (1983) Characterization of sarcomeric myosin heavy chain genes. J Biol Chem 258:670.

13. Umeda PK, Sinha AM, Jakovcic S, Merten S, Hsu HJ, Subramanian KN, Zak R, Rabinowitz M (1981) Molecular cloning of two fast myosin heavy chain cDNA's from chicken embryo skeletal muscle. Proc Natl Acad Sci (USA) 78:2843.

14. Offer G, Moos C, Starr R (1973) A new protein of the thick filaments of vertebrate skeletal myofibrils. J Mol Biol 74:653.

15. Yamamoto K, Moos C (1983) The C-proteins of rabbit red, white and cardiac muscle. J Biol Chem 258:8395.

16. Callway JE, Bechtel PJ (1981) C-protein from rabbit soleus (red) muscle. Biochem J 195:463.

17. Sartore S, Gorza L, Schiaffino S (1982) Fetal myosin heavy chains in regenerating muscle. Nature (London) 298:294.

18. Toyota N, Shimada Y (1984) Isoforms of troponin during regeneration of chicken skeletal muscle fibers after cold injury. Cell Tiss Res 236:549.

19. Matsuda R, Spector DH, Strohman RC (1983) Regenerating adult chicken skeletal muscle and satellite cell cultures express embryonic patterns of myosin and tropomyosin isoforms. Dev Biol 100:478.

20. Saad AD, Fischman DA (1983) Myosin isoform transitions during muscle regeneration in adult chickens. J Cell Biol 97:51a.

21. Dennis JE, Shimizu T, Reinach FC, Fischman DA (1984) Localization of C-protein isoforms in chicken skeletal muscle: ultrastructural detection using monoclonal antibodies. J Cell Biol 98:1514.

22. Bandman E (1985) Continued expression of neonatal myosin heavy chain in adult dystrophic skeletal muscle. Science 227:780.

23. Squire J (1981) The structural basis of muscular contraction. Plenum Press, NY.

24. Bandman E, Matsuda R, Strohman RC (1982) Myosin heavy chains from two different adult fast-twitch muscles have different peptide maps but identical RNAs. Cell 29:645.

25. Saad AD, Fischman DA, Pardee JD (1985) Analysis of thick filament assembly and myosin exchange. This volume.

Molecular Biology of Muscle Development, pages 323–336

CHANGES IN THE SMOOTH MUSCLE MYOSIN ISOFORMS DURING DEVELOPMENT OF CHICKEN GIZZARD[1]

Tomoh Masaki,* Hiromi Takano-Ohmuro,** Reiko Suzuki,* Takashi Obinata,*** Yoshio Hamada,**** Takashi Mikawa,**** Takeshi Tanaka,* Akira Yoshikawa** and Surender S. Katoch*

*Institute of Basic Medical Sciences, University of Tsukuba, Ibaraki 305, **Tokyo Metropolitan Institute of Medical Sciences, Tokyo 113, ***Department of Biology, Chiba University, Chiba 260, ****National Institute of Basic Biology and Institute of Physiology, Okazaki, Aichi 444, Japan

ABSTRACT In chicken gizzard smooth muscle, three myosin isoforms were detected in embryonic and only one isoform in adult gizzard by pyrophosphate gel electrophoresis (PPi-PAGE). The three embryonic isoforms differ from each other in the light chain distribution. Both heavy and light chains of the smooth muscle myosin change during development. A myosin light chain specific for embryo, L_{23}, exists at young stage. Change of S2 domain of myosin heavy chain also occurs during development. At least two myosin heavy chains were immunochemically distingushable. Different isoforms of myosin heavy chain were found in gizzard and vasucular muscles. cDNA sequence of smooth muscle myosin, GMH1 (1.7Kbp long), was shown not to hybridize to the skeletal muscle myosin sequence.

[1]This work was supported by grants from the Ministry of Education and Culture, and Ministry of Health and Welfare of Japan, University of Tsukuba Project Research and Asahi Press Co.

INTRODUCTION

Myosin is one of the major proteins of muscle, and composed of two heavy chain (MHC) and four light chains (MLC) of the different size peptides. Many recent studies demonstrated that different myosin isoforms exist in skeletal, cardiac and smooth muscle cells (1-3), and that these isoforms change during development (4-8).

In skeletal muscle, developmental change of myosin isoform was investigated by chemical, and immunochemical properties of myosin isoforms, i.e. electrophoretic mobilities under non-denaturing conditions (9), light chain compositions (5,9) and their antigenisities (4,5,7,8). In addition, morphological changes in synthetic filaments during development was also reported in skeletal muscle myosin (6).

In contrast to the skeletal muscle myosin, much less is known about the differentiation of smooth muscle myosin during development.

Since the structure of myosin filament in smooth muscle is morphologically distinct from that of striated muscle (10,11), and it appears that one type of myosin isoform forms each own characteristic contractile apparatus, elucidation of structural difference between smooth and striated muscle myosin filaments will give an important clue for solving the mechanism of formation of striation pattern. This idea is strongly supported by the facts that skeletal type α-actinin does not co-exist and has different distribution (12), and that α-actin and β,γ-actin show different distribution.

Therefore, in this report, we will describe the transition of myosin isoform in developing chicken gizzard smooth muscle and compare these isoforms with that of vascular smooth muscle. A part of the present results has already been reported elsewhere (13,14).

MATERIALS AND METHOD

Preparation of gizzard smooth muscle myosin

Myosin was prepared from the gizzards of 9, 10, 13, 15, 17 and 18-day-old chicken embryos, 1-day and 8-day-old post-hatched chickens, and from adult chickens either by modification of Ebashi (15), or by Sobiestzek and Bremel (16).

However, this method yielded a small amount of myosin

from young embryo. Then, in this case gizzard homogenete was extracted with 0.6 M KI in the presence of 5 mM ATP 1 mM $MgCl_2$, 1 mM dithiothreitol (DTT), 0.1 mM diisopropylfluorophosphate (DIFP), 20 mM Tris-HCl (pH 7.0). After centrifugation at 130,000 *g* for 1 hr., the supernatant of the homogenete was layered on the Sephacryl S-400 column (2 cm x 45 cm) and eluted by the same buffer. The fraction containing myosin was dialysed against a buffer containing 12.5 mM Tris-HCl, 0.2 mM DTT, 0.1 mM phenylmethyl sulfonyl fluoride (PMSF), 0.1 mM DIFP, and then spun down at 17,000 *g* for 30 min. Precipitate was dissolved in 0.3 M KCl, 5 mM ATP and centrifuged at 130,000 *g* for 90 min. Thus resulting supernatant was used as myosin (17).

Electrophoresis and immunoblot procedure

Pyrophosphate polyacrylamide gel electrophoresis (PPi-PAGE) was performed essentially by the same method as Hoh et al (18). The amount of myosin applied was in the range of 2 to 5 µg.

SDS-polyacrylamide gel electrophoresis (SDS-PAGE) was performed as described previously (13). Two-dimensional gel electrophoresis of the light chains was carried out according to O'Farrell's method (19) modified by Mikawa (20) as described previously.

One dimensional peptide mapping experiment according to Cleveland's method (21) and immunobloting experiment were performed as described previously (22).

Monoclonal antibody

Monoclonal antibodies (McAbs) against smooth muscle myosin were raised as described previously (22). Three antibodies G3-6, G3-16 and G17-1 among them were used for the present experiment. The former two G3-6, G3-16 reacted to tail region of the heavy chain of myosin molecule and the latter one G17-1 found to S2 region of myosin molecule (14). G3-6 and G17-1 reacted exclusively to a heavy chain of smooth muscle myosin, and the G3-16 reacted to heavy chains of skeletal fast, skeletal slow, cardiac, smooth muscle but not to that of intestinal brush border myosin (14).

Quantitative immunoassay

Antibody binding to myosin was assayed with a solid phase immunoassay. Antigen were absorbed to wells of polyvinyl 96-well microtiter plates as discribed by Winkelmann et al (8). Detecting antibodies were the rabbit anti-mouse IgG labeled by horse radish peroxidase (ELISA).

One dimensional peptide mapping

Peptide mapping analysis was performed according to the Cleveland (21). After PPi-PAGE, each piece of gel containing each myosin isoform was isolated, and then electrophoresis of each isoform was performed in the presence of 0.1% SDS. Resulting heavy chain of each myosin isoform was subjected to the peptide mapping analysis according to Cleveland (21).

Then 200 ng α-chymotrypsin or trypsin was used in each well and reaction was performed for 30 min on gel.

Immunofluorescence microscopy

Frozen gizzard tissue was cut in 6 μm, after fixation with 3% paraformaldehyde, section was immersed in phosphate buffered saline containing 1% bovine serum albumin and exposed to McAbs followed by washing, staining by FITC-second antibody and subjected to the fluorescent microscope as described previously (12).

RESULTS AND DISCUSSION

As reported previously (18), three myosin isoforms were contained in adult chicken breast muscle. Then it is interesting to know how many isoforms of myosin exists in adult smooth muscle. In case of smooth muscle, separation of myosin isoforms was improved by the addition of Mg-ATP in the buffer system (13).

Figure 1 illustrates the PPi-PAGE pattern of myosin from developing chicken gizzard smooth muscle. Three different myosin isoforms were detected in young embryonic chicken gizzard smooth muscle (GM_3, GM_2, GM_1), while only one major myosin isoform was contained in the adult smooth muscle. The electrophoretic mobility of GM_1 in the embryo, which had the fastest electrophoretic mobility, was the same as that of the adult isoform. At 1 day after hatching, one predominant (GM_1) and one minor (GM_2) isoform coexisted. GM_3 was scarecely detectable at this stage. At 8 days after hatching, only GM_1 was present, as in the adult tissue. No other bands corresponding to fast skeletal, slow skeletal, cardiac or embryonic skeletal myosin isoforms was detectable at any point in the development of gizzard muscle. Under the present

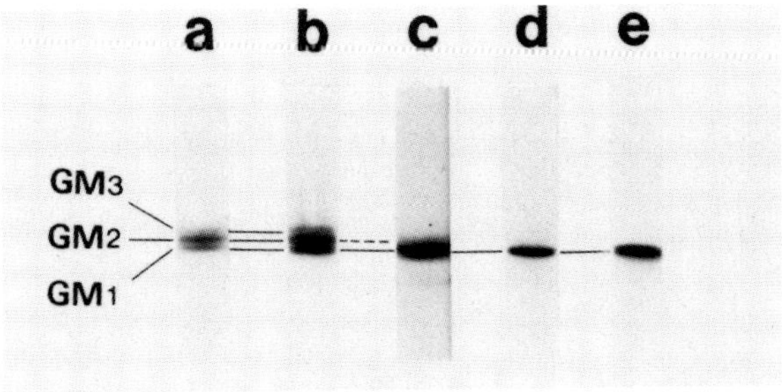

Fig. 1. Pyrophosphate polyacrylamide gel electrophoresis patterns of myosin from gizzard of embryonic and post-hatched chickens. (a) Myosin from 15-day-old embryos. (b) Myosin from 17-day-old embryos. (c) Myosin from 1-day-old post-hatched chicken. (d) Myosin from 8-day-old post-hatched chicken. (e) Myosin from adult chicken. See text.

electrophoretic condition, gizzard myosin isoforms migrated faster than skeletal or cardiac myosin isoforms. However, the mobility of myosin isoform from fibroblasts resembled that of GM_1 isoform.

The light chain distribution in each myosin isoform was observed by SDS-PAGE of individual myosin isoform isolated on PPi-PAGE. Adult myosin isoform (GM_1) contained two kinds of smooth muscle light chains, L_{20} and L_{17}, in descending order of molecular weight according to the nomenclature of Katoh and Kubo (24). The apparent molecular weight of L_{20} and L_{17} in SDS-polyacrylamide gel electrophoresis are about 20,000 and 17,000 respectively as previously demonstrated. The distribution of light chains in each individual embryonic isoform was quite different from each other. The light chain, L_{23}, is specific for embryonic smooth muscle. GM_3 contained both L_{23} and L_{20}, while GM_2 contained L_{23}, L_{20} and L_{17}.

Two-dimensional gel electrophoresis was carried out to compare the light chains present during development of chicken gizzard. The patterns for adult chicken, post-hatched chicken, and embryonic chicken light chains are shown in Fig. 2. The molecular weights and isoelectric points of adult L_{20} and L_{17} light chains were the same as those of embryonic and post-hatched chicken L_{20} and L_{17}. The L_{23} light chain was distinct in both molecular weight

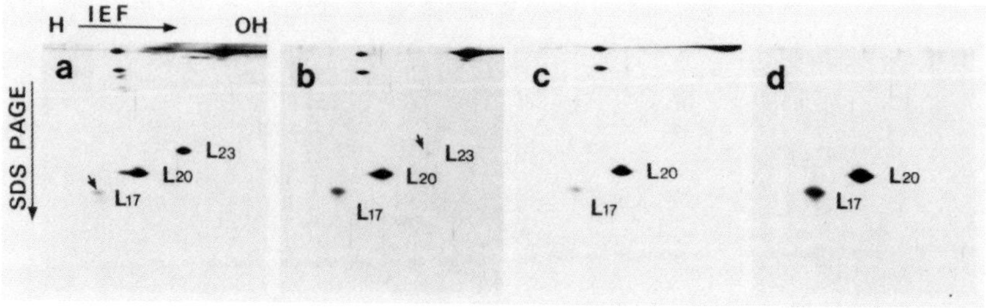

Fig. 2. Two dimensional gel electrophoresis patterns of light chain from 17-day-old embryos (a), 1-day-old post-hatched chicken (b), 8-day-old post-hatched chicken (c), and adult chicken gizzard myosin (d), IEF, isoelectric focusing electrophoresis, L_{23}, L_{20} and L_{17} gizzard myosin light chains.

and isoelectric point from the other light chains. Neither skeletal nor cardiac myosin light chains co-migrated with any of the light chains in the developing gizzard muscle. Katoh and Kubo showed previously that L_{23} was detected only in embryonic or neonatal chicken gizzard (24). However, recently we demonstrated that embryonic chicken skeletal, cardiac and smooth muscle express a common embryo-specific light chain, i.e. L_{23} (23). This light chain was incorporated into embryonic myosin molecules as judged by the results of SDS-PAGE combined with PPi-PAGE.

Then a question arises whether or not the change of heavy chain of myosin occurs during development. The one-dimensional map of α-chymotryptic peptides of myosin in the developing chicken gizzard muscle was performed. The pattern of α-chymotryptic cleavages product of 15-day or 17-day-old embryonic heavy chain was very similar to those of post-hatched or adult heavy chain. However, peptide map of 10-day-old embryonic heavy chain cleaved by trypsin or α-chymotrypsin was similar but definitely different from that of the adult heavy chain. For example, several bands in the lower molecular weight region of adult and embryonic cleavages by α-chymotrypsin are unique in each individual myosin heavy chain.

This difference of myosin heavy chain between adult and young embryonic gizzards was confirmed by immunochemical analysis using monoclonal antibodies as follows.

Solid phase immunoassay using ELISA in polyvinyl

chloride micro-titer plate was performed. A fixed amount of myosin of various developmental stages, was reacted with serial dilutions of the monoclonal antibody (McAb) supernatant. Analysis of McAb G17-1 binding to the myosin from various stages as a function of antibody concentration demonstrated much higher specific binding activity to the myosin of the later stage of embryo or adult myosin than observed with the myosin of younger stage of development. Almost no significant binding of G17-1 was observed with 10-day-old chicken gizzard myosin. Thus the reactivity of G17-1 increased with development.

In contrast, McAb G3-6 or G3-16 reacted equivallently with myosin at all stages of development and no significant difference was seen in plateau values of the peroxidase labelled second antibody bound to the microtiter wells at the saturating levels of G3-6 or G3-16.

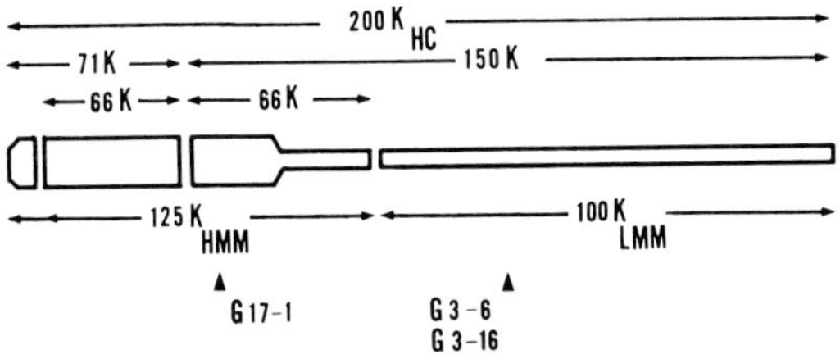

Fig. 3. Schematic representation of the smooth muscle myosin moleculue showing sensitive sites of α-chymotrypsin and epitopes of McAbs. (after Y. Okamoto et al (25))

As shown in Fig. 3, immunoblots with these myosin preparations indicate that G3-6, G3-16 bound exclusively to tail region of myosin molecule, and G17-1 also to S2 of heavy chain of myosin molecule. Thus, with progressive embryonic development the changes in antigenic properties of the S2 region of myosin heavy chain occurs, while the epitope specified by G3-6 or G3-16, i.e. epitope in the tail region of myosin heavy chain molecule, is present in embryonic, post-hatch and adult myosin heavy chains presumably at equivalent molar values. It was also suggested in another paper that chicken skeletal myosin changes epitope in S2 region of heavy chain during development. This change of the myosin heavy chain is slightly ahead of the change in myosin light chain, i.e. disappearance of L_{23} and appearance of L_{17}.

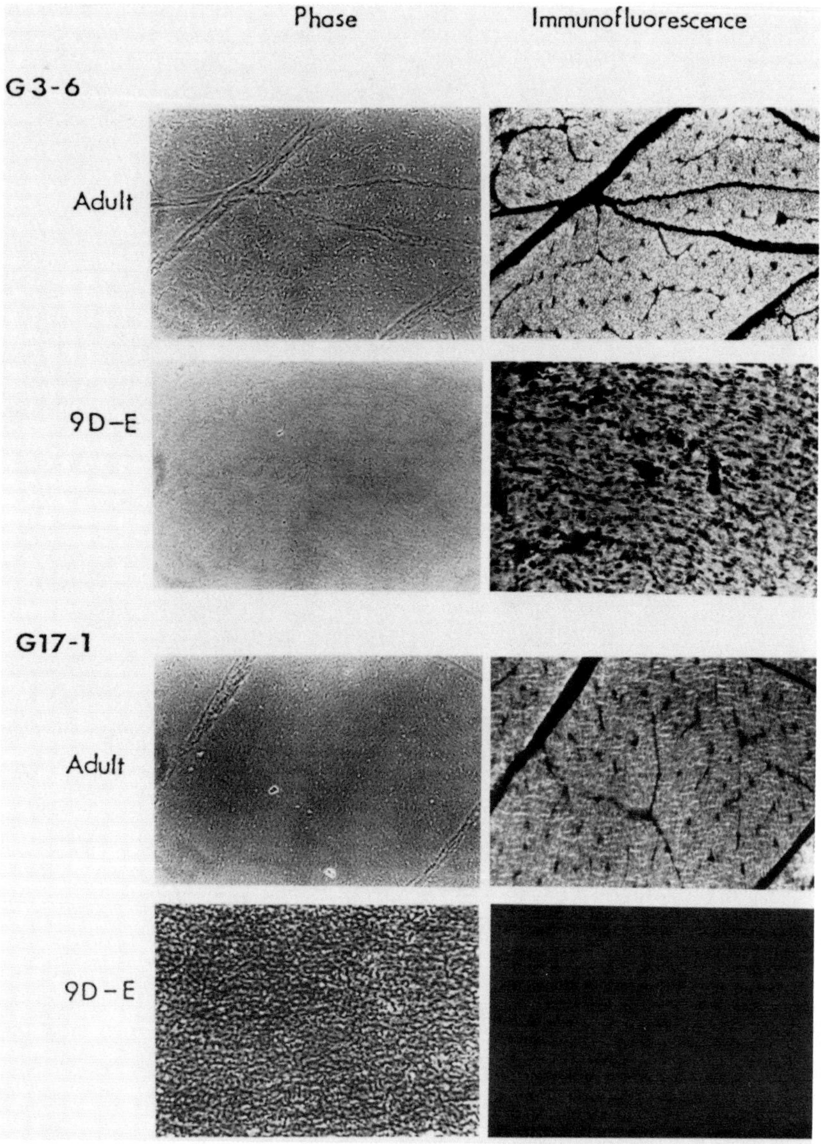

Fig. 4. Immunofulorescence and phase contrast photographs of frozen sections of adult and embryonic (9-day-old) chicken gizzard by McAb G3-6 and G17-1 specific for adult gizzard smooth muscle.

Confirmatory results were also obtained with immunofluorescent analysis of frozen sectioned embryonic and adult chicken gizzard muscle (Fig. 4). G3-6 or G3-16 appeared to stain both of embryonic and adult gizzard muscle, although the staining of embryonic muscle is definitely weaker than that of the adult. However, binding of G17-1 was hardly detected by fluorescent staining at 9-day-old embryo, whereas it was strong in adult smooth muscle.

In addition to this frozen section of gizzard tissue, cultured smooth muscle cell was also stained by these McAbs. Mononucleated cells were isolated from 9-day-old embryonic chicken gizzard in the same way as in the skeletal muscle (26). Polyclonal antibody specific for smooth muscle myosin was previously labeled by tetramethylrhodamine isothiocyanate (RITC). To differentiate smooth muscle cell from fibroblast cell, we used double staining method with this RITC-polyclonal antibody. The cell was stained by RITC-polyclonal anti-smooth muscle myosin plus either of the three McAbs, and then it was stained by FITC-anti-mouse IgG second antibody. If McAb G3-6 was used, all of the embryonic and adult smooth muscle cells were stained both by the polyclonal antibody and by G3-6. However, if McAb G17-1 was used, all adult smooth muscle cells were stained but no embryonic cell was stained even after culture for long period. This result demonstrated that cultured smooth muscle cell does not express adult isoform as in the case of skeletal muscle culture (26). In addition, it was also shown that any of the three McAbs does not stain any other cell or tissue than smooth muscle cell, for example, fibroflast. These results demonstrated that McAb G17-1 stained adult chicken gizzard smooth muscle myosin and distinguished the embryonic and adult myosin. Thus, they suggest that the sequential expression of the epitopes specific for G17-1 occurs in smooth muscle myosin in the same manner as in developing skeletal muscle. However, the transition of heavy chain of smooth muscle myosin is not always parallel to that of the light chain in the same way as that of the skeletal muscle myosin (5,7,9).

Reactivity of McAbs to several smooth muscle tissue was investigated by fluorescent antibody technique. As shown in Table 1, smooth muscle myosin of adult gizzard muscle reacts to both G17-1 and G3-6, whereas adult vascular muscle reacts to G3-6 but not to G17-1. This results suggests that at least two kinds of smooth muscle

TABLE 1

REACTIVITY OF SMOOTH MUSCLE TISSUES
TO MONOCLONAL ANTIBODIES SPECIFIC
FOR SMOOTH MUSCLE MYOSIN HEAVY CHAIN

McAb	Gizzard embryonic	Gizzard adult	vascular muscle
G3-6	+	+	+
G17-1	-	+	-

+: positive, -: negative

myosin exist in adult tissue, i.e. myosin reactive to G3-6 but not to G17-1 and myosin reactive to both McAbs.

To study further the expression of individual myosin heavy chain, and to elucidate structural difference between striated and smooth muscle myosins, we have cloned myosin heavy chain cDNA sequences from chick gizzard smooth muscle.

To obtain recombinant clones for smooth muscle myosin heavy chain cDNA sequences, we first purified poly(A^+) RNA from 13-day-old embryonic chicken gizzard by oligo-dT-cellulose column. Double-stranded cDNAs were inserted into EcoRI site of λgt 11, and y1090 bacteria were transfected (27). Recombinant clones were screened with monoclonal antibodies against chicken gizzard smooth muscle heavy chain, G3-1, G3-16, G-29 and G17-1, by detecting expressed protein. Thus screened cDNA clone, which has 1.7Kbp cDNA insert, was used for detection of other full-length-cDNA. GMH-1, one of them, which was 1.7Kbp long, hybridized at high stringency only to smooth muscle MHC mRNA, but not to RNA from skeletal muscle by Northern blot method (Fig. 5)(28). This result suggests that striated and smooth muscle MHCs are products of different MHC genes.

As reported previously (6), length of the synthetic bipolar myosin filament of rat skeletal muscle under a certain condition, differed for the myosins from different developmental stages. Then a question arises whether or not the difference exists in the length of synthetic filament or in morphological pettern in gizzard smooth muscle myosin from different stages.

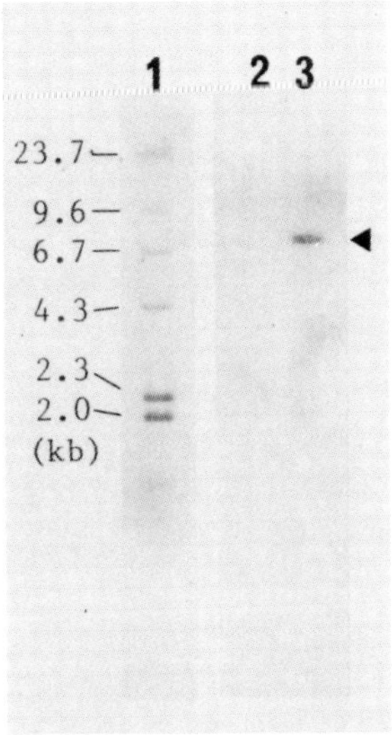

Fig 5. Detection of smooth muscle MHC mRNA sequences by GMHC-1. poly(A^+) RNAs isolated from skeletal (2) and smooth muscle (3) was electrophoresed on denaturating agarose and transferred to nylon paper and hybridized to ^{32}P-labeled GMHC-1 Lane 1 is the Hae III digested λ phage marker.

Thick filament was synthesized by dialysing the embryonic and adult purified myosin dissolved in 0.5 M NaCl, 0.05 M Tris-HCl (pH 6.8), 10 mM $MgCl_2$ and 1 mM ATP, against a large volume of 40 mM NaCl, 30 mM Tris-HCl (pH 6.8). After 3 hrs. the precipitated myosin was collected by centrifubation and resuspended in 0.2 M KCl, 1 mM EDTA, 0.5 mM DTT, 5 mM potassium phosphate buffer (pH 7.0) at concentration of 3 mg/ml. Bipolar filament occurs both at embryonic and adult myosin. However, so far, there was no significant difference in morphological pattern of synthetic filament between adult and embryonic smooth muscle myosin.

In conclusion, in chicken gizzard smooth muscle, PPi-PAGE demonstrated three myosin isoforms in embryonic and only one in adult gizzard muscle. The three embryonic isoforms differ from each other in the light chain distribution. There are three kinds of light chains in embryonic myosin. Two of them exists in adult form. Transition of myosin isoform occurs during development both in heavy and light chains. Embryonic ubiquitous light chain of myosin L_{23} existed abundantly in embryonic smooth muscle myosin. In the heavy chain, change occured at least in S2 region of myosin molecule. This transition does not occur in cultured smooth muscle cell. cDNA sequence of smooth muscle myosin did not also hybridized with skeletal myosin mRNA.

REFERENCES

1. Masaki, T. (1974). Immunochemical comparison of myosins from chicken cardiac, fast white, slow red and smooth muscle. J. Biochem. 76 441.
2. Lowey, S. and Risby, D. (1971). Light chain from fast and slow muscle myosins. Nature 234 81.
3. Obinata, T., Masaki, T. and Takano, H. (1979). Immunochemical comparison of myosin light chains from chicken fast white, slow red and cardiac muscle. J. Biochem. 86 131.
4. Masaki, T. and Yoshizaki, C. (1974). Differentiation of myosin in chick embryo. J. Biochem 76 123.
5. Obinata, T., Masaki, T. and Takano, H. (1980). Types of myosin light chains present during development of fast skeletal muscle in chick embryo. J. Biochem. 87 81.
6. Whalen, R.G., Sell, S.M., Butler-Browne, G.S., Schwartz, K., Bouveret, P., Pinset-Härström, I. (1981). Three myosin heavy-chain isozymes appear sequentially in rat muscle development. Nature 292 805.
7. Bader, D., Masaki, T. and Fischman, D.A. (1982). Isoforms of C-proteins in adult chicken skeletal muscle: detection with monoclonal antibodies. J. Cell Biol. 95 78
8. Winkelmann, D.A., Lowey, S. and Press J.L. (1983). Monoclonal antibodies localize changes on myosin heavy chain isozymes during avian myogenesis. Cell 34 295.
9. Takano-Ohmuro, H., Obinata, T., Masaki, T. and Mikawa, T. (1982). Changes in myosin isozymes during development of chicken breast muscle. J. Biochem. 91 1305
10. Small, J.V. and Sobieszek, A. (1980). The contractile apparetus of smooth muscle, International Review of Cytology 64 241.
11. Craig, R., Megerman, J. (1977). Assembly of smooth muscle myosin into side-polar filaments. J. Cell Biology. 75 990.
12. Endo, T. and Masaki, T. (1984) Differential expression and distribution of chicken skeletal- and smooth-muscle-type α-actinins during myogenesis in culture. J. Cell Biol. 99 2322.

13. Takano-Ohmuro, H., Obinata, T., Mikawa, T. and Masaki, T. (1983). Changes in myosin isozymes during development of chicken gizzard muscle. J. Biochem. 93 903.
14. Suzuki, R., Takano-Ohmuro, H. Tanaka, T. and Masaki, T. (1984). Changes in myosin heavy chain isoform in developing chicken gizzard smooth muscle. Abstract of Annuel Meeting of Jap. Biophysic. Soc. 3D0900.
15. Ebashi, S. (1976). A simple method of preparing actin-free myosin from smooth muscle. J. Biochem. 79 229.
16. Sobieszek, A. and Bremel, D. (1975). Preparation and properties of vertebrate smooth-muscle myofibrils and actioyosin. Eur. J. Biochem. 55 49.
17. Mooseker, M.S., Pollard, T.D. and Fujiwara, K., (1978). Characterization and localization of myosin in the brush border of intestinal epithelial cells. J. Cell Biol. 79 444.
18. Hoh, J.F.Y., McGrath, P.A. and White, R.I. (1976). Electrophoretic analysis of multiple forms of myosin in fast-twitch and slow-twitch muscles of the chick. Biochem. J. 157 87
19. O'Farrell, P.H. (1975). High resolution two-dimensional electrophoresis of proteins. J. Biol. Chem. 250 4007.
20. Mikawa, T., Takeda, S., Shimizu, T. and Kitaura, T. (1981). Gene expression of myofibrillar proteins in single muscle fibers of adult chicken: micro two dimensional gel electrophoretic analysis. J. Biochem. 89 1951.
21. Cleveland, D.W., Fischer, G.S. Kirschner, W. and Laemmli, U.K. (1977). Peptide mapping by limited proteolysis in sodium dodecyl sulfate and analysis by gel electrophoresis. J. Biol. Chem. 252 1102.
22. Fischman, D.A. and Masaki, T. (1982). Immunochemical analysis of myosin with monoclonal antibodies, In Anderson, W. (ed): "Differentiation and Hypertrophy" Amsterdam, Elsevier, p279.
23. Takano-Ohmuro, H., Obinata, T., Kawashima, M., Masaki, T. and Tanaka, T. (1985). Embryonic chicken skeletal, cardiac and smooth muscles express a common embryo-specific myosin light chain. J. Cell Biol. in press
24. Katoh, N. and Kubo, S. (1978). Light chains of chicken embryonic gizzard myosin. Biochim. Biophys. Acta. 535 401.

25. Okamoto, Y., Okamoto, M. and Sekine, T. (1980). Two opposite effect of ATP on the the chymotryptic cleavages in smooth muscle myosin head. J. Biochem. 88 361.
26. Masaki, T., Miyamoto, K. and Obinata, T. (1980). Characterization of myosin heavy chains in cell cultures of embryonic chicken breast muscle. Biomedical Res. 1 443.
27. Reinach, F.C. and Fischman, D.A. (1985). A recombinant DNA approach for defining the primary structure of monoclonal antibody epitopes. The analysis of a conformation-specific antibody to myosin light chain two. J. Mol. Biol. in press
28. Hamada, Y., Mikawa, T. and Masaki, T. (1985). Smooth muscle myosin heavy chain is coded in different gene family from sarcomeric myosin heavy chain gene. Abstract of 17th Jap. Develop. Biologist Soc. in press

Molecular Biology of Muscle Development, pages 337–343

NON-MUSCLE MYOSIN GENES FROM ACANTHAMOEBA

John A. Hammer III, Edward D. Korn and Bruce M. Paterson

Laboratory of Cell Biology, NHLBI and Laboratory of Biochemistry, NCI, National Institutes of Health, Bethesda, MD 20205

ABSTRACT The soil amoeba, *Acanthamoeba castellanii* simultaneously expresses three distinct myosin enzymes, myosin IA, myosin IB and myosin II. We have used a portion of the nematode *C. elegans* myosin heavy chain gene (gift of L. Barnett, MRC, Cambridge), which contains the actin and ATP binding sites and the active thiols, to isolate an amoeba myosin II heavy chain gene. We screened a genomic library of *Acanthamoeba* DNA in phage lambda with this heterologous myosin probe and obtained several groups of overlapping phage. One phage group was shown to contain a myosin II heavy chain gene based on hybrid selection analysis, immunoprecipitation of hybrid selected protein, hybridization to a 5300 bp mRNA, and an exact match between the 3' end nucleotide sequence and a carboxy terminal myosin II chymotryptic peptide sequenced by protein chemical methods (Côté et al. J. Biol. Chem. 259, 12781, 1984). The 5' end of the gene has been tentatively identified and is approximately 5500 bp from the 3' end. We have sequenced the first 20% of the gene. Except for the first ~60 amino acid residues, the amoeba myosin shows nearly 60% exact match with the nematode myosin.

INTRODUCTION

The non-muscle myosin from *Acanthamoeba castellanii*, myosin II, shares many important characteristics with muscle myosins including (i) subunit composition (one pair of 185 kDa heavy chains and two pair of light chains of 17.5 kDa and 17kDa, respectively), (ii) asymmetric structure (2 globular heads and a long -helical coiled-coil tail), (iii) filament formation at low ionic strength, and (iv) Mg^{2+}-ATPase activity which is highly activated by F-actin (1,2).

Myosin II is distinct from muscle myosin, however, in at least several aspects involving enzymatic regulation, structure and self assembly properties. The regulation of myosin II actin-activated Mg^{2+}-ATPase activity occurs via phosphorylation/dephosphorylation of its heavy chains (3). The molecule is phosphorylated by a myosin II-specific heavy chain kinase at 3 serine residues on each heavy chain (4). Sequencing of a 9 kDa carboxy terminal myosin II chymotryptic peptide indicates that the coiled-coil of the myosin tail is disrupted approximately 30 residues from the C-terminus and that the 3 phosphorylated serines reside in this non-helical tailpiece (5). A purified amoeba phosphatase removes these phosphates (6). In this manner the activity of the enzyme can be regulated *in vitro* with the activity being inversely correlated with the phosphorylation state i.e. the fully dephosphorylated enzyme has the highest actomyosin II Mg^{2+}-ATPase activity. Recent evidence suggests that heavy chain phosphorylation exerts its effect via alteration of the filament conformation (7,8). With regard to structure and self assembly, myosin II has a somewhat shorter tail of 90 nm (versus 150 nm for most muscle myosins) and assembles *in vitro* into very small bipolar filaments (9,10). This later characteristic is shared by other non-muscle myosins.

It is our belief that the study of the unique properties of myosin II is essential to understanding the function of myosin II *in vivo*, which probably involves rapid assembly and disassembly of myosin II filaments in the cytoplasm. Towards this end we are attempting to clone and sequence the *Acanthamoeba* myosins and to study myosin structure/function relationships by site-directed mutagenesis.

RESULTS

Our approach was to use a portion of the nematode UNC 54 body wall myosin heavy chain gene (gift of L. Barnett, MRC, Cambridge) to search directly for Acanthamoeba myosin heavy chain genes. The probe used was a 2.7 kb Bam H1 fragment which contains most of the coding information for the nematode myosin head, including the actin and ATP binding sites and the active thiols (11). Initially, we found that this probe hybridized to 8 DNA fragments (totally ∼40 kb of DNA) in Southern blots of Bam H1-digested Acanthamoeba genomic DNA and to two mRNA species (5300 bp and 4250 bp) in Northern blots of amoeba RNA. The fact that this heterologous probe recognized discrete amoeba DNA sequences and RNA transcripts, and that these transcripts were large enough to encode myosin heavy chains (5300 bp is 107% of the size necessary to encode the 185 kDa myosin II heavy chain) strongly suggested that the probe could identify amoeba myosin heavy chain genes. Therefore, we constructed an Mbo 1 partial-digest genomic library of Acanthamoeba DNA in phage lambda 2001 (vector a gift of J. Karn, MRC, Cambridge) and screened 40 genome equivalents (80,000 plaques) with this heterologous probe. We obtained approximately 4 positive phage per genome equivalent. Forty phage clones were purified and rescreened with a 490 bp fragment encompassing the active thiol region of the nematode myosin gene. Restriction mapping of the 7 phage clones which were positive for the thiol probe yielded 2 groups of overlapping phage (3 in one group, 2 in the other) and 2 individual, non-overlapping phage. The group containing 3 members and spanning roughly 21 kb of DNA was identified as containing an amoeba myosin II heavy chain gene based on the following observations: the phage DNA hybrid selected a mRNA which translated in vitro to yield a 185-kDa protein; the 185-kDa protein comigrated in SDS-PAGE with authentic, purified myosin II heavy chain, and was quantitatively and specifically immunoprecipitated by myosin II antiserum; and the phage DNA insert hybridized to a 5300 bp amoeba mRNA.

Our current efforts have been directed towards defining the limits of the myosin II gene. To locate the 3' end of the gene, a mixed 20 bp oligonucleotide probe was synthesized based on a portion of the 58 amino acid residues sequenced from a carboxy-terminal myosin II chymotryptic peptide (5) (amino acid sequence: HEKNKQ; oligomer sequence: CA(TC)GA(AG)AA(AG)AA(TC)AA(AG)CA(AG)(TC)T). The oligomer was end-labelled and hybridized to digests of the myosin II genomic clone. A 1.3 kb Sac I fragment kb which hybridized to the oligomer was subcloned in M13 and sequenced. The sequence contained a stretch of nucleotides encoding exactly the 58 amino acids in the sequenced myosin II peptide. We found, as suspected from the peptide sequencing (5), an additional 8 amino acids at the C-terminus and a termination codon. These results confirmed the identity of the genomic clone, identified the 3' end of the myosin II gene, and provided transcriptional orientation. Approximately 7 kb of DNA exists in the 5' direction in the phage clone group to encode the balance of the gene.

Assuming no introns, the 5' end of the gene should lie approximately 5300 bp from the identified 3' end. We found that DNA fragments from the genomic clone that span the region from the 5' end of the phage insert to within 5800 bp of the identified 3' end failed to hybridize to the 5300 bp myosin II transcript. Furthermore, a 350 bp fragment encompassing the putative ATP binding site in the nematode myosin gene hybridized strongly to a small DNA restriction fragment which mapped to approximately 4800 bp from the 3' end of the myosin II gene. The ATP binding site in myosin II is thought to occur within the amino terminal 15% of the heavy chain (12), similar to other myosins. Recently we have sequenced in these areas and found (i) a stretch of nucleotides about 5500 bp from the 3' end of the gene and 80 bp 5' of an ATG which strongly resembles the "TATA" box sequence described in the _Acanthamoeba_ actin gene (13), and (ii) an open reading frame as far as we have sequenced into the gene (~ 1000 bp), except for a single apparent intron of 177 bp about 500 bp 3' of the ATG. Furthermore, the deduced myosin II amino acid sequence shows highly significant homology with the nematode protein sequence. Except for the first ~ 60 residues of myosin II which are quite different from the nematode, the remainder of the sequence so far completed shows close to 60% exact match with the nematode protein and only a few small gaps in the sequence are necessary to keep the homology strong. Certain areas of sequence are extremely similar. For example, in the area of the putative nematode ATP binding site, 33 of 41 contiguous residues match exactly with the myosin II sequence, and most of the changes are conservative. These results tentatively identify the 5' end of the myosin II gene at approximately 5500 bp from the 3' end and suggest that the genomic phage clone contains the entire myosin II heavy chain gene. The results also suggest that the myosin II gene may contain only the one small 5'-end intron.

DISCUSSION

This is the first report of the isolation and characterization of a non-muscle myosin heavy chain gene. Just as the genes for other highly conserved proteins can be isolated using heterologous gene probes (eg. actin, tubulin), we have been able to isolate the amoeba myosin II heavy chain gene using a nematode myosin gene as a probe. Preliminary sequencing of the amino terminal 20% of the myosin II gene shows areas of strong homology at the amino acid level with the nematode protein and indicate why the heterologous probe worked. By analogy, the myosin II gene could be useful as a heterologous probe to isolate other muscle and non-muscle myosin genes. We are currently sequencing the myosin II gene in its entirety. The sequence should be of great value in furthering our understanding of the unique enzymatic, structural, and regulatory properties of this non-muscle myosin. Our long range goal of studying myosin structure/ function relationships by site-directed mutagenesis should be greatly facilitated by the apparent paucity of intron DNA in the myosin II gene.

ACKNOWLEDGMENTS

J.H. thanks Rob Horlick, Rudi Billeter and Ann Sielers for their advice during this work. J.H. also thanks in particular Carl Schmidt for his excellent advice on technical matters and his helpful discussions during the course of this work.

REFERENCES

1. Maruta H, Korn ED (1977). Acanthamoeba myosin II. J Biol Chem 252:6501.
2. Collins JH, Korn ED (1981). Purification and characteriztion of actin-activatable, Ca^{2+}-sensitive myosin II from Acanthamoeba. J Biol Chem 256: 2586.
3. Collins JH, Korn ED (1980). Actin activation of Ca^{2+}-sensitive Mg^{2+}-ATPase activity of Acanthamoeba myosin II is enhanced by dephosphorylation of its heavy chain. J Biol Chem 255:8011.
4. Collins JH, Cote GP, Korn ED (1982). Localization of the three phosphorylation sites on each heavy chain of Acanthamoeba myosin II to a segment at the end of the tail. J Biol Chem 257:4529.
5. Cote GP, Robinson EA, Appella E, Korn ED (1984). Amino acid sequence of a segment of the Acanthamoeba myosin II heavy chain containing all three regulatory phosphorylation sites. J Biol Chem 259: 12781.
6. McClure JA, Korn ED (1983). Purification of a protein phosphatase from Acanthamoeba that dephosphorylates and activates myosin II. J Biol Chem 258: 14570.
7. Kuznicki J, Albanesi JP, Cote GP, Korn ED (1983). Supramolecular regulation of the actin-activated ATPase activity of filaments of Acanthamoeba myosin II. J Biol Chem 258:6011.
8. Kuznicki J, Cote GP, Bowers B, Korn ED (1985). Filament formation and actin-activated ATPase activity are abolished by proteolytic removal of a small peptide from the tip of the tail of the heavy chain of Acanthamoeba myosin II. J Biol Chem 260: 1967.
9. Collins JH, Kuznicki J, Bowers B, Korn ED (1982) Comparison of the actin binding and filament formation properties of phosphorylated and dephosphorylated Acanthamoeba myosin II. Biochemistry 21:6910.
10. Pollard TD (1982). Structure and polymerization of Acanthamoeba myosin II filaments. J Cell Biol 95: 816.

11. Karn J, Brenner S, Barnett L (1983). Protein structural domains in the Caenorhabditis elegans unc-54 myosin heavy chain gene are not separated by introns. Proc Natl Acad Sci USA 80:4253.
12. Atkinson MAL, Kuznicki J, Korn ED (1984). Physical and enzymatic properties of a soluble head fragment derived from Acanthamoeba myosin II. J Cell Biol 99:35a.
13. Nellen W, Gallwitz D (1982). Actin genes and actin messenger RNA in Acanthamoeba castellanii. J Mol Biol 159:1.

Molecular Biology of Muscle Development, pages 345–361

SARCOMERIC MYOSIN HEAVY CHAIN GENE FAMILY: ORGANIZATION AND PATTERN OF EXPRESSION[1]

V. Mahdavi, E.E. Strehler, M. Periasamy, D. Wieczorek, S. Izumo, S. Grund, M.-A. Strehler and B. Nadal-Ginard

Department of Cardiology, Children's Hospital and Department of Pediatrics, Harvard Medical School, Boston, MA 02115

The regulated expression of structurally distinct, tissue-specific and developmentally regulated protein isoforms is an essential characteristic of cell differentiation, ontogenic development and physiological adaptation. The functional properties of the contractile apparatus in different muscle tissues derive primarily from the composition of its constitutive structural and regulatory proteins. Isoforms of many such sarcomeric proteins are encoded by multigene families, the members of which are subject to tissue- and developmental stage-specific regulation (1). The diversity of these proteins is augmented, in some cases, by alternative RNA splicing of individual gene transcripts (2, and ref in 2). Furthermore, the muscle cells, even after terminal differentiation, are able to reversibly modulate the properties of the contractile apparatus in response to various physiological stimuli, thus providing an additional capacity to generate myofibrillar diversity (3,4).

The plasticity of the muscle fibers is well illustrated by the isoform diversity of myosin heavy chain (MHC), the major determinant of the contractile properties of the muscle. Despite their similarity in primary structure, the expression of the different MHC isoforms is tightly regulated in a tissue- and developmentally-specific manner. In addition, innervation patterns and altered physiological stimuli, such as disuse, electrical stimuli and mechanical overload, are known to cause MHC isoform switches

[1]This work was supported in part by grants from the Muscular Dystrophy Association of America, The American Heart Association and the National Institutes of Health.

(3). Furthermore, various hormones including thyroid hormone, insulin, glucocorticoids and testosterone, can modulate the phenotypic expression of the MHC genes, as best demonstrated by the cardiac ventricular α - and β - MHCs (4 and ref in 4).

In an attempt to understand the molecular basis for the existence of several isoforms of MHC, as well as to elucidate the regulatory mechanisms controlling their tissue-specific, developmental and hormonal regulation, we have cloned the cDNA and genomic DNA sequences coding for the MHCs expressed in several striated muscle tissues. So far seven distinct sarcomeric MHC genes have been isolated and characterized in the rat (4-11). The precise pattern of expression of these seven MHC genes during the normal development of the skeletal and cardiac muscle fibers, as well as in different anatomically defined muscles subjected to different levels of circulating thyroid hormone, has been established.

The overall picture that emerges from these studies is a complex one: a. The MHC genes appear to be linked in tandem in the chromosome and have a highly conserved organization and primary structure. b. In all muscles so far analyzed, more than one — and in the particular case of the extraocular muscle, up to six — MHC genes are expressed. c. All MHC genes -- except one -- are expressed in several different muscle types. d. The developmental pattern of expression and the hormonal regulation of the MHC genes is not intrinsic to any given MHC gene but is determined, at least in part, by the muscle in which a particular gene is expressed.

RESULTS AND DISCUSSION

Structural organization of the MHC multi-gene family. Seven individual MHC gene sequences coding for the different protein isoforms defined as cardiac α -MHC, cardiac β -MHC (=skeletal slow type I MHC, see below), skeletal embryonic, neonatal, fast IIA, fast IIB MHCs and extra-ocular muscle (EOM) MHC are depicted in Fig. 1. The identity of these genomic clones was established by nucleotide sequence comparison with the 3' untranslated and gene-specific sequences of skeletal and cardiac MHC cDNA clones that were characterized previously (4-11, and see below). As seen from the complete genomic map of the cardiac α and β , embryonic and fast IIA skeletal MHC genes, the length of the individual members of the sarcomeric multigene family ranges from 22 to 25kb of DNA. Since the MHC mRNA is 6300 nucleotides long (5) the intron/exon ratio of these genes is relatively high.

Each genomic restriction map was established by the analysis of several overlapping genomic clones. Such analysis revealed that, as previously demonstrated for the α -and β -MHC genes (11), the

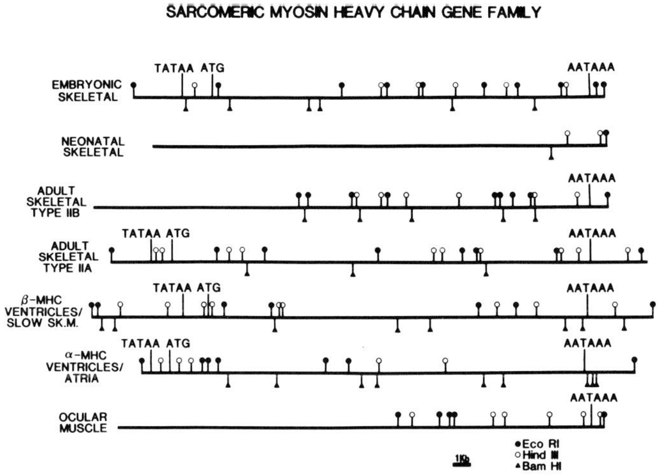

FIG. 1

skeletal MHC genes are also linked in tandem. Indeed, only 2kb of intergenic sequences separate the 3' end of the neonatal MHC from the 5' end of the Fast IIA MHC gene (data not shown). Furthermore, MHC coding sequences have been identified downstream from the 3' end of the embryonic gene and our preliminary data are consistent with the possibility that these sequences represent the 5' end of the neonatal MHC gene. Therefore, consistent with the observation that the sarcomeric MHC genes in mouse (12,13) and human (12) have been mapped onto a single chromosome, and the tandem organization of the cardiac α - and β-MHC genes, it appears that embryonic, neonatal and adult skeletal muscle MHC genes are also organized in a head to tail fashion that reflects the order of their developmental expression (see below).

In order to attain a better understanding of structure-function relationships in the MHC multigene family, the complete nucleotide sequence of the 25kb long embryonic MHC gene was obtained (14,15). Such analysis revealed that this MHC gene is split by 40 introns (Fig. 2). However, the role, if any, of introns in gene structure, expression, and regulation is not yet understood. It has been speculated that introns interrupt protein - coding sequences between structural or functional domains (16). Yet, in the MHC gene introns do not separate the coding sequences at the "expected" positions which would lie between the characteristic 28 amino acid repeats in the rod portion and between rod and S2, and S2 and S1

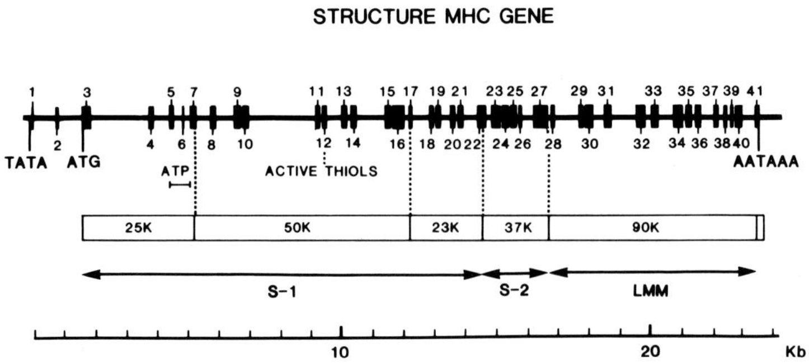

FIG. 2 Structural map of the rat embryonic skeletal muscle MHC gene.

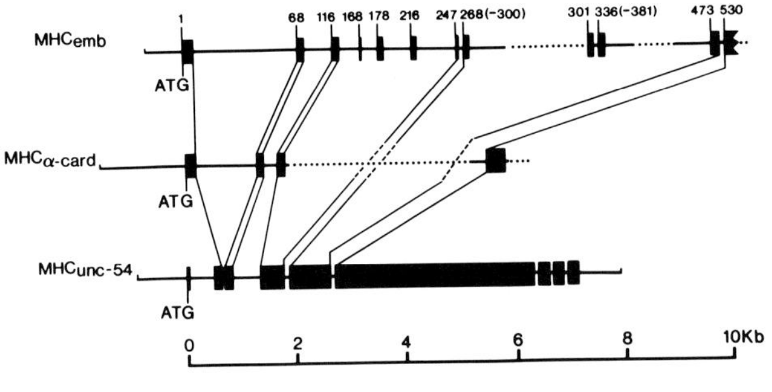

FIG. 3 Structural map of the 5' end region of the rat embryonic skeletal muscle (MHC emb) and α-cardiac (MHC α -card) MHC genes and of the nematode unc-54 gene.

subfragments. However, the fact that all the known positions of introns in other rat MHC genes are fully conserved is striking. Interestingly, most of the intron positions at the 5' end of the gene are also conserved between the MHCs of the rat and nematode _C. elegans_ unc-54 (17) while the latter gene is split by a total of only 8 introns (Fig. 3). The high level of structural conservation of the MHC genes and their close linkage on the chromosome strongly suggest that the sarcomeric MHC genes have originated by duplication of a common ancestral gene. Another feature which seems characteristic of the contractile protein genes is the presence of intervening sequences in the sequences corresponding to the 5' untranslated sequences of the mRNA of several genes analyzed so far. Two introns are found in the embryonic, α -and β -MHC genes, whereas only one intron interrupts the leader sequence of the mRNA of α -actin (18) and troponin T (19) genes. It is therefore possible that these "conserved" interruptions of the sarcomeric genes at their 5' ends are of importance in gene expression or regulation, as it is the case for the immunoglobulin genes (20).

Developmental and tissue-specific pattern of expression of the sarcomeric MHC genes. During the development of cardiac and skeletal muscles, two MHC -isozyme transitions occur involving two MHC isoforms in the ventricle (α -and β -MHC, 4,21) and a minimum of five MHC isoforms in the leg skeletal muscle (9,22). To understand at what level (transcriptional, post-transcriptional or translational) the modulation of expression of the sarcomeric MHC genes occurs and to ascertain the nature and tissue-specificity of the genes involved, we analysed the S_1-nuclease protection pattern of cardiac and skeletal mRNAs obtained from several developmental stages using the seven MHC clones described above. Due to the high degree of sequence homology between the different MHC mRNAs (6), only gene specific probes, most of them containing the untranslated 3' end of the mRNAs, could be used for this analysis.

In the ventricular myocardium at least two MHC mRNAs are detected in the ventricles at almost all stages of development (from 18-day-old fetuses to 6-month-old adults) as shown by the presence of two S1-resistant fragments of the expected sizes, obtained by using a cardiac β -MHC probe generated from pCMHC5 (4,7) (Fig. 4A). The fully protected pCMHC5 probe is more intense than its partially protected counterpart in the ventricular RNAs of 18- and 20-day-old fetuses. After birth, the absolute level of MHC mRNA homologous to the pCMHC5 probe decreases slightly, as visualized by the amount of radioactivity in the 304bp fragment. This MHC mRNA is barely detectable in the 28-day-old ventricle, but reappears in the 3- and 6-month-old animals. The relative

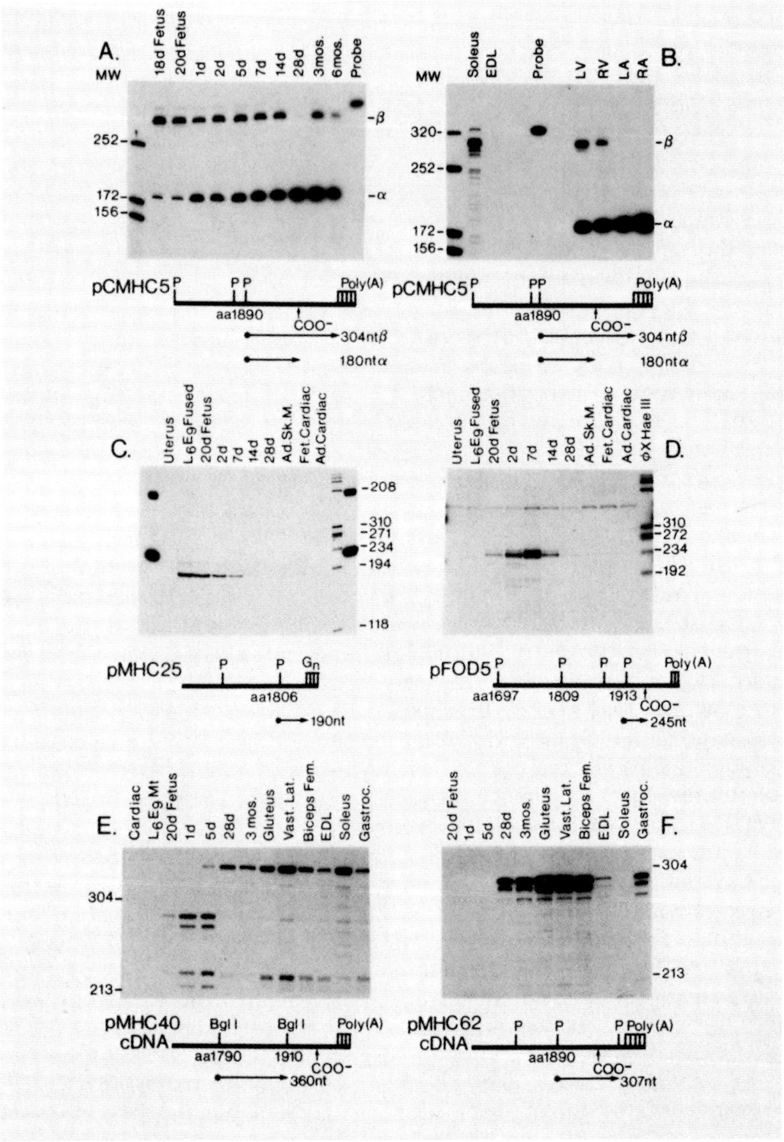

FIG. 4 Developmental and tissue-specific expression of the sarcomeric MHC genes.

abundance of this mRNA correlates very well with the relative levels of β -MHC isoform in the same samples. These results demonstrate that the cardiac MHC gene represented by pCMHC5 codes for the β -MHC.

The 180bp long partially protected fragment shown in Fig. 4A is produced by the hybridization of the portion of the probe common to both cardiac MHC mRNAs and corresponds to the -MHC mRNA (4,7). This mRNA is present in low amounts in the fetus, increases rapidly during the first 3 weeks of postnatal life, and is apparently the only species present at 4 weeks of age. Its pattern of expression correlates very well with the amount of α -MHC mRNA predicted by the accumulation pattern of α -MHC in the same hearts (4).

A pattern very similar to the one shown by 18 day-old fetal mRNA is detected in earlier stages of development starting at day 12 of fetal life when the cardiac tube first becomes apparent in the rat (data not shown). Thus, the α -and β -MHC are the only ones expressed throughout the development of the heart.

It has been previously shown that slow twitch (soleus) skeletal muscle and V3 myosins on one hand and the atrial and ventricular α -MHCs on the other (23) are structurally and immunologically related. These observations suggested that the MHCs present in atrial and soleus muscle might be encoded by the cardiac α -and β -MHC genes, respectively. To test this possibility, S1 nuclease mapping experiments using the gene-specific pCMHC5 probes with atrial, slow and fast skeletal muscle RNAs were performed.

As shown in Fig. 4B, the fully protected fragment, produced by the hybridization of soleus mRNA with the pCMHC5 probe, demonstrates that a MHC mRNA expressed in this muscle is identical to the cardiac β -MHC mRNA and is, therefore, encoded by the cardiac β -MHC gene. Full protection of the pCMHC5 probe was also detected using RNAs from diaphragm and extraocular muscles (see below) which also contain slow fibers. From these results it can be concluded that the slow (type I) MHC and cardiac β -MHC isoforms are encoded by the same gene. Trace amounts of fully protected probe can be also detected in RNAs from right and left atria, suggesting a very low level of expression of the β -MHC gene in this tissue as well.

No partial protection of the pCMHC5 probe was detected with RNA's from any of the skeletal muscles so far tested at any stage of development from 16 day-old fetuses to adult animals, indicating that, in the rat, the cardiac α -MHC gene is probably not expressed in skeletal muscles. In contrast, the α -MHC mRNA is detected in the left and right atria as shown by the presence of a 180bp fragment.

The fact that the α-MHC gene is expressed in both ventricular and atrial myocardium and that the β -MHC gene is expressed in both ventricular myocardium and slow skeletal muscle fibers reveals a new aspect of the tissue-specific expression of the MHC genes and raises important questions related to their developmental stage-mediated regulation. Indeed, the α -MHC gene is expressed in the atria even at its early stage of development when this gene is not expressed in the ventricle. Second, the β -MHC gene is expressed at high levels in the slow muscle fibers of adult but not in fetal stages whereas its pattern of expression is reversed in the ventricles. These results argue strongly against circulating hormonal factors as the main inducer of these genes during development.

Based on biochemical, immunological and peptide mapping techniques, evidence has accumulated to suggest that during the development of skeletal muscle, five types of MHC forms are expressed: a form mainly expressed during fetal life and also during in vitro myogenesis, a neonatal or perinatal form mainly present in the last week of gestation and first weeks of extrauterine life, and three adult forms corresponding to two adult fast (fast oxydative IIA and fast glycolytic IIB) and one slow (slow oxydative IA that, as shown above is identical to the cardiac β -MHC) MHCs. In addition, extraocular muscles express a different MHC as demonstrated at the protein and gene levels (10). Four MHC cDNA clones isolated from fetal (pMHC-25 and FOD5) and adult (pMHC-40 and pMHC-62) muscle cDNA libraries were used for S1 nuclease mapping analysis to determine the temporal and tissue-specific expression of the MHC genes they represent.

A full length protection of the pMHC-25 probe (194bp) is observed with RNA from L_6E_9 myotubes, 20 day fetal, 2 day, 7 day and 14 day old postnatal skeletal muscle (Fig. 4C). Note that this MHC mRNA represented by pMHC-25 is expressed abundantly in early development but decreases gradually to an undetectable level in the postnatal period. The mRNAs from skeletal muscle of 28 day-old and 6 months old animals show little or no full protection of the probe. Furthermore, no protection of the probe is detected either by smooth muscle (uterus) or by fetal or adult cardiac muscle mRNAs. The developmental pattern of expression of the MHC gene represented by pMHC-25 shown above demonstrates that it is in fact the embryonic MHC gene.

As shown in Fig. 4D, when the FOD5 probe is used the full-length protection of the probe (243 bp) is obtained only with skeletal-muscle mRNAs isolated from animals in the perinatal period. The MHC mRNA homologous to the probe is not present in L_6E_9 myotubes, known to express mainly the embryonic MHC mRNA represented by pMHC-25 (6). The MHC mRNA represented by pFOD5 is present in the 20-day-old fetus, and its abundance increases postnatally until the maximum level is reached in the 7-day-old-animals. By 28 days, its concentration has greatly diminished and it can not be detected in adult (3 month-old) mixed skeletal muscle fibers. Furthermore, it is not detected in smooth muscle (uterus) nor in fetal and adult cardiac muscles. The pattern of expression of the MHC gene represented by pFOD5 corresponds very closely with the presence of the MHC protein characteristic of the neonatal period reported previously (22). It can be concluded from these results that the neonatal MHC gene is distinct from the embryonic and adult MHC genes.

The results of similar experiments performed using the MHC cDNA probes isolated from an adult skeletal muscle cDNA library are shown in Fig. 4E and F. The full-length protection of the pMHC-40 probe (Fig. 4E) which is not observed in the RNA isolated from 20 day old fetus, is readily detectable in RNAs from one to five day old animals and becomes more intense in the muscle mRNAs from adult animals. The MHC gene represented by pMHC-40 is expressed in all adult fast muscles containing fast IIA and fast IIB myofibers as well as in slow muscle (soleus) which is constituted by slow type I and fast IIA myofibers. These observations strongly suggest that this gene corresponds to the fast IIA MHC gene, since the slow I MHC gene is identical to the β -MHC gene as detected by the pCMHC-5 probe (see above).

The protection pattern of the pMHC-62 probe obtained with the same RNA samples (Fig. 4F) shows that the corresponding MHC gene is expressed late in development, and only in the adult muscles that have fast IIB fibers. Therefore the identity of this MHC gene can be assigned to fast IIB MHC.

The close correlation between the S_1-nuclease patterns shown above and the previously described pattern of MHC isoform expression at the protein level strongly suggests that the qualitative and quantitative changes in the MHC phenotype observed during development can be entirely accounted for by changes in the level of their respective MHC mRNAs. Translational and post-translational mechanisms, if present, do not appear to play a major role in the production of the MHC phenotype during muscle development (24).

In addition to the MHC genes identified by their corresponding cDNA clones, a MHC genomic clone, λ 10B3 was isolated (10). To study the expression of this gene by S1 nuclease mapping analysis, a genomic DNA fragment that includes the codons for amino acids 1697 to 1736 in MHC and 179bp of intervening sequence was used.

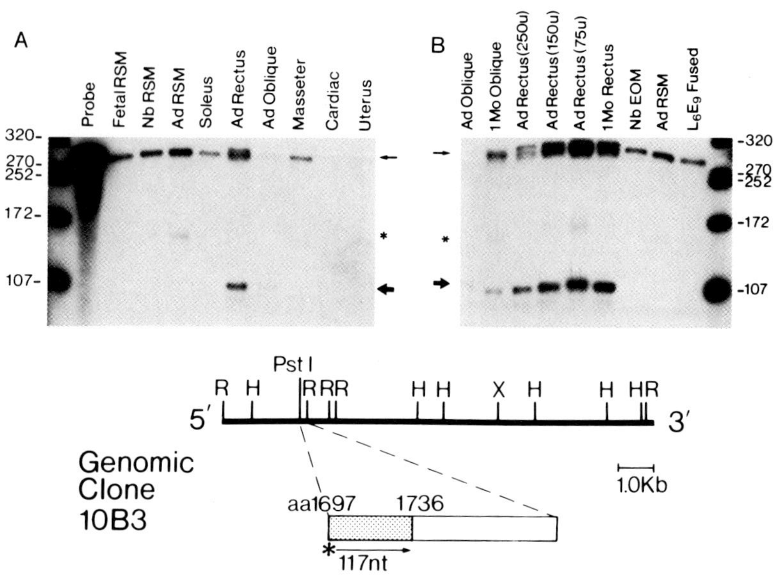

FIG. 5 Developmental and tissue specific expression of the EOM MHC gene.

Hybridization to the homologous mRNA should generate a 117 nucleotides long protected fragment, since at this point an exon-intron splice site occurs and the rest of the sequence in the probe corresponds to intervening sequence (10). From the S1 nuclease protection pattern observed in Fig. 5A, it is apparent that the gene represented by λ 10B3 is expressed only in rat extraocular musculature (recti and obliqui muscles) and not in cardiac or limb skeletal musculature, including the soleus muscle. Furthermore, this probe does not show any protection with RNA isolated from newborn cardiac or adult atrial tissues, fetal skeletal musculature at any stage of development, adult diaphragm or extensor digitorum longus muscles (data not shown). In addition, RNA isolated from the masseter muscle (a muscle which in carnivorous predators displays a hyper-fast contractile myosin) does not show protection when hybridized with the above probe. The band at 296bp seen in Fig. 5A is due to reannealing of the probe itself; the faint band at 155bp is due to incomplete denaturation of the probe because of the high (60%) G-C content in the intervening sequence.

To ascertain the developmental stage specificity of λ10B3 MHC expression, total RNA was isolated from the extrinsic eye muscles at 3 days, 1 month, and 1 year of post-natal life. As seen in

Fig. 5B, the mRNA represented by the λ 10B3 MHC gene is first detected between 3 days and 1 month of post-natal life, and is present throughout adult life. From these results it is reasonable to conclude that the MHC gene represented by λ 10B3 is extraocular muscle-specific.

Multiple MHC isozyme mRNA sequences are expressed in the EOM. To study the developmental expression of MHCs in EOM and to determine whether there are multiple MHC isoforms in these muscles (in addition to the type represented by λ 10B3) we used the probes from rat embryonic (5), neonatal (9), and adult (4,6,7), skeletal and cardiac muscle MHC cDNA clones in S1 nuclease protection analyses. The result of these experiments summarized in Fig. 6 and described in detail elsewhere (10), showed that the adult rat extraocular muscles express all known sarcomeric myosin heavy chain genes (with the possible exception of the α -cardiac gene), including the MHC isoform which is specific for these muscles.

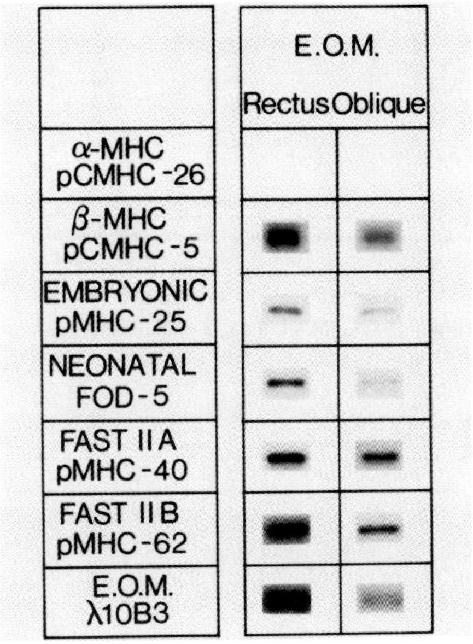

Fig. 6 Expression of the sarcomeric MHCs in extraocular muscle.

During the neonatal stage, 4 different MHC mRNAs are expressed in EOM: embryonic, neonatal, fast IIA, and slow I (β -MHC). However, electrophoresis of native myosin demonstrates that the predominant myosin protein accumulated at this time is the neonatal form (10). In the adult stages, the transcripts from these genes are still retained and supplemented with at least 2 additional ones: the fast IIB and the EOM-specific MHCs.

The expression of a full complement of sarcomeric MHC genes in adult EOM tissue, in addition to embryonic and neonatal isoforms, forces a re-evaluation in defining the developmental and tissue specificity of MHC isoforms. The fact that these genes are expressed at the mRNA and protein levels with regional localization of the embryonic and neonatal isoforms in a particular subset of fibers (10), strongly suggests that the production of these mRNA transcripts is not merely a product of gene transcriptional leakage, but is a controlled and regulated process. Furthermore, the consitutive expression of the embryonic and neonatal myosin isozymes in normal adult muscles clearly demonstrates that their expression is not solely regulated by developmental stage-specific factors. Moreover, the results presented here highlight a novel feature of the MHC multigene family. Although there are predominant MHC isoforms in an isolated muscle at a specific stage of development, with the exception of the EOM-specific MHC, there are not sarcomeric MHC isoforms which exhibit complete developmental- or tissue-"restricted" expression.

Effect of thyroid hormone on the expression of sarcomeric MHC genes. The precise molecular basis for the spacial and developmental pattern of expression of the different MHC genes is not well understood. Among various hormones, thyroid hormone has been shown to cause MHC isoform switches in developing and adult cardiac and skeletal muscle fibers (3,4). In the light of our observation that all MHC genes, with the exception of the EOM specific MHC gene, are expressed in more than one muscle type, the question arised as to whether a given MHC gene will exhibit the same response to thyroid hormone when expressed in different tissues. Using the six different specific MHC cDNA probes described above, we have analysed the MHC mRNAs of seven different muscles obtained from animals with different circulating thyroid hormone levels — euthyroid (normal), hypothyroid (thyroidectomized) and hyperthyroid (normal animals injected with 50 µg/kg of body weight of triiodo-L-thyronine) (25). As seen in Fig. 7, more than one MHC gene is expressed at different levels in each anatomically defined adult muscle and all six genes analyzed are expressed in more than one muscle. However, the "set" of MHC genes expressed is characteristic of each muscle type: α -and β -MHC genes in atrial and ventricular myocardium; β cardiac (=slow I), fast IIA, and IIB, as well as embryonic MHCs in soleus and diaphragm

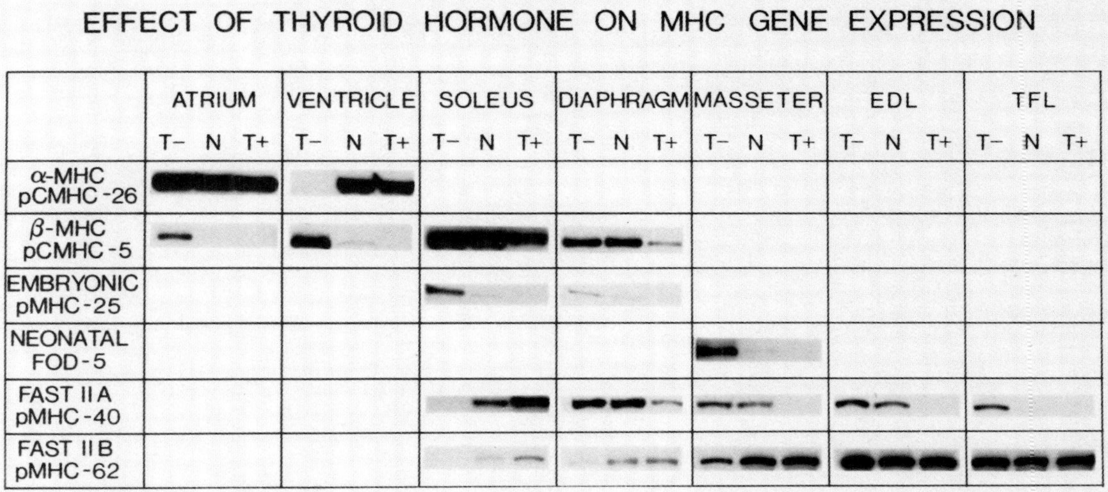

Fig. 7 Effect of thyroid hormone on MHC gene expression. The bands correspond to the full protection of the respective cDNA probes by S1 nuclease mapping analyses. The relative density of each band reflects the relative amount of the specific mRNA coded by the corresponding cDNA probe. T−: Hypothyroid. N: Euthyroid. T+: Hyperthyroid. EDL: Extensor Digitorum Longus. TFL: Tensor Fasciae Latae

muscles; fast IIA and IIB, as well as the neonatal MHC genes in the masseter; fast IIA and IIB MHC genes in EDL and fast IIB MHC gene in TFL. These last three muscles are predominantly composed of fast muscle fibers.

Interestingly, all muscles tested respond to change in thyroid hormone levels. The overall effect of thyroidectomy (T-) is to induce the expression of the MHC gene(s) coding for slow isoforms and/or de-induce the MHC gene(s) coding for fast isoforms. Conversely, hyperthyroidism (T+) has the exact opposite effect. For instance, in TFL muscle, the fast IIA MHC gene is induced in the hypothyroid animals. The product of this gene has a lower ATPase activity than the MHC encoded by the fast IIB gene, the only form normally expressed in this tissue. On the other hand, in soleus and diaphragm muscles of the hyperthyroid animal, the slow I MHC (β) gene is de-induced while expression of the fast IIB gene is induced.

These observations corroborate previous studies that demonstrated the dramatically opposite effect of thyroid hormone on the MHC genes expressed in the ventricular myocardium (4). However, a careful anlaysis of the results demonstrates that the thyroid hormone response of the MHC genes does not appear to be intrinsic to any given MHC gene but is determined by two factors: the muscle in which this gene is expressed and which other MHC genes are expressed in this particular muscle. This phenomenon is more striking in the case of the α -MHC and the fast IIA MHC genes. The level of expression of the α -MHC gene does not change in the atrium in either hypo- or hyperthyroid state whereas, in the ventricle, there is a complete de-induction of this gene by hypothyroidism. The fast IIA MHC gene exhibits two opposite types of response in different muscles of hyperthyroid animals. It is up-regulated in soleus where the fast IIA MHC is the major fast isoform while β -MHC is the major slow isoform. In contrast, this same fast IIA MHC gene is down-regulated in the fast twitch muscles (masseter, EDL, TFL) where the fast IIB MHC is the major fast isoform. To our knowledge, this is the first documented case in which the same genes, expressed in different tissues, have been shown to respond to the same hormonal stimulus in a completely opposite direction. These observations raise significant questions regarding the regulation of the MHC multigene family as well as the molecular mechanisms of thyroid hormone action.

CONCLUSIONS

The results presented here illustrate the highly complex mode of regulation of the expression of the MHC genes. Each member of this multi-gene family can be under different developmental and hormonal regulatory programs, depending on the muscle in which it is expressed. The fact that only certain MHC genes are expressed and/or are inducible in different muscles indicates that these genes are under the regulation of different programs that act at different levels. A basic level of regulation appears to determine which set of MHC genes is made available for expression in a given tissue -- possibly through changes in chromatin conformation, pattern of methylation, etc. -- and become sensitive to inductive factor(s). A second level of regulation, which might involve the presence or absence of certain regulatory sequences in the MHC genes, would determine which genes are sensitive to a set of inductive factors, if these were produced by all muscle cells. This mode of regulation is obviously not sufficient to account for the different developmentally and hormonally regulated patterns of expression exhibited by the MHC genes in different tissues. Therefore, one has to postulate that apart from the intrinsic differences in the primary structure of the different MHC genes that would define their "range" of inducibility, hormonal and/or neural factors, some of them specific for individual muscles, are involved in the regulation of the MHC multi-gene family, either directly or through the availability of the corresponding receptors in particular muscle cells.

REFERENCES

1. Pearson ML, Epstein HF eds. (1982). "Molecular and Cellular Control of Muscle Development." (Cold Spring Harbor, New York: Cold Spring Harbor Laboratory).
2. Nadal-Ginard B, Breitbart RE, Strehler EE, Periasamy M, Ruiz-Opazo N, Mahdavi V (1985). Alternative splicing, a common mechanism for the generation of contractile protein diversity from single genes. In Molecular Biology of Muscle Development UCLA Symposia on Molecular and Cellular Biology, New Series, Volume 29. Emerson C, Fischman DA, Nadal-Ginard, B, Siddiqui MAQ (eds). New York Alan R. Liss.
3. Whalen RG, (1985). Myosin isoenzymes as molecular markers for muscle physiology. J Exp Biol in press.
4. Lompre AM, Nadal-Ginard B, Mahdavi V (1984). Expression of the cardiac ventricular α -and β -myosin heavy chain genes is developmentally and hormonally regulated. J Biol Chem 259:6437.
5. Medford RM, Wydro RM, Nguyen HT, Nadal-Ginard B (1980). Cytoplasmic processing of myosin heavy chain messenger RNA: Evidence provided by using a recombinant DNA plasmid. Proc Natl Acad Sci USA 77:5749.
6. Nadal-Ginard B, Medford RM, Nguyen HT, Periasamy M, Wydro RM, Horning D, Gubits RM, Garfinkel LI, Wieczorek DF, Bekesi E, Mahdavi V (1982). Structure and regulation of a mammalian

sarcomeric myosin heavy-chain gene. In Muscle Development: Molecular and Cellular Control. Pearson ML, Epstein HF (eds). Cold Spring Harbor, NY. p. 46.
7. Mahdavi V, Periasamy M, Nadal-Ginard B (1982). Molecular characterization of two myosin heavy chain genes expressed in the adult heart. Nature 297:659.
8. Wydro RM, Nguyen HT, Gubits RM, Nadal-Ginard B (1983). Characterization of sarcomeric myosin heavy chain genes. J Biol Chem 258:670.
9. Periasamy M, Wieczorek DF, Nadal-Ginard B (1984). Characterization of a developmentally regulated perinatal myosin heavy-chain gene expressed in skeletal muscle. J Biol Chem 259:13573.
10. Wieczorek, DF, Periasamy M, Butler-Browne GS, Whalen RG, Nadal-Ginard B (1985). Co-expression of multiple myosin heavy chain genes, in addition to a tissue-specific one, in extraocular musculature. J Cell Biol, in press.
11. Mahdavi V, Chambers AP, Nadal-Ginard B (1984). Cardiac α - and β -myosin heavy chain genes are organized in tandem. Proc Natl Acad Sci USA 81:2626.
12. Leinwand LA, Fournier REK, Nadal-Ginard B, Shows T (1983). Multigene family for sarcomeric myosin heavy chain in mouse and human DNA: Localization on a single chromosome. Science 221:766.
13. Czosnek H, Nudel U, Mayer Y, Shani M, Barker PF, Pravtcheva DD, Ruddle FH, Yaffe D (1982). The genes coding for the muscle contractile proteins, myosin heavy chain, myosin light chain 2 and skeletal muscle actin are located on three different mouse chromosomes. EMBO J 1:1299.
14. Strehler EE, Mahdavi V, Periasamy M, Nadal-Ginard B (1985). Intron positions are conserved in the 5' end region of myosin heavy-chain genes. J Biol Chem 260:4680.
15. Strehler EE, Strehler M-A, Nadal-Ginard B, in preparation.
16. Gilbert W (1978). Why genes in pieces? Nature 271:501.
17. Karn J, Brenner S, Barnett L (1983). Protein structural domains in the Caenorhabditis elegans unc-54 myosin heavy chain genes are not separated by introns. Proc Natl Acad Sci USA 80:4253.
18. Zakut R, Shani M, Givol D, Neuman S, Yaffe D, Nudel U (1982). Nucleotide sequence of the rat skeletal muscle actin gene. Nature 298:857.
19. Breitbart RE, Nguyen HT, Medford RM, Destree AT, Mahdavi V, Nadal-Ginard B (1985). Intricate combinatorial patterns of exon splicing generate multiple regulated troponin T isoforms from a single gene. Cell 41.
20. Gillies SD, Morrison SL, Oi VT, Tonegawa S (1983). A tissue-specific transcription enhancer element is located in the major

intron of a rearranged immunoglobulin heavy chain gene. Cell 33:717.

21. Schwartz K, Lompre AM, Bouveret P, Wisnewsky C, Whalen RG (1982). Comparisons of rat myosins at fetal stages in young animals and in hypothyroid adults. J Biol Chem 257:1412.
22. Whalen RG, Sell SM, Butler-Browne GS, Schwartz K, Bouveret P, Pinset-Harstrom I (1981). Three myosin heavy-chain isozymes appear sequentially in rat muscle development. Nature 292:805.
23. Gorza L, Sartore S, Schiaffino S (1982). Myosin types and fiber types in cardiac muscle II. Atrial myocardium. J Cell Biol 95:838.
24. Medford RM, Nguyen HT, Nadal-Ginard B (1983). Transcriptional and cell cycle-mediated regulation of myosin heavy chain gene expression during muscle cell differentiation. J Biol Chem 258:11063.
25. Izumo S, Bois D, Nadal-Ginard B, Mahdavi V (1985). All members of the myosin heavy chain multi-gene family respond to thyroid hormone in highly tissue specific manner. Submitted for publication.

Molecular Biology of Muscle Development, pages 363–371

CHARACTERIZATION OF EMBRYONIC MYOSIN HEAVY CHAIN GENES FROM CHICKEN [1]

Gabriele E. Bugaisky, Smilja Jakovcic,
and *Patrick K. Umeda

Department of Medicine, University of Chicago,
Chicago, Illinois 60637

ABSTRACT A chicken genomic library has been screened by recombination in order to isolate and characterize myosin heavy chain genes encoding the embryonic isoforms. Two genomic clones were isolated and one, $\pi\lambda$251, was extensively characterized. By DNA sequencing, $\pi\lambda$251 has been identified as a gene segment encoding an embryonic myosin heavy chain. In the region encoding the final 1.6 kb of mRNA, there are at least 7 introns, one splitting the 3' untranslated region.

INTRODUCTION

In higher vertebrates, the heavy chain (HC) of myosin is encoded by a family of closely related genes. Analyses of various developing and mature striated muscles has shown that individual members of this gene family are differentially expressed (1,2). Since the HC genes are highly conserved, a major problem of identification can arise when a cDNA probe is used to screen a genomic library

[1]This work was supported by USPHS grants HL09172 and HL20592 and a grant from Muscular Dystrophy Association. *This work was done during the tenure of an Established Investigatorship of the American Heart Association and with funds contributed in part by the Chicago Heart Association. G.E.B. is a fellow of the Chicago Heart Association.

by standard protocols. Indeed, many different genomic sequences have been isolated with a single cDNA probe (3).

In order to circumvent the problem of identification, a recently developed screening protocol (4) has been used to directly isolate a unique gene segment encoding the 3' end of an embryonic myosin HC from chicken. Identification of this gene has been verified by DNA sequence analysis. The positions of introns and exons in the portion encoding the final 1.6 kb of mRNA have been determined by electron-microscopic heteroduplex mapping.

MATERIALS AND METHODS

An approximately 300 bp Pst I-Xba I fragment from the 3' end of p251 (5), and an about 300 bp Pvu II-Pst I fragment from the 5' end of p60 (6) were isolated (Figure 1) and subcloned into the microplasmid πAN7. These microplasmids were then independently used to screen a chicken genomic library (7) by recombination (4).

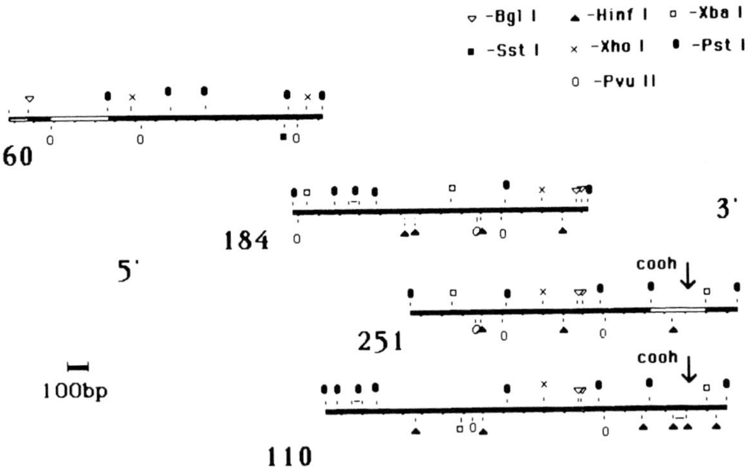

FIGURE 1. Restriction endonuclease maps of embryonic myosin HC cDNA clones. The regions of p251 and p60 that were subcloned into the microplasmid are indicated by the open bars.

The isolated phage were purified by two subsequent platings, and were tested for the appropriate myosin HC gene inserts by Southern hybridizations (8) with the cDNA inserts of the microplasmids. Restriction endonuclease sites were mapped by single and double enzyme digestions. The DNA sequence of various subcloned regions of the phage was obtained by both dideoxy chain termination (9) and chemical cleavage (10) sequencing techniques.

The intron and exon positions in the portion of $\pi\lambda$251 encoding p251 were determined by electron-microscopic heteroduplex analysis (11). Single and double stranded ϕX174 DNA was used as a standard for measurement.

RESULTS AND DISCUSSION

It has been experimentally determined (4) that the recombination frequency drops about 100-fold for a 10% sequence divergence. A highly diverged portion of p251 was therefore used for the screening. The subcloned region of p251 contains 197 bp of the -COOH coding sequence and 96 of 147 bp from the 3'-untranslated sequence. Since p251 is diverged from p110 about 27% in the coding portion and greater than 50% in the untranslated region, the probability that the gene corresponding to p251 would be isolated was fairly high.

For each microplasmid, three separate screenings of 10 genomic equivalents were performed. Only one phage was isolated with each probe. The maps of these genomic clones are presented in Figure 2. The insert of $\pi\lambda$251 is about 16.5kb, and that of $\pi\lambda$60 about 15.5 kb. Initially, the 5'-3' orientations were established by the locations of restriction endonuclease sites in the polylinker region of the microplasmid insert. This was subsequently confirmed by Southern blot analysis with various cDNA probes. The library (7) was constructed from a partial Alu I-Hae III digestion of chicken genomic DNA. It was expected that each probe would yield several unique overlapping clones; however, this was not the case. It may be that incorporation of the microplasmid into other genomic clones resulted in phage DNA too large to be efficiently packaged. In addition, $\pi\lambda$251 and $\pi\lambda$60 contain a large number of Eco RI sites. Thus, these genomic clones would probably be underrepresented in those libraries generated by partial Eco RI cleavage of genomic DNA. This may explain why the genes corresponding to $\pi\lambda$251 and $\pi\lambda$60

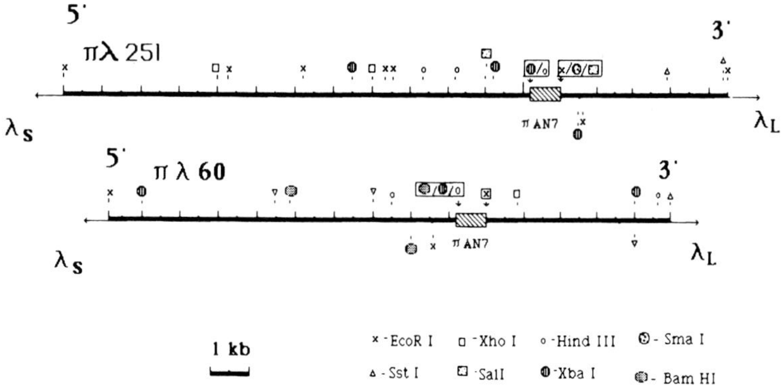

Figure 2. Restriction endonuclease maps of πλ251 and πλ60. The position of the microplasmid is indicated by the hatched box. πλ60 does not contain any Sac II or Sal I sites. The Bam HI sites in πλ251 were not determined.

(without the microplasmid insert) were not isolated from a partial Eco RI library which yielded at least 7 distinct myosin HC genes (12).

Clone p60 was initially characterized as a 5' extension of p184, the latter a 5' extension of p251. Since the restriction maps of the two isolated phage do not overlap, it appeared that either the myosin HC genes in chicken were much larger (>30 kb) than in the rabbit (3), or that the two clones represented portions of two separate genes. Evidence for the latter hypothesis was obtained by sequencing the region (150bp) of overlap between p60 and p184/251. Comparison showed that there were 13 scattered silent nucleotide substitutions (data not presented) and demonstrated that the two cDNA clones code for different mRNAs. Thus, πλ251 and πλ60 probably represent portions of two different genes that are related and not allelic variants of the same gene. The possibility that the genomic clone πλ60 may encode the embryonic HC mRNA represented by p110 is currently being evaluated.

In order to positively identify πλ251, DNA sequencing was performed. We focused on the regions to either side of the microplasmid insert, because the highly diverged 3' portion of the gene was located there. A 1.2 kb Sal I-Hind III fragment lying immediately 5' of the πAN7 insert, and a 0.55 kb Sal I-Eco RI fragment immediately 3' of the microplasmid were subcloned. The 3' end of the larger fragment was identical to the final 200 bp sequence of the cDNA subcloned into the microplasmid (data not presented). The sequence of the smaller fragment was determined from the 3' Eco RI site to about 220 bp upstream (Figure 3). This latter region contains the last 51 nucleotides of the 3' untranslated portion of p251. Since this sequence was not cloned into the microplasmid, it must arise directly from the HC gene, thereby confirming identity of πλ251 as encoding p251. It is intriguing that the 3' 51 bp terminal region is separated from the rest of the 3' untranslated region by an intron. An intron in the 3' untranslated region has only been reported in a few genes from higher vertebrates, including those encoding the thyroid hormone responsive

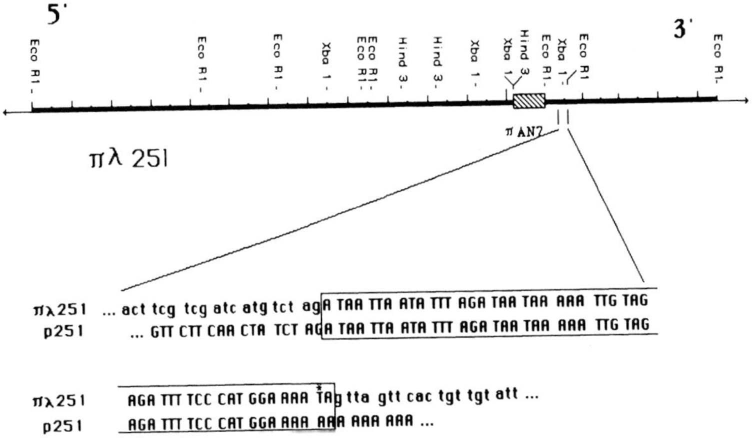

FIGURE 3. Identification of πλ251. The upper line of DNA sequence was obtained from πλ251, and the lower from p251. The highlighted 51 bp sequence is homologous in the 3' terminus of p251, to the poly A+ site. The asterisk marks an ambiguity in the sequence.

spot 14 protein of rat liver (13), and fast myosin light chains 1/3 (14). The presence of an intron could permit alternate splicing in the 3' terminus

The size and number of introns and exons in the portion of πλ 251 encoding p251 has been determined by electron-microscopic heteroduplex mapping. An example of such a hybrid molecule is presented in Figure 4. The plasmid DNA was linearized with Eco RI, resulting in plasmid arms of unequal length. The 5'-3' orientation of the genomic clone was independently confirmed by the location of the plasmid arms in the hybrid molecules. Although this mapping technique may not reliably detect very small introns (<100 bp), two large deletion loops and four smaller ones were observed. The largest deletion loop appears to contain both intervening sequences and the microplasmid insert. By measuring the length of the duplex regions, the positions of the exon boundaries relative to p251 were determined (Figure 5). The 3'-most exon was identified by DNA sequence. The precise genomic location of the sequences encoding the 3' 300 bp Pst I-Xba I

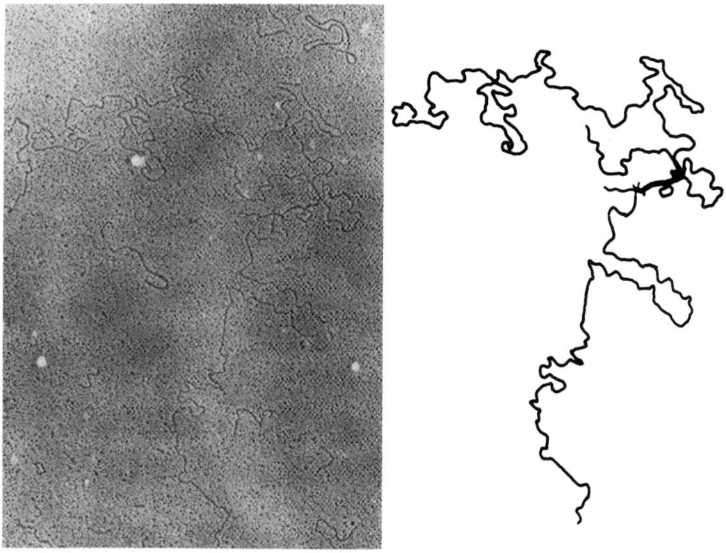

Figure 4. Heteroduplex between πλ 251 and p251. Thick lines represent regions of homology, the flanking thin lines are the phage arms and plasmid sequences. The internal thin lines show regions of non-homology (introns).

fragment contained within the microplasmid could not be determined by heteroduplex mapping. However, in order for the homologous recombination to have occurred, some of this sequence must lie on either side of the inserted DNA. Thus, like other myosin HC genes, the 3' end of the gene represented by πλ251 demonstrates a complex structure of introns and exons.

Preliminary examination of heteroduplexes formed between πλ251 and other cDNA clones indicates that there are additional coding sequences in the 5' portion of the genomic DNA. Once this region is completely characterized, an intron probe will be constructed in order to isolate the 5' end of this gene. Since the mRNA encoded by this HC gene is expressed in early development (5), analysis of the 5' end and flanking regions may elucidate how the myofibrillar protein genes are switched on during terminal differentiation of muscle.

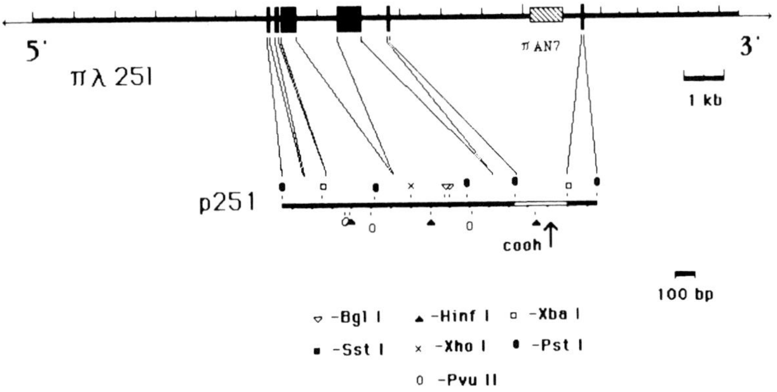

FIGURE 5. Intron-exon positions. Single- and double stranded regions of the heteroduplex molecules were measured, and the lengths (kb) were determined relative to ΦX174 DNA. The approximate positions of each exon relative to p251 are shown.

REFERENCES

1. Masaki T (1974). Immunochemical comparison of myosins from chicken cardiac, fast white, slow red, and smooth muscle. J Biochem (Tokyo) 76:441.
2. Whalen RG, Sell SM, Butler-Browne GS, Schwartz K, Bouveret P, Pinset-Härström I (1981). Three myosin heavy-chain isozymes appear sequentially in rat muscle development. Nature 292:805.
3. Friedman DJ, Umeda PK, Sinha AM, Hsu H-J, Jakovcic S, Rabinowitz M (1984). Characterization of genomic clones specifying rabbit α- and β- ventricular myosin heavy chains. Proc Natl Acad Sci USA 81:3044.
4. Seed B (1983). Purification of genomic sequences from bacteriophage libraries by recombination and selection in vivo. Nucl Acids Res 11:2427.
5. Umeda PK, Sinha AM, Jakovcic S, Merten S, Hsu H-J, Subramanian KN, Zak R, Rabinowitz M (1981). Molecular cloning of two fast myosin heavy chain cDNAs from chicken embryo skeletal muscle. Proc Natl Acad Sci USA 78:2843.
6. Umeda PK, Kavinsky CJ, Sinha AM, Hsu H-J, Jakovcic S, Rabinowitz M (1982). Differential expression of two fast myosin heavy chain mRNAs during chicken skeletal muscle development. in "Muscle Development: Molecular and Cellular Control". Cold Spring Harbor Press, p169.
7. Dodgson JB, Strommer J, Engel JD (1979). Isolation of the chicken α-globin gene and a linked embryonic α-like globin gene from a chicken recombinant library. Cell 17:879.
8. Southern EM (1975). Detection of specific sequences among DNA fragments reported by gel electrophoresis. J Mol Biol 98:503.
9. Sanger F, Coulson AR, Barrel BG, Smith AJH, Roe BA, (1980). Cloning in single stranded bacteriophage as an aid to rapid DNA sequencing. J Mol Biol 143:161.
10. Maxam AM, Gilbert W (1980). Sequencing end-labelled DNA with base-specific cleavages. Methods Enzymol 65:499.
11. Davis, WR, Simon M, Davidson N (1971). Electron microscope heteroduplex methods for mapping regions of base sequence homology in nucleic acids. Methods Enzymol 21:413.

12. Robbins J, Freyer GA, Chisholm D, Gilliam TC (1982). Isolation of multiple genomic sequences coding for chicken myosin heavy chain protein. J Biol Chem 257:549.
13. Liaw CW, Towle HC (1984). Characterization of a thyroid-responsive gene from rat. J Biol Chem 259:7253.
14. Robert B, Doubas P, Akimenko M-A, Cohen A, Garner I, Guenet J-L, Buckingham M (1984). A single locus in the mouse encodes both myosin light chains 1 and 3, a second locus corresponds to a related pseudogene. Cell 39:129.

Molecular Biology of Muscle Development, pages 373–385

STRUCTURAL STUDIES OF THE CHICKEN MYOSIN HEAVY CHAIN GENES[1]

Jeffrey Robbins[2], Keith Kropp, James Gulick, Thomas P. Horan[3], and Nancy Valente

Department of Biochemistry, University of Missouri-Columbia School of Medicine, Columbia, Missouri 65212

ABSTRACT Using a general probe containing nucleotides encoding the amino terminus of a myosin heavy chain (MHC) gene, we have isolated thirty-one unique, non-overlapping clones from chicken genomic libraries which are homologous to MHC structural sequences. High and low stringency hybridizations with probes corresponding to different regions of the genes show that the 5' flanking regions extensively diverge while the structural regions are highly conserved. A subset of seven genes was identified as coding for the fast-white isoforms. Sequencing data shows that the amino acid sequence is highly conserved (in most cases completely conserved) between the members of this group and a canonical exon-intron organization is present at the 5'-ends of the genes.

INTRODUCTION

We have been interested in studying the structure and organization of the MHC gene family in chickens (1,2),

[1]This work was supported by Grant HD13225 to J.R. from the National Institutes of Health

[2]Present Address: Department of Pharmacology and Cell Biophysics, University of Cincinnati Medical Center, Cincinnati, OH 45267

[3]Present address: NPI, Inc., Salt Lake City, UT 84102

the primary purpose being to understand the structural basis for the tissue-specific and developmental-stage specific expression of the different isoforms (3,4,5). In the vertebrates, the number and structure of the MHC genes and their relationship to the different isoforms observed in the various muscle types during the developmental transitions have remained obscure. Using different probes derived from our previously isolated MHC genes (2), we have carried out a number of screenings on different chicken genomic libraries, and have isolated a large number of clones encoding the MHC genes (6,7). We noted, in hybridization experiments, that the 5'-ends of these genes encoding the globular head regions of the different MHC proteins appeared to be highly conserved relative to the sequences which code for the carboxy termini. For example, the 5'-ends of the chicken and nematode genes cross-hybridize with one another, while the 3'-ends are unable to anneal, even at low stringency (Gulick et al, unpublished data).

We describe below the results of screening chicken genomic libraries with a gene probe derived from the 5'-end of an MHC gene which encodes a fast-white isoform expressed in the adult. The structural sequences of the fast-white isoforms are catalogued and their organization compared.

RESULTS

Isolation of an Adult Fast-white Gene

DCM11, a clone containing the 3'-end of a MHC gene has been described (2). N49 and N116 were selected from a chicken genomic library via sequential screenings during "chromosome walks." These three overlapping clones were found to encode a complete MHC gene, consisting of approximately 23 kbp of DNA. The three clones, and their EcoRI sites are shown in Figure 1: the 5' and 3' termini of the gene were confirmed by DNA sequencing.

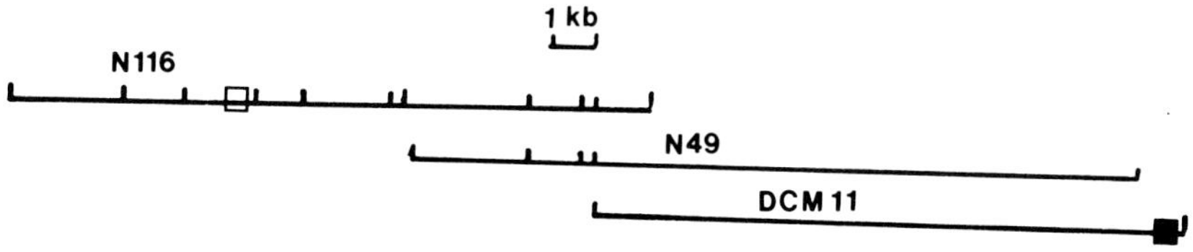

FIGURE 1. EcoRI restriction sites in a MHC gene. Shown is a complete MHC gene containing approximately 23 kbp of DNA. The 5' (□) and 3' (■) termini are indicated.

The isoform identity of this gene was determined by preparing an oligonucleotide unique to this gene, derived from the 5' untranslated (UT) region. This 15-mer, 5'GCTGCGCTCACACAG 3', is complementary to the mRNA transcript; its 5'-end is located one base 5' to the ATG start codon (Gulick et al, manuscript submitted). The oligonucleotide was labelled with ^{32}P-ATP and polynucleotide kinase (8), and hybridized to RNA's derived from the different embryonic and adult muscle sources. The annealed oligomer was then used to prime the synthesis of the strand complementary to the RNA's leader sequence using reverse transcriptase. The data are shown in Figure 2, and indicate that two predominant products, 60 and 58 bases, resulted from the hybridization with the RNA derived from the adult superficial pectoralis. The smaller products observed varied in experimental trials; preliminary sequencing data indicates that they are due to the pausing of the reverse transcriptase.

The data indicate that the RNA encoded by the gene appears to be expressed predominantly in the adult pectoralis; no products were observed in the other lanes even upon prolonged exposure. We estimate that transcription of this gene in the other tissues is negligible (<1%) as compared to the pectoralis. Somewhat surprisingly, few, or no transcripts were present in the RNA derived from the adult gastrocnemius, a muscle which contains both slow-red and fast-red fibers, indicating that these muscles must contain a different population of fast-white transcripts. Hybridization with non-specific probes did indicate that both RNA preparations contained essentially equal amounts of MHC mRNA (data not shown).

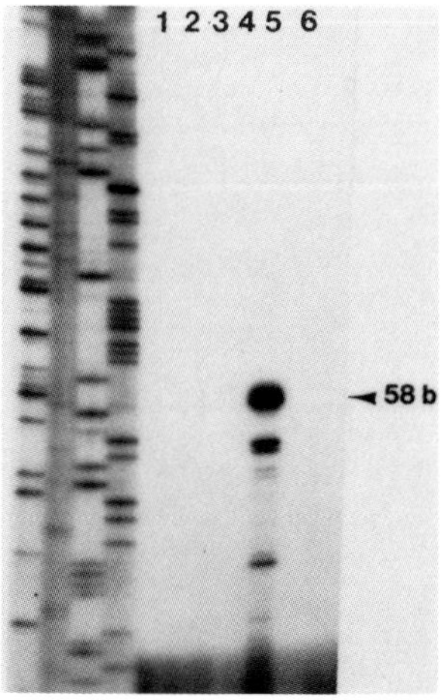

FIGURE 2. Isoform identification of the MHC gene. The hybridizations were driven to completion by annealing 100 ng of oligomer with 50 μg of each of the RNA's. The primer was extended to the 5'-terminus using reverse transcriptase (9) and the products electrophoresed on DNA sequencing gels. The sequencing ladder (left) was used as a marker. The RNA's were derived from the following sources: lane 1; embryonic leg, lane 2; embryonic breast, lane 3; embryonic heart, lane 4; adult gastrocnemius, lane 5; adult superficial pectoralis, lane 6; adult heart.

Screening the Libraries Using the 5' End of the MHC Gene.

The 3050 bp EcoRI fragment containing the 5' end of the gene was subcloned and sequenced; the fragment consists of 1400 bp of flanking sequence containing the promoter elements. Also encoded within this fragment in the remaining 1600 bp are the exons which encode the amino terminus of the globular head region (Robbins et al, manuscript submitted). This fragment was purified and used to exhaustively screen two chicken gemonic libraries at moderate stringency. A large number (92) of isolates were obtained. After duplicates and overlaps were identified, thirty-one unique, non-overlapping clones containing sequences homologous to the probe remained.

The sizes of the EcoRI fragments of the isolates are shown in TABLE 1.

TABLE 1
EcoRI FRAGMENT SIZES

Clone	EcoRI fragments (in kbp)
N29	3.85, 2.80, 2.55, 2.10, 2.05, 0.90
N39	6.10, 4.35, 4.20, 0.80
N50	4.80, 4.00, 3.00, 2.20, 0.68, 0.625, 0.38
N101	6.20, 5.00, 0.10
N109	6.10, 4.20, 3.80, 3.75
N116	3.25, 3.24, 3.05, 2.30, 1.60, 1.15, 1.10, 0.30, 0.20
N118	3.51, 3.50, 3.30, 2.10, 2.09, 1.20, 1.10, 1.00, 0.30, 0.20
N124	5.30, 4.70, 1.51, 1.50, 1.00, 0.80, 0.50, 0.15
N125	3.26, 3.25, 2.60, 2.10, 1.70, 1.10, 1.00, 0.65, 0.55, 0.50, 0.45, 0.20, 0.10
N127	7.00, 2.70, 2.00, 0.90, 0.70, 0.60
N301	15.80
N303	11.80, 2.80, 2.60
N304	10.50, 3.90, 0.90
N305	11.00, 3.00, 2.45
N306	9.00, 3.80, 1.80, 1.55, 0.50
N307	7.70, 2.75, 1.90, 1.37, 1.30, 1.00
N309	7.50, 4.70, 2.15
N310	4.50, 3.80, 1.60, 1.12, 0.94, 0.75, 0.46
N311	9.00, 3.30
N312	5.50, 5.25, 2.80, 0.87
N314	4.50, 4.30, 3.35, 2.20, 1.90, 1.30, 0.50
N315	5.10, 4.50, 4.40, 1.70, 1.00, 0.55
N316	8.30, 4.90
N319	5.10, 4.25, 3.25, 2.15, 1.80, 0.45
N320	13.40, 2.70
N321	6.60, 3.75, 3.60, 1.30, 0.82, 0.74
N322	2.53, 2.48, 2.20, 1.75, 1.50, 1.20, 0.95, 0.35
N324	8.05, 3.05, 1.70, 0.90
N325	9.10, 4.70, 2.70
N326	9.00, 3.80, 1.55, 0.50
N328	8.30, 3.25, 1.40

A Southern transfer (12) of the EcoRI-digested DNAs was made and subsequently hybridized to the probe; the autoradiogram is shown in Figure 3. The DNA from which the probe is derived is in lane 1.

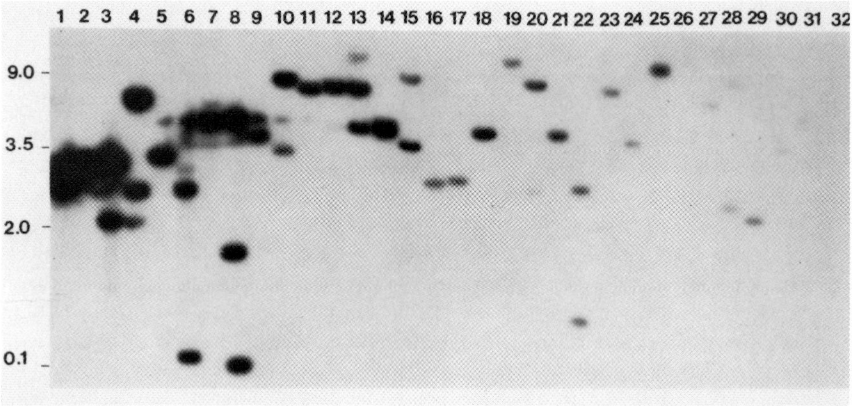

FIGURE 3. Hybridization of the DNA's to the 3050 bp fragment. 0.8 μg of each DNA was digested with EcoRI, electrophoresed in 0.6% agarose, transferred to nitrocellulose and hybridized to the 3000 bp fragment derived from pN116-3.05. Post-hybridization washes were done in 5 x SSC at 62°C. The autoradiogram is a result of a 3 h exposure at -70°C. The clones in lanes 1-32 correspond to the following: N116, N29, N118, N127, N29, N125, N101, N124, N39, N311, N324, N316, N309, N315 N304, N303, N326, N310, N301, N325, N109, N312, N306, N314, N320, N50, N307, N305, N322, N319, N321, N328. The sizes listed to the right are in kilobase pairs. N29 was electrophoresed in duplicate (lanes 2 and 5) in order to monitor the reproducibility of signal intensity.

The DNA's are grouped in order to emphasize the different degrees of homology among the isolates. Shorter exposures showed that six other clones present in lanes 2-8 hybridized to the probe with an intensity approximately equal to that observed in lane 1, which contains the parental DNA. The DNA's of six other isolates, in lanes

9-14 generated a slightly less intense signal, while the DNA's in lane 15-32 were highly variable in their relative abilities to hybridize strongly with the probe. For example, the DNA's derived from clones N109, N310 and N320 (lanes 21, 18 and 25) hybridize strongly while clones N50, N321 and N328 (lanes 26, 31 and 32) generate signals that are barely visible in this exposure. Subsequent analysis of the clones showed (with the exceptions of N50, N319 and N321) that the varying signal intensities were not due to the absence of the (comparable) homologous sequences within the clones.

Conservation of the Structural Sequences at the 5' Ends of the MHC Genes.

In order to define the relationships between the isolates, the 3050 bp EcoRI fragment was cleaved with NcoI. This enzyme cleaves at the site $^{5'}$CCATGG $^{3'}$ within which is contained the codon at which protein synthesis is initiated (Kropp et al, manuscript submitted). The digestion yields a 1400 bp fragment containing the 5' UT nucleotides, as well as the promoter and flanking nucleotides. The 1600 bp fragment contains at least 5 exons and the intervening sequences. Each of these fragments was labelled with [^{32}P] dATP and used to probe Southern transfers of the EcoRI-digested DNA's. After hybridization the filters were subjected to either low stringency (6 x SSC, 62°C) or high stringency (0.1 x SSC, 62°C) post-hybridization washes. The results are shown in Figure 4.

The probe containing the flanking regions hybridized [Panel A] to the DNA's in lanes 1-8 (sequence analysis of the weakly hybridizing clone, N29, in lanes 2 and 5 showed that it did not contain the corresponding region). Little, or no hybridization was observed with the other 24 isolates. However, if a high stringency post-hybridization wash was performed [Panel B], the hybridization was restricted almost entirely to the DNA from which the clone is derived.

The results were strikingly different when the probe containing the structural sequences was used. All of the isolates showed strong or moderate homology with the probe, as evidenced by the signal intensities [Panel C]. The stringent post-hybridization wash had no effect on the overall pattern, or relative signal intensities. We conclude that the sequences encoding the globular head regions of the MHC genes are highly conserved, and that a large

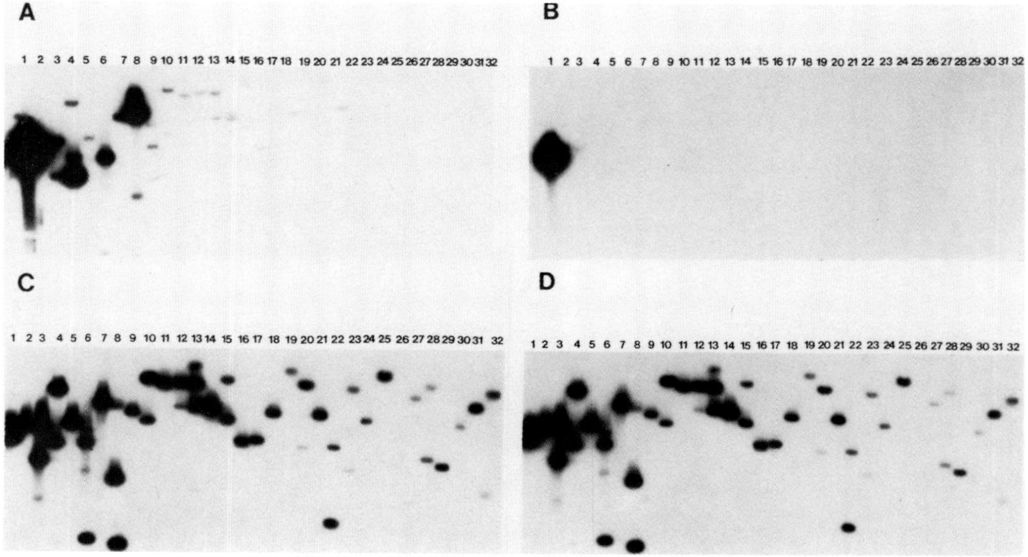

FIGURE 4 Hybridization of the DNA isolates with the 5' flanking and structural MHC sequences. Southern transfers of the EcoRI digested DNA's were carried out as described in Figure 2. Both the 1400 bp and 1600 bp EcoRI-NcoI fragments were labelled with [^{32}P]-dATP to a specific activity of 5 x 10^8 cpm/μg. Hybridizations were carried out in 6 x SSC, 5 x Denhardts solution at 62°C for 12 hr in the presence of 500,000 cpm/ml of the 1400 bp fragment (Panels A and B) or the 1600 bp fragment (Panels C and D). Post-hybridization washes were carried out in 6 x SSC at 62°C (Panels A and C), or in 0.1 x SSC at 62°C (Panels B and D), The order of the DNA's on the autoradiogram is identical to that in Figure 2. Again, N29 was present in lanes 2 and 5.

number of MHC genes (or pseudogenes) are present in the chicken genome.

Sequencing a Subset of the MHC Genes.

The results in Panel A and other data to be presented elsewhere indicated that the seven genes represented in lanes 1-8 could be grouped into a subset, all of which code (or are pseudogenes) for the different fast-white isoforms (Robbins et al, manuscript submitted). Subsequently the first three exons, (as well as the 5'-flanking regions and the introns) of these seven genes were sequenced; the amino acids encoded by the exons are shown in Figure 5.

The sequence conservation between all of the fast-white isoforms is extremely high. Most of the substitutions are conservative, and localized in the first exon. Only a single conservative substitution is present in the second exon, and the third exon, which contains a domain of ATP binding site (10) is completely conserved.

DISCUSSION

These results show that the size of the MHC gene family in the chicken is larger than previously estimated (2). It possible, however, that a number of the isolates correspond to pseudogenes and are not transcribed, or code for "myosin-like" proteins. However, preliminary experiments, in which three randomly selected clones were labelled and hybridized with muscle RNA's indicated that all three did hybridize exclusively to the MHC mRNA. However, unambigious identification of each of these genes will depend upon DNA sequencing.

The relationship between the seven genes whose sequences are partially shown in Figure 5 is interesting. All of the genes share a highly conserved exon structure. Additional sequence information in the 5'-UT regions has shown extensive divergence, and we have exploited this divergence in order to prepare transcript-specific probes (Kropp et al, manuscript in preparation). To date, experiments with these probes have confirmed that the subgroup codes for transcripts that are preferentially expressed in the fast-white fibers. Sequence data has also been obtained

AMINO ACIDS

1st EXON

N101	MASS	DAEM	AVFG	EAAP	YLRK	SEKE	RIEA	QNKP	FDAK	SSVF	VAHP	KESF	VKGT	IQSR	ETGK	VTVK	TEGG	E
N116	MASP	DAEM	AAFG	EAAP	YLRK	SEKE	RIEA	QNKP	FDAK	SSVF	VVHP	KESF	VKGT	IQSK	EGGK	VTVK	TEGG	E
N118	MAT-	DADM	AIFG	EAAP	YLRK	SEKE	RIEA	QNKP	FDAK	SSVF	VVHA	KESY	VKST	IQSK	ESGK	VTVK	TEGG	E
N122			AFG	EAAP	YLRK	SEKE	RIEA	QNKP	FDAK	SSVF	VAHP	KESF	VKGT	IQSR	ETGK	VTVK	TEGG	E
N124	MASP	DAEM	AAFG	EAAP	YLRK	SEKE	RIEA	QNKP	FDAK	SSVF	VVHP	KESF	VKGT	IQSK	ETGK	VTVK	TEGG	E
N125	M-SS	DAEM	AIFG	EAAP	YLRK	SEKE	RIEA	QNKP	FDAK	TSVF	VVHA	KESY	VKST	IQSK	ESGK	VTVK	TESG	E
M127	M-SS	DSEM	AIFG	EAAP	YLRK	SEKE	RIEA	QNKP	FDAK	SSVF	VVHP	KESF	VKGT	IQSK	ETGK	VTVK	TEGG	E

2nd EXON

N101	TLT	VKED	QVFS	MNPP	KYDK	IEDM	AMMT	HLHE	PAVL	YNLK	ERYA	AWMI	Y
N116	TLT	VKED	QVFS	MNPP	KYDK	IEDM	AMMT	HLHE	PAVL	YNLK	ERYA	AWMI	Y
N118	TLT	VKED	QIFS	MNPP	KYDK	IEDM	AMMT	HLHE	PAVL	YNLK	ERYA	AWMI	Y
N122	TLT	VKED	QVFS	MNPP	KYDK	IEDM	AMMT	HLHE	PAVL	YNLK	ERYA	AWMI	Y
N124	TLT	VKED	QIFS	MNPP	KYDK	IEDM	AMMT	HLHE	PAVL	YNLK	ERYA	AWMI	Y
N125	TLT	VKED	QIFS	MNPP	KYDK	IEDM	AMMT	HLHE	PAVL	YNLK	ERYA	AWMI	Y
N127	TLT	VKED	QVFS	MNPP	KYDK	IEDM	AMMT	HLHE	PAVL	YNLK	ERYA	AWMI	Y

3rd EXON

N101	TYS	GLFC	VTVN	PYKW	LPVY	NPEV	VLAY	RGKK	RQEA	PPHI	FSIS	DNAY	QFML	T
N116	TYS	GLFC	VTVN	PYKW	LPVY	NPEV	VLAY	RGKK	RQEA	PPHI	FSIS	DNAY	QFML	T
N118	TYS	GLFC	VTVN	PYKW	LPVY	NPEV	VLAY	RGKK	RQEA	PPHI	FSIS	DNAY	QFML	T
N122	TYS	GLFC	VTVN	PYKW	LPVY	NPEV	VLAY	RGKK	RQEA	PPHI	FSIS	DNAY	QFML	T
N124	TYS	GLFC	VTVN	PYKW	LPVY	NPEV	VLAY	RGKK	RQEA	PPHI	FSIS	DNAY	QFML	T
N125	TYS	GLFC	VTVN	PYKW	LPVY	NPEV	VLAY	RGKK	RQEA	PPHI	FSIS	DNAY	QFML	T
N127	TYS	GLFC	VTVN	PYKW	LPVY	NPEV	VLAY	RGKK	RQEA	PPHI	FSIS	DNAY	QFML	T

FIGURE 5 Amino acid sequences of seven MHC genes. All sequences were obtained by the method of Sanger and Coulson (12) and confirmed by sequencing overlapping fragments or both DNA strands.

for the introns, and shows that extensive sequence conservation is also present in these regions. The second intron, which occurs before the exon containing part of the ATP binding domain is almost completely conserved between the seven genes. This, coupled with the lack of silent site changes in the third exon raises the interesting possibility of a correction mechanism using gene conversion (Robbins et al manuscript submitted).

Our results, concerning the structure of these genes are summarized in Figure 6, and show the exon-intron organization at the 5' ends.

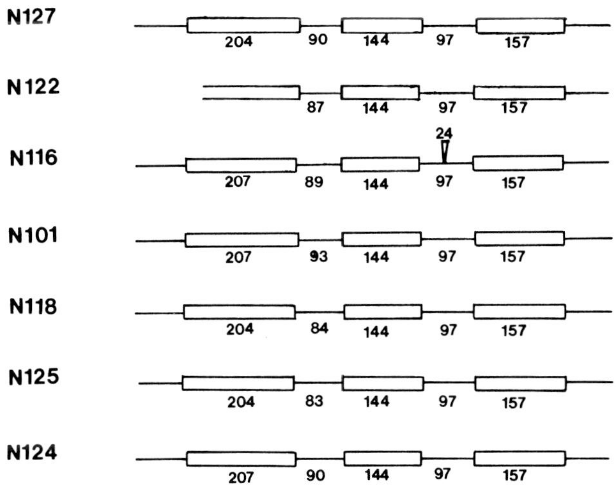

FIGURE 6. Exon-intron organization at the 5' ends. The number of nucleotides in each exon (▭) and intron (—) is indicated. The exact number of nucleotides in the first exon of N122 has not yet been determined.

It is apparent that a canonical exon-intron organization for this subset of MHC genes exists (the 24 bp insert in N116 is a result of a tandem repeat). The basic sequence organization appears to maintained across species and isoforms (11). As sequence data is accumulated, it will be interesting to see if this conservation of organization and structure is maintained within the members of this subgroup throughout the genes' structure, and whether the conservation of internal intron sequence is fortuitous or necessary for expression.

REFERENCES

1. Freyer, G, Robbins, J (1983) The analysis of a chicken myosin heavy chain cDNA clone J Biol Chem 258:7149.
2. Robbins, J, Freyer, G, Chisholm, D, Gilliam, TC (1982) Isolation of multiple genomic sequences coding for chicken myosin heavy chain protein J Biol Chem 257:549.
3. Bandman, E, Matsuda, R, Strohman, RC (1982) Developmental appearance of myosin heavy and light chain isoforms in vivo and in vitro in chicken skeletal muscle Dev Biol 93:508.
4. Gauthier, GF, Lowey, S, Hobbs, AW (1978) Fast and slow myosin in developing muscle fibers Nature 274:25.
5. Gauthier, GF, Lowy, S, Benfield, PA, Hobbs, AW (1982) Distribution and properties of myosin isozymes in developing avian and mammalian skeletal muscle fibers J Cell Biol 92:471.
6. Gulick, J (1983) The Isolation and comparison of normal myosin heavy chain gene clones Master Thesis, University of Missouri-Columbia, Columbia, MO 65202
7. Horan, TP (1985) Myosin heavy chain genes Master Thesis, University of Missouri-Columbia, Columbia, MO 65202
8. Maxam, AM and Gilbert, W (1977) A new method for sequencing DNA Proc Natl Acad Sci USA 74:560
9. Goldberg, DA, Posakany, JW and Maniatis, T (1983) Correct developmental expression of a cloned alcohol dehydrogenase gene transduced into the drosophila germ line Cell 34:59.

10. Okamoto, Y and Yount, RG (1983) Identification of an active site peptide of myosin after photo affinity labeling Biophys J 41:298a
11. Strehler, EE, Mahdavi, V, Periasamy, M and Nadal-Ginard, B (1985) Intron positions are conserved in the 5'-end region of myosin heavy chain genes J Biol Chem 258:468.
12. Sanger, F and Coulson, AR (1975) A rapid method for determining sequences in DNA by primed synthesis with DNA polymerase J Mol Biol 94:441.

Molecular Biology of Muscle Development, pages 387–410

ALTERNATIVE SPLICING: A COMMON MECHANISM FOR THE GENERATION OF CONTRACTILE PROTEIN DIVERSITY FROM SINGLE GENES[1]

B. Nadal-Ginard, R.E. Breitbart, E.E. Strehler, N. Ruiz-Opazo, M. Periasamy, and V. Mahdavi

Department of Cardiology, Children's Hospital, and Department of Pediatrics, Harvard Medical School
Boston, Massachusetts 02115

ABSTRACT Characterization of the detailed genomic structure and mRNA products of three muscle-specific genes reveals that alternative RNA splicing is an important mechanism for the generation of contractile protein diversity subject to developmental and tissue-specific regulation. Transcription of the myosin light chain 1/3 gene, initiated at two different promoters, and of the α -tropomyosin gene, terminated at two different 3' untranslated sequences, yields in each case a structurally distinct primary transcript which may dictate its own splicing pattern in cis. The troponin T gene, however, with a single promoter and one termination site, manifests differential patterns of splicing from structurally identical primary transcripts, necessitating control by trans-acting factors. Implications for the understanding of alternative splicing mechanisms in general, and their role in the generation of isoform diversity in contractile systems in particular, are discussed.

INTRODUCTION

The regulated expression of structurally distinct, tissue-specific and developmentally regulated protein isoforms is an essential characteristic of eukaryotic cell differentiation. The molecular mechanisms responsible for generating this protein

[1]This work was supported by grants from the National Institutes of Health, the Muscular Dystrophy Association, and the American Heart Association.

diversity, while poorly understood, may be broadly categorized into two systems: those that select one gene among the members of a multigene family for expression in a particular setting - - e.g. the myosin heavy chain (1-3) - - and systems that generate protein diversity from a single gene. DNA rearrangement (see 4) and alternative RNA processing (see 5) are mechanisms suitable for differential expression of intragenic sequences leading to the formation of multiple protein isoforms from a single gene. Increasing numbers of genes are now known to use alternative RNA processing to generate protein diversity (see 5-7 for review). Among these genes, most splice exons differentially from primary transcripts that have different primary structures, either because they are transcribed from alternative promoters or because they are terminated at alternative 3' untranslated sequences and/or poly (A) addition sites (see 5-7). In these cases it is possible that the splicing pathway is dependent on the primary structure of the transcripts and, therefore, is cis-regulated. To date, the only well documented example of regulated alternative splicing of transcripts with entirely identical primary structure is the case of the troponin T gene (6). The regulated patterns of splicing of the troponin T gene transcripts require the existence of trans-acting factors that are developmental and tissue-specific.

The generation of protein diversity by alternative splicing appears to be a common mechanism used by muscle cells. The power of this mechanism to produce multiple proteins from a single gene covers a broad range. It is relatively low in the myosin light chain 1/3 gene where two proteins are produced, intermediate in the tropomyosin gene which has the capacity to produce eight different proteins, and extremely high in the troponin T gene which has the capacity to encode 64 different proteins. The coding capacity of these genes raises important but so far unanswered questions about the physiological roles of the different isoforms produced and reveals new, unsuspected and highly complex aspects of the biochemistry of contractility. In a broader sense, these genes might serve as paradigms for the study of the mechanisms involved in the generation of protein diversity by alternative splicing, and might shed light on more general mechanisms of gene regulation relevant to a wide variety of biological systems in addition to muscle.

RESULTS

Utilization of the Two Promoters of the Myosin Light Chain 1/3 Gene Determines the Splicing Pathway of the Two Primary Transcripts.

Organization of the MLC 1/3 gene. In fast skeletal, but not in other muscle types, the alkali light chains exist in two isoforms of different size, termed $MLC1_f$ and $MLC3_f$ (M_r 21,000 and 17,000, respectively). In all vertebrate species so far analyzed (7-10), these two proteins exhibit several interesting features that set them apart from other contractile proteins. $MLC1_f$ (190 amino acids long) and $MLC3_f$ (149 amino acids long) have complete sequence homology for the last 141 amino acids at the carboxyl terminus. The two proteins differ, however, in the length and amino acid sequence of their amino termini. Based on the analysis of cDNA and genomic clones, it has now been documented that in chicken (9), rat (7), and mouse (10), these two proteins are the products of a single gene with an unusual structural organization.

As is clear from Fig. 1, which depicts the genomic organization of the MLC 1/3 gene in the rat compared with those in mouse and chicken, the isoform-specific sequences of $MLC1_f$ (aa 1 to 40) and $MLC3_f$ (aa 1 to 8) are in physical continuity in the genome and located upstream from the those encoding the sequences common to both isoforms. This result conclusively demonstrates that these two isoforms are produced by the same gene.

The organization of the MLC 1/3 genomic sequences in the rat has been further established by a combination of DNA blots, S1 nuclease mapping, DNA sequence analysis, and in vitro transcription experiments. As illustrated in Fig. 1, the sequences encoding the 141 amino acids common to the $MLC1_f$ and $MLC3_f$ proteins, including the common 3' untranslated region, are split in five exons which are grouped in a region of ∿ 5 kb of DNA. Interestingly, the 5' ends specific for $MLC1_f$ and $MLC3_f$ are separated from the common body exons by ∿ 15 and 5 kb of genomic DNA, respectively. Thus, the complete MLC 1/3 gene spans ∿ 25 kb of DNA which, given the size of the mature mRNAs for $MLC1_f$ and $MLC3_f$ (∿ 1.0 and ∿ 0.8 kb without poly (A) tails, respectively) leads to an unusual exon:intron ratio of approximately 1:25. The most, striking aspect of this gene, however, is the organization of the $MLC1_f$- and $MLC3_f$-specific sequences.

As shown in Fig. 1, the 5' untranslated region and amino acids 1-40 of $MLC1_f$ are encoded by the 5' most exon of the gene. Approximately 10 kb downstream lies an exon that contains only the 5' untranslated region of $MLC3_f$, whereas the $MLC3_f$-specific amino-terminal 8 amino acids are located in a separate "mini-exon"

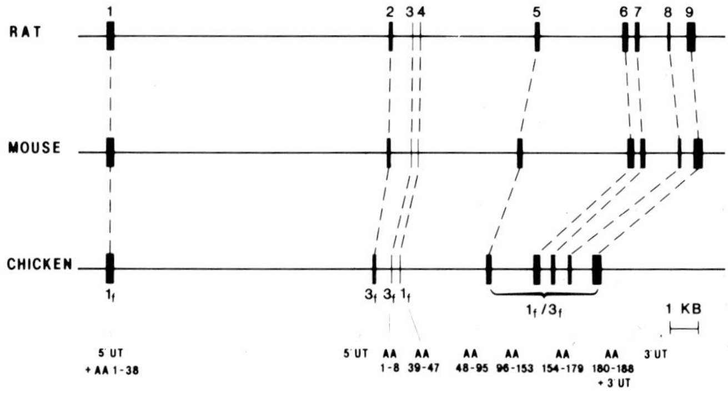

FIGURE 1. Comparison of the $MLC1_f/3_f$ genomic organization in rat, mouse (10), and chicken (9). Exons, numbered 1-9, are represented by boxes, with the encoded rat amino acid sequences indicated below. Exons 1 and 4 are specific to $MLC1_f$, and 2 and 3 to $MLC3_f$, mRNAs.

741 bp downstream. Another 244 bp downstream from this $MLC3_f$ mini-exon lies a $MLC1_f$ specific mini-exon coding for amino acids 41-49. The existence of these two isoform-specific mini-exons in this particular arrangement indicates that differential splicing is involved in generating both $MLC1_f$ and $MLC3_f$ mRNAs (Fig. 2). The $MLC3_f$-specific exons need to be spliced out from the $MLC1_f$ primary transcript in order to generate $MLC1_f$ mRNA, while the mini-exon coding for $MLC1_f$ amino acids 41-49 must be removed in order to generate $MLC3_f$ mRNA, independently of whether the $MLC1_f/3_f$ gene has one or two promoters. The splicing out of the 5' terminal $MLC3_f$-specific exon (exon 2 of Fig. 1) can easily be explained by the fact that this exon lacks the consensus splice

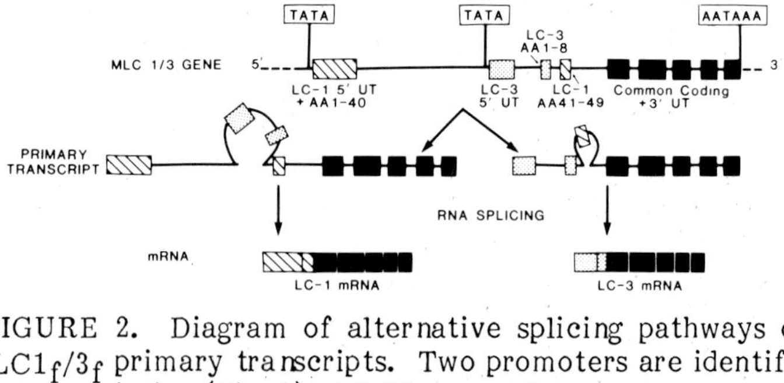

FIGURE 2. Diagram of alternative splicing pathways of the two $MLC1_f/3_f$ primary transcripts. Two promoters are identified by in vitro transcription (Fig. 3). $MLC1_f$-specific exons are represented by striped boxes, $MLC3_f$-specific by gray, and common by solid boxes.

acceptor sequence (7). However, no known mechanism of RNA splicing can account for the alternative and mutually exclusive use of the $MLC1_f$- and $MLC3_f$-specific mini-exons (exon 3 and 4) to form $MLC1_f$ and $MLC3_f$ mRNAs, respectively. These two mini-exons are flanked by functional acceptor and donor sites in agreement with the consensus splice junctions (25). Surprisingly, however, these two exons are not spliced together into the same mRNA molecule, nor are they both spliced out or interchanged in the production of $MLC1_f$ and $MLC3_f$ mRNAs. Yet, the 3' ends of both exons are spliced to the same 5' end of exon 5, the first common exon (see Fig. 2).

$MLC1_f$ and $MLC3_f$ mRNAs are transcribed from two different promoters that are more than 10 kb of DNA apart. The lack of homology between the 5' untranslated sequences of $MLC1_f$ and $MLC3_f$ mRNAs, as determined by the cDNA sequences (7), and the finding of TATAA-like boxes (TATATA and AATAA) 22 and 24 nucleotides upstream from the putative cap sites determined by S1 nuclease mapping, strongly suggested that the MLC 1/3 gene contains two differentially regulated promoters located very far apart. To further investigate this possibility we determined the ability of the corresponding sequences to initiate transcription in

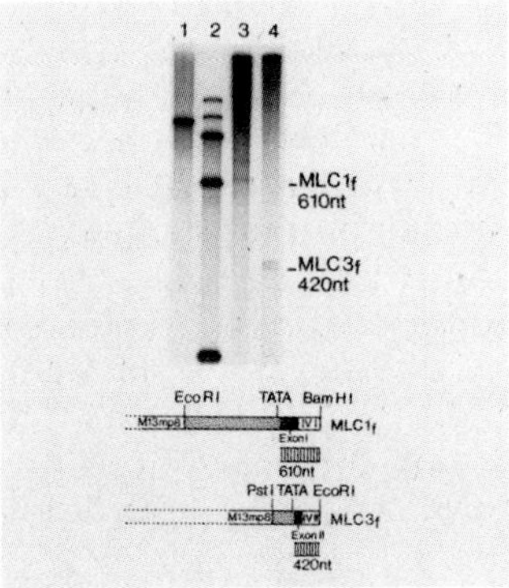

FIGURE 3. Run-off in vitro transcription using $MLC1_f$- and $MLC3_f$-specific DNA templates. A recombinant plasmid containing exon 1 and its immediate 5' flanking genomic DNA was cleaved with BamHI to give a template ($MLC1_f$) which, upon transcription, yields a 610 nucleotide (nt) RNA (lane 3) initiated at the indicated promoter. Similarly, a 420 nt transcript (lane 4) is initiated at the $MLC3_f$ promoter lying in genomic DNA immediately upstream of exon 2 in a second plasmid. Lanes 1 and 2 are a control transcript and size markers, respectively. (Methods per ref. 11.)

vitro using a HeLa cell lysate (11). As shown in Fig. 3, two promoters are identified which, in vitro, initiate RNA transcription at the previously determined cap sites of the two mRNAs. These results taken together with the above data indicate that the generation of $MLC1_f$ and $MLC3_f$ mRNA isoforms involves alternative transcription from two different promoters located more than 10 kb apart. Each one of the transcripts, in turn, is subject to a different splicing pathway as shown in Fig. 2 in order to produce the corresponding mature mRNA. The wide separation of the transcription initiation sites for $MLC1_f$ and $MLC3_f$ mRNAs is a unique feature of the $MLC1_f/MLC3_f$ gene both in rat (7), chicken (9) and mouse (10). It gives rise to two mRNA precursors having signficantly different primary structures, including both size ($\sim$10 and 20 kb) and sequence, which could play a role in determining the pattern of alternative splicing of the isoform-specific exons (see below).

Each of the $MLC1_f$ and $MLC3_f$ primary transcripts has only one accessible splicing pathway. The organization and behavior of the specific mini-exons (exons 3 and 4) strongly suggests that the existence of conformational differences between the $MLC1_f$ and $MLC3_f$ primary transcripts is responsible for generating preferential splicing pathways. According to this hypothesis, the alternative and mutually exclusive use of mini-exons 3 and 4 to generate $MLC1_f$ and $MLC3_f$ mRNAs, respectively, would be a cis-regulated process that may or may not require the additional existence of diffusible trans-acting factors. Although not yet formally proved for the production of any mature mRNA, this concept is appealing because the known behavior of the primary transcripts can be readily explained by the existence of two different promoters that give rise to two mRNA precursors of significantly different size ($\sim$ 10 and 20 kb), primary sequence and, potentially, different conformation. In addition, this hypothesis could easily explain the simultaneous occurance of mini-exon duplication (7) and the appearance of the two splicing pathways specific for each of the transcripts. Moreover, this mechanism requires only a promoter selection in order to produce each of the two mRNAs.

The fact that the relative levels of $MLC1_f$ and $MLC3_f$ mRNAs change, both during development and in different muscle types (7), indicates that the mechanism(s) of alternative promoter utilization and differential splicing of isoform-specific exons involved in the generation of these two mRNAs from the $MLC1_f/3_f$ gene is not stochastic, but responds in a regulated manner to different environmental cues, including the pattern of nerve stimulation. These developmental and tissue-specific switches can now be more easily understood because only one splicing pathway is open to each of the two mRNAs originated from different promoter sites.

Therefore, regulated changes in promoter utilization during development and/or in different tissues, as has been demonstrated for α -amylase (12) and alchol dehydrogenase (13), is the main regulatory event in the expression of the MLC1/3 gene from which the splicing pattern of the transcripts follows in a deterministic manner.

One Smooth and Two Striated, Skeletal and Cardiac, α -Tropomyosin Isoforms are Encoded by the Same Gene.

Tropomyosin (TM), in association with the troponin complex, plays a central role in the Ca^{2+} dependent regulation of vertebrate striated muscle contraction. On the other hand, despite structural similarities with the striated isoforms, the functions of this protein in smooth and non-muscle cells remain undefined, since in these cells contraction is regulated by myosin-linked processes independently of TM.

Using α -TM cDNA clones (14,15) to analyze the structure and expression of the α -TM mRNAs as well as to isolate the corresponding genomic sequences, we have found that three different tissue-specific and developmentally regulated muscle α-TM mRNA isoforms, each encoding a different protein, are produced by a single gene by alternative splicing of "isotype switch exons."

Nucleotide and deduced protein sequences of smooth and striated α -TM cDNA clones define "isotype switch peptide" sequences. The complete nucleotide sequences of a smooth (SM) and a striated (STR) muscle cDNA clones have been determined (15). Examination of these sequences reveals several new structural features of the TM proteins. First, the smooth muscle α -TM is 280 aa long, 4 aa shorter than the rabbit skeletal muscle isoform. Second, the rat striated muscle isoform is identical in aa sequence to its rabbit counterpart, at least throughout the regions which have been compared (aa 81-284 of the rabbit molecule). Third, the comparative analysis between the smooth and striated muscle isoforms uncovers "isotype switch peptides" which reside in different positions along the protein molecule. The striated muscle α -TM is identical in aa sequence to the smooth muscle isoform except for aa 258-284 (STR aa sequence) which defines a carboxyl-terminal "isotype switch peptide." In addition, comparison of the rabbit striated muscle α -TM with the rat smooth muscle isoform shows them to be identical in aa sequence except for two regions: from aa 41-80 (STR aa sequence) which depicts an amino-proximal "isotype switch peptide" and the carboxyl-terminal "isotype switch peptide" discussed above. The interaction of smooth muscle TM with the troponin complex and other components of the contractile system in vitro is substantially different from that of the striated muscle isoform (see 23). Since the only difference in primary sequence between the smooth and striated -tropomyosins resides in the "isotype switch peptides," it is likely that these two peptides specify the troponin I (aa 41-80) and troponin T site 1 (aa 258-284) binding domains, respectively. They can also can explain the observed differences in actin binding and in the actin activation of myosin between the two α -TM isoforms (16).

At the nucleotide level, the most striking observation for the comparative analysis between the STR and SM cDNA clones is their identity in the region coding for aa 81-257 (STR aa sequence). In contrast, sequences located 3' to codon 257 are unique for each cDNA. The fact that in the common aa coding sequence of the two mRNAs there is not a single nucleotide substitution, even at the wobble position, strongly suggests that a single gene specifies the expression of these two tissue-specific α-TM isoforms.

Three different muscle α -TM mRNAs are expressed in a developmental and tissue-specific manner. In order to study the developmental and tissue-specific expression of the α -TM mRNAs, S1-nuclease mapping analysis of RNA isolated from several rat tissues at different developmental stages was carried out using a series of cDNA probes that cover the entire TM mRNA sequence.

The results obtained from this analysis, shown diagramatically in Fig. 4, define the molecular structure of three distinct α -TM mRNAs: STR α_1 and STR α_2 which are abundant in striated muscles, and SM α -TM which is the predominant form in smooth muscle. All three mRNAs are 100% homologous in the region encoding aa 81-257 (77-253 in the SM sequence). The two striated muscle mRNAs continue this identity up to the poly(A) tail, while the SM form contains a different coding sequence after codon 254 and continues with a very long tissue-specific 3' untranslated region. The nucleotide sequence encoding the first 80 aa (77 in the case of the SM and STR α_2-TM sequence) is different in all three mRNAs. The first 40 codons are shared between STR α_1 and SM mRNAs but not with STR α_2 that contains a unique sequence in this region. Interestingly, however, the codons for aa 41 to 76 are shared between STR α_2 and SM mRNAs but not by STR α_1 mRNA that contains a different sequence. This arrangement, as depicted in Fig. 4, suggests that the α-TM gene(s) has an unusual organization with common and exchangeable exons that can be spliced together according to different combinatorial possibilities.

The STR α -TM isoforms are tissue-specific and also developmentally regulated. The STR α_1-TM mRNA is induced early in muscle development and accumulates to high levels in the adult, while STR α_2-TM mRNA can only be detected in adult skeletal and

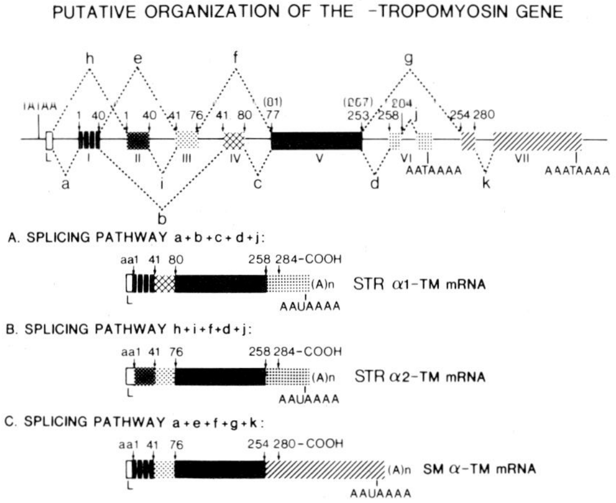

FIGURE 4. Schematic organization and alternative splicing pathways for the origin of three different mRNAs from the α -TM gene. Boxes show the regions of the α -TM gene containing leader (L), common (V), and isotype-specific (I-IV, VI, VII) sequences drawn tentatively as single exons in one of several possible 5' to 3' arrangements. The dotted lines represent proposed splicing events to generate the three different α -TM mRNAs. A, B, and C describe the specific pathways producing the STRα $_1$-TM, STRα $_2$-TM, and SMα-TM mRNAs, respectively.

ventricular tissues. These results demonstrate that, in addition, identical α -TM isoforms are expressed in skeletal and cardiac muscles despite the fact that the two muscle types express different myosin heavy chains, light chains, actins, and troponins (see this volume).

The Three Forms of α -TM mRNA are Encoded by the Same Gene. The results presented so far demonstrate conclusively the existence of three distinct α -TM mRNAs. The fact that the three mRNAs share regions of 100% homology in conjunction with isotype-specific sequences organized in different combinations, raises the question as to whether the different α -TM isoforms are encoded by a single gene or by a highly conserved set of different genes. Blot analysis of rat genomic DNA hybridized to different α -TM cDNA probes demonstrated that the α -TM sequences in the cDNA clones are located in three EcoRI restriction fragments 18, 16, and 8.5 kb long that contain 5' isotype switch-specific, common, and 3' specific sequences, respectively (data not shown). This conclusion has been confirmed by the isolation of three overlapping genomic recombinant clones, spanning ∿ 35 kb of genomic DNA, which include the 16 and 8.5 kb EcoRI α -TM hybridizing fragments detected in the genomic DNA blots. These three genomic clones do not encompass the sequences corresponding to the first "isotype switch peptide" (aa 1-40) which are separated from the common coding region by a minimum of 10 kb of DNA.

Analysis of these genomic clones by Southern blot techniques reveals that the 3' "isotype specific regions" of smooth and striated muscle α -TM mRNAs are in the same 5' to 3' orientation in the gene, physically linked to each other and to the sequences common to all three α -TM isoforms as shown in Fig. 4. Interestingly, these sequences are each interrupted by an intron that separates the codons specifying the carboxyl terminal peptide (codons 258-284 and 254-280) from the respective 3' untranslated sequences. The results shown schematically in Fig. 4 have been confirmed by DNA sequencing analysis (data not shown).

The organization of the α -TM gene shown above rules out the existence of tandemly arranged α -TM genes and conclusively demonstrates that the mRNAs coding for the three α -TM isoforms are produced from this single gene by alternative splicing of a set of exchangeable "isotype switch exons." The α -TM gene contains a single region encoding the sequences common to all three mRNAs -- codons 77 (81) to 253 (257), (see Fig. 4) — while the sequences coding for aa 1 to 40, 41 to 76 (80), 254 (258) to 280 (284) and the 3' untranslated sequences are all duplicated. Independent of the specific order of the different "isotype switch exons" at the 5' end of the gene, it is evident that a complex pattern of alternative splicing pathways (shown schematically in Fig. 4) has to be involved in generating the three documented α -TM mRNAs. In addition, if the "isotype switch exons" can be spliced together in all possible combinations a minimum of eight different mRNAs coding for unique proteins could be generated from this gene. Given the probes presently available, the existence of any of the five mRNAs not yet documented can neither be excluded nor confirmed.

A Combinatorial Pattern of Splicing can Generate up to 64 Different Isoforms From a Single Fast Troponin T Gene.

Troponin T (TnT) is the M_r 30,000 regulatory protein which, in combination with troponins I and C, confers calcium sensitivity to the actomyosin ATPase in the presence of tropomyosin in striated muscles. Antigenically diverse TnT isoforms are present in different muscle types and at various developmental stages. At least three different TnT genes appear to be differentially expressed in vertebrates, specific for fast skeletal, slow skeletal, and cardiac fibers, respectively. In addition to this protein diversity generated by the activation of different genes, we have previously elucidated a novel mechanism of alternative RNA splicing for the developmentally regulated generation of two TnT isoforms from a single rat fast skeletal TnT_f gene (5). It is now clear that the fast TnT gene has a very unusual structural organization. This gene, using a combinatorial pattern of exon splicing in one region in conjunction with the alternative but mutually exclusive use of two other exons, can generate up to 64 different proteins in a developmental and tissue-specific manner. Such extensive coding potential has not been documented for any other gene, with the exception of the immunoglobulins. Recently, it has been demonstrated that the cardiac TnT gene in chicken also uses alternative splicing in order to produce two different proteins (Cooper et al., this volume).

Multiple cDNA clones from transcripts of the unique TnT_f gene. Fig. 5 shows the structural organization of the rat TnT_f gene and four cDNA clones. This organization was determined from the complete nucleotide sequence of two overlapping genomic clones, spanning 21 kb of DNA, as well as the four cDNA clones (6).

While each of the four cDNA inserts represents a unique TnT isoform, all have the sequence encoding aa 37-72 in common. The absolute conservation of this nucleotide sequence, shown to exist in only one copy per haploid genome (5), indicates that all four cDNAs derive from a single gene. Among them, only pTnT-III and IV extend to the translational start codon and into the 5' untranslated sequence. Their corresponding mRNAs remain indentical up to their cap sites, as demonstrated by the 5' extension products of these cDNA species, which gave a single unambiguous sequence comprising 79 nucleotides upstream from the translation start codon. The detection of a single fully-extended 5' untranslated sequence by primer extension assays indicates that the mRNAs corresponding to pTnT-III and IV, as well as any other putative TnT message with 5' sequence complementary to the primer, derive from primary RNA transcripts with identical initiation sites. Since no alternative promoter was identified, nor does the TnT_f gene show any heterogeneity at the 3' end (5), these results demonstrate that structurally identical primary gene transcripts can be alternatively spliced into different mature mRNAs.

The TnT gene has an unprecedented structural organization. The complete nucleotide sequence of the overlapping genomic clones λ TnT-1 and λ TnT-2 (Fig. 5A) has been determined. Eighteen individual exons, together accounting for all the known cDNA sequences, where identified and are diagrammed schematically in Fig. 5B. The protocol used to compose this figure numbers the exons and the amino acids consecutively, based simply on their 5' to 3' location in the genomic sequence, rather than on the four different cDNA sequences. One exception is made for exons 16 (α) and 17 (β), where the original duplicate amino acid number (5) is retained.

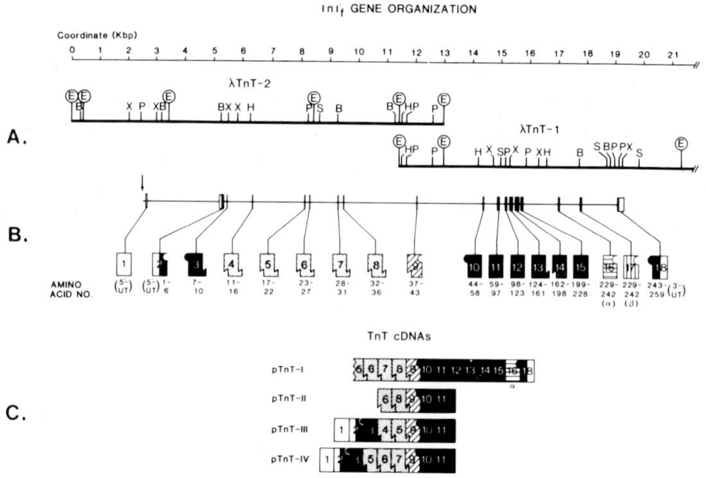

FIGURE 5. Organization of the rat TnT_f gene encoding multiple mRNAs. (A) Partial restriction map of overlapping genomic clones. (B) Diagram of the TnT_f gene structure, aligned with the above map, showing the position and relative sizes of the exons along the upper line and, beneath, schematic representations of the exons: white, untranslated; black, constant; gray, interchangeable and not mutually exclusive; striped, exchangeable but mutually exclusive (see text). (C) Schematic incorporation of the exons in different combinations to form the four cloned cDNA sequences.

The distance from the RNA cap site to the polyadenylation site subtends 17 kb of genomic sequence, as shown in Fig. 5, only 6% of which comprises exons. The exons range in size from a maximum of 243 nucleotides (exon 18) to just 12 nucleotides (exon 7); exons 3 through 9 are among the smallest reported in the literature to date. Exon 1 consists exclusively of untranslated sequence and, as such, represents a spliced leader. Exon 2 and exon 18 each span translated and untranslated sequence at their respective ends of the gene. Of the internal coding exons, a distinct similarity is evident among exons 3, 4, 5, 6, and 8: all begin and end with "split" glutamic acid (E) codons, and glutamic acid as well as glutamine (Q) and valine (V) codons are prevalent. There is particular homology between exons 4 (EQYEEE) and 5 (EEAQEE), and between exons 6 (EVQEE) and 8 (EEVQE). A similar level of homology is evident between exon 16 (α) and 17 (β) (5). The overall redundancy of nucleotide sequence among exons 3 to 8 on one hand, and 16 and 17 on the other, in combination with the matched symmetry of their split codons (see below), suggests that they may have derived from a series of duplication events.

Each exon is flanked by the conventional acceptor and donor splice junction consensus sequences (25) with absolute conservation of the AG and GT dinucleotides, respectively. Close inspection of the individual splice junctions reveals that, while many occur conventionally between two intact triplets (represented schematically as blunt boundaries in the figures), there are some which, instead, fall within a codon; in particular, an exon may terminate lacking either the third nucleotide of its final codon (concave boundaries) or both the second and third nucleotides (sawtooth boundaries). The immediate implication for RNA splicing is that the succeeding exon must, then, commence with a complementary split codon, contributing either one or two nucleotides, respectively, in order to preserve the sense of the reading frame. No anomaly of flanking intron sequences was detected at any of the donor or acceptor sites which might distinguish exons that are subject to alternative RNA splicing from those that are spliced constitutively.

Figure 5C diagrams the schematic pattern of incorporation of the various exons into the four known cDNA sequences. Remarkably, in the upstream region of cDNA divergence, the symmetry of split codons characteristic of exons 4, 5, 6, 7, and 8 permits one or more of these exons, and only these exons, to be excluded with absolute preservation of the translational reading frame. In fact, it is by just such selective exclusion via different pathways of alternative splicing that the mRNA divergence must be generated, as predicted from the cDNA sequences themselves (6). While subtotal 5' extension of pTnT-I and -II leaves their extreme upstream sequences unknown, the following is nevertheless clear: pTnt-I includes each of exons 5, 6, 7, and 8; pTnT-II excludes exon 7; pTnT-III excludes exons 6, 7, and 8; and pTnT-IV excludes exons 4 and 8. Both pTnT-III and -IV include exon 2, containing the 5' untranslated sequence and start codon, and so necessarily include exon 3, which, owing to its unique split codon asymmetry, cannot be ommitted. Exons 9, 10, 11, and a portion of 12, representing aa 37-109, are constant among all four cDNAs. Moreover, it was demonstrated previously (5) that exons 12-15 and exon 18 are also constant in all rat skeletal muscle TnT mRNAs detectable by S1 nuclease analysis with probes spanning this region, and at all developmental stages. Exons 16 (α) and 17 (β), however, are included in a mutually exclusive fashion; no mRNA contains both, as required for preservation of the sense of the reading frame. While pTnT-1 excludes exon 17 (5), a rabbit skeletal muscle cDNA containing a sequence homologous to exon 17 (β), rather than 16 (α), at aa 229-242 has been isolated (17).

Inspection of the genomic sequence immediately upstream of exon 1 reveals cannonical promoter TATAA and CAAT boxes at -24 and -55 nucleotides, respetively, from the RNA cap site, in a guanine-rich region. Since this is the only promoter sequence identified in the gene, it is expected that all TnT_f primary transcripts, while subject to different splicing pathways, are initiated from this unique start site. Moreover, it was shown previously that all transcripts of this gene share the AAATAAA polyadenylation signal in exon 18 and terminate at a unique site 17 nucleotides downstream (5). These data confirm that RNA transcripts completely identical in their primary structure undergo nonidentical combinations of alternative splicing to generate different messages. Factors acting in trans on the primary RNA transcripts must, therefore, be implicated in the control of these processes if they occur in a nonrandom fashion; the common primary structure of these RNA molecules cannot contain sufficient discriminative information to determine regulated differential splicing. However, the possible existence of other promoters further upstream of the one identified, although unlikely, cannot be completely ruled out with the data presently available.

A minimum of ten different TnT_f mRNAs is generated in a developmentally regulated and tissue-specific manner. The incorporation of any one, two three, four, all five, or none of the exons 4, 5, 6, 7 and 8 of the TnT_f gene could produce 2^5, or 32, different mRNA sequences upstream of the codon for aa 37. Since, in addition, exons 16 (α) and 17 (β) are incorporated into each mRNA in an alternative and mutually exclusive manner (5) the coding potential of this gene increases to 64 different mRNAs, each encoding a different protein. Such diverse messages could be generated through an entirely stochastic process of alternative RNA splicing or, rather, through a specifically regulated pattern. In order to distinguish between these two possibilities a series of S1 nuclease analyses was performed using probes derived from the four cDNA clones.

The results shown in Fig. 6 (see ref. 6 for details) and in a series of S1 nuclease analyses of RNAs from different developmental stages (6), demonstrate the existence of fully protected fragments corresponding to each of the respective probes. In addition, partially protected fragments reflecting mRNA species with upstream sequences that diverge from the respective probes at the points corresponding to the boundaries between exons 3/4, 4/5, 5/6, 6/7, 7/8, and 8/9 are also detected. At no point does the divergence occur within an exon, except where two exons share one or more identical 3' terminal nucleotides. In each case the protected portion of the probe is depicted schematically to the right

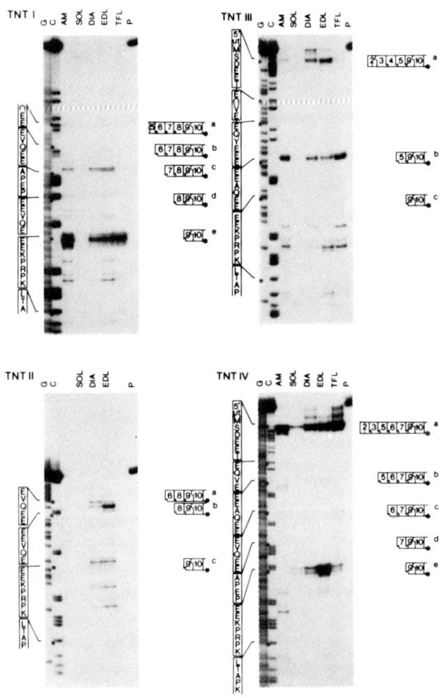

FIGURE 6. Identification of multiple tissue-specific TnT mRNAs. End-labelled (*) probes derived from the four cDNA clones were hybridized individually to rat skeletal muscle RNAs (AM, mixed adult; SOL, soleus; DIA, diaphragm; EDL, extensor digitorum longus; TFL, tensor fasciae latae) and subjected to S1 nuclease digestion. S1 nuclease-resistant fragments were electrophoresed adjacent to the guanine (G) and cytosine (C) base-specific chemical degradation products for the respective probe (P). The corresponding aa sequence for the probe is aligned to the left, and the exon components of the protected bands are diagrammed directly to the right. (Detailed methods and analysis as per ref. 6.)

of the appropriate band. Most significant is that individual muscles and stages of development differ in their pattern of expression of these mRNAs. These results document the expression of a minimum of ten different alternative splicing patterns in the "hypervariable" region of the TnT_f gene, between aa 10 and 37 (exons 4 to 8) depicted schematically in Fig. 7. It is important to emphasize that every pattern potentially recognizable by each of the four probes has, in fact, been documented. While additional combinations may well exist, these ten are the minimum needed to account for all the

S1 nuclease protected fragments. Table 1 shows that each of the ten defined splicing patterns is actually a partial representation of one or more combinations of the five interchangeable exons depending, in particular, on which exon sequences are included 5' to those protecting the probe. Together, these ten classes comprise 22 of the 32 theoretical combinations of the five exons; the remaining ten represent patterns not detectable by the four available cDNA probes. It is likely that their existence will be detected once the necessary probes are available.

Figure 7 diagrams splicing pathways joining either of the two mutually exclusive exchangeable exons 16 (α) and 17 (β) with any of the ten minimum combinations of the interchangeable exons, generating twenty different potential patterns of alternative splicing for the primary RNA transcript of this gene. Indeed, if all of the 32 combinations listed in Table 1 can be spliced downstream to either exon 16 (α) or 17 (β), the theoretical potential doubles to 64. However, as overlapping but, nevertheless, different probes were employed to explore the 3' "isotype switch region" (5) and the 5' hypervariable domain, these data cannot confirm that mature mRNA molecules exist for each of the twenty splicing pathways. For example, the pathway which generates pattern Ia upstream might always include exon 16 (α), excluding 17 (β), while another upstream pattern might invariably include exon 17 (β). While the

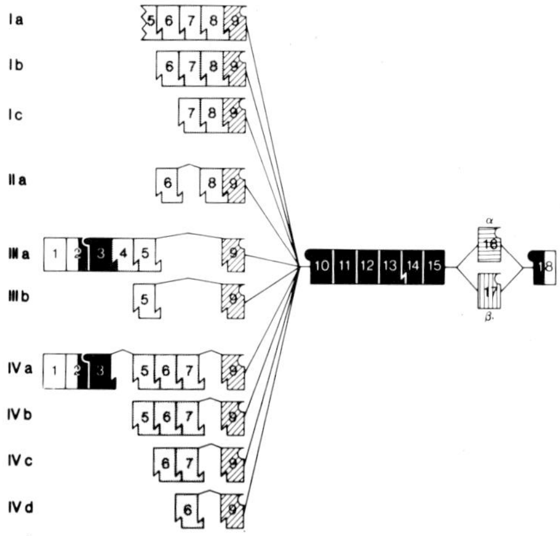

FIGURE 7. A minimum of ten mRNAs is produced by alternative combinations of TnT exons. These ten partial patterns of RNA splicing upstream of exon 9, Ia-IVd, define ten classes of TnT mRNAs as detected by S1 nuclease analysis with the corresponding cDNA probes TNT I-IV. At least one member of each class must exist in vivo (see text). Exons are represented schematically as in Fig. 5.

TABLE 1
THIRTY-TWO COMBINATIONS OF THE FIVE INTERCHANGEABLE TnT EXONS

Ia	4 5 6 7 8	IIIa	4 5 . . .	4 5 . . 8
	. 5 6 7 8	IIIb	. 5 . . .	4 . . . 8
Ib	4 . 6 7 8	IVa	. 5 6 7 .	. 5 . . 8
	. . 6 7 8	IVb	4 5 6 7 .	 8
Ic	4 5 . 7 8	IVc	4 . 6 7 .	4 5 6 . .
	4 . . 7 8		. . 6 7 .	4 . 6 . .
	. 5 . 7 8	IVd	4 5 . 7 .	. 5 6 . .
	. . . 7 8		4 . . 7 .	4 . . . .
IIa	4 5 6 . 8		. 5 . 7 .	. . 6 . .
	4 . 6 . 8		. . . 7 .	. . . . .
	. 5 6 . 8			
	. . 6 . 8			

Numbers refer to the respective exons while dots indicate exclusion of the particular exon by alternative splicing. Of the 32 theoretical combinations of exons 4-8, those 22 underlined fall into 10 distinct mRNA classes detectable by the four cDNA probes in S1 nuclease analyses. The existence of the remaining combinations in the right-most column could not be explored with the available probes. At least one member of each grouping Ia-IVd must exist in the mRNA population, giving an empirical minimum of 10 different species.

experiments reported here do not address whether any or all of these mRNAs are, in fact translated into proteins, the abundance of at least ten distinct isospecies among the steady state RNA populations from the tissues assayed argues strongly for their functional significance. The combinatorial patterns in the 5' region of TnT generate proteins differing not only in amino acid sequence and charge but also in overall size. Resolution of the full spectrum of TnT diversity by protein biochemical methods alone, however, may be limited not only by the particular susceptibility of TnT to proteolysis, but also by the similarity of size and charge of certain of the interchangeable amino-terminal peptides. Those encoded by exons 4 and 5 each comprise six residues with a net charge of -4, and those encoded by exons 6 and 8 each comprise five residues with a net charge of -3.

Wilkinson et al. (18, and this issue) have recently isolated and sequenced three distinct TnT amino-terminal peptides from chicken skeletal muscle that share discontinuous subsegments of amino acid sequence, analogous to the relationships described for the four rat TnT cDNAs and rabbit TnT in this region. These data indicate that mechanisms underlying the complex patterns of alternative RNA splicing documented in the rat are conserved in the avian species. Remarkably, peptides corresponding to all of the rat TnT_f exons are present in one or more of the chicken isoforms. The chicken breast

TnT, however, contains amino-terminal region peptides, composed exclusively of glutamic acid, alanine, and histidine residues, for which there are no counterparts in the four rat cDNAs, nor in the TnT_f genomic sequence, just as the rabbit TnT has two additional short peptides. These data imply a rapid rate of evolution of this gene by exon deletion and/or duplication.

Three types of TnT RNA splicing characterize five protein domains. The degree of diversity among the TnT_f isoforms varies within five defined regions of the primary sequence representing three constant, a variable and one hypervariable domain. The first 10 amino terminal as well as the last 17 carboxyl terminal amino acids appear to be shared among all isoforms. At the mRNA level the homology extends to exons 1 to 3 and 18, respectively. Another constant domain, from aa 44 to 228, generated by the constitutive splicing of exons 10-15, is shared by all the isoforms, as no partial protection of overlapping cDNA probes spanning this sequence was detected in S1 nuclease analyses (5,6). Remarkably, this region of TnT has an extensive predicted α-helical content, in contrast with the remainder of the molecule. It also includes the aa 71-151 and 152-209 peptides responsible, in part, for the binding of TnT to α-tropomyosin and troponin I, respectively (19,20). The helical structure and uniform conservation of this sequence speak to its fundamental role in the interaction of all TnT isoforms with the regulatory complex. The segment of TnT from aa 229 to 242, encoded by each of two alternatively spliced but mutually exclusive exons, 16 (α) and 17 (β) represents a variable domain of the protein. This segment is included in the chymotrypic fragment that binds troponin C and tropomyosin in a Ca^{2+}-sensitive fashion (21-23). Clearly, the isotype switch in this sequence may influence these interactions. The region of TnT manifesting the highest degree of diversity is the hypervariable domain from aa 11 to 36. Encoded by combinatorial patterns of splicing of exons 4-8, the absolute length as well as the sequence of this region is subject to a high degree of variation. A cyanogen bromide fragment of TnT comprising aa 11-43 binds troponin I (20). Again, the diversity in this region may provide means for modulating the interaction of TnT with troponin I.

DISCUSSION

The results here for the MLC 1/3, α-TM and TnT_f genes clearly demonstrate an unsuspected level of complexity in the contractile proteins generated by different modes of alternative splicing of the primary transcripts of the respective genes. In addition, these results raise important questions about the physiological significance of this protein diversity as well as the mechanisms involved in its production.

Current models of processive scanning RNA splice site selection (see 24) cannot explain the observed patterns of splicing of these three genes. Moreover, mechanistic models for alternative RNA splicing must accomodate the existence of at least two types of alternative splicing: a) alternative but mutually exclusive splicing such as demonstrated for exons 3 and 4 of MLC 1/3, the isotype-specific exons of α-TM, and exons 16 and 17 of TnT_f; b) the combinatorially interchangeable exons 4 to 8 of the TnT_f gene. In addition, splicing mechanisms must differentiate between exons that might be selected in an alternative fashion from those included constitutively. Moreover, at least in some cases, such as with the TnT_f gene, they must be capable of generating different mature mRNAs from the same primary transcript in a regulated, that is nonstochasitc, manner.

It is reasonable to expect that the end result of a process that splices an array of exons from an RNA transcript must be a message with an appropriate translational reading frame. Split codons characterize alternatively spliced exons in TnT and other genes (5-10); however, as the putative splicing factors are presumably incapable of recognizing phased tripled nucleotides as such, the split codon symmetries cannot directly control pathways of alternative splicing. Nevertheless, their existence constitutes a strong selective pressure against promiscuous mechanisms that yield incorrect pairings in an evidently intricate combinatorial system. If such incorrectly spliced RNAs are generated, they are unstable, as they do not accumulate in detectable amounts.

It is widely accepted that flanking donor and acceptor site sequences are responsible, at least in part, for the demarcation of exons and introns, resolving the exact points for splicing (25). These consensus sequences are uniformly present in the TnT introns, as are those of most branchpoints for the formation of lariat intermediates (26,27). However, this system, which confers the fidelity of splice site selection common to all exons, must be reconciled with a mechanism for differentiating those few subjects to alternative splicing. No elements of primary intron sequence have been identified that distinguish alternative versus constitutive exons.

Alternative splicing can be mechanistically defined as the ability to selectively mask certain donor and/or acceptor sites in a primary transcript. Although the precise mechanism responsible for this selective masking remains to be elucidated, the data presently available and the behavior of the three genes presented here give some clues as to how this process may be regulated. Among the genes exhibiting regulated alternative splicing, all of those reported to date, with the exception of the TnT_f gene, do so by generating structurally different RNA transcripts, either initiated at different promoters, or terminated at alternative 3' ends (see 5-7 for review). This feature strongly suggests that the primary sequences of the different transcripts of these genes might each contain enough information to govern its own splicing pathway by cis mechanisms alone, for example, different secondary structures; the involvement of trans-acting factors would not be absolutely required, although it would not be ruled out. We have demonstrated here that the TnT_f mRNAs are spliced from entirely identical primary transcripts, sharing both a single promoter and unqiue poly-adenylation site, in a specifically regulated manner. This differential alternative splicing must involve trans-acting factors.

A strict directionality of the splicing process is also suggested by the structural organization of each pair of exons that are spliced in an alternative but mutually exclusive fashion, such as exons 3 and 4 of MLC 1/3, isotype specific exons of α-TM, and exons 16 and 17 of TnT_f. The splicing together of the 3' exon of any one pair to the 3' common exon after the 5' exon of the same pair has already been joined to the 5' common exon would, in all cases, result in an abortive mRNA molecule with a missense of the translational reading frame. Although, due to limitations in the techniques available, the generation of such molecules cannot be excluded, it is clear that they do not accumulate to any significant extent. The data presented here strongly suggest that the first splicing reaction within an isotype switch region determines the second. The demonstration of tissue and developmental specificity in the expression of each of the exons effectively rules out a purely stochastic activation of the first splice site. Once this choice has been made, only the other splice site of the selected alternatively spliced exon can be joined to the remaining common exon, while the equivalent splice site of the non-selected exon is prevented from doing so. Therefore, the exclusion of a given exon from the mature mRNA requires that its flanking splice junctions be overlooked by the splicing machinery. To regulate this process a modified scanning model is conceivable in which the splice sites of the excluded exons are masked and inaccessible to a generic splicing activity that does not by itself discriminate among splice sites.

While no primary intron sequences have been identified to distinguish alternative from constitutive exons, it is possible to envision secondary structures of gene transcripts that may expose the splice sites of particular exons while exlcuding others. In this model, "allowable" splices would be favored by local conformations of the RNA having the least free energy, either alone or stabilized further by additional intramolecular secondary structure (cis), or by interaction with other factors (trans). Regions of the RNA molecule that may be spliced in more than one pattern, then, would assume different conformations subject to such stabilizing influences, whether these derive from structurally different ends of transcripts with alternative initiation or polyadenylation sites, or from specific, coordinately expressed trans-acting factors. Moreover, splices executed at one point in the intermediate can affect splicing elsewhere. In TnT_f, for example, a particular combination of exons 4-8, if spliced first, might dictate the usage of exons 16 (α) or 17 (β), or vice versa. The finding that a mutant Neurospora mitochondrial pre-rRNA with an anomalous 3' extension fails to be spliced (28) lends validity to the concept that distant intramolecular changes might affect local splicing.

The physiological basis for the selective advantage of encoding different protein isoforms in a single gene, rather than in different genes, is not clearly understood. However, the assembly of contractile proteins into a complex organelle such as the sarcomere, the function of which is differentially modulated in a developmental and tissue-specific manner, places stringent requirements on the conservation of the domains involved in that intermolecular assembly. It also necessitates a capacity to generate diversity in the domains involved in the modulation of function. Thus, despite the apparently higher risk for lethal mutations, it might still be advantageous to maintain the constant domain of a protein in a single copy, able to combine with multiple variable domains with specialized characteristics. This arrangement is reminiscent of the immunoglobulin system where diversity is generated by such combination achieved by DNA rearrangement, and in some cases by differential splicing. DNA rearrangement is an irreversible process, well suited to terminally differentiated cells with a short life span. The muscle cell, however, has a very long life span and reversibly modulates the properties of the contractile apparatus in response to developmental and environmental cues. For these cells, alternative RNA splicing is a plastic mechanism, capable of generating diversity in a reversible fashion while maintaining constant the regions involved in conserved intermolecular interactions. This capacity to generate subtly different proteins from a single gene which, in turn, belongs to a multigene family, greatly increases the potential for myofibrillar diversity. The capability of these various protein isoforms to associate in all possible combinations to generate different myofibrils represents an unsuspected complexity in the contractile system which likely plays a significant but, as yet, poorly understood role in determining the physiological properties of the muscle fibers.

ACKNOWLEDGEMENTS

The authors acknowledge the invaluable secretarial assistance of Ms. M. Hager in the preparation of this manuscript.

REFERENCES

1. Lompre AM, Nadal-Ginard B, Mahdavi M (1984). Expression of the cardiac α -and β -myosin heavy chain genes is developmentally and hormonally regulated. J Biol Chem 259:6437.
2. Periasamy M, Wieczorek DF, Nadal-Ginard B (1984). Characterization of a developmentally regulated perinatal myosin heavy chain gene expressed in skeletal muscle. J Biol Chem 259:13573.
3. Wieczorek D, Periasamy M, Butler-Browne GS, Whalen RG, Nadal-Ginard B (1985). Co-expression of multiple myosin heavy chain genes, in addition to a tissue-specific one, in extraocular musculature. J Cell Biol in press.
4. Leder P, Max EE, Seidman JG, Kwan SP, Naw M, Norman B (1981). Recombination events that activate, diversify, and delete immunoglobulin genes. Cold Spring Harbor Symp Quant Biol 45:859.
5. Medford RM, Nguyen HT, Destree AT, Summers E, Nadal-Ginard B (1984). A novel mechanism of alternative splicing for the developmentally regulated generation of troponin T isoforms form a single gene. Cell 38:409.
6. Breitbart RE, Nguyen HT, Medford RM, Destree AT, Mahdavi V, Nadal-Ginard B (1985). Intricate combinatorial patterns of exon splicing generate multiple regulated troponin T isoforms from a single gene. Cell, in press.
7. Periasamy M, Strehler EE, Garfinkel LI, Gubits RM, Ruiz-Opazo N, Nadal-Ginard B (1984). Fast skeletal muscle myosin light chains 1 and 3 are produced from a single gene by a combined process of differential RNA transcription and splicing. J Biol Chem 259:13595.
8. Franks G, Weeds AG (1974). The amino acid sequence of the alkali light chains of rabbit skeletal muscle myosin. Eur J Biochem 44:317.
9. Nabeshima Y, Kurajama-Fuji Y, Muramatsu M, Ogata K (1984). Alternative transcription and two modes of splicing result in two myosin light chains from one gene. Nature 308:333.
10. Robert B, Daubas P, Akimenko MA, Cohen A, Garner I, Guenet JL, Buckingham M (1984). A single locus in the mouse encodes both myosin light chains 1 and 3, a second locus corresponds to a related pseudogene. Cell 39:129.

11. Strehler EE, Periasamy M, Strehler-Page, M-A, Nadal-Ginard B (1985). The myosin light chain 1 and 3 gene has two structurally distinct and differentially regulated promoters evolving at different rates. Submitted.
12. Young RA, Hagenbuchle O, Schibler V (1981). A single mouse α -amylase gene species two different tissue-specifc mRNAs. Cell 23:451.
13. Benyajati C, Spoerel N, Haymerle H, Ashburner M (1983). The messenger RNA for alcohol dehydrogenase in Drosophila melanogaster differs in its 5' end in different developmental stages. Cell 33:125.
14. Garfinkel LI, Periasamy M, Nadal-Ginard B (1982). Cloning and characterization of cDNA sequences corresponding to myosin light chains 1, 2, and 3, troponin-C, troponin-T, α -tropomyosin, and α-actin. J Biol Chem 257:11078-11086.
15. Ruiz-Opazo N, Weinberger J, Nadal-Ginard B (1985). Comparison of α -tropomyosin sequences from smooth and striated muscle. Nature 315:67.
16. Yamaguchi M, Ver A, Carlos A, Seidel JC (1984). Modulation of actin-activated adenosinetriphosphatase activity of myosin by tropomyosin from vascular and gizzard smooth muscles. Biochemistry 23:774.
17. Putney SD, Herliky WC, Schimmel P (1983). A new troponin T and cDNA clones for 13 different proteins found by shotgun sequencing. Nature 302:718.
18. Wilkinson JM, Moir AJG, Waterfield MD (1984). The expression of multiple forms of troponing T in chicken fast skeletal muscle may result from differential splicing of a single gene. Eur J Biochem 143:47.
19. Pearlstone JR, Smillie LB (1977). The binding site of rabbit skeletal -tropomyosin on troponin T. Can J Biochem 55:1032.
20. Pearlstone JR, Smillie LB (1980). The binding sites of rabbit skeletal troponin I on troponin T. Can J Biochem 58:649.
21. Pearlstone JR, Smillie LB (1978). Troponin T fragments: physical properties and binding to troponin C. Can J Biochem 56:521.
22. Pearlstone JR, Smillie LB (1981). Identification of a second binding region of rabbit skeletal troponin T for α -tropomyosin. FEBS Lett 128:119.
23. Pearlstone JR, Smillie LB (1982). Binding of troponin T fragments to several types of tropomyosin. J Biol Chem 257:10587.
24. Sharp PA (1981). Speculations on RNA splicing. Cell 23:643.
25. Mount S (1982). A catalogue of splice junction sequences. Nucl Acids Res 10:459.

26. Keller EB, Noon WA (1984). Intron splicing: a conserved internal signal in introns of animal pre-mRNAs. Proc Natl Acad Sci USA 81:7417.
27. Ruskin B, Krainer AR, Maniatis T, Green MR (1984). Excision of an intact intron as a novel lariat structure during pre-mRNA splicing in vitro. Cell 38:317.
28. Garriga G, Bertrand H, Lambowitz AM (1984). RNA splicing in Neurospora mitochondria: nuclear mutants defective in both splicing and 3' end synthesis of the large mRNA. Cell 36:623.

Molecular Biology of Muscle Development, pages 411–421

ALTERNATIVE SPLICING AS A MECHANISM FOR GENERATION OF MUSCLE PROTEIN DIVERSITY IN DROSOPHILA[1]

Norman Davidson,[2] Scott Falkenthal[3] and Charles E. Rozek[4]

[2]Department of Chemistry, California Institute of Technology, Pasadena, California 91125
[3]Department of Genetics, The Ohio State University, Columbus, Ohio 43210
[4]Department of Biology, Case Western Reserve University Cleveland, Ohio 44106

ABSTRACT Tissue and stage specific diversity of the isoforms of muscle proteins of Drosophila are provided by several mechanisms. There are six separate actin genes, each encoding a single isoform, and each expressed with a unique developmental pattern. However, for both the myosin heavy chain and the myosin alkali light chain, there is only a single-copy gene. However, in each case, two different mRNAs encoding slightly different amino acid sequences at the carboxy termini of the respective proteins are created by alternate RNA splicing of the primary transcript. Surprisingly, the developmental specificities of the splicing patterns appear to be quite different for these muscle specific genes.

---

[1] This work was supported by grants and postdoctoral fellowships from the National Institutes of Health and a grant from the Muscular Dystrophy Association.

INTRODUCTION

It is frequently the case that different isoforms of a particular amino acid sequence are expressed in a stage and tissue specific manner during development of a higher eukaryote. One paradigm for the generation of such diversity is illustrated by the globin gene family. In vertebrates, globins are coded for by an oligogene family with 5 to 10 members. Each gene is expressed in its own unique ontogenetic program and gives rise to a single polypeptide gene product (see for example, 1).

The same mechanism is used by Drosophila for generation of diversity in the actin proteins. Drosophila has six actin genes. Each is expressed with its own unique temporal and spatial program in development, (although the genes fall into three sets of pairs with two members of each pair showing somewhat similar ontogenetic patterns) (2, 3, 4).

However differential regulation of expression of the several members of an oligogene family is not the only mechanism for generation of gene product diversity. An alternative mechanism is developmentally regulated differences in RNA splicing of primary transcripts from a single gene. In the present communication we wish to describe our recent results on how this mechanism is used to generate different isoforms for the myosin heavy chain gene (MHC), and the myosin alkali light chain gene (MLC-ALK) of Drosophila.

RESULTS

A) Myosin Heavy Chains.

Drosophila has one and only one MHC gene expressed in the several muscles of larvae and adults (5, 6). This single copy gene gives rise to several different mRNAs of length ca. 7.2, 8.0 and 8.6 kb. These transcripts arise by alternative splicing of a primary transcript (or transcripts) from a gene of length ca. 19 kb. The several mRNAs differ by alternate splicing patterns and/or polyadenylation at the 3' end of the transcripts.

By a combination of methods involving S1 transcriptional mapping, RNA gel blots, and especially by determination of the DNA sequence of the chromosomal gene and of a number of cDNA clones, the sequence organization of the mRNAs at their 3' end has been deduced. This result is depicted in Fig. 1.

It should be noted that the figure depicts sequences at the 3' terminal 1 to 2 kb of an approximately 8 kb set of mRNAs. Low resolution S1 mapping and R-loop studies have not indicated any evidence for alternative choice of cap sites or splicing patterns in the rest of the gene. However these questions have not yet been studied by the high resolution methods that have been used to determine the structure at the 3' end depicted in Fig. 1.

Figure 1 illustrates the 3' terminal exons denoted C, B and A which are involved in the alternate processing of the MHC mRNAs. The 7.2 and the 8.0 kb mRNAs, present in all stages of development, are formed by the splicing of exon C to exon A. The difference in molecular length results from the alternate selection of poly(A) addition sites within exon A. These two mRNAs provide MHC proteins for larval muscles as well as some adult muscles.

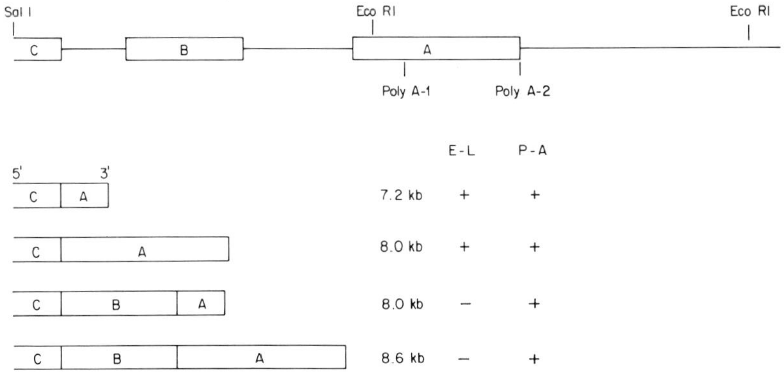

FIGURE 1. Arrangement of the exons C,B,A at the 3' end of the Drosophila MHC gene and in the several resulting mRNAs. Drawing is approximately to scale, the length of exon B is 525 nt, and the length of exon A with the down stream poly(A) site (polyA-2) is 725 nt. For further discussion, see text.

We also have identified two MHC mRNAs which are formed only during pupal and adult stages of development. These mRNAs are formed by the inclusion, by splicing, of exon B between exons C and A. In this case too, two alternate poly(A) addition sites are used within exon A. Thus, choice of polyadenylation sites does not appear to be developmentally regulated. Furthermore, selection of poly(A-1) versus poly(A-2) occurs at approximately equal frequencies and does not alter the amino acid sequence of the MHC polypeptides. Therefore, as shown in Figure 1, four species of mRNA arise from the possible combinations of the two splicing patterns and two poly(A) sites.

DNA sequencing has shown that when exon C is directly joined to exon A, (E-L and P-A stages), the reading frame continues to give 27 carboxyl terminal amino acids in exon A before the termination codon. However for the CBA mRNA (P-A stage only) there is only one amino acid encoded in exon B followed by the termination codon. These data are presented in Fig. 2.

			Exon C		Exon A											
GGT	GCC	AGC	CCA	GCG	CCC	CGT	GCG	ACG	TCC	GTT	AGG	TCA	CAA	TTC	GAC	GGA
Gly	Ala	Ser	Pro	Ala	Pro	Arg	Ala	Thr	Ser	Val	Arg	Ser	Gln	Phe	Asp	Gly
TTG	GCC	TTC	CCA	CCA	ACA	TTC	GAC	CTT	GCT	CCT	GAA	AAC	GAA	TTC	TAA	
Leu	Ala	Phe	Pro	Pro	Thr	Phe	Asp	Leu	Ala	Pro	Glu	Asn	Glu	Phe	*	

GGT	GCC	AGC	CCA	GCG	ATC	TAA
Gly	Ala	Ser	Pro	Ala	Ile	*
	Exon C				Exon B	

FIGURE 2. cDNA sequences and translated amino acid sequences at exon C-exon A and exon C-exon B junctions of the MHC gene.

Thus the overall result is that by alternate splicing of a primary transcript that includes exons C, B, and A and the introns separating them, two different MHC isoforms are produced with different carboxyl terminal amino acid sequences, with a developmentally regulated expression pattern.

b) The Myosin Alkali Light Chain Gene.

The structure of this gene (MLC-ALK gene) is depicted in Fig. 3 (7, 8). The gene begins with a single cap site at the 5' end of exon 1 (Fig. 3). All told there are six exons. Because of alternative selection of poly(A) sites, all within E6, mRNAs of several rather different molecular lengths are formed. However these differences are not developmentally regulated nor do they give rise to any different amino acid sequences for the gene products. The interesting point is that there are two different splicing patterns and mRNA sequence organizations, E1, E2, E3, E4, E5, E6; and E1, E2, E3, E4, E6 with E5 not included in the splicing pattern. The former mRNA is found in both E-L and P-A stages whereas the latter with E5 missing, is found in the P-A stage only.

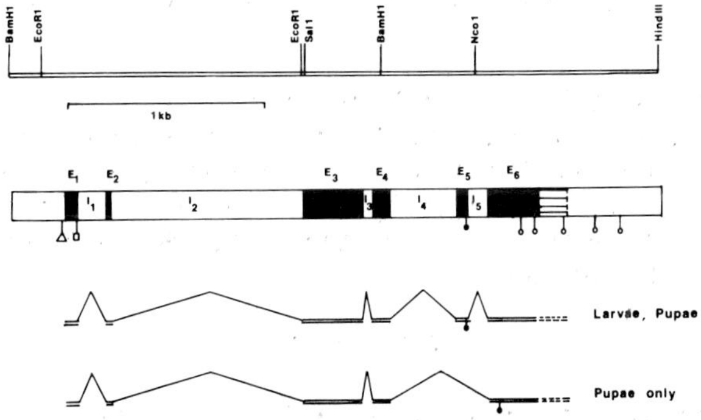

FIGURE 3. Arrangement of introns and exons in the MLC-ALK gene. The cap site is indicated by an open triangle; translation initiation begins at the last 3 nucleotides of E_1. Translation termination occurs at the end of E_5 (filled in circle) and for the E_4E_6 splice, 42 nt into E_6 (not shown). Possible poly(A) sites are shown as open circles. The two splicing patterns are shown below.

Sequence determinations show that the two alternate splicing patterns give rise to the two different translated protein coding sequences depicted in Fig. 4. For the E4, E5, E6 mRNA, reading frame is maintained between E4 and E5 to give 14 amino acids in E5 prior to a termination codon. For the E4, E6 mRNA, there are 14 amino acids encoded in E6 prior to the termination codon. These two carboxyl terminal sequences are quite different from each other, although they have the same length.

```
                                                      Exon 4 |Exon 5
 P   E   D   D   E   G   F   I   P   Y   S                |Q   F   V   Q
CCC GAG GAT GAT GAA GGA TTT ATC CCC TAC TCT C |AG TTC GTC CAG
CCC GAG GAT GAT GAA GGA TTT ATC CCC TAC TCT C |CA TTC CTC GCC
 P   E   D   D   E   G   F   I   P   Y   S                |Q   F   L   A
                                                      Exon 4 |Exon 6

                                                          Exon 5 |Exon 6
 R   L   M   S   D   P   V   V   F   E  TRM         |
CGC CTG ATG AGC GAT CCT GTC GTC TTC GAC TAA ACA | CATTCCTCGCC
AGA ATG TGT GAC AGA CCA GAT CAG CTA AAA TAG TTGAAAACAACAGCTT
 R   M   C   E   R   P   D   Q   L   K  TRM
```

FIGURE 4. Nucleotide sequences and translated amino acid sequences at the E4-E5 and E4-E6 junctions.

Note that the structure of the full gene is known in this case. The cap site has been mapped to the 5' end of exon 1 and there are no developmental differences in splicing nor in transcript initiation in this region, unlike the case of the vertebrate MLC-ALK isoforms (MLC-1/3) (9, 10, 11).

DISCUSSION

A summary of the alternate splicing patterns observed for these two genes is depicted in Fig. 5. We use the notation of the MHC case for convenience of comparison. For both cases there is a penultimate exon B, which is included in one splicing pattern and omitted in the other. For the MLC-ALK case, exon B (E5 in the notation of Fig. 3) is included in all stages of development and spliced out only in the P-A stage. Just the reverse is the case for the developmental variation in the MHC splicing pattern where the middle exon B is included only in the P-A stage.

MLC-3	5' C B A 3'	all stages
	C A	pupal only
MHC	C B A	pupal only
	C A	all stages

5' splice junctions of B are non consensus

MLC	tgaaaacag	AG
MHC	cgaaaacag	AT
consensus	tytyyyncag	G

FIGURE 5. Summary of splicing patterns of MLC-ALK and MHC genes of Drosophila. In this figure, MLC-3 denotes MLC-ALK and CBA of MLC-ALK denotes E4, E5, E6 of Fig. 3.

The only non-consensus splice sequences, according to the usual Chambon rules (reviewed in 12) lie at the 5' end of exon B. In both cases, the consensus CAG at the intron exon junction is present but is preceded by a non-consensus GAAA sequence rather than the usual pyrimidine rich sequence. Presumably, the alternate splicing events for these two genes are occurring in some of the same muscle tissues of the larval and adult muscles. However the same non-consensus splice acceptor sequence is utilized in both stages of development for the MLC-ALK transcript but in only the P-A stage for the MHC transcript. Therefore the molecular mechanism for the developmental differences in splicing with quite different patterns for these two genes must involve sequences within the transcript other than those simply at this single splice junction.

The most important point is that in each case the alternate RNA splicing generates two different isoforms for the carboxyl termini of the proteins, for both the MHC and the MLC-ALK cases. The differences are in one small region involving less than 30 amino acids in polypeptides chains consisting of approximately 2000 and 150 amino acids, respectively. It may be noted however that in the case of the MHC gene of C.elegans, a single missense mutation in this amino acid sequence of length ca. 2000 can prevent assembly of the gene product into functional myofibrils (reviewed in 13). Therefore it is entirely plausible that the amino acid differences between the two isoforms of either protein could have significant effects on the properties or assembly of the resulting myofibrils.

One of us has recently observed that the P-A specific mRNA (E4,E6 of Fig. 3) of the MLC-ALK gene is the only form expressed in the indirect flight muscle, whereas the E-L form of MLC-ALK mRNA (E4,E5,E6 of Fig. 3) is the only form expressed in the musculature of the adult head and abdomen. These latter muscles are capable of supercontraction like larval body wall muscle but ultrastructurally are adult tubular muscle (14).

There are a number of instances known in which a single transcribed region of DNA gives rise to a diverse but related set of polypeptides by a combination of alternative transcription initiation sites, splicing choices, or polyadenylation sites. Examples include immunoglobulin genes, lens crystallin proteins, and endocrine hormones (15, 16, 17). It is striking however that these phenomena seem to be a particularly conspicuous mechanism for generation of isoforms of components of the myofibril, both in Drosophila and in vertebrates (9, 10, 11, 18, 19, 20). If this generalization is confirmed by further studies of other genes, it will raise an interesting question as to the reasons for the special suitability of this mechanism for generating diversity for muscle tissue.

Finally, we note that molecular biologists interested in the problem have frequently discussed the relative merits of two hypotheses as to the driving force in evolution for the existence of an oligogene family coding for very closely related isoforms of actin (for example), that are expressed in a stage and tissue specific manner. One extreme position is that the creation in evolution of separate chromosomal genes provides a convenient mechanism for creating cis acting control sequences to achieve different patterns of regulation, but that the differences in amino acid sequences are the result of neutral mutations that do not have functional significance. The alternative hypothesis is that the slightly different amino acid sequences of the different isoforms have selective value and provide for more effective function of each component in its respective muscle tissue.

It seems implausible that nature would develop gratuitous mechanisms for tissue specific non-consensus RNA processing of one primary transcript. It is much more plausible that the isoforms so generated have specialized functional significance. In any case these questions as to the significance of slightly different amino acid sequences of different isoforms for function are now amenable to experimental study because of our present and growing ability to introduce altered genes into an organism and observe their effects on function.

ACKNOWLEDGMENTS

We acknowledge the collaboration and counsel of our many colleagues at the California Institute of Technology who have contributed to the results and ideas presented here. These include: B.J. Bond, E.A. Fyrberg, N.D. Hershey, K.M. Mixter, W. Mattox and V. Parker.

REFERENCES

1. Efstratiadis A, Posakony JW, Maniatis T, Lawn RM, O'Connell C, Spritz RA, DeRiel JK, Forget B, Weissman SM, Slightom JL, Blechl AE, Smithies O, Baralle FE, Shoulders CC, Proudfoot NJ (1980). The Structure and evolution of the human β-globin gene family. Cell 21:653.
2. Fyrberg EA, Kindle KL, Davidson N, Sodja A (1980). The actin genes of Drosophila: a dispersed
3. Fyrberg EA, Mahaffey JW, Bond BJ, Davidson N (1983). Transcripts of the six Drosopohila actin genes accumulate in a stage- and tissue-specific manner. Cell 33:115.
4. Tobin SL, Zulauf E, Sanchez F, Craig EA, McCarthy BJ (1980). Multiple actin-related sequences in the Drosophila melanogaster genome. Cell 19:121.

5. Rozek CE, Davidson N (1983). Drosophila has one myosin heavy-chain gene with three developmentally regulated transcripts. Cell 32:23.
6. Bernstein S, Mogami K, Donady J, Emerson C (1983). Drosophila myosin heavy chain is encoded by a single gene located in a chromosomal region of other muscle genes. Nature 302:393.
7. Falkenthal S, Parker VP, Mattox WW, Davidson N (1984). Drosophila melanogast has only one myosin alkali light-chain gene which encodes a protein with considerable amino acid sequence homology to chicken myosin alkali light chains. Molec Cell Biol 4:956.
8. Falkenthal S, Parker VP, Davidson N (1985). Developmental variations in the splicing pattern of transcripts from the Drosophila gene encoding myosin alkali light chain result in different carboxyl-terminal amino acid sequences. Proc Natl Acad Sci USA 82:449.
9. Nabeshima Y-I, Fujii-kuriyam Y, Muramatsu M, Ogata K (1984). Alternative transcription and two modes of splicing result in two myosin light chains from one gene. Nature 308:333.
10. Periasamy M, Strehler EE, Garfinkel LI, Gubits RM, Ruiz-Opazo N, Nadal-Ginard B (1984). Fast skeletal muscle myosin light chains 1 and 3 are produced from a single gene by a combined process of differential RNA transcription and splicing. J Biol Chem 259:13595.
11. Robert B, Daubas P, Akimenko M-A, Cohen A, Garner I, Guenet J-L, Buckingham M (1984). A single locus in the mouse encodes both myosin light chains 1 and 3, a second locus corresponds to a related pseudogene. Cell 39:129.
12. Sharp PA (1981). Speculations on RNA splicing. Cell 23:643.
13. Waterston RH, Bolten S, Sive HL, Moerman DG (1982). Mutationally altered myosins in Caenorhabditis elegans. In Pearson ML, Epstein HF (eds): "Muscle Development: Molecular and Cellular Control," New York: Cold Spring Harbor Laboratory, p 119.
14. Dubendorfer A, Blumer A, Deak II (1978). Differentiation in vitro of larval and adult muscles from embryonic cells of Drosophila. Roux Arch Devel Biol 184:233.

15. Early P, Rogers J, Davis M, Calame K, Bond M, Wall R, Hood L (1980). Two mRNAs can be produced from a single immunoglobulin μ gene by alternative RNA processing pathways. Cell 20:313.
16. King CR, Piatigorsky J (1983). Alternative RNA splicing of the murine αA-crystallin gene: protein-coding information within an intron. Cell 32:707.
17. Rosenfeld MG, Mermod JJ, Amara SG, Swanson LW, Sawchenki PE, Rivier J, Vale WW, Evans RM (1983). Production of a novel neuropeptide encoded by the calcitonin gene via tissue-specific RNA processing. Nature 304:129.
18. Basi GS, Boardman M, Storti RV (1984). Alternative splicing of a Drosophila tropomyosin gene generates muscle tropomyosin isoforms with different carboxy-terminal ends. Molec Cell Biol 4:2828.
19. Karlik CC, Coutu MD, Fyrberg EA (1984). A nonsense mutation within the Act88F actin gene disrupts myofibril formation in Drosophila indirect flight muslces. Cell 38:711.
20. Medford RM, Nguyen HT, Destree AT, Summers E, Nadal-Ginard B (1984). A novel mechanism of alternative RNA splicing for the developmentally regulated generation of troponin T isoforms from a single gene. Cell 38:409.

Molecular Biology of Muscle Development, pages 423–436

EXPRESSION OF TROPONIN T VARIANTS IN CHICKEN STRIATED MUSCLE

J.M. Wilkinson, I.M. Bird,[*1] G.K. Dhoot,[†] B.A. Levine[‡] and R.D. Taylor*

Department of Biochemistry, Royal College of Surgeons, Lincoln's Inn Fields, London, WC2A 3PN, Departments of *Biochemistry and †Immunology, University of Birmingham, Birmingham B15 2TT, and ‡Inorganic Chemistry Laboratory, University of Oxford, Oxford OX1 3QR.

ABSTRACT Troponin T is expressed in several closely related variant forms in chicken fast muscle. Sequence analysis of the N-terminal region of three of these variants suggests that they are products of a single gene and that the different forms are produced by a differential splicing mechanism. Investigation of the distribution of these variants in different chicken muscles using monoclonal antibodies shows that there are at least five major variants, two being expressed in breast muscle and three in leg muscles. Only a single form of troponin T is expressed in the adult chicken heart. Proton NMR studies of the N-terminal peptides show that they have a relatively structured conformation which is sensitive to changes in the charge environment.

INTRODUCTION

Contractile activity in striated muscle is regulated via the thin filament by troponin and tropomyosin. Troponin itself is a complex of three proteins, troponin C which binds Ca^{2+} and provides the trigger for muscular

[1]Present address: Courtauld Institute of Biochemistry, The Middlesex Hospital Medical School, London, W1P 7PN.

contraction, troponin I which binds to actin and inhibits the actomyosin ATPase and troponin T which binds to tropomyosin and locates the complex at a repeat of 38.5 nm along the thin filament. The original view of troponin was that of a globular structure located adjacent to every seventh actin monomer (1). More recent work from studies of antibody binding (2), fragment binding (3-7), chemical modification (8) and electron microscopy (9) has suggested a model in which the troponin complex is approximately 26.5 nm long with a head composed of troponin C and troponin I and a tail, 16 nm long, composed of troponin T. It thus spans almost two thirds of the thin filament repeat. A schematic diagram of the complex and its relationship to tropomyosin is shown in Fig. 1. The C-terminal chymotryptic fragment, T2, of troponin T contains binding sites for both troponin I and troponin C and also a site which binds to tropomyosin in the region of Cys-190. In the intact troponin complex this site is Ca^{2+} sensitive. A second binding site for tropomyosin is found in the cyanogen bromide fragment, CB2, from residue 71-181. This region contains a high content of α-helix, binds strongly

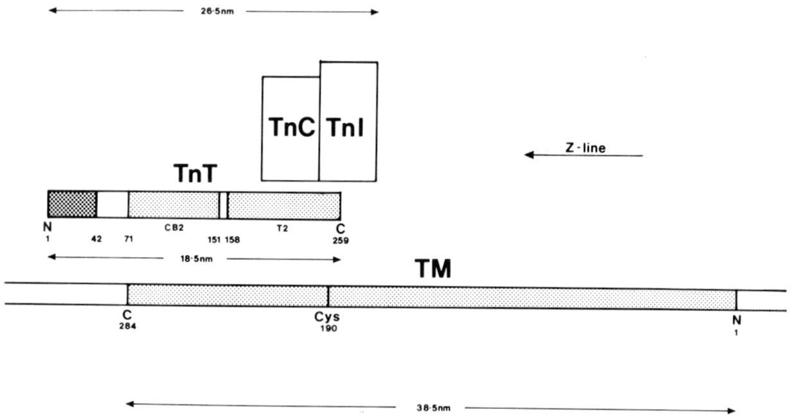

FIGURE 1. Schematic diagram of the interaction of troponin with tropomyosin. The shaded areas in troponin T show the positions of the N-terminal tryptic peptide and the fragments CB2 and T2 from the rabbit protein.

to the C-terminus of tropomyosin and the binding is not Ca^{2+}-sensitive. No direct binding has been shown between the N-terminal region of troponin T and tropomyosin, but fragments which contain this region are able to enhance head to tail aggregation of tropomyosin (10).

In chicken fast skeletal muscle different forms of troponin T have been found in breast and leg muscle (11, 12) and these differences were shown to reside in the N-terminal regions of the proteins (12). Further work (13) has shown that there are two different variant forms present in leg muscle and this paper will discuss the structural relationship to these variants, their distribution in a variety of chicken muscles and their possible functional role in muscle regulation.

N-TERMINAL PEPTIDE SEQUENCES

We have recently described the isolation of different variant forms of troponin T from chicken breast and leg muscle (13). One form, B-TNT, was found in breast muscle while two different forms, $L\text{-}TNT_1$ and $L\text{-}TNT_2$, were prepared from mixed leg muscle. The relationship of these three variants was investigated by the isolation and sequence analysis of their N-terminal tryptic peptides. Peptide mapping and partial sequence analysis of the remaining tryptic peptides (12) shows that the large N-terminal peptides contain all the differences between the parent proteins. The sequences of the three variants are shown in Fig. 2. The sequences from $L\text{-}TNT_1$ and $L\text{-}TNT_2$ were obtained directly from the tryptic peptides but that from B-TNT was obtained from two smaller peptides derived by a minor tryptic cleavage between residues 30 and 31. It is possible therefore that the major tryptic peptide from B-TNT contains some further sequence at this point which has been missed. A second possibility is that the two smaller peptides may be derived from a further variant of B-TNT.

When the three sequences are compared it can be seen that those from B-TNT and $L\text{-}TNT_1$ are the same as that from $L\text{-}TNT_2$ except that each has an insertion of a stretch of unique sequence, these insertions being at different positions in the prototype sequence (Fig. 2). This has led us to propose that chicken fast muscle troponin T is coded for by a single gene which has a number of small exons at the 5' end and that the variant forms of the protein that are expressed are the product of different routes of post-transcriptional modification. A model for such a gene is shown in Fig. 3. A similar means of expressing variant proteins from a single gene has recently been demonstrated in the A1 and A2 myosin light chains of rabbit fast muscle (14, 15) and also in murine αA-crystallin (16).

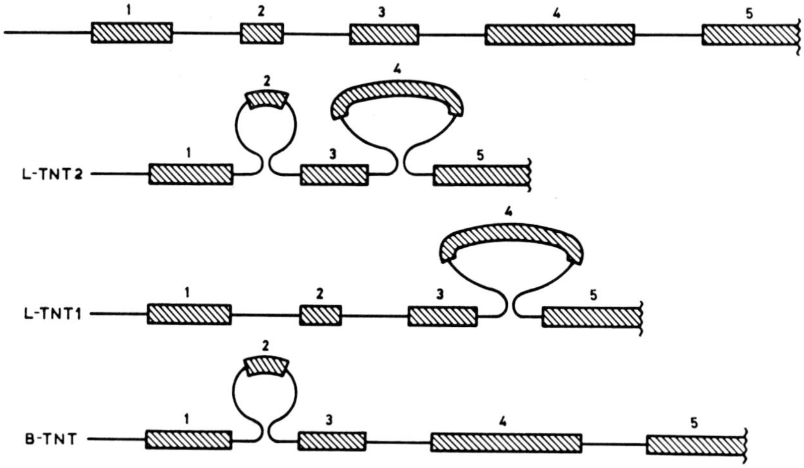

FIGURE 3. Proposed exon structure of the troponin T gene. The hatched bars indicate the exons. The length of the introns is arbitrary. Reproduced from (13) with permission.

MONOCLONAL ANTIBODIES TO TROPONIN T

In order to study the expression of the troponin T variants in chicken muscles, monoclonal antibodies were raised using breast muscle troponin T as antigen. Two antibodies, T1/7 and T1/61, have been prepared (17). Initial assays were carried out using an ELISA with either troponin or troponin T from various sources as the target antigen. No significant difference was observed in the reactivity of the antibodies with either troponin or troponin T from chicken breast or leg muscle. Cross-reactivity with troponin T from rabbit fast and slow muscle

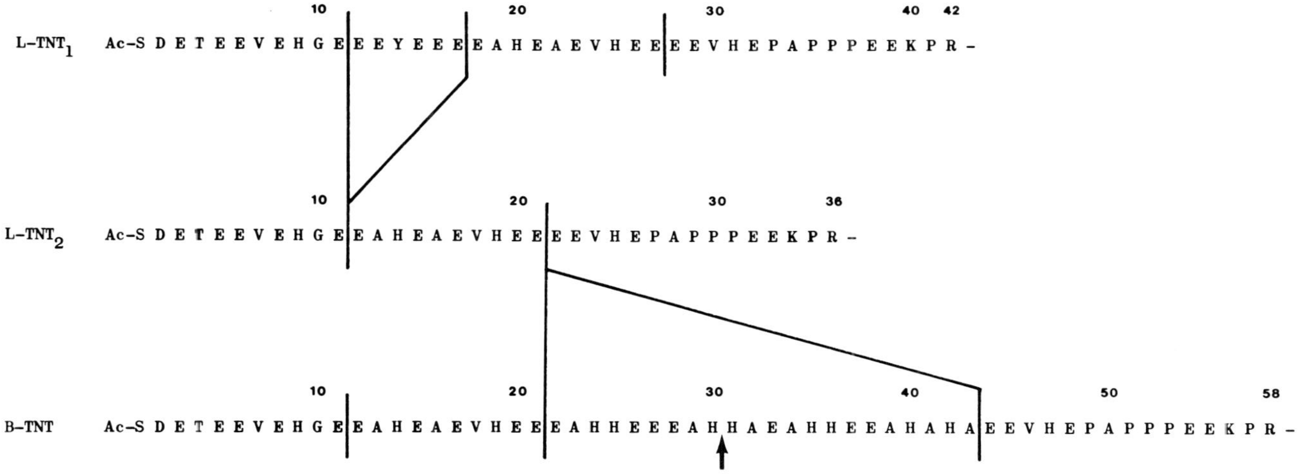

FIGURE 2. Amino acid sequences of the N-terminal tryptic peptides from the three forms of chicken fast troponin T. The vertical lines indicate the proposed exon boundaries. The arrow in the B-TNT sequence shows the junction between the peptides produced by a minor tryptic cleavage.

was also assessed by this method. T1/7 binds weakly to rabbit fast but not slow troponin T while T1/61 does not bind to either.

The antibodies were further assayed by immunoperoxidase staining of thin sections of chicken, human and rabbit muscle. The chicken muscles chosen were the posterior latissimus dorsi (PLD), a typical fast muscle, and the anterior latissimus dorsi (ALD), a typical slow muscle. In the PLD the majority of fibres stained with both antibodies while in the ALD very few fibres were stained. In human biceps a proportion of the fibres were stained and these could be shown to be fast fibres by staining of serial sections for myosin ATPase at pH 9.4. No significant staining of rabbit muscle sections was obtained with either antibody.

These findings were extended by immunostaining of protein blots. Small samples of muscle or muscle acetone powder were homogenised with SDS electrophoresis buffer and the solubilised proteins separated by SDS gel electrophoresis. Protein blots of stained gels were prepared by the method of Jackson and Thompson (18) and immunostained with the two antibodies. Both antibodies stain troponin T bands in a wide selection of chicken muscles, with the exception of the ALD which gave very weak staining. In breast muscle two bands were stained and in leg muscles three bands of staining were generally seen. Staining of appropriate bands in human skeletal muscle samples was obtained with both antibodies. With rabbit muscle T1/7 gave weak staining while T1/61 gave none. Somewhat surprisingly, both antibodies stained a single band in chicken cardiac muscle homogenates but they gave no staining of either rabbit or beef cardiac samples.

Staining of chymotryptic fragments of B-TNT, L-TNT_1 and L-TNT_2 showed that both antibodies bind to epitopes in the C-terminal fragment T2 (Fig. 1). These epitopes are clearly distinct from the binding sites for troponin I, troponin C and tropomyosin which are also in this region of the molecule as the antibodies are able to bind to whole troponin and also to muscle sections and myofibrils.

In summary, the antibodies are specific for different epitopes in the C-terminal region of chicken fast muscle troponin T. They cross-react with chicken cardiac and human fast muscle troponin T but not with that from slow muscle in any species tested. T1/7 cross-reacts weakly with rabbit fast troponin T but T1/61 does not.

DISTRIBUTION OF TROPONIN T IN CHICKEN MUSCLES

The expression of different variants of troponin T in leg, breast and cardiac muscles has been investigated by two dimensional electrophoresis of homogenates of small amounts of tissue. The Basodalt method of Giometti et al. (19) was used for skeletal muscle samples and the Isodalt method of Anderson and Anderson (20, 21) for cardiac muscle. After Coomassie staining, the protein patterns were transferred to nitrocellulose and immunostained with T1/61.

The two dimensional patterns obtained were similar to those reported by others (19, 22) and this allows the positions of a number of contractile proteins and glycolytic enzymes to be identified with some confidence. Fig. 4 shows gels from a number of different muscles with the troponin T spots, detected by immunostaining with T1/61, indicated by arrowheads.

In breast muscle (Fig. 4a) two spots, BT_1 and BT_2, are seen together with minor spots which probably correspond to dephosphorylated forms of the proteins. The finding of two distinct variants in breast muscle was unexpected but found consistently in both pectoralis major and minor. These forms may correspond to the two forms in which the N-terminal tryptic peptide was isolated.

In leg muscles a more complex situation is found. The position of the slow muscle troponin T doublet (22) which does not stain with T1/61, can be determined from gels of ALD muscle (Fig. 4e). The percentage of slow muscle fibres in any particular leg muscle can be determined approximately from the amounts of the slow isoforms of troponin T and other contractile proteins such as α-tropomyosin, troponin I and the myosin light chains which are found on the gels. Using this criterion, fast leg muscles such as the iliotibialis (Fig. 4b) and the semitendinosus express two troponin T variants, LT_1 and LT_2, while slower muscles, such as the adductor superficialis (Fig. 4d) express a third variant, LT_3. Other muscles can be found which express different proportions of these three variants. Thus the biceps femoris (Fig. 4c) expresses mainly LT_1 with smaller amounts of LT_3. On the basis of their isoelectric points and their expression in the major leg muscles, LT_1 and LT_2 probably correspond to the two leg variants which have been sequenced. The third variant, LT_3 has not been described before but is expressed in muscles which contain

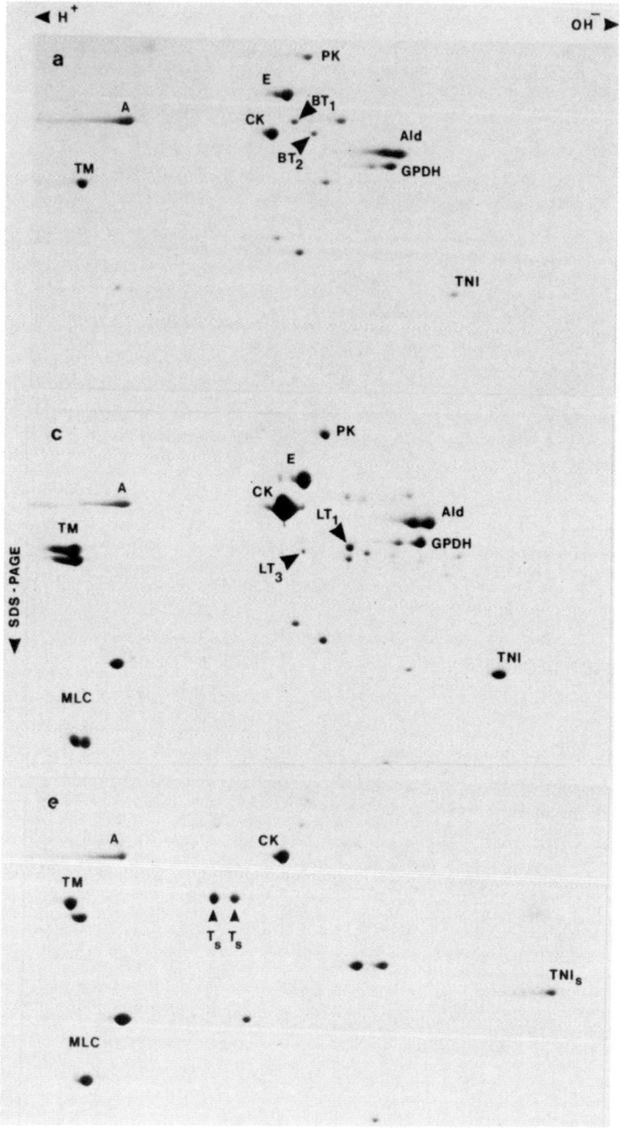

FIGURE 4. Two dimensional gels of chicken muscle homogenates. a, pectoralis; b, iliotibialis; c, biceps femoris; d, adductor superficialis; e, ALD; f, heart. The spots which immunostained with T1/61 are marked by large arrowheads. Other components are identified as:

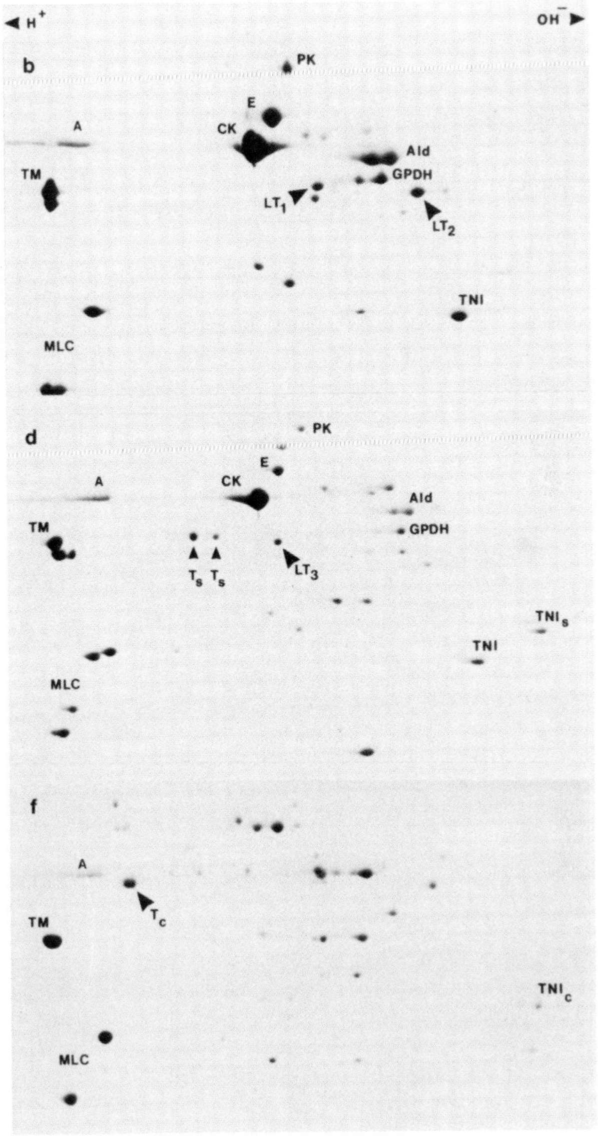

FIGURE 4. (Cont.) A, actin; TM, tropomyosin; MLC, myosin light chains; TNI, troponin I; CK, creatine kinase; E, enolase; PK, pyruvate kinase; Ald, aldolase; GPDH, glyceraldehyde-3-phosphate dehydrogenase. s and c indicate slow and cardiac proteins respectively.

a high percentage of slow fibres. This variant may well be related to the others by variation in the N-terminal sequence.

In addition to the three major variants identified in leg muscles, immunostaining with either antibody revealed the presence of a number of minor spots in the region between LT_1 and LT_2. These spots were only faintly visible by Coomassie staining. This phenomenon was only observed with leg muscle preparations and not with those from breast, ALD or heart muscle. It is unlikely, therefore, to be an artefact due to non specific staining. The possibility of proteolysis cannot be ruled out but is also unlikely considering the method of sample preparation. These extra spots may represent a low level of expression of other variants.

In the chicken heart only a single spot staining with T1/61 was found (Fig. 4f) in agreement with the findings of Mikawa et al. (22) and Murakami and Uchida (23). This result was extended by using tissue from atria and ventricles of adult chicken hearts and using a gel system which gave better resolution in this region of the gel. No significant differences could be found between the gels of atrial and ventricular tissue, in agreement with the finding of Dalla Libera et al. (24) that the myosin light chains are identical in these tissues. In particular only a single major spot of troponin T was found with a minor spot which is probably due to dephosphorylated protein. The chicken heart thus appears to be different from that of the beef where two forms are expressed (25).

STRUCTURAL STUDIES ON THE N-TERMINAL PEPTIDES

As discussed above, current views of troponin structure suggest that the N-terminal region of troponin T may interact with the N-terminal region of tropomyosin. As the differences between the troponin T variants in the chicken reside in this N-terminal region it is possible that they modulate the interaction with tropomyosin and the regulation of contraction. In order to address this question we are investigating the structures of the N-terminal tryptic peptides by proton NMR spectroscopy.

Initial analysis was made using small peptides from the N- and C-termini. Several residues are unique in the peptide sequences, for instance NAc-Ser and Thr-4 at the N-terminus, Lys and Arg at the C-terminus. These provide reporter signals of their environment in solutions of the complete peptides. Analysis of the spectra of the 13 residue peptide from the C-terminus shows the region P A P P P to be in the trans configuration and that it adopts an extended and well defined conformation. The remainder of the sequences contain a high proportion of charged residues which may be expected to produce an extended structure which alternates rapidly between a limited number of conformations of similar free energy. The shortest peptide, from L-TNT_2, contains four His residues which all have abnormally high pK_a values, in one case this is 8.1 and for the other three 7.8. The titration behaviour is influenced by the presence of Glu residues which have an apparent pK_a of 5.0 and also show an inflection with a titration midpoint of about pH 8. The three Val and two Ala side chains in this region of the sequence do not appear to sense these ionisations and must therefore be oriented away from the His and Glu side chains. This region of the sequence thus takes up a relatively structured conformation with the side chain mobility of His and Glu constrained by rapidly exchanging hydrogen bond linkages. Comparison of the peptide from L-TNT_1, with that from L-TNT_2 shows that the introduction of the extra segment E E Y E E E alters the apparent pK_a of all four His residues but still retains the interaction between the His and Glu residues with the His sensing altered micro-environments. The peptide from B-TNT shows a similar phenomenon in that the additional inserted sequence causes a shift in the His pK_a values with virtually all of the twelve His residues having slightly different pK_a's.

This region in rabbit troponin T has been implicated as interacting with the N-terminal region of tropomyosin by Brisson et al. (26). The sensitivity to changes in the charge environment demonstrated by the chicken N-terminal peptides suggests that these changes may modify the interaction with tropomyosin in this region.

CONCLUSIONS

Troponin T is unusal among the contractile proteins of muscle in that it can be expressed in multiple variants which are coded for by a limited number of genes. The only other example of this is the alkali light chains of fast muscle myosin (14, 15). Our studies have been confined to chicken muscle but there is evidence that a similar situation occurs in mammalian muscle. The work of Clarke et al. (27) on beef skeletal and Briggs et al. (28) on rabbit skeletal muscle and of Gusev et al. (29) and Risnik et al. (25) on beef cardiac muscle suggest that variability in the N-terminal region of troponin T is a general phenomenon. There are as yet no data available on troponin T from slow muscle.

The data cited above have all been obtained at the level of protein expression. Recently gene sequences for a number of these proteins have become available. That for chicken cardiac troponin T (30) shows that in the N-terminal region there is an extra exon which is incorporated in some versions of the transcript but not in others in a manner similar to that suggested here for skeletal troponin T. Other gene sequences for the C-terminal region of rat fast troponin T (31) are in good agreement with those from the rabbit (32) and the quail (33). These sequences show that two alternative exons can code for part of the C-terminal sequence of the protein, although in quail breast muscle only one version of the gene is expressed. If these alternative exons are expressed in conjunction with the variability found at the N-terminal end the number of possible variants is considerably increased and may account for the heterogeneity found in the expression of troponin T in chicken leg muscles.

The functional significance of the N-terminal variants is not clear but our preliminary experiments suggest that this region adopts distinct and unusual structures which may interact with tropomyosin. A comparison of the sequence of the N-terminal regions from chicken fast muscles with those from rabbit fast (34), chicken cardiac (30) and beef cardiac (25) shows that there is very little similarity between any of them except for the phosphorylation of the N-terminal serine and the high density of negatively charged residues which suggest that these are all extended structures. It is possible therefore that these sequence differences may provide a means, at the level of gene expression, of modulating the interaction of troponin with tropomyosin and hence the contractile activity of the muscle fibres in which they are expressed.

REFERENCES

1. Ebashi S, Endo M, Ohtsuki I (1969). Quart Rev Biophys 2:351.
2. Ohtsuki I (1975). J Biochem (Tokyo) 77:633.
3. Jackson P, Amphlett GW, Perry SV (1975). Biochem J 151:85.
4. Pearlstone JR, Smillie LB (1978). Can J Biochem 56:521.
5. Pearlstone JR, Smillie LB (1981). FEBS Lett 128:119.
6. Mak AS, Smillie LB (1981). J Mol Biol 149:541.
7. Tanokura M, Tawada Y, Ono A, Ohtsuki I (1983). J Biochem (Tokyo) 93:331.
8. Hitchcock SE, Zimmerman CJ, Smalley C (1981). J Mol Biol 147:125.
9. Flicker PF, Phillips GN, Cohen C (1982). J Mol Biol 162:495.
10. Pato MD, Mak AS, Smillie LB (1981). J Biol Chem 256:602.
11. Perry SV, Cole HA (1974). Biochem J 141:733.
12. Wilkinson JM (1978). Biochem J 169:229.
13. Wilkinson JM, Moir AJG, Waterfield MD (1984). Eur J Biochem 143:47.
14. Nabeshima Y, Fujii-Kuriyama Y, Muramatsu M, Ogata K (1984). Nature (Lond) 308:333.
15. Robert B, Daubas P, Akimenko MA, Cohen A, Garner I, Guenet JL, Buckingham M (1984). Cell 39:129.
16. King CR, Piatigorsky J (1983). Cell 32:707.
17. Bird IM, Dhoot GK, Wilkinson JM (1985). Biochem Soc Trans 13:119.
18. Jackson P, Thompson RJ (1984). Electrophoresis 5:35.
19. Giometti CS, Anderson NG, Anderson NL (1979). Clin Chem 25:1877.
20. Anderson NG, Anderson NL (1978). Anal Biochem 85:331.
21. Anderson NL, Anderson NG (1978). Anal Biochem 85:341.
22. Mikawa T, Takeda S, Shimizu T, Kitaura T (1981). J Biochem (Tokyo) 89:1951.
23. Murakami U, Uchida K (1984). J Biochem (Tokyo) 95:1557.
24. Dalla Libera L, Sartore S, Schiaffino S (1979). Biochim Biophys Acta 581:283.
25. Risnik VV, Verin AD, Gusev NB (1985). Biochem J 225:549.
26. Brisson J-R, Sykes BD, Golosinska K, Smillie LB (1984). Biophys J 45:239a.
27. Clarke FM, Lovell SJ, Masters CJ, Winzor DJ (1976). Biochim Biophys Acta 427:617.

28. Briggs MM, Klevit RE, Schachat FH (1984). J Biol Chem 259:10369.
29. Gusev NB, Barskaya NV, Verin AD, Duzhenkova IV, Khuchua ZA, Zheltora AO (1983). Biochem J 213:123.
30. Cooper TA, Ordahl CP, Personal communication.
31. Medford RM, Nguyen HT, Destree AT, Summers E, Nadal-Ginard B (1984). Cell 38:409.
32. Putney SD, Herlihy WC, Schimmel P (1983). Nature (Lond) 302:718.
33. Emerson CP, Personal communication.
34. Pearlstone JR, Johnson P, Carpenter MR, Smillie LB (1977) J Biol Chem 252:983.

Molecular Biology of Muscle Development, pages 437–444

THE STRUCTURE OF CARDIAC TROPONIN T AS DEDUCED FROM THE NUCLEOTIDE SEQUENCE OF THE GENE

Thomas A. Cooper and Charles P. Ordahl
University of California San Francisco
San Francisco, CA 94143

ABSTRACT

We have solved the primary structure of cardiac troponin T (TnT) by determination of the nucleotide sequence of the gene and its mRNA products. This gene encodes two proteins which differ by the presence, or absence, of ten amino acid residues near the NH-terminus. These two proteins arise via developmentally-regulated, alternate splicing of a single primary gene transcript. The isoform containing the variable ten amino acids is restricted to the early embryonic stages of both cardiac and skeletal muscle development. The embryo-specific expression of this cardiac TnT isoform suggests that it may play a specialized role in early differentiation of all striated muscle.

INTRODUCTION

The development of striated muscle is marked by the sequential, overlapping expression of various myofibrillar protein isoforms which are related but not identical in primary structure. The heterogeneity of the muscle proteins is generated both by differential mRNA splicing (1-8) as well as the regulated expression of multigene family members (9). Most known striated muscle-specific proteins have been found to have more than one isoform. This regulated protein diversity presumably reflects the different functional requirements of different striated muscle fiber types (cardiac, fast, and slow) as well as transient functional requirements of developing muscle fibers (10). At present, very little is known about how cells regulate specific isoform switching by either of these two mechanisms.

Here, we report the structure of two isoforms of cardiac troponin T (TnT) which are encoded by a single gene. The original cardiac TnT cDNA clone (designated 106A4) was isolated from embryonic skeletal muscle where it was found to represent a mRNA that is transiently expressed during development (11). We have recently shown that the same single-copy gene expressed in early embryonic skeletal muscle is expressed throughout cardiac development and produces relatively high abundance mRNA levels in adult heart (12).

RESULTS

cDNA and Genomic Structure Indicates That Two Cardiac Troponin T mRNAs are Generated by Alternate Splicing

To characterize the structure of the cardiac TnT gene and mRNAs, full-length cDNAs were obtained both by oligo-dT priming and a novel primer extension cloning technique (Cooper and Ordahl, submitted) using RNA from day 18 embryonic cardiac muscle. Genomic clones containing the cardiac TnT gene were isolated from a chicken genomic library (13). Cardiac TnT exons were identified by blot hybrization and nucleotide sequence comparisons between cloned cDNA and genomic DNAs. The boundaries of all cardiac TnT exons were determined by nucleotide sequence.

The cardiac TnT gene spans 9.0 kilo-base pairs and is composed of 18 exons which range in size from 6 base pairs to 267 base pairs. All exons are bounded by the appropriate consensus splicing signals (14,15) Exon 17, which is six base pairs in length, is to our knowledge, the smallest exon reported to date. S-1 nuclease mapping detects no cardiac TnT mRNAs lacking this exon (data not shown) indicating that it is spliced efficiently. Exon 17 establishes a new lower limit for the number of nucleotides of a primary transcript that can be recognized as exonic sequence.

As indicated in Figure 1, the nucleotide sequence of several independent cDNA clones spanning almost the entire cardiac TnT mRNA identified two different cardiac TnT mRNAs that are identical except for the presence or absence of an internal 30 nucleotide coding block (shaded box in Figure 1). Both mRNAs contain an identical single open reading frame differing by an internal ten codons encoded by the variable 30 nucleotides.

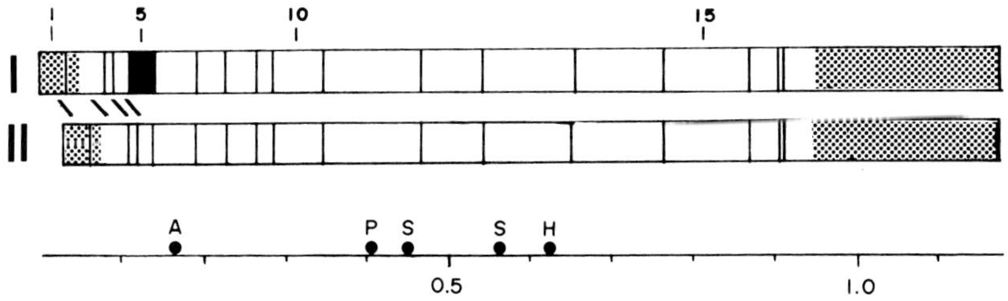

Figure 1: Cardiac troponin T alternate splice products. Exon boundaries are indicated on the mRNAs and every fifth exon is numbered. Untranslated portions of the mRNAs are shaded. The variable region, corresponding to exon 5, is heavily shaded. Restriction endonuclease cleavage sites: A, Acc I; H, Hind III; P, Pst I;, S, Sac I.

Comparison of the cDNA nucleotide sequence to that of the cardiac TnT gene reveals that the two corresponding mRNAs are generated by alternate splicing (inclusion or exclusion) of exon 5 (Figure 1). Furthermore, only these two mRNAs are detected by S-1 nuclease mapping of RNA from cardiac and skeletal muscle (data not shown). In addition, the two mRNAs generated from this gene appear to have the same cap site indicating that a single primary RNA transcript serves as the precursor for both cardiac TnT mRNAs in cardiac and skeletal muscle (Cooper ad Ordahl, submitted). For discussion, we have designated the mRNA and cDNAs containing exon 5 as form I (cardiac TnT-I) and those lacking exon 5 as form II (cardiac TnT-II).

Cardiac TnT Alternate Splicing is Developmentally Regulated in Cardiac Muscle

S-1 mapping experiments also indicate that there is an isoform switch of cardiac TnT in developing cardiac muscle (Figure 2). In this experiment, a probe for the 5' half of the form I mRNA was obtained by 32-P-labeling the complementary strand of a cDNA containing the variable region at the Hind III site (see Figure 1). Cardiac TnT-I mRNAs that anneal to the probe protect the entire 618 base pair cDNA insert from S-1 nuclease digestion. Cardiac TnT-II mRNAs diverge from the form I cDNA probe 476 base

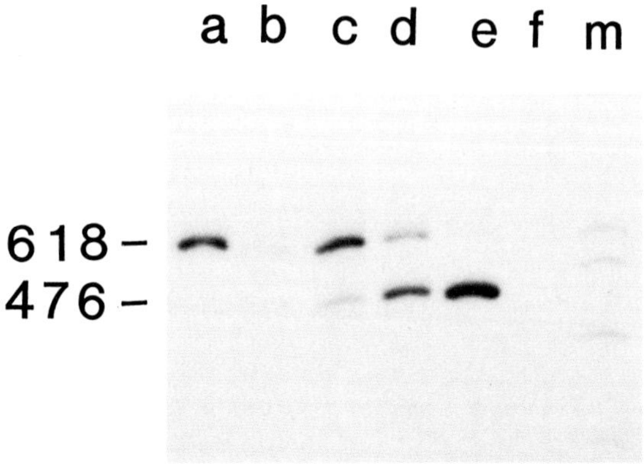

Figure 2: S1 mapping of troponin T mRNA populations. To construct the probe, sp 64 plasmid containing the 5' 618 base pair fragment of a fully extended class I cDNA was linearized at a unique Hind III. The complementary strand was end-labeled with 32-P using polynucleotide kinase (23). S-1 nuclease digestion products were separated on a 6% acrylamide, 8 M urea gel. The size marker is Hpa II digested pBR322 that was 32 P-labelled by reverse transcriptase. The S-1 digestion in lanes a-d contained 5 ug poly(A) RNA; lane e contained 5 ug total RNA. Lanes: m. size markers, a. day 10 embryonic leg muscle, b. day 18 embryonic leg muscle, c. day 10 embryonic cardiac muscle, d. day 18 embryonic cardiac muscle, e. adult cardiac muscle f. day 18 embryonic liver (5 ug poly(A) RNA).

pairs from the labeled Hind III site due to the absence of the variable region. Therefore, the presence and relative abundance of forms I and II cardiac TnT mRNA is indicated by the intensity of bands at 618 and 476 nucleotides, respectively (Figure 2). As indicated in Figure 2, cardiac TnT-I predominates in early cardiac muscle (lane c; day 10 in ovo). In late embryonic cardiac muscle (lane d; day 18 in ovo), form II predominates. In adult cardiac muscle (lane e) form II cardiac TnT mRNA vastly predominates, form II mRNA is barely detectable. This result indicates that during cardiac muscle development, there is a transition from form I to from II cardiac TnT mRNAs.

When skeletal muscle RNA is assayed with the same S-1

nuclease probes, the fragments protected by cardiac TnT are identical to those protected by cardiac muscle RNA (Figure 2 lanes a and b, and data not shown). Figure 2 shows, as reported previously, the abundance of cardiac TnT decreases during late embryonic skeletal muscle development (11,12) (Figure 2, lanes a and b). In addition, the cardiac TnT-I mRNA, which predominates in early embryonic cardiac muscle is also the predominant, if not the sole isoform generated by this gene in early embryonic skeletal muscle.

Cardiac TnT Protein Structures

The above data indicate that cardiac TnT-I serves as an embryonic isoform in skeletal muscle as well as in cardiac muscle. In both cardiac and skeletal muscle, the embryonic cardiac TnT isoform dissappears as development proceeds and is replaced by late-stage TnT isoforms (Figure 3). In cardiac muscle, TnT-I is replaced by the adult isoform (TnT-II) via an alternate splicing switch. In skeletal muscle, however, TnT-I must be replaced by the appropriate skeletal muscle isoform.

Recently, several lines of evidence have demonstrated that the NH-terminal region of TnT interacts with tropomyosin (TM) along the TM COOH terminal one-third and spans the head to tail overlap between adjacent TM molecules (16,17). A direct demonstration of the interaction between the NH-terminus of TnT and the NH-terminus of an adjacent TM has recently been reported (18,19). When the head-to-tail polymerization of TM is interupted by removing the COOH-terminal 11 residues, TM no longer binds to F-actin in vitro (20). The addition of TnT restores the ability of nonpolymerizable TM to bind F-actin (in the absence of calcium) (21). These results indicate that the NH-terminus of TnT can influence the interactions of TM and actin by affecting the stability of TM head-to-tail polymerization.

These kinds of results suggest a more fundamemtal role for TnT, and particularly its NH-terminal domain, in myofibrillar structure and function than has previously been assumed. While it is too early to make any definite predictions, it is an intriguing possibility that the embryo-specific region of cardiac TnT plays a specialized role in early striated muscle differentiation, possibly in nascent sarcomere assembly. It is interesting to consider, in this regard, some differences between nascent sarcomeres

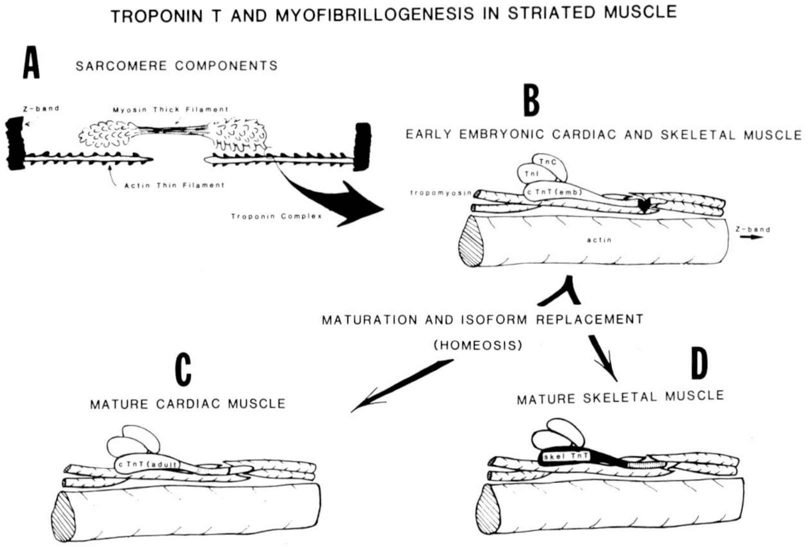

Figure 3: Cardiac TnT Transitions During Striated Muscle Development. Schematic illustration of the embryonic sarcomere (A) in which the thin filament contains the embryonic isoform of cardiac TnT in both cardiac and skeletal muscle (B). During maturation, the embryonic TnT is replaced by the adult cardiac isoform in cardiac muscle (C) via alternate splicing and by the appropriate skeletal isoform in skeletal muscle (D). Isoform replacement is essentally synonymous with homeosis (25) Representation of the thin filament in this figure is adapted from Flicker et al. (17).

and mature sarcomeres which have recently come to light. First, it has been demonstrated that nascent myofibrils form in close association with stress fiber-like bundles of non-muscle actin in cultured cardiac myocytes and in EMS-treated cultured skeletal myocytes (22). Second, newly formed sarcomeres in skeletal and cardiac myocytes may be unable to undergo shortening during contraction even when they are part of a myofibril containing contracting mature sarcomeres (23). Using the information regarding cardiac TnT structure which is now available it should be possible to design gene-specific and protein-specific probes to determine what role cardiac TnT plays in early embryonic muscle.

ACKNOWLEDGMENTS

We would like to thank J.M. Wilkinson for helpful discussions and communication of unpublished results. This work was supported by grants from NIH (R10-GM32018) C.P.O., a postdoctoral fellowship from NICHD (T.A.C.), and a NIH Research Career Development Award (C.P.O.). Some of the work presented here has been submitted for publication elsewhere (Cooper and Ordahl, submitted).

REFERENCES

1. Rozek, C. and Davidson, N. (1983) Cell 32, 23-34
2. Medford, R.M., Nguyen, H.T., Destree, A.T., Summers, E., and Nadal-Ginard, B. (1984) Cell 38, 409-421
3. Basi, G.S., Boardman, M., and Storti, R.V. (1984) Mol. Cell. Biol. 4, 2828-2836
4. Falkenthal, S., Parker, V.P., and Davidson, N. (1985) Proc. Natl. Acad. Sci. U. S. A. 82, 449-453
5. Nabeshima, Y., Fujii-Kuriyama, Y., Muramatsu, M., and Ogata, K. (1984) Nature 308, 333-338
6. Robert, B., Daubas, P., Akimenko, M.A., Cohen, A., Garner, I., Guenet, J.L., and Buckingham, M. (1984) Cell 39, 129-140
7. Periasamy, M., Strehler, E.E., Garfinkel, L.I., Gubits, R.M., Ruiz-Opazo, N., and Nadal-Ginard, B. (1984) J. Biol. Chem. 259, 13595-13604
8. Wilkinson, J.M., Moir, A.J.G., and Waterfield, M.D. (1984) Eur. J. Biochem. 143, 47-56
9. Buckingham, M.E., and Minty, A.J. (1983) in Eukaryotic Genes: Their Structure, Activity, and Regulation (McLean, Gregory, and Flavell, eds., Butterworth, London) p. 1-78
10. Caplan, A.I., Fiszman, M.Y., and Eppenberger, H.M. (1983) Science 221, 921-927.
11. Ordahl, C.P., Kioussus, D., Tilghman, S.M., Ovitt, C., and Fornwald, J. (1980) Proc. Natl. Acad. Sci. U. S. A. 77, 4519-4523
12. Cooper, T.A., Ordahl, C. P. (1984) Science 226, 979-982
13. Dodgson, J.B., Strommer, J., and Engel, J.D. (1979)

Cell 17, 879-887
14. Mount, S.M. (1982) Nucl. Acids Res. 10, 459-472
15. Breathnach, R. and Chambon, P. A. (1981) Rev. Biochem. 50, 349-383
16. Tanokaur, M., Tawada, Y., Ono, A., Ohtsuki, I. (1983) J. Biochem. 93, 331-337
17. Flicker, P.F., Phillips, G.N., Cohen, C. (1982) J. Mol. Biol. 162, 495-501
18. Pearlstone, J.R., and Smillie, L.B. (1984) Biophysical J. 45, 111a
19. Brisson, J.-R., Sykes, B.D., Golosinka, K., and Smillie, L.B. (1984) Biophysical J. 45, 239a
21. Mak, A.S. and Smillie, L.B. Biochem. Biophys. Res. Comm. 101, 208-214
20. Mak, A.S., Golosinska, K., and, Smillie, L.B. J. Biol. Chem. 258, 14330-14334
22. Dlugosz, A.J., Antin, P.B., Nachimias, V.T., and Holtzer, H. (1984) J. Cell Biol. 99, 2268-2278
23. Sanger, J.M., Mittal, B., Pochapin, M.B., and Sanger, J.W. (1984) J. Cell Biol. 99, 26a
24. Chaconas, G. and van de Sande, J. (1980) Methods Enzymol. 65, 75-85
25. Gehring, W.J. (1985) Cell 40, 4-5

Molecular Biology of Muscle Development, pages 445–455

MOLECULAR CHARACTERIZATION OF HUMAN CYTOSKELETAL TROPOMYOSINS

A.R. MacLeod[1], K. Talbot[1], F.C. Reinach[1], C. Houlker[1], L.B. Smillie[1,2], C.S. Giometti[3], and N.L. Anderson[3]

[1]Ludwig Institute for Cancer Research, MRC Centre, Hills Road, CAMBRIDGE CB2 2QH

[2]Department of Biochemistry, University of Alberta, Edmonton, Alberta T6G 2H7, Canada

[3]Molecular Anatomy Program, Division of Biological and Medical Research, Argonne National Laboratory, Argonne, Illinois 60459, U.S.A.

ABSTRACT The heat-stable cytoskeletal protein fraction of human fibroblasts contains five distinct proteins which share the physical and chemical characteristics of tropomyosins. Using a variety of cloned probes we can demonstrate the existence of at least 3 distinct mRNAs encoding different cytoskeletal tropomyosins in human fibroblasts. Structural analysis of these mRNAs indicates that not only do human fibroblast express tropomyosins of the classic non-muscle type but they also express a tropomyosin identical in length and closely related in structure to smooth muscle tropomyosin.

INTRODUCTION

Tropomyosin is a protein which was first isolated from muscle tissue (1). In skeletal muscle, it is associated with actin in the thin filament and serves to mediate the effect of Ca++ on the actin-myosin interaction. This is achieved through interaction with a set of proteins called the troponins.

Tropomyosin is also found in smooth muscle and in the actin-containing microfilaments of the cytoskeleton (2), in which the troponins have not been found. In these tissues the biological role of tropomyosin is obscure.

Early studies showed that tropomyosins from non- muscle sources were consistently smaller than their muscle counterparts (3,4). Sequence analysis of horse platelet tropomyosin has shown it to be 247 amino acids long (5) whereas tropomyosins of 284 amino acids are found in smooth (6,7) and skeletal muscle (8). This difference in structure is commonly thought to represent a fundamental difference in the function of tropomyosin in muscle and non-muscle tissues.

Recent studies have revealed a surprising heterogeneity of non-muscle tropomyosins (9,10,11). As many as five different proteins, varying in apparent molecular weight from 30,000 - 40,000, have been isolated from a single source. This raises a number of questions. How many of these proteins are true tropomyosins? What is the physical nature of the variation in apparent molecular weight between them? What is the biological function of these different tropomyosins in non-muscle tissues? To address these questions, we have undertaken a characterization of human cytoskeletal tropomyosins at the molecular level.

HUMAN FIBROBLAST TROPOMYOSINS

Tropomyosins from both muscle and non-muscle sources are characterized by a striking resistance to denaturation by solvents and heat. This property is due to the α-helical coiled-coil structure of the molecule. To prepare a fraction containing tropomyosin, cytoskeletal extracts of human fibroblasts (MRC-5) were subjected to a solvent precipitation, heat treatment and high speed centrifugation to remove denatured protein (10). Analysis of the supernatant by high-resolution two-dimensional gel electrophoresis revealed the presence of five distinct protein species (Figure 1). We have called these the heat-stable cytoskeletal proteins (hscp) and numbered them according to their apparent molecular weight of 30,000 (hscp30: two components designated hscp30pl and hscp30nm), 32,000 (hscp32), 34,000 (hscp34) and 36,000 (hscp36). This heterogeneity of tropomyosin-like proteins is not due to the heat treatment, since all five proteins can be observed in the untreated cytoskeletal extracts (10). Hscp30pl and hscp30nm correspond to human cytoskeletal proteins identified in a previous study as Cytosk:11 and Cytosk:13

(12). Using the Cleveland method, these were shown to have distinct peptide maps, indicating that hscp30pl and hscp30nm must be the products of different functional genes.

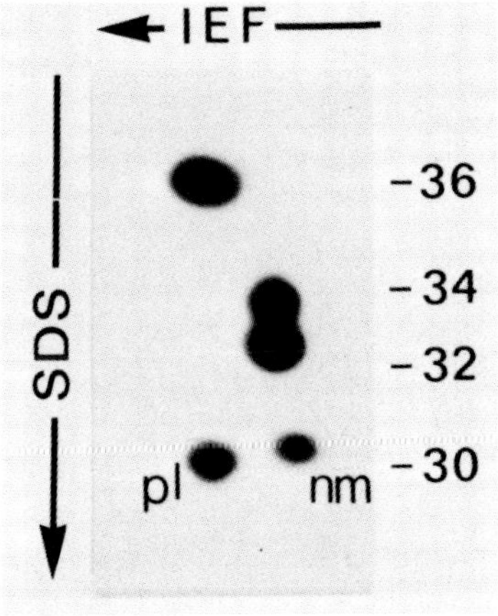

FIGURE 1. Two dimensional gel analysis of the heat-stable cytoskeletal protein (hscp) fraction of human fibroblasts. The numbers in the right hand margin refer to apparent molecular weight x 10^3.

Antibodies to chicken skeletal muscle tropomyosin efficiently immunoprecipitate only hscp36, the other proteins of the hscp fraction cross-reacting very weakly if at all (10). The apparent molecular weight of hscp36 is similar to that of tropomyosins from muscle sources. These characteristics suggest that hscp36 may be more muscle-like than the other proteins of the hscp fraction.

MOLECULAR PROBES FOR HUMAN CYTOSKELETAL TROPOMYOSINS

The original tropomyosin clone used in this study was isolated from a chicken skeletal muscle cDNA library. This clone, designated pTM4 (13), contains the sequence of chicken leg muscle α-tropomyosin corresponding to amino acids 24-284 (14). Rescreening of this library under conditions of very low stringency using the pTM4 clone led to the isolation of the plasmid pβTM7 (10), which contained

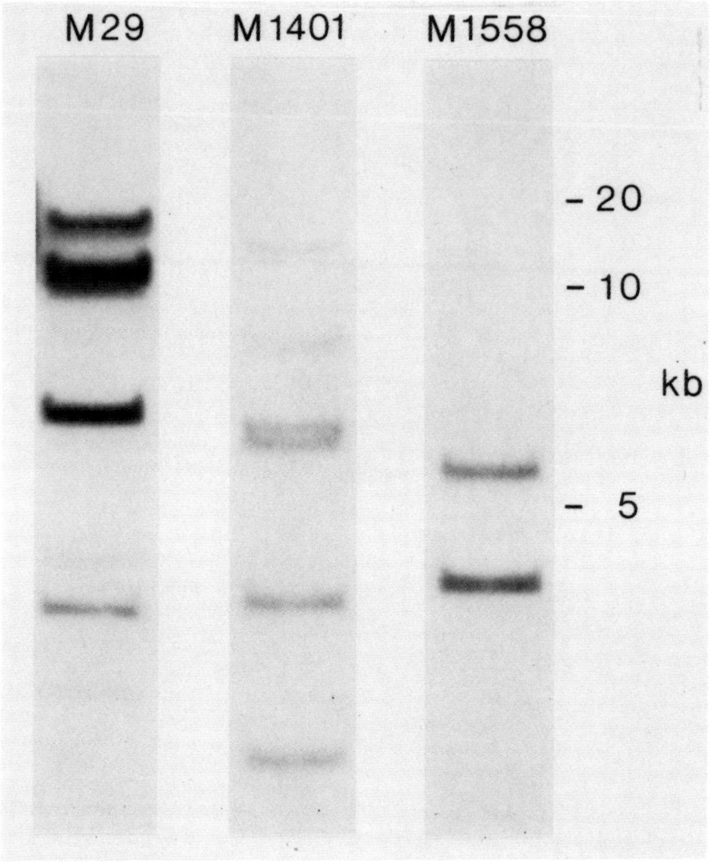

FIGURE 2. Southern blot analysis of Hind III digested human fibroblast DNA using the three different human non-muscle tropomyosin probes. Hybridization conditions were high stringency. Note that the pattern produced by each probe is complex and that the patterns do not overlap.

a small region of the coding sequence of a tropomyosin corresponding to amino acids 207-284. This sequence was similar to that of skeletal muscle tropomyosin except from amino acids 258-284, where very considerable differences were observed. This suggested that pβTM7 was derived from a non-skeletal muscle tropomyosin.

pβTM7 was also found to hybridize to a 1.1 kb mRNA in

human fibroblasts (10) and was therefore used to screen a genomic library of human DNA in order to isolate genomic sequences encoding non-muscle tropomyosins (15). A number of positive clones were obtained and characterized by subcloning and sequence analysis. Subcloned probes containing human non-muscle tropomyosin sequences were used to rescreen the genomic libraries to isolate as many different types of non-muscle tropomyosin sequences as possible. So far, we have identified three separate gene families encoding tropomyosins expressed in human fibroblasts (Figure 2). Each of these gene families is defined by a separate probe subcloned from a member of that family. Each probe contains a distinct and different non-muscle tropomyosin sequence. These probes, M29, M1401 and M1558 each hybridize to multiple restriction enzyme fragments in a digest of human genomic DNA. This complexity is apparent even at very high stringency, indicating that the tropomyosin sequences in any one family are very closely related. Structural analysis of cloned members of each family reveals that this complexity is due to the occurrence of multiple RNA-copy pseudogenes. Each family would appear to be composed of a single functional gene and multiple closely related RNA-copy pseudogenes which are not readily distinguishable by hybridization alone. The probes themselves are derived from RNA-copy pseudogene sequences, since these were the first members of each family to be cloned.

IDENTIFICATION OF mRNAs EXPRESSING CYTOSKELETAL TROPOMYOSINS

Each probe defining a separate family of non-muscle tropomyosin sequences at the DNA level hybridizes to a unique mRNA expressed in human fibroblasts (Figure 3). The nature of the proteins encoded by these mRNAs can be inferred by mRNA selection experiments in which the probe is used to purify the mRNA to which it hybridizes. The mRNA is then translated in vitro and the proteins analyzed by gel electrophoresis. These studies have shown that the 1.1 kb mRNA encodes hscp36 while the 2.5 kb mRNA encodes hscp30nm (10). This procedure failed to yield a result using the M1401 probe so the protein product of the 3.0 kb mRNA is still unidentified.

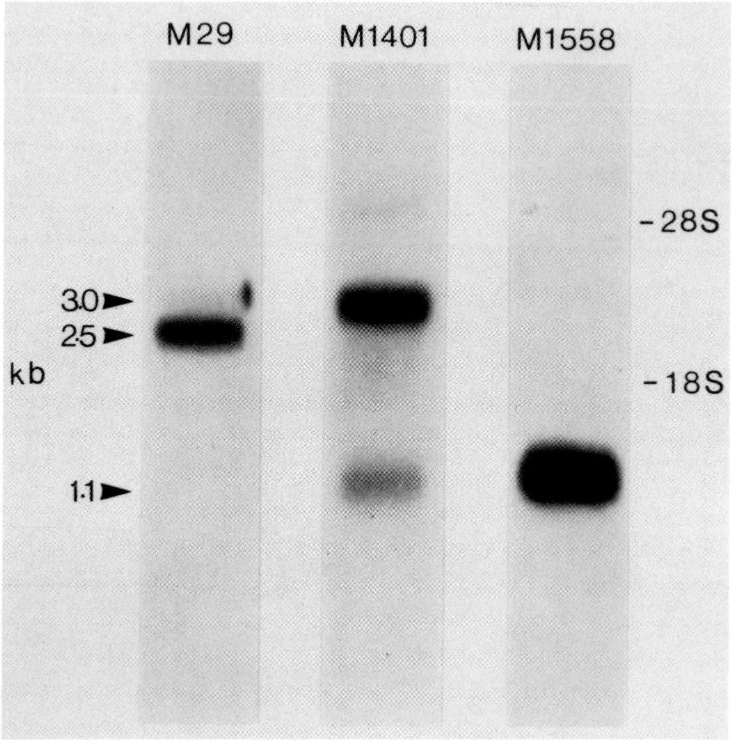

FIGURE 3. Nothern blot analysis of human fibroblast RNA using the three human non-muscle tropomyosin probes. Note that each probe hybridizes strongly to only one mRNA species. Cross-hybridization under these conditions of high stringency is very weak.

In addition to the three mRNAs which can be defined directly, two further mRNAs can be identified by virtue of their weak cross-hybridization to a series of tropomyosin probes (Figure 4).

One of these, a 2.0 kb mRNA, is weakly positive to both pTM4 and M1558, while the other, a 5.0 kb mRNA, is weakly positive to M1558 and M1401. In both cases the intensity of the hybridization signals to these mRNAs appears to be independent of the intensity of cross-hybridization signals of the other tropomyosin mRNAs to these probes. This suggests that the 2.0 kb and 5.0 kb mRNAs are the products of separate functional genes encoding non-muscle tropomyosins. Thus, the complexity of the hscp fraction would appear to be mirrored by a similar complexity of mRNAs encoding non-muscle tropomyosin sequences. However, it remains to be shown by cloning and sequence analysis that the 2.0 and 5.0 kb mRNAs encode genuine tropomyosins.

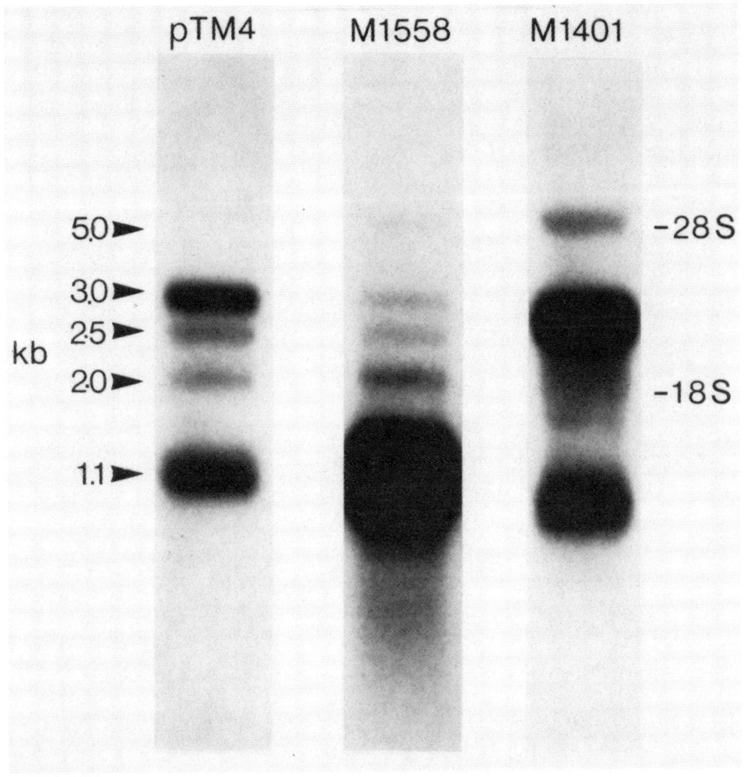

FIGURE 4. Nothern blot analysis of human fibroblast RNA using chicken (pTM4) and human non-muscle tropomyosin probes. Note the appearance of a 2.0 kb and 5.0 kb mRNA species using the pTM4 probe and long exposures of the M1558 and M1401 probes.

STRUCTURAL ANALYSIS OF THE mRNAs ENCODING CYTOSKELETAL TROPOMYOSINS

To characterize the nature of the differences between the cytoskeletal tropomyosins and, in particular, to determine the physical basis for the differences in apparent molecular weight, we have isolated cDNA clones containing cytoskeletal tropomyosin sequences. A cDNA library of human fibroblast RNA was constructed in the vector λgt10 and screened using the M29, M1401 and M1558 probes. This led to the isolation of cDNA clones corresponding to the 1.1, 2.5 and 3.0 kb mRNAs, although attempts to isolate cDNA clones corresponding to the 2.0 and 5.0 kb mRNAs have so far proved unsuccessful. cDNA clones containing the entire protein coding regions of the 1.1 and 3.0 kb mRNAs have been identified and partially characterized. The representation of the 2.5 kb mRNA in the form of cDNA clones was very poor, for reasons which are not understood.

Sequence analysis indicates that the 3.0 kb mRNA encodes a 247 amino acid tropomyosin whose structure is closely related to the known sequence of horse platelet tropomyosin (5), the classic non-muscle tropomyosin. By comparison, the 1.1 kb mRNA was found to encode a 284 amino acid tropomyosin whose sequence most closely resembles the known sequence of chicken gizzard smooth muscle tropomyosin (6). The higher apparent molecular wieght of hscp36 and its greater cross-reactivity with antiserum against skeletal muscle tropomyosin is therefore a consequence of this homology to smooth muscle tropomyosin.

Although a complete cDNA clone corresponding to the 2.5 kb mRNA is not available we can infer a great deal about the structure of the mRNA and nature of the encoded protein from sequence analysis of the cloned RNA-copy pseudogene (15). RNA-copy pseudogenes are sequences which appear to be derived by reverse transcription of a mRNA followed by insertion of the DNA sequence into the genome. Since the sequences are non-functional and therefore not subject to evolutionary constraints on the protein sequence, they incorporate random nucleotide substitutions. Thus, structural analysis of an RNA-copy pseudogene will provide a very useful but not exact representation of the mRNA structure. Sequence analysis of the M29 RNA-copy pseudogene indicates that the 2.5 kb mRNA encodes a 247 amino acid non-muscle tropomyosin (10), and that, like the 3.0 kb mRNA, the large size of the mRNA is due to a very long 3' non-coding region. Our current understanding of human cytoskeletal tropomyosins is summarized in Table 1.

TABLE 1

mRNA (kb)	cDNA	Protein	Size (amino acids)	Pseudo -genes	Functional Gene Isolated
1.1	complete	hscp36	284	2	No
2.5	fragmentary	$hscp30_{nm}$	247	10	Yes
3.0	complete	?	247	5	Yes

It is interesting to note that the largest of the cytoskeletal tropomyosins, hscp36, is encoded by the smallest mRNA. This 1.1 kb mRNA presents the other extreme in having a 3' non-coding region of only 80 base pairs. The significance of this difference in size of the mRNAs is not known, although it is interesting to speculate that 3' non-coding region could contain signals for intracellular compartmentalization.

CONCLUSIONS

From these studies we can begin to answer some of the questions raised earlier. The heterogeneity of the non-muscle tropomyosins is very real and, in the 3 cases for which unique probes are available, is due to the expression of 3 different functional genes. The difference in apparent molecular weight between the 30,000 and 36,000 molecular weight is due to a real difference in size, the longest, hscp36, being identical in length to a tropomyosin of muscle origin. While we have not addressed the question of function, the occurrence of both muscle and non-muscle type tropomyosins in the cytoskeleton is very striking. Non-muscle and muscle tropomyosins differ not only in actin-binding capacity (16) but also in their *in vitro* polymerization properties (17). It is reasonable to expect that these differences will be reflected in different functional roles for the two classes of tropomyosin in the cytoskeleton.

ACKNOWLEDGEMENTS

We thank Mrs Mary-Ann Starkey and Mrs Sue Hamilton for preparation of the typescript.

REFERENCES

1. Bailey K (1948). Tropomyosin: a new asymmetric protein component of the muscle fibril. Biochem J 43:271.
2. Lazarides E (1975). Tropomyosin antibody: the specific localization of tropomyosin in nonmuscle cells. J Cell Biol 65:549.
3. Cohen I, Cohen C (1972). A tropomyosin-like protein from human platelets. J Mol Biol 68:383.
4. Fine RE, Blitz AL, Hitchcock SE, Kaminer B (1973). Tropomyosin in brain and growing neurones. Nature New Biology 245:182.
5. Lewis WG, Coté GP, Mak AS, Smillie LB (1983). Amino acid sequence of equine platelet tropomyosin: correlation with interaction properties. FEBS Lett 156:269.
6. Helfman DM, Feramisco JR, Ricci WM, Hughes SH (1984). Isolation and sequence of a cDNA clone that contains the entire coding region for chicken smooth muscle α-tropomyosin. J Biol Chem 259:14136.
7. Sanders C, Smillie LB (1985). Amino acid sequence of chicken gizzard β-tropomyosin: comparison of the chicken gizzard, rabbit skeletal and equine platelet tropomyosins. J Biol Chem (in press).
8. Mak AS, Smillie LB, Stewart GR (1980). A comparison of the amino acid sequences of rabbit skeletal muscle α- and β-tropomyosins. J Biol Chem 255:3647.
9. Schloss JA, Goldman RD (1980). Microfilaments and tropomyosin of cultured mammalian cells: isolation and characterization. J Cell Biol 87:633.
10. Talbot K, MacLeod AR (1983). Novel form of non-muscle tropomyosin in human fibroblasts. J Mol Biol 164:159.
11. Matsumura F, Lin JJ-C, Matsumura SY, Thomas GP, Topp WC (1983). Differential expression of tropomyosin forms

in the microfilaments isolated from normal and transformed rat cultured cells. J Biol Chem 258:13954.
12. Giometti CS, Anderson NL (1981). A variant of human nonmuscle tropomyosin found in fibroblasts by using two dimensional electrophoresis. J Biol Chem 256:11840.
13. MacLeod AR (1981). Construction of bacterial plasmids containing sequences complementary to chicken α-tropomyosin mRNA. Nucl Acids Res 9:2675.
14. MacLeod AR (1982). Distinct α-tropomyosin mRNA sequences in chicken skeletal muscle. Eur J Biochem 126:293.
15. MacLeod AR, Talbot K (1983). A processed gene defining a gene family encoding a human non-muscle tropomyosin. J Mol Biol 167:523.
16. Coté GP, Smillie LB (1981). The interaction of equine platelet tropomyosin with skeletal muscle actin. J Biol Chem 256:7257.
17. Coté GP, Lewis WG, Smillie LB (1978). Non-polymerizability of platelet tropomyosin and its NH_2- and COOH-terminal sequences. FEBS Lett 91:237.

Molecular Biology of Muscle Development, pages 457–471

A SINGLE GENE CODES FOR GIZZARD AND SKELETAL MUSCLE α TROPOMYOSIN[1]

Marc Y. Fiszman,[2] Elissavet Kardami[3] and Marguerite Lemonnier[2]

[2]Molecular Biology Department, Pasteur Institute, Paris, France and [3]Zoology Department, University California, Berkeley, California 94720

ABSTRACT Affinity purified antibodies against α tropomyosin from chicken Pectoralis major muscle (α_F-TM) recognize specifically the 43 000 daltons tropomyosin from gizzard, hereafter denominated α gizzard tropomyosin (α_{Gz}-TM) . Limited tryptic cleavage provide evidence that these antibodies recognize sites located on the -COOH half of the α_F and the α_{Gz} tropomyosins. Using a cDNA which represents the 3' untranslated region and 90% of the coding sequence of the mRNA coding for α_F tropomyosin , we have characterized the mRNA coding for α_{Gz} tropomyosin which is a 2.0kb molecule. The 1.2kb mRNA (α_F-TM) and the 2.0kb mRNA show greater than 90% homology over 500 nucleotides in their 3' coding sequence. However, the two mRNA have totally distinct 3' untranslated regions. We will show that during skeletal muscle development the two mRNA are being synthesized at the same time. However the rate at which they are expressed is modulated so that in young embryos (5-8 days in ovo) the two messengers are present in equal amount while in older embryos the mRNA corresponding to the skeletal muscle tropomyosin is the major form. It is interesting to note that this phenomenon does not exist during gizzard development since only the 2.0kb mRNA can be detected.

[1]This work was supported by grants from the CNRS, the Association des Myopathes de France and the Muscular Dystrophy Association of America.

On the other hand, the 1.2kb mRNA is the only mRNA to be accumulated when myoblasts differentiate to form myotubes in vitro. By screening a chicken genomic library in λ Charon 4A, we have isolated a recombinant phage which carries a genomic sequence which corresponds to the common region of the two mRNAs and which also corresponds to the two different 3' untranslated regions of the two messengers. Such a structure suggests that a single gene codes for the two messengers by differential splicing.

INTRODUCTION

Tropomyosin, as many other muscle contractile proteins, is found ubiquitously in muscle and non muscle cell types. The skeletal isoforms are composed of two types of subunits, α and β (1-3) which are organized in a dimeric α-coiled-coil structure (4). The gizzard α and β tropomyosins are different in their physico-chemical properties from the striated isoforms.

Here we report that skeletal muscle and gizzard tropomyosins share common antigenic determinants located at the carboxylic end of the molecules, and also that the respective mRNAs have homologous sequence restricted to their coding sequence. Lastly, we will show that the same fragment of genomic DNA carries exons which are common to the two mRNA and codons which are specific for one or the other, therefore implying that the two mRNA are generated from the same gene by differential splicing.

RESULTS

Gizzard and Skeletal Muscle α Tropomyosin Share Antigenic Determinant.

We have raised antibodies against purified α tropomyosin from Pectoralis Major muscle of the chicken. After adsorbtion of the material which cross reacts with the skeletal muscle β tropomyosin and the skeletal slow muscle α and β tropomyosin, the remaining antibodies strongly recognize the α_F-TM and the α_{Gz}-TM (Fig. 1).

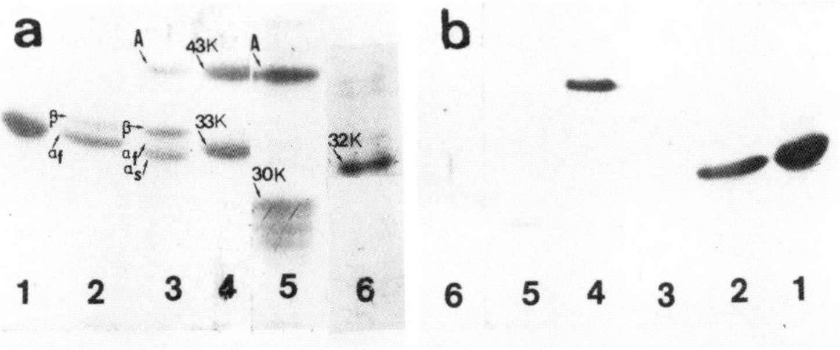

FIGURE 1. Immunological relationships between gizzard and skeletal α tropomyosin. Tropomyosins were separated by SDS/PAGE, transferred onto nitrocellulose and detected by a peroxydase-linked indirect immunoassay (5). a) Amido black stained copy of transferred proteins (6), b) Immunological detection with the purified antibodies at a 1:10 dilution. Lanes 1) Pectoralis Major muscle tropomyosin, 2) PLD tropomyosin, 3) ALD tropomyosin, 4) gizzard tropomyosin, 5) crude preparation of chicken brain and 6) cardiac tropomyosin.

In order to locate the region of the molecule which reacts with the antibodies, we digested the various tropomyosins with trypsin under the conditions described by Pato and Smillie (7). Under these conditions, α_F-TM produces two fragments, T1 and T2, with T2 being the carboxylic end of the molecule since it is sensitive to carboxypeptidase applied before trypsin (Fig. 2a). Gizzard tropomyosin produces several peptides one of which, T'2, comigrate with T2 and also corresponds to the carboxylic end of the molecule (Fig. 2b). When the tryptic peptides are transferred onto nitroelulose and reacted with the purified antibodies, only T2 and T'2 can be visualized (Fig. 2c). This result indcates that the common antigenic determinants are located on the carboxylic end of the two tropomyosins.

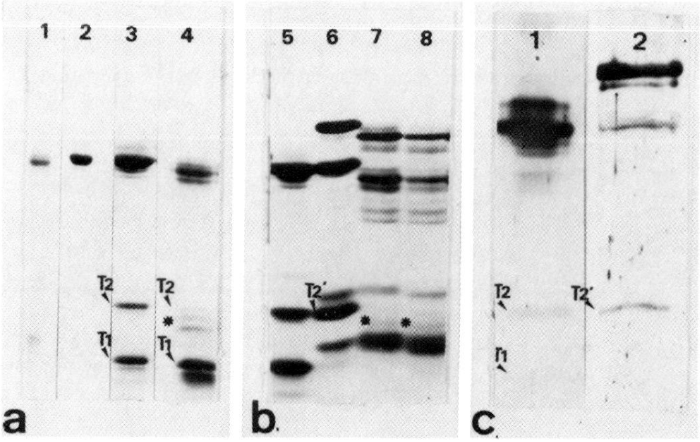

FIGURE 2. Identification of the region which contains the common antigenic sites. Native tropommyosin was treated with trypsin according to Pato and Smillie (7) and in some cases the molecule was treated with carboxypeptidase as described by Mak and Smillie (8) before adding the trypsin. a) and b) SDS/PAGE analysis of the tryptic digests from intact and truncated tropomyosin: 1) Pectoralis Major tropomyosin after carboxypeptidase A treatment, 2) intact Pectoralis Major tropomyosin, 3) and 5) Pectoralis Major tropomyosin after digestion with trypsin for 2 minutes, 4) same tropomyosin but first treated with carboxypeptidase A, 6) gizzard tropomyosin after tryptic digestion for 10 minutes, 7) and 8) same but first treated with carboxypeptidase A. c) Immunological detection: 1) Pectoralis Major tropomyosin after digestion with trypsin and 2) gizzard tropomyosin after digestion with trypsin.

Analysis of the Messenger RNAs.

The use of a recombinant plasmid (pαTM4) which contains a cDNA complementary to the entire 3' non coding region and 90% of the sequence coding for α_E-TM (9) has enabled us to detect homology between the mRNAs coding for the two tropomyosins. When the RNA isolated from leg and gizzard muscles is analyzed by electrophoresis in a 1.3% agarose gel, blotted onto nitrocellulose and hybridized with pαTM4, it can be observed that in the case of leg muscle RNA, pαTM4 recognizes two species, a major 1.2kb mRNA and a minor 2.0kb mRNA, while the same probe recognizes a 2.0kb mRNA species in the gizzard muscle RNA (Fig. 3A).

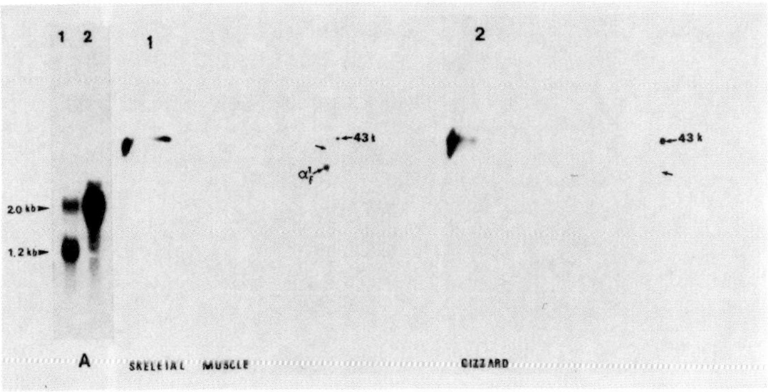

FIGURE 3. Analysis of RNA isolated from gizzard and skeletal muscle. Total RNA was isolated as previously described (10). 2 μg of total RNA was electrophoresed in a 1.3% agarose gel, transferred to nitrocellulose and hybridized with pαTM4 as described by Thomas (11). Lane 1) skeletal muscle, lane 2) gizzard. Hybrid selection of RNA was performed as described by Maniatis (12) using 2 μg of immobilized pαTM4 and 150 μg of total RNA. In vitro translation and analysis of the products by 2D gel electrophoresis was as described in (10).

In vitro translation of the RNA which is specifically retained by PαTM4 in RNA preparations from leg and gizzard muscles shows that, in the former case, the major products are two proteins which comigrate with α_F-TM and α_{Gz}-TM, while in the latter case, the only translation product comigrates with α_{Gz}-TM (Fig. 3). We can therefore conclude that the 1.2kb mRNA codes for α_F-TM while the 2.0kb mRNA codes for α_{Gz}-TM.

Despite the strong homology which exists between the two mRNA species, they nevertheless present distinct differences. We have taken advantage of a Hind III restriction site located at the junction of the coding and non coding sequence of pαTM4 to construct a new recombinant plasmid (pαTM4-200) which contains only the

3' non coding region of pαTM4. When this plasmid is hybridized with total RNA isolated from skeletal and gizzard muscles, only the 1.2kb species is now recognized and no hybridization occurs with the 2.0kb mRNA (Fig. 4 lane 4). Therefore the two mRNA have distinct 3' untranslated regions.

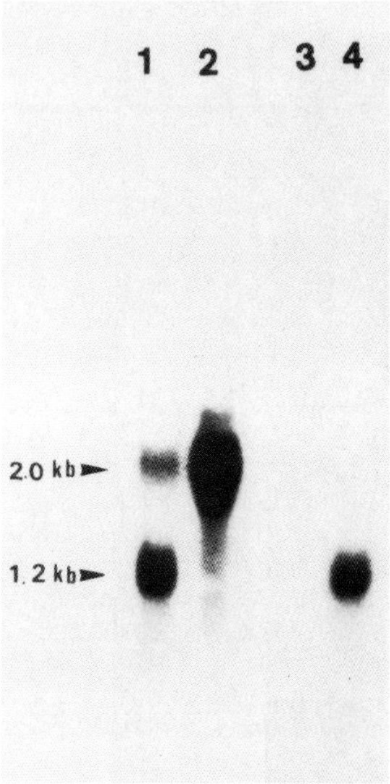

FIGURE 4. Gizzard and skeletal muscle α tropomyosin mRNA have distinct 3' untranslated region. Total RNA was fractionated and analyzed as in Fig. 3. Lanes 1 and 4: skeletal muscle mRNA hybridized with pαTM4 (1) and pαTM4-200 (2), lanes 2 and 3: gizzard mRNA hybridized with the same probes.

The Gene(s) Which Codes for Gizzard and Skeletal Muscle α Tropomyosin.

In order to gain some information on the complexity of the genes which code for α_F-TM and α_{Gz}-TM, Southern blot analysis was carried on chicken genomic

DNA after restriction with Eco RI. Under non stringent coditions of hybridization approximately eight different fragments can be visualized with pαTM4. However when the stringency of hybridization is increased, only three fragments remain (Fig. 5A). It must be noted that if the hybridization is performed with pαTM4-200 only the 1.35kb fragment is vizualized (data not shown), which indicates that this fragment carries the 3' untranslated region of the α_F-TM gene and that there is only one such gene.

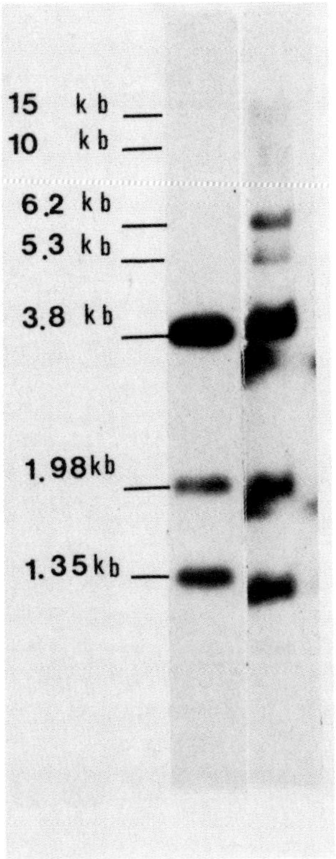

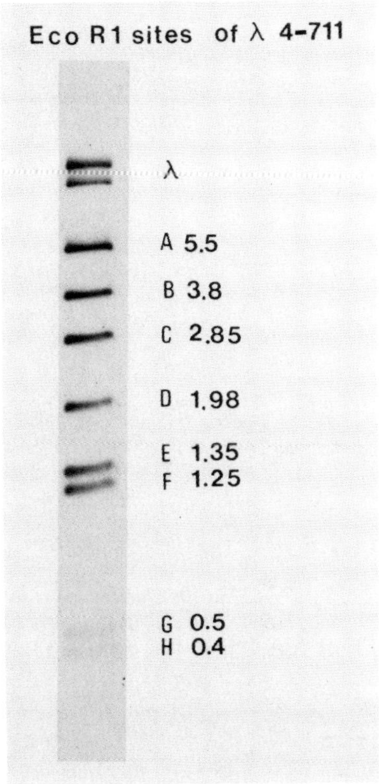

FIGURE 5. Southern blot analysis of chicken genomic DNA. 10 μg of genomic DNA was restricted with Eco R1, fractionated on a 0.8% agarose gel, blotted onto nitrocellulose and hybridized with pαTM4.

FIGURE 6. λ4-711 fragments generated with Eco R1.

By screening a chicken genomic library cloned in λ Charon 4A, several recombinant phages were isolated which hybridized with pαTM4 under low stringency. Under high stringency, only one was still able to hybridize. This phage, called λ4-711, was therefore selected and further analyzed. By restriction with Eco RI, eight fragments are generated (Fig.6) out of which three hybridize with pαTM4 (fragments B, D and E). It is interesting to note that they are identical to the fragments of genomic DNA which hybridize with pαTM4 under stringent conditions.

Figure 7 presents a partial restriction map of λ4-711 as well as its transcriptional polarity. Since some of the restriction enzymes which have been used (Eco R1, Bgl 2, Sst 1) have no sites in the sequence of pαTM4, it can be deduced from the data presented that the gene for

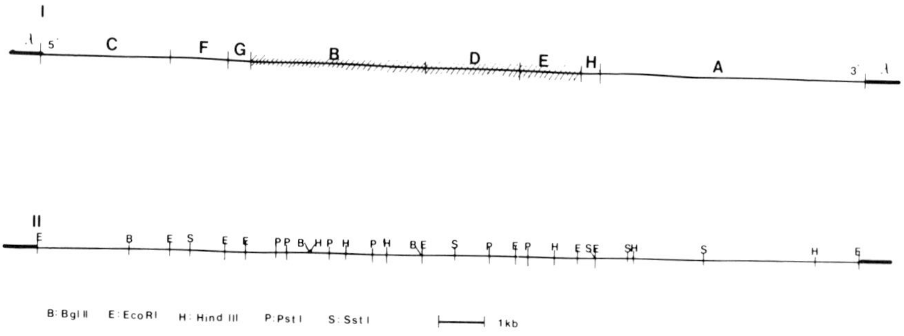

FIGURE 7. Partial restriction map of λ4-711. I) Eco R1 sites (dashed area indicates the fragments which hybridize with pαTM4). II) Partial restriction map using 5 different restriction enzymes.

α_F-TM will be interupted by a serie of introns. To gain more information on the organization of this gene, each Eco R1 fragment was subcloned and used as a probe to hybridize with gizzard and skeletal muscle mRNA (Fig 8). Fragment E, which was the only fragment to hybridize with pαTM4-200 hybridized exclusively with mRNA isolated from skeletal muscle. This result therefore suggests that this fragment contains an exon corresponding to the 3' untranslated region of the α_F-TM mRNA. This was

	C	F	G	B	D	E	H	A
skeletal muscle RNA	–	–		+	+	+		–
gizzard RNA	+	–		+	+	–		+

FIGURE 8. Identification of the genomic fragments which hybridize with gizzard and skeletal mRNA. λ4-711 Eco R1 fragments were subcloned in pBR 325 and used as probes for hybridization with total RNA isolated from gizzard and skeletal muscle and dotted on nitrocellulose

indeed confirmed when the nucleotide sequence of fragment E was determined. Furthermore this sequence analysis has indicated that the entire 3' untranslated region is contained within a single exon (data not shown). Fragments B and D hybridize with mRNA isolated from gizzard and skeletal muscle and therefore they must contain exons which are common to the two types of RNA. No other fragments hybridize with skeletal muscle mRNA which must indicate that the genomic clone which we have isolated is lacking the 5' end of the gene since we know that pαTM4 is 200 nucleotides short for a full length representation of the message. However, two fragments, A and C, did hybridize with gizzard mRNA and Northern blot analysis have shown that this hybridization is with the 2.0kb mRNA species. These fragments must therefore contain exons which are specific for the α_{Gz}-TM mRNA. Such an organization suggests very strongly that there is a single gene which codes for two mRNAs, the 1.2kb α_E-TM mRNA and the 2.0kb α_{Gz}-TM mRNA, and that these mRNAs are generated by differential splicing. Figure 9 is a schematic representation of such a genomic organization. The position of the various exons has been obtained by restriction analysis of the various fragments and hybridization with gizzard and skeletal muscle mRNA.

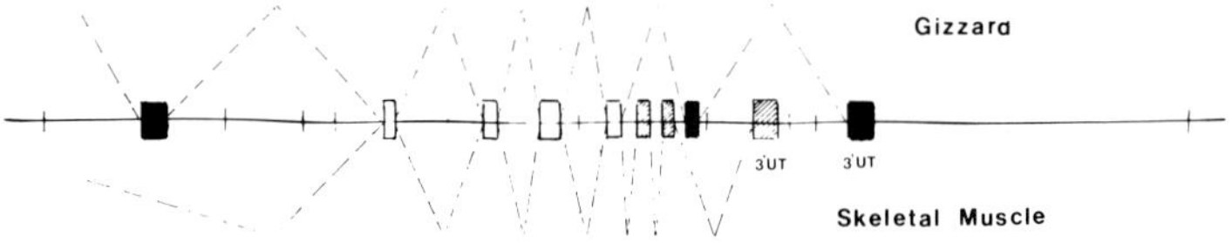

FIGURE 9. Genomic organization of the gene which codes for gizzard and skeletal muscle α tropomyosin. Black area indicates gizzard specific exons, dashed area indicates skeletal muscle specific exons and open area indicates common exons.

Tropomysin Gene Expression During Muscle Development.

When gizzard and skeletal muscle mRNAs were being analyzed, it was striking to find that skeletal muscle mRNA contain both the 1.2kb and the 2.0kb species (see fig. 3). A more detailed analysis was therefore undertaken. Total RNA was isolated from developing limb buds from day 5 to day 19, fractionated on a 1.3% agarose gel, blotted onto nitrocellulose and hybridized with pαTM4. As shown on figure 10, the 1.2kb and the 2.0kb mRNA species are detected at all time. However the two mRNAs are differentially accumulated since they are present in almost equivalent amount between day 5 and 7 but later there is a preferential accumulation of the 1.2kb mRNA while the 2.0kb remains approximately constant. It must be enphasized that we have tested that the 2.0kb mRNA corresponds indeed to the α_{Gz}-TM mRNA both by in vitro translation of hybrid selected mRNA and by hybridization with λ4-711 fragment A (see fig. 3 and data not shown). It is interesting to note that when the same experiment is performed on developing gizzard only the 2.0kb mRNA is detected. Furthermore, when skeletal muscle myoblasts are allowed to differentiate _in vitro_ only the 1.2kb mRNA is detected (data not shown).

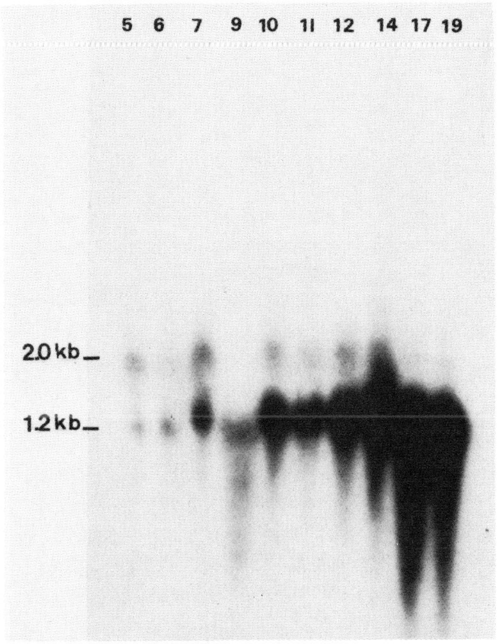

FIGURE 10. RNA expressed by developing skeletal muscle. Total RNA was isolated and analyzed as described for figure 3. The numbers refer to the age of the embryos which were used to prepare RNA.

DISCUSSION

Using affinity purified antibodies prepared against Pectoralis Major muscle α tropomyosin, we have been able to compare the various tropomyosins and to detect a very strong reactivity of these antibodies with gizzard α tropomyosin. It is worth mentionning that there is no reactivity of our antibodies with cardiac α tropomyosin which is a further argument to support our earlier proposition that skeletal muscle α tropomyosin and cardiac α tropomyosin are different proteins in avian (5, 10 and 13) which is a major difference with the situation in mammals(14).

The common epitopes were further localized to the carboxylic end of the two proteins by taking advantage of the observation that trypsin, under mild conditions, cleaves native tropomyosin into two halves (7).

When the mRNAs coding for α_F-TM and for α_{GZ}-TM are analyzed, it was found that α_F-TM is coded for by a 1.2kb mRNA while α_{GZ}-TM is coded for by a 2.0kb mRNA. Furthermore, these two mRNAs show over 90% homology but this homology is restricted to the coding sequence since the 3' untranslated region of one message is different from the 3' untranslated region of the other.

By screening a genomic library of chicken DNA, we have isolated a recombinant phage which contains a 17kb DNA insert which hybridizes both with the 1.2kb and the 2.0kb mRNAs. Further analysis of this phage has allowed us to show that it contains a serie of exons some of which are commonly used for the two messages and some of which are specific for one or the other. Such an organization therefore suggests that a single gene produces the mRNAs coding for the gizzard and skeletal α tropomyosins by alternative splicing of isotype specific exons (see fig. 9). Such an alternative splicing mechanism has also been proposed in the case of the rat tropomyosin (15). However, in this case, three different mRNAs are being produced. As shown in figure 9, it is interesting to note that the two 3' untranslated sequences are contained within single exons while, at least the skeletal isotype specific sequence at the 3' end is contained within two exons. Our result also suggests that gizzard and skeletal α tropomyosin must contain an isotype specific peptide located at the carboxy terminus of the molecules. The same conclusion has been recently obtained for gizzard and skeletal β tropomyosins where it was shown that the two molecules have a divergent sequence over the last 25 amino acids (16). However in this case, the gene(s) has not yet been isolated and it is not known whether these molecules are coded for by two or a single gene.

A major question which remains to be solved concerns the organization of the 5' end of the gene which we have isolated, and specially whether there is a single or a duplicated promoter sequence.

Such a pattern of complex gene sequence and alternative splicing seems to be a common feature of the contractile protein genes since it has been demonstrated for Drosophila myosin heavy chain (17), myosin light chain 1 and 3 (18-20), troponin T (21, 22) and now tropomyosin (15). The mechanism which regulates the splicing pathway has not been elucidated. However, in the case of myosin light chain 1 and 3 and of troponin T, the use of different promoters seems to be involved while, in the case of myosin heavy chain, different polyadenilation sites may play a role in this process (17-22). No information is yet available for tropomyosin. However, our observation that developing skeletal muscle expresses the two messages while developing gizzard or myotubes formed in vitro express only their specific mRNA is of potential interest for the elucidation of the mechanism which regulates this differential splicing.

ACKNOWLEDGMENTS

This work could not have been done without the generous gift of pαTM4 by A. MacLeod.

REFERENCES

1. Sodek J, Hodges RS, Smillie LB (1978). Amino acid sequence of rabbit skeletal muscle α-tropomyosin : The COOH terminal half (residue 142-284). J Biol Chem 253 : 1129.
2. Stone D, Smillie LB (1978). Amino acid sequence of rabbit skeletal muscle α-tropomyosin : the NH_2 terminal half and complete sequence. J Biol Chem 253 : 1137.
3. Mak AS, Lewis WG, Smillie, LB (1979). Aminoacid sequence of rabbit skeletal β and cardiac tropomyosins. FEBS Lett 105 : 232.
4. Smillie LB (1979). Structure and functions of tropomyosins from muscle and non muscle sources. Trends in Biochem Sci 4 : 151.
5. Kardami E, Montarras D, Fiszman M (1983). Fast and slow chicken skeletal muscles contain different α and β tropomyosins. Biochem Biophys Res Commun 110 : 147.
6. Towbin H, Staehelin T, Gordon J.(1979). Electrophoretic transfer of proteins from polyacrylamide gels to nitro cellulose sheets : procedure and some applications. Proc Natl Acad Sci Acad 76 : 4350.

7. Pato MD, Smillie LB (1978). Stability and troponin T binding properties of rabbit skeletal α tropomyosin fragments. FEBS Lett 87 : 95.
8. Mak AS, Smillie LB (1981). Non polymerizable tropomyosin : preparation some properties and F actin binding. Biochem Biophys Res Commun 101 : 208.
9. MacLeod AR (1981). Construction of bacterial plasmids containing sequences complementary to chicken α tropo myosin mRNA. Nucl Acids Res 9 : 2675.
10. Montarras D, Fiszman MY, Gros F (1981). Characterization of the tropomyosin present in various chick embryo muscle types and in muscle cells differentiated in vitro. J Biol Chem 256 : 4081.
11. Thomas P (1980). Hybridization of denatured RNA and small DNA fragments transferred to nitrocellulose. Proc Natl Acad Sci 77 : 5201.
12. Maniatis T, Fritsch EF, Sambrook J (1982). In "Molecular cloning : a laboratory manual" Cold Spring Harbor Lab. p 329.
13. Kardami E, Fiszman MY (1983). Classification of tropomyosin components into an α-like or a β-like family by partial peptide mapping. FEBS Lett 163 : 250.
14. Lewis WG, Smillie LB (1980). The amino acid of rabbit cardiac tropomyosin. J Biol Chem 255 : 6854.
15. Ruiz-Opazo N, Weinberger J, Nadal-Ginard B (1983). Different tissue specific forms of α-tropomyosin encoded by the same gene. J Cell Biol 97 : 329a.
16. Helfman DM, Feramisco JR, Ricci WM, Hughes SH (1984). Isolation and sequence of a cDNA clone that contains the entire coding region for chicken smooth-muscle α-tropomyosin. J Biol Chem 259 : 14136.
17. Rozek CE, Davidson N (1983). Drosophila has one myosin heavy chain gene with three developmentally regulated transcripts. Cell 32 : 23.
18. Nabeshima Y, Fujii-Kuriyama Y, Muramatsu M, Ogata K (1984). Alternative transcription and two modes of splicing result in two myosin light chains from one gene. Nature 308 : 333.
19. Robert B, Daubas P, Akimenko MA, Cohen A, Garner I, Guénet JL, Buckingham M (1984). A single locus in the mouse encodes both myosin light chains 1 and 3, a second locus corresponds to a related pseudogene. Cell 39 : 129.

20. Periasamy M, Strehler EE, Garfinkel LI, Gubits RM, Ruiz-Opazo N, Nadal-Ginard B (1984). Fast skeletal muscle myosin light chain 1 and 3 are produced from a single gene by a combined differential RNA transcription and splicing. J Biol Chem 259 : 13604.
21. Wilkinson JM, Moir AJG, Waterfield MD (1984). The expression of multiple forms of troponin T in chicken fast skeletal muscle. Eur J Biochem 143 : 47.
22. Medford RM, Nguyen HT, Destree AT, Summers E, Nadal-Ginard B (1984). A novel mechanism of alternative RNA splicing for the developmentally regulated generation of troponin T from a single gene. Cell 38 : 409.

Molecular Biology of Muscle Development, pages 473–485

TISSUE-SPECIFIC EXPRESSION OF A PROCESSED CALMODULIN GENE[1]

Joseph P. Stein
Division of Endocrinology, University of Texas
Medical School, Houston, Texas 77030

ABSTRACT Two Calmodulin genes exist in the chicken genome. One, CL1, codes for an authentic calmodulin protein that is expressed in all tissues of the chicken. The other, CM1, appears to be a processed copy of CL1, since it contains no introns and is flanked by 9 bp direct repeats. CM1 codes for a calmodulin-like protein that has 19 amino acid differences from other vertebrate calmodulins. The CM1 gene is expressed only in cardiac and skeletal muscle, albeit at a much lower level than CL1. 500 bp upstream from the initiator AUG is a 200 bp internally repetitive sequence, made up primarily of the tetranucleotide TGGA. This repeat sequence appears to be unique in the chicken genome, and is currently being tested for enhancer activity.

INTRODUCTION

Calmodulin is a Ca^{2+}-binding protein that has been found in virtually every eucaryotic cell type (1,2). It exists as a monomer of molecular weight 16,700 and contains four Ca^{2+}-binding sites (3). Calmodulin functions in a regulatory capacity for a diverse array of enzymatic reactions and biochemical pathways. The pleiotropic nature of calmodulin-regulated events has sparked intensive research efforts into its mechanism of action (for reviews see Refs 1,4-5). Calmodulin is constitutively expressed in target cells for both steroid and peptide hormones (1), but the intracellular concentration is elevated in transformed cells (6) as well as at the G_2/S boundary of the growth

[1]This work was supported grants from the National Institutes of Health and the American Heart Association, Texas Affiliate.

cycle of mammalian cells in culture (7). In order to evaluate the regulatory mechanisms involved in calmodulin function a series of studies was initiated to isolate the chicken calmodulin gene. Initially, Munjaal et al. (8) reported the isolation of a cDNA for eel calmodulin (pCM109) which contained calmodulin-specific sequences coding for amino acid residues 93-148. Legace et al. (9) then used pCM109 to isolate a full-length cDNA for eel calmodulin (pCM116). Finally, these eel cDNAs were used to isolate a cDNA clone for chicken brain calmodulin as well as two different chicken calmodulin genomic clones, CM1 and CL1 (10-11). CL1 is a genomic clone that contains numerous introns and codes for an authentic calmodulin protein (CaM). CM1, on the other hand, contains no introns and appears to be a processed gene derived from CL1. It is unique, however, in that its mRNA, which encodes a variant calmodulin protein (CaML), is expressed in some muscle tissues of the chicken. This manuscript describes the isolation and characterization of this gene, as well as some interesting features of its structure and function.

RESULTS

Organization of the CM1 Gene

750,000 recombinant phage carrying a chicken genomic DNA library were screened for a calmodulin gene using the <u>in situ</u> plaque hybridization technique of Benton and Davis (12). All of the eight positive clones that were found proved to be overlapping fragments containing the CM1 gene. A partial restriction map of the genomic insert of the plaque selected for sequence analysis is shown in Figure 1. The insert was cut by Eco RI into fragments of 11.5, 2.8, and 2.5 kilobases (Kb) as shown. Only the 11.5 Kb fragment hybridized to pCM109, and the calmodulin sequences were localized to its 5' terminus, although the 2.8 Kb fragment has since been shown to contain the 5' coding sequences. DNA sequencing of portions of the cloned DNA established a striking feature of the CM1 gene: a complete lack of intervening sequences. An intact coding sequence, from the ATG initation codon through a termination codon 448 nucleotides downstream, is present. This was unexpected in light of CaM's four domain structure in which the four

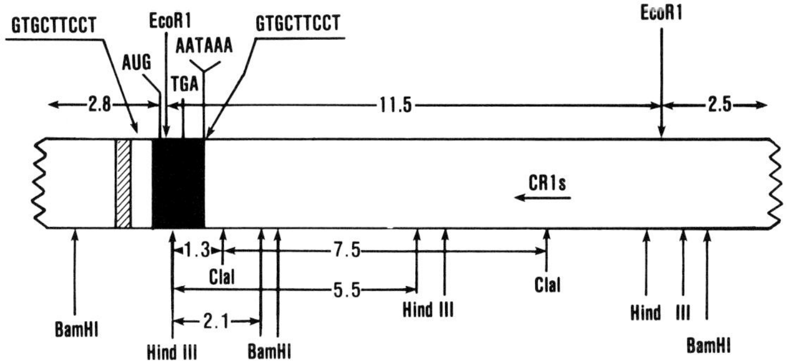

Figure 1. Partial restriction map of the chicken calmodulin gene CM1 and flanking sequences. The calmodulin gene sequence is denoted by the solid box. Distances are given in Kb.

domains are quite homologous. In view of this structure, we had predicted that introns would separate the genomic sequences coding for these peptide domains.

An explanation presented itself with the isolation of a second calmodulin gene, CL1. Initially, 4,500 clones of a chicken brain cDNA library were screened with pCM116, and a calmodulin cDNA isolated whose restriction map differed from that of the CM1 gene. This cDNA was used to isolate a calmodulin gene, CL1, that differed significantly from the CM1 gene: numerous introns were present (10,13). The organization of these two calmodulin genes was markedly different, as demonstrated in Figure 2. Chicken genomic DNA, digested with four different restriction enzymes, was hybridized to labelled probes derived from the CM1 gene (right panel) and a CL1 cDNA clone (left panel), using the standard Southern protocol (14) and stringent hybridization conditions (15). Unique hybridizing fragments were observed in each digest, indicating that the genes are not linked. Furthermore, the lack of multiple hybridizing fragments indicated that CM1 and CL1 were the only calmodulin gene copies in the chicken genome. The nucleotide sequence of the protein coding regions of CL1 was 75% homologous to the corresponding sequences of the CM1 gene, while the 3' nontranslated regions of the two

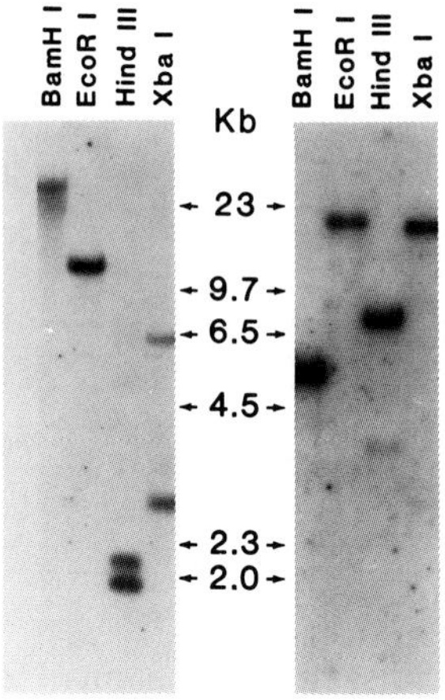

Figure 2. Organization of calmodulin genes CL1 (left panel) and CM1 (right panel) as determined by Southern transfer/hybridization.

genes were only 18% homologous. The exons of the CL1 gene code for a CaM protein that is identical in sequence to bovine, rat and human CaMs (16-18). CM1, on the other hand, codes for a calmodulin-like protein (CaML) that was 87% homologous to these other CaMs. It thus appears that the chicken genome contains two different calmodulin genes whose coding sequences are related but whose organization is considerably different. Furthermore, the sequence homology of CM1 and CL1, despite the lack of introns in CM1, suggests that CM1 might have originated as a processed gene copy derived from CL1.

Expression of Calmodulin mRNAs in Chicken Tissue

In light of the fact that two potential calmodulin genes exist in the chicken genome, it was of interest to examine their expression in different tissues. A cDNA probe for CL1 and a cloned 1.3 Kb fragment of CM1 were both labeled and hybridized to poly(A) RNA separated on 1.5% agarose/formaldehyde gels (19). The results with mRNA from chicken oviduct, brain, breast muscle, and liver are shown in Fig 3. Whereas the CL1 probe hybridizes to one or more mRNA species in all tissues examined, the CM1 probe hybridizes to a mRNA species in muscle only. The lone mRNA species to which CM1 hybridizes is smaller than any of the CL1 mRNAs. These results have recently been extended to other tissues. While spleen and smooth muscle do not contain CM1 trancripts, thigh and cardiac muscle do. Thus, transcripts of the CM1 gene seem to be found only in striated muscle tissue. These data suggest that CL1 is the primary gene responsible for the production of CaM and that CM1 codes for a CaML that has a Ca^{2+} -binding function somehow concerned with striated muscle physiology.

Sequence Organization of Calmodulin Gene CM1

Sequence analysis of the CM1 gene was carried out using the chemical degradation method of Maxam and Gilbert (20) and the thin acrylamide gels of Sanger and Coulson (21). A

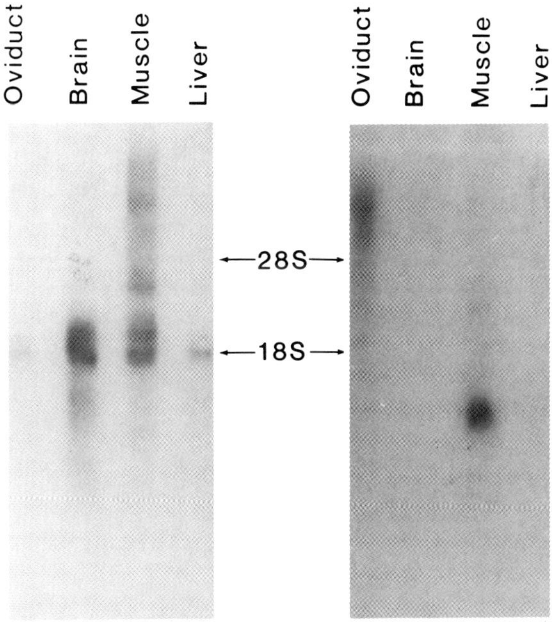

Figure 3. Northern hybridization of chick poly(A) RNA from four different tissues to probes from the CL1 gene (left panel) and the CM1 gene (right panel).

sequence of over 3 Kb of chicken DNA spanning the CM1 gene is presented in Figure 4. An open reading frame coding for CaML begins with an ATG at position 1117, and extends through a TGA at position 1564. Following the termination codon is a 3' non-tranlated region of approximately 300 bp; the polyadenylation site is probably determined by an AATAAA sequence beginning at position 2041. The sequence presented in Figure 4 lends further credence to the hypothesis that CM1 is a processed gene copy derived from CL1. Processed genes normally exhibit three distincitve

```
        10         20         30         40         50         60         70         80         90        100
GGGCTCCCCT CCCCGAAACA CACCGTGCCT CTCGCCAGAC AGGTTTCCAG CATGGCTCAG CCCATTCCTC CCAGCAACGC CTGAGTGTGC CAAGAGGTTT

       110        120        130        140        150        160        170        180        190        200
TTCCAGATGA AGTTTTTCCT GGGCCCCCTC CCCCACCCCC G--------- -TTCCCATTC TATCTCATAA TTTGCTTCTA TTTCGTAGCA TTCCATCCAA

       210        220        230        240        250        260        270        280        290        300
ATCTCATAGC TTTACGTCCT CGCTGGAGTA AAACGGGGAA AAATGGTGGT GTTTTCTTGT GCAGAAGTCA GATTGTATTC TGCTGTAGAG TTTGGTCATC

       310        320        330        340        350        360        370        380        390        400
TGGTCCTATT AGTACCTCGG CCGTTATGGG GACATGGGAA GAAGAGGGTG ATGGATGGAT GGATGGATGG ATGGATGGAT GGATGGATGG ATGGATGAAA

       410        420        430        440        450        460        470        480        490        500
AAATGGATGG ATGGATGGAT GGATGGATGG ATGAAAAAAT GGATGGATGG ATGGATGGAT GGATGGATGG ATGGATGGAT GGATGGAAAA AATGGATGGA

       510        520        530        540        550        560        570        580        590        600
TGGACTGGAT GGATGGATGG ATGGAAAAAA CGGATGGATG GAAAAAATGG ATGGATGGAC GGATGCATGG ATGGTAGGTT GGGCAACTGG GTGGACGTAT

       610        620        630        640        650        660        670        680        690        700
GAGTGAATGG TCAGAAGAAT GAATGGGTGG ATGGTTGATT AGTTGAC--- --GATGAACA TGGACAAGTG GGCAAAGAGC TCATAATCAT GTGCTCCTGA

       710        720        730        740        750        760        770        780        790        800
ACCCCTTGTC ATCATGTGCT TCCTAACTCC CTGATGTAAA AGATGAGCAC AACTTGACAG GGTGATTATC CCTCCCTCAG CCCAGGATTA ACCCCTGATT

       810        820        830        840        850        860        870        880        890        900
AACCAGGTAT AAGCCTCATG ACAAAACTCA CCTGATGTAC AGCCACGTCA GCTCCCTGGG GCCATGCAGC TGTGCTCGGG AAGCCCAGAC AAGGTGTCAT

       910        920        930        940        950        960        970        980        990       1000
GGCTGACTCA GGTTCCCGTC AGCAAGGAAC ACAAGGAGTA CCATCACAGC TGGGCTCTGC TGGACCTGCA CGAGCAAGGA GCAGGAAACC TCCTGCATGT

      1010       1020       1030       1040       1050       1060       1070       1080       1090       1100
CCCAGCTCCT CCTCCTCCAG GAGCAGGAGA ACAGGAAGGA GGAGGCCCCA TCCACACCTA CCCGCCAACG CTGCGTGCTG GGGCAAAGCG TGTGCGAGCG

      1110       1120       1130       1140       1150       1160       1170       1180       1190       1200
GGGAGCGTGG GACGCCATGG CCGAGCGGCT GTCGGAGGAG CAGATCGCTG AATTCAAGGA GGCTTTTTCC CTTTTTGACC GCGACGGCGA TGGCTGCATC

      1210       1220       1230       1240       1250       1260       1270       1280       1290       1300
ACCACGATGG AGCTGGGCAC TGTCATGCGC TCACTGGGGC AGAACCCCAC CGAGGCGGAG CTGCAGGACA TGGTGGGAGA GGTGGACGCG GACGGCAGCG

      1310       1320       1330       1340       1350       1360       1370       1380       1390       1400
GCACCATTGA TTTCCCCGAG TTCCTCTCGC TGATGGCGAG GAAGATGAGG GATTCAGACA GCGAGGAGGA GATCCGTGAA GCTTTCCGTG TCTTCGACAA

      1410       1420       1430       1440       1450       1460       1470       1480       1490       1500
GGACGGTAAT GGTTACATCA GCGCGGCGGA GCTGCGGCAC GTCATGACCA ACCTGGGCGA GAAGCTGACG GACGAGGAGG TGGATGAGAT GATCAAGGAG

      1510       1520       1530       1540       1550       1560       1570       1580       1590       1600
GCAGATTGCA ACAATGACGG GCAGGTCAAC TACGAGGAGT TTGTGAGGAT GATGACGGAG AAGTGACCTT CATTCCCAGC CCACAAGTAA CTCATGCCCG

      1610       1620       1630       1640       1650       1660       1670       1680       1690       1700
ACCCCACGTG CTTTGAAAGC CCCGGGCCAA GGGGTCCGTC CTACCGAATG CAATGGCACC TCGAATGTCT CATTCACCAC CCCTTGGCCA CCTCTTGCAT

      1710       1720       1730       1740       1750       1760       1770       1780       1790       1800
TGCTGCTGAT GGTGGTGGTT GGAAAAAACA GTGTTTAAGG TGCTTTTTAA AATTGTTTCC TTTGTTTTTT CCATAGGGAA AACAGAGACC ACGGCTGCCC

      1810       1820       1830       1840       1850       1860       1870       1880       1890       1900
CCAACGGTCC ACCCCTATCT GCCTTCTCTC TGCCATCCTG GGGTGAAAAA TGATCTTTTA TCCCAACTGG AGCATCGCCA ACCTGACGTC TGTGTTGGAT

      1910       1920       1930       1940       1950       1960       1970       1980       1990       2000
GCAGGACAGG GTCACGACAG CATGGGATGA CGACAGACAG TGGGGGGAGG ACCCTCCTGC TTCCATTTTC AAGATGTGCC TCTTGTGTTC ATAGAGGTGA

      2010       2020       2030       2040       2050       2060       2070       2080       2090       2100
TAGTGGTCGT TGCTCGTCCA TGAGAAATGT GCTGGGGTCC AATAAAATGT GGCCCCTGTG TCTACCAGTG CTTCCTTCTG CTTCACTTTG GGACCACTAC

      2110       2120       2130       2140       2150       2160       2170       2180       2190       2200
TAAGAAACCC CCTCAAAAAA GACACTGCCC CGATGGGGTG ACAGGGCTGG GGATGGGAAC AGAATCACCT CCGTCCCTCT TCTCCACATT TTTTTTCACC

      2210       2220       2230       2240       2250       2260       2270       2280       2290       2300
CCAAAAGATG GAAAAATACT CCCCAGGGCT GCTCAACACT CAGTTCCTGG GTTGTGACCA TCCCCATCCC CAAATCCTAT CCCTTTTGGA ATCTGTTGGC

      2310       2320       2330       2340       2350       2360       2370       2380       2390       2400
TGCACCCATG CTCTGGCTTC TCCCAAAGGA AGGGCACTGA GGGCTCTTTT ATCTCCCTTC CCATCCCAAG AGGAGGGTTG GTAGATCCTG TCTCAAACAA

      2410       2420       2430       2440       2450       2460       2470       2480       2490       2500
GGCAGGCACA GAAATTATCA TCAATCCACC CCACAGTGCT GCCTTGACTC AGTTTCCCCA CAAAGCAGGT GAAACCCCAT TGCAGCAACC CTCCTGCCCT

      2510       2520       2530       2540       2550       2560       2570       2580       2590       2600
GGGGCTGGGA GGATTTCCTT TCTCTGCATG CAGGATTTGG AAAGAGAATT AATTGCAATG GGTCTTATCG ATGCTCCACC ACTACCAAGC AAGCAGAGCA

      2610       2620       2630       2640       2650       2660       2670       2680       2690       2700
GTGGTATGAG CACAACGAAA CCATTCCGAG TGGGAAACAG CAGTTCTGGG GGTGTACVAA CCAACA-AAG GACCTCAACT TTTTGGCAAT CCCAAATAGT

      2710       2720       2730       2740       2750       2760       2770       2780       2790       2800
TTTCCCAGTG TTGGGAAAAG CAGACAATAT A--------- -TGAGCACAC CCACGTTAAG CCCCACCCCA ACCCAACGCA CCAACGTTGG TGCTTTCACC

      2810       2820       2830       2840       2850       2860       2870       2880       2890       2900
ACGTGTTCCC CTCTTGCTCA CACACACAAG AGCAGAAAGC ATTAAAAGAG GCCGTTATAG TGTTGGGGTA CTGTACTT-A C--------- -TGCTGCCCG

      2910       2920       2930       2940       2950       2960       2970       2980       2990       3000
GAGCTTGGCG CAATCCTCGC AAATGGTAGA AACGCTAGAA GGAGAAGGTG AAAGCCGCGG CTTGAGCGCA TCTAGGAGCA AAAAG-GGCT GAAAAGGACG

      3010       3020       3030
GCGCAAAGC CCGCTAAGGA GCGGCTGGAT C
```

Figure 4. Nucleotide sequence of the processed calmodulin gene CM1 and flanking regions.

features: they lack introns, they contain a poly(A) "tail" at their 3' end, and direct repeats of variable length are found at either end. Two of these features are found in the CM1 gene. First, no introns are present, while the "parent" CL1 gene contains numerous introns. Second, a nine base pair direct repeat, GTGCTTCCT, is found in the flanking sequence 5' (position 715) and 3' (position 2,068) of the CM1 gene, as indicated in Fig. 1. Even though no poly(A) tail is present, the 3' 9 bp repeat is found following an A residue 21 bp 3' of the polyadenylation signal AATAAA, which is precisely where the poly(A) tail would normally occur. These features suggest that CM1 is a processed calmodulin gene that originated as a copy of a transcript of the authentic calmodulin gene CL1, and was inserted into a break in a chicken chromosome remote from the position of the CL1 gene.

The CM1 gene sequence presented in Figure 4 encodes a CaML of 148 amino acids that is typical of all CaMs isolated to date. 87% of the amino acid residues are identical to human, rat and bovine CaMs and to the CaM protein encoded by the CL1 gene. The 19 amino acid differences between CaML and these other CaMs are listed in Table 1. 11 of these differences are nonconservative substitutions (indicated by the asterisks between columns). This represents a significant variation because the largest difference among vertebrate CaMs previously analyzed is one amino acid. Even _T. pyriformis_, a protozoan, contains a CaM that differs from human CaM at only 12 positions, 5 of which are nonconservative.

Examination of the amino acid substitutions in the Ca^{2+} - binding subdomains of CM1 suggests the possibility of functional changes in the CaM encoded by this gene. Based on sequence analogies to metal-binding sites proposed by Kretsinger and Barry (22) for parvalbumin and troponin C, Dedman et al. (17) determined the probable metal-binding sites of CaM. These four Ca^{2+} -binding subdomains of CM1 are listed in Table 2. Subdomain 3 is indentical to previously sequenced CaMs, and subdomain 2 is 91% homologous (only one amino acid difference). The single change in subdomain 2 is asp, which is one of the six coordination residues for Ca^{2+}, to a ser, which still could participate in the coordination of a Ca^{2+}. Thus, subdomains 2 and 3 could have fully functional Ca^{2+} -binding sites. Subdomains 1 and 4 are only 75% homologous to the corresponding subdomains in the other vertebrate CaMs, however. Two of the three differences in

TABLE 1
DIFFERENCES IN AMINO ACID SEQUENCE DERIVED FROM CHICKEN CALMODULIN GENES CM1 AND CL1

Pos.	CM1		CL1	Pos.	CM1		CL1
2	GLU		ASP	71	LEU	*	MET
3	ARG	*	GLN	77	ARG		LYS
5	SER		THR	79	SER		THR
21	ARG		LYS	126	LYS		ARG
26	CYS	*	THR	130	CYS	*	ILE
30	MET	*	LYS	131	ASN		ASP
52	VAL	*	ILE	132	ASN	*	GLY
53	GLY	*	ASN	143	ARG	*	GLN
60	SER	*	ASN	147	GLU	*	ALA
70	SER		THR				

subdomain 4 are in critical residues implicated in Ca^{2+}-binding (asp 131 and gly 132), and the third change is the substitution of a bulky cys residue for the adjacent ile 130. In subdomain 1, three residues are replaced by bulkier groups, including the substitution of a cys for thr 26. The presence of a cys residue in both subdomains suggests that a disulfide bond might even be formed that would bridge the central portions of the 1st and 4th Ca^{2+} binding sites. It thus seems possible that this variant CaML might have evolved into a two-domain Ca^{2+}-binding protein.

The nucleotide homologies of the Ca^{2+}-binding subdomains listed in Table 2 also support this hypothesis. Whereas the second and third subdomains of CM1 are very

TABLE 2
CONSERVATION OF Ca^{2+}-BINDING SUBDOMAINS BETWEEN CM 1 AND OTHER CALMODULINS

Subdomain	Amino acids	% Homology Amino acid	Nucleotide
1	20-31	75	64
2	56-67	91	81
3	93-104	100	83
4	129-140	75	61

homologous to the corresponding regions in CL1, subdomains 1 and 4 are considerably less so (64% and 61%, respectively). If these two subdomains no longer functioned as Ca^{2+} - binding sites, then the sequence would be less constrained than subdomains 2 and 3 and evolutionary "drift" would be more rapid. The function of this CaML protein is an intriguing question that is currently under investigation (26).

Repeat Sequences Associated With The CM 1 Gene

Additional interesting features of this gene were noted after close sequence analysis. There is an absence of the sequence elements usually associated with eucaryotic promoters, such as the TATA box. A lack of a functional promoter would partially explain the poor expression of the CM1 mRNA in the muscle tissues in which it is found. We estimate from Northern gels that the CM1 mRNA levels are only about 1/10 the levels of CL1 mRNA in these tissues. This situation resembles the human metallothionine IIB pseudogene (23). Even though this processed gene has an intact coding sequence it is apparently not expressed because it lacks functional promoter sequences near its 5' end. This lack of an effective promoter sequence could perhaps be overridden if a functional enhancer were located nearby. Sequence analysis has revealed a very unusual internally repetitious sequence of approximately 200 bp located 500 bp 5' of the initiator codon ATG. This sequence, which begins at position 352 in Figure 4, is reproduced in Figure 5. The repeat contains over 40 copies of the tetranucleotide sequence TGGA, and can be broken down into the following pattern:

...GAAGAGGGTGATGGATGGATGGATGGATGGATGGATGGATGGATGGATGGATGGATGAAAAAATGGATGGATGGATGG

ATGGATGGATGGATGAAAAAATGGATGGATGGATGGATGGATGGATGGATGGATGGATGGATGGATGGAAAAAATGGATGG

ATGGATGGATGGATGGATGGATGGAAAAAACGGATGGATGGAAAAAATGGATGGATGGACGGATGCATGGATGGTAGGTTG

GGCAACTGGGTGGACGTATGAGTGAATGGTCAGAAGAATGAATGG...(89)...GTGCTTCCT...

Figure 5. The nucleotide sequence of the internally repetitive (TGGA) repeat located upstream of the CM1 gene.

$$\ldots(\text{TGGA})_{12}\text{AAAAA}(\text{TGGA})_{8}\text{AAAAA}$$
$$(\text{TGGA})_{12}\text{AAAAA}(\text{TGGA})_{8}\text{AAAAA}(\text{TGGA})_{3}\text{AAAAA}\ldots$$

Since the trinucleotide TGG is found in the consensus eucaryotic enhancer sequence (24), we are currently testing the possibility that this sequence has enhancer-like activity for transcription of the CM1 gene.

A different class of repeat sequence, a middle-repetitive chicken repeat designated CR1, is found about 6 kb 3' of the CM1 gene (25). Several CR1 sequences are found in tandem surrounding a number of expressed gene families, oriented towards the expressed genes that they surround. These sequences are presumed to play a role in the establishment of an open chromatin region around expressed chicken genes. Three CR1s are located as shown in Figure 1. We have undertaken a 5' genewalk to locate CR1s that might define the 5' boundary of an open chromatin region.

CONCLUSIONS

The data presented here indicates that the chicken calmodulin gene CM1 probably originated as a processed copy of the primordial CL1 gene. We hypothesize that this processed copy was inserted into a distant chromasomal site fortuitously located within a transcriptionally open chromatin domain containing several muscle-specific genes. The CM1 gene has since evolved to code for a CaML protein that has an unknown function in striated muscle tissue.

This hypothesis makes several testable predictions. Obviously, the CaML product must be expressed in cardiac and skeletal muscle, and it must have a demonstrable function. As a first step towards testing this, expression plasmids containing the CM1 coding sequence have been constructed and cloned to produce large amounts of a synthetic CaML protein, which is being functionally characterized in several laboratories (26).

Second, we are presently assessing the enhancing activity of the TGGA repeat. A 583 bp restriction fragment that contains the TGGA repeat has been isolated from the 5' flanking region and cloned into a chloramphenicol acetyltransferase (CAT) expression plasmid that was designed for testing enhancer sequences (24). The protocol for the constructions is outlined in Figure 6. The TGGA repeat has been inserted both 5' and 3' of the CAT gene in

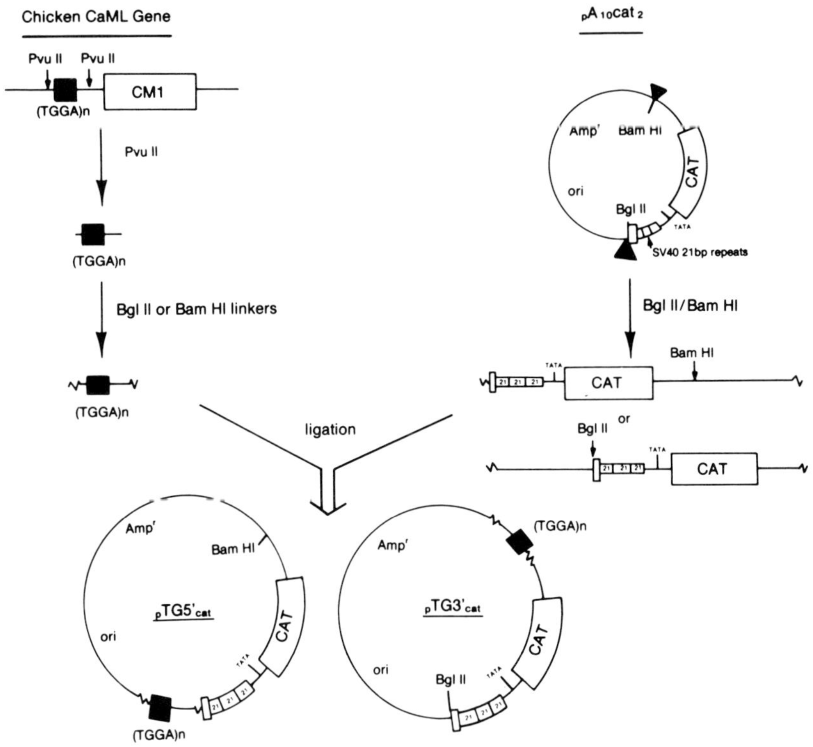

Figure 6. Protocol for the construction of CAT expression plasmids for testing the enhancer activity of the TGGA repeat.

the enhancer-less plasmid, creating two new plasmids designated pTG5'cat and pTG3'cat respectively. Since eucaryotes lack a CAT enzyme, it is a relatively straightforward matter to assay eucaryotic cells containing CAT plasmids for CAT activity. The enhancing activity of the TGGA repeat will be assayed in Xenopus oocytes. If our hypothesis is correct, it will also be necessary to transfect avian muscle cells with the pTGcat plasmids.

Finally, the hypothesis predicts that a set of tandem CR1 repeats are located a finite distance upstream of the CM1 gene, and that these 5' and 3' sets of CR1s then define a transcriptionally open region of chicken chromatin containing several genes that are expressed in chicken muscle. My laboratory is currently engaged in a 5' walk from the CM1 gene, in an attempt to locate a set of 5' CR1 repeats. Once the 5' CR1s are found and mapped, a search will be initiated for other open reading frames between the 5' and 3' CR1s that are expressed in chicken muscle tissue.

ACKNOWLEDGEMENTS

I am grateful to Gerri Hanten and Jamie Allen for excellent technical assistance, and to Dr. Tony Means for his sustained intellectual support.

REFERENCES

1. Means, A.R., Tash, J.S. and Chafouleas, J.G. (1982) Physiol. Rev. 62:1-39.
2. Chafouleas, J.G., Dedman, F.J., Munjaal, R.P. and Means, A.R. (1979) J. Biol. Chem. 254: 10262-10267.
3. Watterson, D.M., Sharief, F. and Vanaman, T.C. (1980) J. Biol. Chem. 255 962-975.
4. Means, A.R. (1981) Recent Prog. Horm. Res. 37:333-367.
5. Klee, C.B. and Vanaman, T.C. (1982) Adv. Protein Chem. 35:213-321.
6. Chafouleas, J.G, Pardue, R.L., Brinkley, B.R., Dedman, J.R. and Means, A.R. (1981) Proc. Nat. Acad. Sci. USA 78:996-1000.
7. Chafouleas, J.G., Bolton, W.E., Hidaka, M., Boyd, A.E. and Means, A.R. (1982) Cell 28:41-50.
8. Munjaal, R.P., Chandra T., Woo, S.L.C., Dedman, J.R. 78:2330-2334.
9. Legace, L., Chandra, T., Woo, S.L.C. and Means, A.R. (1983) J. Biol. Chem. 258:1684-1688.
10. Putkey, J.A., Tsui, K.F., Tanaka, T., Legace, L., Stein, J.P., Lai, E.C. and Means, A.R. (1983) J. Biol. Chem 258:11864-11870.
11. Stein, J.P., Munjaal, R.P., Legace, L., Lai, E.C., O'Malley, B.W. and Means, A.R. (1983) Proc. Nat. Acad. Sci. USA 80:6485-6489.
12. Benton, W.D. and Davis, R.W. (1977) Science 196:180-182.
13. Simmen, R.C.M., Tanaka, T., Tsui, K.F. Putkey, J.A., Scott, M.J., Lai, E.C. and Means, A.R. (1985) J. Biol. Chem. 260:907-912.
14. Southern, E.M. (1975) J. Mol. Biol. 98:503-517.
15. Lai, E.C., Stein, J.P., Catterall, J.F., Woo, S.L.C., Mace, M.D., Means, A.R., and O'Malley, B.W. (1979) Cell 18:829-842.
16. Kasai, H., Kato, Y., Isobe., I., Kawasaki, M. and Okuyama, T. (1980) Biomed. Res. 1:248-264.
17. Dedman, J.R., Jackson, R.L., Schreiber, W.E. and Means, A.R. (1978) J. Biol. Chem. 253:343-346.

18. Sasagawa, T., Ericsson, L.M., Walsh, K.A., Schreiber, W.E., Fischer, E.M. and Titani, K. (1982) Biochem. 21:2565-2569.
19. Thomas, P.S., (1980) Proc Nat. Acad. Sci. USA 77:5201-5205.
20. Maxam, A. and Gilbert, W. (1977) Proc. Nat. Acad. Sci. USA 74:560-564.
21. Sanger, F. and Coulson, A.R. (1977) FEBS lett. 87:107-110.
22. Kretsinger, R.M. and Barry, C.D. (1975) Biochem. Biophys. Acta 405:40-52.
23. Karin, M. and Richards, R. (1982) Nature 299:797-802.
24. Rosenthal, N., Kress, M., Gruss, P. and Khoury, G. (1983) Science 222:797-755.
25. Stumph, W.E., Hodgson, C.P., Tsai, M.J. and O'Malley, B.W. (1964) Proc. Nat. Acad. Sci. USA 81:6667-6671.
26. Putkey, J.A. Slaughter, F.R. and Means, A.R. (1985) Cell (in press).

Molecular Biology of Muscle Development, pages 487–506

THE STRUCTURE AND ORGANIZATION OF ACTIN AND MYOSIN GENES AND THEIR EXPRESSION IN MOUSE STRIATED MUSCLE[1]

Benoît Robert, Paul Barton, Serge Alonso, Arlette Cohen, Philippe Daubas, Ian Garner, André Weydert and Margaret Buckingham

Department of Molecular Biology, Pasteur Institute
75015 Paris, France

ABSTRACT During the initial formation of skeletal muscle fibres and their subsequent maturation, different isoforms of the contractile proteins accumulate. We demonstrate that the foetal myosin light chain $MLC1_{emb}$ is encoded by the same genomic fragment as the cardiac atrial light chain $MLC1_A$, and that the adult slow light chain $MLC1_S$ is a product of the same gene as the cardiac ventricular light chain $MLC1_V$. Analysis of the mRNAs and proteins suggests that these are identical in both cases. We also demonstrate that not only the cardiac actin mRNA, but also the protein accumulates in developing skeletal muscle. Using a genetic approach, we have looked at the linkage relationships and chromosomal localisation of actin and myosin genes in the mouse. Genes expressed in the same muscle phenotype are not linked. The actin and myosin alkali light chains are encoded by dispersed multigene families. Cardiac and skeletal myosin heavy chain genes are also not linked. However the MHC genes expressed sequentially during the development of skeletal muscle are clustered at the same chromosomal locus. $MLC1_F$ and $MLC3_F$ are encoded by a single gene, two mRNAs are generated by initiation at different 5' promoters followed by differential splicing of exons coding for the distinct NH_2 termini of the two

[1]This work was supported by grants from the MRT, CNRS, AGIR and the Muscular Dystrophy Association (MDA) of America. I.G. is the recipient of a grant from the Royal Society of Great Britain, P.B. from the MDA and recently from EMBO.

proteins. Comparison of these genes between chick, rat and mouse indicates the presence of conserved sequences in the 5' promotor region which are probably of functional significance. The 5' promotor region of the α-cardiac actin gene has been sequenced. A $(TG)_{24}$ stretch, potentially capable of forming Z DNA is found upstream of the CAAT box. In some *Mus 1* mouse strains, the whole 5' end of this gene is duplicated.

INTRODUCTION

Myogenesis involves the expression of many genes, which encode the proteins required for muscle activity. Among these are the components of the contractile apparatus (such as actin, myosin, tropomyosin, troponins, etc.), enzymes for the muscle metabolism (such as the M-creatine phosphokinase) and membrane proteins such as the acetyl choline receptor (for review, see (1)). The expression of these different genes must be correlated and finely regulated to insure the accumulation of the muscle proteins in the proper stoechiometry, in order to elaborate a functional sarcomeric structure. The first event of muscle formation is the fusion of myoblasts into multinucleated myotubes. For the skeletal muscle fibre of mouse, this process is first detectable *in utero* at the 10/11th day when limb buds are beginning to form from the mesoderm-derived somites of the embryo (2). This process can take place *in vitro*, where primary myoblasts from foetal muscle or myoblasts from permanent cell lines will fuse when they reach confluence. Therefore, the accumulation of muscle proteins during this initial transition has been studied in great detail. It appears that these proteins are synthesized in a co-ordinate way at the onset of fusion. Evidence has accumulated, especially since cloned probes have been available, that this phenomenon is regulated primarily at the transcriptional level (for review, see (1)). Therefore, myogenic differentiation of cultured myoblast provides a tool to investigate the molecular mechanisms which activate the transcription of different genes to generate a defined phenotype.

In vivo, further transitions are observed. The contractile proteins exist as different isoforms, therefore encoded by multigene families (for review, see (3)). The multiple genes in these families probably arose from repeated duplications of ancestral genes, and have acquired during evolution a tissue specific range of expression. In addition, evidence has accumulated more recently that different isoforms of the same protein may be expressed at specific stages of muscle development (4,5,6). Therefore, after the initial transition from myoblast to myotube, which is accompanied by the expression of the earliest foetal phenotype, the expression of one gene within a multigene family may be substituted by that of another one, in keeping with the stoechiometric requirement of muscle structure.

In order to describe these transitions and to investigate the genes concerned with a view to their manipulation in muscle cells, we have cloned a series of recombinant probes for different myosin and actin sequences (7,8,9). We report here further studies on the interplay in gene expression between the skeletal and cardiac, actin and myosin alkali light chain genes, in striated muscle. Structural features of these genes which may be implicated in their regulated expression are discussed.

THE EXPRESSION OF ACTIN AND MYOSIN GENES IN STRIATED MUSCLE

During the maturation of mammalian cardiac and skeletal muscle tissues, the actin and myosin multigene families follow different developmental strategies (for review, see (1)). Thus specific developmental isoforms of the myosin heavy chains are expressed sequentially in skeletal muscle (e.g. (5)) and in the mouse, for example, transcripts of foetal myosin heavy chain mRNA are replaced by those encoding the protein of adult fast skeletal muscle, at birth (9). The actins and myosin alkali light chains, on the other hand, do not have isoforms that are only accumulated at a specific developmental stage. Instead, in foetal skeletal muscle, proteins of adult cardiac tissue co-accumulate with the adult skeletal isoforms. Thus for the actins, transcripts of the cardiac actin gene are found in late foetal and new born skeletal muscle (6), and conversely transcripts of the skeletal actin gene co-accumulate with those of the cardiac isoform in foetal hearts (e.g. (10)). However there was no evidence that the corresponding proteins were accumulated. It is now clear that this phenomenon pertains also at the protein level. Peptide analyses, which distinguish between the two isoforms, demonstrate that cardiac actin is present in foetal (18 days) mouse skeletal muscle (43% of striated muscle actins), and in myotubes of the mouse muscle cell line, T984 C110 (36% of striated muscle actins) (11). These figures correspond approximately to those estimated for the mRNA (6). Interestingly, significant quantities (16-18%) of skeletal actin protein are also detected in adult pig and bovine hearts (11). This has also been reported at the mRNA level in adult human hearts (ca. 50%) (12), and may be a feature of larger mammals, since in adult rat heart, for example, no significant quantities of skeletal actin mRNA are detected (10). A distinction between the isoforms expressed in hearts of large and small mammals has already been noted for the myosin heavy chains (13).

In the case of the myosin alkali light chains it has been shown that an isoform, $MLC1_{emb}$, found in mammalian

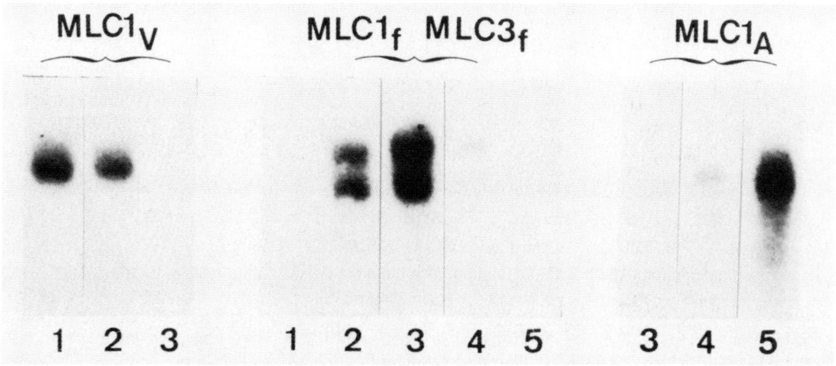

Figure 1. Northern blot hybridization of DNA probes encoding ventricular myosin light chain $MLC1_V$, atrial myosin light chain $MLC1_A$, and the fast skeletal muscle myosin light chains $MLC1_F$ and $MLC3_F$, to RNA from various sources. Track 1; Ventricular muscle RNA, hybridizing to the $MLC1_V$ probe only. Track 2; Soleus muscle, which contains a mixture of fast and slow fibres, hybridizes with $MLC1_V$ and the $MLC1_F/MLC3_F$ probes. Track 3; Adult fast skeletal muscle RNA shows hybridization with the $MLC1_F/MLC3_F$ only. Track 4; 18 day foetal skeletal muscle contains $MLC1_F$ mRNA with no detectable $MLC3_F$ mRNA (centre panel), and contains $MLC1_A$ sequence (right panel). Track 5; Adult atrial muscle RNA hybridizes with the $MLC1_A$ probe only. Note that the mRNA encoding these different isforms are distinguishable on the basis of size.

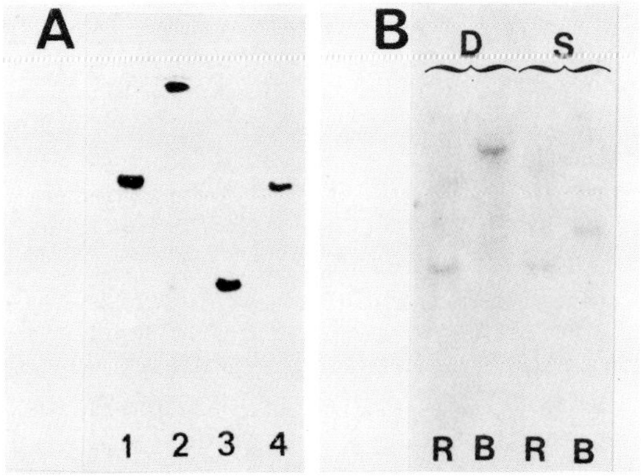

Figure 2. Southern blot hybridization to mouse genomic DNA. (A): Hybridization of $MLC1_V/MLC1_S$ probe to Mus musculus (DBA/2) DNA digested with 1: XbaI, 2: BclI, 3: BamHI, 4: BglI. (B): Hybridization of $MLC1_A/MLC1_{emb}$ probe to DNA from Mus musculus (DBA/2) "D" and Mus musculus "S" digested with EcoRI: R and BamHI: B. Note the restriction fragment length polymorphism for this gene using Bam HI.

foetal skeletal muscle (4) and in foetal hearts (ventricles and atria) (14,15) co-migrates with the major isoform of adult heart atria ($MLC1_A$) on two dimensional gels. Partial peptide analyses also detect no differences between the foetal cardiac and adult atrial isoforms (15,16), although differences have been reported with the foetal skeletal isoform (17) possibly reflecting differences in secondary modification. We have now isolated a recombinant plasmid, containing part of the mRNA sequence of $MLC1_A$, from a cDNA library prepared with mRNA from new born mouse hearts (18). This recombinant has been shown to hybridize to an mRNA encoding a protein that co-migrates on two dimensional gels with the $MLC1_{emb}/MLC1_A$ isoforms. When used as a probe on Northern blots, this plasmid detects an mRNA of 930 nucleotides present in adult cardiac atria and foetal skeletal muscle (Fig.1). Southern blotting experiments with mouse genomic DNA demonstrate the presence of a single restriction fragment, homologous to the recombinant cDNA plasmid. Furthermore, analysis of restriction length polymorphisms for this fragment between two mouse species (Fig.2) and of their segregation in back-crosses (Table I) would suggest that only one locus is present. The recombinant cDNA plasmid, and the mRNA and genomic sequences detected with it, cross-hybridize with the -COOH common coding sequence of the $MLC1_F/MLC3_F$ gene (19), under conditions of low stringency. We therefore conclude that the same genomic locus encodes both $MLC1_A$ and $MLC1_{emb}$ (18). We cannot exclude at present the possibility that slightly different mRNAs are generated from the same genomic fragment, however the size and sequence homology of the $MLC1_A$ and $MLC1_{emb}$ mRNAs suggest this is unlikely. The complete structure of the gene, which we have isolated and are currently sequencing will resolve this question. It will also provide the first amino acid sequence data for the protein. In the region already sequenced (amino acids 74-92 and 140-158 in the $MLC1_F$ sequence), only one amino acid substitution out of thirty-six has occurred between $MLC1_A$ and $MLC1_F$, compared to 6 substitutions over the same regions between the chicken $MLC1_V$ and $MLC1_F$ sequences (20). This, together with the data on cross-hybridization of the coding and 3' non-coding sequences of the mRNAs (18,8) strongly suggests that $MLC1_A$ has evolved recently from $MLC1_F$. Since there is no evidence for an $MLC1_A$ isoform in birds, this may have happened during mammalian radiation.

In late foetal (18 day) skeletal muscle, the mRNAs for $MLC1_A/MLC1_{emb}$ and $MLC1_F$ co-accumulate. The mRNA for $MLC3_F$ is not detectable at this stage in the mouse (Fig.1 and (18)). The absence of the $MLC3_F$ protein in foetal skeletal muscle has been reported in a number of species (e.g. (21,22)). The fact that only $MLC1_F$ transcripts (and protein) accumulate is of particular interest in view of the fact that both $MLC1_F$ and $MLC3_F$ proteins are generated from the same gene (see later).

During skeletal muscle development, the timing of the accumulation of $MLC1_{emb}$ mRNA in relation to $MLC1_F$ mRNA differs from that of the cardiac and skeletal actin mRNAs. Significantly more cardiac actin mRNA is present after birth (6), while the $MLC1_{emb}$ mRNA rapidly disappears (our unpublished observations) suggesting that cardiac actin and $MLC1_{emb}$ genes are responding to different developmental signals. Differences in expression of these isoforms are also seen in myogenic cell lines in culture; in the mouse muscle cell line T980-C110, both cardiac and skeletal actins, and $MLC1_F$ and $MLC1_{emb}$ light chain genes are co-expressed on myotube formation (6,11,23, our unpublished observations), whilst in the L6 rat cell line only the

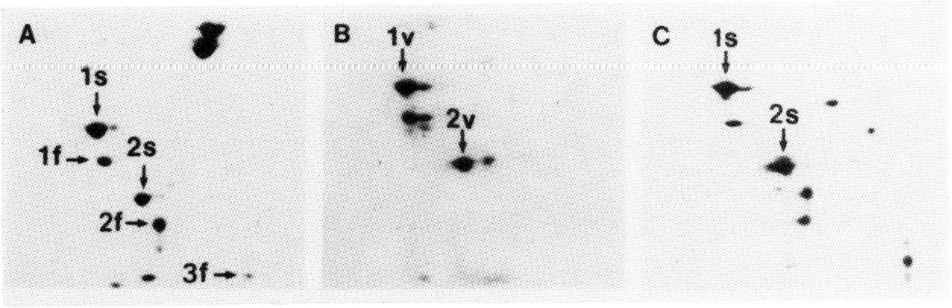

Figure 3. 2-dimensional gel analysis of ventricular and soleus muscle myosin light chain proteins. A: Total extract from adult mouse soleus muscle. Positions of the myosin alkali light chains $MLC1_S$, $MLC1_F$, and $MLC3_F$ are indicated 1s 1f and 3f. The positions of $MLC2_S$ and $MLC2_F$ are also shown (2s and 2f). B and C: Comigration of ventricular and soleus muscle myosin light chains, B: Coomassie blue stained gel showing ventricular proteins, C: autoradiograph of the same gel showing the position of the ^{35}S-labelled proteins produced by *in vitro* translation of soleus muscle RNA.

skeletal actin and $MLC1_{emb}$ gene transcripts are detectable (8,18 and unpublished observations). This raises the intriguing possibility that the L6 line may be analogous to an earlier type of myoblast population (24), while T984-C110 cells would resemble the myoblasts of late foetal muscle.

There is no evidence for the expression of the $MLC1_F$ protein in mammalian cardiac tissue. However the major myosin alkali light chain of adult heart ventricles, $MLC1_V$, has been shown to be very similar by one (e.g. (25) and two (e.g. (26)) dimensional gel electrophoresis and partial peptide analysis (27) to the skeletal slow muscle isoform $MLC1_S$. We have demonstrated that in the mouse $MLC1_V$ and $MLC1_S$ are indistinguishable on two dimensional gels (Fig.3). Cloning of part of the $MLC1_V$ mRNA sequence in a recombinant plasmid has permitted us to investigate the corresponding mRNA and gene. The identity of the recombinant, isolated from a cDNA library prepared with mRNA from new born mouse hearts, was demonstrated in hybridization/translation experiments (46). The cloned cDNA sequence hybridizes to a cardiac ventricular mRNA of 1000 nucleotides. A messenger of the same size and sequence homology is also present in the mouse soleus (Fig.1), a slow skeletal muscle. This muscle contains a mixture of fast and slow fibres and, as expected, mRNA from this source also hybridizes to the $MLC1_F/MLC3_F$ probe (Fig.1). In vitro translation experiments with soleus mRNA demonstrate the synthesis of the $MLC1_S$ isoform, and its co-migration with $MLC1_V$ (Fig.3). In mouse genomic DNA, the $MLC1_V$ probe detects only a single restriction fragment with several enzymes (Fig.2). The analysis of restriction fragment length polymorphisms between mouse species, and their segregation in back-crosses (see Table I), demonstrates that only one genetic locus, homologous to this sequence, is present. These results, together with those on the mRNAs, suggest that only one genomic fragment encodes $MLC1_S$ and $MLC1_V$. We have isolated a recombinant phage containing the $MLC1_V$ gene, and have shown by analysis of R-loops formed between this sequence and $MLC1_V$ mRNA that only a single gene is present in this fragment, thus excluding the possibility of a tandem gene arrangement. We therefore conclude that the same gene encodes $MLC1_S$ and $MLC1_V$ (46). As in the case of the $MLC1_{emb}/MLC1_A$ gene, DNA sequencing should establish whether identical transcription products are generated in cardiac and slow skeletal muscles.

STRUCTURAL ANALYSIS OF MYOSIN AND ACTIN GENES.

The gene encoding myosin light chains $MLC1_F$ and $MLC3_F$.

The two isoforms of the alkali myosin light chains which are expressed in adult fast skeletal muscle, $MLC1_F$ and $MLC3_F$, are encoded by a single gene. This was first suggested by the sequence identity of the 141 COOH-terminal amino-acids of the two proteins (28), which is conserved during evolution (29). This identity is further retained over the 3' non-coding region of the corresponding mRNAs, as shown from cDNA cloning (8,30). In the laboratory mouse, Mus musculus, two loci are detected which are highly homologous to such a cDNA. We have isolated recombinant λ phages representing these two loci (19). One of them is a processed pseudogene, which is absent from some mouse species, such as Mus spretus. Sequencing of the functional locus shows that the common part of the two mRNAs is encoded by five exons, and the specific parts by four exons, which span 21 kb of DNA. The NH_2-terminal peptide for $MLC1_F$ (38 amino-acids) is encoded by two exons, separated by a 10 kb intron within which two exons are found which encode the

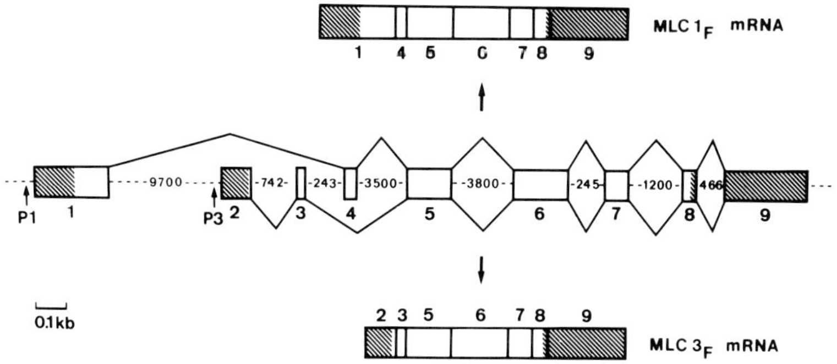

Figure 4. Diagrammatic representation of the $MLC1_F/MLC3_F$ gene, showing the pattern of splicing required to generate the two mRNAs. Exons are numbered and drawn to scale. Intron size is indicated in base-pairs. Dashed areas represent non coding regions of the mRNAs.

NH_2-terminal peptide of $MLC3_F$ (9 amino-acids) (Fig.4). This implies a differential splicing to generate the mRNAs for the two proteins. S1 protection and primer-extension experiments identify distinct cap sites for the mRNAs of $MLC1_F$ and of $MLC3_F$ (19). In the case of the $MLC3_F$, there are, in close proximity to the main cap site, multiple secondary sites for the initiation of the mRNA (31). Upstream of the cap sites, the canonical CAAT and TATA sequences are found. This suggests that transcription is initiated from distinct promotors to generate the pre-messenger RNAs for $MLC1_F$ and $MLC3_F$, each of which would then undergo specific splicing.

As the sequence of the same gene is available for the chicken (32) and for the rat (33), we have undertaken sequence comparison between the three species to detect structural features that have been conserved during evolution and may therefore be involved in the expression of this complex gene (31). The overall structure of the gene, especially the number of introns and their position in the coding sequence is conserved between the three species. Interestingly, intron size is also conserved at the 5' end of the genes, where differential splicing takes place. This is particularly striking in the case of intron III, the size of which (about 240 nucleotides) differs only by three

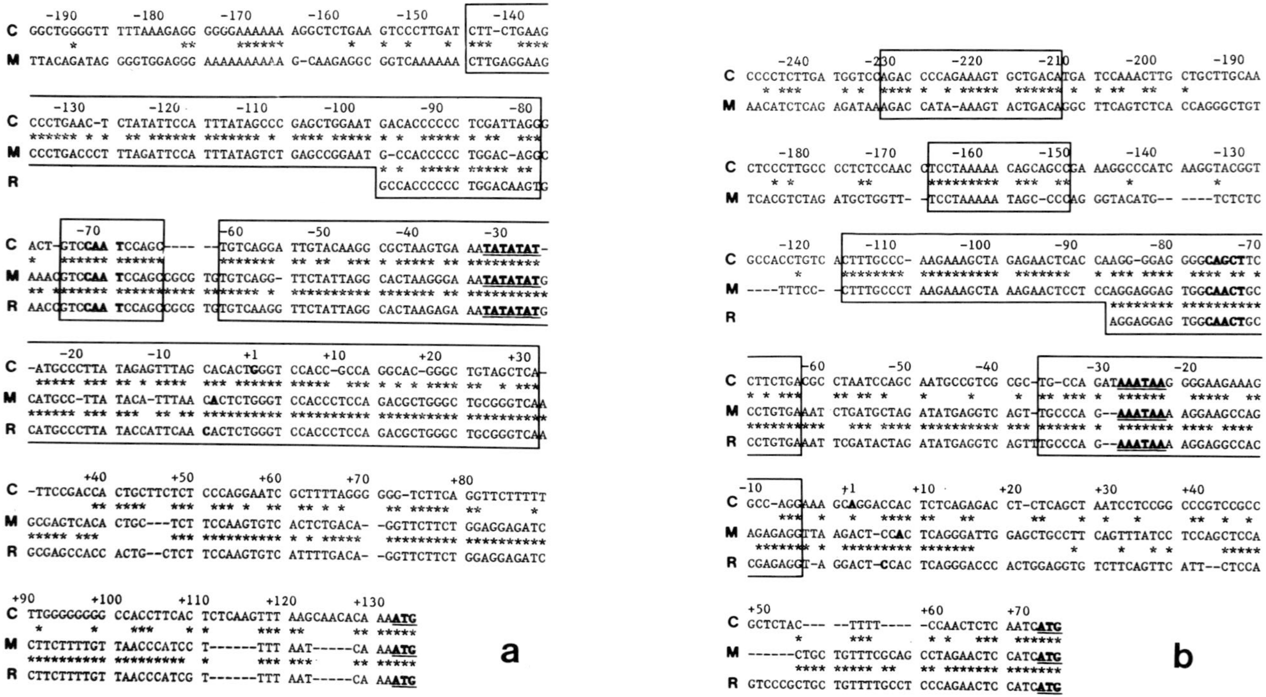

Figure 5. Sequence comparison between the promotor regions of chicken, mouse and rat a) MLC1$_F$ mRNA and b) MLC3$_F$ mRNA. The CAAT boxes, TATA boxes, cap sites and initiation codons are in bold type. Region of homology between the chicken(C), mouse(M) and rat(R) genes are boxed. Numbering starts from the cap sites of chicken mRNAs.

nucleotides between the three genes. This suggests a precise structural requirement for the pre-mRNA in this region for the differential splicing of the exons which generate the two mRNAs. No significant feature which might act as a specific splice signal, in addition to the universal consensus sequences (34), can be detected between genes.

Examination of the potential 5' upstream promotor regions for $MLC1_F$ and $MLC3_F$ transcription indicates that highly conserved sequences are found around the cap sites of the two mRNAs (fig.5a). In the case of the $MLC1_F$ mRNA promotor region, the whole sequence between positions -145 and +30, relative to the cap site, is more than 75% homologous between the mouse and the chicken genes. This is to be compared to the 79% homology noted for the coding regions, and the 30% homology noted for the 3' non coding regions of the corresponding mRNAs. Within this conserved region, three blocks of higher homology can be defined, the first one including the TATA box and the cap site, the second one around the CAAT box, and the third one between positions -80 and -150. The same type of comparison reveals similar, although shorter , blocks of conserved sequences around the initiation region of the $MLC3_F$ mRNA (fig.5b).

These blocks of homology are almost as well conserved as the coding regions of the gene, on which a selective pressure is exerted during evolution. We therefore conclude that they must be of functional importance for the expression of the MLC gene. No significant sequence homology is detected between the 5' regions of the $MLC1_F$/$MLC3_F$ gene and those of other contracile protein genes coding for rat α-skeletal actin (35), rat $MLC2_F$ (36) or chicken α-skeletal actin (37), either within the conserved blocks of homology or outside them. These genes are co-expressed within the same muscle phenotype, and one might have expected a common signal sequence in their 5' promotor regions, which would play a role in co-ordinating their activation. However, one dimensional structural analysis of these genes gives no clear indication for a co-ordinating mechanism.

The α-cardiac actin gene.

In the genome of some inbred strains of mice (e.g. BALB/c), a 3.5 kb DNA fragment, containing the promotor region and the first three exons of the α-cardiac actin gene is duplicated (Fig.6a). In other strains (e.g. C3H), this is not the case, but nevertheless the gene is expressed both in

a

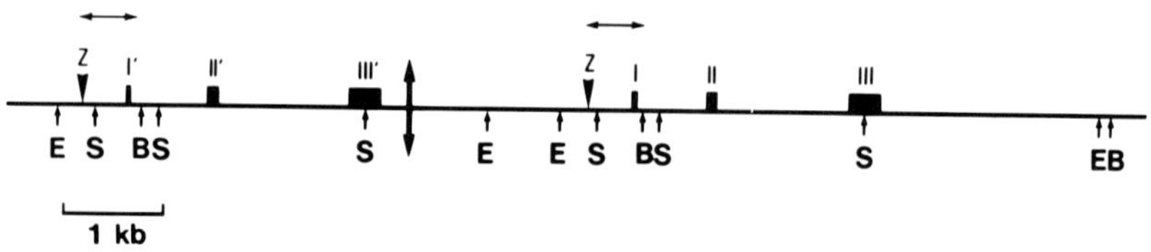

b

GGCGCCAAGTCTTCCGGCATTGTGTGTGTGTGTGTGTGTGTGTGTGTGTGTGTGTGTGTG

TGTGTGTGACTCATTGTCCCTTAGTTTGGAAGGGCTGAAGAGCAATAAGCCCACTCCACA
100

ACTAGGGAGCTCCCCCACCCAAGGGGCGCATTGGCATCACATAGCCTTTCCCCGTCCCCC

ACCCCTTGCTGGCCTGCCCCTCCCTAGCTCCCTATATGGCCATTGCTCTGACTGCCCCCT
200

CCCCTTCCTTACATGGTCTGGGAGCCCCCTGGCTGATCCTCTACCCTGCCCTTGGCTCCA
300

AGAATGGCCTCAGCGGTCCTAGATGGTGCTAAGGCGA**CCAAAT**AAGGCAAGGTGGCAGAT

CAGGGGCCCCCCACCCCTGCCCCCGGCTGCTCCAACTGACCCCGTCCATCAGAGAGC**TAT**
400

AAAGCTGCGCTCCAGGCGACTGAC**ACCCAGTGCCTGCCACCAGCGCCAGCCCAGCTGAAT**

CCAGCCGCCCCTAGCACGGTGAGTCCCAGCCTTGCTCCCTGCAGGACCTTGTCAGCACTG
500

Figure 6. Duplication of the promoting region of the α-cardiac actin gene in BALB/c mouse. a) Structure of the duplicated region. Black boxes represent the exons. The vertical arrow (↕) shows the duplication point, the horizontal arrows (↔) represent sequenced areas. The arrowhead ▼ points to the Z DNA. E=EcoRI; B=BamHI; S=SacI. b) The sequence of the promotor region and of its duplication. The putative CAAT box, TATA box, cap site and splice signal are underlined. The $(TG)_{24}$ sequence, capable of forming Z DNA, is boxed.

cardiac tissue and in developing skeletal muscle. Therefore, the duplication is not functionally significant in this respect. The promotor region and the first three exons have been sequenced both in the actual gene and in the duplication. The first exon encodes a leader sequence for the mRNA, as in the case of the α-skeletal actin gene (35), and has been identified by analogy with the human α-cardiac actin gene, where the cap region of the mRNA is characterised by S1 mapping (Adrian Minty, personal communication). In front of the putative cap site (fig.6b), the canonical sequences CAAT and TATA are present. In addition, 370 bp upsteam of the cap in the mouse cardiac actin gene, a $(TG)_{24}$ stretch is found, which can potentially form a left-handed DNA structure of the Z type. Such a structure may play a regulatory role in the expression of this gene.

The sequence of the promotor region and of the first three exons is identical in the gene and in the duplication. Experiments are in progress to determine whether the duplicated promotor is functional, and what is the fate of a transcript initiated at this promoter. If the duplication is not functional, this might point to factors which are required, in addition to sequence characteristics to define an eukaryotic promotor.

At the duplication point, there is a sequence CACACA which resembles the sequences involved in rearrangement within the immunoglobin genes (38), and the conversion site proposed for the γG-γA globin genes (39).

Genetic organisation of the actin and myosin multigene families.

As already discussed, many genes belonging to different multigene families must interact to achieve a functional muscle fiber; members of different multigene families are activated co-ordinately at the onset of myoblast fusion, while members of the same multigene family, of common evolutionary origin, may be sequentially expressed during developmental transitions. To investigate whether these intricate events necessitate a specific gene organisation, we and others (40,41,42) have undertaken a linkage analysis of the genes for contractile proteins.

Our approach has made use of the Mendelian segregation of alleles of these genes. The alleles were defined as restriction fragment length polymorphisms (RFLPs) detected using our cloned actin and myosin probes between different mouse strains. As such RFLPs are often difficult to detect between inbred strains of the laboratory mouse, Mus musculus, we used interspecific back-crosses between Mus musculus and a distantly related mouse species, Mus spretus. These two species can be interbred under laboratory conditions. Due to the evolutionary distance between them, they exhibit polymorphism at many more loci than do laboratory strains of Mus musculus, for example.

The F1 from a cross between Mus musculus (DBA/2) and Mus spretus (SPE/Pas) were back-crossed against DBA/2, and 42 offspring analysed (43). In addition to those contractile protein genes for which we had cloned probes, 21 previously localised markers, covering half the mouse genome, were followed in the same cross for chromosomal assignment. This analysis enabled us to propose a location for all the genes examined, except the actin genes (43).

Results summarized in table 1 show that there is no linkage between the genes expressed in an adult phenotype (those for $MLC1_F/MLC3_F$, α-skeletal actin, MHC_{Ad}) or in an adult cardiac phenotype (those for $MLC1_V$, α-cardiac actin, MHC_{card}). Similarly, no linkage is observed for genes expressed together in developing striated muscle, such as $MLC1_{emb}$ and α-cardiac actin. The genes from the actin and the myosin alkali light chain multigene families are dispersed in the mouse genome (43), whereas the genes for the myosin heavy chains are clustered at two distinct loci (Weydert et al., manuscript in preparation). At one locus, located on chromosome 11 (43), are found the genes expressed sequentially during skeletal muscle development (see 5,9), namely those encoding the foetal isoform (MHC_{emb}), the perinatal isoform (MHC_{neo}) and the adult isoform (MHC_{Ad}), detected with cloned mouse cDNA probes specific for each of these sequences (Weydert et al., 1983; our unpublished results).

The second myosin heavy chain locus contains a gene homologous to a cloned mouse cardiac (α) myosin heavy chain probe. Given the close physical proximity of the α and β cardiac genes in the rat (44), this locus is likely to contain both cardiac myosin heavy chain genes. No co-segregation is seen between this locus and the skeletal myosin heavy chain genes nor with the genetic markers on mouse chromosome 11. Instead the cardiac myosin heavy chain gene(s) co-segregate with Np1, on mouse chromosome 14, and we therefore conclude that they are unlinked to the skeletal

TABLE I
LINKAGE ANALYSIS OF MYOSIN AND ACTIN GENES IN THE MOUSE

n°	1	2	3	4	5	6	7	8	9	10	11	12	13	14	15	Total n° cosegregant	Chr.
$MLC1_F/MLC3_F$	-	+	+	+	+	-	-	-	-	+	+	-	-	+	+	35/37	1
Idh-1	-	+	+	+	+	-	-	-	-	+	+	-	-	+	+		
$MLC1_V$	-	-	+	-	+	+	+	+	-	+	+	-	-	-	-	37/37	9
Es-14	-	-	+	-	+	+	+	+	-	+	+	-	-	-	-		
$MLC1_{emb}$	-	+	-	-	-	+	-	-	+	-	+	+	+	+	-	34/37	11
Es-3	+	+	-	-	-	+	-	-	+	-	+	+	+	+	-		
MHC_{emb}	N	+	+	-	+	-	+	-	+	+	+	+	+	N	-	40/40	11
MHC_{Ad}	-	+	+	-	+	-	+	-	+	+	+	+	+	-	-	42/42	
MHC_{neo}	-	+	+	-	+	-	+	-	+	+	+	+	+	-	-	21/42	
MHC_{card}	-	-	-	-	+	+	-	-	+	+	+	-	-	+	+	36/37	14
Np-1	-	-	-	-	+	+	-	-	+	+	+	-	-	-	+		
α-skel actin	+	+	-	+	-	-	+	-	+	-	+	+	-	-	-	26/42	(3)
α-card actin	+	-	-	-	+	-	+	-	+	-	+	+	-	+	+		(17)

Forty-two offspring of a back-cross between Mus musculus and Mus spretus were analysed for the segregation of myosin and actin genes, together with previously localised markers. The table represents the results for the first 15 F2 animals. They are scored as homozygous for the musculus allele (-) or heterozygous (+). The results for the total sample are summarised at the end of each line, together with chromosomal assignments. α-skeletal and α-cardiac actin genes do not co-segregate. Chromosomal assignment for them is from (40).

myosin heavy chain genes, and are on a different chromosome. Previous localisation of sarcomeric myosin heavy chain genes in the mouse (40,42) and human (42), carried out with somatic cell hybrids and probes for the skeletal isoforms only, placed them all on the same chromosome (11 in the mouse and 17 in human). From this genetic analysis we conclude that sarcomeric myosin heavy chain genes expressed during the developmental maturation of a muscle are clustered within a single locus and are likely to be physically linked. It would be interesting to establish whether their organisation is like that seen for the α or β globin gene clusters, where the genes are arranged in order of their developmental expression (45).

CONCLUSIONS

Genetic engineering has provided new tools to study the complex interactions that regulate a number of genes implicated in the elaboration of a functional muscle structure. Isolation of gene sequences and genetic analysis point to features which are significant for the expression of these genes. Further progress in the understanding of these mechanisms requires a functional analysis. The reintroduction of cloned genes into differentiating myoblasts, to investigate the function of sequence elements by site-directed mutagenesis, and to isolate the cellular factors that interact with these elements to regulate gene expression, provides such an approach.

ACKNOWLEDGEMENTS

Parts of this work (see reference list) were carried out in collaboration with Drs. Marc Fiszman, Joël Vandekerckhove, Adrian Minty, Jean-Louis Guénet and François Bonhomme. We are grateful to Prof. F. Gros for support and encouragement.

REFERENCES

1. Buckingham ME (1985). Actin and myosin multigene families : their expression during the formation of skeletal muscle. In Campbell PN, Marshall RD (eds):"Essays in Biochemistry", New York : Academic Press, Vol 20, p 77.
2. Ontell M (1982). The growth and metabolism of developing muscle. In Jones CT (ed) "Biochemical Development of the Fetus and Neonate", Elsevier Biomedical Press, p 213.

3. Buckingham ME, Minty AJ (1983). Contractile protein genes. In Maclean N, Gregory SP, Flavell RA (eds): "Eukaryotic Genes, their Structure, Activity and Regulation", London: Butterworths, p 365.
4. Whalen RG, Butler-Browne GS, Gros F (1978) Identification of a novel form of myosin light chain present in embryonic muscle tissue and cultured muscle cells. J Mol Biol 126:415.
5. Whalen RG, Sell SM, Butler-Browne GS, Schwartz K, Bouveret P, Pinset I (1981) Three myosin heavy chain isozymes appear sequentially in developing rat muscle. Nature 292:805.
6. Minty AJ, Alonso S, Caravatti M, Buckingham ME (1982) A fetal skeletal muscle actin mRNA in the mouse and its identity with cardiac actin mRNA. Cell 30:185.
7. Minty AJ, Caravatti M, Robert B, Cohen A, Daubas P, Weydert A, Gros F, Buckingham ME (1981) Mouse actin messenger RNAs. J Biol Chem 256:1008.
8. Robert B, Weydert A, Caravatti M, Minty A, Cohen A, Daubas P, Gros F, Buckingham M (1982) cDNA recombinant plasmid complementary to mRNAs for light chains 1 and 3 of mouse skeletal muscle myosin. Proc Nat Acad Sci USA 79:2437.
9. Weydert A, Daubas P, Caravatti M, Minty A, Bugaisky G, Cohen A, Robert B, Buckingham M (1983) Sequential accumulation of mRNAs encoding different myosin heavy chain isoforms during skeletal muscle development in vivo detected with a recombinant plasmid identified as coding for an adult fast myosin heavy chain from mouse skeletal muscle. J Biol Chem 258:13867.
10. Mayer Y, Czosnek H, Zeelon PE, Yaffé D, Nudel U (1984) Expression of the genes coding for the skeletal muscle and cardiac actins in the heart. Nucl Acids Res 12:1087.
11. Vandekerckhove J, Bugaisky G, Buckingham M (1985) Simultaneous expression of skeletal muscle and heart actin proteins in adult striated muscle, mouse foetal skeletal muscle and fused mouse muscle cells: a quantitative determination of the two striated muscle actin isoforms. J Biol Chem, Submitted.
12. Gunning P, Ponte P, Blau H, Kedes L (1983) Alpha skeletal and alpha cardiac actin genes are co-expressed in adult human skeletal muscle and heart. Mol Cell Biol 3:1985.

13. Lompré AM, Mercadier JJ, Wisnewsky C, Bouveret P, Pantaloni C, d'Albis A, Schwartz K (1981) Species and age dependent changes in the relative amounts of cardiac myosin isozymes in mammals. Dev Biol 84:286.
14. Cummins P, Price KM, Littler WA (1980) Foetal myosin light chain in human ventricle. J Muscle Res Cell Motility 1:357.
15. Whalen RG, Sell SM (1980) Myosin from fetal hearts contains the skeletal muscle embryonic light chain. Nature 286:731.
16. Cummins P (1982) Transitions in human atrial and ventricular myosin light-chain isoenzymes in response to cardiac pressure overload induced hypertrophy. Biochem J 205:195.
17. Dalla Libera L, Carraro U (1983) The suggested identity of myosin light chain of cardiac atrial muscle and embryonic skeletal muscle may be excluded by proteolytic mapping. Cell Biol Int Rep 7:271.
18. Barton PJR, Robert B, Fiszman MY, Leader DP, Buckingham ME(1985a) The same alkali myosin light chain gene is expressed in adult cardiac atria and in foetal skeletal muscle. J Muscle Res Cell Motility. In press.
19. Robert B, Daubas P, Akimenko MA, Cohen A, Garner I, Guénet JL, Buckingham ME (1984) A single locus in the mouse encodes both myosin light chains 1 and 3, a second locus corresponds to a related pseudogene. Cell 39:129.
20. Maita T, Umegane T, Kato Y, Matsuda G (1980) Amino-acid sequence of the L-1 light chain of chicken cardiac-muscle myosin. Eur J Biochem 107:565.
21. Sréter FA, Balint M, Gergely J (1975) Develop Biol 46:317.
22. Gauthier GF, Lowey S, Benfield PA, Hobbs AW (1982) Distribution and properties of myosin isozymes in developing avian and mammalian skeletal muscle fibers (1982) J Cell Biol 92:471.
23. Caravatti M, Minty A, Robert B, Montarras D, Weydert A, Cohen A, Daubas P, Buckingham M (1982) Regulation of muscle gene expression - The accumulation of messenger RNAs coding for muscle-specific proteins during myogenesis in a mouse cell line. J Mol Biol 160:59.
24. Rutz R, Hauschka S (1982) Clonal analysis of vertebrate myogenesis: VII. Heritability of muscle colony type through sequential subclonal passages in vitro. Develop Biol 91:103.

25. Lowey S, Risby D (1971) Light chains from fast and slow muscle myosins. Nature 234:81.
26. Margreth A, Salviati L, Dalla Libera I, Betto R, Biral D, Salvatori S (1980) Transition in membrane macromolecular composition and in myosin isozymes during development of fast-twitch and slow-twitch muscles. In Pette D (ed): "Plasticity of Muscle", Berlin - New York : Walter de Gruyter & Co, p 194.
27. Weeds AG, (1976) Light chains from slow-twitch muscle myosin. Eur J Biochem 66:173.
28. Frank G, Weeds AG (1974) The amino-acid sequence of the alkali light chains of rabbit skeletal muscle myosin. Eur J Biochem 44:317.
29. Matsuda G, Maita T, Umegane T (1981) The primary structure of L-1 light chain of chicken fast skeletal muscle myosin and its genetic implication. FEBS Letters 126:111.
30. Nabeshima Y-i, Fujii-Kuriyama Y, Muramatsu M, Ogata K (1982) Molecular cloning and nucleotide sequences of the complementary DNAs to chicken skeletal muscle myosin two alkali light chain mRNAs. Nucleic Acids Res 10:6099.
31. Daubas P, Robert B, Garner I, Buckingham M (1985) A comparison between mammalian and avian fast skeletal muscle alkali myosin light chain genes: regulatory and evolutionary implications. J Mol Biol, Submitted.
32. Nabeshima Y-i, Fujii-Kuriyama Y, Muramatsu M, Ogata K (1984) Nature 308:333.
33. Periasamy M, Strehler EE, Garfinkel LI, Gubits RM, Ruiz-Opazo N, Nadal-Ginard B (1984) Fast skeletal muscle myosin light chains 1 and 3 are produced from a single gene by a combined process of differential RNA transcription and splicing. J Biol Chem 259:13595.
34. Ruskin B, Krainer AR, Maniatis T, Green MR (1984) Excision of an intact intron as a novel lariat structure during pre-mRNA splicing in vitro. Cell 38:317.
35. Zakut R, Shani M, Givol D, Neuman S, Yaffé D, Nudel U (1982) Nucleotide sequence of the rat skeletal muscle actin gene. Nature 298:857.
36. Nudel U, Calvo JM, Shani M, Levy Z (1984) The nucleotide sequence of a rat myosin light chain 2 gene. Nucleic Acids Res 12:7175.

37. Fornwald JA, Kuncio G, Peng I, Ordahl CP (1982) The complete nucleotide sequence of the chick α-actin gene and its evolutionary relationship to the actin gene family. Nucleic Acids Res 10:3861.
38. Flanagan JG, Lefranc MP, Rabbitts TH (1984) Mechanisms of divergence and convergence of the human immunoglobulin α1 and α2 constant region gene sequences. Cell 36:681.
39. Slightom JL, Blechl AE, Smithies O (1980) Human fetal $^{G}\gamma$- and $^{A}\gamma$-globin genes: complete nucleotide sequences suggest that DNA can be exchanged between these duplicated genes. Cell 21:627.
40. Czosnek H, Nudel U, Shani M, Barker PE, Pravtcheva DD, Ruddle FH, Yaffé D (1982) The genes coding for the muscle contractile proteins, myosin heavy chain, myosin light chain 2, and skeletal muscle actin are located on three different mouse chromosomes. EMBO J 1:1299.
41. Czosnek H, Nudel U, Mayer Y, Barker PE, Pravtcheva DD, Ruddle FH, Yaffé D (1983) The genes coding for the cardiac muscle actin, the skeletal muscle actin and the cytoplasmic β-actin are located on three different mouse chromosome. EMBO J 2:1977.
42. Leinwand LA, Fournier REK, Nadal-Ginard B, Shows TB (1983) Multigene family for sarcomeric myosin heavy chain in mouse and human DNA: Localizatgion on a single chromosome. Science 221:766.
43. Robert B, Barton P, Minty A, Daubas P, Weydert A, Bonhomme F, Catalan J, Chazottes D, Guénet JL, Buckingham M (1985) Use of an interspecific mouse back-cross to investigate genetic linkage between myosin and actin genes. Nature, in press.
44. Mahdavi V, Chambers AP, Nadal-Ginard B (1984) Cardiac α- and β-myosin heavy chain genes are organized in tandem. Proc Natl Acad Sci USA 81:2626.
45. Maniatis T, Fritsch EF, Lauer J, Lawn RM (1980) The molecular genetics of human hemoglobins. Ann Rev Genet 14:145.
46. Barton PJR, Cohen A, Robert B, Fiszman MY, Bonhomme F, Guenet JL, Leader DP, Buckingham ME (1985b) The myosin alkali light chains of mouse ventricular and slow skeletal muscle are indistinguishable and are encoded by the same gene. J Biol Chem, In press.

Molecular Biology of Muscle Development, pages 507–519

REGULATION OF ACTIN GENE EXPRESSION IN MAMMALIAN STRIATED MUSCLE

Adrian J. Minty◇, Edna Hardeman*, Peter Gunning◇, William Bains◇,*, Helen Blau* and Larry Kedes◇

The MEDIGEN Project, Department of Medicine◇ and The Department of Pharmacology*, Stanford Medical School and Veterans Administration Medical Center◇
Palo Alto, California 94305

ABSTRACT The striated muscle actin genes are co-expressed in heart and skeletal muscle. Their levels of expression vary at different stages of development of these tissues. Based on experiments transfecting the human cardiac actin gene and vectors containing the 5′-flanking region of this gene into muscle and non-muscle cells, we propose a two-step regulation of the expression of this gene. One stage comprises a repression mechanism that prevents transcription of this gene in myoblasts and non-muscle cells. This repression is lifted during myoblast differentiation. Subsequently, an upstream region of the gene interacts with muscle-specific factors to modulate its expression. The level of these factors determines the level of expression of cardiac actin.

INTRODUCTION

The idea that sets of genes may be co-ordinately regulated by trans-acting factors which interact with cis-acting regulatory regions adjacent to or within these genes, has long been the preferred model for molecular biologists to explain cellular differentiation (1,2). The development of striated muscles offers an interesting model system to investigate the existence of such regulatory programs which have, as yet, remained somewhat enigmatic. During muscle development many genes, including those encoding the contractile proteins, have to be regulated to establish a complex and changing muscle phenotype of precisely defined stoichiometry. Among these genes, those encoding the striated muscle actins are particularly attractive for study since these proteins constitute major components of striated muscles and have been shown to be developmentally regulated, both *in vivo* and in tissue culture (3,4).

Two major actin isoforms have been detected by peptide mapping and amino-acid sequencing in striated muscles of adult mammals: one in skeletal muscle and one in cardiac muscle (5). These two actins have thus been designated "skeletal" and "cardiac". Subsequent studies on expression of mRNAs for these two proteins have shown that they are co-expressed at varying levels in different striated muscles (3,6,7,8). Fetal and neo-natal rodent striated muscles show a higher degree of co-expression; adult muscles show greater specialization (3,7). Recent work of Vandekerckhove and Weber (5) shows that the muscle-specific actin isoforms found in mammals arose during early chordate evolution. The two striated muscle isoforms first appear in amphibians and Mohun et al (9) have shown that expression of these two proteins is activated simultaneously at early stages of *Xenopus levis* embryonic development. It would thus appear that a gene duplication event generated the two striated muscle actin genes from a primordial muscle actin gene. It seems likely that these two genes have retained a common mechanism of activation and subsequently have evolved separate patterns of modulation.

RESULTS AND DISCUSSION

Organization and Structure of Striated Muscle Actin Genes

The skeletal and cardiac actin genes are encoded by single unlinked genes in both mice and men (10,11,12). The chromosomal dispersion of these two genes may have been involved in the evolution of separate regulatory programs, although it is clear that other developmentally-regulated gene-families have not adopted this evolutionary strategy, for example globin or myosin heavy chain genes (see Gunning et al (12) for discussion). This dispersion might reflect a chromosomal re-grouping of genes from different contractile protein gene families (actin,myosins, tropomyosins, etc.) which are co-expressed in the same myogenic program. However, linkage analysis of cardiac and skeletal actin, myosin light chain and myosin heavy chain genes in the mouse (13) shows that these genes are widely scattered in the genome.

1A. The 5'-end of the human cardiac actin gene

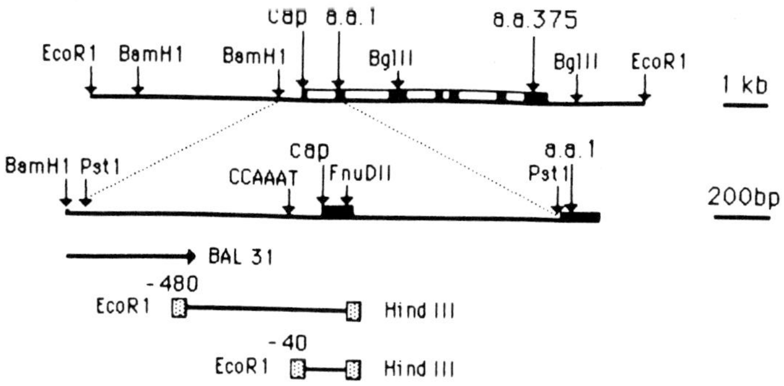

1B. Sequence homologies in the CCAAAT box region

CCAAATAAGGCA.................human cardiac actin [16]

CCAAATATGGC.....................rat and chick skeletal actin [15,20]

CCAAATTTAGGCA............rat αcardiac myosin heavy chain [21]

CCAAAT(A/T)$_{2-3}$ GGC.....CONSENSUS

Figure 1. A. Cloning of 5'-flanking regions of the human cardiac actin gene. The gene was cut at the *Bam*HI site and digested with the nuclease *Bal*31. After addition of *Eco*RI linkers, the gene was recut with *Fnu*DII, which cuts in the 5' leader sequence of the gene. *Hind*III linkers were added and the *Eco*RI-*Hind*III fragments containing different lengths of 5' flanking sequence were ligated to the *Hind*III-*BamH*I fragments containing the bacterial CAT, *neo* or *gpt* genes from the vectors pSV2CAT (26), pSV2*neo* (35), and pSV2*gpt* (36), and cloned in the plasmid pBR322 cut with *Eco*RI and *BamH*I. **B.** Sequences around the CCAAAT box (nucleotides −103 to −98) in the human cardiac actin gene were compared with those of the rat (15) and chicken (20) skeletal actin genes and the rat α-cardiac myosin heavy chain gene (21).

The cardiac and skeletal actin genes have identical exon/intron boundaries in their coding regions, although the introns are larger in size in the cardiac gene (14, 15). They also possess an intron in a similar position in the 5' untranslated region (5' UTR) (15, 16, see Figure 1). The significance of the widespread occurrence of introns in the 5' UTR of actin genes, which contrasts with their absence from 3'

UTRs (17), remains unclear. At least three explanations can be envisaged: that the intron sequence plays a role in regulating gene expression (c.f. immunoglobulin and β-globin genes: 18, 19), that it remains as a relic of the evolution of a new regulatory mode, or that its presence increases the likelihood of such an evolutionary event by allowing the creation of new nucleotide combinations in the vicinity of the promoter. With respect to the second possibility, it is interesting that some strains of mice contain a duplication of the 5′-half of the cardiac actin gene (10; I.Garner and M.Buckingham, personal communication). Such a duplication could create a "pseudo-promoter" which would be free to evolve a new regulatory sequence and subsequently could re-attach to the body of the gene, leaving an intron separating it from the coding region.

Sequence comparisons in the 5′ flanking regions of the chick and rat skeletal actin genes have revealed a 20-nucleotide region around the CCAAAT box showing 19 homologous nucleotides (20). The human cardiac actin gene contains a 11 base-pair region including the CCAAAT box with 10 bases homologous to the skeletal actin sequence (16; see Figure 1). Interestingly, the rat α-cardiac myosin heavy chain gene (21) shows a similar sequence (Fig. 1). The consensus sequence shown in Figure 1 occurred in only one other gene in the mammalian gene sequence data banks— at 1500 base pairs upstream of an interferon gene— indicating that its occurrence in three genes active at high levels in cardiac development may be significant— an assertion that we are currently trying to test experimentally.

Although the CCAAT box has been thought to have a role as part of the basal eukaryotic promoter sequence, this would not seem to be a universal role (for example, see discussion of reference 22). Indeed, we have shown that removal of this region from the cardiac actin promoter has little, if any, effect on its activity in transient expression assays in L-cells (16; see below). Recently Gelinas et al (23) have shown that a single base mutation in the CCAAT box region of a fetal human globin gene is probably sufficient to cause the continued expression of this gene seen in Greek hereditary persistence of fetal hemoglobin. Thus it would seem that the CCAAT box may be important in developmental gene regulation, as opposed to basal gene expression.

Expression of the Striated Muscle Actin Genes

The expression of the striated muscle α-skeletal and α-cardiac actin genes can be summarized as follows:

1. Expression is highly tissue-specific. For example, using dot-blot analyses, we find that both skeletal and cardiac actin mRNAs are expressed at levels at

least 1000-fold greater during skeletal muscle development than they are in adult human liver (Table 1).

TABLE 1

Expression of Sarcomeric Actin Genes During Human Myogenesis

Cell Type[1]	mRNA levels[2] [% skeletal actin mRNA in adult skeletal muscle]	
	Cardiac actin	Skeletal actin
Myoblasts	0.4	0.3
Myotubes [day 1]	33	3
Myotubes [day 4]	100	8
Fetal Skeletal Muscle [24 week]	50	100
Adult Skeletal Muscle	5	100
Adult Liver	<0.1	<0.1

[1]Proliferating human myoblasts were harvested at 20% confluency and human myotubes were harvested after 1 or 4 days in fusion medium (37). Skeletal muscle was obtained from a 24 week human fetus and from an adult human. Liver was also obtained from an adult human. RNA was extracted from the myoblast, myotube, liver and fetal muscle by a modification of the guanidine-HCL method and from adult human muscle as described previously (6).

[2]The levels of cardiac and skeletal actin mRNAs in these RNA preparations were determined by the dot-blot method as previously described (6). The levels were calculated as a percentage of the level of skeletal actin mRNA present in the adult human skeletal muscle RNA preparation.

2. The relative level of expression of these two genes varies during development

and during differentiation of cultured myoblasts. Adult mouse skeletal muscle expresses >95% skeletal and <5% cardiac actin mRNAs (3,4,6). However, during early development, these two actin genes are co-expressed at high levels in rodent and amphibian sarcomeric muscles (3,9). We have measured cardiac and skeletal actin mRNA levels during human myogenesis (Table 1). Human myoblasts grown at low confluence express only trace levels of both mRNAs. Within one day after transfer into fusion medium both mRNAs are induced, with cardiac actin transcripts predominating, as was found in mouse C2 myotubes (4). However, we have found that this is not the case for all cultured myoblast lines: myotubes from the rat L6 and L8 lines express significantly less cardiac actin mRNA (P.Gunning, R.Hickey, A.Minty, unpublished data). Day 4 human myotubes contained a very high level of cardiac actin transcripts: more than the level of skeletal actin transcripts in adult skeletal muscle. Skeletal muscle from a 24 week human fetus contained skeletal and cardiac actin mRNAs in a ratio of about 2:1, a figure comparable to that in late fetal mouse muscle (3). In the mouse, the expression of cardiac actin mRNA persists in new-born muscle indicating that the down-regulation is a relatively late event in myogenic development. As in the mouse, adult human skeletal muscle expresses predominantly skeletal actin (Table 1).

3. The expression of the two actin genes in pure muscle cultures is paralleled in heterokaryons formed between mouse muscle and human fibroblasts, suggesting that muscle trans-acting factors are responsible for the differential expression of these two genes (E. Hardeman, C-P. Chiu and H. Blau, in preparation; see Blau et al, this volume).

4. These two genes are expressed at very high levels. Cardiac and skeletal actins are the most abundant proteins in striated muscle cells, and their mRNAs can constitute up to 5-10% of cellular mRNA (24).Thus unless their mRNAs show a very high stability, a hypothesis which existing data would tend to negate (25), these genes must be subject to very high rates of transcription.

5. The expression of these two genes is activated when cultured myoblasts start to differentiate into myotubes (3,4) and this switch-on can be very rapid: the amount of cardiac actin mRNA reaches a maximum (at a level equivalent to

that found in adult heart) within only 24-36 hours of the onset of differentiation of mouse C2 myoblasts (4).

Human Cardiac Actin Gene Expression in Mouse C2 Myoblasts

In experiments to be reported in detail elsewhere (16) we have introduced vectors including the 5′-flanking region of the human cardiac actin gene attached to the bacterial chloramphenicol acetyl transferase (CAT) gene (26) into C2 myoblasts and non-muscle cells. An upstream region of the gene containing 480 base pairs from the transcription initiation (cap) site (see figure 1) directs high-level transient CAT expression in the C2 muscle cell line, both in myoblasts and in myotubes. The level of transient CAT expression is equivalent to that obtained from the SV40 early promoter in the vector pSV2CAT (26). In contrast, in the non-muscle L- and PCG2- cell lines, the cardiac actin CAT vector shows a 10-20 fold lower level of CAT activity, equivalent to that from the SV40 early promoter without the 72 bp enhancer (26). Deletion of sequence down to 40 base pairs from the cap site reduces transcription >20 fold in C2 cells but has little (<2 fold) effect on the level in L-cells, indicating that we can dissociate a regulatory sequence from the basal gene promoter.

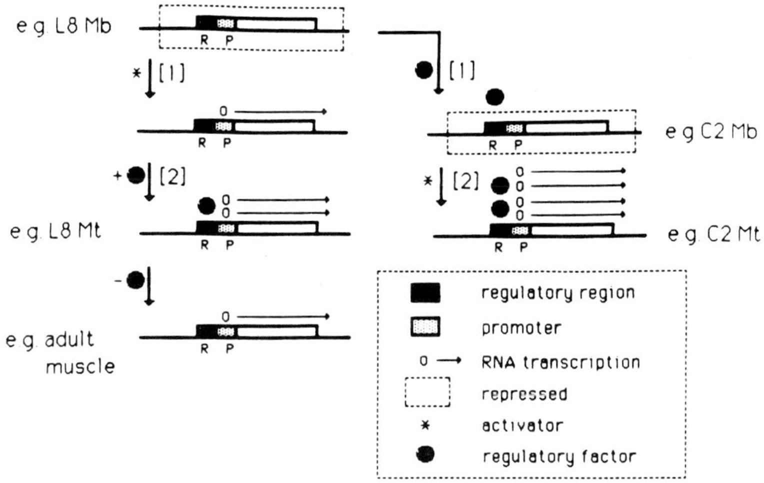

Figure 2. Two-level regulation of cardiac actin gene expression.

We have deduced the following two-level scheme for the regulation of cardiac actin expression (figure 2). A region of the human cardiac actin gene from -480 to -40 interacts with factor(s) present in muscle cells to promote high-level transcription of this gene. The factor(s) are muscle specific and are already present in C2 myoblasts. Why then is the endogenous mouse cardiac actin gene not expressed at maximum levels in myoblasts? Two explanations would be either (a) that the gene is unable to respond to the regulatory factors because it is repressed at the level of chromatin (27), or (b) that it is subject to a post-transcriptional regulation repressing its activity in myoblasts.

We favor the former interpretation since when the *Eco*RI fragment containing the human cardiac actin gene (see figure 1) is stably integrated by transfection into C2 cells it shows a high-level constitutive expression in myoblasts and myotubes. Any post-transcriptional regulation acting on the RNA transcripts of the endogenous mouse gene would also be expected to act on the transcripts of the transfected human gene. However, the transfected gene might escape a repression at the level of chromatin because either (a) this repression acts on a DNA sequence larger than the gene and its immediate surroundings (the sequence contained in the *Eco*RI fragment), or (b) it is imposed earlier in development, and the repression mechanism is no longer active. Experiments measuring the DNase1-sensitivity of the skeletal actin gene in the rat L8 line (28) have suggested that this gene is de-repressed at the level of chromatin during the differentiation of myoblasts into myotubes. Our experiments on C2 cell differentiation indicate that, in addition to this first level of control, the cardiac actin gene, once de-repressed, must interact with a class of regulatory factors which ensure maximal gene transcription. Since these factors accumulate in C2 myoblasts before the de-repression of the endogenous mouse cardiac actin gene, they cannot also be responsible for the de-repression. We have therefore been able to dissociate two separate events in gene regulation: activation and modulation.

Gene Regulation: the Accumulation of Regulatory Factors

We summarize here our current understanding (16) of the regulation of cardiac actin expression. It consists of at least three components:

1. A strong non-cell-specific promoter.

2. An upstream regulatory sequence interacting with factors in muscle cells.

3. A repression mechanism acting in myoblasts and non-muscle cells at the level of chromatin but which does not act on transfected genes.

Aspects (1) and (3) resemble the situation for the α-globin gene which is highly active when introduced into a variety of non-erythroid cells although the endogenous α-globin gene is inactive (19). By contrast, the β-globin gene in Friend erythroleukemia cells requires for maximum activity an erythroid-specific factor, which accumulates only after differentiation (19: c.f. aspect (2) above). Perhaps reflecting the fact that only the β-globin gene requires this factor, transcripts from the α-globin gene begin to accumulate before those from the β-globin gene during Friend cell differentiation (29). For the cardiac actin gene, which requires a muscle-specific factor for maximum expression, the accumulation of this factor *before* gene activation, as in C2 myoblasts, permits rapid synthesis of cardiac actin immediately upon gene activation (4; see also Table 1).

This is an appropriate regulatory strategy when rapid accumulation of large amounts of a particular gene product is required. Is this regulatory strategy general for myoblasts? In contrast to our results discussed above on the introduction of the human cardiac actin gene into C2 myoblasts, and those of Seiler-Tuyns et al (30) who also observed no up-regulation of the expression of transfected chick actin genes in differentiating C2 myoblasts, Melloul et al (31) have shown up-regulation of chimeric genes containing upstream regions of the rat skeletal actin gene in differentiating rat L8 myoblasts. We have also observed up-regulation of the human cardiac actin gene introduced into L8 cells (A. Minty, unpublished results) and rat L6 cells (R. Hickey and P. Gunning, unpublished results). Since the myoblast cells in these experiments are already expressing a significant amount of the human cardiac mRNA, our hypothesis is that this does not reflect gene activation (step 1, fig. 2), but rather that it reflects an increased level of the regulatory factor (step 2, fig. 2) during the differentiation of these cell lines (L6, L8).

The different myoblast cell lines differentiate with strikingly different kinetics. Both by morphological criteria (fusion into myotubes) and biochemical criteria (accumulation of actin mRNAs), the C2 cell line differentiates more rapidly than the L6 or L8 cell lines (4,25,32). It would seem, therefore, that different myoblast cell lines may differ also in the extent to which they express actin gene regulatory factors at the myoblast stage, and that this may reflect or dictate the speed with which they differentiate. It is not clear to what extent the different cell lines mimic different developmental programs *in vivo*. We note, however, that the kinetics of accumulation of actin mRNAs in C2 cells (4) more closely parallel those seen in primary cultures of human (Table 1) or rat (32) myoblasts.

Prospects: a Regulatory Program and a Regulatory Gene?

We propose that the interaction of the -480 to -40 regulatory region with varying levels of muscle-specific factor(s) largely determines the level of cardiac actin mRNA. Absence of these factors, such as in adult skeletal muscle, results in basal level expression (2-5% of maximum; 3,6). Do any other genes respond to the same factors? Two proteins which display similar,though not identical, patterns of accumulation in developing skeletal and cardiac muscle are the embryonic myosin light chain (33) and cardiac troponin T (34). Experiments involving competition of a co-transfected gene by transfection with saturating amounts of this -480 to -40 regulatory region (e.g. see ref.18) could be used to identify co-ordinately regulated genes.

We are currently investigating the use of vectors containing the cardiac actin gene promoter and regulatory region linked to a selectable marker gene (see Figure 1) to isolate the gene(s) encoding the regulatory factor(s). Consistent with our observation that this promoter is relatively strong even in non-muscle cells, these vectors give high transfection efficiencies in L TK^- cells ($\approx 10^{-4}$). However at sub-saturating transfection amounts the efficiencies are about two-fold lower than those obtained with the parental pSV2neo and pSV2gpt vectors (35,36) which contain the SV40 early promoter. L cells containing an integrated cardiac actin promoter linked to *neo* or *gpt*, but sensitive to either neomycin or mycophenolic acid (35,36), have been isolated by co-transfection with the Herpes virus thymidine kinase gene. These cells will provide an assay system for the isolation of the gene(s) encoding the regulatory factor(s). The introduction of such a gene into these cells by subsequent transfection of human genomic DNA should increase transcription from the cardiac actin promoter of the selectable marker gene. Thus those cells that take up a regulatory factor gene in the secondary transfection should now be selectable by virtue of their acquired resistance to the appropriate selection regime. The isolation of such a regulating factor gene should help elucidate the mechanisms involved in the differential expression of genes during myogenesis.

ACKNOWLEDGEMENTS

This work was supported in part by grants from the Muscular Dystrophy Association and the National Institutes of Health to L.K. and to H. B. and by a grant from the Veterans Administration to L.K. A.M. was a Fellow of the Fogarty International Center. E.H. was a Fellow of the NIH. W.B. was a Fellow of the Muscular Dystrophy Association.

REFERENCES

[1] Georgiev G.P. (1969) On the structural organization of the operon and the regulation of RNA synthesis in animal cells. *J. Theor. Biol.* 25: 473.

[2] Davidson E.H., Jacobs H.T., Britten R.J. (1983) Very short repeats and coordinate induction of genes. *Nature* 301: 468.

[3] Minty A.J., Alonso S., Caravatti M., Buckingham M.E. (1982) A fetal skeletal muscle actin mRNA in the mouse and its identity with cardiac actin mRNA. *Cell* 30:185.

[4] Bains W., Ponte P., Blau H., Kedes L. (1984) Cardiac actin is the major actin gene product in skeletal muscle cell differentiation in vitro. *Mol. Cell. Biol.* 4: 1449.

[5] Vandekerckhove J., Weber K. (1984) Chordate muscle actins differ distinctly from invertebrate muscle actins. *J. Mol. Biol.* 179: 391.

[6] Gunning P., Ponte P., Blau H., Kedes L. (1983) α-skeletal and α-cardiac actin genes are co-expressed in adult human skeletal muscle and heart. *Mol. Cell. Biol.* 3: 1985.

[7] Mayer Y., Czosnek H., Zeelon P.E., Yaffe D., Nudel U. Expression of the genes coding for the skeletal muscle and cardiac actins in the heart. *Nucl. Acids Res.* 12: 1087.

[8] Paterson B.M., Eldridge J.D. (1984) α-cardiac actin is the major sarcomeric isoform expressed in embryonic avian skeletal muscle. *Science* 224: 1436.

[9] Mohun T.J., Brennan S., Dathan N., Fairman S., Gurdon J.B. (1984) Cell type specific activation of actin genes in the early amphibian embryo. *Nature* 311: 716.

[10] Minty A.J., Alonso S., Guenet J.-L., Buckingham M.E. (1983) Number and organization of actin-related sequences in the mouse genome. *J. Mol. Biol.* 167: 77.

[11] Czosnek H., Nudel U., Mayer Y., Barker P.E., Pravtcheva D.D., Ruddle F.R., Yaffe D. (1983) The genes coding for the cardiac muscle actin, the skeletal muscle actin and the cytoplasmic β-actin are located on three different mouse chromosomes. *EMBO J.* 2: 1977.

[12] Gunning P., Ponte P., Kedes L., Eddy R., Shows T.(1984) Chromosomal location of the co-expressed human skeletal and cardiac actin genes. *Proc. Natl. Acad. Sci. U.S.A.* 81: 1813.

[13] Robert B., Barton P., Minty A., Daubas P., Weydert A., Bonhomme F., Catalan J., Chazottes D., Guenet J.-L., Buckingham M. (1985) Use of an interspecific mouse back-cross to investigate genetic linkage between myosin and actin genes. *Nature* in press.

[14] Hamada H., Petrino M.G., Kakunaga T. (1982) Molecular structure and

evolutionary origin of human cardiac muscle actin gene. *Proc. Natl. Acad. Sci. U.S.A.* 79: 5901.

[15] Zakut R., Shani M., Givol D., Neuman S., Yaffe D., Nudel U. (1982) Nucleotide sequence of the rat skeletal muscle actin gene. *Nature* 298:85.

[16] Minty A.J., Blau H., Kedes L. (1985) Two-level regulation of cardiac actin expression : an upstream sequence interacting with muscle-specific transcription regulatory factors. In preparation.

[17] Buckingham M.E., Minty A.J. (1983) Contractile protein genes. In Maclean N., Gregory S.P., Flavell R.A. (eds): *Eukaryotic Genes :Their Structure, Activity and Regulation*, London: Butterworths, p.365.

[18] Mercola M., Goverman J., Mirell C., Calame K. (1985) Immunoglobulin heavy-chain enhancer requires one or more tissue-specific factors. *Science* 227: 266.

[19] Charnay P., Treisman R., Mellon P., Chao M., Axel R., Maniatis T. (1984) Differences in human α- and β-globin gene expression in mouse erythroleukemia cells: the role of intragenic sequences. *Cell* 38: 251.

[20] Ordahl C.P., Cooper T.A. (1983) Strong homology in promoter and 3′-untranslated regions of chick and rat α-actin genes. *Nature* 303: 348.

[21] Mahdavi V., Chambers A.P., Nadal-Ginard B. (1984) Cardiac α- and β-myosin heavy chains are organized in tandem. *Proc. Natl. Acad. Sci. U.S.A.* 81: 2626.

[22] Dierks P., van Ooyen A., Cochran M.D., Dobkin C., Reiser J., Weissmann C. (1983) Three regions upstream from the cap site are required for efficient and accurate transcription of the rabbit β-globin gene in mouse 3T6 cells. *Cell* 32: 695.

[23] Gelinas R., Endlich B., Pfeiffer C., Yagi M., Stamatoyannopoulos G. (1985) G to A substitution in the distal CCAAT box of the $^{A}\gamma$-globin gene in Greek hereditary persistence of fetal hemoglobin. *Nature* 313: 323.

[24] Schwartz R.J., Rothblum K.N. (1980) Regulation of muscle differentiation : isolation and purification of chick actin mRNA and quantitation with cDNA probes. *Biochemistry* 19: 2506.

[25] Medford R.M., Nguyen H.T., Nadal-Ginard B. (1983) Transcriptional and cell cycle-mediated regulation of myosin heavy chain gene expression during muscle cell differentiation. *J. Biol. Chem.* 258: 11063.

[26] Gorman C.M., Moffat L.F., Howard B.H. (1982) Recombinant genomes which express chloramphenicol acetyltransferase in mammalian cells. *Mol. Cell. Biol.* 9: 1044.

[27] Weintraub H. (1984) Histone-H1 dependent chromatin superstructures and the suppression of gene activity. *Cell* 38: 17.

[28] Carmon Y., Czosnek H., Nudel U., Shani M., Yaffe D. (1982) DNase sensitivity of genes expressed during myogenesis. *Nucl. Acids Res.* 10: 3085.

[29] Minty A.J., Gros F. (1980) Coding potential of non-polyadenylated mRNA in mouse Friend cells. *J. Mol. Biol.* 139: 61.

[30] Seiler-Tuyns A., Eldridge J.D., Paterson B.M. (1984) Expression and regulation of chicken actin genes introduced into mouse myogenic and non-myogenic cells. *Proc. Natl. Acad. Sci. U.S.A.* 81: 2980.

[31] Melloul D., Aloni B., Calvo J., Yaffe D., Nudel U. (1984) Developmentally regulated expression of chimeric genes containing muscle actin DNA sequences in transfected myogenic cells. *EMBO J.* 3: 983.

[32] Shani M., Zevin-Sonkin D., Saxel O., Carmon Y., Katcoff D., Nudel U., and Yaffe D. (1981) The correlation between the synthesis of skeletal muscle actin, myosin heavy chain, and myosin light chain and the accumulation of corresponding mRNA sequences during myogenesis. *Devel. Biol.* 86: 483.

[33] Whalen R.G., Sell S.M., Erikson A. and Thornell L.E. (1982) Myosin subunit types in skeletal and cardiac tissues and their developmental distribution. *Dev. Biol.* 91: 478.

[34] Toyota N., Shimada Y. (1983) Isoform variants of troponin in skeletal and cardiac muscle cells cultured with and without nerves. *Cell* 33: 297.

[35] Southern P.J. and Berg P. (1982) Transformation of mammalian cells to antibiotic resistance with a bacterial gene under the control of the SV40 early region promoter. *J. Mol. Appl. Gen.* 1: 327.

[36] Mulligan R.C., Berg P. (1981) Selection for animal cells that express the *E. coli* gene coding for xanthine-guanine phosphoribosyl transferase. *Proc. Natl. Acad. Sci. U.S.A.* 78: 2072.

[37] Blau H. and Webster C. (1981) Isolation and characterization of human muscle cells. *Proc. Nat. Acad. Sci. USA* 78: 5623.

Molecular Biology of Muscle Development, pages 521–529

REGULATION OF ACTIN GENE EXPRESSION: ANALYSIS OF POTENTIAL REGULATORY REGIONS

Bruce M. Paterson, Juanita Eldridge and Anne Seiler-Tuyns

Laboratory of Biochemistry, National Cancer Institute
National Institutes of Health, Bethesda, Maryland 20205

SUMMARY

The entire nucleotide sequence of the chicken α-cardiac actin gene has been determined (Eldridge et al., in press, Gene). The gene contains six introns, five of which interrupt the coding region at amino acids 41, 150, 204, 267, and 327. The first intron is in the 5' noncoding region and is 438 bp in length.

The cardiac α-actin gene encodes an mRNA of approximately 1400 bp with 5' and 3' untranslated region of 59 and 184 nt, respectively, and has the codon for the amino acid cysteine between the initiator ATG and the codon for the N-terminal aspartic acid residue of the mature protein.

There are no strong homologies (less than 13 consecutive nt) in the promoter or 3'-untranslated regions between the chicken α cardiac and skeletal α actin genes even though both are expressed in skeletal muscle during development. However, the 3'untranslated region of the chicken cardiac α actin gene demonstrates significant sequence homology (76% over a 200 nt region) with the same region in the partial sequence of the human cardiac gene. The conservation of these sequence homologies between identical isoforms rather than the different α actin genes suggests these conserved regions may have a role in regulation rather than tissue specific expression, as previously proposed.

Preliminary studies on the regulation of β-actin expression during myogenesis suggest sequences 3' to the β promoter mediate the down regulation of expression.

INTRODUCTION

Actin gene expression is tissue specific and is developmentally regulated during the formation and growth of the organism. At least six different actins have been defined by amino acid sequence in vertebrates (1) and include the following isoforms: α-skeletal, α-cardiac, α-smooth muscle, λ-smooth muscle, β-cytoplasmic and λ-cytoplasmic actin. There are four to six aa replacements between the muscle specific isoforms, four replacements between the cytoplasmic isoforms, and 25 replacements between the cytoplasmic and the muscle α-actins. The major sarcomeric actin isoform expressed in developing embryonic avian and murine muscle is cardiac α-actin which is replaced by skeletal α-actin in the early stages of growth (2,3). Here we present the structural organization of the chicken α-actin cardiac gene. The sequence of the promoter and 3' untranslated regions is compared with other vertebrate sarcomeric actin genes.

MATERIALS AND METHODS

Cloning, sequencing and characterization of the chicken α-cardiac actin gene were as described (Eldridge et al., in press, Gene).

All sequence comparisons were done using either the ANALYSEQ computer program of Staden, the NUCALN computer program of Wilber and Lipman with the following parameters: K-tuple size = 3, window size = 20, gap penalty = 7, or SEQA (Kanehisa, NIH) (See Eldridge et al., In press, Gene). Matches of thirteen consecutive nucleotides or more were considered significant.

RESULTS AND DISCUSSION

Organization of the Chicken α Cardiac Actin Gene.

The 8 kb EcoRI insert containing the cardiac α-actin structural gene and flanking DNA is depicted in Fig. 1. The coding portion of the gene is interrupted by five introns. The 5' untranslated region is interrupted by a 438 bp intron and in this respect is similar to all the vertebrate actin gene sequences reported to date. The structural data is tabulated in Table I. Altogether, the cardiac α-actin gene contains seven exons encoding an mRNA of approximately 1400 bp (depending upon the site of poly(A) addition).

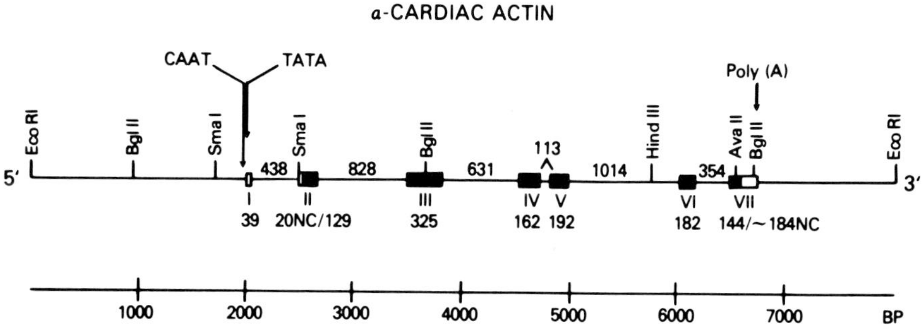

FIGURE 1. Structural map of the chicken cardiac α-actin gene. This is a schematic representation of the 8.0 Kb EcoRI fragment containing the cardiac α-actin gene. The solid bars indicate the coding regions and the open bars mark the untranslated regions. The numbers above the central line refer to the length of the introns in bp; the numbers below the bars indicate the size of each exon in bp; the roman numerals indicate the 5' to 3' order of the exons.

Sequence homologies with other actin genes.

Ordahl and Cooper (4) have reported strong homologies in the promoter and 3' untranslated regions for the chick and rat skeletal α-actin genes and suggested that these sequences may have a role in tissue specific expression of the skeletal α-actin genes. We find no similar homologies when comparing these genes to the CCαA gene in either the CAAT/TATA region, or the 5' untranslated regions. However, the 3'-untranslated region of the chicken and human cardiac α-actin genes shows a significant homology (Fig. 2A) with 76% match in the approximate 200 nt between

TABLE 1
INTRON/EXON PATTERN FOR CHICKEN CARDIAC α ACTIN

Exon/IVS	Positions (Base Pairs)	Size (Base Pairs)	Encoded Region of mRNA
Exon I	1-39	39	39 of 59 nt, 5' untranslated region (ut)
IVS A	40-477	438	---
Exon II	478-626	149	Amino Acids -1 to 41 + 20 nt of 5' ut
IVS B	627-1454	828	---
Exon III	1455-1779	325	Amino Acids 42 to 150
IVS C	1780-2410	631	---
Exon IV	2411-2572	162	Amino Acids 150 t0 204
IVS D	2573-2665	113	---
Exon V	2666-2857	192	Amino Acids 204 to 267
IVS E	2858-3872	1014	---
Exon VI	3873-4054	182	Amino Acids 267 to 327
IVS F	4055-4408	354	---
Exon VII	4409-4708	299	Amino Acids 328 to 374 +) 184 nt in 3' ut.

Abbreviations: ut - untranslated region, nt - nucleotides, IVS - intervening sequence the terminator codon and the polyadenylation signal. The promoter region in the human cardiac gene has not been sequenced (5) so a similar comparison of the promoter regions cannot be made. However, a preliminary sequence has been presented for the mouse cardiac gene (6) and reasonable homology is suggested (Fig. 2B).

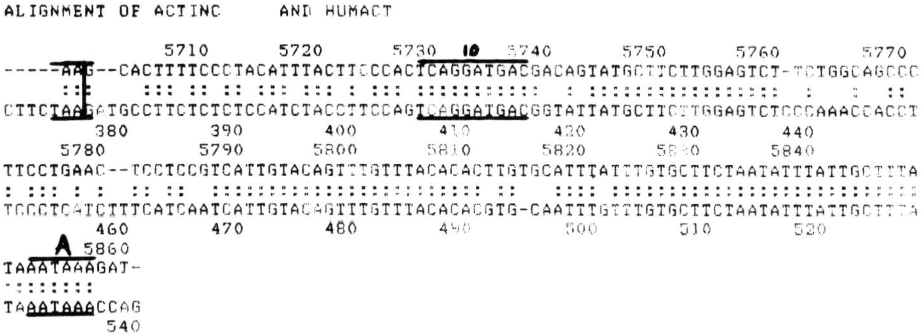

Fig. 2A

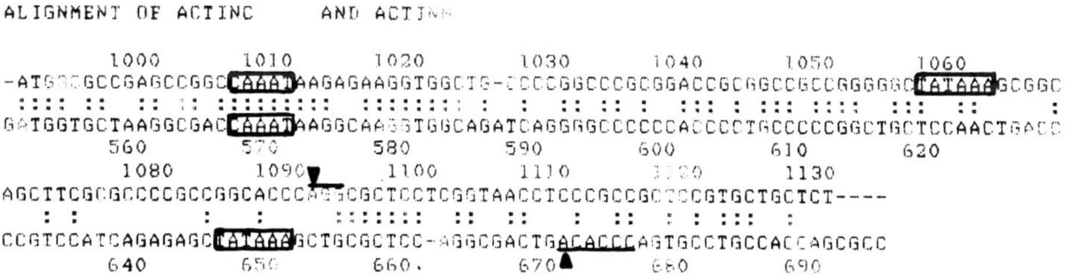

Fig. 2B

```
ALIGNMENT OF ACTINC      AND CHKACT

         5710      5720      5730      5740      5750      5760      5770
CTAAGCACTTTTCCCTACATTTACTTCCCACTCAGGATGACGACAGTATGCTTCTTGGAGTCTTCTGGCAGCCCTTCCTG
 ::: ::  ::: : :  ::    :: :::  ::   : : ::    :  ::: : ::  ::      :       ::::
-TAAACATGTTTACAT-GATCACTTTGCCAACCACACTCAGGATGACAATCTTGTAGG-TTCCAGGCTCCTGAGGACCTG
    2150       2160      2170      2180      2190       2200      2210      2220
         5790      5800      5810       5820      5830      5840      5850
AACTCCTCCGTCATTGTACAGTTTGTTTACACACTTGTGCAT-TTATTTGTGCTTCTAATATTTATTGCTTTATAAATAA
 ::    :   ::   : :  :::  : ::: : :: ::: :  : ::  :::       :: :    :  : ::  :::
CACCAGCCATGCAACTTTCTATTTTGTAACA-ATTTCTGCTTACTGTTGCTGCAAAGCCCATCTCACACAGTGTATGTAA
      2230      2240      2250       2260      2270      2280      2290
5860
AGAT----------------------------------------------------------------------------
::
AGTGTACATAAATTAATTTATTTTACCTCGTTTCGTTTGTTTTTAAAACCAATGCCCTGTGGAAGGAAACATAAAACTTC
       2310      2320      2330      2340      2350      2360      2370

-------------

AAGAAGCATTAAA
       2390
```

Fig. 2C

```
ALIGNMENT OF ACTINC      AND CHKACT

       1000      1010       1020      1030      1040      1050       1060
ATGGCGCCGAGCCGGCCAAATAAGAGAAG-GTGGCTGCCCCGGCCCGCGGACCGCGGCCGCCG-GGGGCTATAAAGCGGC
    : :  :   :::       : :  :  :  ::: : ::: ::: :    ::  ::: ::            : :::
GACACCCAAATATGGCGCGGCCGGGGTCGCATTCCTGTCGCGGGCCGGGCGGTGCTCCCGTCGATAAAAGGCTCCGGGGC
        10        20        30        40        50        60        70        80
         1080       1090      1100      1110      1120      1130
AGCTTCGCGCCCCG-CCGGCACCCAGGCGCTCCTCGGTAACCTCCCGCCGCTCCGTGCTGCTCT------
 :    ::: ::::  :  : :  ::: ::     :     :: ::  ::  :::     :  :
CGGCGGGCGACCCGAGCTACCCGGAGGAGCGGGAGGCGTCTCTGCCAGCGGCCCGACGCCCAGTCAGCAC
        90       100       110       120       130       140       150
```

Fig. 2D

FIGURE 2. Sequence homologies between the chicken and human cardiac α-actin genes in the 3'-untranslated regions. The human cardiac actin sequence data is taken from the partial sequence reported by Hamada et al. (1982). The polyA site and the ten nt consensus region, TCAGGATGAC, are underlined. The termination codon is designated by a bracket. 142 out of 186 bases match for a 76% homology. Sequences have been aligned for optimal homology using the NUCLAN computer program of Wilber and Lipman with the following parameters: K-tuple size = 3, window size = 20, gap penalty = 7 (see Eldridge et al., in press, Gene). A) 3' chicken α cardiac vs 3' human α cardiac actin, B) 5' chicken α cardiac vs 5' mouse α cardiac actin, C) 3' chicken α cardiac vs 3' chicken α skeletal actin, D) 5' chicken α cardiac vs 5' chicken α skeletal actin.

Taken together with our data and that of Kost (1984) and Ordahl and Cooper (1983), sequence conservation in the promoter and 3'-untranslated regions appears to reflect the particular actin isoform rather than sequence elements involved in tissue specific expression. This interpretation is supported by the fact that both the cardiac and skeletal α-actin genes are expressed in developing skeletal muscle (2,3), yet these genes demonstrate no obvious sequence homology in the promoter or in the 5' and 3'-untranslated regions (Fig. 2C and 2D). The ten bp sequence, TCAGGATGAC, found in the 3'-untranslated region of the chick and rat skeletal α-actins, and in the human cardiac α-actin (4) is also present in the same position in the chick cardiac α-actin sequence (Fig. 2). However, it is unlikely that this ten bp homology is muscle specific since the same sequence appears with a 9/10 match in the 5' and 3' flanking regions in several nonmuscle mRNAs (Genetic Sequence Data Bank, SRCHN Program, Lipman and Wilbur, NIH, 1983). These include: the rabbit IG germ line kappa isotype ki (allotype B4) complete J-C and the rabbit IG gamma heavy chain (allotypes VHA2, DE12, 14) mRNA.

Beta Actin Regulation.

Preliminary results with β-actin deletions and promoter exchanges between human α globin and β-actin suggest down regulation of β-actin (7) with cell fusion is mediated by the region of the β-actin gene 3' to the promoter itself. Furthermore, all the sequence upstream from the Xho site in the 5' region of the β-actin gene is not essential for the down regulation of β-actin expression (7) (Fig. 3).

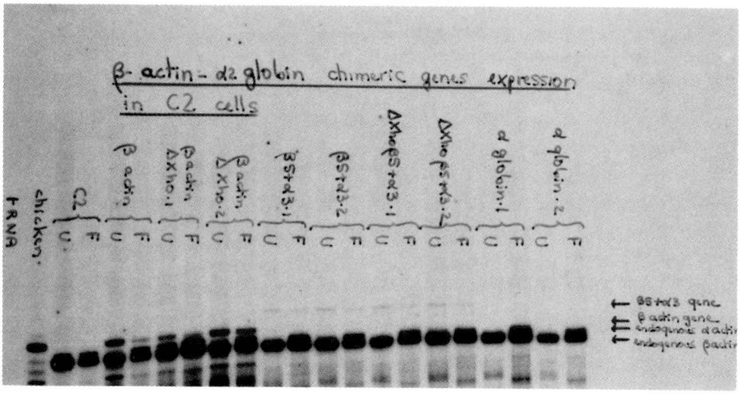

FIGURE 3. Regulation of Deleted β actin and Chimeric α-globin/β-actin genes. Cells were transfected and analyzed as previously described (6).

REFERENCES

1. Vandekerckhove J and Weber K (1979). The complete amino acid sequence of actins from bovine aorta, bovine heart, bovine fast skeletal muscle and rabbit slow skeletal muscle. A protein chemical analysis of muscle actin differentiation. Differentiation 14:123.
2. Minty AJ, Alonso S, Cavavotti M and Buckingham MB (1982). A fetal skeletal muscle actin mRNA in the mouse and its identity with cardiac actin mRNA. Cell 30:185.
3. Paterson BM and Eldridge J (1984). Alpha cardiac actin is the major sarcomeric isoform expressed in embryonic avian skeletal muscle. Science 224:1436.

4. Ordahl CP and Cooper TA (1983). Strong homology in the promoter and 3'-untranslated regions of the chick and rat α-actin genes. Nature 303:348.
5. Hamada H, Petrino MG and Kakunaga T (1982). Molecular structure and evolutionary origin of the human cardiac muscle actin gene. Proc Natl Acad Sci USA 79:5901.
6. Buckingham M, Alonso S, Barton P, Bugiasky G, Cohen A, Daubas P, Garner I, Minty A, Robert B and Weydert A (1984). Actin and myosin genes, and their expression during myogenesis in the mouse. In Davidson E and Firtel R (eds): Molecular Biology of Development, New York: Alan R. Liss, p 275.
7. Seiler-Tuyns A, Eldridge JD and Paterson BM (1984). Expression and regulation of chicken actin genes introduced into normal myogenic and nonmyogenic cells. Proc Natl Acad Sci USA 81:2980.

Molecular Biology of Muscle Development, pages 531–546

REGULATION OF ACTIN GENE EXPRESSION DURING CHICKEN MYOGENESIS[1]

Derk Bergsma, Larry Hayward, Jim Grichnik and Robert Schwartz

Department of Cell Biology, Baylor College of Medicine
Houston, Texas 77030

ABSTRACT A switch in the expression of chicken actin genes occurs during myogenesis, in which the non-muscle actin isoforms such as β and γ are replaced by the α-striated actin isoforms. Measurements of mRNA levels with specific DNA hybridization probes showed that the induction of α-skeletal (10-30 fold) and α-cardiac (10 fold) actin mRNAs was influenced by separate regulatory signals. Results of gene transfection experiments revealed that the DNA regions 5' to the mRNA cap site may play a major role in the transcriptional regulation of the α-skeletal actin gene during myogenesis.

INTRODUCTION

Our laboratory (1) as well as others (2-4) have described a switch in the expression of actin genes during in vitro myogenesis, in which the non-muscle actin isoforms such as β and γ are replaced by the α-striated actin isoforms. We will discuss here our development of specific DNA probes from the chicken actin multigene family to assay the mRNA content of several actin isoforms during myogenesis. In order to understand how these actin genes might be regulated at the transcriptional level, we have recently completed the entire nucleotide sequence of the chicken α-cardiac actin gene (5) and extended the sequence of the chicken α-skeletal actin gene's promoter and its flanking region. This sequence analysis has allowed us to make the first comparison of promoter regions between two different

[1]This work was supported by Public Service grantNS15050, a grant from the Muscular Dystrophy Association and a grant from the American Heart Association, Texas affiliate.

α-striated actin genes. Finally we have constructed hybrid genes with the 5' flanking and promoter sequences of the actin genes fused to the bacterial chloramphenicol acetyltransferase (CAT) gene which were transfected into chicken primary myoblasts. In these transient assays we determined that the α-skeletal actin promoter region placed cis to the CAT gene responded to trans-acting factors present in post-fusion myoblasts. These preliminary experiments suggest that DNA regions 5' to the promoter may play a major role in the transcriptional regulation of the α-skeletal actin gene during myogenesis.

RESULTS AND DISCUSSION

In order to specifically measure the genetic activity of several actin genes during chicken myogenesis, it was necessary to develop probes which would not cross hybridize. We recently isolated 6 different chicken actin genes (6). The α-skeletal, α-cardiac and β-actin genes were used to develop 3'end noncoding hybridization probes which are present in the transcribed portion of one actin mRNA but absent in the other actin mRNA species (Fig. 1).

A short α-skeletal actin cDNA clone called pAC51 (7), which overlaps the last 100 nucleotides of the coding region, was digested with Sau 3A and Pst I to produce a 272 bp fragment which was trimmed with Bal-31. The resulting clone in M13mp19 extends from nucleotides 61 to 272 in the 3' untranslated region, as verified by dideoxy nucleotide sequencing (Fig. 1a).

Construction of the α-cardiac actin probe involved isolation of the Ava II/Bgl II fragment of pAC7.5 and further partial digestion with Dde I to yield a 164 bp insert which was ligated into the Sma I site of M13mp19 (sequence shown in Fig 1b). The 3' end of the β-cytoskeletal actin sequence pAC1 was removed by Pvu II/Pst I digestion, treated with exonuclease III and S1 nuclease to remove the poly(A) region derived from the cDNA, and cloned into the Sma I site of M13mp10 (Fig. 1c). As shown from the sequence data, each 3' noncoding probe contains unique sequences with no significant homology to either of the other two actin mRNA species.

RNA blot analysis was used to provide a relative quantitation of individual actin mRNA species during myogenesis in culture (Fig. 2). Primed extension of labeled single stranded DNA containing specific 3' actin gene sequences served as hybridization probes. We noted that the levels of α-skeletal and α-cardiac actin mRNA in freshly dissociated

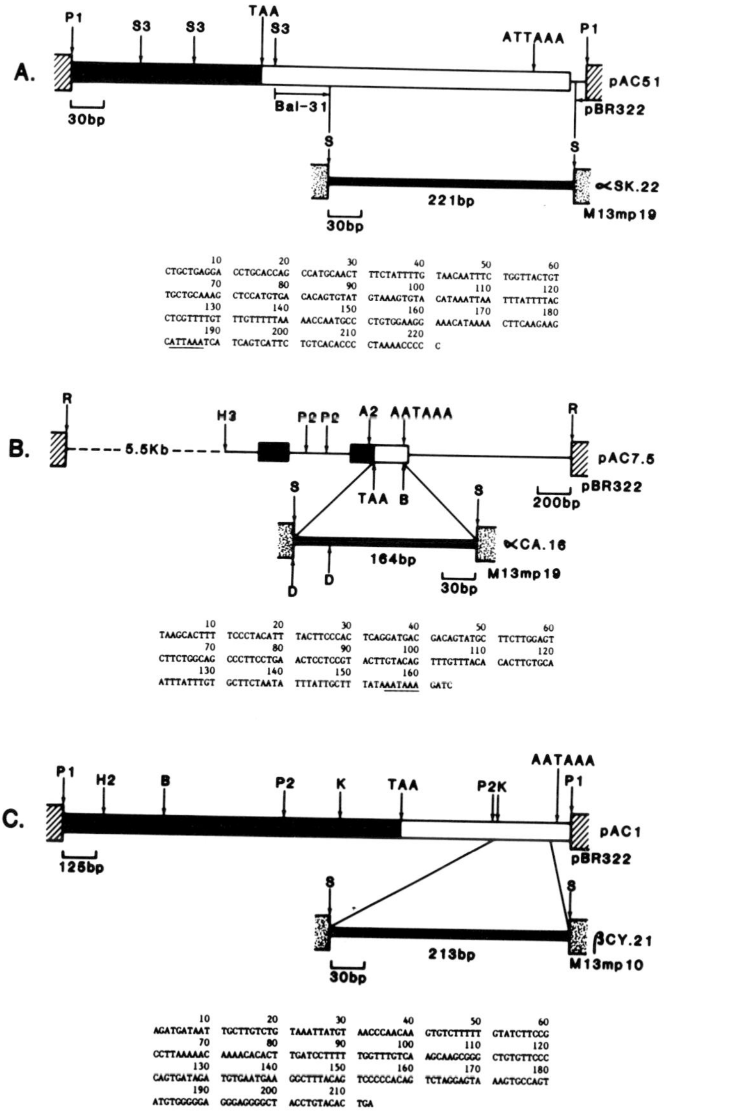

Figure 1. Restriction maps and nucleic acid sequences of the transcribed 3' noncoding regions of the (A) α-skeletal, (B) α-cardiac, and (c) β-actin genes which were subcloned into M13 vectors.

myoblasts (0 h) were lower than amounts in the intact embryonic day 11 hind limb tissue from which the myoblasts were isolated. This result was not unexpected since the dissociation and filtration procedures used in our primary cell cultures discard existing myotubes, which contain any sarcomeric actin mRNAs (8). By 24h in culture (prefusion), α-cardiac actin mRNA was expressed at levels about 10-fold higher than the α-skeletal message. Since recent reports showed the expression of the α-cardiac actin gene in mammalian myoblast cultures and muscle tissues (2,4), it was not surprising for us to observe a similar increase during chicken myogenesis in vitro. The α-skeletal message remained low until 48h (after fusion had begun), suggesting that these two actin genes are influenced by separate control mechanisms. Cytoplasmic β-actin mRNA increased sharply during the proliferative phase prior to fusion, reached a peak by 36h, and then continued to decline as observed at later time points. Cytosine arabinoside (10 uM final concentration) was added to the culture medium at 72h and removed at 120h to selectively kill dividing cells.

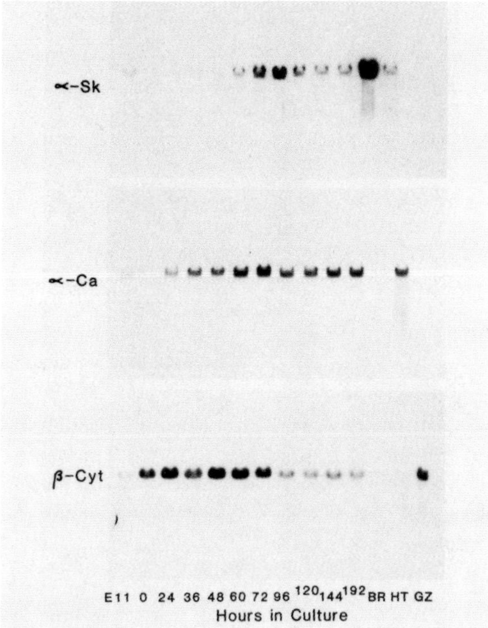

Figure 2. RNA blots of primary myoblst cultures hybridized with ^{32}P labeled 3' noncoding actin DNA probes.

At the completion of fusion (72h-96h), the α-skeletal message had increased 10-30 fold over the initial level in proliferating myoblasts. Under the standard growth conditions used, primary muscle cultures after four days in culture attain α-skeletal actin mRNA levels (on an mRNA/total RNA basis) equivalent to those found in embryonic muscle on day 13 or 14 of incubation. However, at later times in culture, the level of the mRNA declined to less than half its value at 96h. In contrast, α-cardiac actin mRNA was maintained at least through 192h in culture at a level about 50%-75% of that found in a mature heart tissue.

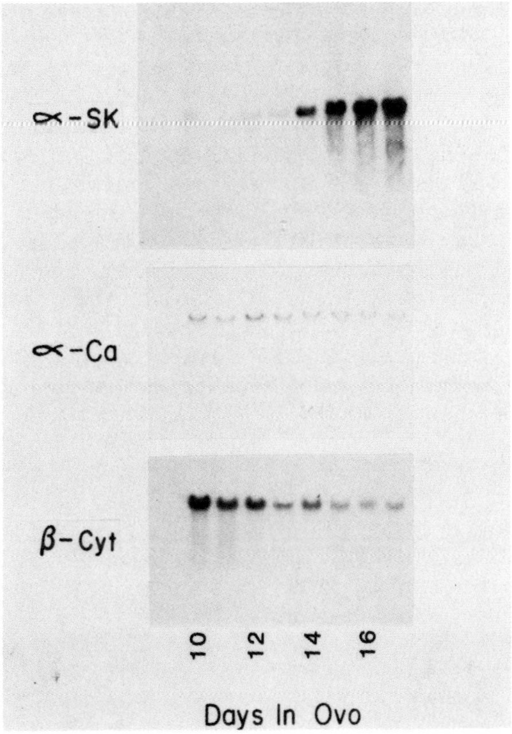

Figure 3. RNA blots of embryonic chicken hind limb stages hybridized with ^{32}P labeled 3' noncoding actin DNA probes.

We then examined changes in steady-state levels of actin mRNA species during chick embryonic thigh and breast development. Total RNA was extracted from these tissues over a period from embryonic day 10 to day 17. As shown in

Fig. 3, β and α-cardiac-actin mRNAs are the predominant actin transcript at embryonic day 10 to day 12. The thigh is mostly composed of replicating myoblasts and fusing myoblasts at these developmental time points (9). RNA products representing the α-skeletal actin gene were at the lowest limit of detectability. However, by the 17th day of embryonic development when the thigh is composed of myotubes, which are polyinnervated by motor neurons and responsive to electrical stimulation (10), the α-skeletal actin mRNA increased over 30-fold in comparison to day 10 values. The major increase in the α-skeletal actin mRNA began at day 14. In contrast, α-cardiac actin mRNA levels had decreased at day 17 by 30% from the levels found at day 10, while the β-actin mRNA was reduced by 70% from its level at day 10. Actin gene expression in embryonic breast muscle indicated a similar increase in the α-skeletal mRNA between day 10 and day 17, with the major increase occurring about one day later (day 15) but was otherwise indistinguishable from expression in thigh (data not shown).

Clearly, two major differences in actin gene expression were observed during myogenesis in culture and in the intact embryo. First, α-cardiac actin mRNA was induced in cultured myoblasts, but failed to be down-regulated in late myogenic cultures in comparison to the reduction of α-cardiac actin mRNA in late stages of embryonic thigh muscle. Second, the content of α-skeletal actin mRNA was approximately one sixth of that observed in 17 day embryonic leg muscle (on mRNA basis per μg total RNA). Thus primary muscle cultures accumulated α-skeletal actin mRNA at levels equivalent to day 13 embryonic muscle. We conclude that primary myoblast cultures partially mimic the in vivo embryonic state. Since myogenesis proceeds in culture in the absence of neurons, the differences in observing the full induction of α-skeletal actin mRNA and the down-regulation of α-cardiac mRNA in the intact animal might be due to neuronal influences. However, this observation does not exclude the possibility of humoral influences which may be involved as well in the switching of the actin gene program. The role of hormones and innervation upon actin gene expression in myoblasts is currently being investigated.

Comparison of the Promoter Regions of the α-Striated Actin Genes

Recent studies have shown that the human α-cardiac actin and the chicken and rat α-skeletal actin share similar gene organization (11-13), and as shown by other investigators (2,4), as well as ourselves in this report, the α-cardiac and α-skeletal actins are expressed during myogenesis. Since these genes also have a common evolutionary history, it seems reasonable to hypothesize that potential regulatory sequences might still be shared by these two striated actin genes. However, the α-skeletal actin gene might also contain regulatory sequences which are unrelated to the α-cardiac actin gene, since the α-skeletal actin mRNA levels during late stage myogenesis in myoblast cultures differ from those and in the intact chicken embryo during muscle development.

```
                                                 CGTGGTTAGG
                                                  ********
TGGAGTACTGCGGGAACATGATTTAACGGAGGGTTGTTAGTTAGAGTA GTATGGTTAGG TCATGGTTGGACTC
                                                 ---------->

GTGTCTTTTAAAGTTTTTTCCAACCTGTGCAATTCTATGATACTAGGCGGACTGCCGCCATGTACTACAGGAAGG

CACACAGCCACAGCTCAGCCTCCAACACACGCTGCCCTTCCCAGCTGCCTGCCCCCTGCTTGGCACAGCCCGTAC

TTGGCCGCACGCTCCCTCACAGGTGAAGCTCGAAAACTCCGTCCCCGTAAGGAGCCCCGCTGCCCCCCGAGGCCT

CCTCCCTCACGCCTCGCCTGCGCTCCCGCTCCCGCACGGCCCTGGGAGAGGCCCCCACCGCTTCGTCCTTAACGG

GCCCGGCGGTGCCGGGGGATTATTTTCGGCCTCTCGGCCCGGGGGGCCCGGCAGACGCTCCTTATACGGCCCGGC

CTCGCTCACCTGGGCCGCGGCCAGGAGCGCCTTCTTTGGGCAGCGCGGGGCCGGGGCCGCGCCGGGCCCGACACC
                                   GCGGGCGG       CCCGCCCAG
                                    *******       ********
CAAATATGGCGACGGCCGGGGCCGCATTCCTGGGG GCCGGGCGG TGCT CCCGCCCGCC TCGATAAAAGGCT
                                   <---------      ---------->

CCGGGGCCGGCGGCGGCCC ACGAGCTACCCGGAGGA..........
                   +1
```

Figure 4. Nucleotide sequence of the chicken α-skeletal actin gene promoter region. Sequences that resemble McKnight's distal signals (14) and the core enhancer (15) are shown above the appropriate DNA sequence.

As shown in Figure 4 we have sequenced the promoter and 5' flanking DNA of the chicken α-skeletal actin gene. Sequencing showed a direct and indirect repeat between the

TATA and CAAT boxes at position -35 and -51. These sequences (GGGCGG or CCCGCC) are identical to the "distal signals" shown by McKnight (14) to potentiate transcription in the thymidine kinase gene. Also 554 nucleotides upstream of the CAAAT box an enhancer core sequence (TGGTTAGG) was detected (15). Enhancer sequences have been shown to independently stimulate transcription of diverse genes relatively independent of position and orientation. Presently, we don't know if this core sequence will potentiate transcription in the chicken muscle system.

Probably, the most exciting observation concerning the α-skeletal actin promoter is the high degree of conserved sequences which are shared between the chicken and rat genes shown in Figure 5. Extensive regions of homology were located between the TATA and CAAAT boxes at -44 to -64 nucleotides, surrounding the CAAAT box at -80 to -100 nucleotides and the long sequence from -110 to -155 5' to the CAAAT box. These homologous sequences provide excellent choices for sites that might bind or interact with muscle specific regulatory factors and direct transcription of the α-skeletal actin genes.

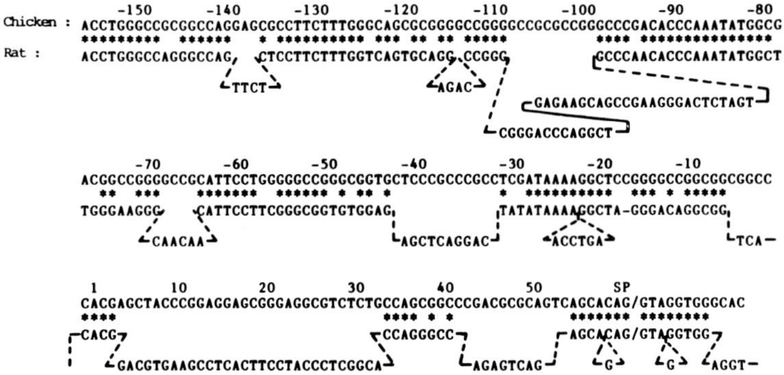

Figure 5. Comparison of promoter regions of the chicken and rat α skeletal actin genes. Regions of homology are shown by a star. Additional sequences in the rat promoter which are not homologous to the chicken sequences are looped out of the primary sequence.

Sequencing of the chicken α-cardiac actin gene has allowed for the first comparison of promoter regions between two different striated actin genes (5). The chicken α-cardiac and the chicken and rat α-skeletal actin genes contain extraordinarily G+C-rich domains which encompass the transcriptional promoter regions. Within the chicken α-cardiac actin gene a region extending a total of 297 nucleotides upstream of the cap site and another region of 480 nucleotides downstream of the cap site is composed of about 70% G+C bases, while the 100 nucleotides immediately preceeding the transcriptional start site was 74% G+C. This G+C richness contributes a striking feature to the striated actin genes: an abundance of potentially methylatable sites.

```
                -280                                                     -240
CCCGGGCCCGACATCTCGCGGACGCACCGCTTGTACATTCCGGAGGGCCATGGAGGGCCACAGAGACTCCGAGCCTCCC
                     -200                                                    -160
                CCGCCC                                                CCGCCC
CTCCCACCACGTCGGCGGGAGGCTCCCTATTTGGCCATGTGGCGGCGGGACCGCTGCTCCTCACCTGCCTTAGATGGCC

                   -120                                                      -80
GGACAGCCGCGCCGGCCTTGGCCATTCATGGCCGCGCTGCGCGCATGGCGAAGCGGCAAATAAGGAGAAGGTGGCTGCC
                      -40
                   GGGCGG                                               GGGCGG      (1)
CGGCCGCGGACGCCGGCCCCGCCGGGGCTATAAAGCGGCAGCTCGCGCCCCGCCGGCACCCAGGCGCTCCTCGGTAACC

 GGGCGG
TCCCGCCGTCCGTGCTGCTCCG/
```

Figure 6. Nucleotide sequence of the chicken α-cardiac actin gene promoter region.

Second, the comparison of the nucleic acid sequences in the promoter regions of the α-cardiac and α-skeletal genes revealed several homologous sequences. The two longest homologous sequences were found at position -296 (CCGGGCCCGACA) and -134 (GCCGCCGCCGG) in the chicken α-cardiac actin gene. In the chicken α-skeletal actin gene these two sequences comprise a single sequence (GCCGCGCCGGGCCCGACA) found between positions -92 and -109. In the chicken α-cardiac gene the homologous sequence located at position -134 was reiterated at positions -17, -35, and -110 as a series of nine nucleotide direct and indirect repeats (consensus sequence of CCGCGCCGG). It is conceivable that these short nine base pair repeat sequences could potentiate transcription in the α-cardiac actin gene by either forming a stem/loop structure (positions -4 to -52) or by acting as separate regulatory elements.

Gene Transfer Experiments with Actin Gene Promoters Fused with the CAT Gene.

In recent years, the function of cloned genes have been studied by transferring them into cells in culture (16,17). In several of these studies, DNA sequences upstream of the cap site were implicated in the regulation of transcription (18-21). Our strategy was to construct a number of hybrid genes with the 5' flanking and promoter sequences of the actin genes fused to the bacterial gene known as CAT (22,23). By transferring plasmids which contain these type of constructs into primary chicken myoblasts it would be possible to correlate promoter function with enzyme activity. In these transient assays the plasmid DNA remains extrachromosomal and is sensitive to potential trans-acting factors. The advantages of this assay system are that CAT is absent from the host genome, is not altered by other enzymatic activities, and the assay is rapid, sensitive and reproducible.

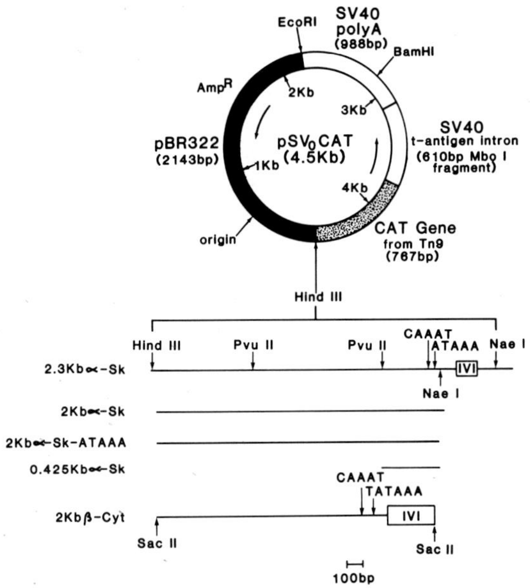

Figure 7. Construction of hybrid chicken actin CAT vectors.

One vector, SVo CAT, contains the CAT gene subcloned into pBR322. Lacking a promoter this construct will not produce assayable CAT. A single Hind III site at the 5' end of the CAT gene provides a cloning site for adding promoter DNA sequences; a hybrid gene may thus be formed. A second vector contains the SV40 early promoter and the 72bp enhancer juxtaposed to the CAT gene. This construction, SV2

CAT, when transfected into most eukaryotic cells produces CAT activity and is subsequently used as a system control. We have made several constructs with the α-skeletal and β-actin gene promoters (Figure 7). The first was made with the 5' flanking and promoter regions of the α-skeletal actin gene in which a 2 kb Hind III/NaeI fragment was blunted with Klenow enzyme and nucleotides, Hind III linkers were added, and then subcloned into the Hind III site of SV_0 CAT. The NaeI site is located 12 nucleotides downstream of the TATA box and recombinant clones were selected for the proper orientation. We also obtained clones that contained the α-actin promoter placed in the opposite orientation.

The same 2 Kb fragment was also treated with an exonuclease to delete the TATA box. The Hind III junctions were then sequenced as a further check on the construction. In a similar manner constructs were made by a Hind III and partial NaeI digestion which provided us with a fragment that contained the promoter, the first intron, the complete 5' leader and 104 nucleotides of the coding region which was fused to the SVo CAT gene. A shorter region of the α-actin promoter was made by a complete PvuII/NaeI digestion, which removed the potential core enhancer sequence and produced a 425 nucleotide fragment inserted next to SVo CAT. Similarly, a 2 Kb SacII fragment from the β-actin gene was cloned into SVo CAT.

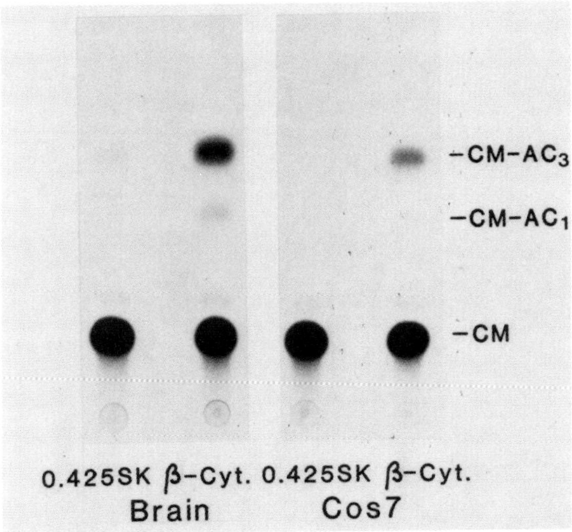

Figure 8. Assay of CAT activity in primary day 11 chick brain cells and Cos7 cells.

The transcriptional activity of these hybrid genes were examined by gene transfection experiments. Initially, the 2Kb α Sk-CAT, 0.42 Kb α Sk-CAT, and 2 Kb β-CAT constructs (10 ug of DNA) were separately transfected with calcium phosphate into primary cultures of chick embryonic day 11 brain cells (3 x 10^6) and mammalian COS 7 cells (Figure 8). These cells were selected because they are non-muscle cell types that express the B-actin isoform but not theα-skeletal actin isoform. Cell supernatants were removed and assayed for CAT enzyme activity with [^{14}C] chloramphenicol and acetyl coenzyme A as described by Gorman et al. (23). The production of acetylated chloramphenicol is monitored by thin layer chromatography on silica gel plates. After 2 days in culture, CAT activity assays revealed, less than 1% of the radioactive chloramphenicol was converted to the acetylated forms in the presence of the alpha skeletal actin promoters in either the chicken or mammalian non-muscle cell type, while the β-actin fusion genes converted 14% and over 20% of the substrate in COS 7 cells and primary brain cell cultures. These preliminary experiments suggest that transcription from the α-actin promoter is tightly restricted in the presence of non-muscle cells. The transcriptional activity of the α-actin promoter was then examined in the homologous chicken myoblast culture system.

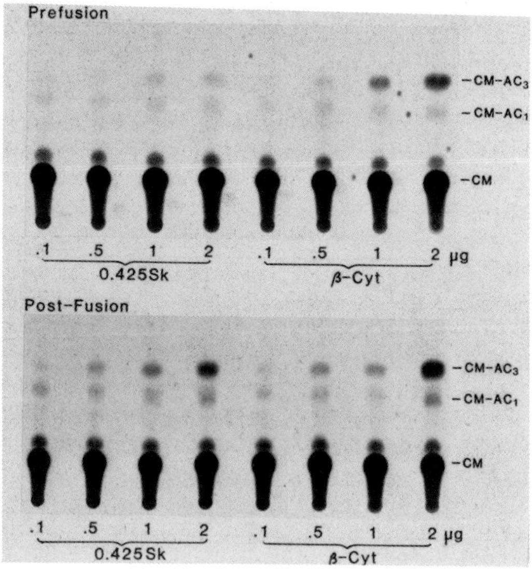

Figure 9. Assay of the α-skeletal β-actin gene promoters fused CAT gene activity during myogenesis in culture.

Dissociated primary myoblasts were plated at 3 x 10^6 cells/100 mm plate and allowed to replicate during the first 24 hours. Myoblasts were then treated with increasing DNA concentrations (0.1 to 2 ug of DNA) of the 0.425 kb α-Sk-CAT and the B-CAT hybrid gene in calcium phosphate precipitates. Cell extracts were made after the 42nd hour (early prefusion) and 74th hour (postfusion) in culture. In these experiments the same number of myoblasts were present as determined by DNA and protein assays. CAT activity appeared to increase in a dose-dependent manner for both the α-skeletal and β-cytoplasmic actin promoter CAT constructs in prefusion myoblasts. Little change (1-2 fold) in CAT activities was detected between the two developmental time points for myoblasts transfected with the β actin hybrid gene. However, the α-skeletal actin CAT construct caused a 10 fold enhancement in CAT activity between the prefusion and postfusion time points. In control experiments, myoblasts transfected with SVoCAT alone did not produce discernable CAT activity, while SV2CAT transfected myoblasts produced CAT activities essentially identical to those of the β-actin gene constructs (data not shown). These results suggest that a portion of an α-skeletal promoter is regulated in a muscle-specific manner.

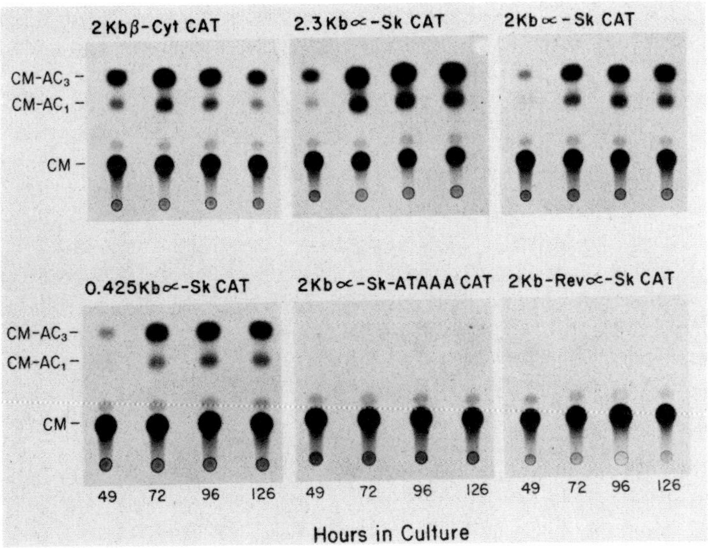

Figure 10. Assay of actin-CAT fused gene activity in primary chicken myoblast cultures.

Experiments were then performed with several other α-skeletal actin hybrids in order to examine the role of DNA sequences flanking the promoter region (Fig. 9). The 2 Kb α-Sk CAT gene, which contains a consensus core enhancer element at -554 nucleotides, was transfected into prefusion chick myoblasts. After 72 hours in culture, CAT activities increased over 9 fold in comparison to prefusion values (48 hours), while by postfusion (96 hours) CAT activity was increased 17 fold. Since the shorter 0.425 α-Sk construct provided a 10 fold stimulation in CAT activity in transfected myoblasts it is possible that additional DNA sequences 5' to the promoter are responsible for potentiating CAT transcription. Whether this increased level of transcriptional activity is due to the core enhancer element or perhaps due to other DNA sequences upstream of the α-skeletal actin promoter will require further experimentation. In control experiments the same 2 Kb α-skeletal DNA placed in the opposite orientation to the CAT gene was transfected into myoblasts. The reversed hybrid gene was ineffective in stimulating any CAT activity. Therefore, we assume that transcription is not initiated in a bidirectional manner within the 2 Kb portion of flanking DNA. In addition, no CAT activity was detected in myoblasts transfected with a properly oriented 2 Kb α-Sk CAT gene deleted at the TATA box. These results are unexpected, since simple deletions at the TATA box in other genes has shown only a minimal effect on reducing transcriptional efficiency. One explanation for these results is that the deleted TATA box altered the spacing of the remaining DNA with the CAT gene, thus causing a defective CAT transcript. This observation is also consistant with transcription being initiated in vivo at a single locus in the α-skeletal actin gene.

This assumption is supported by the detection of a single primer extension product synthesized from a fragment of the 5' α-skeletal actin mRNA leader hybridized to muscle mRNA and our recent experiments in an in vitro transcription Xenopus oocyte system which located a single start site 27 nucleotides downstream of the ATAAAA box in the α-skeletal actin gene (Bergsma, Gossett and Schwartz, in preparation). However, we have not yet demonstrated that the actin-CAT fused genes introduced into cells are initiated at the same start sites. Finally, the construction 2.3 Kb α-Sk-CAT was used to determine a role, if any, for the first intron of the α-skeletal actin gene on CAT activity in transfected myoblasts. In this case, the CAT transcript contains the chicken mRNA leader sequence and a small portion of the actin coding region. A maximum of a 14-15 fold increase in CAT activity was detected at late myogenic stages. Therefore, DNA sequences immediately downstream from the α-skeletal actin promoter did not seem to have an additive effect on potentiating transcriptional activity.

In conclusion, DNA sequences within 425 nucleotides of the promoter region appear to be responsible for the tissue and developmentally specific transcription of the α-skeletal actin gene. This observation is based upon the restricted expression of the actin CAT fused genes in muscle cells versus non-muscle cells and the 10-20 fold induction of CAT activity which coincides with the normal appearance of the α-skeletal actin mRNA during myogenesis in culture. These results support the presence of trans-acting factors in myotubes which stimulate the transcription of the α-skeletal actin. Comparisons of the chicken and rat α-skeletal actin gene promoters within these 425 nucleotides has shown us several regions of homology. We suspect that these conserved regions are the best locations for detecting DNA sequences involved with transcriptional control of a muscle regulated gene.

REFERENCES

1. Schwartz, RJ, and KN Rothblum (1981). Biochemistry 20:4122.
2. Minty, AJ, S Alonso, M Caravatti, and ME Buckingham (1982). Cell 30:185.
3. Shani, M, Nudel, U, Zeuin-Sonkin, R, Zakut, D, Givol, D, Katcoff, Y, Carmon, J. Reter, Frischauf, A and Yaffe, D (1981). Nucleic Acids Res 9:579.
4. Bains, W, Ponte, P, Blau, H, and Kedes, L (1984). Molec and Cell Biol 4:1449.
5. Chang, KS, Rothblum, KN and Schwartz, RJ (1985). Nucleic Acids Res 13:1223.
6. Chang, KS, Zimmer, WE Jr, Bergsma, DJ, Dodgson, JB and Schwartz, RJ (1984). Mol and Cell Biol 4:2498.
7. Schwartz, RJ, Haron, J, Rothblum, KN and Dugaiczyk, A (1980). Biochem 19:5883.
8. Moss, M and Schwartz, R (1981). Mol and Cell Biol 1:289.
9. Hermann, H, Heywood, SM and Marchok, AC (1979). Curr Top Devel Biol 5:181.
10. Oppenheim, R, Pittman, R, Gray, M and Maderdrut, JL (1978). J Comp Neurol 179:619.
11. Hamada, H, Petrino, MG and Kakunaga, T (1982). Proc Natl Acad Sci USA 79:5901.

12. Fornwald, JA, Kuncio, G, Peng, I and Ordahl, CP (1982). Nucleic Acids Res 10:3861.
13. Zakut, R, Shani, M, Givol, D, Neumak, S, Yaffe, D and Nudel, U (1982). Nature 298:857.
14. McKnight, SL and Kingsbury, R (1982). Science 217:316.
15. Laimins, LA, Kessel, M, Rosenthal, N, and Khoury, G in "Enhancers and Eukaryotic Gene Expression" Eds Gluzman, Y, and Shenk, T, Cold Spring Harbor Laboratory (1983), pp. 28-37.
16. Hamer, DH and Leder, P (1979). Nature 281:35.
17. McKnight, SL and Gavis, ER (1980). Nucleic Acids Res. 8:5931.
18. Kurtz, DT (1981). Nature 291:629.
19. Paulakis, G and Hamer, D (1983). Proc Natl Acad Sci USA 80:397.
20. Chao, MV, Mellon P, Charnay, P, Maniattis, P, Grez, M and Schutz, G (1983). Cell 32:483.
21. Penkowitz, R, Bueg, H, Graf, T, Matthias, P, Grez, M and Schutz, G (1983). Cell 32:483.
22. Gorman, C, Merlino, GT, Willingham, MG, Pastan, I and Howard, BH (1982). Proc Natl Acad Sci USA 79:6777.
23. Gorman, CM, Moffat, LF and Howard, BH (1982). Mol and Cell Biol 2:1044.

Molecular Biology of Muscle Development, pages 547–558

GENE TRANSFER INTO MYOGENIC CELLS IN PRIMARY CULTURE

Charles P. Ordahl, Janet Mar, Jennifer Clyne and *Paul Simpson

Department of Anatomy and
*Cardiovascular Research Institute,
University of California at San Francisco,
San Francisco, CA, USA 94143

ABSTRACT

Procedures for transfecting chick muscle cells in primary culture have been studied to permit the use of such cells in analysis of gene regulation during myogenic differentiation. A marker plasmid containing the drosophila alcohol dehydrogenase (dADH) gene was constructed which allows visual identification of transfected cells by virtue of a stain which is specific for the dADH enzyme. The use of this plasmid in transfection experiments demonstrates that primary cultures of chick skeletal muscle and cardiac muscle can be efficiently transfected with appropriate cloned DNAs.

INTRODUCTION

Transfection of cloned DNA into cells in culture has been useful in determining the DNA sequence elements involved in transcription of eucaryotic genes. In some cases it has also been possible to identify DNA sequence elements which are involved in the tissue-specific or hormonally-regulated transcription of genes. Indeed, recent work has indicated that alpha-actin genes are correctly expressed and regulated when transfected into rat myogenic cells in culture (1). These results suggest the feasibility of using DNA transfection as a routine procedure for the study of the DNA sequence elements involved in the regulation of muscle-specific genes.

Most transfection work has involved the use of established cell lines as recipients of the exogenous DNA. Cell lines offer several advantages over primary cell cultures; eg. clonal purity and immortality. The latter property is essential in order to obtain cells which have stably integrated exogenous DNA into their genomic DNA because such cells represent only an extremely low percentage of the total number of transfected cells and must be cloned and expanded to permit analysis.

Primary muscle cell cultures, on the other hand, are never 100% free of non-muscle cells and are not immortal. Nevertheless, there are a number of reasons why it might be advantageous to use primary muscle cell cultures for transfection experiments. First, suitable cell lines for the muscle cell of interest may not exist. For example, there are no avian myogenic cell lines at all nor are there cardiac myocyte cell lines from any species. Moreover, evidence is accumulating to indicate that the precursor cell populations giving rise to embryonic muscle change throughout embryogenesis (see Stockdale, this volume). These different precursor cells may regulate genes differently and it may be impractical, or impossible, to obtain cell lines representing each precursor cell "type". Second, it is likely that some myogenic developmental transitions occur via interactions between the muscle cell and non-muscle cells and/or extracellular molecules produced by non-muscle cells. There is, for example, ample evidence that muscle development and muscle gene expression are influenced by neurotrophic factors and Caplan et al. indicate that extracellular matrix proteoglycans also affect myogenic differentiation (see examples in this volume).

Transfected DNA need not be integrated to be efficiently expressed because non-integrated DNA can be "transiently" expressed for a period of at least 16-72 hours after transfection. Recently, regulated transcription has been demonstrated for several genes using such transient expression transfection experiments (see ref. 2, for example). If cloned muscle-specific genes are also found to be correctly regulated during transient expression transfection experiments it would allow experiments on primary muscle cultures because it would not be necessary to establish stably transfected cell clones.

Primary muscle cultures contain substantial, and variable, numbers of non-myogenic cells such as fibroblasts. As indicated above, it may often be desirable to have such cells in the culture either because they affect the differentiation of the myogenic cells or because they cannot be conveniently eliminated. It would bc undesirable, however, if DNA was preferentially transfected into non-muscle cells because the signal from these cells would dilute or mask that from the muscle cells.

We have been attempting to optimize conditions for performing transient expression transfection experiments with primary muscle cells in culture. To permit visual determination of transfected cell type we sought to develop a transfection vector containing a marker gene which produces a protein for which there is a simple and reliable histochemical stain. We selected the drosophila alcohol dehydrogenase (dADH) gene because the dADH enzyme has been shown to be assayable histochemically in drosophila cells (3). In addition, background staining by endogenous alcohol dehydrogenases in vertebrate cells can be eliminated by the use of butanol as a substrate because the drosophila, but not vertebrate, ADH is capable of using this alcohol as its substrate (4).

In order to ensure efficient expression of the drosophila ADH gene in vertebrate cells we trimmed away essentially all but the coding region of the gene and provided it with a promoter from the Rous Sarcoma virus Long Terminal Repeat (RSV LTR) and SV-40 splice and poly(A) addition signals (5). The inclusion of these viral regulatory sequences ensures that this marker gene will be efficiently expressed in a wide variety of cell types from a wide variety of vertebrate species. Here we demonstrate the efficacy of this marker gene for rapid analysis of transfected cell types in primary muscle cell cultures.

MATERIALS AND METHODS

Preparation of plasmid DNA, restriction enzyme digestions, agarose gel electrophoresis, DNA fragment isolations, ligations and repair synthesis with reverse transcriptase and bacterial transformation have been published and were performed as per standard procedures (6). Deletion using exonuclease III was as described (7).

Cell Culture.

Primary cultures of day 11 chick embryo breast muscle cells were growm on Dulbecco's modified Eagle's medium (DMEM) supplemented with 10% horse serum and 2.5% chick embryo extract (8,9). Cultures were established at 1 million cells per 60 mm gelatinized plate. Medium was changed every 48 hours. Cardiac cell cultures were established using cells from day 8 embryonic hearts by the method of Simpson and Sarion (10).

Cell Transfection.

The DEAE-dextran mediate transfection protocol of Lopata et al (11) was adopted for introduction of DNA into chick primary cell cultures. Conditions for optimal transfection of exogenous DNA into chick primary cells were empirically determined using the plasmid pRSV-CAT (the gift of M. Walker) (2). The following protocol represents the conditions for optimal expression of the transfected CAT gene in chick muscle cells. Cells to be transfected were washed once with serum-free medium and then incubated with 1.5 ml of serum-free medium containing 1 ug/ml of supercoiled plasmid DNA and 200 ug/ml of DEAE-dextran for 30 min at 37°C in a 5% CO_2 incubator. After this time, the transfection medium was removed, the cells washed once with phosphate buffered saline (PBS) and complete medium replaced. Two days after transfection, the cells were assayed for d-ADH activity as described below.

Histochemical Staining for ADH Activity.

Cells were washed with PBS three times and then fixed with 1% formaldehyde in PBS for 2 min at room temperature. This above fixation procedure was arrived at after trying several fixatives including paraformaldehyde, methanol, ethanol and mixtures of methanol and acetic acid. After several washes with PBS to remove formaldehyde, the staining solution (175 mM Tris-Cl, pH 8.2, 0.3 M 2-butanol, 0.44 mg/ml nitro blue tetrazolium, 0.017 mg/ml phenazine nethonsulfate, 0.44 mg/ml NAD) (4) was added and the cells incubated in the dark at room temperature for 2-24 hours. Usually the dark blue colored product appeared within 30-60 min and staining was complete by 2-6 hours. Stained cells were visualized by phase contrast microscopy and photographed.

RESULTS

Construction of the Plasmid pRSV-ADH.

A 4.8 kb Eco RI fragment containing the entire dADH gene and 1-2 kb of its 5' and 3' flanking sequences was isolated from the plasmid aSF-2 (12,13) and subcloned into pBR322 (designated pA4). The majority of the 5' untranslated region, and all of the 3' untranslated region, of the dADH gene was then removed by a combination of exonuclease III deletion and partial restriction endonuclease digestion (Mar et al, submitted). This left the body of the dADH gene intact, including the translation initiation and termination codons, while removing the untranslated and flanking region of the gene. The endpoints of all deletions were confirmed by DNA sequencing.

The truncated dADH gene body was then cloned into a derivatized version of the plasmid pRSV-TG (gift of Dr. H. P. Moore). This plasmid contains the 2295 bp Pvu II-Eco RI fragment of pBR322 (consisting of the origin of replication and the ampicillin resistance gene), the RSV LTR enhanced promoter, the SV40 small t intron and the SV40 early region polyadenylic acid addition site. Thus the transcription of the dADH gene is placed under control of the RSV LTR promoter and polyadenylation of the transcript is directed by sequences contained within the SV-40 segment at the 3' end. The distance between the transcriptional start site of the RSV LTR and the AUG codon of the dADH gene is approximately 116 bp. The resultant plasmid is shown in Figure 1.

Transfection of pRSV-ADH into Chick Primary Muscle Cultures

The ability of the RSV-ADH plasmid to be transfected and expressed in different cell types was tested using primary cultures of chick embryonic skeletal muscle. Leg or thigh muscle tissue from 11 day chick embryos was mechanically disrupted to liberate mononucleate myoblast and fibroblast cells and plated onto gelatinized plates.

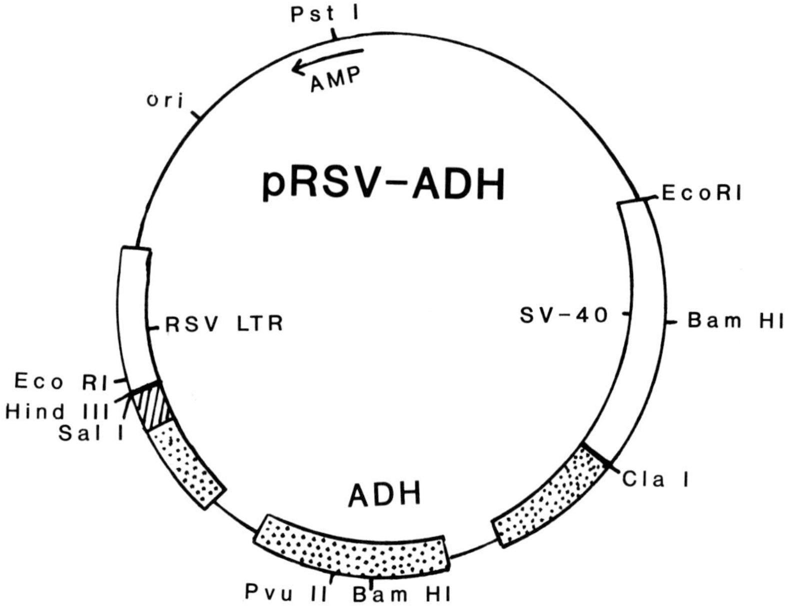

Figure 1. Structure of pRSV-ADH. The general structure of the pRSV-ADH plasmid is shown with pertinent restriction enzyme sites. Thin lines designate pBR322 DNA and dADH introns. The RSV LTR and SV 40 segments are shown as open boxes. The dADH gene exons are stipled and the 5' untranslated region is cross-hatched.

Using a plasmid in which the RSV promoter was linked to the biochemical marker gene chloramphenicol acetyltransferase (RSV-CAT, the gift of M. Walker; ref.2) we had previously established a standardized transfection regime in which cells were transfected 24 hours after plating and then assayed for transient expression 48-72 hours later (although cells can be transfected and/or assayed after much longer periods in culture). We had also optimized the amount of RSV-CAT DNA which yielded maximum levels of CAT activity in these cultures. Thus, for all of these experiments the amount of DNA transfected was well below the saturation level of the system.

Figure 2 (A and B) and Figure 3 (A, B, C) show typical examples of cells in primary cultures of skeletal muscle which have been transfected with RSV ADH 24 hours after plating and then histochemically stained for dADH activity 3 days later when most of the myogenic cells have fused. Large multinucleated muscle cells (Figure 2 A,B; Figure 3 A) and binucleated (Figure 3 B,C) muscle cells are stained. In addition to myogenic cells small fibroblastic cells also stain (Figure 3 C). Preliminary counts (not shown) of stained cell types suggest that the proportion of stained cells is probably proportional to their relative number within the culture. If so, this would indicate that fibroblastic cells and myogenic cells are transfected with approximately equal efficiency.

There are several interesting features of the transfection of skeletal muscle cells with RSV ADH. First, multinucleate muscle fibers tend to stain over their entire length. This is interesting because we do not expect that all of the nuclei within the fiber actually carry transfected DNA. We conclude, therefore, that the dADH protein once synthesized migrates freely through the muscle cell cytoplasm. Second, some fibers stain with a periodic pattern (see, for example, Figure 3A). Careful examination indicates that this does not represent a clustered distribution of the enzyme within the muscle fiber, but is an artifact resulting from the unequal distribution of droplets of non-miscible butanol along the length of the cell. Highly vigorous mixing of the staining mixture before application to the culture can essentially eliminate this artifact. Third, we found that myogenic cells could be successfully transfected both before fusion (as above) and after fusion to form multinucleate myotubes (data not shown). This establishes the feasibility of introducing exogenous DNA into muscle cell nuclei at essentially any stage during their differentiation. Finally, we found that under optimum conditions approximately 200 cells per million (0.02%) of the cells stained for dADH activity. Thus, on a single plate, there are sufficient numbers of stained cells to easily establish the diversity of transfected cell types.

Transfection of Cardiac Myocytes.

We were also interested in determining if chick cardiac myocytes could also be transfected in primary culture because it is of interest to use transfection to study gene regulation in normal and pathological cardiac myocyte differentiation. Chick cardiac myocyte cultures were prepared by procedures developed for the culture of neonatal rat cardiac myocytes for analysis of signal-induced hypertrophy (10). Figure 3D shows an example of a typical branched cardiac myocyte stained for dADH activity indicating that is has been transfected with,

A

B

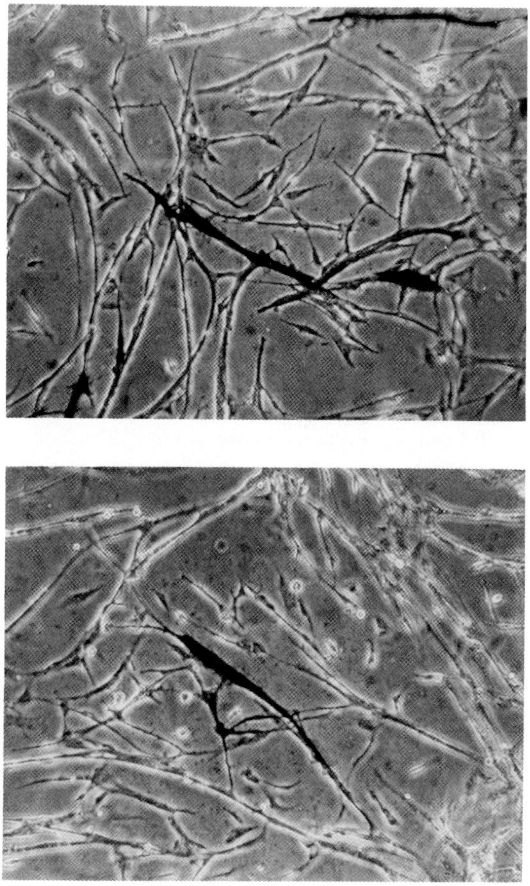

Figure 2 and 3. Phase Contrast micrographs of cultured primary chick skeletal and cardiac muscle cells transfected with pRSV-ADH and histochemically stained for drosophila ADH activity. The black cells are transfected cells expressing the dADH protein. The transfected cells are surrounded by untransfected cells which do not stain. The actual color of the stained product is dark blue.
Panels A and B in Figure 2 (above) and panels A, B and C in Figure 3 (facing) show skeletal muscle primary cultures. Panel D of Figure 3 shows a cardiac myocyte culture. The small size of the cardiac myocytes is attributable to their growth in serum-free medium (see Simpson and Sarion, 1982).

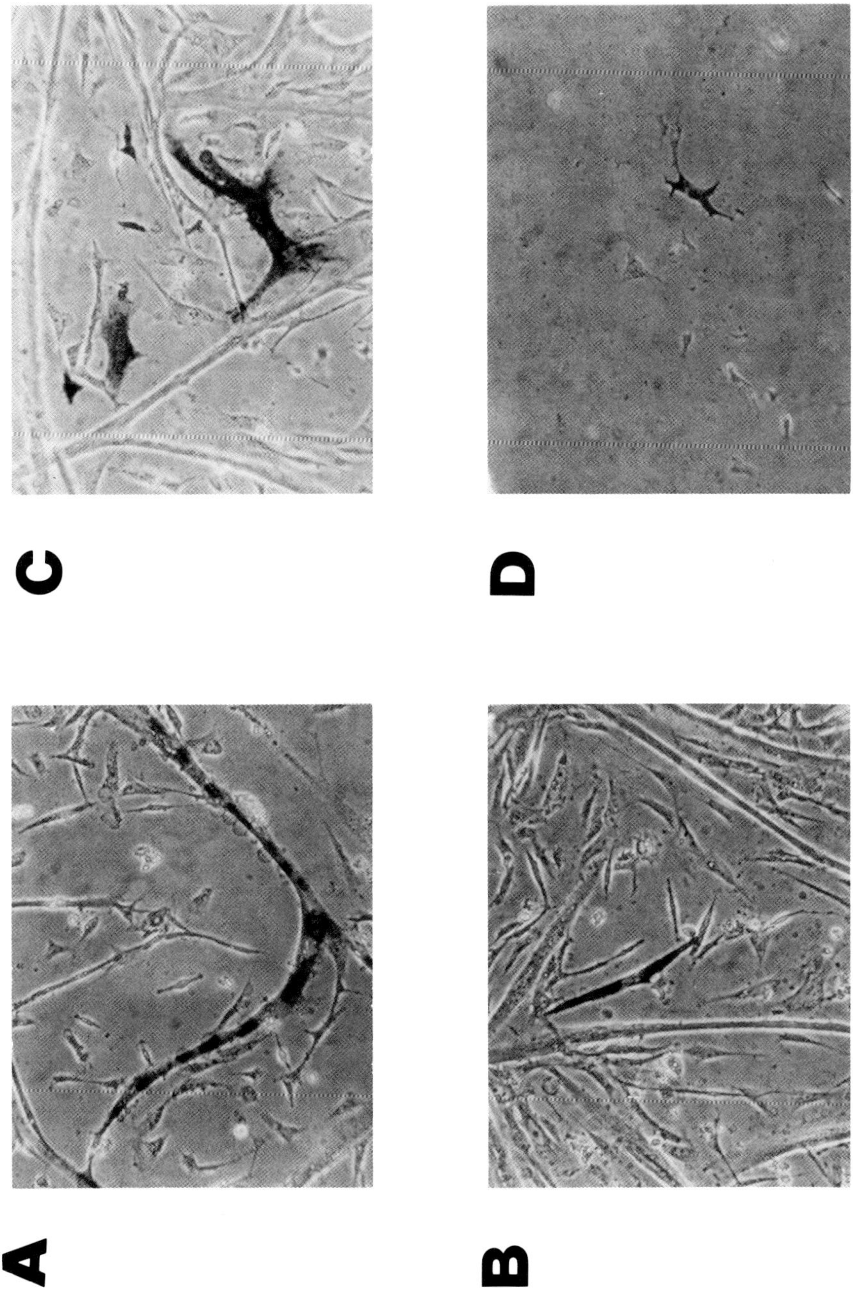
A
B
C
D

and is expressing, the RSV ADH gene. This indicates, therefore, that it is feasible to use the transient expression transfection approach to study gene regulation in cardiac myocyte cultures.

DISCUSSION

We have constructed a plasmid containing a chimeric gene (RSV ADH) in which we have placed the drosophila ADH gene under the transcriptional control of the RSV LTR. When transfected into cells in culture, the transient expression of this gene can be rapidly and easily detected histochemically using a gene-specific stain. Using this plasmid for transfection into primary cultures of skeletal and cardiac muscle we were able to determine; 1) that primary myogenic cells can take up and express exogenous DNA; and 2) that skeletal muscle cells can be transfected both before and after fusion. It remains to be determined whether this basic dADH gene construction will be useful for analysis of cell-specific expression when controlled by muscle-specific promoters.

The RSV LTR promoter used in RSV ADH is highly active in a wide range of mammalian and avian cell types as is the SV-40 splice and poly(A) signals. We expect, therefore, that this chimeric gene can be used for histochemical analysis of transfection in a variety of cell systems. Its principal utility is to determine the relative efficiency of transfection of different cell types within a mixed cell culture to permit rapid optimization of transfection conditions for individual cell types in mixed cell culture.

ACKNOWLEDGEMENTS

We would like to thank M. Walker, H.P. Moore, W. Sofer, and C. Benyajati for generously providing plasmids and advice. A preliminary report of this work has been submitted for publication (Mar et al, submitted). This work was supported by NIH grant GM-32018.

REFERENCES

1. Melloul D, Aloni B, Calvo J, Yaffe D, Nudel U (1984). Developmentally Regulated Expression of Chimeric Genes Containing Muscle Actin DNA Sequences in Transfected Myogenic Cells. EMBO J 3:983-990.
2. Walker MD, Edlund T, Boulet AM, Rutter WJ (1983). Cell-specific Expression Controlled by the 5'-Flanking Region of Insulin and Chymotrypsin Genes. Nature 306:557-561.
3. Goldberg DA, Posakony JW, Maniatis T (1983). Correct Developmental Expression of a Cloned Alcohol Dehydrogenase Gene Transduced into the Drosophila Germ Line. Cell 34:59-73.
4. Ursprung G, Sofer WH, Burroughs N (1970). Ontogeny and Tissue Distribution of Alcohol Dehydrogenase in Drosophila melanogaster. Wilhelm Roux Archiv 164:201-208.
5. Corman CM, Merlino GT, Willingham MC, Pastan I, Howard, BH (1982). The Rous Sarcoma Virus Long Terminal Repeat is a Strong Promoter When Introduced into a Variety of Eukaryotic Cells by DNA Transfection. Proc Natl Acad Sci 79:6777-6781.
6. Maniatis T, Fritsch EF, Sambrook J (1982). Molecular Cloning: A Laboratory Manual. Cold Spring Harbor, Cold Spring Harbor Laboratory, New York.
7. Sakonju S, Bogenhagen DF, Brown DD, (1980). A Control Region at the Center of the 5S RNA Gene Directs Specific Initiation of Transcription: The 5' Border of the Region. Cell 19:13-24.
8. Caplan AI (1976). A Simplified Procedure for Preparing Myogenic Cells for Culture. J Embryol Exp Morph 35:175-181.
9. Konigsberg IR (1979). Skeletal Myoblasts in Culture. Meth Enzymol 58:511-527.
10. Simpson P, Sarion S (1982). Differentiation of Rat Myocytes in Single Cell cultures with and without Proliferating Nonmyocardial Cells: Cross-striations, Ultrastructure and Chronotropic Response to Isoproterenol. Circ Res 50:101-116.
11. Lopata MA, Cleveland DW, Sollner-Webb B (1984). High Level Transient Expression of a Chloramphenicol Acetylatransferase Gene by DEAE-Dextran Sulfoxide or Glycerol Shock Treatment. Nucl Acid Res 12:5707:5717.
12. Benyajati C, Wang H, Reddy A, Weinberg E, Sofer W (1980). Alcohol Dehydrogenase in Drosophila: Isolation and Characterization of Messenger RNA and cDNA Clone. Nucl

Acid Res 8:5649-5667.
13. Goldberg D (1980). Isolation and Partial Characterization of the Drosophila Alcohol Dehydrogenase Gene. Proc Natl Acad Sci:5794-5798.

Molecular Biology of Muscle Development, pages 559-573

THE PROMOTER REGIONS OF TWO MUSCLE-SPECIFIC GENES CONTAIN DNA SEQUENCES WHICH CONFER TISSUE- AND STAGE-SPECIFIC EXPRESSION OF CHIMERIC GENES IN TRANSFECTED MYOGENIC CELLS[1]

Uri Nudel, Danielle Melloul, David Greenberg and David Yaffe

Department of Cell Biology, The Weizmann Institute of Science, Rehovot 76100, Israel

ABSTRACT The expression of muscle-specific genes, and of chimeric genes containing the promoter regions of muscle-specific genes, was analyzed in transfected myogenic L8 and nonmyogenic 3T3 cells. It is demonstrated that the promoter regions of muscle-specific genes contain DNA sequences which confer developmentally regulated expression of fused CAT genes in myogenic L8 cells. The expression of genes which are not muscle-specific was not upregulated following differentiation of transfected myogenic cells.

To test the conservation of regulatory sequences and cellular signals we transfected rat myogenic cells with the chicken skeletal muscle actin gene and found that the expression of this gene was developmentally regulated, thus suggesting that the regulatory cis-acting sequences and the cellular trans-acting factor(s) have been conserved for at least 300 million years.

----

[1]This investigation was supported by the Muscular Dystrophy Association, Inc., USA, by the National Institutes of Health (Grant No. GM-22767) and by the US-Israel Binational Science Foundation. U. Nudel is the incumbent of the A. and E. Bloom Career Development Chair for Cancer Research.

INTRODUCTION

The DNA-mediated gene transfer technique has proved to be a powerful tool for analyzing the control of expression of eukaryotic genes. It has become possible in recent years to isolate eukaryotic genes, to modify them, and to study their expression after re-introduction into cells. This approach has been used to identify and analyze common transcription control sequences such as the CAAT box and ATA box, and cis-acting gene-specific control sequences such as the sequences which regulate induction of transcription by hormones, heavy metals, and heat-shock (e.g., 1-3). The DNA-mediated gene transfer technique is widely used now to identify and to study control sequences involved in the regulation of expression of tissue- and stage-specific genes.

During terminal differentiation of muscle cells in culture, mononucleated myoblasts fuse to form muscle fibers. Cell fusion is accompanied by a rapid accumulation of muscle-specific mRNAs and the corresponding proteins. Earlier studies have demonstrated that: (a) The mRNAs coding for several contractile proteins accumulate just before and during the phase of rapid cell fusion. These mRNAs are not detectable in dividing myoblasts (4, 5). (b) The genes coding for contractile proteins are relatively insensitive to DNAase I in nuclei from dividing myoblasts, and become more sensitive to DNAase I in nuclei from muscle fibers (6). Such a shift in DNAase sensitivity of a gene is associated with its progression from an inactive state to a transcriptionally active one (7). (c) Genes coding for the major skeletal muscle contractile proteins are located on different chromosomes (8).

These findings have suggested that muscle-specific genes are regulated predominantly at the transcriptional level, and that the activation of these genes depends on sequences located near the structural genes. The experiments described in this presentation aimed at identifying DNA sequences involved in the developmentally regulated expression of muscle-specific genes.

RESULTS

DNA Sequences from the 5'-Flanking Region of the Myosin Light Chain 2 Gene Confer Developmentally Regulated Expression of a Fused Bacterial CAT Gene in Stably Transfected Myogenic Cells

We have previously described the construction of two chimeric genes, one containing 730 bp 5' flanking DNA and about two-thirds of the rat skeletal muscle actin gene spliced to the 3' end region of human ε-globin gene; and another one containing 730 bp 5' flanking DNA, the first (untranslated) exon of the skeletal muscle actin gene, and 25 base pairs of the first intron spliced to the bacterial structural gene coding for chloramphenicol acetyl transferase (CAT). These two fused genes were introduced into myogenic L8 cells and their expression was analyzed. We found that in many of the isolated myogenic clones the expression of these fused genes was greatly increased during differentiation of the cultures. In contrast, in clones containing a fused β-actin/CAT gene we observed relatively high levels of CAT activity that did not change upon differentiation of the cultures. We concluded that the promoter region of the skeletal muscle actin gene contains information for tissue- and stage-specific expression of the gene (9, 10).

To examine whether such information was contained in the promoter region of another gene expressed during terminal differentiation of myogenic cells, we constructed a fused gene consisting of the DNA fragment extending from about 1.2 Kbp 5' to the myosin light chain 2 (MLC2) structural gene to the beginning of the first exon, spliced to the CAT gene (Fig. 1). Myogenic L8 cells were cotransfected with this plasmid (pMLCCAT) and the plasmid pIPB1 which contains a neomycin resistance marker. Clones resistant to the antibiotic G-418 were pooled, grown to confluency, and induced to differentiate. Cell extracts were prepared from undifferentiated and differentiated cultures and assayed for CAT activity. In all the pools tested, CAT activity in extracts from differentiated cultures was 20-50-fold higher than in extracts from undifferentiated cultures (Fig. 2). It should be pointed out that the expression of the rat MLC2 gene was tissue-specific in transgenic mice (11, 12).

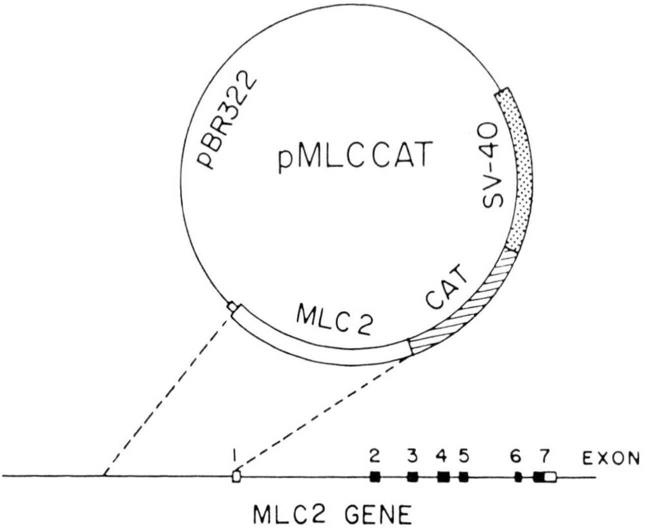

FIGURE 1. The structure of plasmid pMLCCAT. pMLCCAT consists of a 1210 bp long DNA fragment of the 5' flanking sequences of rat MLC2 gene (from the HindIII site at position -1206 to the Hinfl site at position +4 (26)) inserted into the BglII site of the plasmid pCAT-3N (obtained from P. Gruss). The plasmid pCAT-3N is a CAT expression vector containing the coding region of the bacterial CAT gene, the SV40 T antigen intron and poly(A) addition signal, pBR322 sequences and polylinker (thin open bar). In the scheme of the MLC2 gene exons containing untranslated and translated regions are presented by open and black bars, respectively.

In order to identify the DNA sequences in the promoter regions of the skeletal muscle actin gene and MLC2 gene that confer tissue- and stage-specific expression of these genes, we constructed several plasmids in which various parts of the 5' flanking DNA were deleted from the promoter regions of pα-CAT and pMLCCAT and assayed their expression in myogenic cells. These experiments are still in progress, but we have good indications that the regions

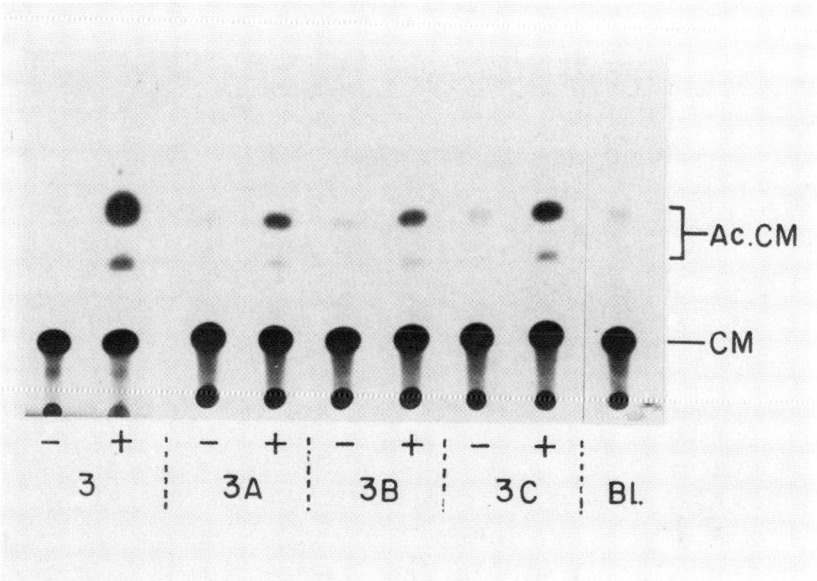

FIGURE 2. CAT activity in pooled clones containing the plasmid pMLCCAT. L8 cells were transfected with pMLCCAT as described previously (9). After the selection with the antibiotic G418, clones were pooled (20-50 clones in each pool) and grown to confluency. Cell extracts were prepared from undifferentiated (-) and differentiated (+) cultures and assayed for CAT activity (9). 3, 3A, 3B, and 3C represent different pools of clones. CM, chloramphenicol; Ac.CM, acetylated chloramphenicol; BL, blank assay to which no extract was added (in 3A, 3B, and 3C the substrate was slightly contaminated with Ac.CM).

between positions +4 and -140 (the cap site is at position +1) are sufficient to confer developmentally regulated expression. However, additional sequences may contribute to the regulation of expression of the genes (data not shown).

TABLE 1
Expression of the chicken skeletal muscle actin gene in transfected rat myogenic cells

Clone No.	Gene copy No.	Induction (fold)
1a	200-400	21.5
7a	3-6	12.7
4c	4-8	7.3
1d	1-3	4.5
2a	40-80	4.4
6b	150-300	3.4
6a	10-20	1.81
1b	4-8	1.44
5b	30-60	1.29
3a	2-4	U.D.*
3b	2-4	U.D.*

*U.D. = chicken actin mRNA was undetectable in undifferentiated and differentiated cultures. The transferred gene copy number was estimated from Southern blot analysis of DNA extracted from each clone. The amount of the chicken actin gene transcript in undifferentiated and differentiated cultures of each of the clones was determined by endonuclease S1 analysis and by dot blot hybridization as described elsewhere (10). Induction (fold) is the ratio between the amounts of chicken skeletal muscle actin mRNA sequences in RNA prepared from differentiated and undifferentiated cultures.

The Expression of the Chicken Skeletal Muscle Actin Gene is Developmentally Regulated in Transfected Rat Myogenic Cells

To test the evolutionary conservation of the DNA sequences specifying the differentiation-dependent expression of a muscle-specific gene, we introduced the chicken skeletal muscle actin gene into rat myogenic cells and analyzed its expression during differentiation of the cell cultures. In these experiments we used the plasmid pGα-actin-1, which contains the chicken skeletal muscle actin gene plus about 2 Kbp 5' flanking DNA and about 2 Kbp 3' flanking DNA (13). We isolated 11 clones containing 1-400 copies of the transferred gene. The expression of the transferred gene was assayed by the S1 mapping technique. Chicken actin mRNA was detected in 9 clones. In 6 of these the amount of chicken actin mRNA increased 3.4-21.5-fold after differentiation of the cultures (Table 1; Ref. 14). These results indicate that DNA sequences specifying differentiation-dependent expression of the skeletal muscle actin gene have been conserved for at least 300 million years.

It was recently reported that the chicken skeletal muscle actin gene was constitutively expressed in transfected mouse myogenic C2 cells and was not upregulated during differentiation of the cultures (15). Possible reasons for the differences between these results and our results were discussed elsewhere (14).

As a control experiment for the specificity of the regulated expression of transferred muscle genes in myogenic cells, we analyzed the expression of an exogenous β-globin gene, which is normally expressed only during terminal differentiation of erythropoietic cells. The plasmid used in these experiments was pMHβ, containing a fused mouse/human β-globin gene. The expression of this gene was developmentally regulated when introduced into mouse erythroleukemic cells (16) and tissue-specific in transgenic mice (17). Myogenic L8 cells were cotransfected with the plasmids pMH and pIPB1, and 11 neomycin resistant clones were isolated. RNA was extracted from undifferentiated and differentiated cultures of these clones, and assayed for the chimeric β-globin mRNA by the S1 mapping technique. In 8 out of 9 clones in which the globin gene was expressed, the amount of the β-globin gene transcripts decreased after differentiation. In one clone a 2-fold increase in the amount of β-globin mRNA sequences was observed (Table 2; Ref. 14). These results indicate that the increase in the expression of the chicken skeletal muscle actin gene following the differentiation of transfected cells is a specific response of a gene normally expressed during myogenesis, and not a consequence of nonspecific changes associated with terminal differentiation.

TABLE 2
Expression of the mouse/human β-globin chimeric gene in transfected rat myogenic cells

Clone No.	Differentiated/ undifferentiated cultures
G1	0.3
G2	0.6
G3	0.5
G4	U.D.*
G5	U.D.*
G6	0.6
G7	0.7
G8	2.0
G9	0.7
G10	0.3
G11	0.2

*U.D. = β-globin mRNA was undetectable in undifferentiated and differentiated cultures. The amount of the chimeric β-globin gene transcript in undifferentiated and differentiated cultures was determined by endonuclease S1 analysis, as described elsewhere (14). The numbers represent the ratio between the amounts of β-globin mRNA sequences in RNA prepared from differentiated and undifferentiated cultures.

DNA Sequences near the Promoter Region of the Rat Skeletal Muscle Actin Gene Confer Developmentally Regulated Expression of a Fused Bacterial CAT Gene in Transiently Transfected Myogenic Cells

The experiments described above were done with stably transfected clones in which the transferred genes were integrated into the host cell genome. In order to exclude possible effects of neighboring host DNA sequences in the site of integration on the expression of the transferred genes, we studied their expression in transient transfection assays. In these experiments we assayed the efficiency of several promoters in nonmuscle cells and in undifferentiated and differentiated L8 cells. As can be seen in Fig. 3, the plasmids pSV2-CAT and pβ-CAT are efficiently transcribed in 3T3 and L8 cells; the plasmid pα-CAT is efficiently transcribed only in differentiated L8 cells. These results demonstrate that the tissue- and stage-specific expression of the transferred muscle genes does not depend on the integration of the genes in specific chromosomal sites.

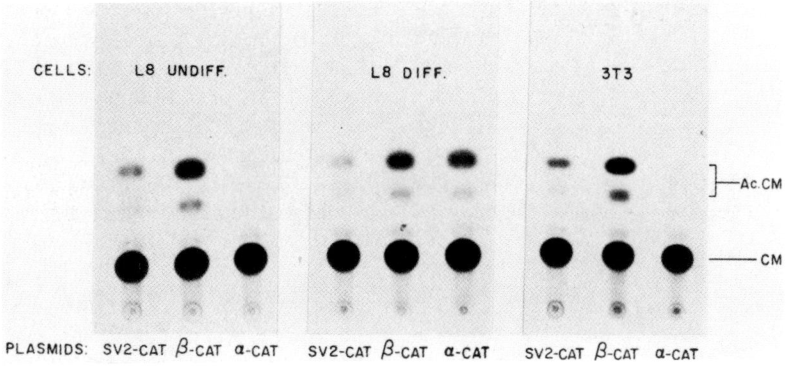

FIGURE 3. CAT activity in 3T3 and L8 cells transiently transfected with the plasmids pα-CAT, pβ-CAT and pSV2-CAT. Cultures of 3T3 cells, proliferating L8 cells, and L8 cells grown for 24 h in the differentiation-promoting H1 medium, were transfected with 10 ug/2.6 x 10^6 cells of the plasmids pα-CAT, pβ-CAT or psV2-CAT. L8 cells and 3T3 cells were harvested 20 and 48 h after transfection, respectively. Aliquots of extracts prepared from the same number of cells (or nuclei in fibers) were assayed. Assay conditions were described previously (9). CM, chloramphenicol; Ac.CM, acetylated chloramphenicol.

DISCUSSION

In this communication and in previous publications (9, 1Ø) we presented data suggesting that information for tissue- and stage-specific expression of the genes coding for skeletal muscle actin and skeletal muscle MLC2 resides in or near the promoter regions of the genes. Since our experiments were done with constructions that do not contain the entire gene sequences, our results do not rule out the possibility that other parts of the genes, such as introns and 3' untranslated regions, are involved in some aspects of tissue-specific control of expression. It should be noted that while the expression of fused genes containing the promoter regions of the skeletal muscle actin gene and MLC2 gene followed the upregulation pattern of the endogenous skeletal muscle actin and MLC2 genes, the expression of the β-actin-CAT fused gene did not follow the downregulation pattern of the endogenous β-actin gene (5, 9, 1Ø). On the other hand, the complete chicken β-actin gene was downregulated in mouse C2 myogenic cells (15). Results presented at this meeting (18) indicate that the downregulation of the chicken β-actin gene in transfected mouse C2 myogenic cells depends on sequences in the 3' half of the gene, and not on sequences in the promoter region.

Since the evolution of the nucleotide sequences of coding regions of genes is dominated primarily by the selective pressure operating at the protein level, it is conceivable that sequences or structures necessary for the regulation of gene expression will be located mainly in noncoding regions (5' flanking, 5' untranslated, introns, and 3' untranslated and flanking DNA). Comparison of the nucleotide sequences of vertebrate actin genes reveals a high degree of sequence homology between the 3' untranslated regions of genes coding for homologous (isotypic) actins in different organisms (e.g., chicken, rat, and human β-actin genes; chicken, rat, and human cardiac actin genes) but not between the 3' untranslated regions of genes coding for actin isoforms that differ in their specificity of expression (e.g., rat skeletal muscle and cardiac actin genes) (19). The sequence conservation in the 3' untranslated regions of homologous actin genes is greater than that found in the silent sites of the coding regions (which are conserved to some extent due to the constraint of the nonrandom codon usage), and certainly greater than expected of neutral sequences. A similar pattern of sequence conservation in the 3' untranslated regions was also found in several other genes (reviewed in Ref. 19). Furthermore, while there is a great variation in the size of the 3' untranslated region of genes coding for different proteins and even for different actin isoforms, mRNAs coding for isotypic proteins in distantly related organisms often have 3' untranslated regions of similar lengths. These observations suggest that the 3' untranslated region may be involved, at least in some genes, in the transcriptional or post-transcriptional regulation of expression of these genes (19). Experiments are currently being carried out to examine whether the 3' untranslated regions of α- and β-actin genes are involved in the regulation of these genes during myogenesis.

In a significant number of myogenic rat clones transfected with the chicken skeletal muscle actin gene, the expression of transferred genes increased during differentiation of the cultures (Table 1). These results suggest that the cis-acting sequences, and probably also the cellular signals conferring tissue-specific expression of the skeletal muscle actin gene, have been conserved for at least 300 million years. Indeed, comparison of the nucleotide sequences in the promoter regions of rat and chicken skeletal muscle actin genes reveals several blocks of highly conserved sequences (14). Also the sequences in the promoter regions of rat and human MLC2 genes are highly conserved. However, no obvious sequence homology between the promoter regions of rat MLC2 gene and skeletal muscle actin gene was detected, except for 100% homology in a block of 12 nucleotides that include the ATA box and 6 nucleotides upstream (U. Nudel and C. Cisar, unpublished observations).

Our experiments (9, 10) as well as experiments done by others (e.g., 16, 20-23) have demonstrated that the regulation of expression of transferred genes is less stringent than the regulation of expression of the endogenous genes. Usually, transferred tissue- and stage-specific genes are expressed at low levels in tissues or stages of development in which the homologous endogenous genes are not expressed (e.g., low levels of expression of muscle-specific genes in dividing myoblasts and low levels of expression of the β-globin gene in myoblasts and myotubes). A model based on gene transfer experiments, implying two steps in the activation of β-globin gene, was proposed by Charnay et al.

(24). According to the model, the first step of activation of the endogenous β-globin gene is a tissue-specific derepression step which may be due to a change in the chromatin structure, removal of a repressor, or demethylation of DNA. This is followed by positive activation or induction by a tissue-specific trans-acting factor(s).

It has been shown that transferred DNA usually integrates into a single or very few chromosomal sites; the sites of integration are different in each clone (25). Since selection of stably transfected myogenic cells depends on the expression of the selective marker, we selected only clones of myogenic cells in which the cloned genes were inserted into a chromatin domain permissive for expression. The low background activity of these transferred muscle-specific genes in undifferentiated cells may be attributed, at least in part, to the effect of the site of integration. Similarly genes located on episomes (in the transient transfection experiment) are in a chromatin environment favoring expression. In order to be fully expressed the transferred muscle-specific genes require activation by a tissue-specific factor(s). On the other hand, the endogenous muscle-specific genes are probably located in a chromatin domain which is inactive in nonmyogenic cells and in undifferentiated myogenic cells. Thus, in order to be fully expressed they require a tissue-specific derepression step followed by activation by a tissue-specific transacting factor(s) (see Refs. 9 and 12 for further discussion). Experiments are currently being carried out aiming at the identification of the trans-acting regulatory protein(s) and the control sequences.

ACKNOWLEDGMENTS

The excellent technical assistance of Ms. Sara Neuman and Ms. Ora (Saxel) Fuchs and Ms. Zehava Levy is greatly appreciated. We thank Dr. T. Maniatis for plasmid pMH , Dr. P. Gruss for the plasmid pCAT-3N, Dr. C. Ordahl for the plasmid pG -actin-1 Drs. C.M. Gorman and B.H. Howard for plasmids pSVØ-CAT and pSV2-CAT, Dr. S. Silverstein for plasmid pIPB1.

REFERENCES

1. Dean DC, Knoll BJ, Riser ME, O'Malley BW (1983) A 5'-flanking sequence essential for progesterone regulation of an ovalbumin fusion gene. Nature 305:551
2. Karin M, Haslinger A, Holtgreve H, Richards RI, Krauter P, Westphal HM, Beato M (1984) Characterization of DNA sequences through which cadmium and glucocorticoid hormones induce human metallothionein IIA gene. Nature 308:513
3. Pelham HRB, Bienz M (1982) A synthetic heat shock promoter element confers heat-shock inducibility on the herpes simplex virus thymidine kinase gene. EMBO J 1:1473
4. Yablonka Z, Yaffe D (1977) Synthesis of myosin light chains and accumulation of translatable mRNA coding for light chain-like polypeptides in differentiating muscle cultures. Differentiation 8:133
5. Shani M, Zevin-Sonkin D, Saxel O, Carmon Y, Katcoff D, Nudel U, Yaffe D (1981) The correlation between the synthesis of skeletal muscle actin, myosin heavy chain, and myosin light chain and the accumulation of the corresponding mRNA sequences during myogenesis. Dev Biol 86:483
6. Carmon Y, Czosnek H, Nudel U, Shani M, Yaffe D (1982) DNAase I sensitivity of genes expressed during myogenesis. Nucl Acids Res 10:3085
7. Mathis D, Oudet P, Chambon P (1980) Structure of transcribing chromatin. Cohn WE (ed) "Progress in Nucleic Acids Research and Molecular Biology," New York: Academic Press, p 1
8. Czosnek H, Nudel U, Shani M, Barker PE, Pravtcheva DD, Ruddle FH, Yaffe D (1982) The genes coding for muscle contractile proteins, myosin heavy chain, myosin light chain 2 and skeletal muscle actin are located on 3 different mouse chromosomes. EMBO J 1:1299
9. Melloul D, Aloni B, Calvo J, Yaffe D, Nudel U (1984) Developmentally regulated expression of chimeric genes containing muscle actin DNA sequences in transfected myogenic cells. EMBO J 3:983
10. Nudel U, Melloul D, Aloni B, Greenberg D, Yaffe D (1984) Developmentally regulated expression of chimeric muscle genes transferred into myogenic cells. In Davidson E, Firtel R (eds) "Molecular Biology of Development," Alan R Liss, Inc, New York, p 647

Molecular Biology of Muscle Development, pages 575–590

EXPRESSION OF RAT SKELETAL MUSCLE GENES IN TRANSGENIC MICE[1]

Moshe Shani and David Yaffe

Department of Cell Biology
The Weizmann Institute of Science
Rehovot, 76100, Israel

ABSTRACT To study the control elements involved in tissue- and stage-specific expression of genes during myogenesis, the rat skeletal muscle myosin light chain 2 (MLC2) gene and an actin/globin chimeric gene were introduced into mice by microinjection into fertilized eggs. Four transgenic mice carrying the rat MLC2 gene and four transgenic mice carrying the chimeric gene were produced. In the offspring of two of the three transgenic mice carrying the rat MLC2 DNA sequences that were analyzed for expression, transcripts of the introduced gene were detected in skeletal muscle, but not in several other tissues tested. Thus, the injected rat DNA fragment containing the MLC2 gene included cis-acting sequences that were sufficient to specify the correct developmental expression of this gene in transgenic mice. None of the offspring of the transgenic mice carrying the actin/globin chimeric gene expressed the inserted DNA sequences in skeletal muscle. In one of them the introduced gene was expressed in testes, thymus and lung.

----

[1]Supported by the Muscular Dystrophy Association, Inc., USA, and by the Israel Academy of Sciences and Humanities.

INTRODUCTION

Transfection of cell cultures is now widely used as a means to study the regulation of cloned genes. Investigations using a number of developmentally regulated genes have shown that the DNA sequences in the cloned DNA fragment include cis acting control sequences required for tissue- and stage-specific expression of the genes (1-8). However, several important differences between the expression of the native and cloned genes have been noted. For example, low levels of expression of the foreign gene are often detected also in undifferentiated cells or in cell types in which the expression of the native gene is undetectable (e.g., Refs. 2, 7, 8). It is possible that by introducing cloned genes into somatic cells, important processes related to the programmed control of expression of tissue-specific genes (e.g., DNA methylation) are circumvented. It is, therefore, important to investigate also the control of expression of cloned genes introduced into the germ line. Such genes would be present in every cell of the organism and its offspring throughout development and would, therefore, be exposed to the same developmental signals as the endogenous genes. This approach had been applied mainly in Drosophila and in mice. In Drosophila, single copy genes have been inserted using the specific transposable P-element as a vector (9). In every case reported, the expression of the transferred gene was tissue-specific (10-14). In mice, several genes were inserted into the germ line (15-27); however, expression was reported in only a few cases (17, 19, 21, 23-27).

In the present study the rat skeletal muscle myosin light chain 2 (MLC2; 28) and an actin/globin chimeric gene (7) were injected into mouse fertilized eggs. We have shown previously that expression of MLC2 and skeletal actin genes is undetectable in proliferating myoblasts as well as in nonmuscle tissues (29-31). Expression of these genes is associated with changes in their chromatin organization, as measured by DNAase I sensitivity (32) and by sensitivity to restriction enzymes (Einat et al., unpublished). This indicates that the transcriptional activation of these genes occurs during the transition to the stage of formation of multinucleated fibers.

We report here on the results obtained with four transgenic mice carrying the rat MLC2 DNA sequences and four transgenic mice carrying the actin/globin chimeric DNA sequences. In the offspring of two out of the 3 transgenic mice carrying the rat MLC2 gene, the introduced gene was expressed specifically in skeletal muscle. None of the offspring of the transgenic mice that carry the actin/globin chimeric gene expressed the inserted DNA sequences in skeletal muscle. In one of them, the introduced gene was expressed in inappropriate tissues such as testes, thymus and lung.

RESULTS

Expression of the Rat MLC2 Gene in Progeny of Transgenic Mice

A recombinant plasmid containing the entire rat skeletal muscle MLC2 gene, in a 4.7 Kb DNA fragment (pRH3MLC; 28) was used for injection into fertilized eggs of (BALB/c x C57BL/6J)F1 mice. Male pronuclei were microinjected with 1-2 pl DNA solution containing about 500 molecules. Spleen DNA samples from born mice were analyzed by Southern blot hybridization. Four out of 26 born mice containing the rat MLC2 DNA sequences were identified. Figure 1 shows Southern blot analysis of DNA isolated from four positive mice and a control mouse. As expected, hybridizable DNA fragments of 4.7 and 4.3 Kb were generated in the HindIII digest of spleen DNA of transgenic mice carrying the MLC2 gene (Fig. 1). In addition to the expected DNA fragments, several minor DNA fragments hybridized to the probes. The latter could be either junction fragments with cellular DNA or rearranged sequences. Southern blots of DNA digested with restriction enzymes which cut the plasmid only once indicated that the majority of the injected DNA is probably integrated tandemly in a single site or a few sites in the host genome. The number of copies of the inserted DNA was estimated to be 20, 10, <1, and 16 copies per cell for transgenic mice MLC/1, MLC/2, ML/3 and MLC/4, respectively.

Except for mouse MLC/3, the three transgenic mice transmitted the introduced genes to about 50% of their progeny, indicating the integration of the rat gene into a single chromosomal site. Mouse MLC/3, containing less than 1 copy per cell of the rat MLC2 gene, did not transmit the inserted DNA to any of its first 60 offspring, indicating that this mouse is probably a mosaic in which the injected DNA was integrated in only a small fraction of its somatic cells.

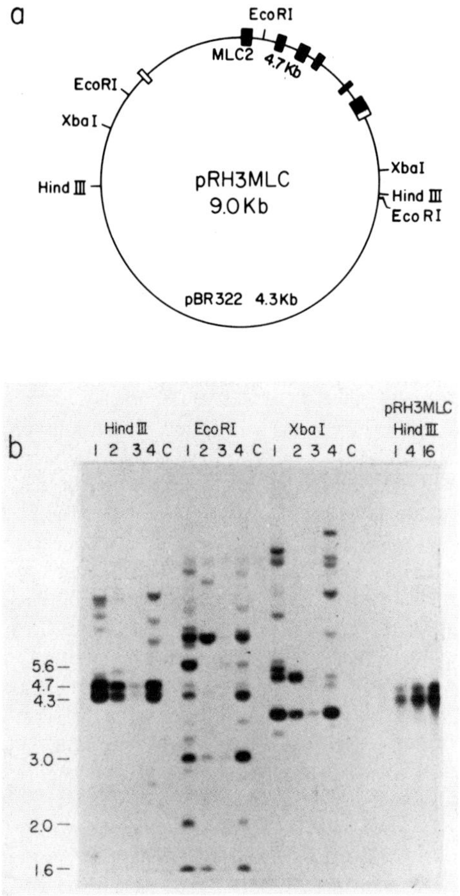

FIGURE 1. Map of plasmid pRH3MLC and Southern blot analysis of DNA from four transgenic mice. (a) The 4.7 Kb HindIII DNA fragment containing the rat MLC2 gene was cloned into the HindIII site of pBR322 (28). Empty boxes represent the 5' and 3' untranslated regions; solid boxes represent coding sequences. (b) Total spleen DNA (10 ug) from the four transgenic mice (1 to 4) and a control mouse DNA (C) was digested with HindIII, EcoRl or XbaI. The DNA fragments were blot hybridized with ^{32}P-labeled pRH3MLC DNA. Plasmid pRH3MLC DNA samples digested with HindIII corresponding to 1, 4 and 16 copies per diploid mouse genome were included to estimate the copy number of the inserted DNA in transgenic mice. Molecular weights are shown in kilobases.

Recent reports from several laboratories have indicated that in about 1/3 of the cases tested, attempts to produce homozygous strains of transgenic mice resulted in lethality, due to insertional mutagenesis (33, 34). However, we were able to produce homozygous strains of transgenic mice carrying the MLC2 gene.

To determine whether the expression of the rat MLC2 gene in transgenic mice is tissue-specific, RNA was extracted from muscle as well as from seven other tissues of progeny of transgenic mice MLC/1, MLC/2 and MLC/4. The presence of rat MLC2 transcripts in the RNA preparation was determined by S1 endonuclease mapping analysis, using as a probe a DNA fragment derived from the 5' untranslated exon of the rat gene. No rat transcripts could be detected in RNA prepared from skeletal muscle or any of the other tissues of transgenic mouse MLC/1. In contrast, in the offspring of mice MLC/2 and MLC/4, rat transcripts were detected in skeletal muscle but not in RNA prepared from thymus, testes, brain, heart, lung, kidney and liver. Figure 2 shows the S1 analysis with RNA prepared from several tissues of two progeny of the transgenic mouse MLC/4.

The relative levels of the rat transcripts and their site of initiation was then determined. The noncoding strand of a 289 bp Sau3A-EcoRI DNA fragment, ^{32}P-labeled with polynucleotide kinase, was hybridized to various RNA preparations, as indicated in Fig. 3. RNA from rat skeletal muscle or from differentiated cultures of the rat myogenic cell line L8 protected a 50 nucleotide DNA fragment (Fig. 3a). The appearance of a doublet spaced one nucleotide apart is most likely due to S1 nibbling. RNA from control and transgenic mouse skeletal muscle protected a 76-nucleotide DNA fragment (Fig. 3b). This unexpected finding suggests that there is a great sequence homology between the 5' region of the mouse mRNA and the 5' exon plus the proximal 20 nucleotides of the flanking region of the rat MLC2 gene. This implies that the endogenous mouse mRNA is initiated about 20 nucleotides upstream to the cap site of the rat gene. Under more stringent hybridization conditions this cross reactivity almost disappeared (Fig. 3a). RNA of 3 out of 4 progeny of transgenic mouse MLC/2 and 2 progeny of mouse MLC/4 protected the diagnostic 50 nucleotide DNA fragment. It was estimated that the levels of expression were about 1 to 2% and 2 to 4 times those found in rat skeletal muscle, in transgenic mice MLC/2 and MLC/4, respectively.

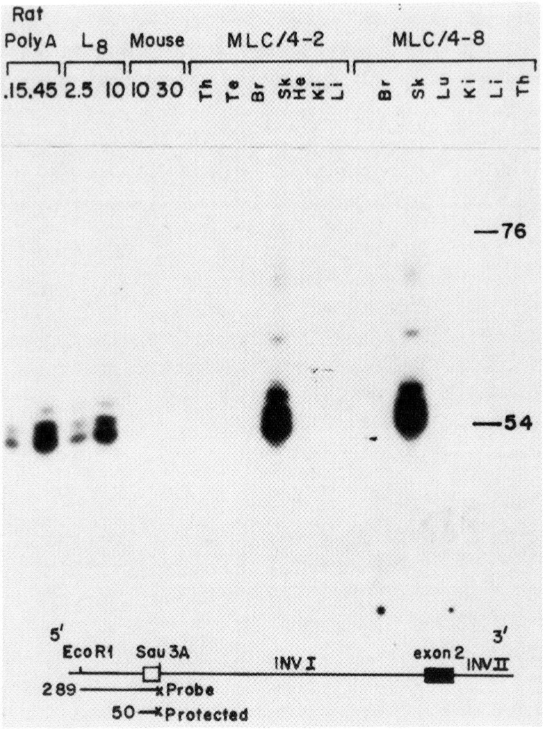

FIGURE 2. Expression of the rat MLC2 gene in skeletal muscle of progeny of transgenic mouse MLC/4. Total RNA (20 ug) from the indicated tissues, and the indicated amounts of total RNA from mouse skeletal muscle, differentiated cultures of the rat myogenic cell line L8, or polyA$^+$ RNA of rat skeletal muscle were hybridized with 5'-end labeled Sau3A-EcoR1 DNA fragments (see scheme, bottom part). Th, thymus; Te, testes; Br, brain; Sk, skeletal muscle; He, cardiac muscle; Ki, kidney; Li, liver; Lu, lung. After treatment with S1 nuclease, the resistant hybrids were sized on 6% polyacrylamide/7 M urea gel. Bottom part: Map of the 5' region of the rat MLC2 gene and the probe used for the hybridization. Open box: The 5' untranslated exon. Solid box: The first coding exon. The 5' label is indicated by an asterisk and the length of the probe and the protected fragment are also indicated.

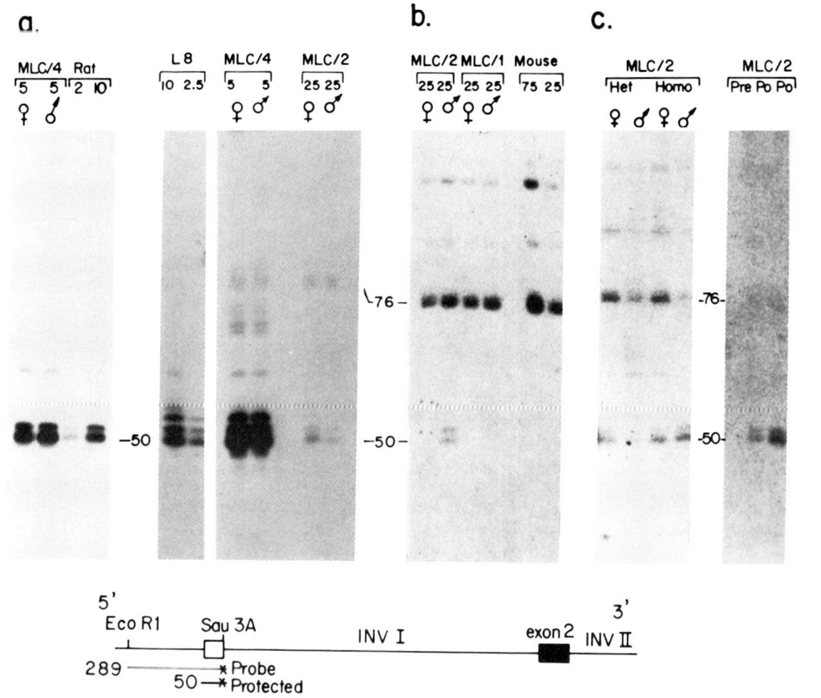

FIGURE 3. The initiation site of transcription and the relative levels of the rat MLC2 gene transcripts in progeny of transgenic mice MLC/1, MLC/2 and MLC/4, and in primary muscle cultures of mouse MLC/2. Total RNA was isolated from the following sources: Skeletal muscle of male (♂) and female (♀) progeny of transgenic mice MLC/1, MLC/2 and MLC/4, rat (Rat) and mouse (Mouse) skeletal muscle, differentiated cultures of the rat myogenic cell line L8 and proliferating mononucleated myoblasts (Pre) or multinucleated fibers (Po) from primary muscle cultures prepared from homozygous progeny of mouse MLC/2. The amounts of RNA (ug) used in (a) and (b) are indicated. In (c) 20 ug of RNA were used. In (a) hybridization was performed at 42°C and S1 digestion at 37°C. In (b) and (c) hybridization was at 37°C and S1 digestion at 32°C. For details see legend to Fig. 2. Het and Homo - heterozygous and homozygous progeny of the transgenic mouse MLC/2.

The expression of the rat MLC2 gene in the offspring of transgenic mouse MLC/2 is a stable heritable characteristic. Figure 3c shows that muscle RNA from the third generation offspring contained rat MLC2 mRNA in amounts comparable to those found in the first generation.

We have previously shown that in proliferating rat myoblasts there are undetectable levels of MLC2 mRNA, and that the main increase in the amount of this mRNA occurs during the stage of cell fusion (30, 31). To test whether the introduced gene is expressed specifically in differentiated cells, primary skeletal muscle cultures were prepared from the homozygous progeny of transgenic mouse MLC/2. RNA was isolated from proliferating mononucleated myoblasts and from differentiated cultures harvested at early and advanced stages of formation of multinucleated fibers. Figure 3c demonstrates that in cultures consisting mostly of mononucleated cells, rat MLC2 transcripts are not detectable, while increasing amounts of this mRNA is found in differentiating cultures containing multinucleated fibers. These results show that the expression of the introduced rat gene, similar to the endogenous gene, is stage-specific.

Expression of an Actin/Globin Chimeric Gene in Transgenic Mice

The actin/globin chimeric gene (comprised of about 2/3 of the 5' region of the rat skeletal muscle actin gene, plus 730 bp upstream from the cap site and about 1/3 of the 3' region of the human ε-globin gene, including the polyadenylation site, and 172 bp 3' flanking region (Fig. 4), introduced into the rat myogenic cell line L8, was developmentally regulated in most transfected clones. The expression of the transfected gene greatly increased during differentiation, and in some clones, the level of expression exceeded that of the endogenous actin gene (7). Furthermore, the temporal relation between the differentiation of the transfected cells and the rate of accumulation of the chimeric gene transcripts was very similar to that of the native skeletal muscle actin gene. These results suggested that the transfected gene and the endogenous counterpart responded to similar control signals.

This gene construct was injected into mouse fertilized eggs, as described above. Out of 36 born mice,

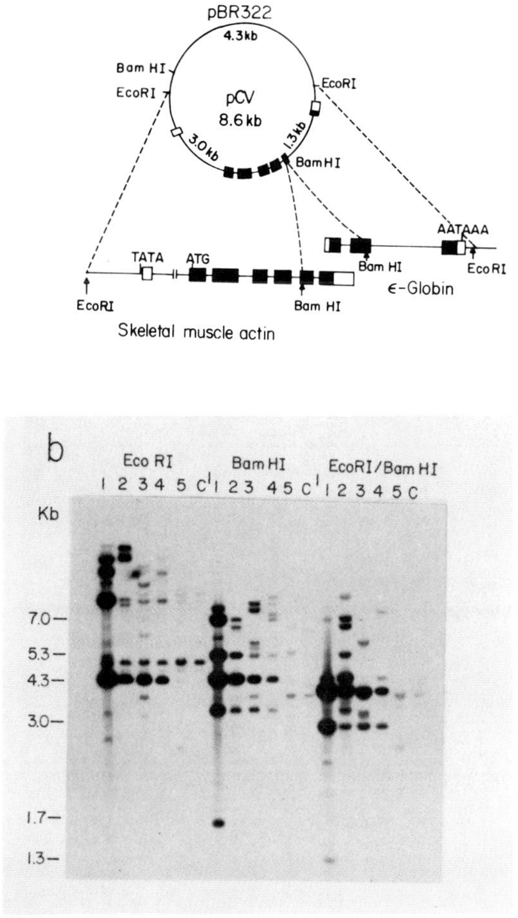

FIGURE 4. Map of plasmid pCV and Southern blot analysis of DNA from four transgenic mice. (a) The chimeric rat actin/globin gene was constructed as described in the text (7). Empty bars: 5' and 3' untranslated regions. Solid bars: Coding regions. (b) Total spleen DNA (10 ug) from transgenic mice (1 to 5) and a control mouse DNA (c) was digested with EcoRI, BamH1 or EcoR1/Bam/H1. The DNA fragments were blot hybridized with ^{32}P-labeled pCV DNA. Molecular sizes are shown in kilobases.

4 transgenic mice were identified. Figure 4 shows Southern blot analysis of DNA isolated from these mice. Mouse CV/5 carried an incomplete plasmid. Instead, a small DNA fragment (the structure of which was not determined) was integrated. The number of copies of the inserted DNA was estimated at 15, 8, 8 and 4 copies per cell for mice CV/1, CV/2, CV/3 and CV/4, respectively.

The presence of the chimeric gene transcripts in the four transgenic mice was determined by the S1 endonuclease mapping, using as a probe a 187 bp DdeI DNA fragment derived from the human ε-globin region of the construct. The labeled DNA probe was hybridized with RNA of skeletal muscle and other tissues of these mice (as indicated in Fig. 5), as well as with RNA of differentiated cultures of a stably transfected clone of the rat myogenic cell line (L8-46; Ref. 7). RNA from differentiated L8-46 cells protected a DNA fragment of about 121 nucleotides. No detectable signal could be seen with skeletal muscle RNA of transgenic mice CV/1, CV/3 and CV/4. Similar results were obtained with RNA from skeletal muscle of mouse CV/2 (data not shown).

In three of the transgenic mice no transcripts could be detected in any other of the tissues examined. However, transcripts were detected in RNA from testes, thymus and lung of mouse CV/3. Such transcripts were most abundant in testes and at a lower level in thymus and lung. No transcripts were detected in the liver or kidney RNA. These transcripts protected the same size fragment of the probe as that protected by RNA from differentiated cultures of L8-46, indicating that they possess the correct 3' end. No information is yet available on the initiation of transcription of the chimeric gene in these tissues.

DISCUSSION

In the present study a cloned rat MLC2 gene and a chimeric actin/globin gene were inserted into the germ line of mice via microinjection into fertilized eggs. About 1-2% of the injected eggs gave rise to transgenic mice, in which multiple copies of the injected DNA had integrated in tandem into the host genome.

In the offspring of one transgenic mouse carrying the rat MLC2 gene no rat transcripts could be detected in skeletal muscle or in any of the seven other tissues tested.

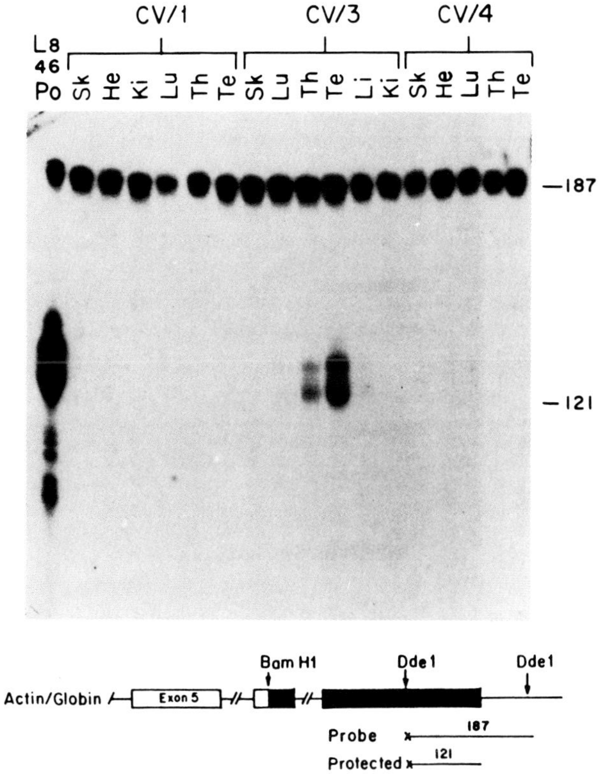

FIGURE 5. S1 analysis of actin/globin chimeric gene transcripts in various tissues of transgenic mice CV1/, CV/3 and CV/4. Total RNA (20 ug) from the indicated tissues and total RNA from differentiated cultures of a transfected clone of the rat myogenic cell line (L8-46) were hybridized with a 187 bp DdeI DNA fragment derived from the human ε-globin region of the construct. Open box: Actin coding sequences. Solid box: Human ε-globin coding sequences. The 3' label is indicated by an asterisk and the length of the probe and the protected fragment are also indicated.

In the offspring of the second transgenic mouse, low levels of expression were detected only in skeletal muscle. In the offspring of the third transgenic mouse, very high levels of the rat MLC2 transcripts were detected in skeletal muscle, but not in the other tissues which were examined. These results, indicate that the 4.7 Kb rat DNA fragment, containing the rat MLC2 gene, includes sufficient genetic information to direct tissue-specific expression. The more than a hundred-fold difference in the level of expression observed could be due to integration into different chromosomal domains.

The mode of expression of the introduced gene in transgenic mice seems to be stably inherited. Moreover, expression of the rat gene is also developmentally regulated in primary muscle cultures derived from transgenic mice. We have established myogenic cell lines from mice MLC/2 and MLC/4, and are presently comparing the control of expression of the introduced gene to that of the endogenous counterpart, at the RNA, chromatin and DNA methylation levels.

In the four transgenic mice carrying the chimeric actin/globin gene, no detectable levels of transcripts could be found in skeletal muscle RNA. Moreover, in one transgenic strain the chimeric gene mRNA was detected in several nonmuscle tissues. If these results are reproducible, they would suggest a difference in the control of expression of the 2 cloned genes, the control of the MLC2 gene construct being more autonomous and less influenced by the chromosomal site of integration. The lack of expression of the chimeric gene is of special interest, since the same construct was developmentally regulated in stably transfected myogenic cells (7). This may imply that there are different sets of control elements: those that are sufficient to direct expression in somatic cells and those that are required for developmentally regulated expression in transgenic mice. Several possible reasons for the apparent lack of expression of the chimeric gene in skeletal muscle of transgenic mice should be considered:

(a) The 5' flanking region included in the inserted rat DNA is too short and does not contain all control elements required for tissue-specific expression in transgenic mice.

(b) The different results may reflect an intrinsic difference between the genes in the strength of the control DNA sequences and their interaction with the chromatin domain into which the cloned gene became integrated. Surveying the available data on other genes introduced into transgenic mice indicates considerable variability in the mode of expression of these genes (16, 18, 19, 22-26, 33, 34).

(c) The substitution of the authentic rat skeletal muscle actin 3' region by the human globin 3' end may interfere with the regulation of expression of the gene. It has been shown recently that DNA sequences 3' to the site of initiation of transcription of the globin gene are involved in the regulation of expression of this gene (8, 35). Moreover, sequence comparison has indicated a very high degree of homology in the 3' untranslated region between mRNA coding for isotypic actins in distantly related organisms (e.g., rat and chick β-actin (36) and rat and human cardiac actin (37)) but not between 3' untranslated regions of mRNA coding for very similar actins differing in tissue specificity (e.g., skeletal and cardiac muscle). This suggests that this region of the gene is involved in transcriptional or posttranscriptional control of expression of the genes (38; Nudel et al., this volume).

More experiments are needed to examine these possibilities.

ACKNOWLEDGMENTS

We thank Dr. Uri Nudel for fruitful collaboration, Ms. Sara Neuman, Ora (Saxel) Fuchs and Olla Yoffe for their excellent technical assistance, and Dr. J. Calvo and X. Young for constructing the plasmid pCV.

REFERENCES

1. Renkawitz R, Beng H, Graft T, Mathias P, Grez M, Schutz G (1982). Expression of a chicken lysozyme recombinant gene is regulated by progesterone and dexamethasone after microinjection into oviduct cells. Cell 31:167-176.
2. Chao MV, Mellon P, Charnay P, Maniatis R, Axel R (1983). The regulated expression of β-globin genes introduced into mouse erythroleukemia cells. Cell 32:483-493.

3. Kondoh H, Yasuda K, Okada TJ (1983). Tissue-specific expression of a cloned chick -crystallin gene in mouse cells. Nature 301:440-442.
4. Rice D, Baltimore D (1982). Regulated expression of an immunoglobulin k gene introduced into mouse lymphoid cell line. Proc Natl Acad Sci USA 79:7862-7865.
5. Oi Vt, Morrison SL, Herzenberg LA, Berg P (1983). Immunoglobulin gene expression in transformed lymphoid cells. Proc Natl Acad Sci USA 80:825-829.
6. Gillies SD, Morrison SL, Oi VT, Tonegawa S (1983). A tissue-specific enhancer element is located in the major intron of a rearranged immunoglobulin heavy chain gene. Cell 33:717-728.
7. Melloul D, Aloni B, Calvo J, Yaffe D, Nudel U (1984). Developmentally regulated expression of chimeric genes containing muscle actin DNA sequences in transfected myogenic cells. EMBO J 3:983-990.
8. Charnay P, Treisman R, Mellon P, Chao M, Axel R, Maniatis T (1984). Differences in human α- and β-globin gene expression in mouse erythroleukemia cells; the role of intragenic sequences. Cell 38:251-263.
9. Spradling AC, Rubin GM (1982). Transposition of cloned P-elements into Drosophila germ line chromosomes. Science 218:341-347.
10. Rubin GM, Spradling AC (1982). Genetic transformation of Drosophila with transposable element vectors. Science 218:348-353.
11. Scolnick SB, Morgan BA, Hirsh J (1983). The cloned dopa decarboxylase gene is developmentally regulated when reintroduced into the Drosophila genome. Cell 34:37-45.
12. Goldberg DA, Posakony JW, Maniatis T (1983). Correct developmental expression of a cloned alcohol dehydrogenase gene transduced into the Drosophila germ line. Cell 34:59-73.
13. Spradling AC, Rubin GM (1983). The effect of chromosomal position on the expression of the Drosophila xanthine dehydrogenase gene. Cell 34:47-57.
14. Richards G, Cassab A, Bourouis M, Jarry B, Dissons C (1983). The normal developmental regulation of a cloned sgs3 'glue' gene chromosomally integrated in Drosophila by P-element transformation. EMBO J 2:2137-2142.
15. Gordon JW, Ruddle FH (1981). Integration and stable germ line transmission of genes injected into mouse pronuclei. Science 213:1244-1246.

16. Wagner E, Stewart T, Mintz B (1981). The human β-globin gene and a functional viral thymidine kinase gene in developing mice. Proc Natl Acad Sci USA 78:5016-5020.
17. Wagner TE, Hoppe PC, Joklick JD, Scholl DR, Hudinka RL, Gault JB (1981). Microinjection of a rabbit β-globin gene into zygotes and subsequent expression in adult mice and their offspring. Proc Natl Acad Sci USA 78:6278-6380.
18. Lacy E, Roberts S, Evans EP, Burtenshaw MD, Costantini FD (1983). A foreign β-globin gene in transgenic mice: Integration at abnormal chromosomal positions and expression in inappropriate tissues. Cell 34:343-358.
19. Brinster RL, Chen HY, Trumbauer M, Seneor AR, Warren R, Palmiter RD (1981). Somatic expression of Herpes thymidine kinase in mice following injection of a fusion gene into eggs. Cell 27:223-231.
20. Jahner D, Stuhlmann H, Stewart CL, Harbers K, Lohler J, Simon I, Jaenisch R (1982). De novo methylation and expression of retroviral genomes during mouse embryogenesis. Nature 298:623-628.
21. Palmiter RD, Brinster RL, Hammer RE, Trumbauer ME, Rosenfeld MG, Birnberg NC, Evans RM (1982). Dramatic growth of mice develop from eggs microinjected with metallothionein-growth hormone fusion genes. Nature 300:611-615.
22. Burki K, Ullrich A (1982). Transplantation of the human insulin gene into fertilized mouse eggs. EMBO J 1:127-131.
23. McKnight GS, Hammer RE, Kuenzel EA, Brinster RL (1983). Expression of the chicken transferrin gene in transgenic mice. Cell 34:335-341.
24. Brinster RL, Ritchie KA, Hammer RE, O'Brien RL, Arp B, Storb U (1983). Expression of a microinjected immunoglobulin gene into the spleen of transgenic mice. Nature 306:332-336.
25. Grosschedl R, Weaver D, Baltimore D, Costantini F (1984). Introduction of a -immunoglobulin gene into the mouse germ line: Specific expression in lymphoid cells and synthesis of functional antibody. Cell 38:647-658.
26. Swift CH, Hammer RE, MacDonald RJ, Brinster RL (1984). Tissue-specific expression of the rat pancreatic elastase I gene in transgenic mice. Cell 38:639-646.

27. Stewart TA, Pattengale PK, Leder P (1984). Spontaneous mammary adenocarcinomas in transgenic mice that carry and express MTV/myc fusion genes. Cell 38:627-637.
28. Nudel U, Calvo J, Shani M, Levy Z (1984). The nucleotide sequence of a rat skeletal myosin light chain 2 gene. Nucl Acids Res 12:7175-7186.
29. Katcoff D, Nudel U, Zevin-Sonkin D, Carmon Y, Shani M, Lehrach H, Frischauf AM, Yaffe D (1980). Construction of a recombinant plasmids containing rat muscle actin and myosin light chain DNA sequences. Proc Natl Acad Sci USA 77:960-964.
30. Shani M, Zevin-Sonkin D, Saxel O, Carmon Y, Katcoff D, Nudel U, Yaffe D (1981). The correlation between the synthesis of skeletal muscle actin, myosin heavy chain and myosin light chain and the accumulation of the corresponding mRNA sequences during myogenesis. Dev Biol 86:483-492.
31. Yablonka Z, Yaffe D (1977). Synthesis of myosin light chains and accumulation of translatable mRNA coding for light chain-like polypeptides in differentiating muscle cultures. Differentiation 8:133-143.
32. Carmon Y, Czosnek H, Nudel U, Shani M, Yaffe D (1982). DNAase I sensitivity of genes expressed during myogenesis. Nucl Acids Res 10:3085-3098.
33. Schnieke A, Harbers K, Jaenisch R (1983). Embryonic lethal mutation in mice induced by retrovirus insertion in the Alphal(1) collagen gene. Nature 304:315-320.
34. Wagner EF, Covarrubias L, Stewart TA, Mintz B (1983). Prenatal lethalities in mice homozygous for human growth hormone gene sequences integrated in the germ line. Cell 35:647-655.
35. Wright S, Rosenthal A, Flavell R, Grosveld F (1984). DNA sequences required for regulated expression of β-globin genes in murine erythroleukemia cells. Cell 38:265-273.
36. Nudel U, Zakut R, Shani M, Neuman S, Levy Z, Yaffe D (1983). The nucleotide sequence of the rat cytoplasmic β-actin gene. Nucl Acids Res 11:1759-1771.
37. Mayer Y, Czosnek H, Zeelon PE, Yaffe D, Nudel U (1984). Expression of the gene coding for the skeletal muscle and cardiac actins in the heart. Nucl Acids Res 12:1087-1100.
38. Yaffe D, Nudel U, Mayer Y, Neuman S (1985). Highly conserved sequences in the 3' untranslated region of mRNAs coding for homologous proteins in distantly related species. Nucl Acids Res, in press.

Molecular Biology of Muscle Development, pages 591–603

CYTOPLASMIC TRANSLATION INHIBITORY RNA OF CHICK EMBRYONIC MUSCLE: POSSIBLE ROLE IN MYOGENESIS AS ANTIMESSENGER RNA[1]

Shashiprabha Dasgupta[2], Mark S. Eller, Stephen Zolnay, Chellah Jayabaskaran, Hansell Stedman, and Satyapriya Sarkar.

Department of Muscle Research, Boston Biomedical Research Institute and Department of Neurology, Harvard Medical School, Boston, Massachusetts 02114.

ABSTRACT A cytoplasmic 10 S ribonucleoprotein (iRNP) isolated from chick embryonic muscle is a potent inhibitor of mRNA translation in vitro and contains a 4 S translation inhibitory RNA species (iRNA). The iRNA inhibits the binding of mRNA to the 43 S pre-initiation complex during mRNA translation. Several subspecies of iRNA have now been resolved with retention of biological activity. The sequence of the largest iRNA (134 nucleotides) indicates that it is unrelated to other eukaryotic small RNAs described in the literature. Several iRNA subspecies contain 5'-ppp termini and are therefore, bonafide gene transcripts. The iRNA shows differential inhibition of in vitro translation of different mRNAs. The iRNA subspecies hybridize with muscle poly$(A)^+$ mRNA and globin mRNA in dot-blots but not with 28 S, 18 S, 4 S, and 5 S RNAs. These results strongly suggest that iRNA may act as a cellular "anti-messenger RNA" in a manner similar to the prokaryotic micRNA (mRNA-interfering complementary RNA). A possible in vivo function of iRNA accordingly is to regulate efficiently the intracellular mRNA levels, thus augmenting the transcriptional controls which operate during myogenesis.

[1]This work was supported by NIH (HL-07266, AM-13238) and MDA.
[2]Recipient of a postdoctoral fellowship from MDA.

INTRODUCTION

It is currently accepted that myogenesis *in vitro* and *in vivo* is primarily regulated by transcriptional control (reviewed in references 1-3). The transition from the myoblast to myotube stage in cultured muscle cell is associated with the appearance of a new abundant class population of mRNAs consisting of the mRNA sequences coding for the myofibrillar proteins (4). The increased synthesis of myofibrillar proteins at or after fusion is correlated with the appearance of this new class of mRNAs (4,5). The genes coding for the myofibrillar proteins are coordinately regulated during terminal differentiation (6,7,8) resulting in accumulation of the protein products in stoichiometric amounts. It has been shown that the increase in the accumulation of cytoplasmic myosin heavy chain mRNA during myogenesis is due to a combination of two cellular processes, transcriptional induction of the myosin heavy chain gene and the cell cycle-mediated increase in stability of the mRNA (9). While these studies clearly stress the importance of transcriptional control in myogenesis, the possibility of additional cytoplasmic processes which may regulate the translation process in conjunction with the transcription of the myofibrillar genes, have been suggested by several studies. For example, during chick pectoralis muscle development *in ovo*, it has been observed that the mRNA level for skeletal fast myosin light chain LC_3 shows a noncoordinate relationship to the corresponding protein level (10). This suggests that the translation of LC_3 mRNA may be regulated by additional cytoplasmic control during myogenesis *in vivo*.

We have previously described the isolation and characterization of a cytoplasmic 10 S ribonucleoprotein (iRNP) particle containing a low molecular weight heterogeneous RNA (iRNA) species in the 70-130 nucleotide size range from chick embryonic muscle (11,12,13). Both the 10 S iRNP and the deproteinized 4 S iRNA are potent inhibitors of mRNA translation *in vitro*. While these results suggest that iRNA and iRNP may be involved in regulation of translation in embryonic muscle, their biological role *in vivo* remains to be understood. We report here the resolution of iRNA into biologically active single subspecies. The nucleotide sequence of iRNA subspecies and their specific interaction with mRNA suggest that they are bonafide RNA transcripts unrelated to other eukaryotic nuclear and cytoplasmic small RNAs. The significance of these results in relation to a physiological role of iRNA and iRNP is discussed.

RESULTS

The iRNA and iRNP Represent a Family of Cytoplasmic Macromolecular Entities.

The biochemical properties of iRNA and iRNP which indicate that they represent a novel class of cytoplasmic entities present in embryonic muscle cells, are briefly summarized. Both 10 S iRNP and the 4 S iRNA are potent inhibitors of *in vitro* mRNA translation (12). The 10 S iRNP contains a complex set of proteins (M_r 12,000-100,000) which are electrophoretically distinct from those of eukaryotic mRNP particles (11,12). The proteins present in iRNP do not show any immunoreaction with six monospecific autoimmune antibodies (RNP, Sm, PM-1, SSA, SSB and ScL-70). This indicates that the protein moieties of iRNP are distinct from those of small nuclear RNP (SnRNP) and other nuclear RNA-containing particles (14), which show characteristic immunoreaction with the autoimmune antibodies. The iRNA consists of a class of RNA species in the 70-130 nucleotides in size, which have very little double-stranded structure (11,12,14). The 10 S iRNP particle can be specifically reconstituted from the dissociated RNA and protein components (13). The proteins of iRNP do not show any translation inhibitory activity. Whereas iRNA is RNase-sensitive, both native and reconstituted iRNP are RNase-resistant (13). This suggests that a possible biological function of the proteins of iRNP is to provide a RNase-resistant structure to iRNA, which would otherwise be sensitive to nucleolytic cleavage.

With regard to the translation inhibitory activity of iRNA and iRNP, they have no effect on the elongation phase of peptide synthesis in an *in vitro* assay system (15). The iRNA inhibits mRNA binding to the 43 S pre-initiation complex (Met-$tRNA_f$-40 S ribosome-eIF_2) and thus prevents the assembly of the fully functional 80 S initiation complex during mRNA translation. Interestingly, iRNA does not affect the formation of the 43 S complex. Furthermore, iRNA has no effect on the phosphorylation of the alpha subunit of the initiation factor eIF_2 in an in vitro assay system (14). This mode of inhibition distinguishes iRNA and iRNP from other well-known eukaryotic cellular inhibitors such as the hemin-controlled repressor, which block the initiation step by stimulating phosphorylation of the eIF_2 alpha subunit.

Biologically active single iRNA subspecies, their nucleotide sequence and specific interaction with mRNAs.

In view of the electrophoretic heterogeneity of iRNA (12; see also Fig. 1, Lane A), the question of whether or not all components of iRNA are biologically active and whether different iRNA subspecies may show differential inhibition of translation of different mRNAs, was considered. It should be stressed that the biochemical properties of iRNA and its mode of inhibition of mRNA translation indicate that iRNA is clearly distinct from tcRNA, another small RNA of chick embryonic muscle which has been reported to inhibit myosin heavy chain mRNA translation in a discriminatory manner (16; reviewed in ref. 12,14). By using a combination of high performance liquid chromatography (HPLC) and polyacrylamide gel electrophoresis (PAGE) in nondenaturing gels, we have recently resolved iRNA into several highly purified subspecies with full retention of biological activity (17,18). The 134 nucleotide size $iRNA_I$ subspecies, the largest component present in unfractionated iRNA is conveniently resolved as a single peak by HPLC gel filtration on a TSK 4000 SW column at 24°C using 50 mM ammonium acetate, pH 6.8 as the eluant (17). This HPLC-resolved peak shows a single RNA band on PAGE in 99% formamide (Fig. 1, Lane B). The other iRNA components are only partially resolved by HPLC (17). However, PAGE on 8% gels using nondenaturing conditions (19) resolved some iRNA subspecies as single bands and others as partially resolved mixed bands. The resolved iRNA subspecies were eluted from the gel slices, labeled at the 3'-end with $[^{32}P]pCp$ and then analyzed by PAGE on a 48 cm sequencing gel (20). The autoradiograms (Fig. 1) show that band II iRNA is resolved into two subspecies $iRNA_{IIa}$ (Lane F) and $iRNA_{IIb}$ (Lane E) in the 105-110 nucleotide size range; band III iRNA

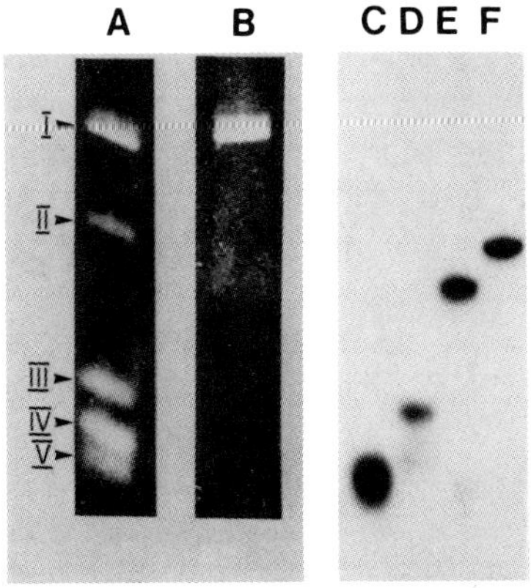

Fig. 1. Electrophoretograms of iRNA subspecies resolved by HPLC and PAGE. For details see also text. Lanes A and B, ethidium bromide staining of unfractionated iRNA and HPLC purified $iRNA_I$ subspecies respectively. Lanes C-F, autoradiograms of [^{32}P]pCp-labelled iRNA subspecies resolved by PAGE. Lane F, $iRNA_{IIa}$; Lane E, $iRNA_{IIb}$; Lane D, $iRNA_{III}$; and Lane C, mixture of incompletely resolved $iRNA_{IV}$ and $iRNA_V$.

is resolved into one major, $iRNA_{IIIa}$, and two minor subspecies (Lane D) in the 85-90 nucleotide size range; and bands IV and V iRNA are present as a mixture which shows a diffuse band (Lane C).

All of the resolved iRNA subspecies (Fig. 1) were found to be potent inhibitors of in vitro mRNA translation in incubations programmed with globin mRNA and muscle poly(A)$^+$ mRNA. However, different subspecies gave different degrees of inhibition with respect to the same mRNA species suggesting that the resolved iRNA components on a molar basis showed considerable differences in biological activities (results not shown). Interestingly, $iRNA_I$ subspecies showed differential inhibition of translation of different mRNA's (Fig. 2) in the order: muscle mRNAs > globin mRNA > poliovirus mRNA.

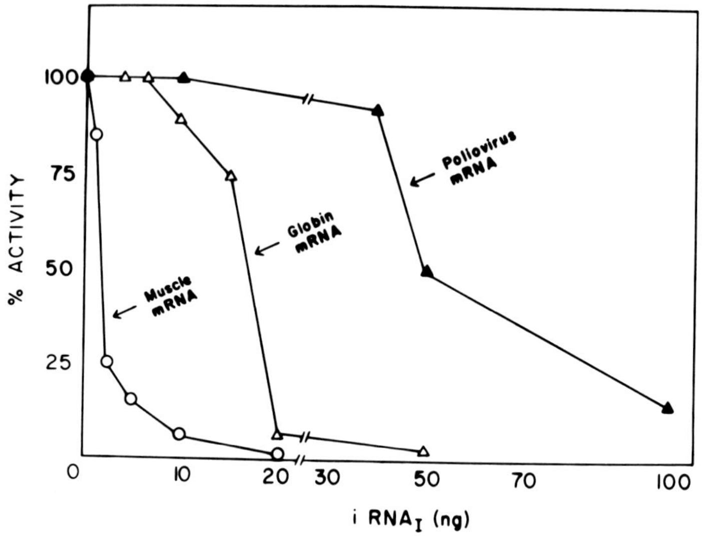

Fig. 2. Effect of purified $iRNA_I$ subspecies on in vitro translation of different mRNAs. Translation assays were carried out as previously described (12) using saturating amounts of muscle poly(A)$^+$mRNA (0.3 ug; open circles); globin mRNA (0.25 ug; open triangles); and poliovirus mRNA (0.5 ug; closed triangles).

The possbility that a single iRNA subspecies may inhibit selectively the translation of mRNA coding for a specific myofibrillar protein, was also tested. The translation products of embryonic chick muscle poly(A)$^+$ mRNA directed incubations in the absence and presence of various concentrations of HPLC purified $iRNA_I$ were examined by fluorography. The fluorogram (Fig. 3) shows that the majority of the bands, appear to be inhibited in a nonselective manner. Furthermore, this mode of concentration-dependent inhibition was observed with all levels of iRNA tested (lanes 2-5). Although these results suggest that iRNA may affect the translation process in a non-discriminatory manner, the possibility that a single iRNA subspecies may exert a pronounced activity to a specific mRNA, cannot be strictly ruled out. Using a number

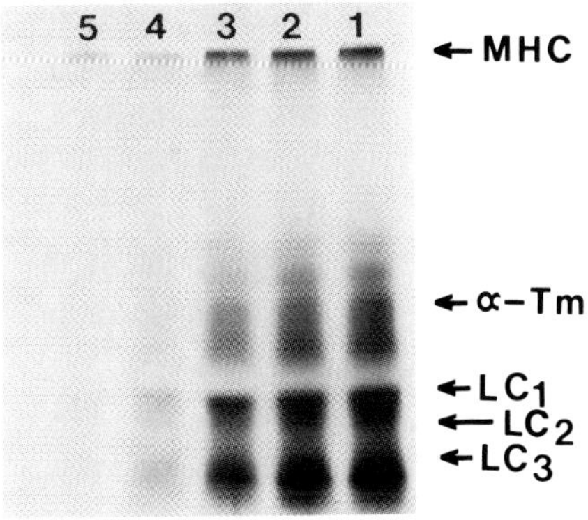

Fig. 3. Fluorography of in vitro translation products obtained with muscle poly(A)$^+$ mRNA in the presence and absence of HPLC purified $iRNA_I$ subspecies. The translation assays and fluorography were carried out as previously described (11,13). The arrows indicate the positions of the following myofibrillar protein bands run as markers: myosin heavy chain (MHC); alpha tropomyosin (alpha-Tm); skeletal myosin light chains 1, 2 and 3 (LC_1, LC_2 and LC_3) and troponin C (TnC). Lane 1: chick embryonic muscle poly(A)$^+$ mRNA (0.3 ug); lanes 2-5: incubations containing 0.3 ug of muscle poly(A)$^+$ mRNA and 1,2.5, 5 and 50 ng of $iRNA_I$ respectively. The control incubation containing nuclease-treated lysate alone did not show any radioactive band.

of hybrid-selected specific myofibrillar mRNAs, we are now screening each of the resolved iRNA subspecies for such inhibitory effect.

The nucleotide sequence of HPLC purified $iRNA_I$ subspecies was also determined. The RNA was labelled at the 5'-end with gamma [^{32}P]ATP and polynucleotide kinase, then cleaved with base-specific RNases, and the resulting oligonucleotides were analyzed using the RNA sequencing gel system (20). Several features of the nucleotide sequence of $iRNA_I$ (Fig. 4) should be noted. We observed microheterogeneity at the 5'-end (nucleotides 1-10) and 3'-end (nucleotides 106-134) by examining the sequence ladder and by homochromatography analysis. The 5'-triphosphate terminus was established by in vitro capping

```
               10        20        30
PPP-A AGUUUUAUUGCGGCGGUUCACUCGGCUGG
    G
        40        50        60
GGAACACUGGCUUGUUGCCAAGAAGAAGCC

        70        80        90
GCCACCUAGAAAAAGGCCGGAUUACAGGAU

       100       110       120       130
CCUCUUAUGGCCUAUUCCAGCAUGGUCUUGUUGUGGCGAGGCAG
```

Fig. 4. The nucleotide sequence of HPLC purified $iRNA_I$ subspecies. For details see also text. The sequences underlined from positions 1-10 and 106-134 show microheterogeneity.

of the $iRNA_I$ subspecies and analysis of the capped product after PAGE and RNase digestion (20,21). Two 5'-ppp termini, pppG and pppA were obtained. These results indicate that purified $iRNA_I$ subspecies represent multiple molecules of the same size arising from closely related gene transcripts. A computer search for homologous sequences of known RNA molecules indicates that $iRNA_I$ represents new RNA species. Furthermore, its sequence is quite distinct from other eukaryotic nuclear and cytoplasmic small RNAs such as Ro, La and the SnRNAs. It should be noted that $iRNA_I$ has no sequence homology to the U-rich tcRNA (16), which has been reported to inhibit myosin heavy chain mRNA translation in a discriminatory manner.

Preliminary sequence analysis of $iRNA_{IIB}$ and $iRNA_{IIIA}$ subspecies indicates that they also contain 5'-ppp termini, and their sequences are distinct from that of $iRNA_I$. It appears that iRNA represents a family of small cytoplasmic RNAs previously not reported in the literature. Furthermore, the 5'-ppp termini present in several purified iRNA subspecies indicate that they are bonafide gene transcripts. Their cytoplasmic occurence is not due to degradation of other cellular RNAs. The biogenesis of these RNAs does not involve processing at the 5'-end. Further studies on sequencing the multiple components present in the

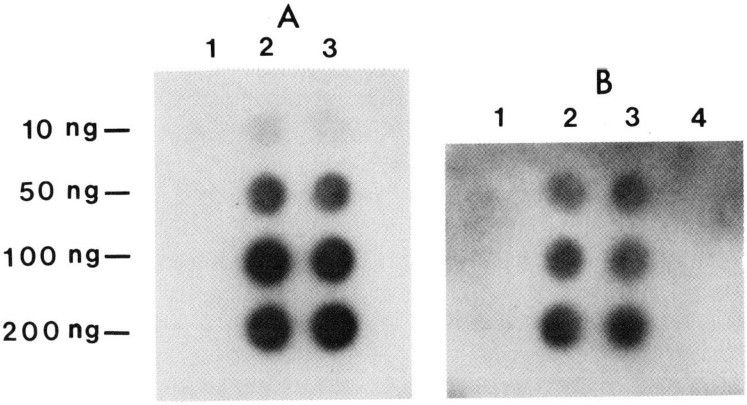

Fig. 5. Analysis of iRNA-mRNA interaction by dot-blot hybridization. Different RNAs in indicated amounts were immobilized on nitrocellulose paper (22). Hybridization of 3'-end labeled iRNA subspecies ($1\text{-}5\times10^5$ cpm/blot) was carried out in a solution containing 5xSSC, 50% formamide, 0.1% SDS and 0.1 mM EDTA. The reactions were carried out at 50°C for 2 hr, gradually cooling to 37°C over 6 hr and incubation at 37°C overnight. The blots were washed five times, 15 min each, with 2XSSC, 0.1% SDS at room temperature, and five times with 0.2XSSC, 0.1% SDS at room temperature. The filters were then dried and autoradiographed. Panel A: hybridization with ^{32}P-labelled $iRNA_{IIa}$ subspecies. Lane 1, chick muscle tRNA; Lane 2, globin mRNA; Lane 3, embryonic chick muscle poly(A)$^+$ mRNA. Panel B: hybridization with ^{32}P-labelled $iRNA_{III}$ subspecies. Lane 1, chick muscle 5 S RNA; Lane 2, embryonic chick muscle poly(A)$^+$ mRNA; Lane 3, globin mRNA; Lane 4, chick muscle 28 S and 18 S rRNAs.

mixture of bands IV and V iRNA are now in progress using recombinant DNA technology to construct cloned probes to individual iRNA species.

The possibility that iRNA interacts with mRNA and this interaction leads to the inhibition of mRNA translation by iRNA, was tested by dot-blot hybridization experiments (22). [^{32}P]pCp-labelled various iRNA subspecies were hybridized with nonradioactive mRNAs immobilized on nitrocellulose paper. The autoradiograms obtained in experiments using

$iRNA_{IIa}$ and $iRNA_{III}$ subspecies are presented as representative results (Fig. 5). Both globin mRNA and chick muscle poly(A)$^+$ mRNA hybridized with the iRNA subspecies. The intensity of the hybrid spots was dependent on the amount of mRNAs used. Other cellular RNAs such as chick muscle tRNA, 5 S RNA, and 18 S and 28 S rRNAs did not show any detectable hybridization. Also, prokaryotic RNAs such as E. coli tRNA, 5 S RNA and 16 S and 23 S rRNAs were unable to interact with iRNA (results not shown). Similar results were obtained with $iRNA_{IIb}$ and the mixture of iRNA bands IV and V. It should be noted that so far we have been unable to show any interaction of iRNA with other components of the translation system such as binding of iRNA to 40 S ribosomal subunit and initiation factors. Specific RNA-RNA interaction involving mRNA and iRNA, as shown by these experiments, presumably occurs at the 5'-untranslated region of the mRNAs which contain the ribosome binding site. Currently we are engaged in defining the binding sites involved in interaction of purified iRNA subspecies with specific myofibrillar mRNAs.

DISCUSSION

Several features of the present study indicate that iRNA and iRNP are true cellular entities and represent a family of cytoplasmic macromolecules which may have a regulatory role in cellular mRNA metabolism. These features are: the resolution of iRNA into multiple biologically active subspecies; the sequence of the 134 nucleotide size iRNA which shows that it is distinct from other eukaryotic small RNAs; the presence of 5'-ppp termini in several iRNA subspecies indicating that they are bonafide gene transcripts; the ability of iRNA to hybridize with mRNA; and the differential degree of inhibition of different mRNA translation by a purified iRNA subspecies. Although the observed interaction of iRNA with mRNA may account for the inhibitory effect of iRNA on mRNA translation, it should be pointed out that these interactions are likely to produce a quantitative modulation rather than an absolute mRNA-discriminatory effect. However, since mRNA binding to the 43 S pre-initiation complex is a rate-limiting step in polypeptide synthesis (23), quantitative modulation of this step by iRNA can lead to striking effects on cellular translation patterns. The iRNA-mRNA interaction is reminiscent of the translational inhibition by a recently reported prokaryotic low molecular weight RNA transcript, referred to as mic RNA (mRNA-interfering complementary RNA) (24). This RNA regulates efficiently the expression of a specific bacterial gene (ompC) at the level of translation by hybridizing with the complementary sequence present at the 5'-translational start region of the mRNA (24). It is conceivable that iRNA acts in an analogous manner. Thus, the in vivo function of iRNA may involve efficient regulation of the intracellular levels of functional mRNA which is actively translated. The iRNA, acting as "anti-messenger RNA" may not only modulate the translation process, but may also lead to destabilization of the untranslated mRNAs. Thus, iRNA may act in conjunction with transcriptional control which is known to operate during myogenesis (1-9).

REFERENCES

1. Pearson ML, Epstein HF (1982). Regulatory mechanisms in muscle development: A persepctive. In Pearson ML, Epstein HF (eds): "Muscle Development: Molecular and Cellular Control", Cold Spring Harbor Laboratory, p. 559.
2. Buckingham ME, Minty AJ (1983). Contractile protein genes. In MacLean N, Gregory S, Flavell R (eds): "Eukaryotic Genes: Their Structure, Activity, Regulation", London: Butterworths, p. 365.
3. Saidapet CR, Munro HN, Valgeirsdottir K, Sarkar S (1982). Quantitation of muscle-specific mRNAs by using cDNA probes during chicken embryonic muscle development in ovo. Proc Natl Acad Sci USA 79:3087.
4. Paterson BM, Bishop JO (1977). Changes in the mRNA population of chick myoblasts during myogenesis in vitro. Cell 12:751.
5. Devlin RB, Emerson CP (1979). Coordinate accumulation of contractile protein mRNA during myoblast differentiation. Dev Biol 69:202.
6. Garfinkel L, Periasamy M, Nadal-Ginard B (1982). Cloning, identification and characterization of alpha-actin, myosin light chain 1,2 and 3, alpha-tropomyosin, and troponin C and T. J Biol Chem 257:11078.

7. Hastings KE, Emerson CP (1982). cDNA clone analysis of six co-regulated mRNAs encoding skeletal muscle contractile proteins. Proc Natl Acad Sci USA 79:1553.
8. Affara NA, Rober B, Jacquet M, Buckingham ME, Gros F (1980). Changes in gene expression during myogenic differentiation. 1. Regulation of mRNA sequences expressed during myotube formation. J Mol Biol. 140:441.
9. Medford RM, Nguyen HT, Nadal-Ginard B (1983). Transcriptional and cell cycle-mediated regulation of myosin heavy chain gene expression during muscle cell differentiation. J Biol Chem 258:11063.
10. Roy RK, Sarkar S (1982). Correlation between the protein and mRNA levels for myosin light chains and tropomyosin subunits during chick fast muscle development in vivo. FEBS Lett 149:22.
11. Mukherjee AK, Guha C, Sarkar S (1981). The translational inhibitory 10 S cytoplasmic ribonucleoprotein of chicken embryonic muscle is distinct from messenger ribonucleoproteins. FEBS Lett 127:133.
12. Sarkar S, Mukherjee AK, Guha C (1981). A ribonuclease-resistant cytoplasmic 10 S ribonucleoprotein of chick embryonic muscle: A potent inhibitor of cell-free protein synthesis. J Biol Chem 256:5077.
13. Mukherjee AK, Sarkar S (1981). The translational inhibitor 10 S cytoplasmic ribonucleoprotein of chick embryonic muscle: Dissociation and reassociation. J Biol Chem 256:1103.
14. Sarkar S (1984). Translational control involving a novel cytoplasmic RNA and ribonucleoprotein. In Cohn WE, Moldave K (eds): Progress in Nucleic Acid Research and Molecular Biology", New York: Academic Press 31:267.
15. Winkler MM, Lashbrook C, Hershey JWB, Mukherjee AK, Sarkar S (1983). The cytoplasmic 4 S translation inhibitory RNA species of chick embryonic muscle: Effect on mRNA binding to 43 S initiation complex. J Biol Chem 258:15141.
16. McCarthy TL, Siegel E, Mroczkowski B, Heywood SM (1983).Characterization of translational-control ribonucleic acid isolated from embryonic chick muscle. Biochemistry 22:935.
17. Dasgupta S, Pluskal MG, Sarkar S (1984). The

cytoplasmic 4 S translation inhibitory RNA species of chick embryonic muscle: Fractionation of biologically active subspecies by high performance liquid chromatography. Prep Biochem 14:331.

18. Sarkar S, Jayabaskaran C, Dasgupta S (1985). A novel cytoplasmic translation inhibitory RNA and RNP of chick embryonic muscle: Properties and possible role in myogenesis. J Cell Biochem (supplement 9B). Abstract number 0586, p. 69.
19. Guthrie C, McClain WH (1973). Conditionally lethal mutants of bacteriophage T4 defective in production of a transfer RNA. J Mol Biol 81:137.
20. Silberkland M, Gillum AM, Rajbandary UL (1979). Use of in vitro ^{32}P labelling in the sequence analysis of non radioactive tRNAs. In Moldave K, Grossman L (eds): "Methods in Enzymology", New York: Academic Press 59:58.
21. Martin SA, Moss B (1975). Modification of RNA by mRNA guanyl transferase and mRNA (guanine-7-)methyl transferase from vaccinia virions. J Biol Chem 250:9330.
22. White, BA, Barcroft FC (1982). Cytoplasmic dot hybridization. J Biol Chem 257:8569.
23. Lodish HF (1976). Translational control of protein synthesis. In Snell EE, Boyer PD, Meister A, Richardson CC (eds): "Annual Review of Biochemistry", Palo Alto, Annual Reviews Inc 45:39.
24. Mizuno T, Chou M-Y, Inouye M (1984). A unique mechanism regulating gene expression: Translational inhibition by a complementary RNA transcript (micRNA). Proc Natl Acad Sci USA 81:1966.

Molecular Biology of Muscle Development, pages 605–617

MUSCLE GENES AND PROTEINS IN *CAENORHABDITIS ELEGANS*[1]

R.H. Waterston, D.G. Moerman, G.M. Benian, R.J. Barstead, I. Mori and R. Francis

Department of Genetics, Washington University
St. Louis, MO 63110

ABSTRACT Our efforts to understand the role of each of the muscle genes in *C. elegans* myogenesis and muscle contraction have followed three different but interrelated approaches. Antibodies have been developed and used to define additional components of *C. elegans* body wall musculature. Studies of gene interactions have revealed epistatic relationships that help to define the function of some gene products. Transposition of the repetitive element *Tc1* into the muscle gene *unc-22* has allowed us to clone this gene and begin the dissection of its function in the nematode.

INTRODUCTION

Mutational analysis in the nematode *Caenorhabditis elegans* has identified 20 genes required for normal muscle assembly, and another half-dozen specific suppressor genes (1-3). The products of four of these have been determined: *unc-54* encodes one myosin heavy chain of the body wall muscle; *unc-15* probably encodes paramyosin; and *act-1* and *act-3* encode two identical actins. Further studies of these genes, their mutants forms and their protein products are providing insight into the function of these components in

[1]This work was supported by grants from the Public Health Service (GM23883) and the Muscular Dystrophy Association of America.

muscle assembly and contraction (4-8).

The products of the remaining 20 or more genes are as yet unknown. These genes might encode other critical components of thick or thin filaments; elements of the dense body or M-line; or other proteins not incorporated into the mature lattice structure, but used in assembly, in regulation of contraction, or in control of gene expression. To take full advantage of these mutants to increase our understanding of muscle assembly and function will require that their products be defined at a molecular level. In genetics, the association of a specific product with a gene has historically been a difficult step. The few gene:protein assignments made so far in C. elegans muscle have required prior knowledge of the protein. A major strength of genetics however is its potential to reveal processes or components not readily studied through biochemical techniques. To bridge the gap between gene and proteins so that we can take advantage of the existing genetic analysis, we have pursued three complementary approaches. At one level, we have used polyclonal and monoclonal antibodies to define new components of C. elegans muscle (9, unpublished results). These antibodies are proving useful both in the analysis of mutant muscle structure and in the isolation of DNA sequences encoding these antigens. At another level, we have examined gene interactions using mutants in genes whose product is known with other mutants to develop more specific hypotheses about the function altered by these other mutants. Finally, we have discovered a mutator system in one strain of C. elegans (10). We and others have shown that this mutator activity is the result of transposition of the 1.6 kb repetitive element Tc1 (11,12). This has allowed us to clone the unc-22 gene by tranposon tagging and to show that it encodes a mRNA of about 14kb. Current studies are directed at determining the unc-22 protein product, its site of expression and its function.

ISOLATION OF ANTIBODIES TO MUSCLE COMPONENTS

The myofilament lattice of the body wall muscle cells in C. elegans is attached to the overlying cuticle throughout the entire length of the animal. Our studies suggest this attachment is mediated through the dense body for thin filaments and the M-line for thick filaments, which in turn are anchored in the plasma membrane. The muscle cell is attached to the underlying hypodermal cell through the basal lamina. Within the hypodermal cell are intermediate filaments stretching between half-desmosomes facing the basal lamina and the cuticle. This chain of components functions to transmit the tension developed in a muscle contraction to the cuticle. As these components and the myofilament lattice remain attached to the cuticle in low salt washes, we were able to obtain a preparation enriched for them. Various fractions of this preparation were used as immunogens for the production of hybridoma cell lines. The hybridoma culture supernatants were assayed by indirect immunofluorescence against cuticular muscle fragments, and those identifying interesting patterns were subcloned and analysed further. These monoclonal antibodies have identified components of the M-line, intermediate filament subunits both within the muscle cell and in the hypodermal cell, and the dense body (R.F and R.H.W., unpublished results). Studies of the latter set of antibodies have been particularly useful in furthering our understanding of nematode muscle organization (1,9).

To date antibodies revealing the periodic distribution of the dense body in the I-band can be subdivided into 3 groups based on the patterns of fluorescent staining of the muscle. Three antibodies (MH31, MH35, MH40) react with the full length of the dense body, but only react with the complete dense bodies found in symmetrical I bands (Fig. 1a; compare with Fig. 1b). In contrast a second group of antibodies reacts only with the base of the dense bodies, and also reacts with the shorter plaque-like dense bodies found in the half I-bands found at cell margins (Fig. 1b). (To distinguish these structures from the more developed dense bodies found in symmetrical I-bands, we have called these structures dense plaques). Both dense bodies and plaques serve as thin filament attachment points, but only the dense bodies are bipolar. Patterns of fluorescent staining with phalloidin after certain fixation procedures suggest that actin is also a major component of the dense bodies. On Western blots the antibodies to the dense plaques and the bases of the dense bodies react with a polypeptide of 107,000 daltons and a pI of about 6.6 (p107b). The antibodies exclusively reacting with dense bodies react with a polypeptide also of 107,000 daltons but with a pI of 6.1 (p107a). Immunological cross-reactivity of anti-p107a antibodies with chicken smooth muscle alpha-actinin suggest that p107a is the nematode analog of vertebrate alpha-actinin. Comparison of partial proteolysis of p107b and p107a confirm that they are different polypeptides, but some partial proteolysis bands are shared between the two polypeptides, suggesting that they may be related.

The third group of antibodies, represented by MH25 (Fig. 1c), reacts with the dense bodies, dense plaques and the M-lines. Studies similar to those with p107a and p107b show that the epitopes for these antibodies are restricted to the base of both the M-line and dense body. Studies of wild type and cleavage arrested embryos suggest that this antigen(s) is membrane associated, rather than intracellular as has been found for p107a and p107b. High magnification electron micrographs show that similar structures are present on the extracellular side of both dense bodies and M-lines (2), and perhaps the antigen recognized by these antibodies is part of this structure. As yet we have not been able to locate the corresponding protein on Western blots.

The differential localization of these three groups of antibodies within the dense body led to the simple model of the dense body shown in Fig. 1d. Perhaps assembly of the

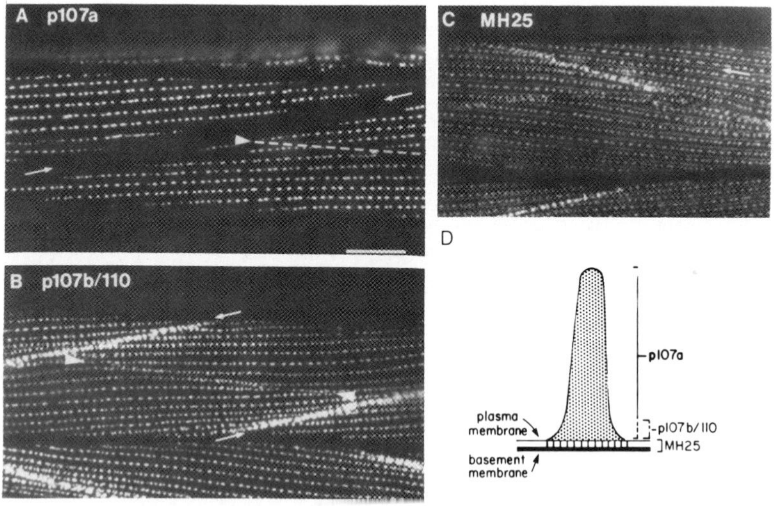

FIGURE 1

Fluorescent micrographs of body wall muscle cells stained with monoclonal antibodies reacting with (a) 107,000 dalton, pI 6.1 protein related to α-actinin, (b) 107,000 dalton, pI 6.6 protein, and (c) an undetermined antigen (recognized by monoclonal MH25). (d) Model for the organization of the components stained in a, b, and c, in the dense-body.

dense body begins with the aggregation of a membrane protein which acts as a nucleation center for deposition of first p107b, then p107a. The existence of the dense plaques containing p107b, but no detectable p107a, suggests that this second addition step is separately controlled. The bipolarity of dense bodes as opposed to dense plaques is a major functional distinction between the two structures and p107a deposition could either confer this bipolarity or be a consequence of the acquisition of it.

These antibodies provide us with a means of recovering the DNA sequences encoding the antigens through the use of expression libraries of either C. elegans genomic DNA or cDNAs (13). These DNA sequences can be positioned on the genetic map with in situ hybridization to metaphase chromosomes as developed by Albertson (14), or in greater detail via direct genetic mapping of associated restriction fragment length polymorphisms (RFLP's; 15). Examination of known muscle mutants in the region for alterations in the DNA, RNA or protein product of the gene, may lead eventually to the assignment of additional genetically defined genes to specific products, as was the case for the actin genes (4,5).

GENE INTERACTIONS

Mutations in genes with known products can be combined in the same animal with mutations in genes whose products are not yet established. Analysis of the resulting phenotypes can suggest functions for the latter gene products. In previous work we have demonstrated that specific unc-54 alleles alleviate the phenotype of unc-22 mutants (16). By direct DNA sequencing (7; J. E. Kiff and R.H.W. unpublished results), these unc-54 alleles were found to be missense mutations in the myosin head in regions implicated in the mysoin ATPase activity (17, 18). Since unc-54 missense mutants are the only indirect suppressors of unc-22 mutants, this result suggests that the unc-22 protein may regulate myosin ATPase activity.

More recently we have examined the interactions of unc-45 mutants with unc-54 and sup-3, an indirect suppressor of unc-54 null alleles (19). The four known alleles of unc-45 are temperature sensitive. Electron micrographs of homozygotes for two unc-45 alleles grown at the restrictive temperature show that these mutants have a reduced number of thick filaments, comparable to the reduction seen in unc-54 null mutants (20, R.H.W., unpublished data). Double mutants of unc-54; unc-45 are no worse phenotypically than the unc-54 mutants alone. This result contrasts with the results seen with other double mutants of unc-54 nulls with other muscle mutants such as unc-15 or unc-82 mutants, where the double mutant is more severe than either mutant alone (2). A more striking result is obtained using combinations of unc-45 and unc-54 mutants with sup-3. The strong suppressor derivative of sup-3, e1407 st90 st92, results in strong suppression of unc-54 null alleles such as e190, but only weak suppression of dominant missense alleles. The same sup-3 derivative shows only weak suppression of unc-45(m94). However the triple mutant unc-54(e190); unc-45(m94); sup-3 moves as well as the unc-54(e190); sup-3 double mutant(R.H.W. and A.M. Curry, unpublished data). Thus in both a wild type background, and more obviously in a sup-3 background, unc-54 is epistatic to unc-45. This result leads to the hypothesis that unc-45 encodes a product which modifies or obligatorily interacts with the unc-54 myosin heavy chain.

TRANSPOSABLE ELEMENTS IN C. ELEGANS

A second approach for identifying products of genetically defined genes makes use of the repetitive element Tc1. Tc1 is present in only 30 copies in the standard Bristol strain (N2) used in most genetic studies, but is present in 500 or more copies in a related variety, Bergerac (strain BO) (21). These copies are dispersed in the genome, and give rise to RFLP's on average of every half map unit between the two strains (21). These RFLP's can be recovered for a given region of interest by constructing congenic lines, and the entire region can be cloned by established techniques. In this instance, however, the approach is made more powerful by the availability through John Sulston of a large set of cosmid clones already organized into groups of overlapping clones (personal communication).

An alternative use of Tc1 stems from the observation that it is an active transposable element in the Bergerac strain of C. elegans. Insertion of Tc1 into a locus may disrupt gene function and in turn "tag" the gene for cloning by recombinant DNA techniques. We have in a single series of experiments both demonstrated Tc1 transposition and cloned the muscle gene unc-22. Results obtained in our studies of Tc1 induced unc-22 mutants have suggested new strategies for making the approach more powerful.

The Bergerac strain BO has mutator activity: During studies of the BO strain, we noted that unc-22 mutations occurred spontaneously at unusually high frequencies-- 5×10^{-4} as compared to 6×10^{-7} for the Bristol strain (10). Further, 33 out of 34 spontaneous mutants reverted spontaneously to wild type. For one mutant studied in detail, 3 revertants were found in 1300 animals and, more generally, revertants of the mutants occurred at a frequency of $\sim 10^{-3}$.

The reversion frequency of unstable BO mutations was decreased by replacing the Bergerac genome with increasing amounts of Bristol genome. The replacement of the Bergerac genome was accomplished by a series of crosses of Bristol males into the mutant strain, with reisolation of the mutant, allowing random assortment of other chromosomes. Activity decreased significantly after four rounds, and after 8 rounds of back crossing, reversion frequency had dropped to $< 10^{-6}$ for each of two different unstable mutations. We attempted to locate a region of the genome responsible for the mutator activity using markers for each of the chromosomes separately, but found that mutator activity was not restricted to a single region linked to any of the markers.

Mutator activity is reflected in Tc1 transposition: The difference in Tc1 copy number between the Bristol (30 copies) and Bergerac (>400) lead to the to the hypothesis that transposition of this element might be the molecular correlate of the mutator activity. Using Tc1 as a probe against a Southern blot of restricted DNA from wild type and the backcrossed stabilized unc-22 mutation showed that several different Tc1's were present in the mutant but absent in wild type (11). To determine which fragments were simply closely linked RFLP's that had been carried through the crosses, and which fragment, if any might represent a Tc1 insertion into unc-22, we isolated a rare revertant of the stabilized strain and compared the Tc1 hybridizing fragments between mutant and isogenic revertant. A variety of restriction enzymes revealed a major band different between the two strains. Of the several different enzymes producing a detectable difference, we elected to clone a 3.8 kb BglII fragment, as it was particularly well separated from other common fragments.

Restriction enzyme digestion of the resultant clone with EcoRV, which cuts Tc1 in the 54 bp inverted repeats, revealed a 1.6kb Tc1 fragment, along with a 1.1 kb insert derived fragment. This fragment hybridized primarily to a single band of 2.2kb in BglII digests of wild type DNA from either

Bristol or Bergerac, and weakly to a second band of 2.8kb. In two of the spontaneous mutants the primary hybridizing band was 3.8kb, in both the original Bergerac background or in the congenic Bristol strains. In derived strains wild type for unc-22 as a result of either reversion or intragenic recombination, the primary band was 2.2 kb. This absolute correlation of the size of this fragment with the state of the unc-22 gene regardless of background demonstrated conclusively the transposition of Tc1 and suggested strongly that this sequence derived from the unc-22 gene.

Recovery of unc-22 genomic sequences: As the next step we recovered lambda clones from a genomic library containing hybridizing sequences. We recovered clones containing both strongly hybridizing and weakly hybridizing sequences. These clones were mapped by restriction endonucleases. To our surprise the two sets of clones overlapped, and the two hybridizing sequences were separated by about ~4 kb (Fig. 2). Using these lambda clones as probes against Southern blots of DNA from 12 additional spontaneous mutants we found that all of them contained an anomalous hybridizing band in one or more digests that could be accounted for by Tc1 insertions. Mapping indicated that all 14 insertions were distributed at different positions over more than 20 kb of

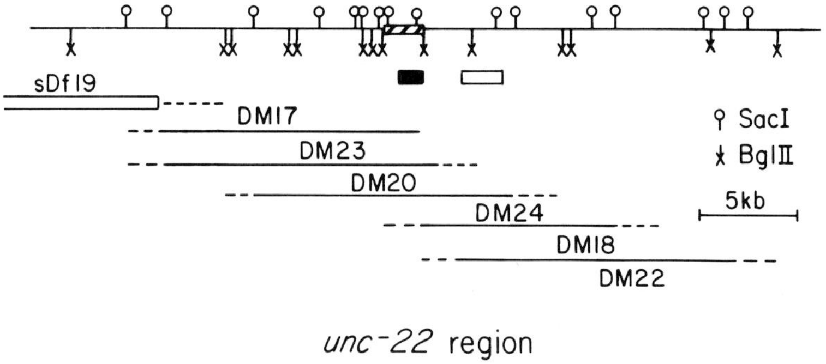

FIGURE 2
The cloned region containing unc-22 is diagrammed. The hatched region on the line represents the initial cloned BglII fragment, while the filled box below the line shows the region used as a unique probe. The open box represents the second region of homology.

the cloned segment. Southern blots of DNA from heterozygous animals carrying a deficiency known to end within unc-22 show a rearranged DNA fragment as well as bands which are present in reduced amounts. The deficiency extends left of unc-22 on the genetic map and alters fragments located only at the left end of the cloned region, thus establishing the orientation of the cloned sequence on the genetic map.

Two RNA transcripts are produced from the cloned region (G.M.B. and R.H.W., unpublished results). One of 1.6 kb hybridizes only to probes from the left most region. A second transcript of 14kb is recognized by probes from throughout the region containing Tc1 insertions. This and the following observations leave little doubt that this is the unc-22 transcript: 1) Accumulated transcript is reduced in animals containing the unc-22(s32) amber allele. In other loci in C. elegans and in other animals, chain terminating mutations are often associated with reduced levels of steady state message. 2) Of 4 Tc1 induced alleles tested, all 4 have a larger hybridizing transcript. Hybridization with Tc1 probes shows that Tc1 is incorporated into these mutant messages, but is absent in the wild type transcript.

In further studies, we have determined using single stranded probes that the direction of transcription is from right to left on the genetic map. Subcloning of a 1.7 kb SstI fragment into the lac Z gene of a modified lambda gt11 vector with a unique Sst I site called lambdaRB1, results in a large fusion polypeptide of a size suggesting that an open reading frame extends through most, if not all, of the fragment (R.J.B. and L. Schrieffer, unpublished results). This fusion protein has been used as an immunogen in rabbits and initial results indicate that the antibody recognizes a protein of more that 400,000 daltons on Western transfers of nematode proteins. Further studies with the antiserum are in progress.

Factors responsible for Tc1 transposition: Insertion of Tc1 into a gene of interest may provide a general method for recovering DNA sequences for that gene. The presently available Bergerac mutator strain however has several disadvantages for applying this strategy, particularly its very high Tc1 copy number. In addition the Bergerac strain is mildly uncoordinated, making it difficult to recognize several types of mutants. The spontaneous mutation frequency, though high for unc-22, is lower for other genes of interest. Lastly most of the genetics done in C. elegans has been done in the Bristol strain N2, in which Tc1 is inactive(10,12).

As part of our studies of Tc1 transposition we have attempted to define the factors responsible for its transposition and to derive more suitable strains for study. As described above mutator activity was altered in hybrids between Bergerac and Bristol strains. To test the hypothesis that this activity was related to overall Tc1 copy number *per se*, we constructed F1 hybrids between Bristol and Bergerac strains, and allowed the chromosomes to assort randomly while deriving 14 independent isogenic lines (I.M. and D.G.M., unpublished data). These strains were then tested for Tc1 activity by measuring the reversion frequency of a known Tc1 induced allele included in all the strains; the same strains were also evaluated by dot blot analysis for the overall Tc1 content. As expected the various hybrids differed in their Tc1 content, none having as many as the Bergerac strain, and none having as few as the Bristol strain (Fig. 3). The Tc1 mutator activity also differed from strain to strain, but there was no correlation between Tc1 copy number and reversion frequency. For example the strain RW7037 has relatively few copies (<90) but has only an order of magnitude reduction in reversion frequency. This strain has

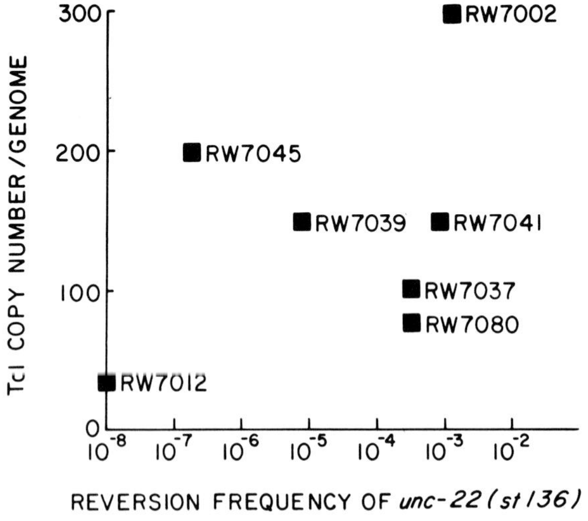

FIGURE 3
The number of copies of Tc1 elements in the strains indicated is plotted against the reversion frequency of the *unc-22(st136)* allele.

also lost the mild uncoordinated phenotype. A further cross of this strain with Bristol followed by isolation of independent lines has led to the recovery of a strain with still fewer copies of Tc1, with little or no further decrease in mutator activity. These results suggest that only a fraction of the Tc1 elements are primarily responsible for mutator activity or that factors that do not cross-hybridize to Tc1 control its transposition. These two strains no longer possess the disadvantageous traits of very high copy number and mild uncoordination. If mutator activity can be raised either by mutation or other means these strains or their derivatives may facilitate the application of Tc1 tagging to other genes.

New strategies for cloning genes using Tc1 induced mutants: The observation that Tc1 is incorporated into the unc-22 mRNA of at least four Tc1 induced alleles is similar to the results of others with other transposable element systems (22). Although how this occurs is puzzling, this feature might be very useful. First it suggests that Tc1 induced alleles in other genes may also contain Tc1 in an elongated mRNA. The detection of a new Tc1 hybridizing RNA species on a Northern transfer of the mutant RNA could help verify that a given spontaneous allele was indeed Tc1 induced. One might even be able to estimate the size of the normal transcript. More speculatively, perhaps a cDNA for the RNA of interest could be derived by priming with an oligonucleotide complementary to the inverted repeat of Tc1.

CONCLUSION

Each of the three different series of experiments described in this paper contributes to our understanding of how muscle structure and function is dictated by the genome. The antibodies obtained define the components of each of the major structures of the body wall muscle and can be used to recover the genes specifying these proteins. The analysis of mutants can reveal interactions between gene products that would be hard to demonstrate by other means. When the function of one of the components of an interacting pair of genes is known it can be used to predict, or at least to limit the range of, possible functions. Finally the transposable element, Tc1, makes it possible to clone genes and determine their protein products which have only been defined genetically. Once cloned the function of the gene product can be further defined using antibodies and genetic tests. A combination of all three approaches should lead to detailed molecular descriptions of the process of muscle development and function in C. elegans.

REFERENCES

1. Waterston, R.H., and Francis, G.R. (1985) Genetic analysis of muscle development in Caenorhabditis elegans. Trends in Neurosciences, in press.
2. Waterston, R. H., Thomson, J. N., and Brenner, S. (1980) Mutants with altered muscle structure in Caenorhabditis elegans. Develop Biol 77, 271-302.
3. Zengel, J. M. and Epstein, H. F. (1980) Identification of genetic elements associated with muscle structure in the nematode Caenorhabditis elegans. Cell Motility 1, 73-97.
4 Waterston, R.H., Fishpool, R.M., and Brenner, S. (1977). Mutants affecting paramyosin in Caenorhabditis elegans. J Mol Biol 117, 679-697.
5. Waterston, R. H., Hirsh, D., and Lane, T. R. (1984). Dominant mutations affecting muscle structure in Caenorhabitis elegans that map near the actin gene cluster. J Mol Biol 180, 473-496.
6. Landel, C. P., Krause, M., Waterston, R., and Hirsh, D. (1985). DNA rearrangements if the actin gene cluster in C. elegans accompany reversion of three muscle mutants. J Mol Biol 180, 497-513.
7. Dibb, N. J., Brown, D. M., Karn, J., Moerman, D. G., Bolten, S. L., and Waterston, R. H. (1985). Sequence analysis of mutations that affect the synthesis, assembly, and enzymatic activity of the unc-54 myosin heavy chain of Caenorhabditis elegans. J Mol Biol, in press.
8. Miller, D. M., Ortiz, I., Berliner, G. C., and Epstein, H. F. (1983). Differential localization of two myosins within nematode thick filaments. Cell 34, 477-490.
9. Francis, G.R. and Waterston, R.H. (1985). Localization of muscle proteins of Caenorhabditis elegans implicated in thin filament attachment and I-band organization. Manuscript in revision.
10. Moerman, D.G and Waterston, R.H. (1984). Spontaneous unstable unc-22 mutations in C. elegans var. Bergerac. Genetics 108, 859-877.
11. Moerman, D.G., Benian, G.M., and Waterston, R.H. (1985). Cloning of the unc-22 gene of Caenorhabditis elegans by Tc1 transposon tagging. Manuscript submitted.

12. Eide, D.J. and Anderson, P. (1985). Transposition of Tc1 in Caenorhabditis elegans. Proc Natl Acad Sci USA, in press.
13. Young, R.A. and Davis, R.W. (1983). Efficient isolation of genes using antibody probes. Proc Natl Acad Sci USA 80, 1194-1198.
14. Albertson, D.G. (1984). Localization of the ribosomal genes in Caenorhabditis elegans chromosomes by in situ hybridization using biotin-labeled probes. EMBO J 3, 1227-1234.
15. Files, J.G., Carr, S., and Hirsh, D. (1983). Actin gene family of C. elegans. J Mol Biol 164, 355-375.
16. Moerman, D.G., Plurad, S., Waterston, R.H., and Baille, D.L. (1982). Mutations in the unc-54 myosin heavy chain gene of Caenorhabditis elegans that alter contractility but not muscle structure. Cell 29, 773-781.
17. Walker, J.E., Sarate, M., Runswick, M.J., and Gay, N.J. (1982) Distantly related sequences in the alpha- and beta-subunits of ATP synthase, myosin, kinases and other ATP-requiring enzymes and a common nucleotide binding fold. EMBO J 1, 945-951.
18. Yount, R. G., Ojala, D. and Babcock, D. (1971). Interaction of P-N-P and P-C-P analogs of adenosine triphosphate with heavy meromyosin, myosin and actomyosin. Biochemistry 10, 2490-2496.
19. Riddle, D.L. and Brenner, S. (1978). Indirect suppression in Caenorhabditis elegans. Genetics 89, 299-314.
20. Epstein, H. F., and Thomson, J. N., (1974). Temperature sensitive mutation affecting myofilament lattice assembly in Caenorhabditis elegans. Nature(Lond) 250, 579-580.
21. Emmons, S.W., Yesner, L., Ruan, K., and Katzenberg, D. (1983). A transposon in Caenorhabditis elegans. Cell 32, 55-65.
22. Doring, H.P., Freeling, M., Hake, S., Johns, M.A., Kunze, R., Merckelbach, A, Salamini, F., and Starlinger, P. (1984). A Ds-mutation of the Adh1 gene in Zea mays L. Mol Gen Genet 193, 199-204.

Molecular Biology of Muscle Development, pages 619–628

SUP-3 SUPPRESSION AFFECTS MUSCLE STRUCTURE AND MYOSIN HEAVY CHAIN ACCUMULATION IN CAENORHABDITIS ELEGANS[1]

Anthony J. Otsuka

Department of Genetics, 345 Mulford Hall,
University of California, Berkeley, California 94720

ABSTRACT Suppression of paramyosin mutations by sup-3 has been shown to result in the formation of thick filaments of normal appearance distributed in regions of partially organized muscle lattice structure. In the presence or absence of muscle mutations, sup-3 results in the elevation of MHCA, one of the two known body-wall myosin heavy chains. The correlation of sup-3 mutations with MHCA increase suggests that sup-3 may regulate the expression or accumulation of MHCA.

INTRODUCTION

The genetic dissection of muscle development in the nematode Caenorhabditis elegans has been facilitated by the study of interacting muscle mutations (1-3). The interaction of genes can often be detected by genetic suppression (1), in which mutations in one gene can be ameliorated by changes in another gene. Although the most familiar examples of genetic suppression involve altered tRNA's, sup-3 is unlikely to use this mechanism because suppression is specific for particular genes rather than for alleles.

sup-3 suppresses mutations in two genes that are known to encode components of muscle thick filaments (unc-15 and unc-54) as well as defects in a gene (unc-87) of unknown function (1). sup-3 was originally isolated (1) as a

[1]Supported by funds from the National Institutes of Health (GM06709), the National Science Foundation (PCM83-16126) and the State of California (SCAR).

suppressor of unc-54, which encodes the predominant body-wall myosin heavy chain (MHCB). sup-3 was also found to suppress certain defects in paramyosin, which is encoded by unc-15 and is found in the core of wild type thick filaments (1).

R. Waterston and his colleagues have demonstrated that the level of a second, less abundant, body-wall myosin heavy chain (MHCA) was elevated 1.7 to 2.2 fold in suppressed unc-54 mutants (3). It was of interest to determine whether suppressed paramyosin (unc-15) mutants displayed an analogous increase in MHCA levels. Elevation of MHCA levels in both suppressed unc-15 and unc-54 mutants could indicate that the MHCA increase may be important to the suppression mechanism. It is also possible that sup-3 directly affects the accumulation of MHCA, independently of the state of the muscle. It is reported here that mutant muscle structure is improved and body-wall myosin heavy chain A accumulates as a direct result of sup-3.

It is desirable to determine the physiological mechanisms, whether similar or different, for suppression of paramyosin and myosin heavy chain mutants in addition to characterizing the genetic nature of sup-3 and its biochemical effect on MHCA accumulation. Although, it is not possible to unambiguously explain the physiological basis for suppression at the present time, electronmicroscopic studies have shown that thick filaments are more prevalent in suppressed strains.

MATERIALS AND METHODS

Electronmicroscopy.

Animals at the L4 to young adult stage were sectioned and electronmicrographed (4).

Growth of Synchronized Cultures of Nematodes.

Nematodes were cultured on *Escherichia coli* by standard methods (5). Adult worms were repeatedly settled by centrifugation to remove laid eggs, killed with 2% hypochlorite, rinsed, and transferred to seeded plates under sterile conditions. Larvae hatched from eggs *in utero*, representing individuals synchronized to within three hours, were grown at 20 C to the L3 stage (identified by the developing vulva). The worms were collected and bacteria were removed by several water washings.

Preparation of Nematode Homogenates.

Worm homogenates were prepared by combining equal volumes of nematodes and water, adding a volume of 2X loading buffer (6) equal to the nematode suspension, boiling for three minutes, vortexing, and boiling for an additional minute.

Polyacrylamide Gel Electrophoresis.

The gel buffer system (6) was that used by Waterston et al. (3). Homogenate aliquots of 10, 20 and 30 ul were loaded on 3% polyacrylamide gels. Gels were silver stained (7) and scanned with a computerized Zeinoh soft laser densitometer. At least two lanes from each of two gels were averaged for each experiment. The amount of protein was maintained within the linear range of silver staining.

RESULTS

Suppressed Paramyosin Mutants Have Improved Muscle Structure.

The muscle structure of unc-15 (e73) is highly disorganized with large spike-like paracrystals of paramyosin generally localized at the ends of the rhomboidal muscle cells (2). The suppressed unc-15 (e73) sup-3 (e1407) animal displays a pattern that is similar to that in suppressed unc-54 mutants (1), with weak birefringence and a feather-like appearance under polarized light due to local regions of muscle fiber organization (data not shown). Paramyosin paracrystals are still evident in suppressed paramyosin mutants. Strains containing only sup-3 (e1407) have wild type appearance under polarized light (1).

Suppressed Animals Have Thick Filaments of Normal Appearance.

In order to determine how sup-3 might affect muscle structure, transverse sections of muscle from suppressed paramyosin mutants were studied by electron microscopy. As controls, sections were prepared from wild type animals, paramyosin mutants, and sup-3 worms.

In the wild type muscle, the thick and thin filaments form a fairly regular array (Figure 1a). The vertical granular band at the left is the M-line which appears as a distinct band due to the oblique striation of nematode muscle.

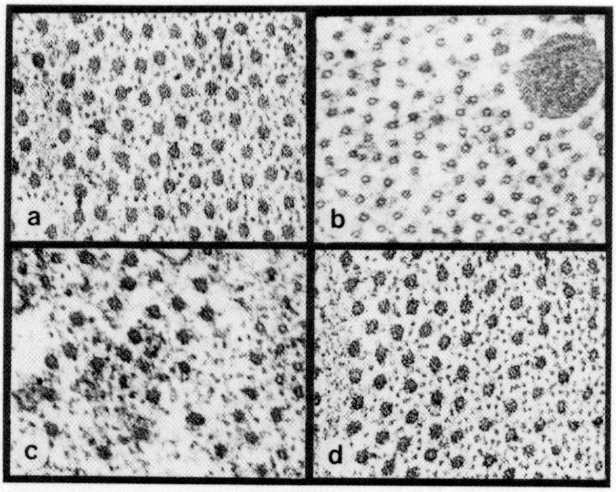

FIGURE 1. Cross sections of muscle filament lattices. The sections are: a) wild type (N2); b) unc-15 (e73); c) sup-3 (e1407) unc-15 (e73) double mutant; and d) sup-3 (e1407).

In sections of unc-15 (e73) animals (Figure 1b), there are hollow filaments about three-fourths the diameter of normal thick filaments (4). In addition, a paramyosin paracrystal is present in the upper right corner. The hollow filaments and paracrystals are localized at the anterior and posterior ends, with respect to the length of the worm, of the elongate muscle cells (4). It is possible that the hollow filaments are thick filaments lacking the paramyosin core (4) or they may be part of a core structure of thick filaments (8).

In addition to the hollow filaments, there are filaments that are of larger diameter than normal thick filaments (9). In preparations of filaments isolated from unc-15 (e1214) (9) or (e73), these are short abnormal filaments with tassle-like ends (9). The short filaments may be rudiments of normal thick filaments. In wild type thick filaments, MHCA is found in the H-band region of the sarcomere and MHCB is found in the A-band region (10). In agreement with this observation, isolated thick filaments show a localization of MHCA to a region midlength along the filament (10). MHCB slightly overlaps the MHCA region and continues outwards to both ends of the filament. In the absence of functional paramyosin, the short filaments formed in unc-15 mutants may represent MHCA-containing abortive intermediates of normal thick filament assembly.

In the suppressed strain, sup-3 (e1407) unc-15 (e73), the muscle is clearly disorganized (Figure 1c). However, there are a large number of thick filaments of normal diameter. These "normal" thick filaments show local areas of lattice formation and probably provide the motive capability of the suppressed animals. Filament preparations isolated from the same suppressed mutant contain many filaments of normal length and appearance as well as intermediate length filaments in addition to the short filaments (A. Otsuka, D. Miller, I. Ortiz and H. Epstein, unpublished results).

The structure of sup-3 (e1407) muscle is essentially identical to wild type (Figure 1d). Again, the vertical strip at the left is the M-line. At the right are a few hollow filaments that represent the very ends of the normal thick filaments.

sup-3 Directly Affects MHCA Accumulation.

We have been able to show that, in the presence or absence of muscle mutations, sup-3 results in the increase of MHCA. In top to bottom order, Figure 2 displays the two body-wall-specific myosins (MHCA and MHCB) and the two pharyngeal myosins (MHCD and MHCC). Qualitatively, the level of MHCA increases from wild type (lane a) to unc-15 (lane b) to sup-3 unc-15 (lane c) and sup-3 (lane d).

These results are expressed more quantitatively in Figure 3. The MHCA/MHCB ratio increases more than two-fold from 0.20 in wild type (Figure 3, line 1) to 0.46 in animals that contain sup-3 (e1407) alone (line 3). MHCA is accumulating because a parallel increase is seen in the MHCA/(MHCC+MHCD) ratio, but amount of the predominant MHCB, and therefore the MHCB/(MHCC+MHCD) ratio, is not dramatically affected. The increase in the MHCA/MHCB ratio is slightly smaller for the weaker sup-3 (e1390) allele.

In suppressed paramyosin mutants, MHCA levels are also

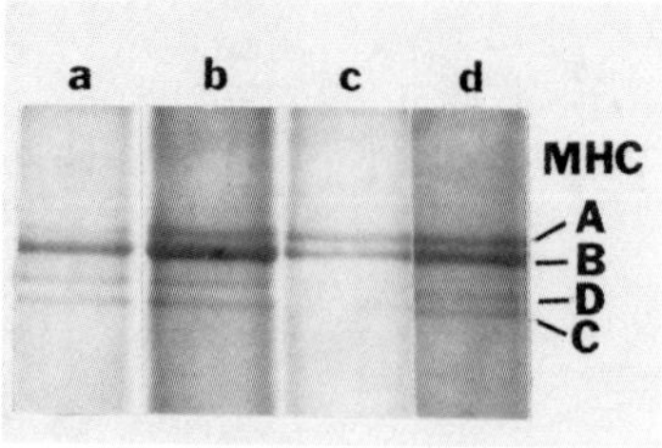

FIGURE 2. Electrophoretic separation of myosin heavy chains. Lanes are as follows: a) wild type; b) unc-15 (e73); c) sup-3 (e1407) unc-15 (e73); and d) sup-3 (e1407).

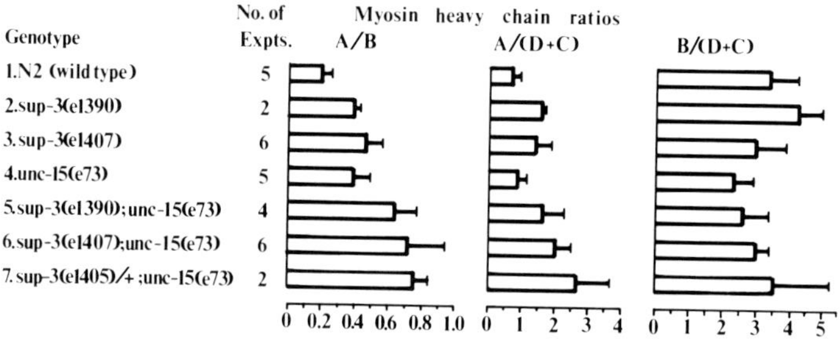

FIGURE 3. Ratios of myosin heavy chains.

increased. Even in the absence of sup-3, the MHCA/MHCB ratio is higher in unc-15 (e73) than in wild type (Figure 3, lines 4 and 1). This may be due to the presence of many shortened thick filaments containing predominantly MHCA. The MHCB level may be lower because of the lack of filament elongation and consequent MHCB degradation. In combination with a decrease in body-wall muscle mass, this results in a lowering of the MHCB/(MHCD+MHCC) ratio from 3.4 in wild type to 2.3 in unc-15 (e73) (Figure 3, line 4), although the level of MHCA does not change considerably.

In the case of suppressed unc-15 (e73) mutants, the MHCA/MHCB ratio increases from 0.39 in unc-15 (e73) to 0.72

for the double mutant sup-3 (e1407) unc-15 (e73) or 1.8 fold. A similar comparison yields a 1.6-fold increase for the weaker sup-3 (e1390) allele. The marked elevation of MHCA can be seen from the relative increase of MHCA/(MHCD+MHCC) ratios by 2.4 and 1.9-fold compared to unc-15 for these same strains. Thus, there is a clear elevation in MHCA in suppressed paramyosin mutants which is correlated with the formation of normal appearing thick filaments. A smaller recovery in levels of MHCB can be seen in suppressed paramyosin mutants because of the assembly of MHCB into thick filaments and an increase in the mass of body muscle compared to pharyngeal muscle.

The semidominance of the sup-3 mutations (1) leads to the prediction that even in heterozygotes the level of MHCA should be elevated. The presence of a single copy of the wild type allele does not block MHCA increase as can be seen from the high MHCA levels in sup-3 (e1405)/+;unc-15/unc-15 animals (Figure 3 line 7).

DISCUSSION

It has been shown that sup-3 suppression results in the formation of thick filaments of normal diameter and length which are able to form localized regions of functional muscle lattice. These regions of partial lattice structure lead to a similar muscle appearance in both suppressed myosin and paramyosin mutants. It has been shown here that MHCA increases not only in suppressed myosin mutants but in suppressed paramyosin mutants as well. Further, a similar MHCA increase was found in strains that contain sup-3 but do not contain any muscle mutations.

It is desirable to incorporate these observations into a model that explains the physiological basis for sup-3 suppression and the genetic nature of sup-3 mutations. One simple model is that the limiting factor in the two different mutants is the number of normal length thick filaments. Elongation of the thick filaments may be an important aspect of the suppression mechanism because MHCA is normally limited to the midlength region of the thick filament. Because there is no MHCB in suppressed myosin null mutants, MHCA must be used to extend the filament to normal length. In the case of unc-54 null mutants, thick filaments can be formed that contain MHCA along their entire length (10). Thus the weak sup-3 suppression of unc-54 may be due to a mass effect resulting from the increase of MHCA.

In paramyosin mutants, a structure is formed that may be a MHCA-containing nucleation site for thick filament elongation. Again, MHCA could play a role in the extension of this structure to form a normal length thick filament. In the case of unc-15 mutations, strong sup-3 suppression may be linked to a change in thick filament assembly as much as to an increase in MHCA. Specifically, though MHCA is present in unc-15 mutants, regions of normal thick filament organization are much less prevalent than in suppressed strains. In spite of only a two-fold increase in MHCA, obvious areas of filament organization are present in suppressed animals. This observation suggests that additional factors are involved.

One possibility for the formation of normal length thick filaments is the delocalization of MHCA throughout the length of the filament. sup-3 may code for a protein that limits the boundaries of MHCA, i. e., a type of capstone that prevents MHCA incorporation towards the ends of the thick filament. Unless the amount of such a capstone had to be stoichiometrically produced or unless the capstone was made of several interacting subunits, the wild type product should serve to prevent the semidominance seen with sup-3. However, if the mutant product could act as a "poisonous subunit" then it could interefere with the formation of the capping structure.

Another possibility is that sup-3 is a thick filament protein that alters the packing of proteins into the thick filaments. For example, a protein substratum on which MHCA and paramyosin assemble could be altered to allow MHCA to assemble into thick filaments even though paramyosin is defective.

In addition to the possibility that transcription of MHCA is altered, the accumulation of MHCA may result from a change in the MHCA protein itself. The altered protein may be more efficiently incorporated into thick filaments and thereby protected from degradation or the protein monomer itself may be more resistant to degradation. An altered protein structure that was more able to form thick filaments than the wild type protein would explain the semidominance of sup-3. Another way that MHCA could be more efficiently incorporated into thick filaments is a change in a factor involved in muscle assembly.

The genetic nature of sup-3 and its relationship to the MHCA structural gene needs to be considered. Recent hybridization studies of cloned MHCA DNA to chromosomes *in situ* showed that MHCA is located on linkage group V in the general vicinity of sup-3 (D. Albertson and D. Miller, personal communication). Further, it has been shown by Southern blot analysis that the coding region for MHCA does not contain any detectable aberrations in a strain containing the eDf1 (e1405) deficiency which produces a very strong sup-3 phenotype (D. Miller, I. Greenwald, D. Albertson, A. Coulson, and J. Sulston, personal communication). However, restriction fragments were found indicating that the end point of the eDF1 deletion is very near the MHCA gene. It can be hypothesized that sup-3 may be a cis element that alters MHCA expression although other mechanisms have not been ruled out.

There are several possibilities for increased transcription of the MHCA gene: 1) insertion of a promoter on a transposable element, 2) fusion of MHCA to a stronger promoter, 3) loss of a cis-acting element that limits transcription, 4) loss of a negative regulator or alteration of a positive regulator, 5) amplification of the number of MHCA gene copies.

Several possible roles have been considered for sup-3. Three of these consider the sup-3 product as an integral part of the thick filament and seek to explain the presence of "normal" length thick filaments in suppressed muscle mutants as well as the semidominance of sup-3. These three possibilities are that sup-3 produces a capstone protein that normally limits MHCA to the midlength of the thick filament, that it encodes a thick filament protein substratum that interacts with paramyosin and MHCA, or that it encodes MHCA. A variation on the third possibility is that sup-3 affects the regulation of MHCA in either a cis or trans manner. At the present time it is not possible to exclude any of these mechanisms, however, we favor the simple idea that sup-3 affects a structural component of the thick filament, for example MHCA.

ACKNOWLEDGEMENTS

We would like to thank M. Chalfie and N. Thompson for help with the electron microscopy, P. Anderson, E. Hedgecock, A. MacLeod, W. Sharrock, H. Epstein, D. Miller, S. Brown, D. Riddle, and R. Waterston for stimulating discussions and unpublished results.

REFERENCES

1. Riddle DL, Brenner S (1978). Indirect suppression in Caenorhabditis elegans. Genetics 89:299.
2. Waterston RH, Thomson JN, Brenner S (1980). Mutants with altered muscle structure in Caenorhabditis elegans. Develop Biol 77:271.
3. Waterston RH, Moerman DG, Baillie D, Lane TR (1982). Mutations affecting myosin heavy chain accumulation and function in the nematode Caenorhabditis elegans. In Shotland DL (ed): "Disorders of the Motor Unit," John Wiley and Sons, p 747.
4. Waterson RH, Fishpool RM, Brenner S (1977). Mutants affecting paramyosin in Caenorhabditis elegans. J Mol Biol 117:679.
5. Brenner S (1974). The genetics of Caenorhabditis elegans. Genetics 77:71.
6. Neville DM Jr (1971). Molecular weight determination of protein-dodecyl sulfate complexes by gel electrophoresis in a discontinuous buffer system. J Biol Chem 246:6328.
7. Merril CR, Goldman D, Sedman SA, Ebert MH (1981). Ultrasensitive stain for proteins in polyacrylamide gels shows regional variation in cerebrospinal fluid proteins. Science 211:1437.
8. Epstein HF, Miller DM, Ortiz I, Berliner, GC (1985). Myosin and paramyosin are organized about a newly identified core structure. J. Cell Biol. 100:904.
9. Mackenzie JM Jr, Epstein HF (1980). Paramyosin is necessary for determination of nematode thick filament length in vivo. Cell 22:747.
10. Miller DM III, Ortiz I, Berliner GC, Epstein HF (1983). Differential localization of two myosins within nematode thick filaments. Cell 34:477.

Molecular Biology of Muscle Development, pages 629–638

THE SUP-3 LOCUS IS CLOSELY LINKED TO A MYOSIN HEAVY CHAIN GENE IN CAENORHABDITIS ELEGANS

David M. Miller[1] and Ichiro Maruyama[2]

[1]Department of Zoology, North Carolina State University, Raleigh, NC 27695 and [2]MRC Centre, Laboratory of Molecular Biology, Cambridge, CB2 2QH, England

ABSTRACT *C. elegans* body wall muscle thick filaments are assembled from two different types of myosin heavy chain, MHC A and MHC B. The structural gene for the MHC A protein has been identified from the reaction of bacterial fusion peptides with specific monoclonal antibodies. We have also shown that a mutation in the *sup-3* locus that enhances the accumulation of the MHC A protein is closely linked to the MHC A gene. A model of *sup-3* action is proposed.

INTRODUCTION

Two different types of myosin heavy chain, MHC A and MHC B, are expressed in the body wall muscle of the nematode, *Caenorhabditis elegans* (1, 2). Immunomicroscopy of muscle tissue and isolated thick filaments has shown that both myosin isoforms are incorporated into a single kind of body wall muscle thick filament. In the proposed model, MHC A occupies the central portion of the filament whereas MHC B is restricted to the lateral regions (3). Two additional minor isoforms, MHC C and MHC D, are produced in the pharynx but their locations within thick filaments from this muscle have not been determined (2,4).

[1]To whom correspondence should be addressed. Supported by grants from the Burroughs Wellcome Fund and from the European Molecular Biology Organization.

The MHC B protein is encoded by the unc-54 gene (5). Null mutations such as unc-54 (e190) that completely eliminate MHC B result in muscle disorganization and uncoordinated movements (6). Reversion analysis has revealed compensatory mutations in a second locus, sup-3, which partially suppress this "unc" phenotype (7). Sup-3 mutations also suppress the paramyosin defective mutant, unc-15(e73). (7,8). The improved muscle structure and movement are correlated with a 2-3 fold increase in the accumulation of MHC A in sup-3 mutants (4).

The complete nucleotide sequence of the MHC B gene has been determined (9). Three additional genes, myo-1, myo-2, and myo-3, have been isolated from a lambda phage genomic library by hybridization with fragments of the MHC B gene (9, 10). In this paper, we describe experiments demonstrating that myo-3 encodes the MHC A protein and that a mutation in the sup-3 locus is closely linked to the MHC A gene. Based upon these data and other considerations, a model of sup-3 action is proposed.

METHODS

Expression Vector Cloning

DNA from lambda phage bearing the unc-54, myo-1, myo-2, or myo-3 genes (10) were cut with Hind III and ligated into the plasmid expression vector, pUR 288 (11). Recombinant clones expressing fusion peptides with β-galactosidase were identified by colony immunoassay with either rabbit antimyosin or with monoclonal antibodies to C. elegans myosin (10, 12). Bacterial homogenates were separated by SDS-PAGE and the reactions of monoclonal antibodies with specific hybrid proteins were determined by immunoblotting (3).

Southern Blotting

Nematodes were grown in liquid culture and harvested by the sucrose method (13). DNA was prepared according to Emmons et al. (1979) (14) and digested with various restriction enzymes. Agarose gel electrophoresis and Southern blotting were conducted according to standard procedures (15). The MHC A cosmids, C02B11 and C14G1, were radiolabeled by nick translation (15). High stringency hybridization and washing conditions were exactly as described by Maniatis et al. (1982) (15).

RESULTS

Identification of the MHC A Gene

The unc-54 gene has been shown to encode the MHC B protein from the analysis of mutants affecting MHC B structure (5,6,10). Similar mutations in the additional MHC genes have not been isolated, however. Thus, other methods were needed to correlate the myo-1, myo-2, and myo-3 genes with the isoforms (MHC A, MHC C, and MHC D) that they encode. Our strategy was to express fragments of each gene in E. coli (12) and then to screen the resultant fusion peptides with a panel of monoclonal antibodies to identify epitopes specific to each MHC isoform (3,16). Several features of the C. elegans MHC genes (9, 10) facilitated this approach. Immunoblots of chymotryptic digests have revealed that most of the available monoclonal antibodies are directed toward epitopes in the myosin rod (3) (Miller, Berliner, and Epstein, unpublished data). This finding is consistent with the DNA sequences of these genes which are less conserved in the rod than in the 5' terminal myosin head domain. In addition, the myosin rod region of each gene, with the exception the 3' end, is encoded by a single long exon (3-5 kb). Finally, due to the frequent occurrence of the amino acid pair, lys leu, in the periodic rod sequence, there are a number of Hind III sites in this region all with the same translational phasing of AAG CTT (9, 10). Thus, it was possible to clone these Hind III fragments into a single expression vector with the matching reading frame, pUR 268 (11), to produce fusion peptides spanning most of the immunogenic rod portion of each gene (12).

The immunoblotting experiment shown in Fig. 1 indicates that myo-3 corresponds to the MHC A gene. The MHC A specific monoclonal antibody, 5-6 (3), reacts exclusively with the fusion peptide encoded by the 460 bp fragment from the S-2 or "hinge" region of the myo-3 gene but not with fusion peptides from the homologous regions of the other three genes. This finding is consistent

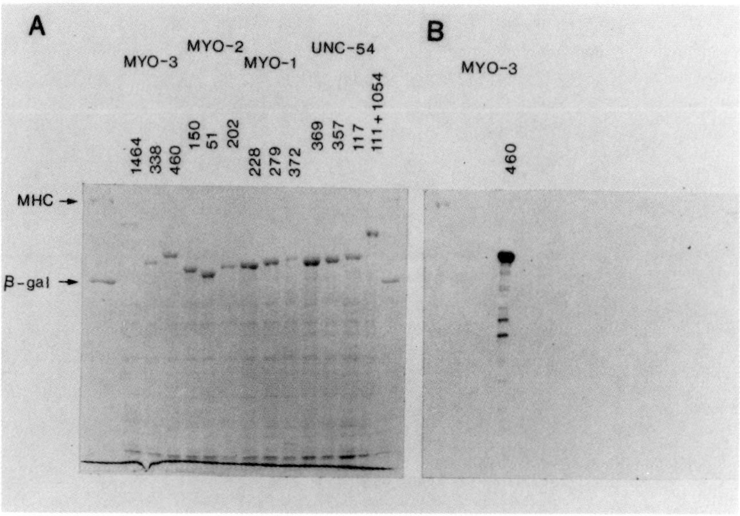

Figure 1. Reaction of the MHC A specific monoclonal antibody, 5-6, with a fusion peptide encoded by the myo-3 gene. Bacterial clones were induced with IPTG, solubilized in SDS, and separated on a 6% SDS-polyacrylamide gel. A) Coomassie stained gel of bacterial homogenates. The fusion peptides correspond to the prominent bands in the high molecular weight range bounded by nematode myosin heavy chain (MHC M_r ~226kd) and by β-galactosidase (β-gal, M_r 116 ~kd) standards in the outside lanes. The numbers across the top of the gel indicate the size of the corresponding Hind III inserts in bp. B) Immunoblot of the gel shown in panel A reacted with monoclonal antibody 5-6. Note specific reaction with the fusion peptide encoded by the 460 bp fragment of myo-3. The 1464 bp and the 1054 bp fragments encode portions of the LMM domain, the remaining Hind III fragments in this experiment are derived from the S-2 regions of the respective MHC genes (10,12).

with the previous observation that antibody 5-6 is specific to the chymotryptic S-2 fragment of MHC A (3). Similar experiments with other monoclonal antibodies have verified that the unc-54 gene encodes MHC B and that myo-1 and myo-2 encode MHC D and MHC C, respectively (10, 12).

Sup-3 (eDf1) is Closely Linked to the MHC A Gene

Recent in situ hybridization data have mapped the MHC A gene to a region on chromosome V that includes sup-3 (17) (D. Albertson, personal communication). We have now used the Southern blotting technique (15) to identify restriction fragment differences between sup-3 and wild type DNA in a region adjacent to the MHC A gene.

The sup-3 allele, eDf1 (formerly, e1405), is a strong suppressor of unc-15(e73) (7) and is correlated with increased accumulation of MHC A (A. Otsuka, personal communication). eDf1 is a homozygous lethal mutation and exhibits a null phenotype for several genes in the vicinity of sup-3 which suggests that the lesion is a DNA deletion or rearrangement. The eDf1 allele can be maintained as a heterozygote in the semi-balanced strain unc-15(e73); sup-3(eDf1)/sma-1(e30). (The homozygous progeny unc-15(e73); sma-1(e30) are "unc" and "small" and do not reproduce well.) (7)

Wild type (N2) and sup-3 mutant DNA were digested with several different restriction enzymes, blotted, and hybridized with radioactive cosmids containing MHC A gene inserts. The probes in this instance span 30-40 kb of sequence in the vicinity of the MHC A gene (J. Sulston, personal communication) and thus produce a complex pattern of bands. As shown in Fig. 2, eDf1/+ DNA exhibits ~15 kb Pst I and ~8 kb Bcl I fragments that are not present in N2 DNA. The polymorphic bands are not detected in homozygous sma-1(e30) DNA (data not shown) nor is it likely that the restriction fragment differences are due to unc-15(e73) since the unc-15 gene is located on chromosome I (8). Additional control experiments have shown that the new bands are not partial digestion products nor do they hybridize with cosmid vector sequences (data now shown). The precise location of the eDf1 lesion has not been determined but recent results have excluded the possibility that the lesion is 3' to the MHC A gene (Miller and Maruyama, unpublished data).

DISCUSSION

The C. elegans body wall muscle thick filament is composed of two types of myosin. The different locations of myosin A and myosin B in the thick filament are correlated with distinct structural domains and suggest

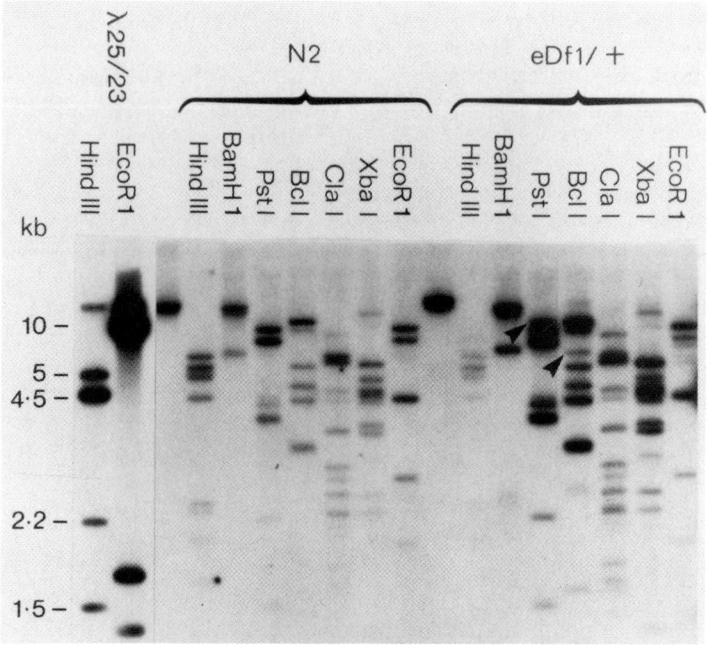

Figure 2. Southern blot comparing hybridization of MHC A gene probes with wild type (N2) and eDf1/+ DNA. Arrows point to restriction fragment differences in the Pst I and Bcl I lanes. Standard size markers on the left were cut from a phage clone (λ25/23) bearing the MHC A gene.

unique roles in assembly. Myosin A occupies a 1.8 μ long central region that spans the antiparallel packed domain of the structural bare zone and also includes the flanking regions in which adjacent myosin rods are in parallel orientation. Myosin B is exclusively incorporated into the 4.5 μ long parallel-packed polar regions of these 9.7 μ filaments. Myosin A would appear to participate in the initiation of filament assembly whereas myosin B polymerizes during filament elongation only (3).

The myosin heavy chains, MHC A and MHC B, are coordinately expressed at a constant ratio of A/B ~ 1/4 throughout the period of myogenesis (4, 18). This finding is consistent with the relative fractions of the body wall thick filament occupied by myosin A (18%) and myosin B (90%) and rules out the possibility that sequential assembly of the myosins is dependent upon temporal MHC gene expression (3). It seems more likely that this mechanism is determined by structural differences between the myosin A and myosin B rod domains.

A model of the myosin B rod has been deduced from the sequence of the MHC B gene. The periodic array of amino acid side chains is correlated with the coiled coil structure of the myosin molecule and with overlap distances between adjacent myosins that occur in native thick filaments (19). In the experiments described in this paper, we have identified the MHC A gene. It is hoped that structural analysis of the MHC A rod sequence (19) will reveal the molecular basis of discrimination between these two closely related myosin isoforms.

Mutations in the sup-3 locus that suppress the MHC B null and paramyosin defective phenotypes are correlated with increased amounts of MHC A (4, 20). Improved movement is clearly related to the assembly of more thick filaments (7). It is conceivable that the elevated MHC A levels are secondary to some other sup-3 mediated event that facilitates thick filament assembly in these mutants. One may suppose, in this instance, that a higher rate or extent of myosin A polymerization would protect a larger fraction of the nascent MHC A polypeptide chain from proteolytic degradation. However, since MHC A is increased 2-3 fold in sup-3(e1407) (4, 20), this mechanism would require that at least 50% of the MHC A synthesized in the wild type is degraded rather than incorporated into filaments. This seems unlikely.

We favor the alternative hypothesis that the assembly of more thick filaments is a direct consequence of the increased amount of MHC A in sup-3 mutants. In the mutant unc-54(e190), myosin A is incorporated into both the central portion of the filament and into the polar regions normally occupied by myosin B (3). Increasing MHC A presumably results in more of these unusual filaments which in turn mediate muscle contraction. Filament preparations from the paramyosin mutant, unc-15(e73), include large paracrystalline arrays of the mutated paramyosin, short stubby filaments, hollow tubes, and a few normal looking thick filaments (Otsuka, Miller, Ortiz and Epstein, unpublished data). The hollow tube structures and short filaments may correspond to intermediates in C. elegans thick filament assembly (21). The occurence of these structures presumably reflects the distribution of e73 paramyosin between these different complexes. In this scenario, the additional MHC A drives further assembly of the defective paramyosin into abnormal but at least partially functional thick filaments.

We have now shown that the sup-3 mutant, eDf1 (7), is closely linked to the MHC A gene. The juxtaposition of these two loci suggests that sup-3 may in fact correspond to the MHC A gene and that sup-3 mutations affecting MHC A protein levels alter DNA sequences that normally regulate MHC A gene expression. Preliminary results suggest that eDf1 is 5' to the MHC A gene (Miller and Maruyama, unpublished data). This finding is consistent with the hypothesis that eDf1 deletes a sequence that normally regulates transcription of the MHC A gene. A large number of sup-3 alleles exhibiting varying degrees of suppression have been isolated (7, 20). Revertants of the sup-3 phenotype that restore MHC A to normal levels have also been identified (20). These effects might correspond to offsetting mutations in a promoter sequence (22). Other mechanisms including post-transcriptional events or the alteration of a nearby locus encoding a protein that normally regulates MHC A expression may also account for the sup-3 effect, however. Our current approach is to clone and sequence these sup-3 alleles as a first step towards distinguishing between these possibilities.

ACKNOWLEDGEMENTS

We thank F. E. Stockdale and J. Karn for their contributions to the expression vector cloning. We also thank D. Albertson, A. Otsuka, H. F. Epstein, R. Waterston, D. Riddle, and S. Brown for communicating unpublished results, I. Greenwald for suggesting the experiment with eDf1, R. Fishpool for technical assistance, and A. Coulson and J. Sulston for providing the cosmid probes. We gratefully acknowledge the MRC Laboratory of Molecular Biology and Sydney Brenner for providing the resources with which to perform this work.

REFERENCES

1. Schachat FH, Harris HE, Epstein HF (1977). Two homogeneous myosins in body-wall muscle of Caenorhabditis elegans. Cell 10:721.
2. Mackenzie JM, Schachat FH, Epstein HF (1978). Immunocytochemical localization of two myosins within the same muscle cells in Caenorhabditis elegans. Cell 15:413.
3. Miller DM, Ortiz I, Berliner GC, Epstein HF (1983). Differential localization of two myosins within nematode thick filaments. Cell 34:477.
4. Waterston RH, Moerman DG, Baillie D, Lane TR (1982) Mutations affecting myosin heavy chain accumulation and function in the nematode Caenorhabditis elegans. In Schotland DL (ed): "Diseases of the Motor Unit," New York: John Wiley and Sons, p. 747.
5. Macleod AR, Karn, J, Brenner S (1981). Molecular analysis of the unc-54 myosin heavy chain gene of Caenorhabditis elegans. Nature 291:386.
6. Epstein HF, Waterston RH, Brenner S (1974). A mutant affecting the heavy chain of myosin in Caenorhabditis elegans, J Mol Biol 90:291.
7. Riddle DL, Brenner S (1978). Indirect suppression in Caenorhabditis elegans. Genetics 89:299.
8. Waterston RH, Fishpool RM, Brenner S (1977). Mutants affecting paramyosin in Caenorhabditis elegans. J Mol Biol 117:825.
9. Karn J, Brenner S, Barnett L (1983). Protein domains in the Caenorhabditis elegans unc-54 myosin heavy chain gene are not separated by introns. Proc Natl Acad Sci USA 80:4253.
10. Karn J, Dibb N, Miller DM (1984). Cloning nematode myosin genes. In Shay JW (ed):"Cell and Muscle Motility," New York:Plenum Press, vol. 6, p 185.
11. Rüther U, Muller-Hill B (1983). Easy identification of cDNA clones. EMBO J 2:1791.
12. Miller DM, Stockdale FE, Karn J (1985). The myosin heavy chain isoforms of Caenorhabditis elegans are encoded by separate genes. Proc Natl Acad Sci USA:in press.
13. Harris HE, Epstein HF(1977). Myosin and paramyosin of Caenorhabditis elegans: biochemical and structural properties of wild type and mutant proteins. Cell 10:709.

14. Emmons SW, Klass MR, Hirsh D(1979). Analysis of the constancy of DNA sequences during development and evolution of the nematode _Caenorhabditis elegans_. Proc Natl Acad Sci USA 76:1333.
15. Maniatis T, Fritsch EF, Sambrook J(1982). In "Molecular Cloning," New York:Cold Spring Harbor Laboratory.
16. Epstein HF, Miller DM, Gossett LA, Hecht RM (1982). Immunological studies of myosin isoforms in nematode embryos. In Pearson M, Epstein HF(eds):"Muscle Development:Molecular and Cellular Control," New York: Cold Spring Harbor Laboratory, p 7.
17. Albertson D (1984). Location of the ribosomal genes in _Caenorhabditis elegans_ chromosomes by _in situ_ hybridization using biotin-labeled probes. EMBO J 3:1227.
18. Garcea RL, Schachat FH, Epstein HF(1978). Coordinate synthesis of two myosins in wild type and mutant nematode muscle during larval development. Cell 15:421.
19. MacLachlan AD, Karn J (1983). Periodic charge distributions in the myosin rod amino acid sequence match cross-bridge spacings in muscle. Nature 299:226.
20. Brown SJ (1984). Genetic interactions affecting muscle organization in the nematode. Ph.D. Dissertation. Columbia:University of Missouri.
21. Epstein HF, Miller DM, Ortiz I, Berliner GC (1985). Myosin and paramyosin are organized about a newly identified core structure. J Cell Biol 100:904.
22. McKnight SL, Kingsbury RC, Spence A, Smith M(1984). The distal transcription signals of the herpesvirus tk gene share a common hexanucleotide control sequence. Cell 37:253.

Molecular Biology of Muscle Development, pages 639–652

GENETIC AND MOLECULAR ANALYSES OF DROSOPHILA ACTIN GENE FUNCTION[1]

Eric A. Fyrberg, James W. Mahaffey,[2]
Christine C. Karlik and Michelle D. Coutu

Department of Biology, The Johns Hopkins University
Baltimore, MD 21218

ABSTRACT To further elucidate how synthesis and assembly of myofibrillar actin is regulated we have undertaken genetic investigations of muscle development in Drosophila melanogaster. *Drosophila* lends itself well to both classical and molecular genetic approaches, and possesses a set of muscle fibers, termed indirect flight muscles (IFM) which offers a number of experimental advantages. For technical reasons our genetic investigations have focused on one particular actin gene, act88F, expressed only within developing flight muscles. To date, we have characterized four act88F alleles, all of which are associated with profound flight muscle abnormalities. We review this data, and discuss strategies for further investigating actin synthesis and assembly in *Drosophila*.

INTRODUCTION

The myofibril is one of the more completely understood eucaryotic organelles, and in this sense represents an apt subject for investigating epigenesis of cytoarchitecture. Myofibril assembly has been intensively investigated for more than twenty years, during which time several

[1]This work was supported by grants from the National Institute of Health and Muscular Dystrophy Association.
[2]Present address: Department of Biology, Indiana University, Bloomington, Indiana 47405

principles have been defined, at least at the morphological or phenomenological levels. Briefly, the program of myofibril assembly seems to be triggered when mononucleated stem cells (myoblasts) undergo a transition to "committed" postmitotic myocytes[1] and begin fusing to form syncitial muscle fibers. Thereafter, steady state levels of myofibrillar proteins (and the mRNA's which encode them) are elevated approximately 500 fold[2], and assembly of sarcomeres is initiated. Skeins of thin (actin-containing) filaments first seen at the cell periphery become integrated with thick (myosin-containing) filaments to form nascent myofibrils, which are subsequently displaced into the cell interior[3]. During their inward migration thick and thin filaments become interdigitated in a precise fashion, and anchored within electron dense Z-discs. Concomitant with myofibril assembly, levels of enzymes required for muscle metabolism accumulate to high levels.[4]

My premise is that mechanisms which regulate synthesis and assembly of myofibrillar components can best be elucidated using a combination of classical and molecular genetic approaches. Purely genetic investigations have been remarkably successful in defining loci which transiently facilitate protein synthesis and assembly in bacterial and bacteriophage systems.[5] Application of these methods is considerably more tedious in metazoans because their biological complexity impedes recognition of mutants which primarily affect a specific event (for example regulation of actin synthesis in flight muscles). By combining genetic and molecular approaches, however, one can systematically isolate mutants affecting a particular class of structural or regulatory genes.

For several reasons we have chosen to investigate muscle development in the fruit fly, Drosophila melanogaster. In addition to lending itself well to both classical and molecular genetic techniques, Drosophila provides an advantageous system of muscle fibers for molecular genetic studies. Approximately 50% of the adult thoracic volume is occupied by a series of 26 muscle fibers referred to as the indirect flight muscles.[6] These large fibers are not required for viability or fertility, a factor which facilitates characterization of mutant homozygotes. Finally, Drosophila offers the capability to extend conventional genetic analysis using the technique of germ-line transformation,[7] which facilitates studies of noncomplementing dominant or semi-dominant alleles.

RESULTS AND DISCUSSION

Organization of Myofibrillar Genes

During the previous five years, several of the genes which encode Drosophila myofibrillar proteins have been cloned and characterized, and some of the pertinent data are summarized in Figure 1. Actins of the fly are encoded by a dispersed family of six genes, which are named according to their chromosomal locations (act5C, act42A, act57A, act79B, act87E and act88F).[8] In contrast, myosin heavy chains are encoded by a single gene, which is differentially spliced so as to generate multiple mRNA's and proteins.[9] Other contractile protein genes employ a combination of these two strategies. For example, both tropomyosins and myosin light chains are encoded by pairs of genes. In the case of tropomyosins, both genes are differentially spliced, and collectively encode four tropomyosin isoforms and two tropomyosin-related proteins.[10] In the case of myosin light chains, the gene encoding alkali light chains has been recently shown to be differentially spliced ,[11] while the second gene (which likely encodes the EDTA light chains) has not yet been characterized.

Figure 1 illustrates also the map locations of fifteen semi-dominant mutations which disrupt the integrity of flight muscles.[12,13] A total of fourteen of these clearly fall into two clusters, one in the location of the myosin heavy chain gene, the other that of the tropomyosin and act88F actin genes. The molecular bases of eleven of these mutants are presently reasonably well defined: MHC 1, 2, 3 and 4 all appear to be defective alleles of the myosin heavy chain gene.[14] In contrast, Ifm(3)1, (3)2, (3)4, (3)7, and raised all appear to be act88F actin gene alleles, while Ifm(3)3 and (3)5 are tropomyosin gene mutants.

Our studies have concentrated on the actin genes, and in particular, the act88F actin gene. Our previously published dot blot and northern blot experiments have shown that poly(A)$^+$ mRNA complementary to this gene accumulates only within indirect flight muscle fibers.[8] Since the gene is expressed specifically within this expendable class of muscle fibers we predicted that mutations within it would confer a flightless phenotype

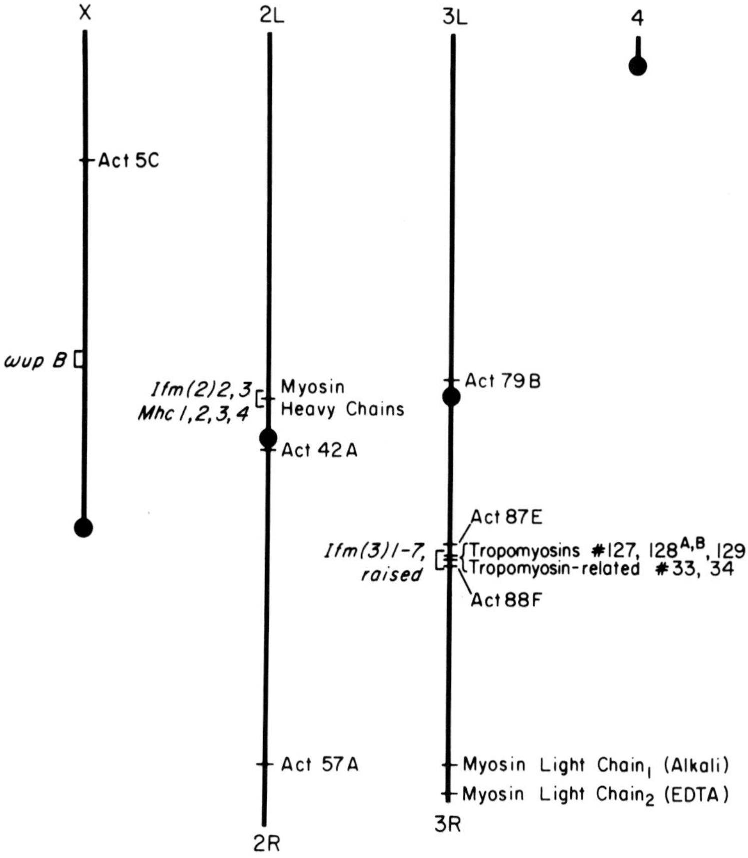

Figure 1. Chromosomal Location of Drosophila Contractile Protein Genes. The diagram schematically represents the four chromosomes (X,2,3, and 4) of diploid Drosophila cells. Locations of particular contractile protein genes are indicated along the right side of each chromosome. On the left sides of the respective chromosomes are indicated approximate locations of fifteen semi-dominant mutations which disrupt flight muscle development.

but not affect viability. As described below, several of the act88F actin gene mutants have precisely this phenotype.

The first mutant to be correlated with a defective act88F allele was Ifm(3)7.[15] Unambiguous association of the mutant phenotype with a defective actin gene was facilitated by the fact that mutant homozygotes synthesize an aberrant actin isoform. Two dimensional electrophoretic patterns of wild type (Canton-S) and Ifm(3)7 thoracic proteins, shown in Figure 2, reveal a novel protein (labeled A') having an isoelectric point and molecular weight close to those of the normal Drosophila isoforms (actins III, II and I). Hybrid selection and translation experiments have revealed that the abnormal isoform is specified by the Ifm(3)7 act88F gene, and antibody binding experiments have unambiguously demonstrated that the protein is actin. The novel isoform has a slightly lower molecular weight than does wild type actin, but we have not noted a corresponding reduction of mRNA size. Taken together, these observations suggested that the mutation was attributable to a nonsense codon near the 3' end of the protein coding region. To demonstrate that this was the case, the mutant act88F allele was isolated and sequenced. Results of this work showed that a single G > A transition within codon 355 converts the tryptophan-specifying triplet (TGG) to an opal (TGA) terminator, thus deleting the carboxy-terminal 20 amino acids.

We have documented two unexpected properties of the truncated polypeptide. The first is that the mutant isoform is stable, accumulating to high levels within affected muscle fibers. Based on this observation we have suggested that the mutant isoform assembles with normal actin monomers, and thus disrupts myofibril assembly. A second, more remarkable aspect of the truncated isoform is that it is a strong inducer of the synthesis and accumulation of several Drosophila heat shock proteins. We do not as yet understand the molecular basis of this induction. Perhaps the most plausible explanation is that the aberrant isoform improperly assembles with cytoskeletal actin isoforms, thus disrupting elongating muscle fibers to the extent that heat shock protein synthesis is induced.

A second act88F allele which has proven both interesting and useful was recovered from the raised mutant.[12] raised is a flightless Drosophila mutant having

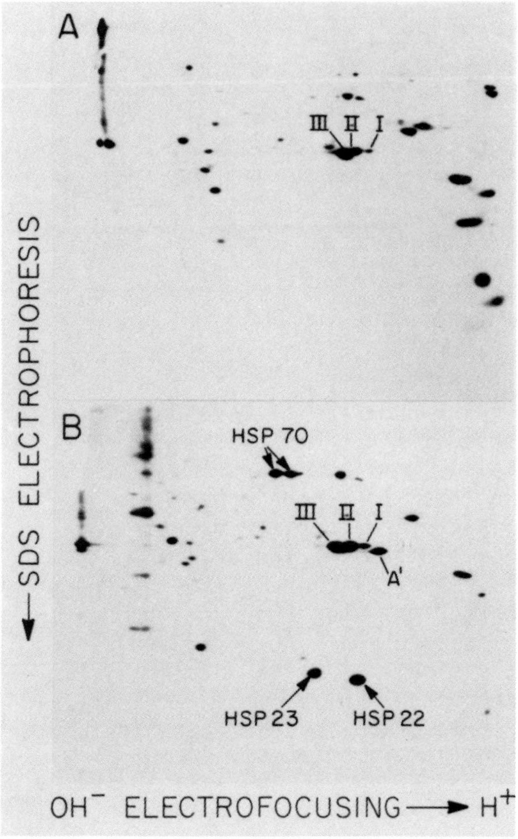

Figure 2. Two Dimensional Electrophoretic Analysis of Ifm(3)7 Thoracic Proteins. Thoraces of wild type (Panel A) and Ifm(3)7 homozygotes (Panel B) were injected with ^{35}S-methionine and allowed to incorporate the isotope for 2 hr. Comparison of accumulated proteins reveals that Ifm(3)7 homozygotes accumulate the normal actin isoforms, labeled III, II and I, as well as a truncated isoform lacking the carboxy-terminal twenty amino acids. The truncated isoform is caused by a nonsense mutation within codon 355 of the act88F gene (Karlik *et al.*[15]). Inspection of Panel B reveals also that the mutation causes several heat shock proteins (in particular, HSP 70, HSP 23 and HSP 22) to accumulate, for as yet unknown reasons.

aberrantly positioned wings and grossly abnormal indirect flight muscles. The latter point is illustrated in Figure 3. Panel A illustrates a longitudinal section of wild-type myofibrils. The precise interdigitation of thick and thin filaments can be seen, as can the regularity of Z-disc organization and spacing. Panels 3B and 3C show analogous sections of raised homozygotes. Morphologically normal but highly disorganized thick filaments are apparent in these sections, but thin filaments and Z-discs are completely lacking. Panel 3D shows a micrograph of a raised/Canton-S heterozygote. Sarcomeres, while present, have a fragile, weaving appearance. The length of these sarcomeres is approximately 50% greater than that of wild type, for as yet unknown reasons.

Our analyses of raised have revealed that most of the observed abnormalities are attributable to a defective act88F actin gene. The initial indication was based upon the observation that muscle defects could be rescued by introducing the wild type act88F allele into germline chromosomes according to the scheme illustrated in Figure 4. The wild type act88F allele was first cloned into a

---

Figure 3. Electron Microscope Analysis of Adult raised Indirect Flight Muscles. (A) shows an electron micrograph of myofibrils in 4 day adult Canton-S indirect flight muscles. Interdigitating thick and thin filaments are apparent, as are Z discs and M lines. (B) and (C) illustrate analogous sections of 4 day adult homozygous raised muscles. In (B), only disorganized thick filaments and mitochondria are seen. (C) shows electron-dense bodies, which are occasionally seen in raised homozygotes. The electron-dense material is associated with a filamentous network and resembles the "Z bodies" found during normal Drosophila muscle development. (D) illustrates a longitudinal section of thoracic muscles of a 4 day adult raised/Canton-S heterozygote. Although sarcomeres are apparent, they appear more fragile than those of Canton-S and approximately 50% longer. Sacomeres having an identical appearance and length are seen during earlier stages of flight muscle development in wild-type flies. We do not know whether this disparity in observed sarcomere lengths is attributable to corresponding in vivo length differences or is an artifact resulting from some aspect of tissue preparation. Bar represents one micron.

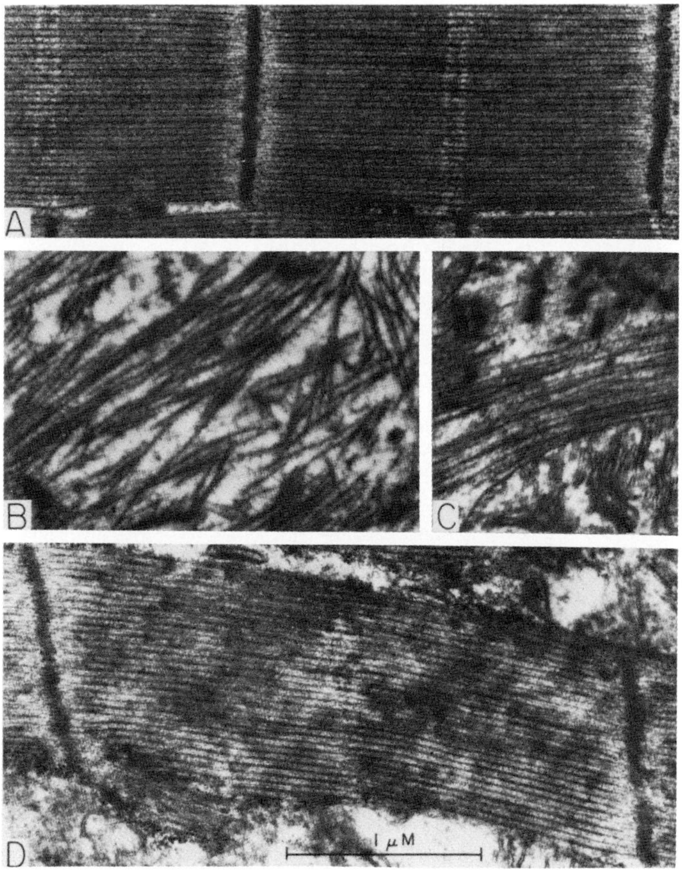
A
B
C
D
1 μM

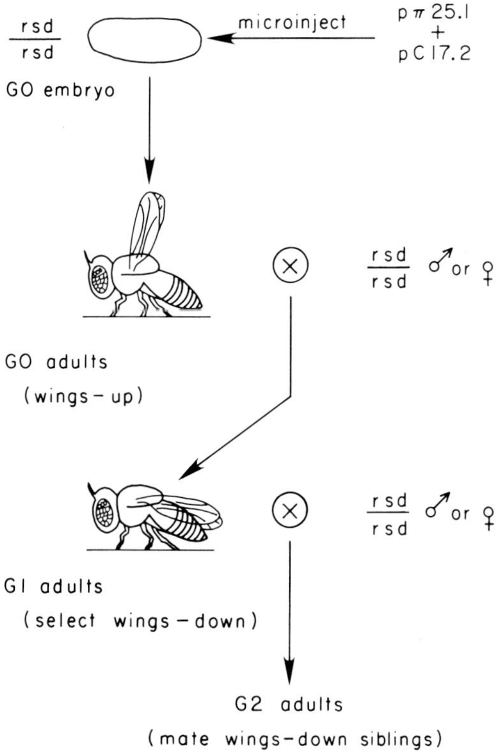

Figure 4. Scheme for Detection of Transformed raised Flies. Homozygous raised embryos were injected with a mixture of pC17.2 and pπ 25.1 plasmids (100 μg/ml and 50 μg/ml, respectively) prior to pole cell formation. The injected embryos that survived to become adult (GO) were mated to homozygous raised flies and their progeny (G1) were screened for wing position. G1 flies that held their wings in the normal position were back-crossed to raised homozygotes in order to establish stocks, as shown in the diagram.

p-element vector,[7] thus creating plasmid pC17.2. Transformants were ultimately recognizable by virtue of the fact that the wild type allele will complement the diploid mutant actin allele of raised, and thus confer normal wing position. Southern blot-hybridization of act88F - derived probes to genomic DNA of transformed stocks confirmed that their chromosomes contain supernumerary copies of this particular actin gene.

The precise molecular nature of the defect within the raised act88F gene is not known, but we have made three relevant observations. The first is that sequences of protein coding regions and splice junctions are normal. We have failed to detect any clearly deleterious nucleotide changes within these regions. Second, in the course of RNA blot-hybridization experiments we discovered that the allele specifies only 5-10% of wild type steady state RNA levels. Finally, we have preliminary evidence indicating that the defect resides 5' to the transcribed region. This latter point is based upon functional tests of Canton-S/raised act88F gene chimeras. Collectively, these observations imply that the allele has a defect within 5' "regulatory" sequences, but the precise nature of the lesion remains obscure (P.K. Geyer, unpublished observations).

We have characterized two additional act88F actin gene alleles which are associated with (respectively) mutants Ifm (3) 2 and Ifm (3) 4. Both alleles appear to have missense mutations, and both block sarcomere formation in flight muscles. In summary, then, we have so far characterized four defective act88F actin gene alleles; two missense, one nonsense, and one presumed "regulatory" lesion. Despite the diverse nature of the defects, all block sarcomere formation within flight muscles, directly demonstrating the critical role of actin in myofibril assembly and/or integrity.

One final mutant which we have characterized proves to be attributable to a defective tropomyosin gene. Both Ifm (3)3 and Ifm (3)5 have identical insertions of an 8.8 kb copia-like mobile element into the more distally located tropomyosin gene.[16] The insertion blocks accumulation of mRNA's which encode an IFM-specific tropomyosin isoform, but does not affect production of another, non-IFM, isoform by the same gene. This particular mutation does not block myofibril formation in homozygotes, although it does disrupt sarcomeres to the extent that they are nonfunctional.

EXTENSION OF THE CLASSICAL GENETIC APPROACH USING P-ELEMENT MEDIATED TRANSFORMATION

The work described above has answered several long standing questions concerning the relationship of actin gene organization and structure to the regulated synthesis of the corresponding isoforms, and has additionally allowed correlation of particular defects with specific muscle abnormalities. It is possible to extend these investigations to address more precisely formulated questions using the DNA transformation approach. This strategy permits one to systematically mutagenize cloned actin gene copies and subsequently to assess their ability to direct myofibril assembly and function *in vivo*. In the remainder of this paper I will discuss application of this general approach to three important areas of muscle biology.

Further definition of cis and trans acting effectors of myofibrillar gene activity

Obviously, to understand the genetic program which directs synthesis and assembly of myofibrillar components it is necessary to learn more about the regulatory cues. It is now clear that eucaryotic gene regulation frequently involves both cis acting and trans acting effectors.[17] The germline transformation method is presently facilitating definition of cis acting regulatory elements of a variety of genes. The approach involves deleting or altering particular gene regions and subsequently assaying the ability of the allele to function properly. We have used this method to demonstrate that only 1400 bp of 5' flanking sequence is required for the wild type act88F actin gene to rescue the *raised* mutant (P.K. Geyer, unpublished results). Presumably, any cis-linked sequences required for regulated expression reside either within this DNA, or within introns and untranslated regions.

It is equally important to define the nature of trans acting effectors. Mutant alleles of genes which encode such effectors can be isolated by classical genetic screening techniques, but usually only after incorporation of a specific and easily recognized phenotypic marker into the protocol. The most straightforward manner in which to proceed would seem to be by fusing the promoter and regulatory region of a particular myofibrillar gene to an easily recognized catalytic function. For example, we have recently begun construction of fusions between the act88F actin gene and the structural portion of the gene encoding *Drosophila* alcohol dehydrogenase. Were such a fusion gene introduced into germline chromosomes of an Adh-null strain we would expect all Adh synthesis to be directed by the act88F regulatory region. Mutant derivatives lacking Adh activity would then potentially be attributable to defective alleles of trans acting effectors. Evidence that such an approach is technically feasible has come from recent experiments of J. Bonner and colleagues. In addition to using the approach to isolate regulatory mutants affecting synthesis of heat shock proteins,[18] he has demonstrated also that alcohol dehydrogenase synthesized only within flight muscles will confer tolerance (by the whole fly) to 6% ethanol. To demonstrate this he constructed a fly strain in which Adh synthesis was directed by the HSP-70 heat shock gene promoter, and which was also homozygous for the defective act88F allele recovered from the *Ifm(3)7* strain. Since the defective actin polypeptide induces HSP-70 synthesis in flight muscles, these fibers contain high levels of the enzyme, as visualized by staining protocols. How enzyme synthesized exclusively in flight muscles is able to metabolize ethanol remains to be determined.

Definition of the functional domains of myofibrillar proteins

Much the same approach can be used to examine in detail the functional significance of particular polypeptide sequences of myofibrillar proteins for generation of contractile force. *In vitro* mutagenesis protocols can be used to engineer particular amino acid replacements, and their effect on contractility documented after germ line transformation. Recently we have demonstrated that both of the aforementioned missense alleles of the act88F actin gene will induce muscle abnormalities after introduction into germ line chromosomes (C.C. Karlik and D.L. Saville, unpublished results). By testing a variety of *in vitro* mutagenized alleles in this fashion we hope ultimately to define actin domains requisite for assembly and/or contractility.

ACKNOWLEDGMENTS

I am indebted to Drs. Jose Bonner, Sandy Bernstein, Scott Falkenthal, and Chuck Rozek for discussions of their published and unpublished results.

REFERENCES

1. Nadal-Ginard, B. (1978). Commitment, fusion, and biochemical differentiation of a myogenic cell line in the absence of DNA synthesis. Cell 15, 855-864.
2. Devlin, R.B., and Emerson, C.P. Jr. (1978). Coordinate regulation of contractile protein synthesis during myoblast differentiation. Cell 13, 599-611.
3. Peng, H.B., Wolosewick, J.J., and Cheng, P-C. (1981). The development of myofibrils in cultured muscle cells: A whole-mount and thin-section electron microscopic study. Develop. Biol. 88, 121-136.
4. Shainberg, A., Yagil, G., and Yaffe, D. (1971). Alterations of enzymatic activities during muscle differentiation In Vitro. Develop. Biol. 25, 1-29.
5. King, J. and Casjens, S. (1974). Catalytic head-assembling protein in virus morphogenesis. Nature 251, 112-119.
6. Deak, I.I. (1977). Mutations of Drosophila melanogaster that affect muscles. J. Embryol. exp. Morph. 40, 35-63.
7. Spradling, A.C., and Rubin, G.M. (1982). Transposition of cloned p elements into Drosophila germline chromosomes. Science 218, 341-347.
8. Fyrberg, E.A., Mahaffey, J.W., Bond, B.J., and Davidson, N. (1983). Transcripts of the six Drosophila actin genes accumulate in a stage and tissue specific manner. Cell 33, 115-123.
9. Rozek, C.E. and Davison, N. (1983). Drosophila has one myosin heavy chain gene with three developmentally regulated transcripts. Cell 32, 23-34.
10. Karlik, C.C., Mahaffey, J.W., Coutu, M.D., and Fyrberg, E.A. (1984a). Organization of contractile protein gene within the 88f subdivision of the D. melanogaster third chromosome. Cell 37, 469-481.

11. Falkenthal, S., Parker, V.P., and Davidson, N. (1985). Developmental variations in the splicing pattern of transcripts of the Drosophila myosin alkali light chain gene result in different carboxy-terminal amino acid sequences. Proc. Natl. Acad. Sci., 82, 449-553.
12. Mahaffey, J.W., Coutu, M.D., Fyrberg, E.A., and Inwood, W. (1985). The flightless Drosophila mutant raised has two distinct genetic lesions affecting accumulation of myofibrillar proteins in flight muscles. Cell 40, 101-110.
13. Mogami, K., and Hotta, Y. (1981). Isolation of Drosophila flightless mutants which affect myofibrillar proteins of indirect flight muscles. Mol. Gen. Genet. 183, 409-417.
14. Mogami, K., O'Donnell, P.T., Bernstein, S.I., Wright, T.R.F., and Emberson, C.P. Jr. (1985). Myosin heavy chain mutants in Drosophila melanogaster. Submitted for publication.
15. Karlik, C.C., Coutu, M.D., and Fyrberg, E.A. (1984b). A nonsense mutation within the act88F actin gene disrupts myofibril formation in Drosophila indirect flight muscles. Cell 38, 711-719.
16. Karlik, C.C., and Fyrberg, E.A. (1985). An insertion within a variably spliced Drosophila tropomyosin gene blocks accumulation of only one encoded isoform. Cell, in press.
17. Parker, C.S. and Topol, J. (1984). A Drosophila RNA polymerase II transcription factor contains a promoter-region-specific DNA-binding activity. Cell 36, 357-369.
18. Bonner, J.J., Parks, C., Parker-Thornburg, J., Mortin, M.A., and Pelham, H.R.B. (1984). The use of promoter fusions in Drosophila genetics: Isolation of Mutants affecting the heat shock response. Cell 37, 979-991.

Molecular Biology of Muscle Development, pages 653–666

DIFFERENTIAL ROLES OF MYOSIN ISOFORMS IN FILAMENT ASSEMBLY[1]

Henry F. Epstein

Department of Neurology, Biochemistry, and
Cell Biology
Baylor College of Medicine
Houston, Texas 77030

ABSTRACT Wild-type body wall muscle cells of *Caenorhabditis elegans* produce at a constant ratio two myosin heavy chain isoforms, A and B, that form homodimeric myosins. Electron microscopy of negatively stained complexes of isoform-specific antibodies with isolated thick filaments shows that the surface of the 9.7 μm long filament is differentiated with respect to myosin content: a medial 1.8 μm zone contains myosin A and two polar 4.4 μm zones contain myosin B.
Biochemical and electron microscopic studies show that at 0.45 M KCl, pH 6.35, myosin B and paramyosin are solubilized. The medial all-myosin A region with novel core structures extending in a polar manner remain. These dissociation experiments suggest a sequential model for wild-type thick filament assembly in which myosins A and B would participate in the initiation and termination of assembly, respectively. Analysis of mutant thick filaments clarifies the relationship of the myosin isoforms. CB190 (*unc*-54 I) thick filaments contain myosin A only and have normal length. CB1214 (*unc*-15 I) mutants produce no paramyosin, and their thick filaments are composed of a medial myosin region

[1]This work was supported by grants from the National Institute of Aging, National Institute of General Medical Sciences, Muscular Dystrophy Association, National Foundation-March of Dimes and The Welch Foundation.

with protruding polar core structures. Mutants in the unc-52 II and sup-3 V loci have increased ratios of myosin A to B. The medial myosin A regions increase in length while the overall filament lengths are similar to wild-type. The CB1407 mutant has normal myofibrillar structure as well. Therefore, the specific pattern of molecular sorting of myosin isoforms does not appear necessary for filament assembly or organization. The distribution of the two myosins appears determined by their concentrations, mole ratios, and the underlying substructure. Gene-specific hybridization experiments indicate that mRNA levels for A and B chains differ, suggesting potential mechanisms for the regulation of isoform biosynthesis. The biochemical and structural results suggest that the regulated biosynthesis of the two heavy chain isoforms controls the initiation and termination of thick filament assembly by the differential interactions of the myosins.

INTRODUCTION

Specific component proteins of cytoplasmic structures are known to exist in multiple forms within the same cell. Examples of this phenomenon include myosins (1-3), actins (4,5) and the α-and β-tubulins (6,7). Characteristic forms of these proteins are found in particular cell types and tissues. Two or more closely related forms of these proteins have been demonstrated in specific cells.

Protein molecules of any of these classes assemble into polymers that exhibit class-specific structural motifs. For example, nearly all purified myosins can aggregate in bipolar filaments, the actins can associate into F-actin filaments, and the tubulins can form microtubules. The close similarities of the class-specific structural properties in these synthetic polymers of proteins from diverse biological sources correlate with the highly conserved amino acid sequences and molecular structures of the component proteins over a wide phylogenetic and histological range.

Despite these general structural similarities, many uniquely characteristic cell structures are derived from these proteins. Myosins and actins are components of the thick and thin filaments of striated muscles. These filaments have exactly determined dimensions, regular internal orders and symmetrical relationships within higher order arrays. All of these properties are muscle-type specific. Tubulins are components of the microtubules in distinct arrangements within mitotic spindles, flagellae and variously modified cilia. Two hypotheses may explain the commonality and diversity observed in these examples. One explanation is based upon the occurrence of different variants or isoforms of these structural proteins. The distinct intrinsic molecular properties of the isoforms would be sufficient to explain their assembly into different structures. The second hypothesis proposes that additional protein components and enzymes interact with the fundamental structural proteins to produce diverse molecular arrangements. These additional components may interact differentially with specific isoforms of the major structural proteins, but it is the interactions between the different kinds of proteins that are critical to the differentiation of assembly. Most recent evidence favors the latter hypothesis (8,9).

The occurrence of structural protein isoforms having very similar or identical amino acid sequences particularly in the case of actins has led to a radically different hypothesis concerning these multiple gene products. With respect to actin isoforms, Fyrberg et al. (10) have proposed that different actin-coding sequences became associated evolutionarily with specific regulatory elements in the genome. These regulatory DNA sequences would in turn link the biosynthesis of actin isoforms to particular inductive patterns in the differentiation of specific cells. In this model, the actin isoforms themselves would not have intrinsic properties distinguishing their function or assembly.

In this paper, a model is presented for a coordinate regulatory mechanism in which the differential synthesis of myosin heavy chain isoforms is linked to their distinct roles in the initiation and termination of thick filament assembly. Intrinsic properties of these myosin heavy chain forms probably determine their exclusive association into homodimeric myosin isoenzymes. The coupling of the regulation of biosynthesis and formation of these homodimers to their different behavior during filament assembly is dependent upon additional filament-associated proteins. This model was derived from studies of myosin isoforms in the body wall muscle cells of Caenorhabditis elegans. In these nematodes, genetical as well as biochemical and structural experiments permit analysis of the relationship between the assembly and synthesis of myosin isoforms.

Two myosin heavy chain isoforms A and B, are coded by different genetic loci, _myo_-3 V and _unc_-54 I (11) respectively. The synthesis of the A and B isoforms may be modified by mutation in the _sup_-3 V (12) and _unc_-52 II (13) loci, respectively. The synthesis of the two forms produces a constant ratio through most of the nematode life-cycle (14). Only homodimeric myosin molecules are detected (15, 16). The myosin isoenzymes, A and B, are localized to the medial and polar regions of the thick filaments (17). These locations appear to reflect different interactions of the myosins with respect to non-myosin internal structures within the thick filaments and the behavior of the myosins during processive dissociation of the filaments (18).

RESULTS

Different Locations of the Body Wall Myosins.

This section reviews published work of Miller et al. (17). Monoclonal antibodies that discriminate between myosins A and B have been produced and characterized. In immunoblot titrations, antibody 28.2 reacts preferentially with the product of the _unc_-54 gene, myosin heavy chain B; antibody 5-6 reacts preferentially with the product of the _myo_-3 gene, the more minor myosin heavy chain A. The weaker reactions of these monoclonal antibodies with additional myosin heavy chain isoforms were reduced effectively to an undetected level by applying low concentrations of antibody to nematode muscle and isolated thick filaments.

By use of these antibodies under such specific conditions, the myosin isoforms have been localized within thick filaments of body muscle. The patterns of immunofluorescence produced by these two antibodies are different and appear complementary. Each A-band reacted with antibody 28.2 appears as a pair of parallel stripes; a small central region within each of these A-bands remains unlabeled. Conversely, reaction with the 5-6 antibody produces a narrow fluorescent stripe running down the center of each A-band while the peripheral regions remain unlabeled. Isolated nematode thick filaments were decorated with these antibodies separately and examined by electron microscopy. The antibody reactions under these conditions corroborate and extend the immunofluorescence analysis. Antibody 28.2 reacts with the full lengths of isolated filaments except with a medial region about 0.9 µm long. Antibody 5.6 reacts with a central 1.8 µm segment.

From these antibody localization results and assuming a standard thick filament length of 9.7 µm (19), we propose a model for the locations of myosins A and B as indicated in Figure 1. Myosin A occupies a medial segment about 0.9 µm long. On each side of that zone is a 0.45 µm long segment which contains myosins A and B. The major remaining, most polar regions of the thick filaments about 4.0 µm long are composed of myosin B.

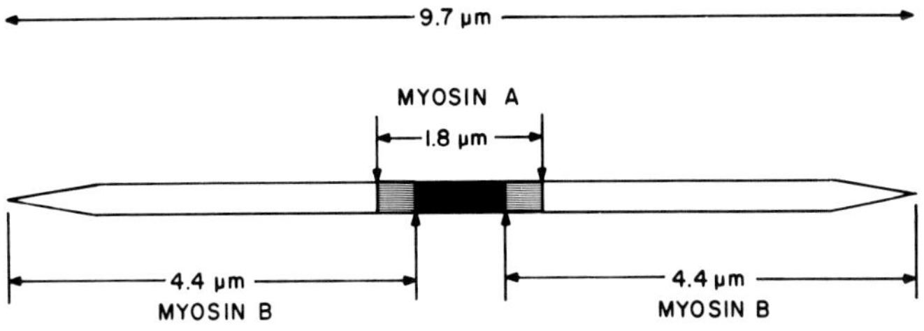

FIGURE 1. Model of Nematode Thick Filament With Distribution of Myosins A and B Suggested by Specific Antibody Reactions. Reprinted with permission from Cell (MIT Press) (17).

The Two Myosins and Paramyosin Are Organized about a Newly Identified Core Structure.

The section reviews published material of Epstein et al. (18). Body muscle thick filaments can be dissociated processively by increasing concentrations of KCl. The previously established major proteins of the thick filaments, myosins A and B and paramyosin, are solubilized to ∿80 and 95%, respectively, by the KCl dissociation. Under these conditions, smooth-appearing structures of 15 nm diameter are observed. These structures are morphologically distinct from the rough myosin or paracrystalline surfaces and are not labeled with monoclonal antibodies specific to either myosin isoenzyme or paramyosin. By their diameter and conditions required for isolation, these structures are termed polar core structures. They appear to extend along the lengths of the thick filaments from the polar tips to the central 0.9 µm regions and possibly through the latter zones. In the polar zones, paramyosin forms an intermediate domain between the core structure and the surface myosin. The thick filament would consist of three coaxial domains. (Figure 2).

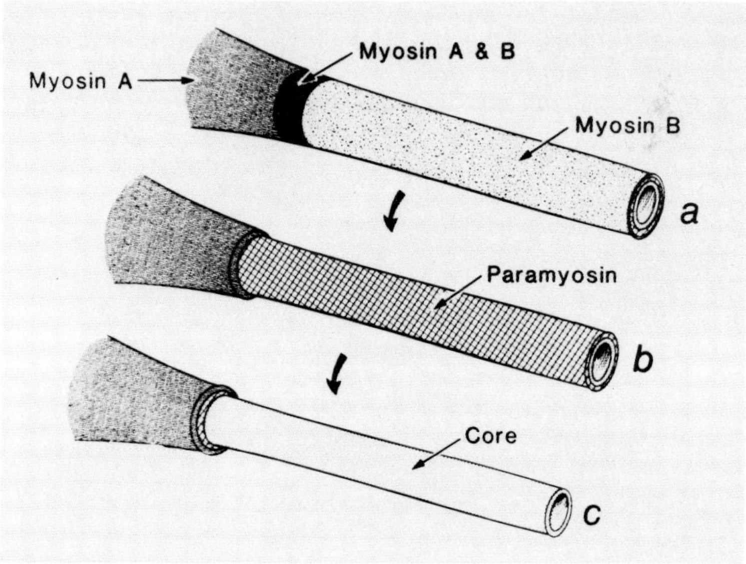

FIGURE 2. Schematic diagram of thick filament domains. In a, a portion of the intact native filament is depicted with the subdomains of differing myosin content indicated. In b, myosin B has been removed, and the paramyosin domain is exposed. In c, paramyosin as well as myosin B have been removed and the polar core structure is exposed. Reproduced from The Journal of Cell Biology, 1985, 100, 904-915 by copyright permission of The Rockefeller University Press.

The structure of these supramolecular domains is not constant. The thick filaments taper in overall diameters from 33.4 to 14.0 nm as measured by electron microscopy of fixed, embedded, sectioned and positively stained muscle. This tapering has important implications for any scheme of packing of the myosin or paramyosin molecules. Since the smallest diameters of filament cross-sections (14.0 nm) are similar to the width of negatively stained polar core structures, depletion of the myosin and paramyosin domains must occur. Preliminary results in our laboratory suggest that the paramyosin domain may not extend to the polar ends of the filaments. Therefore, the exact packing of myosin and paramyosin molecules must change significantly from the central to polar regions.

The internal density of the core changes its appearance along the filament length. The central density of thick filaments in cross-sections appears consistently solid in the central regions, intermediate in most of the polar regions, and hollow near the ends. The internal molecular components of the core structures that change to produce this variable electron density may serve as a template for the exact determination of length of the polar core structures and, in turn, of the tapering paramyosin and myosin domains. RNA and proteins have been implicated as macromolecular templates for length determination in tobacco mosaic virus and in bacteriophages λ and T4, respectively.

Genetic Dissection of an Assembly Pathway and its Regulation.

A pathway for the assembly of nematode body wall thick filaments may be postulated on the basis of the dissociation experiments and related morphological observations. A schematic diagram outlining this postulated pathway is shown in Figure 3. The pathway would represent a sequence of reactions and intermediate structures. Myosin A would be restricted to the rate-limiting nucleation step in wild-type muscle. The next two steps would involve non-myosin proteins. Myosin B would participate in the termination of assembly. The two myosin isoenzymes would participate in kinetically and structurally distinct roles.

Studies of appropriate muscle-defective mutants in C.elegans support the plasibility of this model and may extend its implications to the regulated coupling of assembly and biosynthesis. The CB190 (e190, unc-54 I) mutant produces no myosin B. Only 20% of the wild-type number of thick filaments are present in its body-wall muscles; this fraction correlates with the fraction of body wall myosin that is myosin A in wild-type (20). No structures resembling any of the postulated intermediates have been observed in sections or isolated preparations. The isolated filaments are of normal length, but in contrast to wild-type, myosin A is present in both polar and central regions. In the proposed model, nucleation is critically dependent upon myosin A concentration, and once nucleation has occurred, assembly continues to termination,

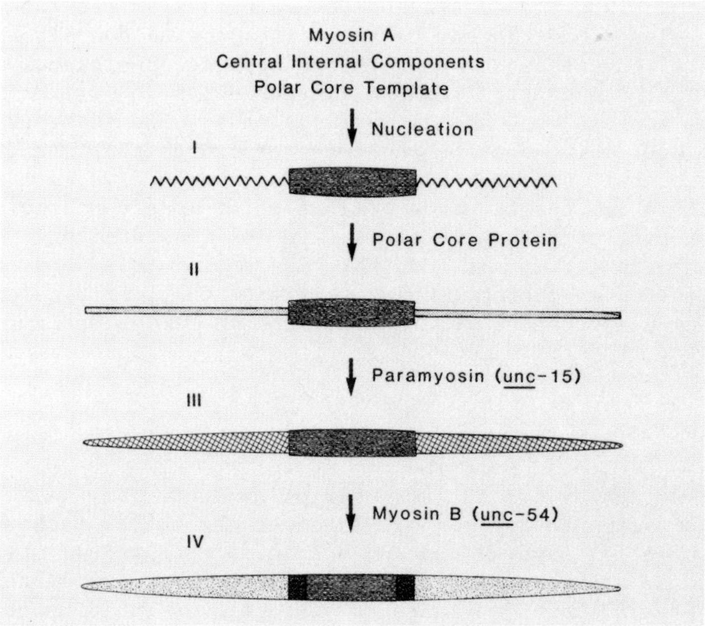

FIGURE 3. Pathway of thick filament assembly, Myosin A, postulated polar core template molecules, and postulated central internal components are required for nucleation (I). Postulated polar core protein polymerizes about the polar core template to form the core structure (II). Paramyosin polymerizes about the polar core structure (III). Myosin B polymerizes about paramyosin and the core structure, and assembly terminates (IV). Reproduced from The Journal of Cell Biology, 1985, 100, 904-915 by copyright permission of The Rockefeller University Press.

if possible. In this case, most of the myosin A is used to form the polar subdomains of initiated filaments. Fewer filaments are formed because the myosin A concentration becomes too low for additional nucleation. CB190 filaments can be dissociated with KCl and exhibit core structures under these conditions. The core structures do not require myosin B for their formation and are likely to be composed of a different protein.

Filament-like structures can also be isolated from the paramyosin-deficient mutant CB1214 (e1214, unc-15 I). These structures morphologically resemble intermediate II of the above scheme except that some myosin B can bind to the polar core surface. Both myosins A and B are located in 1.5-2.0 μm long, tapering, bipolar structures out of which the hollow core structures protrude. This density must be due to protein other than to myosin or paramyosin. The postulated central internal components may be responsible. These observations indicate that the polar core structure does not require paramyosin for its formation and is likely to be composed of protein other than myosin or paramyosin.

The most revealing mutant observations concern the effects of alterations in the relative mole ratios and concentrations of myosins A and B upon the distribution of myosin A is isolated thick filaments. These results are the unpublished work of Dr. Laurel A. Traeger in my laboratory. In these experiments, the wild-type strain, N2, and CB190 may be considered as end points in the genetic titration. N2 represents the normal ratio of A:B, about 1:4, and normal concentrations; CB190 produces only A at about normal concentration (14). Mutant strains due to mutations in the unc-52 II locus (HE200, CB669) and in the sup-3 V locus (CB1407) produce intermediate situations due to distinct mechanisms (12, 13). This comparison is in Table 1. The length of myosin A-containing central regions were determined by localization of monoclonal antibody 5-6 reaction with isolated thick filaments as determined by electron microscopy. The lengths of these zones increased as the mole fraction of myosin A increased. Independent of this variable, increased myosin A concentration was a more significant factor than decreased myosin B relative concentration (viz HE200 vs. CB1407). Two tentative conclusions may be drawn from these experiments. The amounts and ratios of the myosin isoenzymes appear to be coupled to their distribution with the thick filaments. The non-myosin substructures of the filaments appear to create a gradient of affinity for myosin A that is highest in the central region and decreases toward the poles. All regions of these filament substructures appear to have some affinity for A. Experiments to determine the lengths of myosin B containing zones by reaction with monoclonal antibody 28.2 in these mutants are in progress. Mutants that increase myosin B levels or which reduce myosin A levels have not been isolated.

It is noteworthy that the *unc*-52 mutants, CB669 and HE200, and the *unc*-54 mutant, CB190, that are associated with decreased production of myosin heavy chain B also have decreased number of thick filaments and varying levels of filament lattice disorganization. The *sup*-3

TABLE 1
MYOSIN A TITRATION

Strain	MHCA/MHCB[a]	Mechanism[b]	Myosin A[c] Zone Length
N2	1/4	Normal	1.74±0.08μm
CB669	1/3	Decreased MHC B	2.05±0.06μm
HE200	1/2	Decreased MHC B	2.18±0.01μm
CB1407	1/2	Increased MHC A	3.26±0.18μm
CB190	1/0	No MHC B	(9.7μm)[d]

[a]Abstracted from References 12, 13, 14.
[b]As indicated in References 12, 13, 14.
[c]Determined by monoclonal antibody 5-6 reaction.
[d]Myosin A is distributed along full length of thick filaments.

strain, CB1407, that is associated with increased production of myosin heavy chain A, has a normal number of thick filaments and wild-type sarcomere organization. Therefore, the wild-type distribution of myosins A and B along thick filaments does not appear to be necessary for thick filament assembly and organization *per se*. The disruptions in the other mutants are related to insufficient amounts of myosin B and the resulting decrease in thick filament number.

DISCUSSION

Why do the body wall muscle cells of wild-type *C.elegans* produce two myosin isoenzymes at a constant ratio and sort them in a highly specific manner in distinct zones of the thick filaments? The answer may be derived from a consideration of the processes required for the production at a constant ratio and sorting in a highly specific manner of the two myosins. The production at a constant ratio must be based upon the regulation of the biosynthesis of the A and B myosin heavy chain isoforms. It is not known whether the responsible mechanisms operate at transcriptional, post-transcriptional processing or translational steps. Dr. Sandra Honda in my laboratory is quantitating *in vivo* steady-state mRNA levels and the *in vitro* nuclear runoff transcriptions of the myosin heavy chain genes in wild-type and mutant strains in order to answer these questions. In this regard, the *sup*-3 V and *unc*-52 II loci have been suggested to be regulatory with respect to the relative synthesis of the A and B isoforms, respectively (12, 13). The sorting in a highly specific manner can be related to the proposed pathway of assembly of thick filaments and to the differential affinity of non-myosin components for the two isoenzymes. Myosin A would interact with the central internal components to form a nucleation structure as an initial step in assembly. Myosin B at usual wild-type concentrations would interact preferentially with paramyosin and polar core structure to terminate assembly. I propose that the presence of the two myosin isoenzymes and their differential interactions couples the initiation and termination of thick filament assembly to the regulated biosynthesis of the distinct myosin heavy chains. The major function of the two isoforms is to permit regulation of different steps in thick filament assembly. The rate-limiting step requires myosin A which is in limiting quantities in wild-type. Filaments once initiated are assured of completion by the availability of the more abundant myosin B. Dr. Honda finds that the mRNA for myosin heavy chain B is in at least ten-fold excess over the mRNA for the A form.

This teleonomic model may have relevance for other situations involving isoforms of cell structural proteins. The most general principle may be that the existence of isoforms and the differential regulation of their synthesis in turn drives the formation of related but distinct assemblies of these proteins. In the case of identical or very similar products of distinct genes, the temporal regulation of synthesis may provide the critical distinction. Striking examples of this case include the cytoskeletal and sarcomeric actins which are ordinarily compartmentalized but can behave identically under specific experimental conditions (21, 22). A second principle is that the different assemblies of structural protein isoforms are usually correlated with the specific presence of other kinds of proteins. There may be quantitative differences in the interactions between specific isoforms and these assembly-specific proteins. The phenomenon of isoforms provides a mechanism for utilizing the control of gene expression and protein synthesis to regulate the assembly of proteins into specific cellular structures.

ACKNOWLEDGEMENTS

I wish to thank Dr. Laurel A. Traeger for the use of her unpublished data and Gary Berliner and Irving Ortiz for their invaluable assistance in this research.

REFERENCES

1. Starr R, Offer G (1973). Polarity of the myosin molecule. J Mol Biol 81:17.
2. Epstein HF, Waterston RH, Brenner S (1974). A mutant affecting the heavy chain of myosin in _Caenorhabditis elegans_. J Mol Biol 90:291.
3. Chizzonite RA, Everett AW, Clark WA, Jakovic S, Rabinowitz M, Zak R (1982). Isolation and characterization of two molecular variants of myosin heavy chain from rabbit ventricle. J Biol Chem 257:2056.
4. MacLeod C, Firtel RA, Papkoff J (1980). Regulation of actin gene expression during spore germination in _Dictyostelium discoideum_. Dev Biol 76:263.
5. Garrels JI, Gibson W (1976). Identification and characterization of multiple forms of actin. Cell 9:793.
6. Fulton C, Simpson PA (1975). Selective synthesis and utilization of flagellar tubulin. The multi-tubulin hypothesis. In Goldman R, Pollard T, Rosenblum J

(ed): "Cell Motility", Cold Spring Harbor Laboratory, p 987.
7. Cabral F, Sobcl ME, Gotlesman MM (1980). CHO mutants resistant to colchicine, colcemid or griseofuluin have an altered β-Tubulin. Cell 20:29.
8. Korn ED, (1982). Actin polymerization and its regulation by proteins form nonmuscle cells. Physiol Rev 62:672.
9. Kuczmarski ER, Spudich JA (1980). Regulation of myosin self-assembly: phosphorylation of Dictyostelium heavy chain inhibits formation of thick filaments. Proc Natl Acad Sci USA 77:7292.
10. Fyrberg EA, Bond BJ, Hershey ND, Mixter KS, Davidson N (1982). Structural studies of the Drosophila melanogaster actin genes. In Pearson ML, Epstein HF (ed): "Muscle Development: Molecular and Cellular Control", Cold Spring Harbor Laboratory, p 87.
11. Karn J, Dibb NJ, Miller DM, personal communication.
12. Waterston RH, Moerman DG, Baillie D, Lane TR (1982). Mutations affecting myosin heavy chain accumulation and function in the nematode Caenorhabditis elegans. In Schotland DL (ed): "Diseases of the Motor Unit", John Wiley and Sons, p 747.
13. Zengel JM, Epstein HF (1980). Mutants altering coordinate synthesis of specific myosins during nematode muscle development. Proc Natl Acad Sci USA 77:852.
14. Garcea RL, Schachat F, Epstein HF (1978). Coordinate synthesis of two myosins in wild-type and mutant nematode muscle during larval development. Cell 15:421.
15. Schachat FH, Harris HE, Epstein HF (1977). Two homogeneous myosins in body-wall muscle of Caenorhabditis elegans. Cell 10:721.
16. Schachat F, Garcea RL, Epstein HF (1978). Myosins exist as homodimers of heavy chains: demonstration with specific antibody purified by nematode mutant myosin affinity chromatography. Cell 15:405.
17. Miller DM, Ortiz I, Berliner GC, Epstein HF (1983). Differential localization of two myosins within nematode thick filaments. Cell 34:477.
18. Epstein HF, Miller DM, Ortiz I, Berliner GC (1985). Myosin and paramyosin are organized about a newly identified core structure. J Cell Biol: 100:904.

19. Mackenzie JM, Epstein HF (1980). Paramyosin is necessary for determination of nematode thick filament length *in vivo*. Cell 22:747.

20. MacLeod AR, Waterston RH, Fishpool RM, Brenner S (1977). Identification of the structural gene for a myosin heavy chain in *Caenorhabditis elegans*. J Mol Biol 114:133.

21. Gunning P, Ponte P, Kedes L, Hickey RJ, Skoultchi AI (1984). Expression of a human cardiac actin in mouse L cells: a sarcomeric actin associates with a nonmuscle cytoskeleton. Cell 36:709.

22. McKenna N, Meigs JB, Wang Y-L (1985). Identical distribution of fluorescently labeled brain and muscle actins in living cardiac fibroblasts and myocytes. J Cell Biol 100:292.

Molecular Biology of Muscle Development, pages 667–682

ANALYSIS OF THICK FILAMENT ASSEMBLY AND MYOSIN EXCHANGE BY FLUORESCENCE ENERGY TRANSFER

A. D. Saad, D. A. Fischman and J. D. Pardee

Department of Cell Biology and Anatomy
Cornell University Medical College
New York, New York, 10021

ABSTRACT A fluorescence energy transfer (FET) assay has been established to examine myosin assembly into thick filaments and its exchange between preassembled filaments. Purified pectoralis muscle myosin was covalently labeled with donor (IAENS) or acceptor (IAF) fluorochromes and assembly of myosin into thick filaments monitored continuously by quench of IAENS fluorescence. Complete assembly results in 40-45% quench of IAENS fluorescence. After full assembly 15-20 ug/ml of unassembled myosin remains in equilibrium with filaments. Immunolabeling of thick filaments demonstrated that both fluorescently labeled and unlabeled synthetic filaments retain the 14.5 nm repeat structure characteristic of native filaments. Exchange of myosin molecules between synthetic thick filaments was quantified by mixing separately labeled donor and acceptor filaments and measuring quench of IAENS fluorescence. Rapid ($t_{1/2}$ <5 min) and extensive exchange (80% complete) of myosin between synthetic thick filaments has been detected.

This work was supported by research grants from NIH (AM32147) and Muscular Dystrophy Association. Dr. Saad was a Revson Fellow for part of the tenure of this research.

INTRODUCTION

There is inadequate understanding of the mechanisms regulating the assembly, remodeling and disassembly of thick myofilaments in striated muscles (1-10; reviewed in 11). Progress in this field has been limited, in part, by the inavailability of simple, direct and unambiguous assays of myosin polymerization. To approach this problem a fluorescence energy transfer (FET) assay has been developed to study the assembly of myosin into thick filaments and the exchange of myosin between filaments. FET is a rapid, sensitive and quantitative technique that permits continous observation of the polymerization reaction. In principle, the FET assay is based on transfer of resonance energy from an excited "donor" fluorochrome to an "acceptor" fluorochrome. Transfer occurs only when the two fluorochromes are less than 7.0nm apart (12). Consequently, FET has been used to examine spatial relationships between sites on the same protein (13-15). The technique has also been used to analyze polymerization (16-19) and exchange (16,17) of actin, and to characterize the properties of a recently discovered Ca^{++}-regulated protein (severin) that clips F-actin filaments into oligomers (20). To our knowledge, this technique has not been used to analyze the polymerization and exchange properties of the myosin molecule. In this report we present our initial applications of FET to the phenomenon of thick filament assembly and exchange. Although preliminary in nature, it is apparent that the approach offers great promise and should permit a detailed analysis of the assembly properties of myosin isoforms and the putative regulatory functions of thick filament accessory proteins.

MATERIALS AND METHODS

Fluorescent Labeling of Myosin

Myosin was isolated from the pectoralis major muscle of adult White Leghorn chickens using previously described procedures (21,22). Purified myosin was dialyzed against a low salt buffer (0.1M KCl, 10mM potassium phosphate, pH 6.9) to form synthetic thick filaments. Intact filaments were then fluorescently labeled with either 5-(iodoacetamidoethyl)-aminonapthalene-1-sulfonic acid (IAENS; donor) or iodoamido-fluorescein (IAF; acceptor). The reaction was terminated by addition of 1.0mM dithiothreitol. Unreacted fluorochromes were removed by dialysis against a high salt buffer (0.5M KCl, 10mM potassium phosphate, pH 7.2) which also disassembled the filaments. The stoichiometric ratio of covalently attached fluorochrome to myosin was determined by spectrophotometry. Thick filament morphology was assessed by electron microscopy after negative staining with uranyl acetate. Distribution of the fluorochromes was established by limited proteolysis with chymotrypsin under low ionic strength (23), followed by SDS-PAGE (24), and visualization of fluorescently labeled peptides under ultraviolet illumination.

Assembly Assays

Acceptor and donor labeled myosin molecules were mixed at a 5:1 ratio in a high salt buffer (0.5M KCl, 5mM $MgCl_2$, 5mM ATP, 0.2mM DTT, 10mM Tris-maleate, pH 6.9). Baseline fluorescence of the mixture was measured for 10 minutes and assembly initiated by dilution to lower ionic strength (final buffer = 0.1M KCl, 5mM $MgCl_2$, 5mM ATP, 0.2mM DTT, 10mM Tris-maleate, pH 6.9; myosin concentration= 0.8mg/ml). Filament assembly was continously monitored by the decrease in donor fluorescence at 470nm (fluorescence quench). When the assembly reaction reached equilibrium, the extent of thick filament formation was determined by sedimentation, and filament morphology assessed by electron microscopy. All assembly and exchange assays were performed at constant temperature (22°C) in a Perkin-Elmer 650-40 fluorescence spectrophotometer.

Exchange Assays

Synthetic thick filaments were separately labeled with either donor or acceptor fluorochrome, disassembled by dialysis in high salt buffer, and reassembled by dialysis against low salt buffer (0.1M KCl, 5mM $MgCl_2$, 5mM ATP, 0.2mM DTT, 10mM Tris-maleate, pH 6.9). Donor and acceptor filaments were then mixed at a 5:1 molar ratio of acceptor to donor and exchange monitored by following quench of donor fluorescence at 470nm. Filaments were visualized by electron microscopy before and after exchange.

For exchange assays using ^{125}I-myosin, isotope labeled myosin was prepared by the Biorad enzymo-bead procedure using 1mCi Na-^{125}I with 200ug of myosin. For this assay 5 or 15 ug of unassembled ^{125}I-myosin was added to a 1 ml solution of synthetic filaments (1mg/ml) and incubated for 3 hr. Filaments were sedimented, and the extent of myosin exchange calculated from the specific activities of pelleted thick filaments and unassembled myosin in the supernatant. Percent exchange

$$= \frac{\text{Specific activity myosin in filaments}}{\text{Specific activity unassembled myosin}} \times 100$$

$$= \frac{\text{CPM in filaments/[myosin in filaments]}}{\text{CPM unassembled/[unassembled myosin]}} \times 100$$

TABLE 1

FLUORESCENT LABELING OF MYOSIN

Fluorochrome	Labeling Site	Dye/Protein (mol/mol)
IAENS (donor)	-SH	9-10
IAF (acceptor)	-SH	9-10

RESULTS

Characterization of Fluorescent Labeled Myosin

The fluorochromes reported in these studies were IAENS (donor) and IAF (acceptor). Other fluorochromes, including the amino specific label fluorescein-isothiocynate, were also used with equivalent results but will not be discussed here. Both IAENS and IAF bound to myosin with a dye to protein molar ratio of 9-10 fluorochromes per intact myosin molecule (Table 1). Intramolecular distribution of fluorescence was established by chymotryptic digestion of labeled myosin (Fig. 1). Both rod and head portions of the myosin molecule were labeled, even though coupling was performed when myosin was in a polymerized form. Our aim was to label the entire myosin molecule; in future studies we will attempt site specific labeling to define the juxtaposition of specific regions of myosin during thick filament formation.

Thick Filament Assembly

To establish the sensitivity of FET for detecting thick filament formation, coassembly of donor and acceptor labeled myosin molecules was performed. A 5:1 ratio of acceptor to donor myosin was mixed in high salt. Baseline fluorescence was monitored for 10 min to insure that no fluorescence changes occur in unassembled myosin mixtures. Assembly was initiated by dilution and fluorescence quench followed continously. As shown in Fig. 2, 43.3% quench of donor fluorescence was observed during assembly. To confirm that complete assembly had occurred, thick filaments were pelleted at 120,000 x g for 15 min. Approximately 98% of the myosin sedimented as thick filaments. The critical concentration of myosin, i.e. that concentration of soluble myosin present at equilibrium with assembled filaments, was 15-20ug/ml at a filament concentration of 0.8mg/ml.

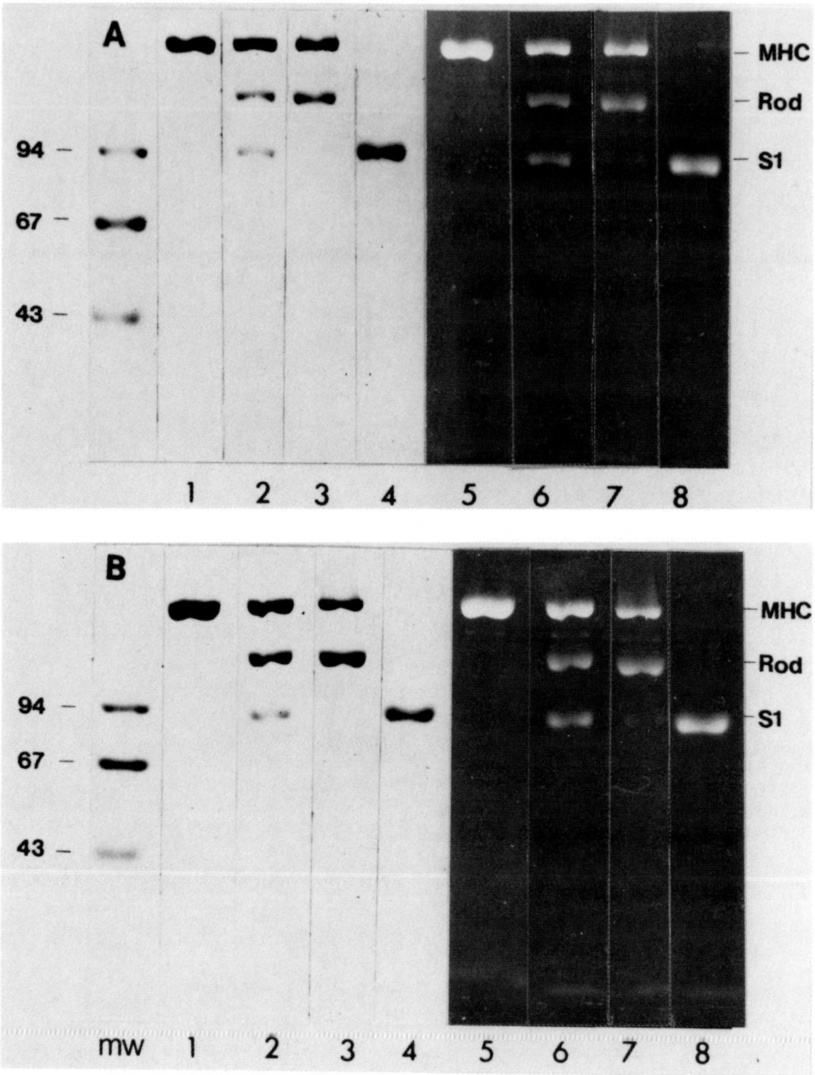

FIGURE 1. Chymotryptic digest of IAENS (A) or IAF (B) labeled myosin. Lanes 1-4, Coomassie stained gel; 5-8 gel visualized under ultraviolet light. Lanes 1 and 5, undigested myosin; lanes 2 and 6, total digest; lanes 3 and 7, low salt insoluble fraction after digestion; lanes 4 and 8, low salt soluble fraction after digestion.

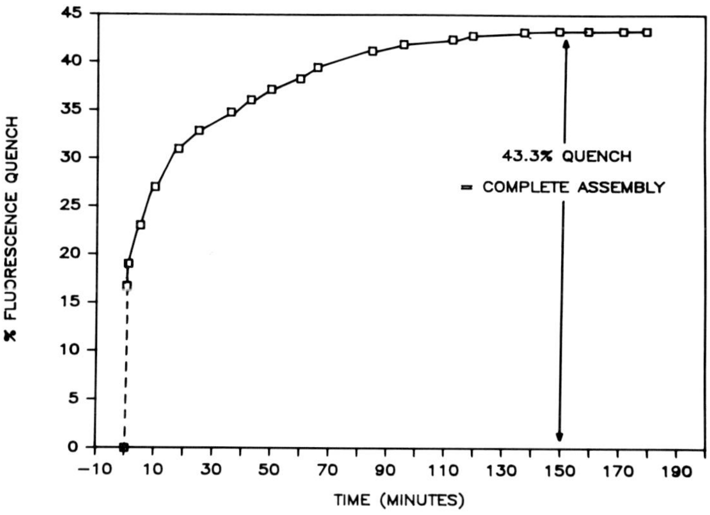

FIGURE 2. Detection of myosin thick filament assembly by fluorescence energy transfer. Coassembly of IAENS labeled myosin (donor) with IAF labeled myosin (acceptor) monitored by increase in donor fluorescence quench at 470nm.

Filament Morphology

By negative staining transmission electron microscopy the synthetic thick filaments formed from fluorescently labeled myosin appeared identical in morphology to those filaments prepared from unlabeled myosin (Fig. 3). Both exhibited wide distributions in length with mean values of ~3.0-3.5um (Fig.4).

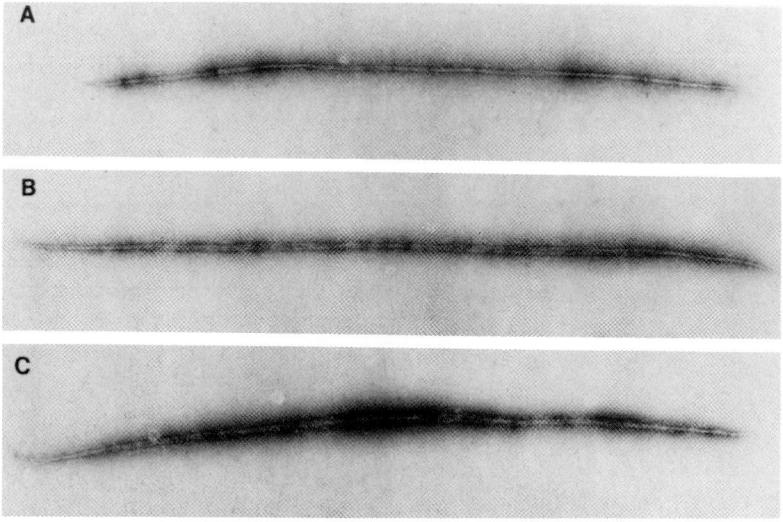

FIGURE 3. Morphology of unlabeled and fluorochrome labeled synthetic thick filaments. (A) unlabeled filaments, (B) IAENS labeled filaments, (C) IAF labeled filaments.

Furthermore, the structure of synthetic filaments was compared with native filaments using an anti-LMM monoclonal antibody, MF20 (21). MF20 has previously been shown to bind native thick filaments at 14.5nm intervals which defines the axial repeat of rods in these structures (25). When MF20 was reacted with synthetic filaments an identical periodicity of 14.5 nm was observed (Fig. 5). Thus, the axial packing of myosin in synthetic filaments matches that of native filaments. Similar results have been obtained using fluorescently labeled filaments (data not shown) indicating that fluorescent labeling of myosin does not appear to alter thick filament structure.

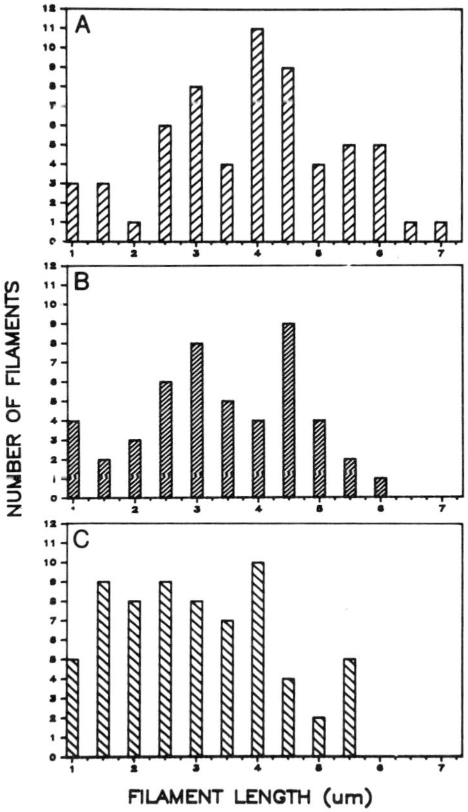

FIGURE 4. Length distribution of unlabeled and fluorochrome labeled synthetic filaments. (A) unlabeled filaments, (B) IAENS labeled filaments, (C) IAF labeled filaments.

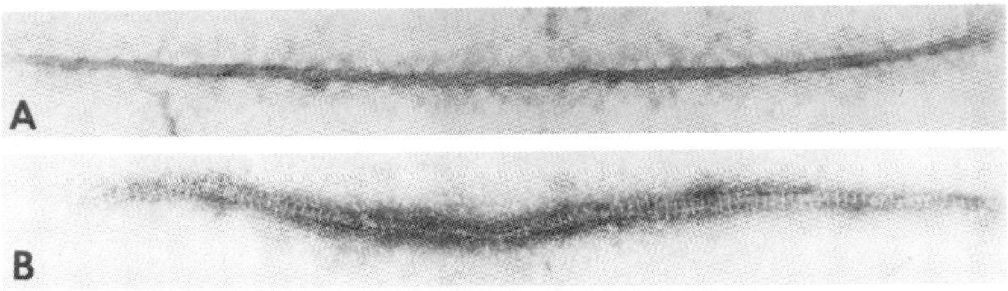

FIGURE 5. Axial periodicity of synthetic thick filaments. (A) filaments + non-immune mouse IgG, (B) filaments + MF20.

Myosin Exchange Between Thick Filaments

The exchange of myosin molecules between thick filaments at equilibrium was also examined using FET. For this assay, unlabeled, donor-labeled and acceptor-labeled filaments were separately prepared by dialysis. Baseline fluorescence, i.e. zero quench, was determined by mixing unlabeled and donor-labeled filaments. Acceptor and donor-labeled filaments were then mixed at a 5:1 ratio of acceptor to donor. Extent of exchange was monitored by measuring donor fluorescence quench as a function of time. Within 180 min, a 32% quench in donor fluorescence was observed (Fig. 6). Since complete exchange (i.e. total randomization of donor and acceptor myosin) would result in the

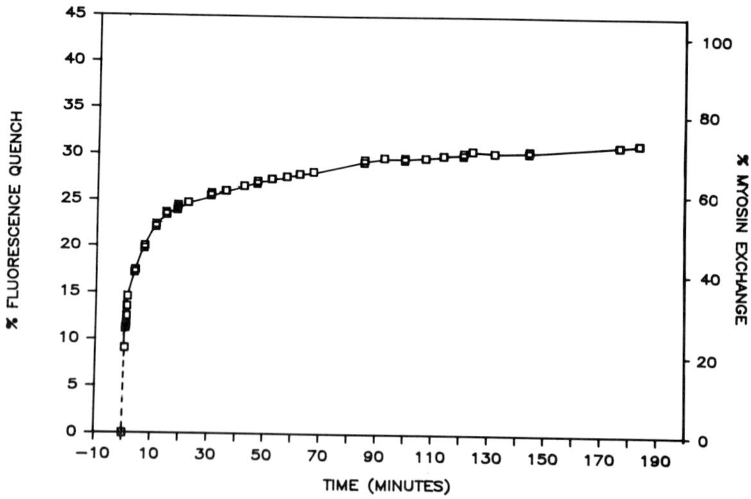

FIGURE 6. Detection of myosin exchange between thick filaments. Exchange between IAENS labeled filaments (donor) and IAF labeled filaments (acceptor) measured by increase in donor fluorescence quench at 470nm.

43% quench observed during filament assembly, 32% quench represents 75% exchange of myosin molecules between thick filaments. The $t_{1/2}$ of the exchange reaction was < 5 min. Sedimentation of thick filaments after the exchange period demonstrated that filaments remain fully assembled during exchange. The critical concentration of myosin after exchange was 15-20 ug/ml, in close aggrement with the critical concentration measured after myosin filament assembly. In addition, SDS-PAGE of exchanged filaments indicated that myosin had not undergone proteolysis during the assay (data not shown).

To determine if myosin exchange occurs between thick filaments and the 15-20 ug/ml pool of soluble myosin in equilibrium with filaments, another type of exchange assay was performed. A trace amount (5 or 15 ug/ml final concentration) of ^{125}I-myosin was added to preformed synthetic thick filaments (1mg/ml) and the incorporation of ^{125}I-myosin into filaments determined (Table 2). The specific activity of ^{125}I-myosin in the pelleted thick filaments was compared to the

TABLE 2

EXCHANGE OF ^{125}I-MYOSIN INTO THICK FILAMENTS

^{125}I-myosin	Soluble Myosin		Myosin in Filaments		Percent Exchange
$\frac{ug}{ml}$	$\frac{ug}{ml}$	$\frac{CPM}{ml}$	$\frac{ug}{ml}$	$\frac{CPM}{ml}$	
5	17.5 $\pm$3	3000	980 $\pm$10	137,000	81.0 $\pm$12
15	18.0 $\pm$2	5000	980 $\pm$10	242,000	86.0 $\pm$11

specific activity remaining unassembled in the supernatant. By this method, the extent of exchange was determined to be 81-86%. ^{125}I-myosin remaining in the supernatant was assembly competent as evidenced by coassembly with added unlabeled myosin.

DISCUSSION

A FET assay system has been developed to follow the assembly of myosin into thick filaments and myosin exchange between filaments. Conditions for fluorescent labeling of myosin molecules have been established. The labeled myosin assembles into synthetic thick filaments with no observable alterations in filament morphology compared to unlabeled filaments. Furthermore, using an anti-LMM monoclonal antibody the packing of myosin molecules into synthetic thick filaments appeared to be similar, if not identical, to that of isolated, native thick filaments. Since sulfhydryl directed fluorescent probes were used in these experiments, it is presumed that the reactive SH_1 and SH_2 sites on the myosin head have been labeled. Athough we have not examined ATPase activity (26), fluorescent labeling of myosin of had no apparent effect on its ability to assemble into thick filaments.

To detect exchange of myosin between thick filaments two different assay systems were employed: FET and ^{125}I-myosin incorporation. Both methods indicated that synthetic myosin thick filaments are in dynamic equilibrium with a small pool (15-20 ug/ml) of soluble, unassembled myosin. Whether extensive exchange occurs _in vivo_ remains to be established. Nevertheless, these observations suggest a possible mechanism for myosin turnover in preexisting myofibrils. Future studies will focus on the exchange properties of native thick filaments, the role of myosin isoform variants in assembly and exchange reactions, and the role of myosin binding proteins in the regulation of thick filament structure.

REFERENCES

1. Bernengo JC, Cardinaud R (1982) State of myosin in solution. J Mol Biol 159:501.
2. Burke M, Harrington WF (1972) Geometry of the myosin dimer in high salt media. II. Hydrodynamic studies on macromolecules of myosin and its rod segments. Biochem 11:1456.
3. Davis JS (1981) Pressure jump studies on the length regulation kinetics of the self-assembly of myosin from vertebrate skeletal muscle into thick filaments. Biochem J 197:309-314.
4. Godfrey JE, Harrington (1970) Self-association in the myosin system at high ionic strength. II Evidence for the presence of a monomer-dimer equilibrium. Biochem 9:894.
5. Herbert TJ, Carlson FD (1971) Spectroscopic study of the self association of myosin. Biopoly 10:2231.
6. Josephs R, Harrington WF (1966) Studies on the formation and physical chemical properties of synthetic myosin filaments. Biochem 5:3474.
7. Josephs R, Harrington WF (1968) On the stability of myosin filaments. Biochem 7:2834.
8. Kaminer B, Bell AL (1966) Myosin filamentogenesis: Effect of pH and ionic concentration. J Mol Biol 20:391.
9. Katsura I, Noda H (1973) Further studies on the formation of reconstituted myosin filaments. J Biochem 73:245.
10. Reisler E, Smith C, Seegan G (1980) Myosin minifilaments. J Mol Biol. 143:129.
11. Harrington WF, Rodgers ME (1984) Myosin. Ann Rev Biochem 53:35.
12. Stryer L (1978) Fluorescence energy transfer as a spectroscopic ruler. Ann Rev Biochem. 47:819.
13. Haugland RP (1975) Myosin structure: Proximity measurements by fluorescence transfer. J Supramol Struct 3:338.

14. Marsh DJ, Lowey S (1980) Fluorescence energy transfer in myosin subfragment-1. Biochem 19:774.
15. Takashi R (1979) Fluorescence energy transfer between subfragment-1 and actin points in the rigor complex of actosubfragment-1. Biochem 18:5164.
16. Pardee JD, Simpson PA, Stryer L, Spudich JA (1982) Actin filaments undergo limited subunit exchange in physiological salt conditions. J Cell Biol 94:316.
17. Pardee JD, Spudich JA (1982) Mechanism of K^+-induced actin assembly. J Cell Biol 93:648.
18. Taylor DL, Reidler J, Spudich JA, Stryer L (1981) Detection of actin assembly by fluorescence energy transfer. J Cell Biol 89:362.
19. Wang Y, Taylor DL (1981) Probing the dynamic equilibrium of actin polyemrization by fluorescence energy transfer. Cell 27:429.
20. Yamamoto K, Pardee JD, Reidler J, Stryer L, Spudich JA (1982) The mechanism of interaction of Dictyostelium severin with actin filaments. J Cell Biol 95:711.
21. Bader D, Masaki T, Fischman DA (1982) Immunochemical analysis of myosin heavy chain during myogenesis in vivo and in vitro. J Cell Biol 95:78.
22. Reinach FC, Masaki T, Shafiq S, Obinata T, Fischman DA (1982) Isoforms of C-protein in adult chicken skeletal muscle: Detection with monoclonal antibodies. J Cell Biol 95:78.
23. Weeds AG, Pope B (1977) Studies on the chymotryptic digestion of myosin. Effects of divalent cations on proteolytic susceptibility. J Mol Biol 111:129.
24. Laemmli UK (1970) Cleavage of structural proteins during the assembly of the head of bacteriophage T4. Nature (Lond.) 227:680.
25. Shimizu T, Dennis JE, Masaki T, Fischman DA (1985) Axial arrangement of the myosin rod in vertebrate thick filaments: Immunoelectro microscopy with a monoclonal antibody to light meromyosin. (J Cell Biol - In Press).

26. Sekine T, Yamaguchi M (1963) Effect of ATP on the binding of N-ethylmaleimide to SH groups in the active site of myosin ATPase. J Biol Chem. 54:196.

Acknowledgements

The authors express their appreciation to James Dennis for his assistance with the electron microscopy.

Molecular Biology of Muscle Development, pages 683-692

DISTRIBUTION OF MYOFIBRILLAR M-BAND PROTEINS

Hans M. Eppenberger, Theo Wallimann, Monika Eppenberger, and Martin Bähler[2]

Department of Cell Biology, Swiss Fed. Institute of Technology, 8093 Zurich Switzerland

INTRODUCTION

The contractile elements of cross-striated muscle fibers consist of myofibrils which are assembled from arrays of thick myosin filaments interdigitating with thin actin filaments. The thick filaments form the so-called A-band. The center of the A-band, the cross-bridge free area of thick myosin filaments ("bare zone"), is traversed by the so-called M-band which consists of at least three non-myosin proteins: MM-creatine kinase (MM-CK), M-protein, myomesin (1-3). Other non-myosin components form regularly striped patterns perpendicular to the longitudinal axis of the A-band extending from the bare zone in both directions, e.g. C-protein, H-protein, 86 kD protein (4-6). It must be assumed that some of these 'minor proteins' are involved in the assembly and maintenance of the thick filament array of the A-band and/or may have enzymatic and regulatory functions within the myofibrils. Here we describe our attempt to study the nature and possible function of the electrondense material which we frequently observe in the bare zone region of isolated native thick filaments (7). It is of interest to find out if this material is identical with one

This work was supported by SNF grant no 3.491-0.83

[2]Present address: Lab. Molec.Cellular Neurosci., Rockefeller Univ.New York, N.Y. 10021-6399

of the known M-band proteins. The binding is presumably rather strong since the material remains attached when native thick filaments are furthermore disassembled into so-called "bare zone assemblages" (8).

METHODS

Myofibrils were prepared from chicken pectoralis muscle according to a described method (9), while native thick filaments, A-segments and bare zone assemblages were prepared according to Niederman and Peters (10). For a complete disassembly native thick filaments were dialysed against relaxing buffer containing 0.6 M KCl as described by the same authors. Polyclonal antibodies against the M-band proteins came from laboratory stocks and were affinity purified (11,12). Monospecific polyclonal antibodies against M-protein (165 kd) and myomesin (185 kd) were prepared from antiserum, containing

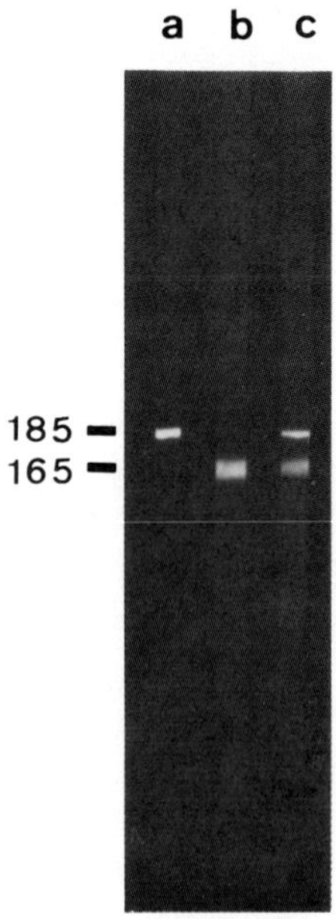

FIGURE 1.

Immunofluorescence staining of myofibrillar proteins after electrophoretical transfer from SDS-PAGE to nitrocellulose. a) Incubation with affinity purified rabbit anti-chick myomesin (185 kd); b) anti-M-protein (165 kd); c) anti-serum cross-reacting with M-protein and myomesin (before affinity purification). Second antibody: FITC-labeled goat anti-rabbit IgG.

antibodies against both high molecular weight M-band proteins, by an affinity purification procedure described by Olmsted (13). The specificity is shown in fig. 1; thus the earlier identification by means of monoclonal antibodies of the 185 kd myomesin as a new M-band protein besides the 165 kd M-protein (3) has here been confirmed. By the use of the monospecific polyclonal antibodies the application of the colloidal gold labeling technique to specimens prepared for electronmicroscopy was possible.

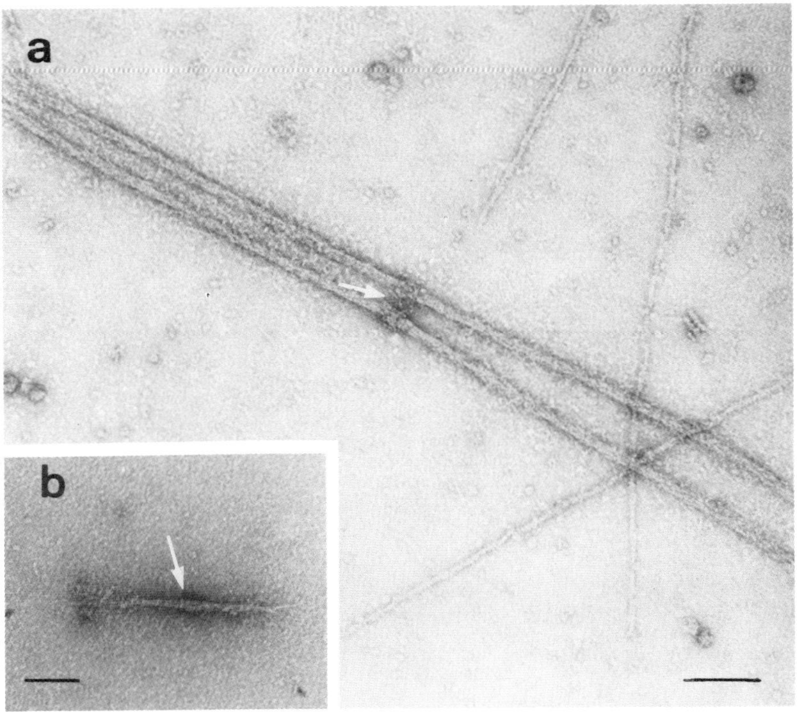

FIGURE 2. Negatively stained preparations of a) native thick filaments with cuff-like extra material in the M-region of the bare zone (arrow); b) bare zone assemblage formed at pH 8.0 (see ref. 10). Extra material indic. by arrow Bar: 0.1 µm.

RESULTS AND DISCUSSION

We could show that the electron-dense cuff-like material shown in fig. 2a in the middle of the bare zone of isolated native thick filaments remains attached to single bare zone assemblages (fig. 2b). This indicates a very strong attachment of the cuff-like extra material. In order to find out more about the nature of this material we used antibodies against the 3 known M-band proteins MM-CK, M-protein and myomesin in connection with the colloidal gold labeling technique to investigate the kind of proteins involved (8). Heavy gold labeling could be observed within the bare zone region of native thick filaments after incubation with antibodies against 165 kd M-protein and 185 kd myomesin increasing the cuff-like appearance of the electron-dense material in the middle of the bare-zone (fig. 3a,b). Both antigens were also found to be present in single bare zone assemblages. Incubation with antibodies against the third known M-band protein, MM-CK, resulted in a less intense decoration of the M-band region of native thick filaments (fig. 3c) and only rarely was labeling observed in the middle of bare zone assemblages. In addition, the number of filaments labeled by anti-MM-CK antibodies was reduced compared to the labeling frequency obtained with anti-M-protein and anti-myomesin antibodies. This was of particular interest since in M-bands of myofibrils or isolated A-bands which remained intact in our thick filament preparations a strong staining of MM-CK antigen could be observed when labeled by the indirect immunofluorescence technique (not shown). MM-CK is the major component of the electrondense material which makes up for the M-band substriations (m-bridges) M4M4' and possibly M1 (14,15) which interconnect the thick filaments. These A-band substructures remain rather intact in isolated A-segments indicating that MM-CK stays bound to isolated A-segments as well. This finding may be an indication that the structural role of MM-CK in the M-band concerns more the proper interconnection and maintenance of

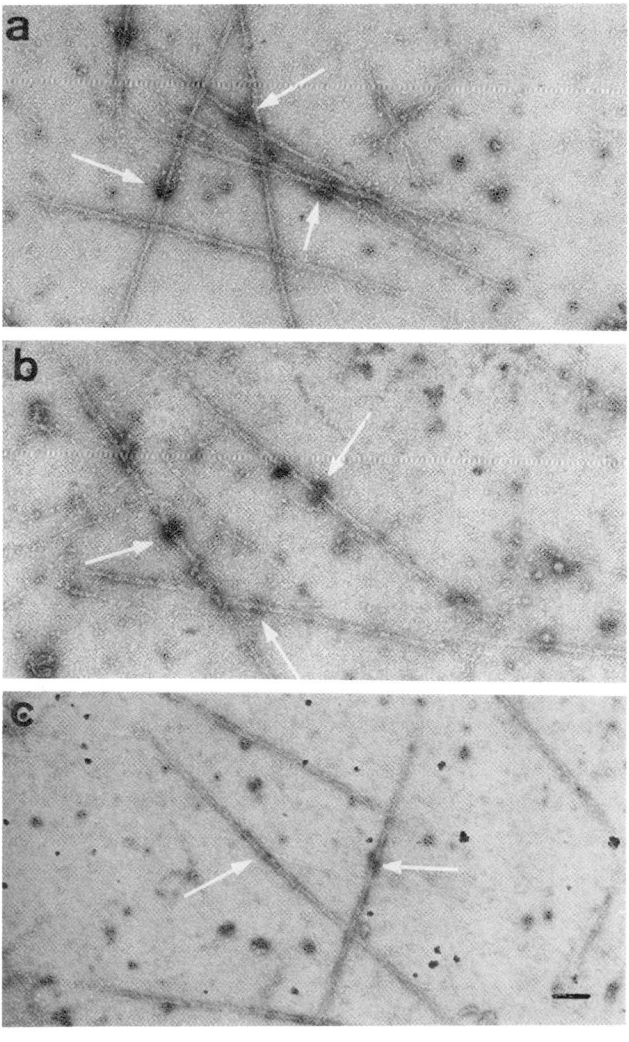

FIGURE 3. Indirect gold labeling for M-band proteins of isolated native thick filaments (negative stain): a) Incubation with anti-M-protein IgG; b) anti-myomesin IgG; c) anti-MM-CK IgG. Arrows indicate gold deposit at bare zones. Bar: 0.1 µm.

register of thick filaments within the whole A-segment than the direct interaction with the myosin array within a single thick myosin filament. Structural data obtained by Luther et al. (16) also support such a model. MM-CK is rather easily and selectively removed from myofibrils either by low salt (5 mM Tris, pH 7.8) treatment or incubation with an excess of monovalent anti-MM-CK Fab'fragments (1). While this enzyme apparently detaches rather easily from the bare

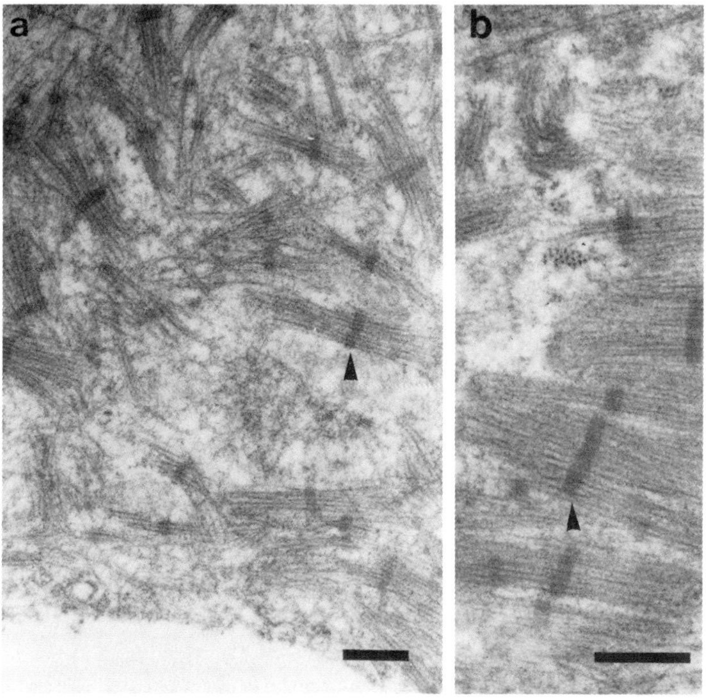

FIGURE 4. Presence of myomesin in "floating" A-segments of moyfibrils in skeletal muscle cell cultures after 2 day phorbol ester treatment. b) higher magnification of a). Arrows indicate anti-myomesin IgG decorated M-bands. For details see ref.17. Bar 0.5 µm. (Figure taken from ref. 17 with permission).

zone of isolated native thick filaments, M-protein and myomesin, under comparable conditions, remain attached and adhere strongly to this region even in partially depolymerized thick filaments or bare zone assemblages.

Two lines of experiments present evidence of a regulatory or organizational role of M-band proteins. In experiments with differentiating myogenic cells in culture it has been shown that after incubation for 24 hrs with phorbol ester myofibrils start to break up along the Z-lines leaving isolated, of actin-filaments stripped, A-segments. These 'floating' A-segments contain still MM-CK within most of the M-bands (17) while the high molecular weight M-band proteins can be chown in all A-segments by antibody decoration (fig. 4). Only after prolongued phorbol ester incubation break down of A-segments and disappearance of M-band proteins has been shown (17). As can be seen from fig. 5 a somewhat similar behaviour is observed in the skeletal muscle of dystrophic Syrian hamster (strain BIO 14.6/U). Here again the sarcomeres of myofibrils in the affected muscle fibers seem to break down along the Z-lines while the A-segments and their thick filament arrays remain fairly intact (18). Antibody decoration clearly demonstrates presence of M-protein and myomesin in the M-bands of the disintegrating sarcomeres, thus suggesting a role of these proteins in stabilizing thick filament arrays. The two high molecular weight M-band proteins may indeed be responsible for the proper initial packing of myosin into thick filaments and the subsequent anchoring of other M-band proteins e.g. MM-CK which then further stabilize the complete A-bands and/or provide as in the case of MM-CK a convenient ATP regenerating system right on the spot (19). The early appearancc during dcvelopment of myomesin and M-protein (20) as well as the relatively late integration of MM-CK into the M-band structure during heart muscle differentiation (21) support the model of an interdependence of the M-band proteins in the assembly and maintenance of those complete A-bands which contain thick filaments of the proper size, while preliminary experiments allow the assumption that no M-band protein binds to synthetically formed thick filaments of irregular size (8).

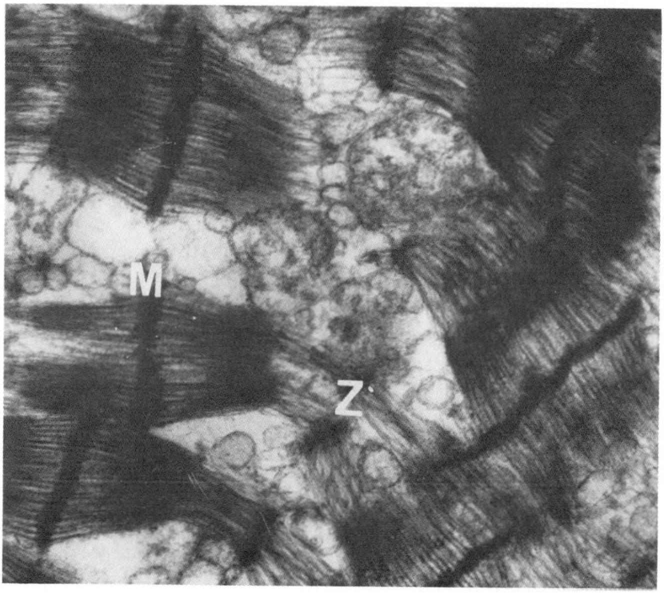

FIGURE 5. Longitudinal section of adductor muscle fibers from a 101 day old dystrophic hamster (strain BIO 14.6/U). Immunolabeling with affinity purified rabbit-anti chicken M-protein/ myomesin IgG. Heavy decoration of M-band region (M). Notice weak or broken Z-band structure (Z).

ACKNOWLEDGMENT

We thank Dr. R. Krauer for providing the necessary photographs for figure 5 and Mrs. E. Abächerli for writing the manuscript.

REFERENCES

1. Wallimann T, Pelloni G, Turner D, Eppenberger HM (1978). Monovalent antibodies against MM-cratine kinase remove the M-line from myofibrils. Proc Natl Acad Sci USA 75:4296.
2. Masaki T, Takaito O (1974). M-protein. J biochem 75:367.
3. Grove BK, Kurer V, Lehner C Doetschman TC, Perriard JC, Eppenberger HM (1984). A new 185,000-dalton skeletal muscle protein detected by monoclonal antibodies. J Cell Biol 98:518.
4. Starr R, Offer G (1978). The interaction of C-protein with heavy meromyosin and subfragment-2. Biochem J 171:813.
5. Starr R, Offer G (1983). H-protein and X-protein. J Mol Biol 170:675.
6. Bähler M, Eppenberger HM, Wallimann T (1985). A novel thick filament protein of chicken pectoralis muscle: 86 kd protein. J Mol biol submitted.
7. Bähler M, Wallimann T, Eppenberger HM (1984). Idication for a role of M-line proteins in assembly and structure of thick filments. J Muscle Res Cell Mot 5:194.
8. Bähler M, Wallimann T, Eppenberger HM (1985). Myofibrillar M-band proteins represent constituents of native thick filaments, frayed filaments, bare zone assemblages and repolymerized bare zone assemblages. J Muscle Res Cell Mot submitted.
9. Wallimann T, Turner DS, Eppenberger HM (1977). Localization of creatine kinase isoenzymes in myofibrils. I. Chicken skeletal muscle. J Cell Biol 75: 297.
10. Niederman R, Peters LK (1982). Native bare zone assemblage nucleates myosin filament assembly. J Mol Biol 161: 505.
11. Perriard JC, Caravatti ER, Perriard E, Eppenberger HM (1978). Quantitation of CK isoenzyme transitions in differentiating chicken embryonic heart muscle and myogenic

cell cultures by immunoadsorption. Arch Biochim Biophys 191:90.
12. Stehler EE, Pelloni G, Heizmann C, Eppenberger HM (1980). Biochemical and ultstructural aspects of M_r 165 000 M-protein in cross-striated chicken muscle. J Cell Biol 86:775.
13. Olmsted JB (1981). Affinity purification of antibodies from diazotized paper blots of heterogeneous protein samples. J Biol Chem 256:11955.
14. Strehler EE, Carlsson E, Eppenberger HM, Thornell LE (1983). Ultrastructural localization of M-band proteins in chicken breast muscle as revealed by combined immunocytochemistry and ultramicrotomy. J Mol Biol 166:141.
15. Wallimann T, Doetschman TC, Eppenberger HM (1983). Novel staining pattern of skeletal muscle M-lines upon incubation with antibodies against MM-creatine kinase. J Cell Biol 96:1772.
16. Luther PK, Crowther RA (1984). Threedimensional reconstruction from tilted sections of fish muscle M-band. Nature 307:566.
17. Doetschman TC, Eppenberger HM (1984). Comparison of M-line and other myofibril components during reversible phorbol ester treatment. Eur J Cell Biol 33:265.
18. Eppenberger M, Schoenenberger R, Eppenberger HM, (1984). Myofibrillar M-line Structure in normal and dystrophic hamster muscle. Muscle Nerve 7:304.
19. Wallimann T, Schlösser T, Eppenberger HM (1984) Function of M-line-bound CK as intramyofibrillar ATP regenerator at the receiving end of the CP shuttle. J Biol Chem 259: 5238.
20. Eppenberger HM, Perriard JC, Rosenberg U, Strehler EE (1981). The M-protein myomesin: a specific protein of cross-striated muscle cells. J Cell Biol 89:185.
21. Anversa P, Olivetti G, Bracchi PG, Loud AV (1981). Postnatal development of the M-band in rat cardiac myofibrils. Circ Res 48:561.

Molecular Biology of Muscle Development, pages 693–707

EXPRESSION OF M-BAND PROTEINS DURING MYOGENESIS[1]

J.C. Perriard, U. Achtnich, L. Cerny[2], H.M. Eppenberger, B.K. Grove[3], H.P. Hossle, B. Schäfer

Institute for Cell Biology, Swiss Federal Institute of Technology, Hönggerberg, CH-8093 Zurich, Switzerland

ABSTRACT

We have found expression of both myomesin and M-protein was muscle specific. Slow muscle tissue (ALD) contained only traces of M-protein, but the myomesin content was about twice as high as in fast skeletal muscle. Electrophoretic analysis of M-protein did not show any variations in various muscles and heart, but heart myomesin behaved as a different isoprotein. ELISA measurements of these proteins in developing breast muscle as well as in myogenic cell cultures showed myomesin to be detectable at early stages while presence of M-protein lagged behind. In myogenic cell cultures the antigens, especially myomesin, were found to be associated with the M-band of myofibrils as soon as they were expressed.
In addition the gene of M-creatine kinase (M-CK) another component of the M-band was partially characterized and all available evidence points to existence of a rather large (> 40kb) gene. The cDNA of its non muscle isoprotein B-CK was identified by immunological screening of a λ gtll gizzard cDNA library with anti B-CK antibody. A surprisingly high homology of the amino acid sequence of B-CK with

[1] This work was supported by Swiss National Science foundation and a grant from the MDA.

[2] Present address: Department of Food Science, University of California, Davis.

[3] Present address: Department of Anatomy, University of Umea, Umea, Sweden

various CK's from other species and with chicken M-CK was observed.
Furthermore experiments are reported investigating the isoprotein specificity of M-CK association with the H-zone. Chicken heart cells not expressing M-CK were microinjected with purified MM-CK, M-CK mRNA enriched RNA and "in vitro" prepared M-CK mRNA transcribed by SP-6 RNA polymerase from full length M-CK cDNA cloned into plasmid pSP6. It appears that the RNA is translated in heart cells and indications of proper association with the myofibrils were observed.

INTRODUCTION

The mechanisms of interdependence of the regulated expression of new genes during development and the utilization of the newly produced (iso)proteins as building blocks for the formation of functional organelles is still poorly understood. Differentiating myogenic cells are a very suitable experimental system to study these problems. In the course of differentiation, myogenic cells undergo multiple changes at the morphological and the biochemical level. Mononucleated cells fuse to form myotubes containing contractile organelles, the myofibrils. These morphological changes are coordinated with the expression of a set of muscle specific genes and are accompanied by many isoprotein switches.

Myofibrillogenesis may depend on the formation of two essential structural elements:

The z-disk delineates the sarcomere, is involved in transmitting the tensile force, in the anchorage of myofibrils in the cells and also in keeping thin filaments in the ordered register.

The M-band forming a transverse structure in the middle of the A-band is probably important to organize the thick filaments into the well ordered, perfectly registered A-bands and may have a organisational role in the formation of thick filaments. Both the M-band and the z-disc are made up in part of minor myofibrillar components, some have been isolated and partially characterized, others are probably still elusive.

Many contractile proteins are first expressed as non-muscle isoproteins. As the cells undergo the differentiative transition, they express a new set of muscle specific isoproteins, which make up the myofibrils (e.g. isoproteins for actin, myosin, tropomyosin, α-actinin etc.) or participate in the energy metabolism like the creatine kinases (CK). A isoprotein switch for CK is observed as B-CK, the non skeletal muscle form, which is expressed among others in embryonic cells, becomes repressed as cells differentiate and start expression of the M-CK isoprotein, characteristic for adult skeletal muscle (1). A fraction of this newly appearing isoenzyme MM-CK (active dimer) was found to be bound to the M-band of skeletal muscle myofibrils and appears to be responsible for the electron dense structure observed by electron microscopy (2). Other minor proteins like myomesin and M-protein localize within the same region of skeletal and heart myofibrils (3,4).

There are several problems we will address within this paper which are open questions of muscle cell biology. First it is of interest to learn more about the minor proteins that may be indispensable for proper myofibril formation and maintenance. In this context we wanted to investigate when and how Myomesin, M-protein and MM-CK are first expressed, then accumulated and incorporated in the regions of the myofibrils like the M-band. We believe this will lead to a better understanding of the interplay of newly expressed proteins and their utilization for organelle formation.

Since the elaboration of the muscle phenotype depends on the expression of a set of muscle specific genes it will be important to learn more about the mechanisms of muscle specific gene activation. Many isoprotein switches are good examples since the comparison of the regulation of a non skeletal muscle protein gene (like e.g. B-CK) to the muscle specific isoprotein gene (M-CK) will help to reveal the mechanism of tissue (muscle) specific gene expression.

The isoprotein switches observed in muscle development lead to the further question of the significance for myofibrillogenesis of the biochemical differences between the non muscle forms and the muscle forms of such proteins. We wanted to investigate this question first using the isoprotein set of B-CK and M-CK, since only the M-CK isoprotein was found to be associated with the M-band of skeletal sarcomeres (5).

RESULTS AND DISCUSSION

Myomesin and M-Protein Expression

Myomesin and M-protein are both located in the M-band of adult skeletal and heart muscle and shown to be different proteins by several criteria. Two classes of monoclonal antibodies generated against a purified M-band protein fraction were found which do not share epitopes as shown by competition experiments (3). The same results were obtained with an affinity purified polyclonal antibody (6) and no crossreaction was observed between the purified antibodies and the denatured antigens (7, see Eppenberger et al. this vol.). In addition the skeletal muscle form of myomesin migrates with an apparent M_r of 185 kd while M-protein has a M_r 165 kd as detected by immunoblots (3).

Both of the proteins are muscle specific. Using a highly sensitive, double antibody sandwich ELISA (DAS-ELISA) technique, concentration of these two proteins and the muscle specific form of creatine kinase, M-CK, were determined in a variety of tissue extracts (8). Relative concentrations were expressed in % of the concentration in pectoral muscle (TABLE I). From these results it is evident that both Myomesin and M-protein are expressed specifically in cross striated muscle. In tissues like tongue and thymus that also contain muscle cells, they are also present. This pattern of expression is shared by M-CK although there are traces of M-CK detectable in other tissues but at very low levels (less than 1/1000 of the levels in pectoral muscle).

A different pattern of expression of M-band proteins in slow tonic muscle (Anterior lat. dorsi, ALD, to a lesser extent in A.profundus) is also evident. M-protein levels are over 10 fold lower while values for myomesin are 2 fold higher than those in fast muscle (TABLE 1 and FIG.1). The results could be interpreted that in these muscle types myomesin replaces the M-protein and thus results in myomesin levels twice as high, which would require a compensatory activation of the myomesin gene or an increased stability of the protein. This hypothesis gains further support from the analysis of ALD at the cellular level by immunohistochemical staining of cryosections. While all fibers stain positive for myomesin there are only a few cells positive for M-protein (9). It is therefore possible that the majority of cells accumulate myomesin at a level twice as high and that the residual M-protein comes

TABLE 1
RELATIVE CONCENTRATIONS OF MYOMESIN, M-PROTEIN, AND M-CK[a]

Tissue	Myomesin	M-Protein	M-CK
Pectoralis major	100	100	100
Post.latissimus dorsi	116	107	b
Adductor profundus	250	10	b
Ant. latissimus dorsi	225	2	b
Heart	110	13	1
Gizzard	c	c	0.024
Gut	c	c	0.032
Tongue	22	14	37.6
Thymus	12	5.7	45.5
Lung	c	c	0.14
Brain	c	c	0.05
Liver	c	c	0.05
Spleen	c	c	0.04
Kidney	c.	c	0.03

[a] A double antibody sandwich ELISA (DAS-ELISA) assay was used to determine relative concentrations of myomesin and M-protein. M-CK concentrations were determined by a direct competition ELISA assay. All determinations are standardized against pectoral muscle (100%) (8).
[b] not done.
[c] not detectable.

from the few cells expressing both types of M-band proteins at levels characteristic for fast muscle, in the case of ALD they are of type IIA (fast oxydative fibers).

Then the M-band proteins of various muscle types were compared to each other by the immunoblotting technique with monoclonal antibodies specific for myomesin and M-protein as shown in FIG. 1. In all muscles examined M-protein appears to be the same as judged by its electrophoretic mobility with the exception of anterior lat. dorsi ALD (FIG. 1B, lane d) where the traces of M-protein escaped detection. On the basis of these results it cannot be rigorously ruled that a slow isoform of M-protein would not react with the monoclonal antibodies (mAb) made against pectoral M-band proteins.

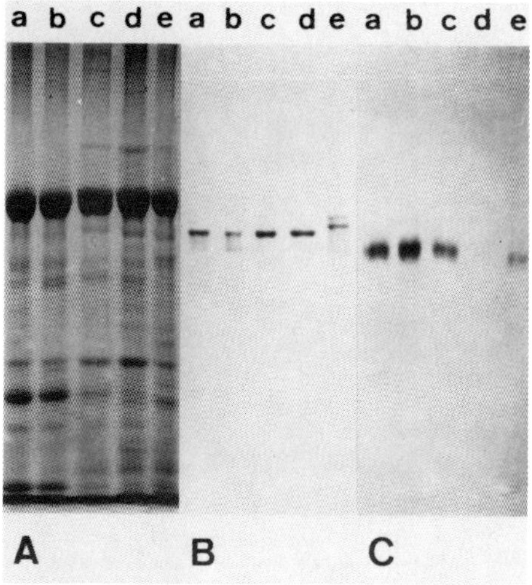

FIGURE 1. Myomesin and M-protein in various muscle types. SDS extracts of muscle tissue containing each 40 ug of protein were electrophoresed on a 5% SDS-polyacrylamide gel and the resolved proteins were electrophoretically transferred onto nitrocellulose paper. The samples of section A were stained for protein, the blot of B was stained with mAb B5 (anti-myomesin) or C mAb A6 (anti-M-protein) and bound antibody was made visible with horse radish peroxidase labeled rabbit anti mouse antibody, chloronaphtol and hydrogen peroxide as substrates (3). a:Pectoralis; b:PLD; c:Adductor profundus; d:ALD; e:heart.

On the other hand such an interpretation is not likely since similar result are obtained with polyclonal antibodies.

As indicated in TABLE 1 and in FIG. 1C (lane e) the amount of M-protein in heart muscle is also much lower but appears to be distributed over all heart cells at lower concentrations. It appears that heart muscle has a different ratio of M-protein to myomesin. In addition the blots stained for myomesin show bands reacting with anti-myomesin antibodies in all extracts as expected from DAS-ELISA determinations (TABLE 1). While the myomesins of the different types of skeletal muscle are qualitatively uniform (FIG. 1B, a-d) heart type myomesins migrate as two protein species with higher M_r (FIG. 1B, lane e) and suggest that heart myomesins are distinct isoproteins.

The M-band proteins are thus also a set of proteins that are differentially regulated and display distinct heart muscle specific isoproteins. In order to investigate the involvement of myomesin and/or M-protein in the early phases of myofibril formation, the pattern of their expression during muscle development was followed. In the embryonic pectoral muscle and in the heart, myomesin is expressed already during early periods while M-protein lags behind 6-7 days in the pectoral muscle. In the heart both proteins appear earlier (8). The accumulation of the two proteins was also investigated in differentiating myogenic cell cultures because they offer the possibility to follow easily the accumulation at the biochemical level by DAS-ELISA assays and the incorporation of the newly expressed proteins into the M-band of nascent myofibrils.
As shown in FIG. 2 a similar pattern as in the developing embryonic pectoral muscle was observed. Myomesin can be detected and accumulates during the first days of culture while M-protein was detectable 1 1/2 days later. Both proteins reached maximal accumulation on day 6.
Sister cultures were subjected to double immunofluorescence analysis for myomesin and M-protein using mAb B4 for the detection of myomesin and a monospecific polyclonal rabbit antibody against M-protein (8). In young cultures most cells were negative, few cells were positive for both antigens or for myomesin alone. As differentiation proceeded, many cells became positive for myomesin and displayed at marginal zones a punctate fluorescence along nascent myofibrils, while M-protein was only readily detected when myomesin appeared in a typical crosstriated pattern. In all the stages of developing cells tested, both antigens were associated with myofibrillar structures and very seldom showed a diffuse pattern, suggesting that the proteins are immediately incorporated into nascent organelles or that they might even be involved in the organelle formation in helping to create nucleation structures for further polymerization.

Myomesin and M-protein appear as many other muscle proteins to be subjected to differential regulation during development. Myomesin may be associated with early stages of myofibrillogenesis since it is always present prior to M-protein and is expressed, in contrast to M-protein, in all muscle fiber types so far examined. The elucidation of the differential genetic expression and the primary amino acid structures will be achieved with more ease once the corresponding cDNAs are cloned.

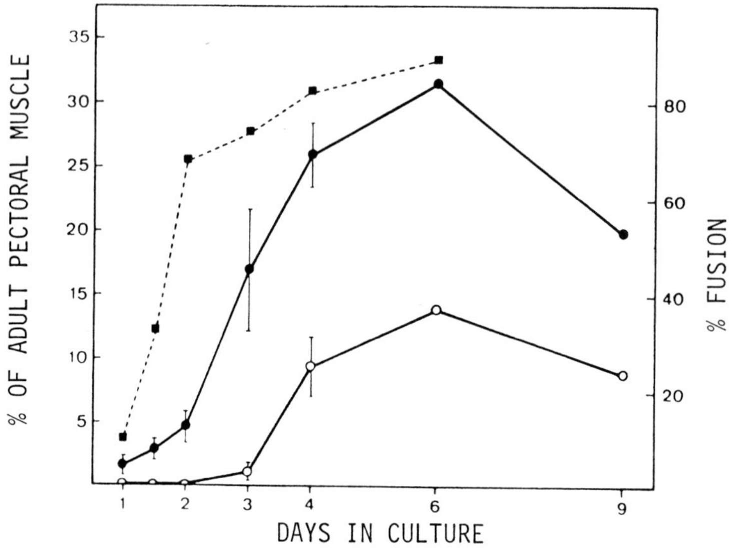

FIGURE 2. Accumulation of myomesin and M-protein during myogenesis "in vitro". Extracts from myogenic cell cultures were prepared in the course of differentiation. Myomesin (●—●) or M-protein (○—○) were determined by DAS-ELISA (see TABLE 1) and data are given as relative values in percent of the ones observed in pectoral muscle. Morphological differentiation is reflected in the increasing proportion of nuclei in multinucleated cells (■- -■). The values represent several independent experiments.

The Creatine Kinase Isoenzyme Switch

On the basis of these results it is likely that minor proteins like myomesin or M-protein are newly expressed gene products necessary for myofibril formation. The mechanisms of activation can only be studied if the corresponding gene structures are known. If, in addition, the genes were studied that undergo isoenzyme switching like the creatine kinases, comparison of the nonmuscle isoprotein gene to its muscle counterpart should help to recognize the muscle typical components of the regulatory mechanisms.

B-CK is expressed in many adult non skeletal muscle tissues and also temporally in embryonic muscle cells (10). As the myogenic cells undergo differentiation, B-CK is repressed and the muscle specific M-CK gene is activated. The CK switch involves dual regulation, first temporal activation and later repression of B-CK concomittant with the activation of M-CK. Previous studies have shown that the switch is very likely controlled at the transcriptional level (11,12).

Although the M-CK gene structure in the chicken is still elusive, more information has become available. Genomic blots have indicated that the M-CK gene of the chicken genome appears to be unusually big. When genomic DNA was doubly digested with the restriction enzyme Eco Rl and several other enzymes (with 6 base pair recognition sites) like Hind III, DNA bands were detected hybridizing with M-CK cDNA of the size of more than 36 kb. The corresponding gene must be even bigger since an EcoRl site was found some 120 bp from the 5'end of the cDNA. Earlier attempts to clone the M-CK gene have failed, possibly because all the λ -phage libraries were screened in not absolutely recombination deficient host bacteria. Only occasional clones including only a short segment of the 3'end of the cDNA were obtained. Recently chicken genomic libraries utilizing the vector λ charon 35 growing in recombination deficient hosts were constructed by J. Mar and Ch. Ordahl (UC. San Francisco) which we partially characterized. This preliminary characterization demonstrated some of the already anticipated peculiarities. The clones are very seldom cut by hexanucleotide recognizing restriction enzymes and the size of the inserted sequence hybridizing with a large part of M-CK cDNA appears to be in the range of 40 kb. Further work is required to elucidate the M-CK gene structure. In this context it will be interesting to learn more about the B-CK gene, since it is phylogenetically closely related to M-CK. Studies on B-CK primary protein structure, gene structure and its expression require the molecular cloning of B-CK cDNA, since the preliminary crosshybridization studies using M-CK cDNA and B-CK mRNA gave inconclusive results.

Molecular Cloning of B-CK cDNA

An expression cDNA library in λgt 11 phages was constructed (by U. Rosenberg; 13) with cDNA derived from chicken gizzard poly A^+

RNA and plaques were screened using a polyclonal antibody against chicken B-CK. Some 25 000 recombinant phages gave 5 truly positive phage isolates identified by the production of a fusion protein of β-galactosidase with anti B-CK reactive peptides. These fusion proteins could be found in lysates of IPTG induced phages having β-galactosidase with a higher M_r as the vector β-galactosidase. They also reacted with anti B-CK antibody on immunoblots. The cDNA

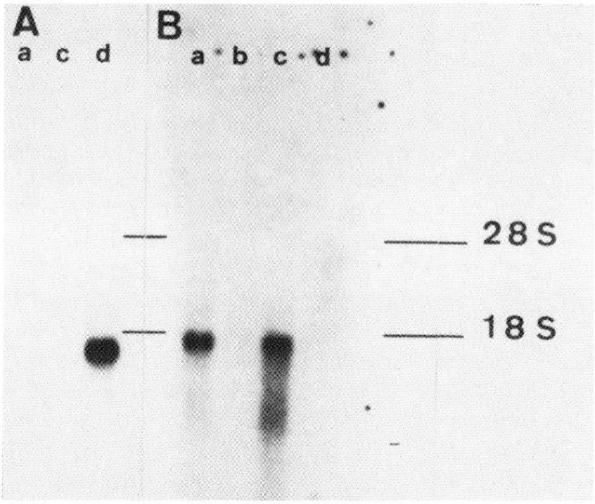

FIGURE 3. Northern hybridization of RNA from different chicken tissues with cDNA from clone 13a.
RNA was resolved on 1.2 % agarose gels, blotted onto nitrocellulose and hybridized with nick-translated cDNA containing M-CK sequences (A) or clone 13a (B), RNA was from heart (a), breast muscle (b), gizzard (c) or leg muscle (d). The bands of the ribosomal RNA are indicated as size markers.

inserts of the phages were subcloned into the plasmid pUC8 and used as probes in northern blot analysis of RNA prepared from various chicken tissues like heart, gizzard, leg muscle, and breast muscle. In FIG. 3 such an experiment is shown for the clone 13a. The size of the hybridizing bands probed with cDNA of clone 13a with approximately 1550 bp (FIG. 3B lanes a,c) is slightly bigger than the M-CK mRNA from leg or breast muscle (FIG. 3A lane d) of some 1350 bp (12,14). The B-CK cDNA probe hybridized to RNA from heart and gizzard but not to the M-CK containing RNA preparations from leg and breast muscle. This pattern of hybridization is the one expected for B-CK cDNA. Further proof that the cloned cDNA was creatine kinase came from sequencing experiments. The cDNA inserts of the original gt 11 clones were subcloned into M13 phages and subjected to the procedure of dideoxy sequencing. For 4 out of the 5 isolates the open reading frames were translated into amino acid sequences which were homologous over wide stretches to the primary M-CK sequences and especially matched the so called active center peptide almost perfectly. As shown in TABLE 2, the CK amino acid sequences known and published to date of M-CK from chicken (14), rabbit (15), rat (16) and the sequences for the fish creatine kinases of Torpedo marmorata (17) and Torpedo californica (18) show extensive homologies among each other as well as with the newly obtained sequence of B-CK from chicken gizzard cDNA.

Significance of Isoprotein Differences of M-CK and B-CK for Myofibril Association

An interesting situation exists in chicken heart cells. Although the M-protein and the chicken heart specific myomesin are localized in the H-zone of heart myofibrils (4) no electron dense M-band structure was observed (19) and no M-CK is expressed in chicken heart. In mammalian heart cells, however, M-CK is expressed and an electron dense M-band structure exists (20). Thus in principle it should be possible to reconstitute also the chicken M-band if MM-CK could be transferred into chicken heart cells.

To this end purified MM-CK was microinjected into chicken heart cells which were cultured for an additional period. The cells were then fixed and stained with MM-CK specific antibody. Only injected cells stained positive, while all heart cells showed the counterstain with anti-myomesin. In some areas the fluorescence of M-CK showed a typical crosstriated pattern superimpossible with the antimyomesin pattern. If instead of purified MM-CK, muscle poly A^+ RNA enriched on a sucrose gradient for M-CK mRNA was microinjected into heart cells, the injected cells became positive for M-CK and some of the fluorescence showed again the crosstriated pattern. Obviously there are no powerful restrictions of translation of muscle specific proteins in heart cells. Of course, the interpretation of

TABLE 2
PRIMARY CK STRUCTURE DERIVED FROM cDNA NUCLEOTIDE SEQUENCES[a]

```
----NS--LL-M-Y-VDD-Y----V----------LD---K---RQ-S------------
MPFSSTHNKHKLKFSAEEEFPDLSKHNNHMAKVLTPELYKRLRDKETPSGFTLDDVIQTG
---GN----Y--NYKS---Y----------------D---K-------------------
---GN----F--NYKSQ--Y----------------D--NK-------------------
---GN----W--NY--A---------------A--LDI--K-------------------
---GN----W--NY--A---------------A--LDI--K--------------I----

--------------------------E------E----------E-K----AD--Q----
VDNPGHPFIMTVGCVAGDEESYEVFKDLFDPVIQDRHGGYKPTDKHRTDLNHENLKGGDD
----------------------T--------I-N-----F------K-------------
----------------------T--------I-N-----F------K-------------
--------------------C------------E------------K----QD-------
--------------------C------------E------------K----Q--------

---N-------------R-FC------------I--------G--G-DL--K--A-RN--
LDPKYVLSSRVRTGRSIKGYSLPPHCSRGERRAVEKLSVEALNSLEGEFKGRYYPLKAMT
---H----------------T------------------------T-----K-----S--
---N----------------T------------------------T-----K-----S--
---N---------------IA-----------L----CI-G-AT-T---Q-K----ST-S
---N---------------IA-----------L----CIDG-AT-T---Q-K----SS-S

DA----------------------------------------------I-----------
EQEQQQLIDDHFLFDKPVSPLLLASGMARDWPDARGIWHNDNKTFLVWVNEEDHLRVISM
-------------------------------------------S----------------
-------------------------------------------S----------------
DA---------------I---------------G------ND-S----------------
DA---------------I---------------G------ND------------------

Q---------T---T--TQ--TL--SKNYE---NP-----------------A---I---
EKGGNMKEVFRRFCVGLKKIEEIFKKAGHPFMWTEHLGYILTCPSNLGTGLRGGVHVKLP
-----------------Q---------------N-----V-------------------A
-----------------Q---------------N-----V-------------------A
Q-----------------------V---RG---N-----V------------------I-
Q--------------------D--V---RG---N----TV------------------I-

N-GK-E--G-V-K------------------G---V------------L----------L
KLSQHPKFEEILHRLRLQKRGTGGVDTAAVGAVFDISNADRLGFSEVEQVQMVVDGVKLM
H--K--------T------------------S-----------S-------L--------
N--K--------T------------------------------S-------L--------
H-CK-E--S-V-K-T------------E---SIY--------------------------
H-CK-E--S-V-K-T----------------SIY--------------------------

I----R--KG-S---LM----   1.: B-CK CHICK
VEMEKKLEQNQPIDDMIPAQK   2.: M-CK CHICK
--------KG-S---------   3.: M-CK RABBIT
--------KG-----------   4.: M-CK RAT
-----R--NGKS---L-----   5.: CK   TORPEDO MARMORATA MUSCLE
-----R--NGKS---LM----   6.: CK   TORPEDO CALIFORNICA
```

[a] The chicken M-CK sequence is listed in full in the one letter code. In the other sequences, corresponding amino acids are indicated by -, amino acid substitutions are listed with symbols. The chicken B-CK amino acid sequence is not fully verified by sequencing a full length cDNA clone (13). For references see text.

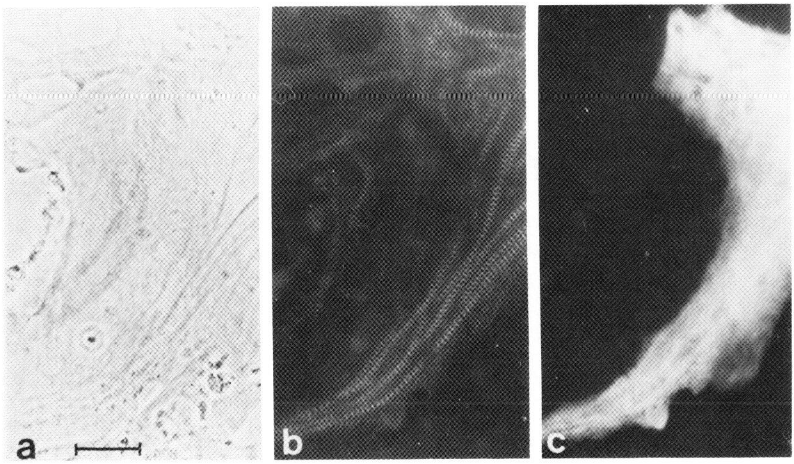

FIGURE 4. Microinjection of "in vitro" produced M-CK mRNA into cultured chicken heart cells. After injection, cells were cultured for further 4 hrs and then subjected to double immunofluorescence staining. A: phase contrast; B: anti-myomesin, C: anti MM-CK. Bar = 20 um

these experiments still suffers from the use of heterogeneous RNA.

Homogeneous mRNA for M-CK was produced from a full length M-CK cDNA cloned into the plasmid pSP6 (21). The transcript was injected into the heart cells. Although after immunological staining the injected cells have indicated the presence of M-CK (FIG.4), the conditions of the procedure still need to be optimized, in order to assess the association of M-CK with the myofibril. Some faint indications of myofibril association of the immuno-reactive material could be seen. Careful extraction of these injected cells should help to increase the signal to noise ratio.

ACKNOWLEDGEMENTS

We are very grateful to Dr.J. Mar and Dr. Ch. Ordahl (U.C. San Francisco) for the sharing of the new λ charon 35 chicken genomic libraries with us, to Dr. U. Rosenberg (MPI Tübingen) for the construction of the chicken cDNA libraries in λ gtll. The technical assistance of Ms. G.Kunz is gratefully acknowledged. JCP is very grateful to

Dr. B. Nadal-Ginard (Harvard Med. School, Boston) for the lab facilities and encouragement during a sabbatical when part of the B-CK cDNA work was done. Drs. F. Chytil and R. Billeter are gratefully acknowledged for the critical reading of the manuscript.

REFERENCES

1. Perriard JC, Caravatti M, Perriard ER, Eppenberger HM (1978). Quantitation of creatine kinase isoenzyme transitions in differentiating chicken embryonic breast muscle and myogenic cell cultures by immunoadsorption. Arch Biochem Biophys 191:90.
2. Wallimann T, Pelloni G, Turner DC, Eppenberger HM (1978). Monovalent antibodies against MM-creatine kinase remove the M-line from myofibrils. PNAS USA 75:4296.
3. Grove BK, Kurer V, Lehner C, Doetschmann T, Periard JC, Eppenberger HM (1984). A new 185,000-dalton skeletal muscle protein detected by monoclonal antibodies. J Cell Biol 98:518.
4. Strehler EE, Pelloni G, Heizmann CW, Eppenberger HM (1979). M-protein in chicken cardiac muscle. Expt Cell Res 124:39.
5. Wallimann T, Moser H, Eppenberger HM (1983). Isoenzyme-specific localization of M-line bound creatine kinase in myogenic cells. J Muscle Res Cell Mot 4:429.
6. Olmsted JB (1981). Affinity purification of antibodies from diazotized paper blots of heterogenous protein samples. J Biol Chem 256:11955.
7. Cerny L, Eppenberger HM, Perriard JC (1985). Immunological crossreactivities among creatine kinase isoenzymes of different species. (in preparation)
8. Grove BK, Cerny LC, Perriard JC, Eppenberger HM (1985). Myomesin and M-protein: Expession of two-M-band proteins in pectoral muscle and heart during development. (submitted).
9. Thornell LE, Grove BK, Cerny L, Perriard JC, Eppenberger HM (1985). Fibre type specific distribution of myomesin and M-protein in chicken muscle (in preparation).
10. Caravatti M, Perriard JC, Eppenberger HM (1979). Devel. regulation of creatine kinase isoenzymes in myogenic cell cultures from chicken. J Biol Chem 254:1388.

11. Perriard JC (1979). Developmental regulation of creatine kinase isoenzymes in myogenic cell cultures from ckicken. J Biol Chem 254:7036.
12. Rosenberg UB, Kunz G, Frischauf A, Lehrach H, Mähr R, Eppenberger HM, Perriard JC (1982). Molecular cloning and expression during myogenesis of sequences coding for M-creatine kinase. PNAS 79:6589.
13. Hossle HP, Rosenberg U, Perriard JC. Molecular cloning and complete nucleotide sequence of chicken B-CK cDNA (in preparation).
14. Ordahl CP, Evans GL, Cooper TA, Kunz G, Perriard JC (1985). Complete amino acid sequence of chick muscle creatine kinase. J Biol Chem 259:15224.
15. Putney SD, Herlihy W, Royal N, Pang H, Aposhian HV, Pickering L, Belagage R, Biemann K, Page D, Kuby S, Schimmel P (1984). Rabbit muscle creatine phosphokinase: cDNA cloning, primary structure and detection of human homologues. J Biol Chem 259:14317.
16. Benfield PA, Zivin RA, Miller LS, Sowder R, Smythers GW, Henderson L, Oroszlan S, Pearson ML (1984). Isolation and sequence analysis of cDNA clones coding for rat skeletal muscle creatine kinase. J Biol Chem 259:14979.
17. Giraudat J, Devilliers-Thierry A, Perriard JC, Changeux JP (1984). Complete nucleotide sequence of Torpedo marmorata mRNA coding for the 43,000 dalton protein: muscle-specific creatine kinase. PNAS 81:7313.
18. West BL, Babbitt PC, Mendez B, Baxter JD (1984). Creatine kinase protein sequence encoded by a cDNA made from Torpedo california electric organ mRNA. PNAS 81: 7007.
19. Wallimann T, Kuhn R, Pelloni G, Turner DC, Eppenberger HM (1977). Localization of creatine kinase isoenzymes in myofibrils. II. Chicken heart muscle. J Cell Biol 75:318.
20. Wallimann T (1985). Localization and function of M-line bound-creatine kinase: M-band model and creatine phosphate shuttle. In: Cell and Muscle Motility, Vol. VI (J.W. Shay ed.) in press.
21. Krieg PA, Melton DA (1984). Functional messenger RNAs are producedby SP6 "in vitro" transcription of cloned cDNAs. Nucleic Acids Res 12:7057.

Molecular Biology of Muscle Development, pages 709–723

USE OF A COCARCINOGEN (TPA) AND A CARCINOGEN (EMS) TO PROBE MYOFIBRILLOGENESIS

P. Antin, S. Forry-Schaudies, S. Tokunaka, A. Duran, J. Eshleman, and H. Holtzer

Department of Anatomy, School of Medicine
University of Pennsylvania, Philadelphia, PA 19104

ABSTRACT The cocarcinogen TPA (12-O-tetradecanoyl-phorbol-13-acetate) and the carcinogen EMS (ethyl methanesulfonate) selectively and reversibly alter the morphology and lineage-specific synthetic program of definitive myogenic cells in culture. Myogenic cells maintained in these agents are transformed into large isodiametric multinucleated myosacs and myosheets, respectively. Cytoimmunofluorescent and electron microscopic analysis reveals that both agents block the emergence of striated myofibrils. In contrast, microtubules, and desmin and vimentin intermediate filaments are present throughout the myosacs and myosheets. Prominent, well separated stress fiber-like structures (SFLS) subtend the upper plasma membrane of these multinucleated structures and can exceed 0.5mm in length. These SFLS bind rhodamine labelled phalloidin (Rho-phalloidin), anti-alpha-actinin, and anti-chick brain myosin (anti-CBM) in patterns similar to stress fibers in normal fibroblasts. Biochemical analysis reveals that the rates of protein synthesis and accumulation of mRNAs coding for myosin heavy chain, LC1, LC2, alpha and beta tropomyosins and troponin C are greatly reduced in TPA and EMS treated myogenic cultures. The synthesis of desmin, vimentin and tubulin, and the accumulation of mRNAs coding for these proteins is unaffected.

When TPA or EMS are removed from the culture medium, the arrested myogenic cells synthesize muscle-specific proteins and assemble them into functional myofibrils.

1. This work was supported by HD07152, CA18194, HL18708, HL15835 and the Muscular Dystrophy Association.

There is a one-on-one spatial relationship between an individual preexisting SFLS and a single nascent myofibril, as indicated by the colocalization of fluorescent reagents that stain SFLS and those that specifically stain myofibrils. The early forming SFLS-myofibril complexes rapidly lose their association with the upper plasma membrane and are displaced into the sarcoplasm where newly assembled thick and thin filaments are added peripherally to the enlarging myofibrils. With further maturation, the previously prominent SFLS disappear over the belly of the myotube and remain only as relatively inconspicuous structures in the growth tips and pseudopodial extensions of the myosacs and myosheets. A similar temporal and topographical relationship between SFLS and emerging myofibrils has been observed in normal, untreated skeletal and cardiac myoblasts. Based on these data, a multistage model is presented for myofibrillar formation, which distinguishes between the forces ordering the nascent myofibril from those that control subsequent growth in diameter.

INTRODUCTION

Definitive muscle cells selectively synthesize a set of contractile protein isoforms that interact in a precise fashion to form the invariant myofibril. Although many of the proteins comprising the myofibril have been well characterized, little is known about the mechanisms that regulate their interaction to form a highly ordered structure. Light and electron microscopic observations of both *in vitro* and *in vivo* skeletal muscle have suggested that other components of the cytoskeleton, including microtubules (1,2), intermediate filaments (3-5) and stress fibers (6-8) influence myofibril formation. However, the contributions of these cytoskeletal elements to the process of myofibrillogenesis have proven difficult to examine microscopically, since normally developing myotubes are relatively thick and cylindrical and myofibril assembly is a surprisingly rapid event.

We have found that the cocarcinogen TPA and the carcinogen EMS reversibly alter myogenesis in a striking and reproducible manner (9-16). Young developing myotubes exposed to these agents form large, isodiametric, multinucleated "myosacs" or "myosheets", respectively.

While maintained in the presence of TPA or EMS, these multinucleated structures fail to synthesize many of the myofibrillar proteins and consequently fail to form ordered myofibrils. When shifted to normal medium, these myofibril-depleted structures begin to synthesize their myofibrillar proteins and assemble them into long, narrow, well-separated, typically-striated myofibrils. The synthetic and optical properties of these drug treated, reversibly arrested, myogenic cells has provided an ideal system for the study of myofibrillogenesis in vitro. We have taken advantage of these properties to study the temporal and topographical relationships between preexisting SFLS, intermediate filaments, microtubules and the emergence of nascent and maturing myofibrils.

RESULTS

Effects of TPA on Developing Myogenic Cultures.

Myotubes in control day 4 cultures may be over 0.5mm in length, occasionally branch, and have cross-sectional diameters of 15-30um. Their numerous nuclei tend to line up in long rows and are equally distributed throughout the myotube. Most myotubes contract spontaneously. Cytoimmunofluorescent staining with antibodies to light meromyosin (LMM), alpha-actinin, C protein or myomesin (gift of Dr. H. Eppenberger) reveals hundreds of mature, striated myofibrils in every myotube and one or two myofibrils in every elongated postmitotic mononucleated myoblast (1).

When day 4 cultures were exposed to TPA (0.5uM) for an additional 72 hours, 100% of the numerous multinucleated structures formed isodiametric myosacs. Electron microscopic and immunofluorescence studies have demonstrated that striated myofibrils were virtually absent in TPA-induced myosacs (9-16). Pulse labelling of cultures with ^{35}S-methionine, followed by one and two dimensional gel electrophoresis and fluorography, revealed that the synthesis of many myofibrillar proteins was greatly reduced following exposure to TPA, although the synthesis of desmin, vimentin and tubulin was unaffected (fig 1). During the 72 hours of exposure to TPA, the nonmuscle fibroblasts and presumptive myoblasts replicated at a rate four fold higher than in control cultures. Thus, after TPA treatment,

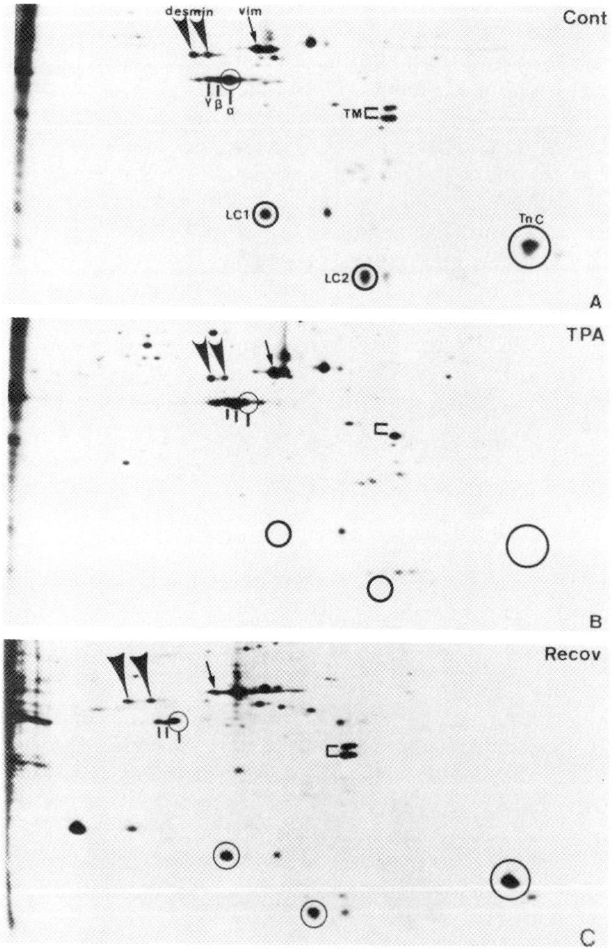

Figure 1. Fluorographs of two-dimensional polyacrylamide gels showing the proteins synthesized during a 1 hour pulse of ^{35}S-methionine. A) Day 7 control myogenic culture; B) Day 7 culture treated with TPA from days 4-7; and C) TPA-treated culture recovered in normal medium for 4 days. Note the loss of most muscle-specific proteins in B) and their synthesis in C). vim = vimentin; α, β, γ, = actin isoforms; TM = muscle-specific tropomyosins; LC1 and LC2 = fast myosin light chains; TnC = troponin C. The arrows and circles in B) and C) correspond to proteins labelled in A). Gels contain equal numbers of counts per minute.

myosacs were often overgrown with layers of mononucleated cells and extracellular matrix. Clearly, the selective inhibition of TPA on the synthesis of myofibrillar isoforms cannot be due to a generalized "cytotoxic" property.

If left in TPA for 5 or more days these myosacs invariably degenerated. If, however, they were shifted to normal medium after 3 days in TPA, they rapidly recovered. Elongation was prominent within 15 hours, and by 48 hours numerous, normally striated myofibrils hundreds of microns in length were evident. The influence of microtubules and intermediate filaments on myofibrillogenesis - particularly the unexpected interaction between thick filaments and microtubules - has been followed by allowing TPA myosacs to recover in normal medium plus Colcemid and normal medium plus taxol (2,13,17).

Because recovering TPA-myosacs assembled nascent myofibrils rapidly, and were relatively thick, they were less favorable for the study of the cytological relationship between SFLS and myofibrillogenesis than were inordinately flattened EMS-myosheets.

Effects of EMS on Developing Myogenic Cultures

We deliberately used a concentration of EMS (7.5mM) that was cytotoxic to replicating presumptive myoblasts and fibroblasts, but spared postmitotic myogenic cells. By exposing very young myogenic cultures to EMS, we obtained populations 4-6 days later which were highly enriched for postmitotic myogenic cells. When day 1 cultures were treated with EMS for the next 72 hours, the number of surviving mononucleated cells was 10-20% of that in day 4 control cultures. Despite this cytotoxicity for replicating cells, EMS did not inhibit fusion of the postmitotic myoblasts that emerged during the period of culture. However, as with TPA treated cultures, the resulting multinucleated structures displayed a highly abnormal morphology. EMS induced the formation of strikingly flattened, immense isodiametric myosheets that often exceeded 0.5mm in diameter yet were rarely greater than 1.0um in thickness. EMS-myosheets failed to bind anti-LMM (fig 2b), anti-MM-CPK, anti-myomesin, or anti-M-band protein (data not shown). Antibodies to alpha-actinin revealed an extensive network of SFLS in these myosheets that were similar to stress fibers in fibroblasts (Fig 2a). At higher magnification, a punctate anti-alpha-actinin

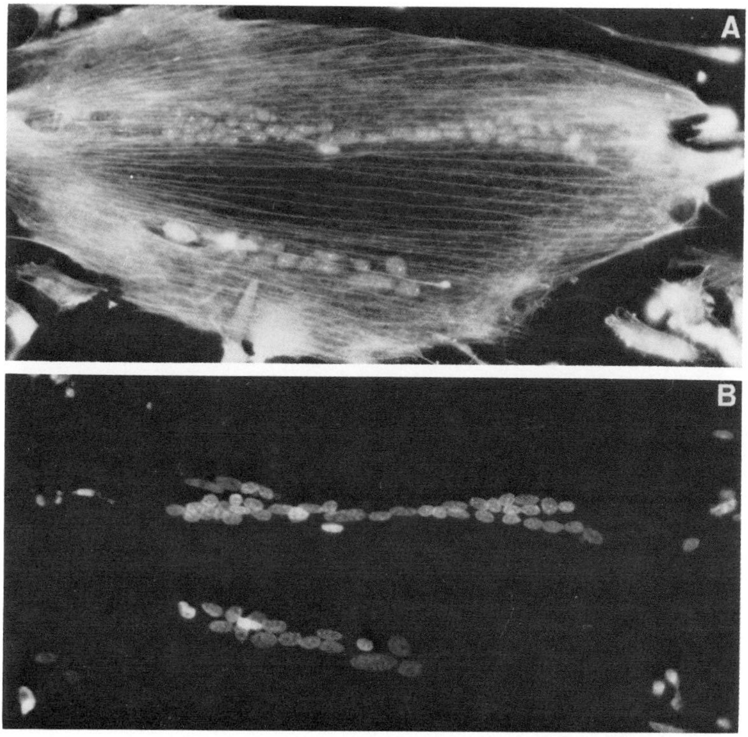

Figure 2. Double labelled cytoimmunofluorescent micrographs of a typical myosheet from a 4 day culture exposed to EMS for the last 72 hours: A) stained with anti-alpha-actinin plus propidium iodide nuclear dye and visualized in the rhodamine channel. B) The same microscopic field viewed in the fluorescein channel, stained with anti-LMM plus bisbenzamide H nuclear dye. Note the extremely long, fine stress fiber-like structures (SFLS) that bind anti-alpha-actinin, and the total absence of staining with anti-LMM. A small number of mononucleated alpha-actinin positive cells are also visualized.(see ref. 7 for antibody characterizations and staining procedures).

staining pattern of 0.6-1.3um periodicity was discernable in both SFLS and fibroblast stress fibers. Through focusing suggested that individual SFLS subtended only the upper plasmalemma. Both SFLS and stress fibers in fibroblasts bound antibodies against chick brain myosin (anti-CBM) and tropomyosin in a punctate fashion. Stress fibers in fibroblasts and SFLS in EMS myosheets bound Rho-phalloidin in a continuous fashion (7).

EM studies confirmed the submembranous localization of the SFLS. SFLS were never observed more than 0.2um from the upper plasma membrane. It is of interest that in myosheets SFLS occupy the "upper" subcortical domain, whereas stress fibers in spread fibroblasts commonly occupy the "lower" or substrate associated domain. Consistent with the immunofluorescence data, EM sections also failed to reveal any suggestion of thick filaments, Z-bands or other evidence of disrupted or degenerating myofibrillar components.

In agreement with the immunofluorescence and EM findings, analysis of protein fractions by SDS PAGE showed that both the accumulation (fig 3a) and synthesis (fig 3b) of myosin heavy chain was greatly reduced following EMS treatment. In contrast, accumulation and synthetic rates of desmin, vimentin and tubulin were unaffected or slightly increased. Immunoblot analysis of these samples, using anti-LMM and anti-CBM, verified the large reduction in muscle-specific myosin heavy chain in EMS treated cultures, and also the concommitant increase in levels of the nonmuscle myosin isoform (data not shown).

The selective effects of EMS on myofibrillar protein accumulation and synthesis was reflected in reduced levels of translatable mRNAs coding for these proteins. Compared to 4 day controls, greatly reduced amounts of myosin heavy chain, myosin fast light chains 1 and 2 , tropomyosins and troponin C were translated in rabbit reticulocyte lysates charged with total mRNAs from EMS cultures (eg. Fig. 4). It was of interest that several unidentified proteins were translated at higher rates in these lysates than lysates from controls.

Myosheets Recovering From Exposure to EMS

To follow the emergence of myofibrils in recovering EMS myosheets we used double label immunofluorescence with antibodies to several myofibrillar and cytoskeletal proteins. By using a variety of double staining permutations, we were able to correlate the earliest appearance and organization of muscle-specific myofibrillar proteins with preexisting cytoskeletal structures.

For 72 hours following removal of EMS from the culture medium, myosheets continued to spread and flatten on the

Figure 3

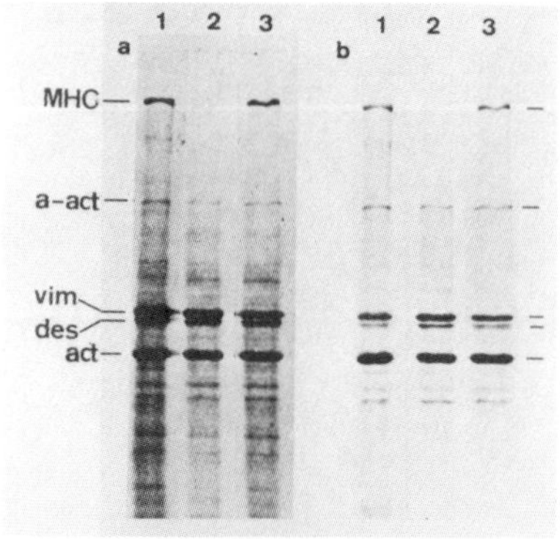

Figure 4

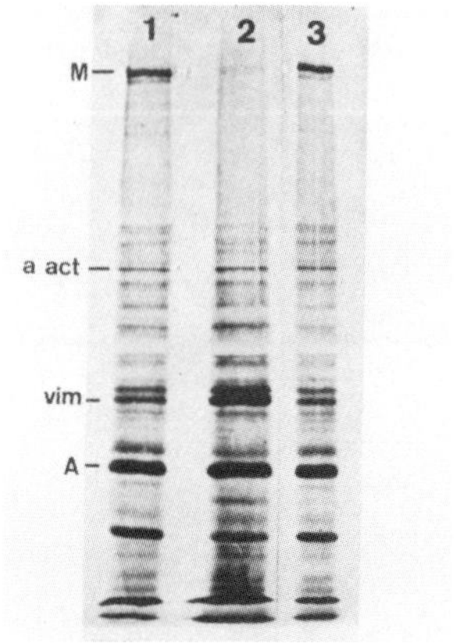

Figure 3. A) Coommassie blue stained 10% polyacrylamide gel of triton-X 100 insoluble proteins from: 4 day control myogenic cultures (lane 1): 4 day myogenic cultures exposed to EMS for the last 72 hours (lane 2), cultures exposed to EMS and allowed to recover in normal medium for 4 days (lane 3). Cultures were pulsed for 1 hour with ^{35}S-methionine prior to sacrifice. Sample loads were standardized to 0.5ug of DNA. B) Fluorograph of the gel shown in A). MHC = myosin heavy chain; a-act = alpha-actinin; des = desmin; vim = vimentin; act = actin. Both the accumulation (panel A, lane 2) and synthesis (panel B, lane 2) of MHC was greatly reduced following exposure of cultures to EMS. Cultures allowed to recover in normal medium for 4 days accumulated and synthesized MHC at control levels (panels A and B, lane 3). Other cytoskeletal proteins were apparently unaffected by EMS treatment.

Figure 4. Fluorogram of a 10% SDS gel revealing <u>in vitro</u> translation products of total cellular RNA isolated from: 4 day control myogenic cultures (lane 1); 4 day myogenic cultures exposed to EMS for the last 72 hours (lane 2): EMS treated cultures allowed to recover in normal medium for 4 days (lane 3). 1.0 ug of total RNA was translated in a rabbit reticulocye lysate containing ^{35}S-methionine. Following incubation, lysates were analyzed by electrophoresis and fluorography. M = myosin heavy chain; a act = alpha-actinin; vim = vimentin; A = actin.

substrate, sometimes forming multinucleated structures greater than $0.2mm^2$. Myofibrillar proteins remained immunologically and biochemically undetectable. During this "silent" period, SFLS, intermediate filaments and microtubules were conspicuous. Approximately 4 days after removal of EMS (exact times varied slightly between experiments), staining with anti-LMM revealed extremely fine, continuous, parallel fibrils of various lengths spaced throughout the myosheets. Double labelling, using anti-LMM and anti-CBM, revealed that these first detectable anti-LMM positive fibrils colocalized along the more extensive, preexisting network of SFLS. During the next 24 hours, fibrils that bound anti-LMM increased in number, became thicker and longer, and the colocalization along the SFLS more striking. Eventually anti-LMM stained fibrils became coextensive over most of the length of SFLS that also bound anti-CBM. Double stains using anti-CBM (fig. 5a) and anti-LMM (fig. 5b), anti-alpha-actinin, or Rho-phalloidin demonstrated that anti-LMM staining localized exclusively with the SFLS. By day five of recovery, the emergence of a striated myofibrillar pattern with a distinct 2.3-2.7um periodicity became evident along portions of these fluorescent fibrils. Concurrently, in these areas, SFLS staining was gradually lost. By day seven of recovery, anti-LMM staining was localized exclusively to striated myofibrils.

A similar transition from SFLS to striated myofibril was observed using fluorescent reagents that recognized both muscle and nonmuscle protein isoforms. An antibody against smooth muscle alpha-actinin stained the Z-bands of striated myofibrils with a periodiciy of 2.3-2.7um whereas it stained stress fibers of spread fibroblasts in a punctate pattern of 0.6-1.3um periodicity. At intermediate times of recovery, anti-alpha-actinin myofibrillar staining patterns were often interspersed with SFLS staining patterns along a single fluorescent fibril. With the loss of anti-CBM staining of SFLS, anti-alpha-actinin localized exclusively to the Z-bands of definitive nascent myofibrils. A similar transition of staining pattern along a single filament was also observed with an antibody recognizing both muscle and nonmuscle tropomyosin, and with Rho-phalloidin. Rho-phalloidin stains both stress fibers of fibroblasts and SFLS in EMS treated myosheets as uniformly stained structures. The complex, striated binding pattern of Rho-phalloidin to myofibrils varies

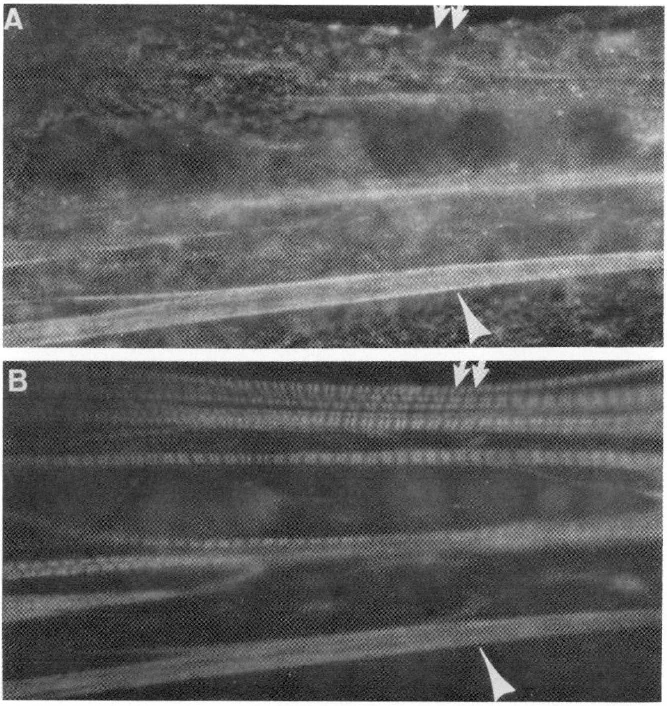

Figure 5. Double label cytoimmunofluorescence micrographs showing the staining patterns of antibodies to myofibrillar and nonmyofibrillar myosin during "intermediate" and "late" stages of early myogenesis. A) a small portion of a recovering EMS-myosheet stained with anti-CBM (chick brain myosin) visualized in the rhodamine channel. B) The same microscopic field visualizing anti-LMM staining in the fluorescein channel. Single arrows point to a SFLS that stains coextensively with anti-CBM and anti-LMM. Within the same mysheet, myofibrils at a later stage of assembly that show a striated anti-LMM staining pattern (Fig. 5B, double arrows) no longer bind the anti-CBM (Fig. 5A, double arrows).

with the extent of sarcomere contraction (7). Nevertheless, its continuous staining of stress fibers in fibroblasts and SFLS in myosheets can be readily distinguished from its staining pattern of definitively-striated myofibrils.

Using antibodies to desmin and vimentin, we failed to detect any obvious morphological correlation between these intermediate filaments and emerging, nascent myofibrils. On the other hand there was a positive correlation between the overall orientation of long microtubules as revealed by anti-tubulin staining and the overall orientation of SFLS and nascent myofibrils. Not uncommonly, a single microtubule in these myosheets could be followed for many hundreds of microns (see also 1,2,17). Following the displacement of the nascent striated myofibrils from their subcortical domain, the characteristic relocalization of desmin intermediate filaments was observed in recovering EMS myosheets. Long after the definitive Z-bands had been assembled, anti-desmin was observed to shift from a predominantly longitudinal orientation to an I-Z band location. This shift in desmin localization long after Z-bands have assembled is similar to that observed during normal myofibrillogenesis (4,5,18).

Discussion

Stress fibers have been studied largely in flattened cells. Little is known of their function. We have found that; 1) although beta-and gamma-actins, non-myofibrillar myosins, etc. are normally present in polygonal chondroblasts, red blood cells, liver cells, and nerve cells, they do not form morphologically typical stress fibers; and 2) pp60V-src, TPA and heat shock rapidly disassemble typical stress fibers in flattened fibroblasts. We suspect a better understanding of the relationship between stress fibers, SFLS and nascent myofibrils requires a better understanding of the function(s) of non-myofibrillar contractile protein isoforms and the rules governing the interconversion of ordered stress fibers into the more "randomized" microfilaments.

It is of interest to compare the emergence of individual striated myofibrils in 1) postmitotic mononucleated myoblasts, 2) early multinucleated myotubes, and 3) recovering EMS-myosheets. Anti-desmin and anti-LMM first localize in a diffuse fashion throughout the postmitotic mononucleated myoblast. As the cell elongates, the anti-desmin remains diffuse, whereas the anti-LMM becomes confined to what under the fluorescence microscope appears as a long, non-striated myofibril. EM sections often reveal that such a fibril consists of overlapping, poorly aligned, but fully assembled thick filaments, 1.5um in length. This unstriated LMM-positive fibril rapidly undergoes a transition into typical tandem sarcomeres of interdigitating thick and thin filaments. Each A-band in the earliest sarcomeres is 1.5um long (1,19-21). First detected subcortically in the vicinity of the nucleus, the single striated myofibril rapidly extends in both directions, terminating as an ill-defined, nonstriated fibril. It will be of interest to determine whether the early LMM-positive unstriated fibril that first appears in postmitotic mononucleated myoblasts assembles in proximity to a single overlying SFLS.

As the 20-30um myoblast lengthens to a mononucleated bipolar cell over 400um long, new, definitive-sized sarcomeres add to both ends of the preexisting myofibril. Mini-sarcarmeres are never observed, nor is there any evidence that sarcomeres intercalate into preexisting myofibrils. 10-30 new sarcomeres can be added to a single elongating myofibril in 24 hours (unpublished data). Several long myofibrils can be observed in a definitive myoblast. Fusion of such myoblasts results in myotubes with multiple, but well separated, myofibrils. The "ends" of these are not readily followed. We believe they remain attached to the inner surface of the cell membrane - possibly linked to the latter by microfilaments and vinculin - and there continue to insert new, full-sized sarcomeres. Clearly, this must be further studied. Some evidence for bidirectional addition of new sarcomeres at each end of a myofibril was reported for maturing cardiac myoblasts (7). However, the most convincing evidence for bidirectional addition of new sarcomeres is derived from double-staining experiments of recovering EMS-myosheets. During early stages of recovery, strings of several hundred tandem, definitive-sized sarcomeres appear in the body of the EMS-myosheet. Both ends of such a nascent striated myofibril are continuous with, and merge into, a SFLS. As recovery proceeds, not only do all long SFLS disappear, but virtually every myofibril now extends from one end of the myosheet to the other. After recovery, stretches of several hundred sarcomeres tapering off at both ends into ill-defined fibrils are never observed.

The addition of new sarcomeres at the ends of striated myofibrils probably involves different mechanisms than growth in diameter of a definitive striated myofibril (7,19,21,22). The diameter of a single myofibril either in a normal myotube or in a recovering EMS-myosheet is remarkably uniform over distances exceeding hundreds of microns. Furthermore, all along this length their profiles or edges are crisp and well demarcated. Overlapping or poorly aligned thick and thin filaments - so characteristic of nascent myofibrils (8,22) - are never observed along the periphery of an enlarging striated myofiber. The entire complex of thousands of tandem sarcomeres enlarges manyfold in diameter and does so as a single unit. During this process the "first generation" sarcomeres must be preserved in the core of the enlarging myofibril. Efforts to demonstrate pools of soluble MHC or alpha-actin in the sarcoplasm at sites distant to the myofibrils have thus far been negative. These observations have suggested, though not proven, that the circumferential growth of tandem sarcomeres involves newly synthesized and newly polymerized thick and thin filaments, and that these were synthesized and polymerized in close proximity to each myofibril (7,19-21). No obvious spatial relationship between nuclei and emerging nascent or elongating or enlarging myofibrils has been detected. If a gradient of mRNAs for myofibrillar isoforms from nuclei to cytoplasm exists, this is not obviously reflected in the deployment of the myofibrils. In situ hybridization experiments to localize the various myofibrillar transcripts should make a significant contribution to this problem. It will be interesting to determine if the cDNA-RNA complexes for the myofibrillar proteins aggregate in proximity to the enlarging myofibrils.

REFERENCES

1. Antin P, Forry-Schaudies S, Friedman T, Tapscott S, Holtzer H (1981). Taxol induces postmitotic myoblasts to assemble interdigitating microtubule-myosin arrays which exclude actin filaments. J Cell Biol 90:300.
2. Toyama Y, Forry-Schaudies S, Hoffman B, Holtzer H (1982). Effects of taxol amd colcemid on myofibrillogenesis. Proc Natl Acad Sci USA 79:6556.
3. Fellini SA, Bennett GS, Holtzer H (1978). Selective binding of antibody against gizzard 10-nm filaments to

different cell types in myogenic cultures. Am J Anat 153:451.
4. Bennett GS, Fellini SA, Toyama Y, Holtzer H (1979). Redistribution of intermediate filament subunits during skeletal myogenesis and maturation in vitro. J Cell Biol 82:577.
5. Gard D, Lazarides E (1980). The synthesis and distribution of desmin and vimentin during myogenesis in vitro. Cell 19:263.
6. Fischman D (1970). The synthesis and assembly of myofibrils in embryonic muscles. Curr Topics Devel Biol 5:235.
7. Dlugosz AA, Antin PB, Nachmias VT, Holtzer H (1984). The relationship between stress fiber-like structur+es and nascent myofibrils in cultures cardiac myocytes. J Cell Biol 99:2268.
8. Peng HB, Wolosewick JJ, Cheng P-C (1981). The development of myofibrils in cultured muscle cells: a whole mount and thin-section electron microscopic study. Devel Biol 88:121.
9. Croop J, Toyama Y, Dlugosz AA, Holtzer H (1980). Selective Effects of phorbol 12-myristate 13-acetate on myofibrils and 10 nm filaments. Proc Natl Acad Sci USA 77:5273.
10. Croop J, Dubyak, Toyama Y, Dlugosz A, Scarpa A, Holtzer H (1982). Effects of 12-O-tetradecanoyl-phorbol-13-acetate on myofibril integrity and calcium content in developing myotubes. Devel Biol 89:460-474.
11. Antin P, Holtzer H (1982). Reversible blocking of myogenesis by a chemical carcinogen. Anat Rec 202:8a.
12. Dlugosz AA, Tapscott SJ, Holtzer H (1983). Effects of phorbol-12-myristate 13-acetate on the differentiation program of skeletal myoblasts. Cancer Res 43:2780-2789.
13. Holtzer H, Forry-Schaudies S, Antin P, Dubyak G, Nachmias V (1984). The tumor promoter TPA and the carcinogen EMS induce incoordinate synthesis of muscle proteins. In Wolf S, Strohman R (eds): "Gene Expression in Muscle", New York: Plenum Press.
14. Cohen R, Pacifici M, Rubinstein N, Biehl J, Holtzer H (1977). Effect of a tumor promoter on myogenesis. Nature 222:538.
15. Toyama Y, West C, Holtzer H (1979). Differential response of myofibrils and 10-nm filaments to a carcinogen. Am J Anat 156:131.
16. Doetschman TC, Eppenberger HM (1984). Camparison of

M-line and other myofibril components during reversible phorbol ester treatment. Eur J Cell Biol 33:265.
17. Holtzer H, Forry-Schaudies, Dlugosz A Antin P, Nachmias V, Dubyak G (1985). Interaction between IFs, microtubules and myofibrils in fibrogenic and myogenic cells. In Burrell CD, Kelly DD, Pagels HR, Markes EA (eds): "Intermediate Filaments", New York, New York Acadamy of Sciences, In Press.
18. Tokuyasu KT, Maher PA, Singer SJ (1984). Distribution of vimentin and desmin in developing chick myotubes in vivo I. immunofluorescence study. J Cell Biol 98:1961.
19. Holtzer H, Marshall J, Finck H (1957). An analysis of myogenesis by the use of fluorescent antimyosin. J Biophys Biochem Cytol 3:705.
20. Okazaki K, Holtzer H (1965). an analysis of myogenesis in vitro using fluorescein-labelled antimyosin. J Cytol Histochem 13:726.
21. Holtzer H, Sanger J, Ishikawa H, Strahs K (1972). Selected topics in skeletal myogenesis. Cold Spring Harbor Symposia on Quantitative Biology 37:549.
22. Shimada Y, Obinata J (1977). Polarity of actin filaments at the initial stage of myofibril assembly in myogenic cells in vitro. J Cell Biol 72:777.

Molecular Biology of Muscle Development, pages 725–739

CYTOSKELETAL ORGANIZATION IN EMBRYONIC CHICK SKELETAL MUSCLE CELLS IN VITRO REVEALED BY THE DETERGENT-EXTRACTION, FREEZE-DRY METHOD[1]

Yutaka Shimada and Yuji Isobe

Department of Anatomy, School of Medicine
Chiba University, Chiba 280, Japan

ABSTRACT Spatial cytoskeletal organization in embryonic chick skeletal muscle cells in vitro was studied by the Triton-extraction, freeze-dry, and rotary-shadowing method. After the detergent-extraction and freeze-dry (without cryofracture), a dense and highly interconnected cytoskeletal network was exposed in young myoblasts at 6 hr and myotubes at 6 days in vitro. The most predominantly found components in this network were identified as actin filaments by their ability to be decorated with S-1. Fusion competent myoblasts at 24-48 hr in vitro and young myotubes shortly after fusion did not possess such a subsarcolemmal network, but exhibited a loose network in which actin filaments were found as small bundles. In cryofractured myotubes at advanced stages, the cytoskeletal network could clearly be compartmentalized into the dense subsarcolemmal and loose endoplasmic regions. While actin filaments were abundant in the subsarcolemmal region, nearly all the filaments in the endoplasmic intermyofibrillar area were intermediate filaments. In the latter domain, nascent myofibrils were intimately associated with intermediate filaments at various levels of myofibrils. Some of these intermediate filaments were anchored to the subsarcolemmal actin filament network. The possible roles of the cytoskeletons observed in developing skeletal muscle cells are discussed.

[1]This work was supported by grants from Japanese Ministry of Education, Science and Culture; National Center for Nervous, Mental and Muscular Disorders and Cardiovascular Diseases of Japanese Ministry of Health and Welfare; Muscular Dystrophy Association of America; and Kazato Foundation.

INTRODUCTION

The characteristic cross-striated appearance of myofibrils in skeletal and cardiac muscle cells arises as a result of the regular assembly of contractile and regulatory proteins. Major parts of these muscle proteins are now recognized as the most commonly encountered elements in nonmuscle cytoskeletons and, further, both the skeletal and cardiac muscle cells at their early developmental stages also have cytoskeletal components similar to those of nonmuscle cells. Thus, a detailed analysis of the spatial organization of cytoskeletal filaments during muscle cell development appears to be essential for a better understanding of the phylogeny of the cellular motile systems.

Recent studies on cultured muscle cells by whole mount electron microscopy have revealed the interactions between cytoskeletal components and their associations with developing myofibrils (1, 2). In addition, detergent extraction has been applied to muscle cells with the effect of producing simplified cytoskeletons that retain many structural characteristics of filaments (3-5). Although such studies have revealed new details about the cytoskeletal organization in muscle cells, these methods have the disadvantage that whole mount transmission electron microscopy, whether by high voltage or standard instruments, and whether cells were extracted or not, provides images in which many structures are superimposed, making interpretations difficult. Scanning electron microscopy of detergent-extracted samples has also been used to observe cytoskeletons in muscle cells in vitro (3, 4). Although this procedure produces three-dimensional images, its resolution is insufficient to reveal individual filamentous structures.

A different approach for studying the spatial organization of cytoskeletons has been introduced by Heuser and Kirschner (6), who examined platinum replicas of detergent-extracted and freeze-dried cells. This procedure gives images of outstanding quality and high information value because it provides excellent three-dimensional resolution at the supramolecular level. In the present study, using the platinum replica method combined with the freeze-drying or the freeze-fracturing and dry procedure of detergent-extracted samples, we have examined the organization of cytoskeletons in embryonic chick skeletal muscle cells in vitro.

MATERIALS AND METHODS

Primary embryonic skeletal muscle cell cultures were made from 11-day old chick embryo pectoral muscles. The cells were plated in 35-mm plastic culture dishes, each containing several fragments of carbon-coated glass coverslips on the bottom. The cultures were lysed with 0.5% Triton X-100 (Triton) in KMP buffer (50 mM KCl, 5 mM $MgCl_2$, 2 mM EGTA and 10 mM $K\text{-}PO_4$ buffer [pH 7.5]) and then fixed with 2% glutaraldehyde in the same buffer. For myosin subfragment-1 (S-1) decoration of actin-containing structures, Triton-extracted cells were treated with S-1 at a concentration of 1 mg/ml in KMP buffer before the fixation step. (S-1 was provided by Dr. N. Shimizu.)

The procedures for preparing replicas were as follows: After experimental treatment and fixation, the cultures grown on coverslips were immersed in 70% ethanol in water as a vola tile cryoprotectant. On these coverslips, other coverslips without cells were placed. The entire "sandwiches" were frozen by plunging into Freon 22 cooled by liquid nitrogen and then were peeled off under liquid nitrogen. Some samples were similarly frozen in Freon 22 without making "sandwiches". The cryofractured as well as non-fractured cells on coverslips were transferred into the freeze-fracture device, and warmed to -95°C in order to sublime away all the ice from the surface of the cover glasses. Replicas were made by rotary shadowing with platinum at a shadowing angle of 20-60° and with carbon at 60-90°. Electron microscope negatives of the replicas were photographically reversed and printed as negative images.

For thin sectioning, cells grown directly on the portions of plastic dishes were postfixed in 1% OsO_4, dehydrated in ethanol and embedded in Epon 812.

RESULTS

Developmental Changes of Overall Cytoskeletal Organization in Myogenic Cells

By the freeze-dry (without cryofracture) and rotary-replication method in combination with the detergent-extraction of the cells, we examined the overall organization of their cytoskeletons during development.

Shortly after plating, myogenic cell cultures contained mononucleated myoblasts. A considerable number of myoblasts at 6 hr possessed a dense and highly interconnected filament network appearing like a sheath enveloping the entire cell

(Fig. 1a). In sharp contrast, at 24-48 hr in vitro when many of the cells left the proliferative cycle and prepared for cell fusion, mononucleated cells had a loose cytoskeletal network. Through this network, a nucleus and some filament bundles arranged longitudinally were seen (Fig. 1b).

Slender myotubes were formed by fusion of mononucleated myoblasts. In these young myotubes, cytoskeletal filaments were still sparse and formed a loose network (Fig. 1c). Individual elements of this cytoskeleton exhibited a gently curvilinear and undulating pattern, and were generally oriented parallel to the long axis of the cell. As myotubes matured in vitro, their cytoskeletal network became dense and intricate. That is, the cytoskeletal network of myotubes of a 6 day culture had dense and anastomotic webs of filaments extended just beneath the extracted cell membrane (Fig. 1d). This subsarcolemmal cytoskeletal network resembled that of 6 hr myoblasts in gross morphology. Through this dense network, some bundles of filaments (stress fibers and/or myofibrils) that lay under the network were scarcely visible.

Localization of Actin Filaments in Subsarcolemmal Domain of Myogenic Cells

We used S-1 to decorate actin filaments with periodic arrowheads to allow an easier distinction from the other filament types within the cytoskeletal network.

In mononucleated myoblasts at 6 hr after plating and multinucleated myotubes at 6 days in culture, the subsarcolemmal cytoskeleton was seen to be composed mainly of 8-10 nm filaments. Since, after treatment with S-1, these filaments were thickened to 20-25 nm and looked to be studded along their length with numerous globular structures (Fig. 2a), they were identified as actin filaments. These filaments formed a dense and highly interconnected network together with a few undecorated intermediate filaments, diameters of which ranged

---

FIGURE 1. Rotary-deposited platinum replicas of Triton-extracted, freeze-dried, but not cryofractured myogenic cells at 6 hr (a), 24 hr (b and c) and 6 days (d) in vitro. A 6-hr myoblast (a) and a 6-day myotube (d) possess a dense network extending just under the plasmalemma. A mononucleated myoblast (b) and a young myotube (c), both from 24 hr cultures, exhibit a loose cytoskeletal network. Through this network, bundles of filaments (arrows) and nuclei (N) are discernible. Lipid droplets (L). Bar, 2 µm. X 7,000.

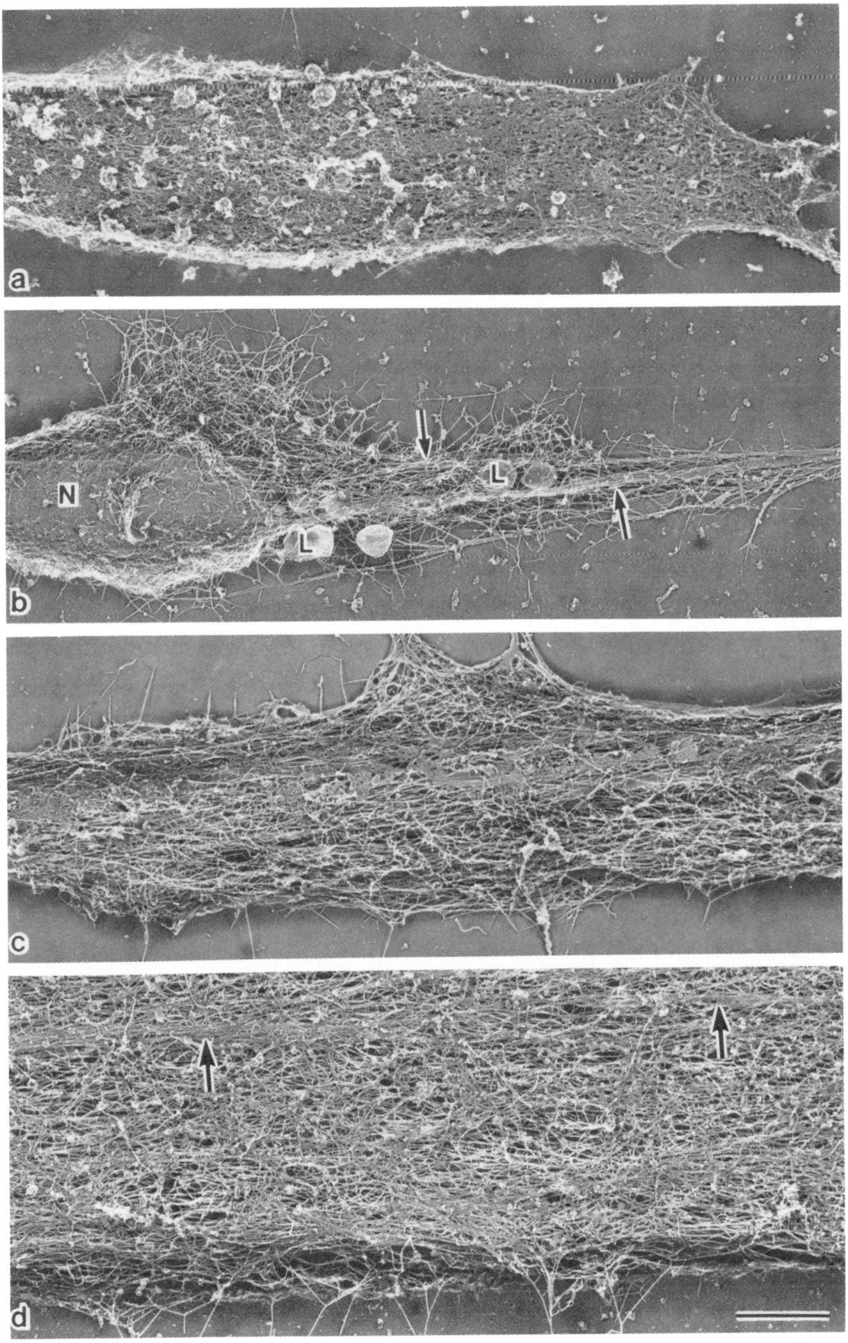
a
b
N
L
L
c
d

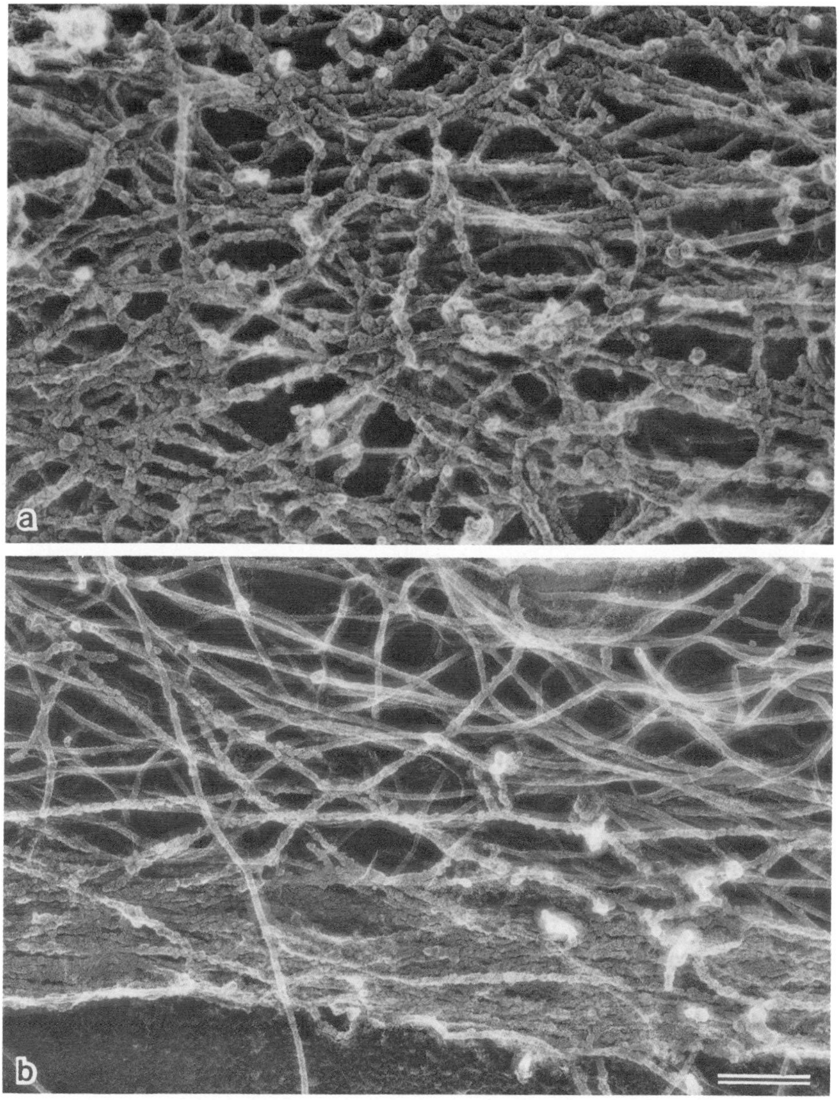

FIGURE 2. Higher magnification views of cytoskeletal networks of cells extracted with Triton and treated with S-1. (a) Under the sarcolemma of a 6-hr myoblast a dense network of filaments is seen. This network consists mainly of S-1 decorated actin filaments. (b) A 24-hr myoblast in the same culture possesses a loose cytoskeletal network, in which actin filaments form bundles. Bar, 0.2 μm. X 70,000.

from 12 to 14 nm in replicas, as well as thinner filaments of 3-5 nm in diameter.

In fusion competent myoblasts seen at 24 hr and young myotubes at 24-48 hr in culture, actin filaments did not form a dense network enveloping the entire cell, but were mainly restricted to bundles which were oriented along the long axis of the cell (Fig. 2b). A few individual or small groups of actin filaments as well as many intermediate filaments were generally aligned along the cell axis. These cytoskeletal elements were associated with each other to form looser and anisotropic networks than seen in younger myoblasts and more mature myotubes.

Cytoskeletal Domains in Myotubes

With Triton-extraction alone, it was difficult to examine the cytoskeletal architectures that were located in the deep endoplasmic area of advanced myotubes, because the replica image includes only those structures that are accessible to the evaporated metal. In order to gain further insight into the spatial distribution of cytoskeletal structures in the deep endoplasmic domain, we exposed such a region by the cryofracturing procedure after the detergent extraction and S-1 treatment.

Fig. 3 illustrates an oblique "cut end" of a myotube at 6 days in culture. The cytoskeleton of myotubes at this stage of development could be compartmentalized into subsarcolemmal and endoplasmic regions by the differences of arrangement and predominant species of the component filaments. As seen in detergent-extracted cells without cryofracturing (Fig. 2a), many S-1 decorated actin filaments were seen to form a dense network in the subsarcolemmal area. Only a few actin filaments were detectable in a relatively loose network of the endoplasmic domain. Most of the filaments in the latter region were intermediate filaments. Microtubules with a diameter of 22-24 nm were easily distinguished from these filaments. They were relatively sparse and ran a rather straight course through the endoplasmic webs of intermediate filaments. In addition, thick bundles of filaments were seen to run along the longitudinal axis of the cells. Since many of the constituent filaments within these bundles were decorated with S-1, they were regarded as nascent myofibrils.

Association between Cytoskeletal Filaments and Myofibrils

In a 6-day old myotube extracted with Triton, cryofractured, but not treated with S-1, developing myofibrils were

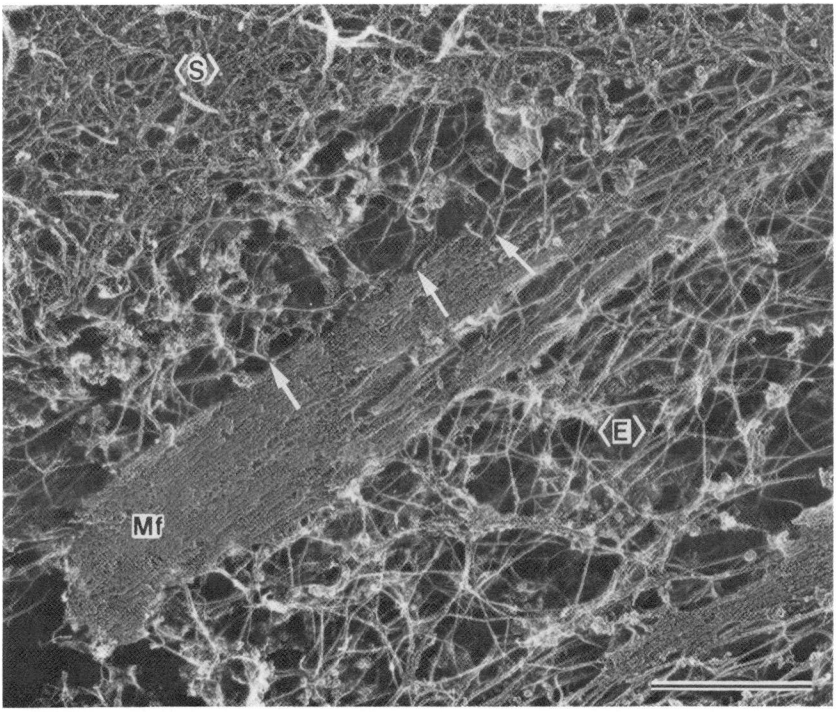

FIGURE 3. Inner view of a cryofractured myotube treated with S-1 after Triton-extraction. The fracture traveled through the endoplasm from the upper cell border, so both the subsarcolemmal (S) and the endoplasmic (E) cytoskeletal domains are seen. While most filaments in the subsarcolemmal region are decorated, very few in the endoplasmic region are decorated except for those that compose myofibrils (Mf). Myofibrils are connected to each other and to subsarcolemmal network by undecorated intermediate filaments (arrows). Bar, 1 μm. X 24,000.

observed as dense bundles composed of thin actin and thicker, knobby myosin filaments within the loose endoplasmic network. A, I and Z bands were discernible but presented a jagged or irregular appearance. The surface of these nascent myofibrils appeared to be intimately associated with the endoplasmic intermediate filaments at various levels (Fig. 4).

Occasionally, near the irregularly shaped Z bands, many intermediate filaments were seen to connect myofibrils at var-

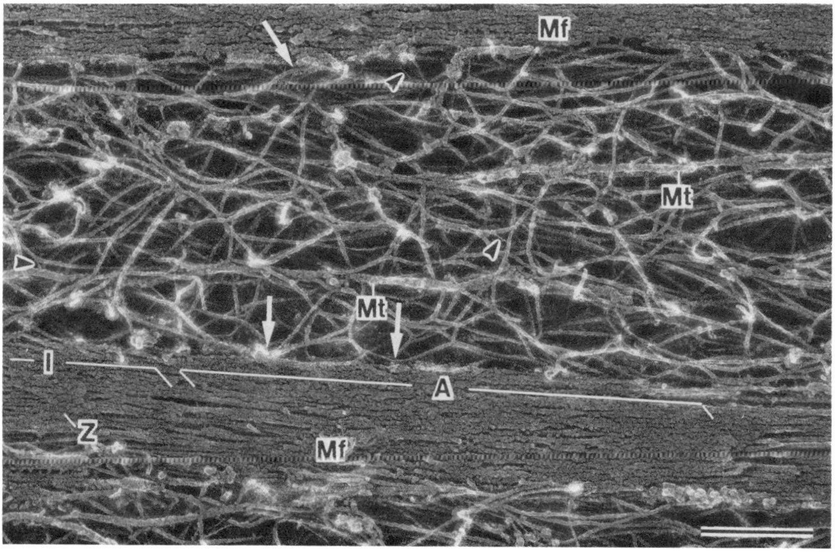

FIGURE 4. Survey view of the endoplasmic cytoskeletal region of a cryofractured myotube. The fracture traveled through this myotube almost parallel to its long axis. Running between myofibrils (Mf) and making a complex but loose network are nearly all intermediate filaments with a diameter of 12-14 nm. The surface of myofibrils is intimately associated with intermediate filaments (arrows). Also visible in this network are thicker and straighter structures thought to be microtubules (Mt). These filaments and tubules are linked to each other with short 3-5 nm filaments (arrowheads). A bands (A); I bands (I); Z bands (Z). Bar, 0.5 μm. X 38,000.

ious angles, as if these filaments radiated out from these bands (Fig. 5a). These radiating filaments were in turn linked to adjacent myofibrils at or near their Z bands. The connections of the myofibrils to the intermediate filaments were either direct at the lateral surfaces or the ends of the filaments, or by 3-5 nm filaments (Figs. 4 and 5a).

A thin section image of such interfibrillar bridges was presented in Fig. 5b. While some intermediate filaments were observed to interlink neighboring myofibrils transversely or obliquely at or near the Z band level, other filaments ran rather longitudinally in the intermyofibrillar spaces. Some of these longitudinal filaments were seen to connect myofibrils at various levels.

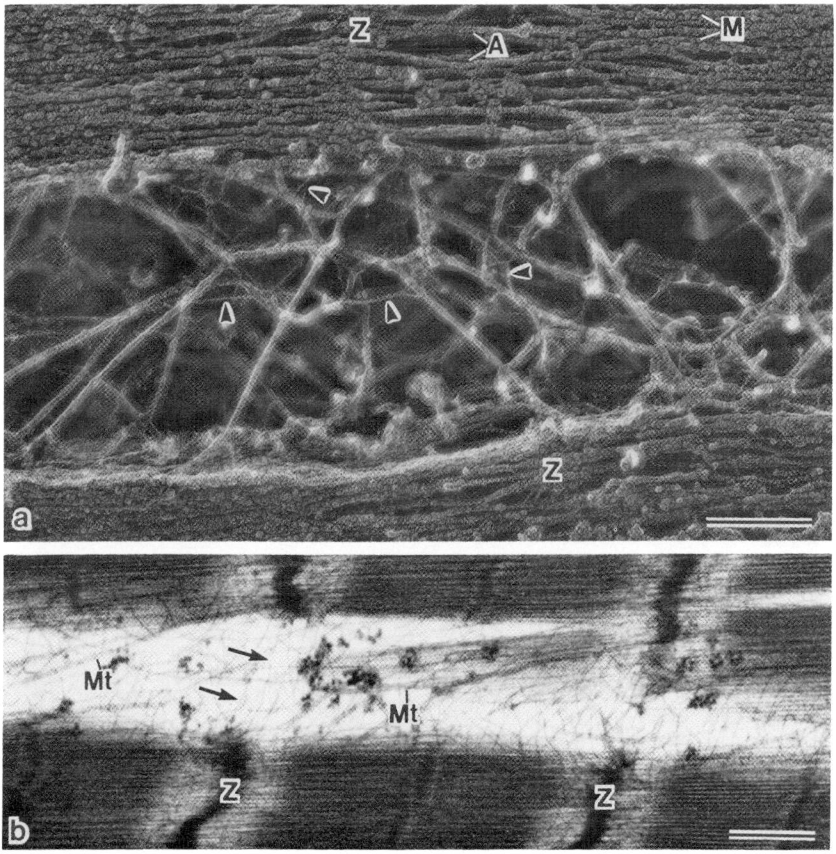

FIGURE 5a. Close-up view of myofibrils and the endoplasmic cytoskeleton. Myofibrils are composed of thin actin (A) and thicker knobby myosin (M) filaments. Many intermediate filaments appear to radiate out from Z bands (Z) of myofibrils to connect adjacent myofibrils at and near their Z bands. Note that some 3-5 nm filaments (arrowheads) act as cross-links between intermediate filaments and between those and myofibrils. Bar, 0.2 μm. X 80,000.

FIGURE 5b. Thin section of an extracted myotube. The section was cut along a plane equivalent to the fracture in Fig. 5a to permit a comparison. Note that some intermediate filaments run along myofibrils, while others (arrows) connect myofibrils at and near Z bands (Z). Microtubules (Mt). Bar, 0.5 μm. X 26,000.

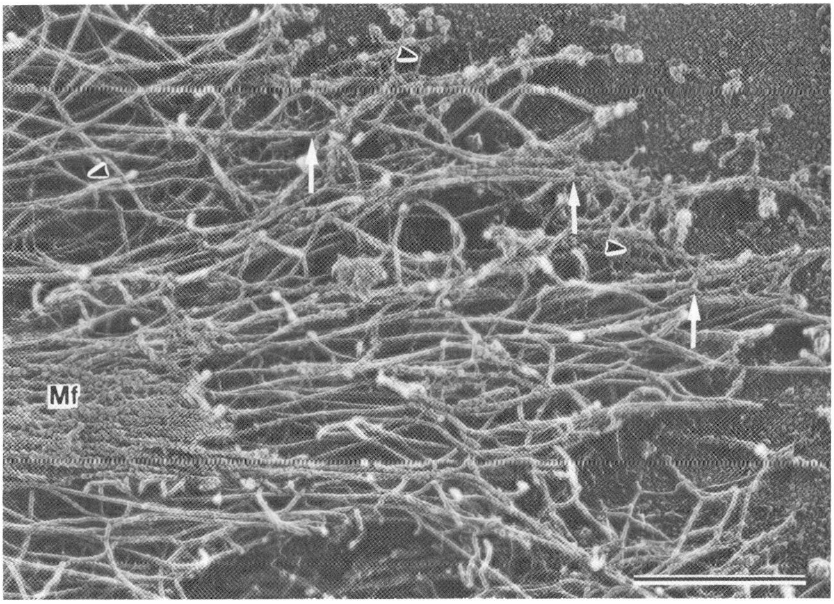

FIGURE 6. High magnification view of the endoplasmic cytoskeleton and a myofibril (Mf) in a Triton-extracted, S-1 decorated and cryofractured myotube. The fracture traveled through the endoplasm of the myotube near the substrate. While nearly all the thin filaments in this myofibril are decorated, most endoplasmic filaments are not. Some of these undecorated filaments, which connect to the surface or the fractured end of the myofibril, are seen to associate with decorated actin filaments just above the substrate (arrows). 3-5 nm filaments (arrowheads) are also observed to cross-connect to actin filaments. Bar, 0.5 μm. X 52,000.

Association of Cytoskeletons between Subsarcolemmal and Endoplasmic Domains

In obliquely fractured myotubes, some intermediate filaments which were connected with myofibrils on one hand appeared to enter into and end at the inner portion of the subsarcolemmal cytoskeletal network on the other (Fig. 3). Such connections between myofibrils and the subsarcolemmal network via intermediate filaments were also clearly discernible in S-1 treated myotubes fractured near the substrate (Fig. 6). From the obliquely fractured myofibril, whose actin filaments were decorated, many undecorated intermediate filaments were seen to fan out and to interlink with decorated actin filaments that were positioned just above the substrate (since the cell membrane had been removed with Triton, the subsarcolemmal cytoskeletal network became to locate just above the substrate). Consequently, myofibrils were linked to the subsarcolemmal cytoskeletons through endoplasmic intermediate filaments.

DISCUSSION

The technique of Triton-extraction, freeze-dry rotary replica, coupled with the use of ethanol as a volatile cryoprotectant (7, 8), has allowed routine investigation of the complex spatial disposition of networks of filamentous proteins in chick skeletal muscle cells developing in vitro. With this high resolving power method, we confirmed our previous work with the scanning electron microscope (9) that the extracted myogenic cells display an extensive network of nonmyofibrillar cytoskeletal components. We further demonstrated that these cytoskeletal networks, especially those beneath the sarcolemma, undergo profound changes in their filament arrangement during development. Three distinct stages of overall cytoskeletal organization are discerned. Myoblasts at 6 hr in vitro (presumably at the proliferative stage) possess a dense, highly interconnected filamentous network which resembles that of fibroblasts and other cell types (6, 10). Myoblasts at 24 hr in culture (fusion competent cells) and myotubes shortly after fusion have sparse cytoskeletal networks with small filament bundles aligned along the long axis of cells. At later stages of development, a dense interconnect filament network reforms.

Observing myogenic cells treated gently with detergent by the whole mount method, Fulton et al. (4) have previously reported that the filament networks in these cells undergo two significant alternations during development, i.e., at the times when cells are in preparation for, and at, fusion. The first change of the overall filament patterns found between the proliferative and postproliferative myoblasts shown in their work appears to correspond well with that presented here in myoblasts between 6 and 24 hr in culture. They suggest the occurrence of the second change at the time of cell fusion, since fusion competent myoblasts and young myotubes immediately after fusion have different patterns in their filament arrangement. Our study showed, however, that this distinct change takes place during the maturing process of myotubes. This difference between their and our observations as to the time when the latter change occurs is likely due to the different extraction conditions, electron microscopical methods used, and areas of myogenic cells observed. That is, their extraction is very mild and there is a possibility that some portions of the plasma membranes are left behind; their whole mount method provides images of superimposition of many structures including the subsarcolemmal and endoplasmic cytoskeletal networks, while our platinum replication permits us to examine the overall architecture of the entire surface cytoskeletons; further, they had to observe the flattened cell tips for the whole mount transmission work, while we studied the cell bellies where fusion appears to occur most frequently (11) and which appear to represent the most uniformly differentiated area whithin a given cell.

By using the S-1 decoration technique, it becomes clear that the distibution and arragement of actin filaments vary according to the stages of the development. While these filaments are found as a major component element of the dense and anastomotic webs at the subsarcolemmal domain of 6 hr myoblasts and 6 day myotubes, they are restricted to the bundles in the loose cytoskeletal networks of fusion competent myoblasts and young myotubes shortly after fusion, both seen in 24-48 hr cultures. It has not been clarified how the actin filaments in the network form seen in 6-hr myoblasts are reorganized into the bundle form in 24-hr myoblasts. It is probable, however, that actin in the network form is first depolymerized into monomers or oligomers, and then they are polymerized some bundle forms, because it has been reported that the protein content of the Triton-insoluble cytoskeleton relative to total cytoplasmic protein is lowest at this time (4).

The cryofracturing procedure allows a clear demonstration of the difference in the cytoskeletal architecture between the subsarcolemmal and endoplasmic regions within a single myotube at advanced stages. By the treatment with S-1 during the sample preparation, it is clear that this regional difference of cytoskeletal organization is due to the disparity of the filament components and density. In the periphery of myotubes close to the sarcolemma, cytoskeletons are condensed and arranged randomly. Actin filaments are the most predominantly found component in this domain. On the other hand, in more central areas, cytoskeletons are sparse and many of them are aligned along the longitudinal axis of the cells. In the latter domain, intermediate filaments are enriched and less actin filaments are found.

The association of intermediate filaments with myofibrils in adult and embryonic muscles has been reported previously in conventional thin section studies (12-14) and in a whole mount study with a high voltage transmission electron microscope (1). These studies have shown that intermediate filaments are connected with myofibrils at the Z band level. Using a high resolution three-dimensional technique, the present study with embryonic skeletal muscle cells and our previous study with cardiac myocytes (15) have revealed the intimate association of intermediate filaments with myofibrils not only at Z bands but also at various other levels of myofibrils. It appears that, as muscle cells mature and/or the irregularity of the Z band structure decreases with increasing differentiation, intermediate filaments converge on Z bands to connect neighboring myofibrils. Our data also indicate that some intermediate filaments which link myofibrils are associated also with actin filaments in the subsarcolemmal cytoskeletal networks. These results suggest that, during muscle development, intermediate filaments might act as an anchoring apparatus of the myofibrils to the cell membrane.

Recently, Wang and Ramirez-Mitchell (16) demonstrated, in KI-extracted adult muscle, the existence of an extensive network of longitudinal intrafibrillar bridges of intermediate filaments that connected adjacent sarcomeric structures of the same myofibrils and transverse interfibrillar bridges that linked neighboring myofibrils. They suggest that these intermediate filament bridges would be capable of transmitting tension in both directions, to myotendinous junctions and to the sarcolemma. Thus, intermediate filaments associated intimately with developing myofibrils observed in the present study may also represent an early stage of such interconnections. It is possible that as myofibrils increase in diameter, these filaments come to be incorporated within them.

REFERENCES

1. Peng HB, Wolosewick JJ, Cheng P-C (1980). The development of myofibrils in cultured muscle cells: a whole-mount and thin-section electron microscopic study. Dev Biol 88: 121.
2. Porter KR, Anderson K (1980). The morphogenesis of myofibrils from the microtrabecular lattice as observed in cultured myoblasts. In Ebashi S, Maruyama K, Endo M (eds): "Muscle Contraction: Its Regulatory Mechanisms." Tokyo: Japan Sci Soc Press/Berlin: Springer-Verlag, p 527.
3. Pudney J, Singer RH (1979). Electron microscopic visualization of the filamentous reticulum in whole cultured presumptive chick myoblasts. Am J Anat 156: 321.

4. Fulton AB, Prives J, Farmer SR, Penman S (1981). Developmental reorganization of the skeletal framework and its surface lamina in fusing muscle cells. J Cell Biol 91: 103.
5. Ip W, Danto SI, Fischman DA (1983). Detection of desmin-containing intermediate filaments in cultured muscle and nonmuscle cells by immunoelectron microscopy. J Cell Biol 96: 401.
6. Heuser JE, Kirschner MW (1980). Filament organization revealed in platinum replicas of freeze-dried cytoskeletons. J Cell Biol 86: 212.
7. Schiller A, Taugner R (1980). Freeze-fracturing and deep-etching with the volatile cryoprotectant ethanol reveals true membrane surfaces of kidney structures. Cell Tissue Res 210: 57.
8. Sawada H, Yamada E (1981). A freeze-fracture deep-etching replica method with volatile cryoprotectant. J Electron Microsc 30: 341.
9. Isobe Y, Shimada Y (1983). Myofibrillogenesis in vitro as seen with the scanning electron microscope. Cell Tissue Res 231: 481.
10. Bell PBJr (1981). The application of scanning electron microscopy to the study of the cytoskeleton of cells in culture. Scanning Electron Microscopy/1981 II: 139.
11. Shimada Y (1971). Electron microscope observations on the fusion of chick myoblasts in vitro. J Cell Biol 48: 128.
12. Ishikawa H (1983). Fine structure of skeletal muscle. In Dowben RM, Shay JW (eds): "Cell and Muscle Motility." New York: Plenum Press, Vol 4, p 1.
13. Price MG, Sanger JW (1983). Intermediate filaments in striated muscle: a review of structural studies in embryonic and adult skeletal and cardiac muscle. In Dowben RM, Shay JW (eds): "Cell and Muscle Motility." New York: Plenum Press, Vol 3, p 1.
14. Price MG, Sanger JW (1984). Reduction of density and anisotropic distibution of intermediate filaments occur during avian skeletal myogenesis. Am J Anat 171: 427.
15. Isobe Y, Shimada Y, Manasek FJ (1984). Cytoskeletons and myofibrils in cardiac myocytes in vitro observed by scanning and transmission electron microscopy. In Nora JJ, Takao A (eds): "Congenital Heart Disease: Causes and Processes." New York: Futura, p 145.
16. Wang K, Ramirez-Mitchell R (1983). A network of transverse and longitudinal intermediate filaments is associated with sarcomeres of adult vertebrate skeletal muscle. J Cell Biol 96: 562.

Molecular Biology of Muscle Development, pages 741–748

INTERMEDIATE FILAMENTS IN DEVELOPING SKELETAL MYOTUBES IN VIVO[1]

K.T. Tokuyasu, P.A. Maher,
A.H. Dutton and S.J. Singer

Department of Biology,
University of California, San Diego,
La Jolla, CA 92093

ABSTRACT We studied the organization and composition of intermediate filaments (IF) in developing chick skeletal muscle myotubes *in vivo* using immunofluorescent and immunoelectron microscopy. With polyclonal anti-vimentin and anti-desmin antibodies cross-absorbed with the heterologous antigen, we showed that vimentin is gradually reduced in concentration as the embryonic myotube matures *in vivo* and becomes undetectable at the time of hatching. We also showed that vimentin and desmin coexist in the same IF in developing myotubes. In developing myotubes of the 11 day chick embryo, IF are wavy and, as a whole, longitudinally directed. As the myotubes mature, the myofibrils come into lateral registry while desmin and vimentin are still widely distributed around the Z line. Desmin is eventually confined to the Z line in the postnatal chick. The labeling of desmin and vimentin occurs exclusively in filaments throughout all stages of development. These findings indicate that IF change their direction rather than their form during myotube development and that the confinement of IF to the Z line occurs subsequent to the lateral registration of myofibrils *in vivo*.

[1]This work was supported by Grants HL-30282, GM-15971 and CA-38006-01 from the U.S. National Institutes of Health.

INTRODUCTION

We began our studies on intermediate filament (IF) proteins in skeletal muscle in an attempt to answer three specific questions. 1. What is the protein composition of IF in skeletal myotubes during myogenesis in vivo? Two earlier in vitro studies (1,2) had shown that both desmin and vimentin were present in developing myotubes in culture. In one study (1) it was found that vimentin remained throughout myogenesis whereas in the other study (2) it was seen to disappear as the myotubes matured. 2. What is the temporal relationship between the appearance of cross-striations in myotubes and the confinement of IF proteins to the Z line in vivo? It had been proposed (3), previously, that IF were responsible for the development of lateral registry of the myofibrils in the myotubes. 3. How do the IF proteins reorganize during myogenesis in vivo? The aforementioned in vitro studies (1,2) had shown that IF fill the embryonic myotubes in a longitudinal pattern. However, it was also known that in adult skeletal muscle IF proteins were confined to the level of the Z line (1-4). In addition, we had shown that the IF proteins were still in a filamentous form in adult skeletal muscle (4).

We employed affinity purified, cross-absorbed polyclonal antibodies to chicken erythrocyte vimentin and chicken gizzard desmin in combination with immunofluorescent and immunoelectron microscopic analysis of frozen sections of 11 day, 17 day and 21 day embryonic chicken thigh muscle to answer these questions. The results are described herein and have been reported elsewhere (5,6).

MATERIALS AND METHODS

All techniques and reagents used in these studies have been described elsewhere (5,6).

RESULTS

Antibody Preparation.

In order to study the organization of IF proteins during myogenesis in vivo we needed antibodies to the IF proteins that were directed solely to the antigen of interest. Because of the 65% sequence homology between chicken desmin and vimentin (7) polyclonal antibodies directed against a pure preparation of one of the proteins have a high probability of exhibiting some cross-reaction. This is exactly what we found. For our source of vimentin we used chicken erythrocytes (8). Polyclonal antibodies generated to electrophoretically-purified vimentin were isolated by affinity chromatography and used for immunofluorescent staining. The antibody labeled not only lysed chicken erythrocytes but also the Z line of adult iliotibialis muscle and gizzard smooth muscle cells. When this antibody was used in immunoblotting experiments with extracts from chicken erythrocytes, adult iliotibialis muscle and adult gizzard smooth muscle a positive reaction was observed with all three tissues. However, after this antibody to vimentin was passed over a desmin affinity column, the column run-through still immunoblotted the chicken erythrocyte extract but no longer showed any reaction with either the skeletal muscle or gizzard extracts. In addition, the cross-absorbed anti-vimentin did not stain frozen sections of either adult skeletal muscle or gizzard.

A similar problem was encountered with our affinity-purified desmin antibody produced against chicken gizzard desmin. When this antibody was used in immunoblotting experiments with chicken erythrocyte, gizzard and adult skeletal muscle extracts a positive reaction was seen with all three samples. Since chicken erythrocytes do not contain desmin (8), it was apparent that the desmin antibody was cross-reacting with vimentin. After passing the desmin antibody over a vimentin affinity column the antibody still reacted strongly with adult skeletal muscle and gizzard in both immunoblotting and immunofluorescence experiments but no longer immunoblotted chicken erythrocytes.

11 Day Myotubes.

The organization of the myofibrils in myotubes of 11 day embryonic thigh muscle could be readily observed in Nomarski images of semi-thin frozen sections. At this stage of development the myofibrils do not fill the myotubes and are generally well separated. Although the individual myofibrils are striated most of them are not in lateral registry and so cross-striations are only infrequently observed. The antibody to vimentin labeled the myotubes, myoblasts, and the non-muscle cells in the intermyotube spaces whereas the anti-desmin reacted only with the myotubes and presumptive post-mitotic myoblasts. Both desmin and vimentin appeared to be present uniformly throughout the cytoplasm of the myotubes even in areas where no myofibrils were present. The immunofluorescent labeling of the myotubes for vimentin was however weaker than that for desmin.

Occasionally, we did observe several myofibrils in lateral registry in the 11 day embryonic thigh muscle. When these myotubes were immunofluorescently labeled for vimentin or desmin a striated pattern of fluorescence was observed. However when micrographs of these sections were examined closely using a myotube nucleus as a point of reference it became apparent that the dark cross-bands corresponded to the in-register Z lines. This is not what one would observe if the IF proteins were beginning to concentrate at the Z lines. In order to clarify what was occuring in these myotubes we examined similar sections by immunoelectron microscopy. In all cases, whenever we saw several myofibrils in lateral registry we found the Z band to be covered by sarcoplasmic reticulum. Immunolabeling for both desmin and vimentin was virtually absent in the areas which were occupied by sarcoplasmic reticulum, but was uniform in adjacent areas. Thus, the cross-striated immunofluorescent pattern seen in some 11 day embryonic myotubes was due to the exclusion of desmin and vimentin from the Z line by sarcoplasmic reticulum rather than the confinement of IF proteins to the Z line.

Immunolabeling of ultrathin frozen sections of 11 day embryonic thigh muscle with anti-desmin and anti-vimentin made it possible to observe the IF at low power in the electron microscope so that relatively large areas of a myotube could be examined. Thus, we were able to confirm that all the IF run longitudinally at this stage of development. These filaments showed no particular relationship with any level of the sarcomeres of the myofibrils, either in orientation or distribution. At higher power the immunolabeling for desmin and vimentin could be clearly seen and was localized exclusively to filaments. The labeling for desmin was fairly uniform along the filaments whereas that for vimentin was distributed in patches. The overall level of anti-vimentin antibody labeling was substantially lower than that of the anti-desmin antibody, however. These results suggested that desmin and vimentin existed in the same IF *in vivo* and this was confirmed by double immunolabeling experiments (Fig. 1 and 2). With the appropriate combinations of primary and secondary antibodies it was apparent that all the IF contained both desmin and vimentin. No extensive lengths of one or the other protein were seen.

FIGURES

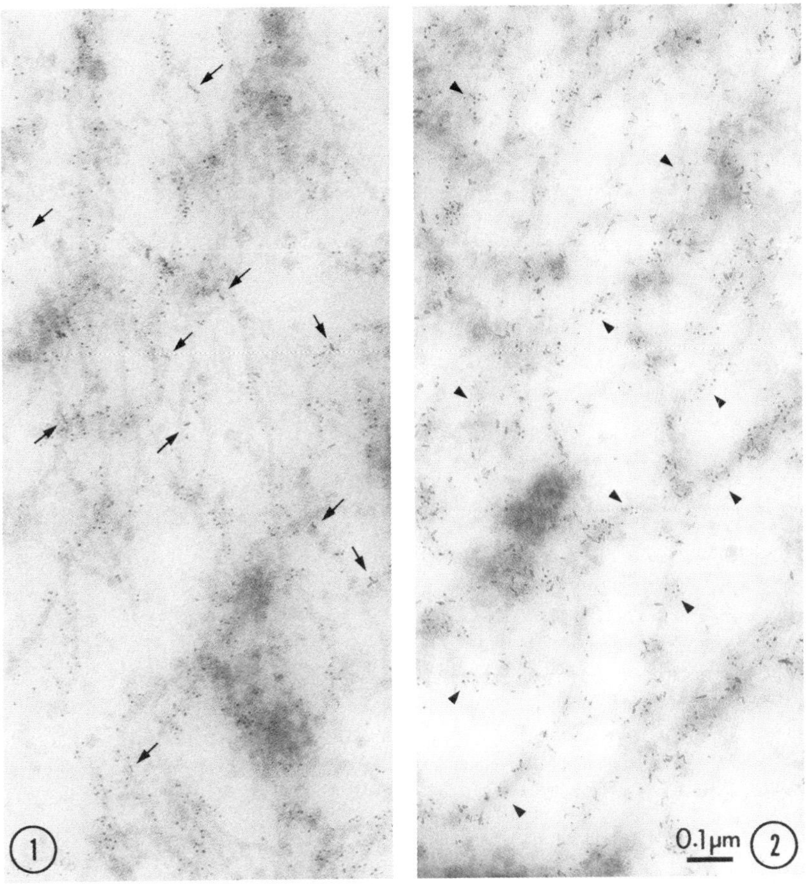

FIGURE 1. A portion of a myotube of an 11 day embryo double immunolabeled for desmin (ferritin) and vimentin (Imposil). All the labeling is localized to filaments. Although vimentin labeling is sparse (arrows), a few individual filaments can be seen that are labeled with

both markers. Bar, 0.1 µm. x 60,000.

FIGURE 2. A portion of a myotube of an 11 day embryo double immunolabeled for desmin (Imposil) and vimentin (ferritin). The exchange of the markers makes the vimentin labeling (arrowheads) relative to the desmin labeling much more abundant than in Fig. 1. Many individual filaments can now clearly be seen to be labeled with both markers in a clustered manner. Bar, 0.1 µm. x 60,000.

---

17 Day Myotubes.

Examination of semi-thin frozen sections of 17 day embryonic muscle by light microscopy showed the myotubes to be packed with myofibrils which were nearly in lateral registry. Immunofluorescent labeling for vimentin was variable among the myotubes and appeared to be strongest in the less well-developed myotubes whereas immunofluorescent labeling for desmin was equally strong in all the myotubes. Within the individual myotubes the fluorescence was less uniform than in the 11 day embryo and appeared as alternating bands of dark and light. When micrographs of sections were enlarged and examined with respect to a reference point it became obvious that the mid-levels of the light bands corresponded to the Z lines. Studies in the electron microscope confirmed and extended these immunofluorescence results. As compared with the 11 day embryonic muscle, the intermyofibrillar spaces were greatly reduced and thus IF were difficult to discern in the frozen sections. However, in tangential plastic sections of 17 day thigh muscle IF were readily visible in the intermyotube spaces and appeared to be more concentrated at the level of the Z band. These same areas could be shown to contain linear arrays of ferritin particles representing desmin labeling in ultrathin frozen sections indicating that desmin was still confined to IF. Labeling for vimentin was very low but in double labeling experiments with desmin the two proteins appeared to be confined to the same IF.

21 Day Myotubes.

Light microscopic examination of semi-thin frozen sections of 21 day embryonic thigh muscle showed the myotubes to be completely filled with myofibrils that were as precisely aligned as in adult skeletal muscle. When sections were immunolabeled for vimentin very little fluorescence was seen. Antibodies to desmin still reacted strongly with the myotubes and the fluorescence was more or less confined to the Z line. The degree of this confinement did not correlate with the degree of lateral registry of the myofibrils. Thus, even myotubes where the fluorescent labeling for desmin was much wider than the Z line and fluorescent filaments spanning the sarcomeres were still apparent, had precisely aligned myofibrils. Examination of anti-desmin labeled ultrathin frozen sections still revealed linear arrays of ferritin particles which represented the desmin antibody. The confinement of this labeling to the Z band was variable although in no case was it as spread out as in the 17 day myotubes.

DISCUSSION

We have examined the organization and distribution of the IF proteins desmin and vimentin in 11 day, 17 day and 21 day embryonic chicken thigh muscle and correlated these results with the myofibrillar organization at each of these stages of development. In 11 day myotubes the myofibrils were single, sparse and individually striated but not in lateral registry while the IF were longitudinally arrayed and contained both desmin and vimentin. By 17 days of development the myotubes were densely packed with myofibrils which showed substantial lateral registry. Although the IF were still longitudinally arrayed they were more concentrated at the Z line. These filaments contained large amounts of desmin and a small amount of vimentin. The myofibrils in 21 day myotubes showed complete lateral registry whereas the IF were still not completely confined to the Z line. These filaments appeared to contain desmin almost exclusively. The conclusions to these studies can be summarized as follows. 1. Desmin and vimentin are in the same IF in developing skeletal muscle myotubes. 2. During the course of myogenesis in skeletal muscle myotubes vimentin is gradually reduced in concentration and becomes undetectable at the time of hatching. 3. Desmin and vimentin remain in a filamentous form throughout skeletal muscle myotube development but change their arrangement from longitudinal throughout the myotube to transverse at the level of the Z line. 4. The confinement of IF proteins to the Z line is subsequent to, not simultaneous with or prior to, the lateral registration of myofibrils in skeletal muscle myotubes *in vivo*.

ACKNOWLEDGMENTS

We gratefully acknowledge the excellent technical assistance of Mr. J. Michael McCaffery and Mrs. Margie Adams.

REFERENCES

1. Gard DL, Lazarides E (1980). The synthesis and distribution of desmin and vimentin during myogenesis *in vitro*. Cell 19:263.
2. Bennett B, Fellini SA, Toyama Y, Holtzer H (1979). Redistribution of intermediate filament subunits during skeletal myogenesis and maturation *in vitro*. J Cell Biol 82:577.
3. Lazarides E, Granger BL, Gard DL, O'Connor CM, Breckler J, Price M, Danto SI (1982). Desmin- and vimentin-containing filaments and their role in the assembly of the Z disk in muscle cells. Cold Spring Harbor Symp Quant Biol 46:413.
4. Tokuyasu KT, Dutton AH, Singer SJ (1983). Immunoelectron microscopic studies of desmin (skeletin) localization and intermediate filament organization in chicken skeletal muscle. J Cell Biol 96:1727.
5. Tokuyasu KT, Maher PA, Singer SJ (1984). Distributions of vimentin and desmin in developing chick myotubes *in vivo*. I. Immunofluorescence study. J Cell Biol 98:1961.
6. Tokuyasu KT, Maher PA, Singer SJ (1985). Distributions of vimentin and desmin in developing chick myotubes *in vivo*. II. Immunoelectron microscopic study. J Cell Biol 100:1157.
7. Zehner ZE, Paterson BM (1985). The Chicken vimentin gene: Aspects of organization and transcription during myogenesis. Annals NY Acad Sci (in press).
8. Granger BL, Repasky EA, Lazarides E (1982). Synemin and vimentin are components of intermediate filaments in avian erythrocytes. J Cell Biol 92:299.

Molecular Biology of Muscle Development, pages 749–772

THE STRIATED MUSCLE CYTOSKELETON: EXPRESSION AND ASSEMBLY IN DEVELOPMENT

Elias Lazarides and Yassemi G. Capetanaki

Division of Biology, California Institute of Technology
Padadena, CA 91125

ABSTRACT Muscle cells express an extramyofibrillar cytoskeleton, whose molecular composition is, at least in part, remarkably analogous to that of nucleated avian erythrocytes. Two domains of this cytoskeleton, a spectrin based membrane skeleton domain and an intermediate filament domain, have been established as model systems to analyze muscle morphogenesis. Each domain undergoes a characteristic molecular remodelling in postmitotic cells, regulated predominantly at the transcriptional level, and through a series of posttranslationally regulated events assembles into the adult muscle cytoskeletal phenotype following a well-defined temporal order.

INTRODUCTION

The formation of a muscle fiber represents one of the most dramatic examples of macromolecular assembly in eukaryotic cells. This cell is composed of multiple structural units or domains, with the myofibril being the most predominant one, all organized in a characteristic striated appearance. The mechanism by which all these structural domains assemble into a muscle fiber has been the subject of many earlier light and electron microscopic studies, but the molecular details of this process are mostly unresolved. If we are to elucidate the mechanisms that underlie muscle morphogenesis we need to define first the various cytoskeletal domains and their constituent polypeptides, then the mechanism of assembly of each domain and subsequently the mechanism of joining of these domains to each other to bring about the ultimate supramolecular phenotype of the muscle cell. Over the past several years it has become apparent that muscle cells contain several structural domains in addition to myofibrils. Here we summarize the evidence for the existence of two such cytoskeletal domains, one composed of intermediate filaments, and the other composed of polypeptides characteristic of the red blood cell plasma membrane skeleton. These two domains exhibit, both in terms of disposition and molecular composition, a striking homology to those present in nucleated red blood cells. We describe studies on the regulation of their expression, assembly and topogenesis during myogenesis and use them as a model system to illustrate some of the levels of regulation that are operative in the structural morphogenesis of muscle cells.

The Existence of a Cytoskeleton in Striated Muscle and Implications for Its Physiological Function

Light and electron microscopists have long suspected that muscle cells contain a framework transverse to their long axis and responsible for maintaining the striated appearance of these cells during cycles of contraction and relaxation (reviewed in 20, 51). Microscopic studies of contracting muscle cells suggested that the sarcolemma is physically coupled to the underlying myofibrils by inextensible bridges at periodic intervals along the plasma membrane that coincided with the Z lines and occasionally also with the M lines of myofibrils. Invariably these sites of attachment of myofibrils to the plasma membrane appear under the electron microscope as electron dense plaques. A result of this inextensible connection of the sarcolemma with the underlying myofibrils is the "scalloping" or "outpocketing" of the former in the areas overlying the sarcomeres in between Z lines or in between Z lines and M lines (48,50,51). The constituent filaments of these bridges are approximately 100 Å in diameter and belong to a class of cytoplasmic filaments known as intermediate filaments (48). These observations have provided firm morphological grounds that striated muscle cells do indeed contain a structural network disposed at right angles to the long axis of the cells and have given rise to the hypothesis that this network is responsible for mechanically integrating myofibrils laterally to each other and to the plasma membrane as well as to cytoplasmic organelles such as mitochondria, nuclei and glycogen particles (20,29). In this fashion all organelles, the sarcolemma and myofibrils are physically coupled to each other by inextensible filamentous linkages at the area of the Z line and has been further hypothesized that they additionally contribute to the resting tension of muscle fibers (50). Attachment of this network to the membrane occurs at electron dense plaques associated with the sarcolemma, subjacent to the myofibril Z lines. However, the molecular composition of either these intermediate filaments or the plasma-membrane associated attachment sites remained undefined.

Since the late 1960's it was established that intermediate filaments in muscle cells were distinct from actin and myosin filaments (26). Even though their relationship to the filaments found traverse to the long axis of adult muscle cells was not known, electron microscopic studies on their cytoplasmic disposition during myogenesis had provided the first evidence of their involvement in the final stages of myofibril assembly and overall muscle morphogenesis during differentiation. Kelly had suggested that these filaments are probably involved in the lateral alignment of myofibrils and the maintenance of the striated phenotype of muscle cells by virtue of the fact that during myogenesis, well after myofibrils have assembled and sarcomeres have formed, the cytoplasmic distribution of these filaments changes and they become associated gradually with the peripheries of Z discs (28).

The existence of such a myofiber network in all types of muscle cells has been established in molecular terms recently through the biochemical and immunological identification and characterization of intermediate filament proteins in myogenic cells. The potential mechanism of attachment of these filaments to the plasma membrane has been advanced further by the identification of components of the red blood cell membrane skeleton as well as components of adhesion sites of actin filaments to the plasma membrane in areas of the sarcolemma that coincide with the electron dense areas of attachment of Z discs and intermediate filaments to the sarcolemma. Together these studies have established the existence of a myogenic cytoskeleton. From parallel studies with nucleated erythrocytes, in particular chicken erythrocytes, a striking similarity in the molecular composition and structure of the cytoskeleton of muscle and nucleated erythroid cells has emerged, supporting the notion that, from a cytoskeletal point of view, the avian erythrocyte can be viewed as a vestige of muscle cells (30,32). Analysis of the expression and assembly of these polypeptides during muscle differentiation have allowed muscle morphogenesis to be approached from a new and different point of view.

Cytoskeletal Domains in Avian and Mammalian Erythrocytes

The membrane-skeleton. The structure and properties of the red blood cell membrane skeleton in both mammalian and avian red blood cells have been reviewed extensively in recent years and no attempt will be made here to review this literature. Instead we will present a general outline of the molecules involved as background of the studies with muscle cells.

Mammalian and avian red blood cells contain a network of proteins composed of spectrin, actin and several associated polypeptides (for reviews see 6,8,30). This network lines the inner surface of the erythrocyte's plasma membrane and serves as its cytoskeleton, contributing to the shape, integrity, and elasticity of the cell. Principal component of this membrane-skeleton is spectrin, a protein composed of two nonidentical polypeptides termed α-spectrin and β-spectrin. The formation of this skeleton involves the self-association of spectrin predominantly into $(\alpha,\beta)_2$ tetramers and to a certain extent $(\alpha,\beta)_{2n}$ oligomers (34,35) and the crosslinking of actin by the spectrin. The association of this skeleton with the membrane is mediated, at least in part, by the protein ankyrin (variably termed goblin in avian erythrocytes (1,4,41)), which binds both to the β subunit of spectrin and the cytoplasmically oriented NH_2-terminal segment of the transmembrane anion transporter (band 3) (6,8,12,27,35). The resulting molar ratio of the complex between the anion transporter, ankyrin, β-spectrin, and α-spectrin is 1:1:2:2. Protein 4.1 strengthens this network by forming a stable ternary complex with spectrin and actin (for a review see 35). In addition, protein 4.1 may be involved in the anchorage of the spectrin-actin skeleton to the lipid bilayer, since it associates at least with the cytoplasmic domain of glycophorin, a transmemberane sialoglycoprotein (2) and can potentially also bind to the cytoplasmic domain of the anion transporter close to the binding site for ankyrin

(35). Mammalian and avian erythrocytes also contain a protein structurally analogous to skeletal muscle tropomyosin. It is present in this cell in approximately the same ratio to actin as in skeletal muscle (7 actins:1 tropomyosin), suggesting a structural role (14,18).

The intermediate filament skeleton. During mammalian erythroid development, erythroblasts cease synthesizing the intermediate filament subunit vimentin and intermediate filaments are gradually diluted out by subsequent cell proliferation and turnover prior to enucleation (44,45). Thus adult mammalian red blood cells do not contain intermediate filaments. However, unlike mammalian erythrocytes, the nucleated avian erythrocytes possess, in addition to the spectrin-based membrane skeleton, intermediate filaments which form a three-dimensional network in these cells, interlinking the nucleus to the membrane-skeleton on opposite sides of the cell (23,24,54,55). Avian erythrocyte intermediate filaments are composed of two polypeptides: vimentin (Mr 52,000) and a high molecular weight protein (Mr 230,000) known as synemin (22). Vimentin functions as the main core subunit of the filaments, while synemin is associated along the vimentin filament backbone with a characteristic periodicity and may function to crosslink them. Immunoelectron microscopic studies have suggested that synemins of adjacent filaments can self-associate and thus mediate lateral crosslinking of the filaments (23). Depending on the stage of development and maturation of the chicken, red blood cells may also express the 70,000 dalton subunit of neurofilaments (NF70; 29) which copolymerizes in an apparently random fashion with vimentin within single filaments (25a).

A summary of the known protein constituents of the spectrin network and the intermediate filament network in adult avian erythrocytes is given in Table 1. A hypothetical diagram depicting the interrelationship of the spectrin network and the intermediate filament network in avian erythrocytes is shown in Figure 1. The two interconnected cytoskeletal domains link specific transmembrane polypeptides and the nucleus thereby proving an inextensible linkage between the two. This system may function to mechanically integrate the cell and contribute to its inflexibility and its overall rigidity and resistance to gross shape changes. This is unlike enucleate mammalian red blood cells which lack

CHICKEN RED BLOOD CELL CYTOSKELETON

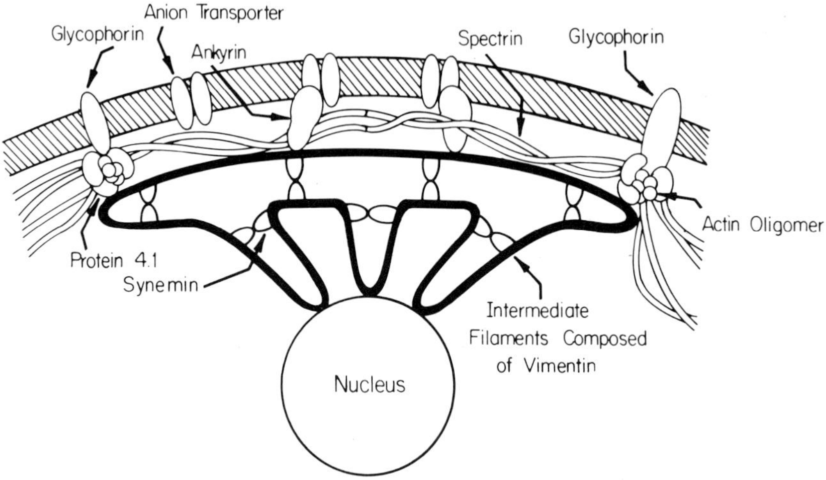

Fig. 1. A highly schematic arrangement for the major polypeptides of the chicken erythrocyte cytoskeleton. The position of the marginal band of microtubules is not shown (see 23).

TABLE I. Cytoskeletal domains in the avian erythrocyte.

Membrane Skeleton	Intermediate Filament Skeleton
α-spectrin (Mr 240,000)	Vimentin (Mr 52,000)
β-spectrin (Mr 220,000)	Synemin (Mr 230,000)
Ankyrin (Goblin; Mr 260,000)	Neurofilament
Actin (Mr 42,000)	Subunit NF70 (Mr 70,000)
Protein 4.1 (family of polypeptides)	
Tropomyosin (family of polypeptides)	
Anion Transporter	
Glycophorin*	

*Not yet identified in avian erythrocytes

intermediate filaments and which are substandially more deformable.

Cytoskeletal Domains in Striated Muscle Cells

Analogous studies to those carried out with avian erythrocytes have established that muscle cells contain an extramyofibrillar cytoskeleton whose molecular composition is strikingly similar to that of avian erythrocytes. At least two distinct cytoskeletal domains have been shown to constitute the myogenic cytoskeleton (Figure 2). The first is a transcytoplasmic domain composed predominantly of the muscle specific intermediate filament subunit desmin, lesser amounts of vimentin depending on the type of muscle, the filament crosslinking protein synemin and small amounts of spectrin and actin which form a scaffold around the periphery of individual Z discs (20-22,29). It has been proposed that this Z-disc cytoskeletal scaffold mechanically integrates and unifies the contractile actions of the myofiber by physically linking together myofibrils laterally at their Z discs (29). Substantial immunoelectron microscopic evidence indicates that these proteins are in the form of intermediate filaments at these sites (53). The second cytoskeleton domain is composed of the erythroid form of spectrin (αβ-spectrin) and ankyrin (goblin) and is located predominantly on the cytoplasmic surface of the sarcolemma. In relaxed skeletal muscle, the distribution of spectrin and ankyrin has been shown to be as an intricate gridlike network of rings surrounding the myofiber, which are localized on the sarcolemma directly adjacent to the Z lines and, to a lesser extent, the M lines. These rings are interconnected by regularly spaced lines of spectrin that extend along the sarcolemma in the long axis of the myofiber (Figure 2) (38,41,42,49). In contracted muscle, the network of spectrin is still apparent, although the interval between the rings adjacent to the Z lines appears to decrease (38,49).

The distribution of spectrin on the sarcolemma, together with the fact that in erythrocytes it binds actin filaments and can become extensively cross-linked indicates that in muscle cells spectrin plays an important role in the maintenance of the shape and structural integrity of the sarcolemma during contraction and relaxation. We have hypothesized that intermediate filaments emanating from the Z lines are bound to spectrin, which, through its attachment to ankyrin and to transmembrane proteins in discrete areas of the sarcolemma

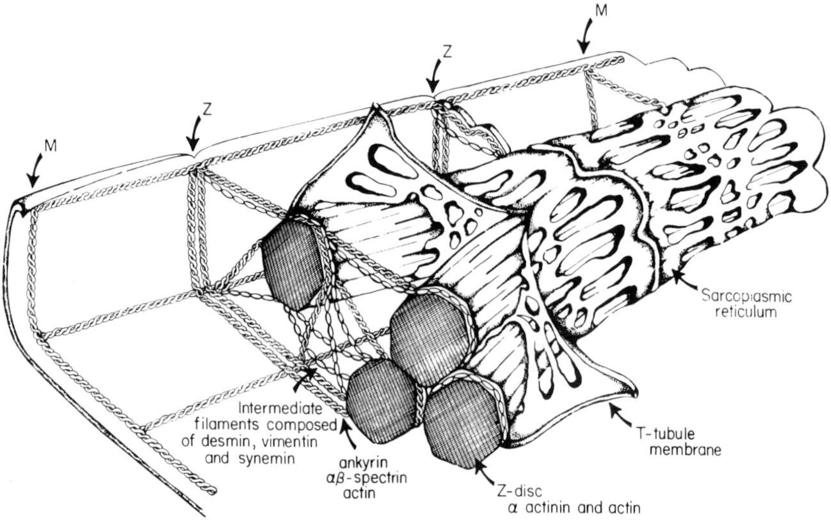

FIG. 2. Schematic representation of a skeletal myofiber with the interior exposed to show the distribution of the intermediate-filament proteins and the membrane skeleton proteins in relationship to the Z discs and the sarcolemma.

adjacent to the Z lines, results in the establishment of a topologically invariant and inextensible anchorage site for the transcytoplasmic Z-disc cytoskeletal scaffold on the sarcolemma (42). As the myofiber shortens during muscle contraction and the interval between Z lines decreases, areas of the sarcolemma that contain little spectrin pucker to form outfoldings, whereas spectrin-containing foci directly adjacent to the Z lines remain attached to the myofiber (Figure 3). Thus spectrin may be involved in the control of membrane deformability and in establishing a molecular continuity between the myofiber at the level of the Z line and the surrounding sarcolemma (42). These studies have provided a basis for understanding the molecular mechanism involved in the mechanical integration of the contractile actions of the muscle fiber and a new approach to elucidate the mechanism of tension development and membrane deformability in

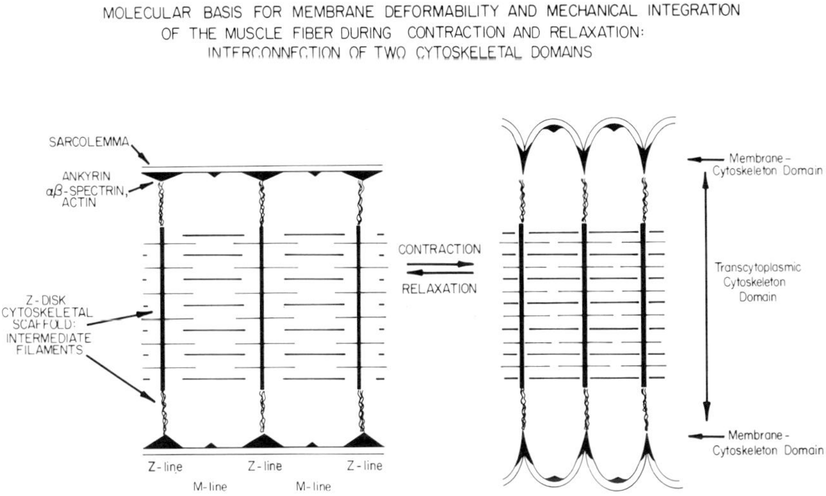

FIG. 3. Molecular basis for membrane deformability and mechanical integration of the muscle fiber during contraction and relaxation (for details, see text).

contracting muscle. Furthermore, they have provided a firm molecular basis for the century-old light microscopic observations that the sarcolemma is coupled to the underlying myofibrils at the level of the Z line.

Recent studies by Craig and collaborators have indicated that vinculin, a protein that associates specifically with the adhesion sites of actin filaments to the plasma membrane in non-muscle cells (17,52), is also expressed in muscle cells and has a distribution at the sarcolemma analogous to that of spectrin and ankyrin (46,47). It appears therefore that multiple actin-binding proteins may mediate the attachment of actin and the whole cytoskeletal framework of myofibrils with the sarcolemma. Vinculin is also a component of the fascia adherens areas of intercalated disks in cardiac muscle (52) and may thus as well mediate the attachment of the ultimate sarcomeric actin filaments to the plasma membrane. The vinculin, spectrin and ankyrin containing sites on the sarcolemma have been referred to as costameres (47) and coincide with the electron dense adhesion sites of Z-discs and intermediate filaments with the sarcolemma. The existence of this cytoskeletal network at the peripheries of Z-discs, at right angles to sarcomeres has been referred to as a holomere (41) to emphasize the existence of a new extra-sarcomeric structure in muscle cells.

TABLE II. Protein composition of Z disc-cytoskeletal domains.

Peripheral Z-disc domain		Membrane-cytoskeletal domains
intermediate filaments	desmin	erythroid (αβ-spectrin)
	vimentin	ankyrin
	synemin	vinculin
		β,γ-actin

Studies on the expression of other polypeptides of the red blood cell membrane skeleton, such as the anion transporter and protein 4.1, in muscle cells have thus far failed to reveal any such analogues (12,25). Therefore at present the mechanism of attachment of the ankyrin-spectrin network to the muscle sarcolemma remains unknown.

Expression and Assembly of Cytoskeletal Domains during Muscle Differentiation

Analysis of the expression and assembly of these two cytoskeletal domains during myogenesis has revealed different mechanisms operative for each domain. The polypeptide composition of each domain is remodelled when myoblasts become postmitotic. Both domains are expressed in the proliferative myoblasts but upon terminal differentiation specific subunits of each domain begin to be expressed when the cells become postmitotic. In this manner the subunit composition of each domain is gradually changed to that observed in the adult. All the remodelling components for both domains are expressed concurrently upon the onset of terminal differentiation; however their segregation and assembly into their respective domains occurs independently for each domain. Subsequently to the production of each subassembly the domains associate with each other into the final structural phenotype seen in the adult cell. This occurs in a strict temporal order and well after the assembly of contracting sarcomeres and represents one of the final morphogenetic stages in muscle fiber assembly.

The morphogenetic pathway of intermediate filaments. Immunological (5,15,22) and electron microscopic (28,53) analysis of myogenic cells at various stages of differentiation have shown that after cell fusion intermediate filaments accumulate in the cytoplasm and exhibit no apparent association with newly assembling myofibrils. However several days later and well after the assembly of fully differentiated sarcomeres and Z disc, intermediate filaments exhibit a gradual change in their distribution and ultimately assemble at the peripheries of Z discs at right angles to the long axis of the cell. Analysis of the expression of the two filament subunits, desmin and vimentin, using cDNA probes and metabolic labeling has indicated that their expression is regulated independently (10,11,15). Vimentin is expressed constitutively in myoblasts and throughout differentiation while desmin accumulates specifically when the cells become postmitotic and thereafter. Both desmin (10) and vimentin (11a,56) are encoded by single copy genes in the haploid chicken genome. The desmin gene is transcribed into a single mRNA species (10), which accumulates specifically in cells of muscle origin; the vimentin gene is transcribed into two mRNA species which vary by approximately 300 bp at their 3' untranslated regions (11a,56). While both vimentin mRNA species accumulate constitutively in myoblasts and during myogenesis *in vitro* (11,11a) and *in vivo* (56), the desmin mRNA accumulates specifically when myoblasts become postmitotic and thereafter (11,11a). Depending on the type of muscle the two vimentin mRNAs either gradually decline to levels lower than those of desmin (e.g. skeletal muscle) or to basal levels well below those of desmin (e.g. smooth and cardiac muscle). The mechanism behind this muscle specific differential accumulation of these two mRNA species is not known; it may be due to a reduction in the transcription of the vimentin gene, differential stabilization of the vimentin and desmin mRNAs, by a dilution of the preexisting vimentin mRNA by relative increase in the transcription of the desmin gene or a combination of the above. The net effect is that the ratio of desmin to vimentin polypeptides in adult muscle is variable from almost exclusive accumulation of desmin (e.g. smooth and cardiac muscle) to a variable accumulation of both desmin and vimentin (e.g. skeletal muscle).

Both immunofluorescence (15) and immunoelectron microscopy (53) have indicated that, upon the onset of accumulation of desmin, the two intermediate filament subunits copolymerize into single filaments. Studies on the assembly of vimentin into filaments in chicken embryo erythroid cells have indicated that assembly is not limited posttranslationally. Assembly of vimentin occurs rapidly and completely from a small soluble pool of vimentin polypeptides (7a). When assembly of vimentin is inhibited, its synthesis is not altered suggesting that synthesis and assembly are not coupled events (36a). From these observations three major conclusions have emerged. First, the levels of each subunit (desmin and vimentin) synthesized or accumulated are regulated at the transcriptional or posttranslational, and not at the translational level. Second, the composition of each intermediate filament and the population of these filaments as a whole gradually change during differentiation as the amount of desmin mRNA and protein synthesized increase during this time. Finally the assembly of each subunit (desmin or vimentin) is stochastic and the ratio of the two subunits in each filament is variable and dependent on the amount of each subunit available for assembly at any instance in development.

The mechanism of assembly of the crosslinking protein synemin is currently unknown. Synemin expression is initiated concurrently with that of desmin when myoblasts become postmitotic and thereafter accumulates on intermediate filaments indistinugishably from desmin and vimentin (21,22). By analogy to desmin it is presumed that the levels of synemin mRNA at any instance in development may determine directly the level of synemin available for polymerization, and consequently, the level of synemin in association with the filaments. This is probably the case since the ratio of newly synthesized synemin to vimentin in erythroid cells and synemin to desmin in myogenic cells is approximately the same as that at steady state, namely 1:20-1:50 (22,37a). The process of synemin assembly has been analyzed in more detail during avian erythroid development. In these cells, and presumably also in muscle cells, synemin associates with the core of vimentin filaments at periodic intervals of approximately 180 nm (23). Most of the newly synthesized synemin assembles onto the growing filaments, similarly to vimentin (37a). However, the kinetics of assembly of synemin are substantially slower than those of vimentin, which has given rise to the hypothesis that the rate of vimentin polymerization and filament growth limits the rate and extent of synemin binding and assembly (37a). The mechanism by which synemin binds to vimentin is currently unknown, nor is it known how the periodicity of synemin binding along the backbone of the vimentin filaments is determined. The data obtained from the analysis of the kinetics of vimentin and synemin coassembly have been consistent with the hypothesis that polymerization of vimentin generates synemin-binding sites through some long-range suprastructure of the filament as a result of a higher-order change in the structure of vimentin upon polymerization (37a). Perhaps this higher-order change in vimentin structure may also dictate the periodic binding sites for synemin (for discussion see 23,37a). Unlike the regulation of expression and assembly of intermediate filaments which takes place primarily at the transcriptional level, the redistribution of the filaments to the peripheries of Z discs observed late in differentiation appears to be regulated post-translationally. Immunofluorescence (5,15,22) and immunoelectron microscopic (53) studies have indicated that this association occurs by the redistribution of intact filaments; all three polypeptides desmin, vimentin and synemin redistribute concurrently to the peripheries of Z-discs (15,22). The exact mechanism that governs this gradual redistribution is currently unknown. Desmin and vimentin may associate preferentially with Z discs due to the de novo appearance of binding sites at these structures. Another possibility involves the developmentally regulated expression of Z-disc-specific desmin and vimentin mRNAs, whose products would only localize at Z discs. Posttranslational modifications of desmin and vimentin also may affect their subcellular localizations and interactions. Lastly, the ratio of vimentin to desmin may influence intermediate filament distribution.

Neither the amount nor the size of the desmin mRNA changes significantly between 4 and 8 days of culture, a time at which the rearrangement of intermediate filaments is observed (10–11a). Since the desmin gene appears to exist as a single copy in the haploid chicken genome (10), it therefore seems unlikely that a switching of desmin mRNAs occurs. However, we cannot rule out the possibility that desmin transcripts are differentially processed during differentiation to yield distinct mRNAs with indistinguishable electrophoretic mobilities. Unlike many other myofibrillar structural proteins which exhibit multiple protein isoforms, thus far desmin has not been shown to exhibit such diversity. Rather, all available evidence is consistent with a single gene giving rise to one size mRNA and polypeptide in all muscle tissues examined.

As the single vimentin gene gives rise to two mRNAs with the same amino acid coding region (11a,56), the redistribution of vimentin-containing filaments cannot be caused by the appearance of a Z-disc-specific vimentin gene product. Furthermore both vimentin mRNAs are present throughout myogenesis, with no significant change in their abundance through 8 days of culture, at a time when the redistribution is well under way (11,11a).

The relatively constant levels of vimentin and desmin expression during the reorganization of intermediate filaments (at ~8 days) renders unlikely also the possibility that changing ratios of subunit composition affect localization. Thus the transition of vimentin- and desmin-containing filaments to Z discs is effected either by the appearance of specific Z-disc receptors or filament-binding proteins, or by modification of the subunits themselves. Earlier studies have indicated that changes in the phosphorylation of specific sites on desmin and vimentin, mediated by cAMP-dependent kinases, may be one of the posttranslational mechanisms that regulate the redistribution of the filaments. The phosphorylation of specific peptides on desmin and vimentin can be modulated only at a time that coincides with the onset of the redistribution of the filaments and thereafter (16) and can be blocked reversibly if the cells are maintained in high exogenous cAMP levels (31). Furthermore certain monoclonal antibodies to desmin recognize desmin only after its transition to the Z lines is initiated and thereafter into adulthood (13). The differential reactivity to this antibody presumably reflects conformational differences, a change in chemical modification or differential masking which only allows this epitope to be reactive upon the association of desmin with Z discs and thereafter. All this evidence points to one or more posttranslational events regulating this step in muscle morphogenesis.

Expression and Assembly of the Membrane Skeleton; Remodelling of the Plasma Membrane during Myogenesis.

The spectrin phenotype expressed in fully differentiated muscle cells has been shown to be exclusively αβ-spectrin, which is similar to that expressed in the erythrocyte, although there are slight differences in the relative electrophoretic mobilities of erythrocyte β-spectrin and the antigenically related proteins in muscle (38). The analysis of the spectrin phenotype of mitotic myoblasts reveals that these cells, on the other hand, express αγ-spectrin and little or no β-spectrin (39), which appears to be similar to the phenotype expressed in other mitotic cells and also in brain tissue (7,9,19,33,39,40). The expression of the fully differentiated spectrin phenotype is developmentally regulated and requires the gradual switch in the subunit composition of spectrin from αγ- → αβ-spectrin during myogenesis (39).

Pulse-labeling experiments followed by immunoprecipitation of newly synthesized spectrin subunits and ankyrin have revealed that as myogenic cells become postmitotic and enter the phase of terminal differentiation the synthesis of β-spectrin and ankyrin is initiated (39,41,43). Concurrently there is a dramatic increase in the levels of α-spectrin, β-spectrin and ankyrin mRNA suggesting that the expression of these two spectrin subunits and ankyrin are regulated by the mRNA abundance (37). During terminal differentiation, the amounts of β-spectrin, α-spectrin and ankyrin in the cell gradually increase whereas that of γ-spectrin decrease, even though the synthesis of γ-spectrin is continuous throughout myogenesis. This results eventually in the predominance of the αβ-spectrin phenotype in the adult muscle cell (39,43). Pulse-chase experiments have shown that this differential accumulation of β-spectrin over γ-spectrin appears to result from the gradual stabilization of αβ-spectrin and the relatively rapid degradation of newly synthesized γ-spectrin. α-Spectrin does not appear to play a significant role in the switch of the polymorphic spectrin subunit, since its rate of synthesis and stability are similar initially to those of γ-spectrin in the myoblast, and later to those of β-spectrin in the mature myotube (39). The accumulation of ankyrin follows that of α-spectrin and β-spectrin suggesting that the three subunits assemble with each other and accumulate gradually as a complex (43). Studies with fusion blocked cells have indicated that even though expression of these three membrane-skeleton polypeptides is regulated mainly at the transcriptional level or by RNA stabilization, their extent of assembly and accumulation at the plasma membrane is regulated posttranslationally. Blocking cell fusion with a Ca^{2+} chelator reversibly inhibits the accumulation of these three subunits, but has no effect on their rates of synthesis, suggesting that expression and assembly are independently regulated (43). From these data we have postulated that the assembly and differential stabilization of the two spectrin phenotypes is regulated at the posttranslational level by the availability on the plasma membrane of receptors specific for γ- and β-spectrin, respectively (39,42). Thus in the myoblast, the αγ-spectrin phenotype is expressed as a result of the coordinate synthesis of γ-spectrin and its fusion-dependent specific membrane receptor which allows the assembly and stabilization of newly synthesized αγ-spectrin on the membrane. Subsequently, the synthesis of β-spectrin and its specific membrane receptor at the onset of and during, terminal differentiation results in the gradual assembly and stabilization of αβ-spectrin. During the same period, the amount of the γ-spectrin-specific membrane receptor declines thus causing newly syunthesized, but unassembled, γ-spectrin to be degraded relatively rapidly. An alternative to this hypothesis is that both γ- and β-spectrin utilize the same receptor but that β-spectrin has a higher affinity for this receptor than γ-spectrin. Therefore when the synthesis of β-spectrin is initiated after the cells become postmitotic, it gradually competes off newly synthesized γ-spectrin from binding to this membrane receptor. An essential feature of this model is that the assembly and stabilization of αβ-spectrin or αγ-spectrin is independent of the relative rates of synthesis and the size of the pools of unassembled γ- and β-spectrin and is driven solely by the affinity of the membrane receptor for β or γ-spectrin, respectively; α-spectrin is not limiting since it is present in sufficient amounts to saturate all available β and γ-spectrin on the membrane. Hence the cell can assemble αβ-spectrin and begin to switch its spectrin phenotype, even though αγ-spectrin is present initially in vast excess over αβ-spectrin (39,42). The synthesis and accumulation of ankyrin in postmitotic cells concurrently with β-spectrin and the inhibition of the accumulation of both in fusion blocked cells, argues that their extent of assembly is regulated at a step beyond that of ankyrin, possibly by the expression of the ankyrin binding site in the membrane (43). Studies with cDNA probes and antibodies to the avian erythrocyte anion transporter have thus far failed to reveal the expression of a homologous mRNA or polypeptide in myogenic cells (12). Therefore the possibility exists that the ankyrin-αβ-spectrin binding site at the plasma membrane in muscle cells is distinct from that in erythroid cells.

The expression and accumulation of the αβ-spectrin phenotype during myogenesis coincides with a dramatic reorganization of the myogenic cytoskeleton. Initially, this takes the form of a reorganization of spectrin as elements of the spectrin matrix begin to coalesce to form a grid-like network on the plasma membrane (39). This may be the result of increased cross-linking of adjacent spectrin tetramers by actin as more αβ-spectrin accumulates on the sarcolemma, which may also cause a gradual decrease in the observed mobility and concomitant localized clustering of other proteins in the plane of the membrane (e.g., acetylcholine receptors; 3). Eventually, the spectrin network forms ring structures around each Z line, which are interconnected by lines of spectrin in the long axis of the myofiber (38,41). This reorganization of spectrin occurs concurrently with the reorganization that intermediate filaments undergo at the periphery of Z discs.

Thus, the temporal and spatial assembly of the spectrin and intermediate filament cytoskeleton domains have several similar characteristics: (1) there is a gradual switch of the predominant subunits upon terminal differentiation from the phenotype expressed in the mitotic cells to that of the postmitotic, fully differentiated cells; and (2) during terminal differentiation there is a reorganization of the networks from a random distribution in the mitotic cell into a highly ordered interconnected structure in the fully differentiated cell. The molecular trigger(s) that initiates these changes in subunit composition and distribution upon terminal differentiation are unknown but as we discussed above the former is regulated at the pretranslational level, while the latter is predominantly posttranslational.

CONCLUSIONS

The studies summarized here indicate that striated muscle cells contain a multidomain cytoskeletal structure that is superimposed onto myofibrils. Undoubtedly the domains described to date may only represent a subset of more to be discovered. Their identification and molecular characterization, however, point to the complexity of the system and how multilevel the regulation of muscle morphogenesis is as exemplified from the analysis of the expression and assembly of these cytoskeletal domains. Although it is well established that expression of many myofibrillar proteins is regulated primarily at the transcriptional level their assembly and segregation to their various sarcomeric substructures may be also regulated by a number of posttranslational events analogous to those described above for the cytoskeletal domains.

ACKNOWLEDGEMENTS

The research summarized here was supported by grants from the National Institutes of Health, the National Science Foundation, and the Muscular Dystrophy Association to E.L. Y.G. was supported also by a Senior Research Fellowship from the American Cancer Society California Division.

REFERENCES

1. Alper, LS, Beam KG and Greengard P (1980). Hormonal control of $Na^{+}K^{+}$ cotransport in turkey erythrocytes. Multiple site phosphorylation of goblin, a high molecular weight protein of the plasma membrane. J Biol Chem 255:4864-4871.
2. Anderson RA and Lovrien RE (1984). Glycophorin is linked by band 4.1 protein to the human erythrocyte membrane skeleton. Nature 307:655-658.
3. Axelrod D, Ravdin P, Koppel DE, Schlesinger J, Webb WW, Elson, EL and Podleski, TR (1976). Lateral motion of fluorescently labeled acetylcholine receptors in membranes of developing muscle fibers. Proc Nat Acad Sci 73:4594-4598.
4. Beam KG, Alper SL, Palade GE and Greengard P (1979). Hormonally regulated phosphoprotein of turkey erythrocytes: localization to plasma membrane. J Cell Biol 83:1-15.
5. Bennett GS, Fellini SA, Toyama Y and Holtzer H (1979). Redistribution of intermediate filament subunits during skeletal myogenesis and maturation in vitro. J Cell Biol 82:577-584.
6. Bennett V (1982). The molecular basis for membrane-cytoskeleton association in human erythrocytes. J Cell Biochem 18:49-65.
7. Bennett V, Davis J and Fowler WA (1982). Brain spectrin, a membrane-associated protein related in structure and function to erythrocyte spectrin. Nature (London) 299:126-131.
7a. Blikstad I and Lazarides E (1983). Vimentin filaments are assembled from a soluble precursor in avian erythroid cells. J Cell Biol 96:1803-1808.
8. Branton D, Cohen CH and Tyler J (1981). Interaction of cytoskeletal proteins on the human erythrocyte membrane. Cell 24:24-32.
9. Burridge K, Kelly T and Mangeat, P (1982). Nonerythrocyte spectrin: actin-membrane attachment proteins occurring in many cell types. J Cell Biol 95:478-486.
10. Capetanaki YG, Ngai J and Lazarides E (1984). Characterization and regulation in the expression of a gene coding for the intermediate filament protein desmin. Proc Nat Acad Sci 81:6909-6912.
11. Capetanaki YG, Ngai J and Lazarides E (1984). Regulation of the expression of the genes coding for

the intermediate filament subunits vimentin, desmin and glial fibrillary acidic protein. In Borisy GG, Cleveland D and Murphy D (eds): "Molecular Biology of the Cytoskeleton," New York: Cold Spring Harbor Laboratory, pp. 415-434.

11a. Capetanaki, YG, Ngai J, Flytzanis CN and Lazarides E (1983). Tissue-specific expression of two mRNA species transcribed from a single vimentin gene. Cell 35:411-420.

12. Cox JV, Moon RT and Lazarides E (1985). Anion transporter: highly cell-type specific expression of distinct polypeptides and transcripts in erythroid and non-erythroid cells. J Cell Biol 100:1548-1557.

13. Danto SI and Fischman D (1984). Immunochemical analysis of intermediate filaments in embryonic heart cells with monoclonal antibodies to desmin. J Cell Biol 98:2179-2191.

14. Fowler V and Bennett V (1984). Erythrocyte membrane tropomyosin. Purification and properties. J Biol Chem 259:5978-5989.

15. Gard DL and Lazarides E (1980). The synthesis and distribution of desmin ad vimentin during myogenesis in vitro. Cell 19: 263-275.

16. Gard DL and Lazarides E (1982). Cyclic AMP-modulated phosphorylation of intermediate filament proteins in cultured avian myogenic cells. Mol Cell Biol 2:1104-1114.

17. Geiger B (1979). A 130K protein from chicken gizzard: its localization at the termini of microfilament bundles in cultured cells. Cell 18:193-205.

18. Georgatos SD and Lazarides E (1983). Tropomyosin in mammalian and avian erythrocytes. J Cell Biol 97:279a.

19. Glenney JR, Glenney P and Weber K (1982). Erythroid spectrin, brain fodrin ad intestinal brush border protein (TW260/240) are related molecules containing a common calmodulin-binding subunit bound to a variant cell type-specific subunit. Proc Nat Acad Sci USA 79:4002-4005.

20. Granger BL and Lazarides E (1978). The existence of an insoluble Z disc scaffold in chicken skeletal muscle. Cell 15:1253-1268.

21. Granger BL and Lazarides E (1979). Desmin and vimentin coexist at the periphery of the myofibril Z disc. Cell 18:1053-1063.

22. Granger BL and Lazarides E (1980). Synemin: a new high molecular weight protein associated with desmin and vimentin filaments in muscle. Cell 22:727-738.
23. Granger BL and Lazarides E (1982). Structural associations of synemin and vimentin filaments in avian erythrocytes revealed by immunoelectron microscopy. Cell 30:263-275.
24. Granger BL, Repasky EA and Lazarides E (1983). Synemin and vimentin are components of intermediate filaments in avian erythrocytes. J Cell Biol 92:299-312.
25. Granger BL and Lazarides E (1984). Membrane skeletal preotein 4.1 of avian erythrocytes is composed of multiple variants tha exhibit tissue-specific expression. Cell 37:595-607.
25a. Granger BL and Lazarides E (1983). Expression of the major neurofilament subunit in chicken erythrocytes. Science 221:553-556.
26. Ishikawa H, Bischoff R and Holtzer H (1968). Mitosis and intermediate-sized filaments in developing skeletal muscle. J Cell Biol 38:538-555.
27. Jay DG (1983). Characterization of the chicken erythrocyte anion exchange protein. J Biol Chem 258:9431-9436.
28. Kelly DE (1969). Myofibrillogenesis and Z-band differentiation. Anat Rec 163:403-426.
29. Lazarides E (1980). Intermediate filaments as mechanical integrators of cellular space. Nature 283:249-256.
30. Lazarides E (1984). Assembly and morphogenesis of the avian erythrocyte cytoskeleton. In Borisy GG, Cleveland D and Murphy D (eds): "Molecular Biology of the Cytoskeleton," New York: Cold Spring Harbor Laboratory, pp. 131-150.
31. Lazarides E and Gard DL (1982). Modulation of intermediate filament organization by cyclic nucleotide analogues and β-adrenergic agonists during myogenesis in vitro. In O'Malley BW (ed): "Gene Regulation," UCLA Symposia on Molecular and Cellular Biology: Academic Press, pp. 343-358.
32. Lazarides E and Granger BL (1983). Transcytoplasmic integration in avian erythrocytes and striated muscle. Modern Cell Biol 2:143-162.
33. Levine J and Willard M (1981). Fodrin: axonally transported polypeptides associated with the internal periphery of many cells. J Cell Biol 90:631-643.

34. Liu S-C, Windisch P, Kim S and Palek J (1984). Oligomeric states of spectrin in normal erythrocyte membranes: biochemical and electron microscopic studies. Cell 37:587-594.
35. Marchesi VT (1985). Stabilizing infrastructure of cell membranes. Ann Rev Cell Biol (in press).
36. Moon RT and Lazarides E (1984). Biogenesis of the avian erythroid membrane skeleton: receptor-mediated assembly and stabilization of ankyrin (goblin) and spectrin. J Cell Biol 98:1899-1904.
36a. Moon RT and Lazarides E (1983). Canavanine inhibits vimentin assembly but not its synthesis in chicken embryo erythroid cells. J Cell Biol. 97:1309-1314.
37. Moon RT, Ngai J, Wold BJ and Lazarides E (1985). Tissue-specific expression of distinct spectrin and ankyrin transcripts in erythroid and non-erythroid cells. J Cell Biol 100:152-160.
37a. Moon RT and Lazarides E (1983). Synthesis and posttranslational assembly of intermediate filaments in avian erythroid cells: Vimentin assembly limits the rate of synemin assembly. Proc Nat Acad Sci USA 80:5495-5499.
38. Nelson WJ and Lazarides E (1983). Expression of the β-subunit of spectrin in nonerythroid cells. Proc Nat Acad Sci USA 80:363-367.
39. Nelson WJ and Lazarides E (1983). Switching of the subunit composition of muscle spectrin during myogenesis in vitro. Natue (London) 304:364-368.
40. Nelson WJ, Granger BL and Lazarides E (1983). Avian lens spectrin: subunit composition compared with erythrocyte and brain spectrin. J Cell Biol 91:1271-1276.
41. Nelson WJ and Lazarides E (1984). Goblin (ankyrin) in striated muscle: identification of the potential membrane receptor for erythroid spectrin in muscle cells. Proc. Nat. Acad. Sci. USA 81:3292-3296.
42. Nelson WJ and Lazarides E (1984). Assembly and establishment of membrane-cytoskeleton domains during differentiation. Spectrin as a model system. In Elson E, Frazier W and Glaser L (eds): "Cell Membranes,: Plenum Publ. Corp. 2:219-246.
43. Nelson WJ and Lazarides E (1985). Post-translational control of membrane-skeleton (ankyrin and αβ-spectrin) assembly in early myogenesis. J Cell Biol 100:1726-1735.

44. Ngai J, Capetanaki YG and Lazarides E (1984). Differentiation of murine erythroleukemia cells results in a rapid repression of vimentin gene expression. J Cell Biol 99:306-314.
45. Ngai J, Capetanaki YG and Lazarides E (1984). Expression of the genes coding for the intermediate filament proteins vimentin and desmin. In Fischman D, Wang E and Liem R (eds): "Intermediate Filaments," New York Academy of Sciences.
46. Pardo JV, Siliciano JD and Craig SW (1983). Vinculin is a component of an extensive network of myofibril-sarcolemma attachment regions in cardiac muscle fibers. J Cell Biol 97:1081-1088.
47. Pardo JV, Siliciano JD and Craig SW (1983). A vinculin-containing cortical lattice in skeletal muscle: transverse lattice elements ("costameres") mark sites of attachment between myofibrils and sarcolemma. Proc Nat Acad Sci 80:1008-1012.
48. Pierbon-Bormioli S (1981). Transverse sarcomere filamentous systems: "Z- and M-lines." J Musc Res Cell Motil 2:401-408.
49. Repasky EA, Granger BL and Lazarides E (1982). Widespread occurrence of avian spectrin in nonerythroid cells. Cell 29:821-833.
50. Street SF (1983). Lateral transmission of tension in frog myofibers: A myofibrillar network and transverse cytoskeletal connections are possible transitters. J Cell Physiol 114:346-364.
51. Tiegs OW (1954). The flight muscles of insects - their anatomy and histology; with some observations on the structure of striated muscle in general. Roy Soc Lond Phil Trans Ser B 238:221-348.
52. Tokuyasu KT, Dutton AH, Geiger B and Singer SJ (1981). Ultrastructure of chicken cardiac muscle as studied by double immunolabeling in electron microscopy. Proc Nat Acad Sci 78:7619-7623.
53. Tokuyasu KT, Maher PA and Singer SJ (1985). Distribution of vimentin and desmin in developing chick myotubes in vivo. II. Immunoelectron microscopic study. J Cell Biol 100:1157-1166.
54. Virtanen I, Kurkinen M and Lehto V-P (1979). Nucleus-anchoring cytoskeleton in chicken red blood cells. Cell Biol. Int. Rep. 3:157-162.
55. Woodcock CLF (1980). Nucleus-associated intermediate filaments from chicken erythrocytes. J Cell Biol 85:881-889.

56. Zehner ZE and Paterson BM (1983). Characterization of the chicken vimentin gene: single copy gene producing multiple mRNAs. Proc Nat Acad Sci USA 80:911-915.

Molecular Biology of Muscle Development, pages 773–791

CONTROL OF MUSCLE GENE EXPRESSION: TRANSCRIPTION OF MYOSIN LIGHT CHAIN GENE AND THE ROLE OF REPETITIVE ELEMENTS IN GENE REGULATION

M.A.Q. Siddiqui, Ana Maria Zarraga, Diarmuid Nicholson, Amrut Deshpande, Chandrika Saidapet and Charmaine Mendola

Roche Institute of Molecular Biology,
Roche Research Center, Department of Biochemistry,
Nutley, New Jersey 07110

ABSTRACT The chicken cardiac muscle myosin light chain (MLC_{2A}) gene segment (λLC5) containing an active promoter was isolated and partly characterized. A MLC_{2A} gene construct was introduced into an expression vector and used for transfection of appropriate host cell lines. The transcription activity was screened by CAT assay and S1-nuclease protection of MLC_{2A} gene specific RNA transcripts. It appears that the transcription of MLC_{2A} gene was dependent upon the presence of an unknown avian specific factor. Recombinant plasmids, pSS48 and pSS116, containing the cDNA and the genomic DNA sequences respectively, for a chicken repetitive DNA transcript, 7S RNA, were constructed and characterized. Hybridization data and base sequence analysis of these clones indicated that a short sequence, located in proximity to the MLC_{2A} gene promoter, is homologous to 7S RNA. The specific recognition by 7S RNA of sequences in the functionally significant domain of MLC_{2A} gene, and the appearance of 7S-like RNA transcripts in chick blastodermal cells prior to the transcription of genes for major muscle proteins, suggest that the 7S sequence might serve as a transcriptional signal element involved in control of MLC_{2A} gene expression.

INTRODUCTION

Recent studies on structural and functional analysis of eukaryotic genes have led to the understanding that the mechanism(s) underlying control of gene transcription is quite complex and might involve both cis- and trans-acting regulatory factors. It is known that short conserved nucleotide sequences located in the flanking regions of structural genes are important for gene functions (1-5). Some of these sequences are repeated in the respective genomes. Eukaryotic genomes are made up of both single copy and repetitive sequences, with the repetitive DNA interspersed among the single copy DNA. Several classes of repetitive DNA have been described in recent years (6-11). Although the functional significance of these DNAs is not clear, several features of the repeat sequences suggest that these elements might have dispersed in the genome via RNA intermediates (12). Distinct low molecular weight RNA that can form strong hybrids with the dominant families of repetitive DNA have been observed in many organisms (13-15). Thus, from the stand point of identifying gene controlling elements, the potential role of repetitive sequences in control mechanisms is of importance.

Muscle cells in culture and their progenitors provide a useful system to investigate the mechanism(s) involved in gene activation and control. Cardiac muscle development is particularly attractive in this regard, since it constitutes an early event during fetal cell development (16) and offers a unique advantage in the availability of a population of blastodermal cells which are predetermined and undifferentiated. The contractile proteins of the cardiac muscle, like in other muscles, are members of large families of iso-proteins whose expression is precisely regulated and undergoes gene switching during muscle development. It is of interest, therefore, to investigate the molecular events responsible for the changes in cardiac muscle gene expression from one developmental stage to another and also during various physiological and pathophysiological states that affect the cardiovascular system.

We recently described the construction and characterization of several recombinant plasmids containing the cDNA sequences for chick heart myosin light chain (MLC_2) (17,18), myosin heavy chain (MHC) (19) polypeptides, the major constituents of cardiac muscle proteins, and for a representative non-myofibrillar chick muscle protein, glyceraldehyde-3-phosphate dehydrogenase (20). We also described the properties of a recombinant plasmid (pSS48) containing the sequence of a low molecular weight RNA (7S RNA) isolated from the embryonic chick cardiac muscle tissue (21). The hybridization pattern of pSS48 to the chick genomic DNA indicated that 7S RNA is the product of the repetitive portion of the chick genome. 7S RNA is preferentially transcribed in chick blastodermal cells earlier than the transcription of genes for major muscle proteins. Of particular interest is the fact that pSS48 DNA identifies with specificity the single copy genes for MLC_2 and MHC. In order to establish the relationship between the 7S specific gene sequence and the myosin genes and to examine whether 7S-like sequences are indeed present in the functionally significant domain of myosin genes, we isolated and characterized a genomic clone, λLC5, for MLC_{2A}. In this communication, we describe the structural and biological properties of the MLC_2 gene and the potential role of the repeat sequences in its regulation.

RESULTS

Myosin Light Chain (MLC_2) Gene: The genomic clones for MLC_{2A} were isolated by screening the chicken genomic library constructed in λ charon 4A with a previously identified and characterized cDNA clone, pML10, for chick cardiac MLC_{2A} (17,18). Of the several isolates, two clones, λLC13, λLC5, were selected for further characterization. λLC13, which hybridized to both the coding and 3'-non-coding segments of pML10 DNA, was truncated at the 5'-end (Fig. 1). On the other hand, λLC5, which contains a large, approximately 7.0 kb, DNA segment of the upstream region of MLC_{2A} gene, was truncated at the 3'-end. A detail structural characterization of these clones isolated independently by Arnold and coworkers has recently been reported (22). Fig. 1 shows the partial sequence analysis of λLC 5

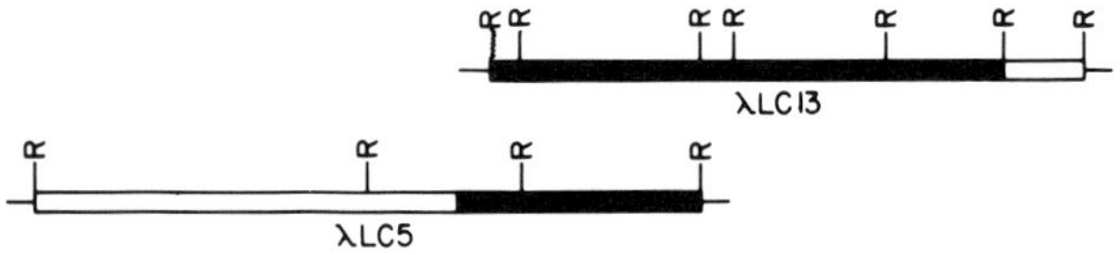


```
                                                  GGGATAGGGGGAGCTAATGGCNG
AGGGCTGAGCAAGTGCTTGGCTTAAGTGGGTGACAGCTGCCCCAGTGGGGGTGATAGAGAGGATCTCTGTGCTGT
GGAGCTCAGCATTCACCCCCAGCCTCATCCAGCCCCTTTTGGATGGATTCTGGCTCTTGGTCCCTGTGGGATGCT
CCTGCCAGCTCACCCCAATGTAGGGCTTACCCCTGCTTACCCCTTGGCCATCAGACCCCGACGCTTCCCTTGCCT
CTGCCCATTAATGCCATCCCAATGAG..CTACCCTGAAGCAAAAGGGAGGCACTGCACCGTGACGTGGGGGACAG
TGTTGGGCCAGATAACCCCATTGCAGCCCAGCAGAGGGGAAGAGCACAGCCTCTGCCCACTCGAAGTGACAGCTC
GTGGGAGCTGCAGGGCCTCCTGGTGTGCTGCTAACCTTGAAAGCCTCTGTGTCTCGCATGTCACGTCGGCCAGT
GACAACAGCAAGCTTACTGCAGATTAGAGGAGGAGAAAGTCAGAGACAAGAGGGTCGGGGGTGGCTGGTTTACC
TATTACAGCCAAAAGTGGACATGGGGTTATTTTTAGCCTGGAATGGGGTGTATTTATTGTTTCAGCCCAGGGAG
GCAGAGGGACTGCCTCTGAGATTCCTCCTGCAGCACCATCAGCACCTCTGCGAAGACATGGTGAGTGCTGGGAG
                                                          MeT
CCACAAGGGGGGCAGCAGGCCTGTGGCACTGCGTTCTCATTGGGATGCCATTAATGGGCAGAGAGCAAGGGAAG
CTTCGGGGTCTGATGGCCAAGGGGTAAGCAGGGGTAAGCCCTACATTGGGGTGAGCTGGCAGGGAGCATCCCAC
AGGGACCGGTGGGGACAGGTGAGGGCAGAGAGGTGTCTCCTGTGTTGGTAGCACTCAGCAGAACTTCCTGACGG
GGAGAGGTCATTTGGGGGCAAATGCTGTCTTTTTCTTATCCTTTCCCCCGTGAAGACACGGAAAGGACAGTTCT
GTCCCTGAAGCCCTCTGGCTGTGTTTCCTCTTGTTAGAATCCTAACTTTTGGAAGAGAAAAAGCTGTATGGATG
CGTTGTGACCAAAAGTGAGGGCGTTTCTGGTCGATTCAAACCTCAGTGCAAGGAGACGGCATGTTTTAAAATAT
GCAAATGCATTACTTAGCACTATGTTCTTAGTGAGAGGCTCAGAGCAATAAGGAGACATCCAATGACCACAGCC
TTATGTGTAATCCCAGCAGCGTGCTCCTGACTCTGACCCTGAGCAGAACTCAGCTGGCCTCGACTGTTTTATCC
ACCGCAGCTGAGAAAGCAGAAATGGGAAATCCAGCTCCTTTCTGCTCCGACCTATTTGGCAAAATTCGTCAGGA
AAGCAAAGGCAGAGCTGTACTCTGAGCAGAAATCTTAACGCTCCCAGCACACGGGGAACCTCAGGCACCCAGGG
ACTGAGCAGGCACCGGTTCCATGCTGTCATGAAGGACTTATTGCTAAGCACTGTCATGAAGGACTTATGACTAA
GCACAGGGTGGTTCAATGCTTTTATTCCACACCATAAAAAGACAGTAAGTGAGTGCTACATACTTGAACCCCGC
AG GCA CCC AAG AAA GCC AAG AAG AGG ATT GAA GGT GCG AAT TCC AAT GTC TTC TCC
   ALA PRO LYS LYS ALA LYS LYS ARG ILE GLU GLY ALA ASN SER ASN VAL PHE SER
ATG TTC GAG CAG GCC CAG ATC CAG GAA TTC AAA GAG GT AGGTGGATT
MET PHE GLU GLN ALA GLN ILE GLN GLU PHE LYS GLU
```

FIGURE 1. Partial sequence of MLC_{2A} gene. Recombinant phages, λLC5 and λLC13 were isolated by screening the chicken genomic library with a previously identified MLC_{2A} cDNA clone, pML10, (17,18). λLC5, which contains a large 5'-flanking region was digested with Eco RI and the resulting 6.0, 2.8, and 3.2 kb fragments were subcloned into pBR322. Plasmid pλLC5.2 containing the 2.8 kb middle DNA fragment was sequenced according to Maxam and Gilbert (38). A portion of the sequence showing the first exon containing the 30 N-terminal amino acids (see text), the intron, and the 5'-flanking region is shown. The TATA-like sequence (TATTTA) is boxed and the CAT sequence is underlined.

representing the first exon, the first intron, and the putative promoter region. As per the sequence, there are two additional N-terminal amino acids, Met and Ala, which are not found in the mature MLC_{2A} protein (23). The first amino acids, Met is separated from Ala by a large intron. The codons for amino acids, Met, Ala, preceding the codon for the N-terminal Pro in mature MLC_{2A} are also present in rat and chicken skeletal muscle MLC_{2A} genes (24,25). Seventy five bp upstream to the first Met codon, is the TATA-like sequence (TATTTA) characteristic of eukaryotic promoters (26) and 28 bp further upstream is the CAT box (27).

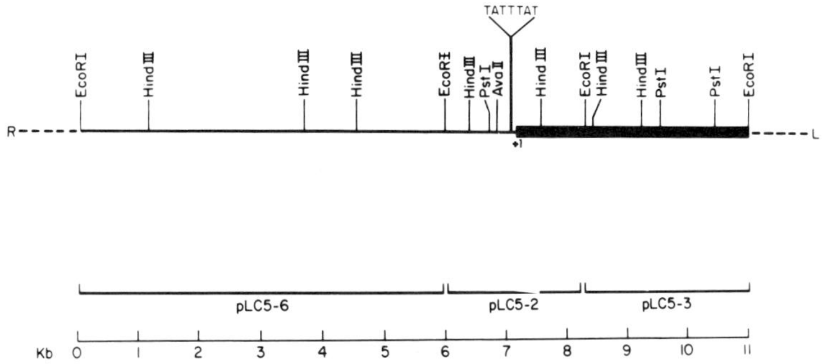

FIGURE 2. Physical map of λLC5 DNA. λLC5 DNA was sub-cloned into pλLC5.2, pλLC5.3, pλLC5.6 as indicated in Fig. 1. Solid bar denotes the MLC_{2A} gene truncated at the 3'-end. +1 indicates the putative transcription start site (see text). R and L are the right and left λ arms.

Evidence for Promoter Activity in MLC_{2A} Gene: λLC5 DNA was digested with restriction enzyme Eco Rl and the resulting fragments were sub-cloned into pBR322 by ligation at the Eco Rl site. pλLC5.3, pλLC5.2, pλLC5.6 are the three sub-clones containing the entire 3.2 kb, 2.8 kb and 6.0 kb DNA sequences respectively with the exception of a 50 bp segment located between the 3.2 and 2.8 kb fragments which was lost due to an additional Eco Rl site in λLC5 DNA. As indicated in Fig. 2, the sub-clone pλLC5.2 contains the putative promoter sequence TATTTA and the transcription initiation site (see below). In order to ascertain whether pλLC5.2 DNA contains a functionally active promoter(s), several constructs were made using DNA fragments obtained by restriction digestion of pλLC5.2 and inserted at the Hind III site of the expression vector pSV0-Cat. The chimeric vectors were then used to transfect the rat myogenic cell line L6, and a Japanese quail fibroblast line QT35 (28). The transient expression of the gene was monitored by assaying for the CAT (chloramphenicol acetyl transferase) activity described earlier (29). The transcription activity was also assayed for the accumulation of gene specific transcripts by Sl nuclease protection method (30). As shown in Fig. 3, the construct pL_o5-2iCat contains a Hind III-Hind III fragment of about 1.6 kb encompassing the putative promoter-like region, the codon for the first amino acid, Met, and a portion of the first intron. When assayed for CAT activity, the pLC_o5-2iCat transformed QT35 cell extract promoted the formation of 3-acetate chloramphenicol, whereas the activity with the mock transfected or control (pSV0Cat) cell extracts was negligible. No activity was found when the same vector was used to transfect L6 cell line under identical conditions. Thus, it would appear that an avian specific factor might be required for transcription of chick MLC_{2A} gene.

To tentatively map the transcription initiation site, total RNA was extracted from the 48 hr old transfected cells and analyzed for Sl nuclease protection with a 5'-end labeled double-stranded Pst I-Pst I DNA probe (Fig. 3). The 5'-end label nucleotide in the probe DNA is 28 bp upstream to Met codon and it was assumed that the transcript for MLC_{2A} gene begins with a leader region longer than 28 nucleotides. As shown in

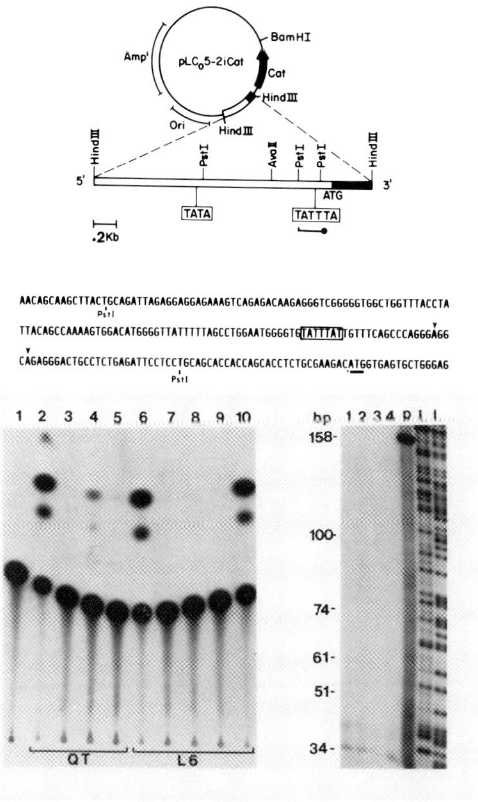

FIGURE 3. Transient expression and identification of transcription initiation site. A Hind III-Hind III fragment of about 1.6 kb isolated by digestion of pλLC5.2 DNA with Hind III was cloned into the expression vector, pSV0Cat at the Hind III site. The resulting plasmid $pLC_0$5.2i Cat was used for transfection of L6 and QT35 cells (see text). After 48 hr, the extracts were assayed for CAT activity (29) (bottom left) and the RNA was used for S1 protection assay (30) using a Pst I-Pst I fragment of about 160 bp labelled at 5'-end (bottom right). Bottom left: Lanes 1, ^{14}C-chloramphenicol; 2, ^{14}C-chloramphenicol after incubation with CAT; 3, extract from QT35 cells transfected with pSV0Cat (control); 4, extract from QT35 cells transfected with

$pLC5_05.2iCat$; 5, same as in 4, except the transfection was with $pLC5_02Cat$ (data not shown); 6, extracts from QT35 cells transfected with pRSV2Cat; 7-10 are same as lanes 3-6, except extracts from L6 cells were used. Bottom right, Lanes 1, 2 and 4 contain 2, 5 and 10 µl of protected DNA sample; lane 3, same as in lane 4 except the RNA was from pSV0Cat transfected cells; P, control unprotected Pst I-Pst I fragment (10,000 cpm); L, ladder of Pst I-Ava II DNA fragment. Sequence in the center of the figure shows the promoter region containing the TATA-like sequence (boxed), ATG sequence (underlined) and the transcription initiation site indicated by arrows.

Fig. 3 (bottom right side), the size of the protected fragment is about 30 nucleotides, indicating that the transcription start site is about 20 bp downstream to the TATTTA region. These results agree with the primer extension data of Arnold *et al.* (31). The fragment was not protected when RNA from cells transfected with control plasmid pSV0 Cat was used for hybridization.

Properties of plasmid pSS48 (7S RNA). In order to study the structure, function and biogenesis of 7S RNA, we constructed a recombinant plasmid (pSS48) containing the entire cDNA sequence for 7S RNA. The construction of the plasmid and its characterization has been described (21). The pattern of hybridization of pSS48 [^{32}P] DNA to the chick genomic DNA indicated that 7S RNA is derived from the repetitive chick genome. A rough estimate of repetition frequency, based on a reconstruction experiment, indicated that there are about 300 copies of 7S-like sequences per haploid chick genome.

It is believed that at 24 hr of development, the presumptive heart-forming cells, which are undifferentiated and morphologically indistinguishable from other cells of the embryo, can be identified based on their location in the blastoderm. These cells are apparently highly primitive in character but have the potential to undergo heart formation. At 40-50 hr of development, depending upon the conditions of incubation of the fertilized eggs, the myofibrillar ultrastructure begins to appear. When total RNA isolated from different stages of development was examined by fractionation on formaldehyde-agarose gel and hybridization with the appropriate DNA probes for actin, MHC and MLC, no hybridization signal was detectable up to about 60 hr of development. On the other hand, 7S RNA-like transcripts are already present in 40 hr blastoderms(Fig.4). Indeed,

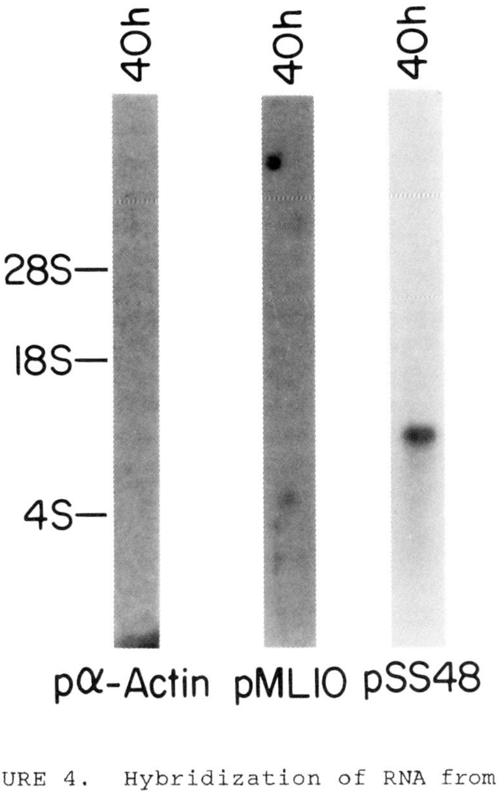

FIGURE 4. Hybridization of RNA from 40 hr chick blastoderms. RNA was isolated from 40 hr chick blastoderms as described earlier (21) and hybridized with the nick translated plasmid DNAs. pα-actin, chick α actin clone (a kind gift of C. Ordahl); pML10, chick MLC_{2A} cDNA clone (17,18); pSS48, 7S RNA specific clone (21).

a faint hybridization signal can be discerned as early as 24 hr (21). Thus, 7S RNA appears in chick embryonic cells preferentially earlier than the onset of transcription of genes for MLC, MHC and actin, the major constituents of cardiac muscle proteins. Of particular interest is that pSS48 DNA (or 7S RNA) under low stringency of hybridization recognizes with specificity both pML10 and pMHC8 DNAs (21). Base sequence analysis of these DNAs revealed that DNA sequences homologous to pSS48 DNA, or 7S RNA, are located in the 3'-non-coding regions of the respective plasmids. There are examples in literature where repeat sequences are present in 3'-untranslated regions of several developmentally regulated poly(A) RNAs (4,5). Indeed, a repetitive element isolated from the mouse genome appeared to identify a family of mRNAs both during development and during oncogenic transformation (5).

Localization of 7S-Like Element in 5'-Flanking Region of MLC_{2A} Gene: In order to search for the 7S-like sequence(s) in the 5'-flanking region of MLC_{2A} gene (λLC5) the plasmids pλLC5.2, pλLC5.3, pλLC5.6 were digested with Eco R1 to release the respective inserts and hybridized with pSS48 (-dA) [^{32}P]DNA. The latter probe was prepared by digestion of pSS48 DNA insert with restriction enzyme DdEI (which removes the poly (dA) tail from pSS48 DNA) in order to avoid the non-specific poly dA/dT hybridization. As shown in Fig. 5, the pSS48 (-dA) [^{32}P]DNA hybridized only to pλLC5.2 DNA insert. The hybridization to high molecular weight DNA bands is due to contaminant plasmid pBR322 DNA present in the probe preparation. Under elevated stringency of washings, the hybridization to pλLC5.2 disappeared preferentially, suggesting that the sequence(s) complementary to pSS48 DNA are short.

Since 7S sequence is repeated in the chick genome, we attempted to ascertain whether the 29 bp sequence in the pλLC5.2 promoter region is also repetitive. For this purpose, the total chick heart repetitive DNA fraction (21) was used as probe. As shown in Fig. 6, the 29 bp sequence does not hybridize to the repetitive DNA probe and thus is not a part of 7S sequence which is repeated in the chick genome. On the other hand, a repetitive sequence was located in pλLC5.6 DNA, but not in pλLC5-2 or pλLC5-3 DNAs. pSS48 DNA insert used as control readily hybridized with the probe, as expected. Thus, although 7S contains a repetitive sequence(s), the short nucleotide stretch homologous to pλLC5-2 promoter region is not repeated. Whether this sequence in 7S RNA is flanked by a repeat sequence is yet to be determined. That 7S-like sequence in MLC_{2A} gene indeed plays a role in MLC_{2A} gene transcription is not known. Experiments are in progress where artificial gene constructs free of

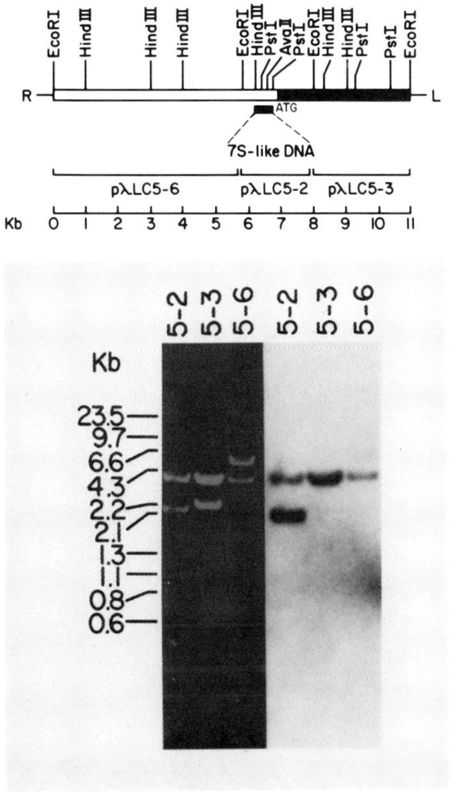

FIGURE 5. Localization of 7S-like sequence in λLC5 DNA. The sub-clones pλLC5.6, pλLC5.2, pλLC5.6, were digested with Eco RI and hybridized with nick-translated pSS48 (-dA) [^{32}P] DNA. The top figure shows a partial restriction map of λLC5 DNA and localization of 7S-like sequence based on further restriction digestion and hybridization with the probe above. Bottom portion shows the ethidium bromide profile (left) and the autoradiograph (right). The size markers are indicated on left.

this sequence are being tested for transcriptional activity by transfection into appropriate host cells.

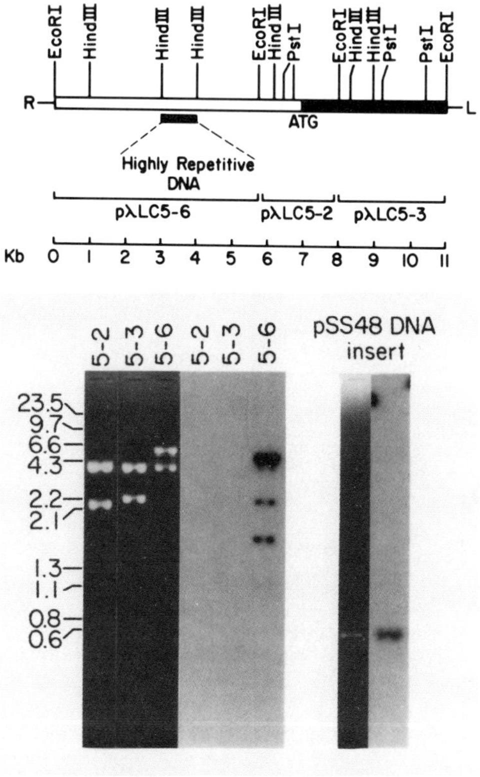

FIGURE 6. Localization of highly repetitive DNA sequences in λLC5 DNA. The sub-clones, pλLC5.6, pλLC5.3, pλLC5.3 were hybridized as above with the sheared repetitive DNA fraction of chick embryonic heart tissue (21). The top portion of the figure shows a partial restriction map of λLC5 DNA and localization of highly repetitive sequence by further restriction digestion and hybridization with the probe DNA. The bottom portion shows the hybridization pattern of the digested DNAs of λLC5 sub-clones and pSS48 DNA, used as control.

Isolation of 7S Gene: The genomic hybridization experiments with pSS48 DNA as probe showed that 7S-like sequences are repeated but not widely dispersed in the chick genome. Under high stringency, distinct bands of strong intensity were visible within 24 hr of exposure of the film. For the purpose of 7S gene isolation, a DNA band of about 2.8 kb that hybridized to pSS48

```
                                             -300
AAGCTTCAGGAGGCGGTGGTGGTGGGGCAAGAGTCCGCCA
                  -250
CCACCACCCGTTCTC..GCCCACTCTCCGAACCGATGGAT
                                  -200
CAAACAACCGCCTCTCCACCATCCAACTCTGATTCACCCA
                                      -150
CCTCCTCACAAAACAACTAATAACCCCCCTAAACAAGGCT

GGTCACAATGAGCCCTCCTACTCACCTCACTTATCCTAAG
        -100
ACTCCTCTTCCATTAACCTCCTAGGCCTTCCTTCCCCTAC
                  -50
ACCTTCACCCAACTACCCAGCTATCAATAAACATGGCCTT
                                  +1
AGCCCTGCCACTATGACTAGCCACCTTACTTAACAGGCCT
                                       50
ACGAAACCAACCCTCCGCCTCCTTAGGACACCTACTCCCT

GAAGGCACCCCCACCCCACTGATTCCAGCCCTAATCATAA
       100
TCGAAACAACCAGCCTACTTATTCGGCCATTAGCCCTAGG
                 150
AGTACGCCTAACAGCAAACCTCACAGCTGGTCACCTACTT
                                 200     BstE II
ATCCAACTTATCTCTACAGCCACAATCGCCCTACTACCAA
                                       250
TAATACCATCAATCTCCGCCCTAACGGCACTCATCCTATT

CCTACTAACCATCCTAGAAGTGGCAGTAGCAATAATCCAA
       300
GCCTACGTCTTCGTCCTCCTCCTAAGCCTCTACTTACAAT
                 350
GAAAATATTTAATGGCACACCAAGCACACTCCTACCACAT
                                  400
AGTTGACCCAAGCCCATGACCAATCTTCGGCGCAGGAGCA
                                      450
CTACTAACCACCTCTGGCCTAATCATATGGTCCACTACAG

CTCGACCACCCACTGACAATAGG
```

FIGURE 7. Partial sequence of 7S genomic DNA (pSS116). The genomic DNA containing 7S sequence was isolated as described in text. The entire sequence in the cDNA clone for 7S RNA (pSS48) with the exception of 3'-poly(dA) stretch (see text) is arbitrarily numbered +1 through 344 and underlined. The inverted repeats are indicated by arrows.

[^{32}P]DNA was recovered from the gel and the DNA was cloned into pBR322 at the Hind III site. On the basis of restriction mapping and hybridization, a prototype plasmid, pSS116, was chosen from a group of putative 7S genomic clones for further characterization. The base sequence data of this clone is shown in Fig. 6. The DNA contains the entire sequence of the cDNA plasmid, pSS48, (with the notable exception of the poly(dA) tract) and the flanking sequences of about 30 bp at 5'- end and about 1.5 kb region at the 3'-end. The lack of poly(dA) sequence in pSS116 would suggest that poly(A) is added to 7S RNA post transcriptionally, a feature characteristic of polymerase II transcription products. But previous experiments (unpublished data) and other structural features of 7S RNA (21) suggested that 7S RNA is transcribed by polymerase III. An examination of the sequence also revealed the presence of a consensus polymerase III promoter-like sequence and the T-rich termination signal, characteristic of polymerase III genes, in pS116 DNA. It is known that pseudogenes of many small RNAs, transcribed by polymerase III, are truncated at the 3'-end. However, unlike in pSS116, where only the polydA segment is deleted, the truncations in the pseudogenes occur at random. Whether or not the pSS116 DNA is a pseudogene is yet to be resolved. Preliminary experiments utilizing a construct containing pSS116 DNA in pSV0-Cat vector in a transient expression assay yielded an RNA product that hybridized to pSS48 DNA.

DISCUSSION

The primary sequence data for the eukaryotic genes have made available many interesting structural features which might be directly involved in regulatory interactions. Short, conserved, and repetitive sequence elements present in the flanking regions of many genes are apparently required for regulation (1-3; see also ref. 33). The sequences shared between human and rodent genes responding to glucocorticoids, for example, are 24 nucleotides in length, whereas repeats of 9 nucleotides in 5'-flanking regions of four genes are involved in inductive control in yeast. A conserved sequence, probably an octanucleotide, was reported to be a regulatory element present in the upstream region of V

CHARACTERISTICS OF SMALL RNAs

Property RNA	Alu-like	7S RNA (7S K, 7S L)	U-RNA	Chick Cardiac 7S
5' Ends	GGCPyPuGG	Some capped	Capped	Not capped
3'-oligo dA	Yes	Yes	No	Yes
Polymerase	Pol III	Pol III	Pol II	Pol III (?)
Size	300	300-400	100-200	400
Inverted Repeats	Yes	Some	No	Yes
Subcellular Localization	Nucleus/ cytoplasm	Nucleus/ cytoplasm	Nucleus	Cyto- plasm(?)
Ribonucleo- protein Complex	Some tran- scripts in RNPs	Yes	Yes	N.d.
Genome Copy	300,000	Middle repetitive	10-1000	Repeti- tive
Tissue Specificity	No	No	No	Yes
Conserved in Evolution	Yes	Yes	Yes	No
Function	?	Protein transport Gene regu- lation?	RNA pro- ces- sing	?

(variable) gene of immunoglobulin light chain (2) and the mutation in a 12 bp repetitive sequence in metallothionein gene promoter severely affected the metal regulation (3).

In the case of MLC_{2A} gene, we found that the 5'-flanking region of the gene shares homology with 7S sequence. Although we do not, at present, know whether this sequence is required for MLC_{2A} gene function, the key feature in this case is that a novel and physiologically relevant RNA species, product of the middle repetitive chick genome, identifies this sequence in the single copy MLC_{2A} gene and probably in other developmentally expressed muscle genes as well. Because of the specificity with which the 7S sequence recognizes the cardiac muscle gene sequences, the majority of which are expressed in a coordinate fashion, we speculate that 7S RNA is involved in coordinate control of these genes. We envision that the putative recognition and interaction of unrelated but coordinately induced genes with a regulatory factor(s), presumably a protein, is mediated by a common tissue specific element, such as 7S RNA. Such an interaction may result in disruption of the chromatin structure in a way to propagate signals to polymerase II and/or other proteins involved in initiation of gene transcription. We do not, at present, know what role, if any, the sequences present in the 3'-non-coding region of muscle genes, also homologous to 7S RNA, might play in this model. It is conceivable that the two regions might be recognized in a cooperative manner to fully provide the necessary activity. There are examples where multiple conserved elements are associated with efficient induction of gene activity (34-36). Implicit in this model of gene induction, however, is the availability of 7S RNA before the start of gene transcription for muscle proteins. Indeed, in early chick blastodermal cells (24 hr), 7S RNA is synthesized prior to the transition of chick myoblast to myocytes and this occurs before the mRNAs for the major proteins, MHC, MLC, actin, are actively transcribed.

The structural and biological properties of 7S RNA differ from those of other known small RNAs. The properties of these RNAs and their comparison with the chick cardiac 7S RNA are summarized in Table I. A notable difference is the high degree of tissue and species specificity in the 7S sequence of cardiac muscle. Although the three major small RNA species, namely 7S K and L RNAs and Alu-like RNAs share extensive homology with similar RNAs of other species, the pattern of repetitive DNA transcription, in general, appears to be tissue specific and is apparently regulated during development (32,37). The structural features of 7S RNA are similar to those transcribed by polymerase III. Although the putative 7S gene is amputated at the interjunction of poly(dA) stretch and the 5'-sequence, the presence of flanking repeats, the polymerase III consensus sequences and the 3'- poly(A) tail in the RNA are common to genes of many other small RNAs. These properties led to the hypothesis that a mechanism of reverse transcription might be responsible for the insertion of these sequences and their dispersion in the genome. The chick cardiac 7S sequence(s) is, however, less dispersed and repetitious than other small RNA genes which are apparently distributed widely and randomly and a heterogeneity is apparent in their genomic hybridization pattern. Thus, the 7S RNA appears to be a different but physiologically relevant entity and for further understanding of its role in gene regulation, the distribution and location of 7S-like sequences in many more developmentally regulated genes would need to be explored.

REFERENCES

1. Donahue TF, Davis RS, Lucchini G, Fink GR (1983). Cell 32:89.
2. Bergman Y, Rice D, Grosschedl R, Baltimore D (1984). Proc. Natl. Acad. Sci. USA 81:7041.
3. Stuart GH, Searle PF, Chen YH, Brinster RL, Palmiter RD (1984). Proc. Natl. Acad. Sci. USA 81:7318.
4. Tchurikov H, Naumova A, Zalentsova E, Georgiev E (1982). Cell 28:365.
5. Murphy D, Brickell PM, Latchman DS, Willison K, Rigby PWJ (1983). Cell 35:865.
6. Singer MF (1982). Cell 28:433.
7. Shimizu Y, Yoshida K, Ren CS, Fujinaga K, Rajagopalan S, Chinnadurai G (1983). Nature 302:587.

8. Digiovanni L, Haynes SR, Misra R, Jelinek WR (1983). Proc. Natl. Acad. Sci. USA 80:6533.
9. Houck CM, Reinhart FP, Schmid CW (1979). J. Mol. Biol. 132:289.
10. Ullu E, Tschudi C (1984). Nature 312:171.
11. Glundelfinger ED, DiCarlo M, Zopf D, Melli M (1984). EMBO Journal 3:2325.
12. Jagadeosawaran P, Forget B, Weissman SM (1981). Cell 26:141.
13. Jelinek W, Leinwand L (1978). Cell 15:205.
14. Weiner AM (1980). Cell 22:209.
15. Harada F, Kato N (1980). Nucleic Acids Res. 8:1273.
16. Romanoff AL (1960). In The Avian Embryo, MacMillan Press, New York.
17. Arnold HH, Siddiqui MAQ (1979). Biochemistry 18:5641.
18. Arnold HH, Krauskopf M, Siddiqui MAQ (1983). Nucleic Acids Res. 11:1123.
19. Jakowlew SB, Khandekar P, Datta K, Arnold HH, Narula SK, Siddiqui MAQ (1982). J. Mol. Biol. 156:673.
20. Arnold HH, Domday H, Weibaur K, Datta K, Siddiqui MAQ (1982). J. Biol. Chem. 257:9872.
21. Khandekar P, Saidapet C, Krauskopf M, Zarraga AM, Lin WL, Mendola C, Siddiqui MAQ (1984). J. Mol. Biol. 180:417.
22. Winter B, Klapthor H, Weibaur K, Delias H, Arnold HH (1985). J. Biol. Chem. (In Press).
23. Matsuda G, Maita T, Kato Y, Chen JI, Umegawa T (1981). FEBS Lett. 135:232.
24. Nudel U, Calvo JM, Shani M, Levy Z (1984). Nucleic Acids Res. 12:7175.
25. Nabeshima Y, Fuji-Kuriyama Y, Muramatsu M, Ogata K (1984). Nature 308:333.
26. Breathnach R, Chambon P (1981). Ann. Rev. Biochem. 50:349.
27. Efstratiadis A, Posakony JW, Maniatis T, Lawn RM, O'Connell C, Spritz RA, DeRiel JKD, Forget BG, Weissman SM, Slighton JL, Blechl AE, Smithies O, Baralle EE, Shoulders CC, Proudfoot NJ (1980). Cell 21:653.
28. Moscovici C, Giovannella-Moscovici M, Jimenez H, Lai MC, Hayman MJ, Vogt P (1977). Cell 11:95.

29. Gorman C, Moffat L, Howard B (1982). Mol. Cell Biol. 2:1044.
30. Weaver RF, Weissman C (1979). Nucleic Acids Res. 7:1175.
31. Arnold HH (In this issue).
32. Murphy S, Altrada F, Ullu E, Tripodi M, Silengo L, Melli M (1984). J. Mol. Biol. 177:575.
33. Davidson EH, Jacobs HJ, Britten RJ (1983). Nature 301:468.
34. Pelham HRB (1982). Cell 30:517.
35. Hinnebirsch AL, Firk ER (1983). J. Biol. Chem. 258:5238.
36. Payvar F, DeFranco D, Firestone GL, Edger B, Wrange D, Oknet S, Gustafson JA, Yammamoto Y (1983). Cell 35:381.
37. Constantini FD, Britten RJ, Davidson EH (1980). Nature 287:111.
38. Maxam AM, Gilbert W (1980). Methods in Enzymology 65:499.

Molecular Biology of Muscle Development, pages 793–807

STRUCTURE AND EXPRESSION OF MUSCLE GENES IN CHICKEN: DNASE I SENSITIVITY AND HYPOMETHYLATION

B. Winter, K. Wiebauer, P. Lohse, and H.H. Arnold

Department of Toxicology, Medical School,
University of Hamburg, Grindelallee 117
D-2000 Hamburg 13, FRG

ABSTRACT The genes coding for cardiac myosin light chain (MLC2-A) and for cytoplasmic ß-actin were isolated from chicken by molecular cloning techniques. These genes were characterized by nucleotide sequence analysis and their expression pattern. To study the regulation of muscle gene expression, DNAse I hypersensitive sites and the state of methylation were determined for the MLC2-A gene in expressing and nonexpressing tissues. A DNA segment located about 1.5 kb upstream of the promoter exhibits a heart-tissue specific DNAse I-hypersensitive site and a region of hypomethylation. The correlation of these two parameters with the expression of the gene suggest a functional involvement of this upstream DNA element in gene regulation. Transfection of the cytoplasmic ß-actin gene into myogenic cell lines results in its down-regulation when cells are induced to differentiate. Since this resembles the regulation of the ß-actin gene *in vivo* , the transfection system seems suitable to investigate the molecular basis for its expression in myogenic cells.

INTRODUCTION

The contractile proteins of the striated muscles are encoded by multigene families and constitute an excellent system to investigate differentiation and developmental control of gene expression. As muscle cells differentiate a set of genes including those coding for the contractile apparatus are being activated, whereas other genes, like the cytoplasmic ß-actin are down-regulated at the same time. To study the gene structure and the molecular mechanisms underlying differential gene expression the cardiac myosin light chain gene (MLC2-A) (1,2) and the cytoplasmic ß-actin gene were isolated from a genomic chicken library. The nucleotide sequences of both genes were determined (2,3). It has been shown for a number of genes that transcriptional activity is associated with alterations in chromatin structure (4,5). This is revealed by the preferential sensitivity of DNA in intact nuclei to digestion with DNAse I. Moreover, so-called DNAse I-hypersensitive sites, particularly in actively expressed genes have been described (6,7). These hypersensitive regions represent defined local alterations of the chromatin, detected by the ability of DNAse I to introduce cuts into the DNA double-strand (8,9). The precise nature of the DNAse I-hypersensitive sites is not known but they seem to be associated with specific DNA secondary structure (10) and are suggestive for interactions of specific factors with distinct gene regions. Many hypersensitive sites have been found near the 5'end of a gene, frequently in close association with the transcriptional promoter (11-13) but also in regions located further upstream (14-16).

Based on the analysis of several genes in which a reproducible correlation between the degree of hypomethylation and the level of expression has been found (17), the general hypothesis has emerged that expressed genes are hypomethylated whereas silent genes are fully modified.

To identify DNA segments of the MLC2-A gene which might be essential for its tissue specific transcription we examined the chromatin for DNAse I-hypersensitive sites and naked DNA for its pattern of methylation. Several tissue-specific DNAse I-hypersensitive sites were detected in heart muscle. One site located about 1.5 kb upstream of the promoter is particularly interesting since the same DNA region is markedly and specifically undermethylated in heart tissue. The strict correlation between a DNAse I-hypersensitive site and corresponding hypomethylation on one hand and expression of the gene on the other hand suggests a functional involvement of the 5'flanking DNA segment in the regulation of MLC2-A gene expression.

Regulated expression of the cytoplasmic ß-actin could be achieved in differentiating myogenic cell lines which had been transfected with the gene cloned into the selectable vector pSV-neo 2.

EXPERIMENTAL PROCEDURES

Isolation of Chicken Genes from a Genomic Library.

$1x10^6$ recombinants of a chicken library (20) were screened with nick translated pML10 DNA, a cDNA clone for the MLC2-A gene (1) and also with cloned cDNA coding for cytoplasmic ß-actin (21). Four overlapping recombinant phages were isolated that contained the structural gene for MLC2-A and two clones containing the cytoplasmic ß-actin gene. The identity of the clones was confirmed by direct nucleotide sequence analysis (22).

Determination of DNAseI-hypersensitive Sites

Intact nuclei were isolated by the procedure of Hewish and Burgoyne (23) using polyamines as a stabilizing agent in the homogenization buffer. Nuclei were digested with DNAse I at concentrations from 0.1 to 16 units per mg of DNA for 30 min at 15°C. DNA isolated from DNAse I treated nuclei was digested with appropriate restriction endonucleases, separated on agarose gels, and blotted onto nitrocellulose filters. For the indirect end-labelling technique described by Wu (8), suitable hybridization probes were subcloned, labelled by nick translation, and hybridized to the filters.

Determination of the Methylation Pattern in the MLC2-A Gene.

DNA was isolated from embryonic heart and liver and digested with either Eco RI, Hind III or BAM H1. Samples of these digests were additionally digested with either Hpa II, Hha I or Msp I. To determine methylated sites in the gene specific fragments the digested DNA was run on agarose gels, blotted on nitrocellulose filters, and hybridized to either cloned cDNA or genomic subclones of the MLC2-A gene.

Transfection of Cloned Genes into Myogenic Cell Lines.

The ß-actin Gene. A 10.5 kb BAM H1 fragment of the recombinant phage -Ac31 was subcloned in the single BAM H1 site of plasmid vector pSV-neo 2 (24) conferring geneticin (G 418) resistance to transfected cells. RNA was isolated from stably transfected C2 mouse myoblasts and also L8 rat myoblasts before and after fusion to differentiated myotubes. Nuclease S1 protection experiments were performed using a single strand specific probe from the 3' untranslated region of the ß-actin gene.

RESULTS

Restriction Maps and Partial Sequences of Cloned Chicken Genes.

A chicken library constructed from partial Hae III and Alu I digested DNA ligated with Eco RI linkers into -charon 4A vector (20), was screened with cloned chicken cardiac myosin light chain cDNA probe (pML10; ref.1) and cloned ß-actin cDNA (21). Partial restriction maps and exon-intron organization of the MLC2-A gene and of the ß-actin gene are shown in Fig.1. The complete sequence of the MLC2-A gene has been determined and will be published elsewhere. The sequence at the 5'end including exons 1 and 2 is shown for the unambiguous identification of the gene (Fig.2). Exon 1 contains the 5' leader sequence plus the AUG start codon and is separated from exon 2 by the largest intron of the entire gene. It is interesting to note that this intervening sequence contains a direct repeat of 28 nucleotides in lenght (interrupted by 2 nucleotides) close to the 3' splice junction. The ß-actin gene was partially sequenced and compared to the nucleotide sequence published by Kost et al. (3). Except for minor variations both independently isolated genes exhibit the same sequence in noncoding regions. The identity of the gene was also established by direct sequence analysis of exon sequences.

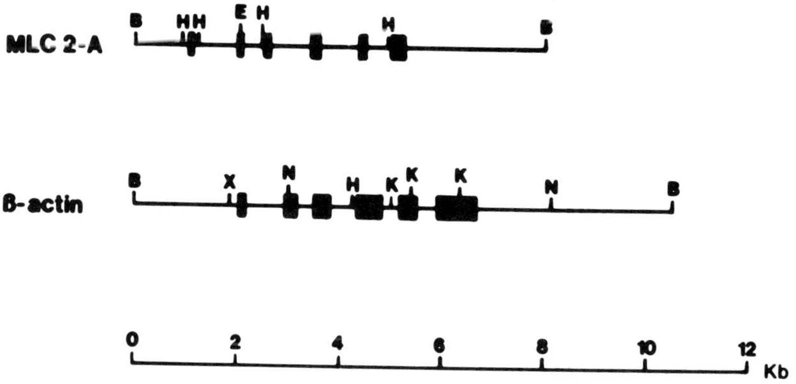

FIGURE 1. Physical Maps of the Genes Coding for MLC2-A and ß-Actin.

The exon-intron organization was derived from restriction mapping, heteroduplex-analysis, and nucleotide sequence analysis.

Down-regulation of the Cytoplasmic Actin Gene in Differentiated Myocytes.

The steady-state levels of mRNA coding for the muscle specific -actin and the cytoplasmic ß-actin was determined in skeletal muscle of developing chicken embryos from days 5 to 20 in ovo. Whereas the concentration of -actin mRNA gradually increases during this time period the concentration of ß-actin mRNA drops to almost undetectable levels by day 20 when most cells are fully differentiated (Fig. 3). To study the down-regulation of the cytoplasmic mRNA, we cloned a 10.5 kb BAM H1 fragment containing the complete ß-actin structural gene plus 5' and 3'flanking DNA into the dominant selection vector pSV-neo 2 that confers resistance against geneticin (G 418) to successfully transfected cells. For the determination of chicken ß-actin mRNA nuclease S1 protection experiments were carried out with the 3'untranslated sequence of a chicken ß-actin clone (21) that had been introduced into M13 phage mp8

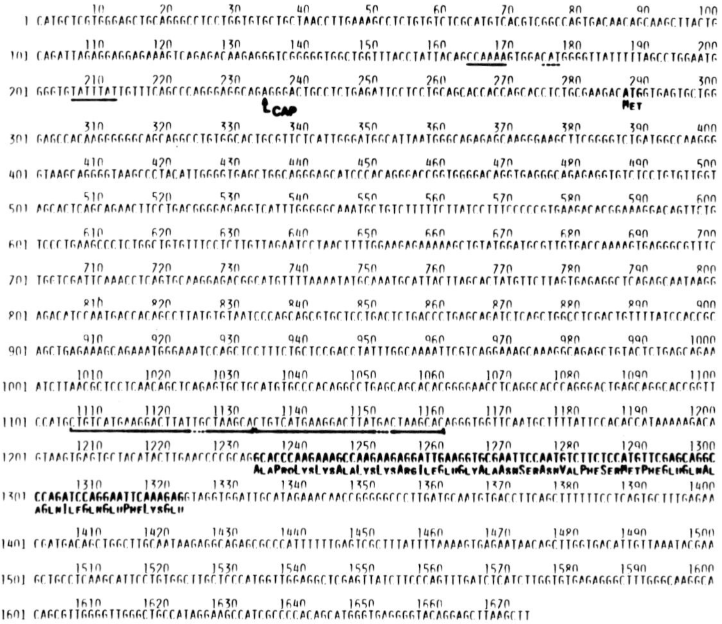

FIGURE 2. Nucleotide Sequence of the 5'End of the MLC2-A Gene.

The cap site was determined by primer-extension and S1 protection experiments. The putative promoter elements and the long direct repeats are underlined.

generating a single stranded, specific probe. When C2 mouse myoblasts or L8 rat myoblasts were transfected with the pSVneo-actin hybrid vector numerous stable transfectants could be isolated under selective conditions. The expression of chicken ß-actin mRNA was measured in RNA from proliferating undifferentiated myoblasts and also from differentiated myotubes. As shown in Fig. 4, in all isolated clones the chicken ß-actin gene is readily expressed in undifferen- tiated myoblasts, but appears to be down-regulated after the cells were shifted to conditions in which they differentiate and form multinucleated myobtubes. This result resembles the in vivo situation of the ß-actin gene in myogenic cells and therefore the transfected gene constitutes a suitable system to investigate

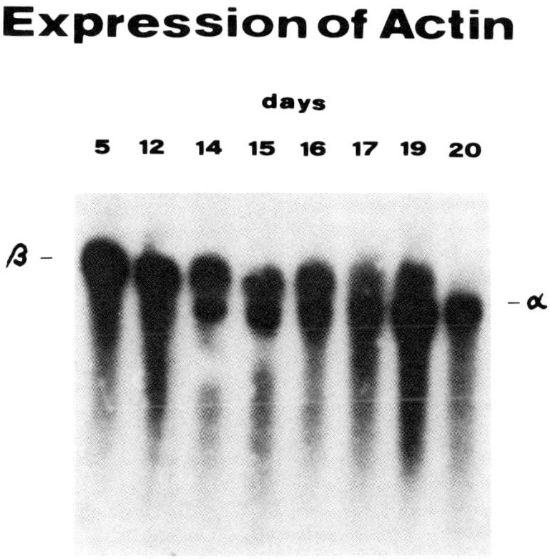

FIGURE 3. Steady-State Levels of Actin mRNAs in Developing Chicken Skeletal Muscle.

RNA was isolated from embryonic chicken leg muscle at days indicated, blotted onto filters and hybridized to nick-translated (^{32}P) actin cDNA.

---

further which parts of the gene might be essential for the down-regulation.

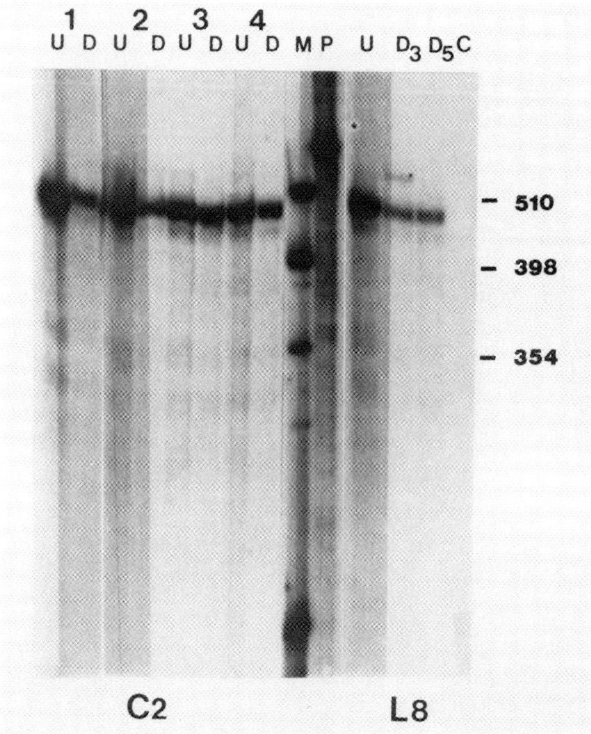

FIGURE 4. Nuclease S1 Analysis of RNA from L8- and C2 Myogenic Cell Lines Containing the pSVneo 2-Actin Construct.

RNA from undifferentiated (U) and differentiated (D) C2 myoblasts and L8 myoblasts (U: undiff.; D_3 = 3 days diff.; D_5 = 5 days diff.) were hybridized to a radioactive labelled single strand ß-actin probe (P) and challenged with nuclease S1. The protected fragments were separated on a 6% PAA-gel. M = size marker; C = RNA from untransfected C2 cells.

Tissue Specific DNAse I-Hypersensitive Sites and Hypomethylation in the Cardiac MLC2-A Gene.

As shown in Fig. 5 the MLC2-A gene in chicken is

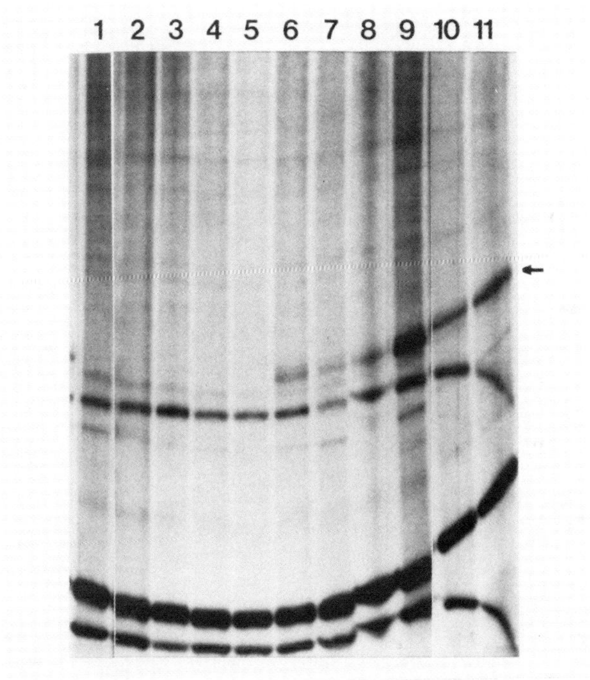

FIGURE 5. Analysis of RNA from Several Chicken Muscles by Primer Extension.

No RNA (lane) or RNA from chicken ALD muscle (lane 2), pectoralis muscle (lane 3), embryonic leg muscle (lane 4), adult leg muscle (lane 5), embryonic heart days 12-16 (lanes 6-10) and adult heart (lane 11) were hybridized to a 5'labelled 42 bp Eco RI fragment of MLC2-A exon 2 and transcribed with AMV reverse transcriptase. The 140 bp extension product indicative for the MLC2-A mRNA is marked by the arrow.

specifically expressed only in heart muscle. Using a 42 bp primer from exon 2 and RNAs from several different skeletal muscles as well as from embryonic and adult heart the properly extended fragment can only be detected in heart. In order to identify DNA segments of the MLC2-A gene which might be essential for the tissue specific expression we examined the methylation pattern and the DNAse I-hypersensitivity in chromatin. This analysis was extended into the 5' and 3' gene flanking sequences. The general strategy and the results of the experiments are shown in fig. 6.

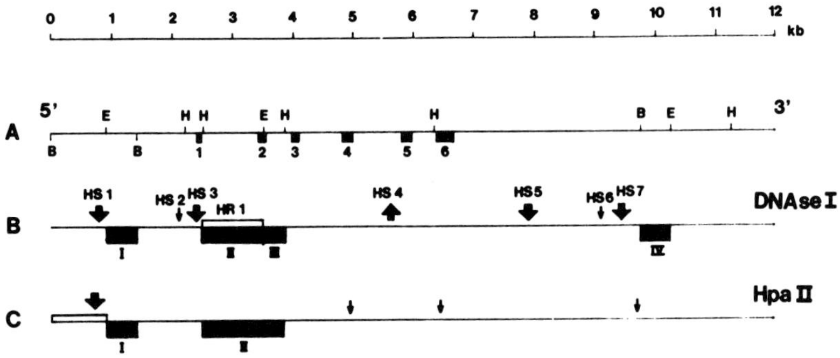

FIGURE 6. Map of DNAse I-Hypersensitive Sites and Hypomethylated Regions of the Chicken MLC2-A Gene.

A) Partial restriction map with exons 1-6 indicated by solid boxes. B = BAM HI; E = Eco RI; H = Hind III.
B) Map of localized DNAse I cleavage sites. Tissue specific hypersensitive sites of heart chromatin are indicated by broad vertical arrows. Thin vertical arrows indicate DNAse-I-hypersensitive sites found in nuclei of heart and liver. The black boxes underneath the map represent the different fragments (I to IV) used as hybridization probes.
C) Map of hypomethylated regions as detected by DNA restriction analysis with the enzyme Hpa II. The thin arrows indicate sites of hypomethylation common to heart and liver. The thick arrow shows a region of hypomethylation found only in DNA from heart.

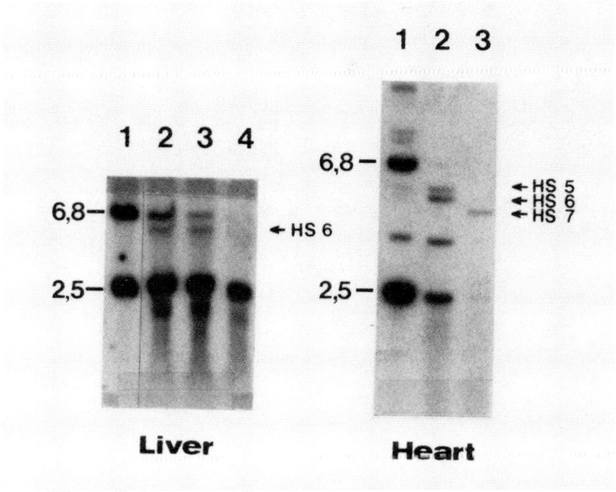

FIGURE 7. DNAse I digestion pattern of the MLC2-A gene region.

Liver and heart nuclei were treated with increasing concentrations of DNAse I as described in Materials and Methods.

The following concentrations of DNAse I were used: For liver: no DNAse I (lane 1); 1 unit/ml (lane 2); 2 units/ml (lane 3); 4 units/ml (lane 4). For heart: no DNAse I (lane 1); 0.5 units/ml (lane 2); 1 unit/ml (lane 3). For each lane 15 ug DNA isolated from nuclei were digested with Eco RI to completion and hybridized to probes II and III.

The positions of DNAse I cleavage sites were determined in intact nuclei from heart and liver tissues by the indirect end-labelling technique. As shown in Fig. 7, when nuclei from liver and heart were treated with increasing concentrations of DNAse I and DNA isolated from these nuclei was digested with Eco RI, blotted on filters, and hybridized with probe II and III (see Fig. 6), one DNAse I-hypersensitive site was found in liver nuclei but two additional sites were found in heart. Both tissue-specific sites HS 5 and HS 7 are located in the 3'flanking DNA. A similar experiment now using probe I for DNA digested with BAM HI is shown in fig. 8. In

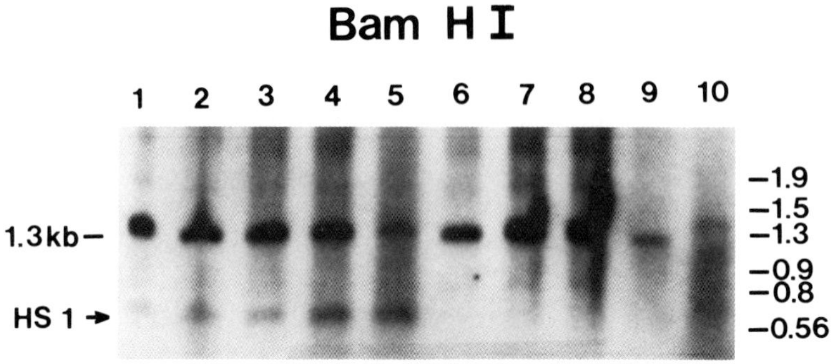

FIGURE 8. DNAse I-specific cleavage site in the 5'flanking gene region.

The gel contains BAM HI digested DNA from DNAse I treated heart nuclei in lanes 1 through 5, and DNA from liver nuclei in lanes 6 through 10. Probe I was used for hybridization. The hypersensitive site HS 1 appears only in heart DNA.

---

this experiment a heart tissue specific DNAse I-hypersensitive site (HS1) could be detected which is located about 1.5 kb upstream of the gene promoter.

The extent of DNA methylation was analyzed throughout the gene. DNA was isolated from chicken heart and liver and digested with the methylation-sensitive restriction enzymes Hpa II and Hha I, followed by digestion with either Hind III (Fig. 9 A), Eco RI (Fig. 9 B) or BAM HI (Fig. 9 C). The isoschizomere Msp I served as control for the presence of Hpa II recognition sites. When Hind III digested DNA was hybridized to cloned radioactive cDNA or Eco RI digested DNA was hybridized to probe II, the pattern obtained for heart and liver DNA were identical indicating the same degree of methylation in both tissues (Fig. 11 A and B). In contrast, when BAM HI digested DNA was hybridized to probe I, the 1.35 kb 5'flanking DNA fragment appeared markedly hypomethylated in DNA from heart tissue but completely modified in DNA from liver (Fig. 11 C). This hypomethylated region is the only tissue-specific undermethylation of the complete gene and it coincides with the DNAse I hypersensitive site HS 1.

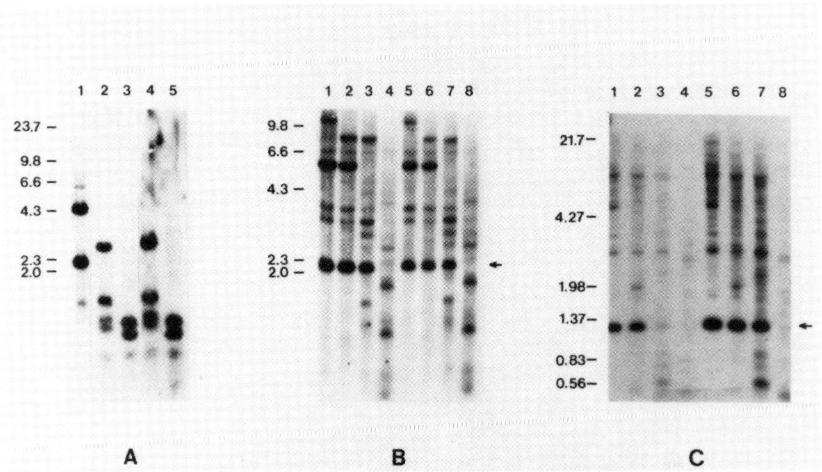

FIGURE 9. The Methylation Pattern of the MLC 2-A Gene.

A) DNA from heart and liver was digested with Hind III and subsequently either with Hpa II or Msp I. Lanes 1 to 3 contain DNA from heart and lanes 4 and 5, DNA from liver. The blot was hybridized to radioactive cDNA clone pML10 (13). Digestion was carried out with Hind III (lane 1), Hind III plus Hpa II (lanes 3 and 4), Hind III plus Msp I (lanes 3 plus 5).
B) DNA from heart (lanes 1-4) and liver (lanes 5-8) were digested with Eco RI (lanes 1 and 5), Eco RI plus Hha I (lanes 2 and 6), Eco RI plus Hpa II (lanes 3 and 7), and Eco RI plus Msp I (lanes 4 and 8). The subclone pMLC 17-I (genomic 1.35 kb Hind III fragment) was used as hybridization probe. The 2.5 kb genomic fragment indicative for the MLC2-A gene is indicated by the arrow.
C) DNA from heart (lanes 1-4) and liver (lanes 5-8) was digested with BAM HI (lanes 1 and 5) and in addition with Hha I (lanes 2 and 6), Hpa II (lanes 3 and 7) and Msp I (lanes 4 and 8). Probe I which detects a 1.3 kb fragment of gene flanking DNA (see arrow) was used for hybridization.

DISCUSSION

We have studied the regulation of the cytoplasmic actin gene and the cardiac myosin light chain 2 gene from chicken. Whereas the ß-actin gene is ubiquitously expressed in all tissues except in differentiated muscle cells where it is down-regulated, the MLC2-A gene is expressed specifically only in terminally differentiated heart muscle. The mechanism(s) underlying the down-regulation of the ß-actin gene are not understood and could include repression of transcriptional activity, increase of mRNA degradation or both. The fact that the transfected ß-actin gene also shows decreased steady-state concentrations of mRNA in differentiated muscle cells should facilitate future experiments designed to determine which parts of the gene participate in its regulation.

Our approach to determine DNAse I-hypersensitive sites and hypomethylation in the MLC2-A gene was an attempt to assigne cis-acting gene elements which are associated with tissue-specific gene expression. The detection of a 5'upstream DNA segment that shows tissue-specific DNAse I-hypersensitivity and hypomethylation will open up further investigations regarding the structure and function of this DNA.

ACKNOWLEDGEMENTS

We thank I. Humburg, H. Glaser and K. Söhren for their skillful technical assistance and Chr. Mink for typing the manuscript. The work was supported by grants from Deutsche Forschungsgemeinschaft.

REFERENCES

1. Arnold, H H, Krauskopf, M, Siddiqui, MAQ (1983). Nucleic Acids Res 11: 1123-1131.
2. Winter, B, Klapthor, H, Wiebauer, K, Delius, H, Arnold, H-H (1985).J Biol Chem, in press.
3. Kost, Th, Theodorakis, N, Hughes, SH (1983). Nucleic Acids Res 11: 8287-8300.
4. Weintraub, H, Groudine, M (1976).Science 93: 848-858.
5. Mathis, D, Oudet, P, Chambon, P (1980). Proc Nucleic Acid Res Mol Biol 24: 1-556.
6. Wu, C, Bingham, PM, Livak, K, Holmgren, R, Elgin, SCR (1979).Cell 16: 797-806.
7. Stadler, J, Larsen, A, Engel, JD, Dolan, M, Groudine, M, Weintraub, H (1980).Cell 20: 451-460.
8. Wu, C (1980). Nature (London) 284: 856-860.
9. Cartwright, JL, Abmayr, SM, Fleischmann, G, Lowenhaupt, K, Elgin, SCR (1982).CRC Crit Rev Biochem 13: 1-86.
10. Weintraub, H (1983). Cell 32: 1191-1203.
11. Keene, MA, Elgin, SCR (1981). Cell 27: 57-64.
12. McGhee, JD, Wood, WJ, Dolan, M, Engel, JD, Felsenfeld, G (1981). Cell 27: 45-55.
13. Sweet, RW, Chao, MV, Axel, R (1982). Cell 31: 347-353.
14. Chung, SY, Folsom, V, Wooley, Y (1983). Proc Natl Acad Sci USA 80: 2427-2431.
15. Burch, JBE, Weintraub, H (1983). Cell 33: 65-76.
16. Fritton, HP, Sippel, AE, Igo-Kemenes, T (1983). Nucleic Acids Res 11: 3467-3484.
17. Razin, A, Riggs, AD (1980). Science 216: 604-610.
18. Ehrlich, M, Wang, RY-H (1981). Science 212: 1350-1357.
19. Wigler, M (1981). Cell 24: 285-286.
20. Dodgson, JB, Strommer, J, Engel, JD (1979). Cell 17: 879-887.
21. Cleveland, DW, Lopata, MA, MacDonald, RY, Cowan, NJ, Rutter, WJ, Kirschner, MW (1980). Cell 20: 95-105.
22. Maxam, AM, Gilbert, W (1977). Proc Natl Acad Sci USA 74: 560-564.
23. Hewish, DR, Burgoyne, LA (1973). Biochem Biophys Res Commun 52: 475-481.
24. Subramani, S, Mulligan, RS, Berg, R (1981).Mol Cell Biol 1: 854-864.

Molecular Biology of Muscle Development, pages 809–823

MOLECULAR ANATOMY OF CARDIAC MYOSIN HEAVY CHAIN GENES[1]

Patrick K. Umeda,[†¶] James E. Levin,[§2] Achyut M. Sinha,[†] Leanne L. Cribbs,[†] Douglas S. Darling,[†] David J. Ende,[†] Huey-Juang Hsu,[†] Evelyn Dizon[†] and Smilja Jakovcic[†]

Departments of [†]Medicine, and [§]Molecular Genetics and Cell Biology, University of Chicago
Chicago, Illinois 60637

ABSTRACT We have analyzed the structure of the 5' regions of the rabbit α- and β-ventricular myosin heavy chain (HC) genes. In both genes, the N-terminal exon encodes the first 67 amino acids of HC, and also contains 12 base pairs (bps) of 5' untranslated mRNA sequences in the HCα gene, but only 8 bps in the HCβ gene. This exon, however, does not include the cap site of the gene, since primer extension analysis indicates that the 5' untranslated region of the HCα and HCβ mRNAs are 68 and 99 nucleotides (nts) in length, respectively. Sequence analysis of the HCα primer extension cDNA localizes the remaining 5' untranslated sequences to two separate exons, and identifies the cap site approximately 920 bps upstream of the N-terminal exon. Comparisons of the 5' flanking regions of the rabbit HCα gene with those of the rat HCα, reveals a 400-bp region of 80% sequence homology immediately upstream of the cap site of both genes. The strong conservation of non-coding sequences suggests a functional role of this region in regulating the expression of the cardiac myosin HCα gene. In constrast, only

[1]This work was supported by in part by USPHS Grants HL09172 and HL20592 from the National Institutes of Health, a grant from the Muscular Dystrophy Association, and an [¶]Established Investigatorship of the American Heart Association and with funds contributed in part by the Chicago Heart Association.
[2]Recipient of the NIGMS-NRSA Grant GM07281 from the National Institutes of Health.

limited homology was noted between the 5'non-coding sequences of the rabbit α- and β-HC genes. The divergent 5' seqences have been used to directly monitor transcription of the respective genes in nuclear run-off experiments.

INTRODUCTION

Cardiac and skeletal muscles contain multiple molecular forms of myosin HCs specifying myosins having different ATPase activities (1-7). The expression of these isomyosins follows a defined developmental pattern that can be altered by physiologic stimuli. Since myosin ATPase activity has been directly correlated with the velocity of fiber shortening (3), the changes in isomyosin expression may alter the contractile properties of the muscle.

In the mammalian ventricle, there are two major forms of myosin HCs, HCα and HCβ, specifying myosins having a relatively high and low ATPase activity, respectively (4-6). In addition, minor HC variants have also been described in selective regions of the myocardium (7). The expression of the α- and β-HCs follows a distinct developmental program in both atria and ventricles (8,9), and can be altered as a result of increased hemodynamic load due to either physiologic stimuli (4-6), or pathologic conditions (10,11). In particular, the administration of large doses of thyroid hormone to adult rabbit expressing primarily the β form of the protein, results in complete conversion to the α form in 2 to 3 weeks (6,8). Conversely, administration of the anti-thyroid drug propylthiouracil (PTU) leads to the expression of only the β isoform. Comparisons of the relative synthesis rates of the α- and β-HCs with the relative amounts of the respective mRNAs indicate that these changes of myosin HC expression are regulated at the pretranslational level (12). Thus, thyroid hormone may be directly regulating the transcription of the cardiac myosin HC genes, similar to its action on the genes for growth hormone (13,14) and malic enzyme (15).

In order to examine the molecular mechanisms responsible for the differential expression of cardiac myosin HC during development and after thyroid hormone administration, we have isolated the genes for the rabbit cardiac myosin HCs (16). Here we report the characterization of the 5' portions of the α- and β-HC genes, and the delineation of the transcriptional initiation site of the HCα gene. The structural analysis reveals a unique genomic organization where the 5' untranslated sequences are interrupted by two introns, and also identifies non-coding sequences that are highly conserved in both the rabbit and rat myosin HCα genes. The characterization of the 5' regions of the cardiac myosin HC genes enables the direct analysis of their transcription using nuclear run-off experiments.

RESULTS AND DISCUSSION

Structure of the Rabbit Cardiac Myosin HC Genes.

We have previously reported the isolation and characterization of genomic clones encoding the two major forms of cardiac myosin HCs in the rabbit (16). The organization of the coding and intervening sequences of the cloned gene regions is shown in Figure 1. Within two overlapping clones, λMHCα12/1 and λMHCα8/1, approximately 25 kbps of genomic DNA encoding the HCα gene have been reconstructed. The HCα gene is divided into at least 37 exons ranging in size from 30-300 bps. The 3' terminal exon of the gene encodes 7 amino acids at the C-terminus of the protein, and all of the untranslated sequences of the HCα mRNA. The latter sequences establish the identity of the HCα gene, since S1 nuclease mapping analyses (17) demonstrate distinct differences in the 3'untranslated sequences of the HCβ mRNA.

Within another genomic clone, λMHCβ6/1, we have reconstructed the 5' half of the HCβ gene. DNA sequence analysis of the 5' and 3' exons indicates that this region encodes

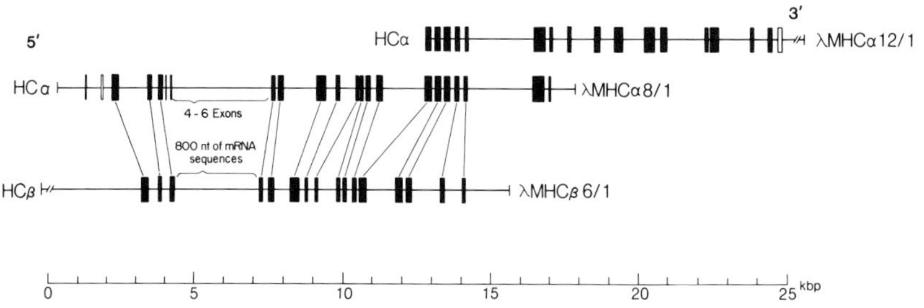

FIGURE 1. Structure of the rabbit α- and β-ventricular myosin HC genes. Shaded boxes designate the protein coding exons, while the open boxes represent the 5'- and 3'-untranslated mRNA sequences.

the entire S1 and 275 amino acids of the S2 portions of the HC. As shown by electron-microscopic heteroduplex analysis (16) and DNA sequencing (see below), the corresponding exons of both genes are similar in size, but are interrupted by introns that differ markedly in both their length and sequence. The differences in the introns show that the α- and β-genomic clones represent two distinct genes.

Thus far, DNA sequence analysis of exons encoding the N-terminal portions of the protein confirms the conservation of intron positions in the protein coding regions of the rabbit α- and β-HC genes suggested by the electron-microscopic heteroduplex mapping. The locations of introns are also the same for the 5' regions of two rat sarcomeric myosin HC genes (18,19), and for the 3' portion of a human myosin HC gene (unpublished) that have been sequenced. The conservation of intron positions in the protein coding regions may be a general feature of all mammalian sarcomeric heavy chain genes, similar to the structure of other gene families, such as globin (20) and vitellogenin (21).

Structure of the 5' Coding and Flanking Regions.

Although myosin HC genes of several different species have been isolated and characterized (16,18,19,22-27), the transcriptional initiation (cap) site of the gene has not been rigorously identified. Therefore, to characterize the 5' region of the cardiac myosin HC genes, we have subcloned and sequenced the corresponding portions of the genomic clones.

Figure 2 shows the sequences upstream from the N-terminal exons of the cardiac myosin HC genes. The high degree of homology between the derived amino acid sequence in this region and the directly determined sequence of rabbit skeletal myosin HC (61% for HCα and 70% for HCβ) (28), identifies the N-terminus of the protein. In both genes, this exon encodes only the first 67 amino acids of the HC. The 3' boundary of the exon is flanked by a GT dinucleotide that is characteristic of 5' splice junctions (29), and is confirmed by the identification of downstream exons in both genes encoding the next 48 amino acids of the HC. There is over 80% homology in the nucleotide sequences of the α and β N-terminal coding regions, while the flanking and non-coding regions are highly divergent.

The HCα gene sequences also contain consensus CAT and TATA boxes characteristic of eukaryotic promoters (30) approximately 920 bps upstream of the initiation codon. To

```
                                PstI PvuII
Rabbit HCα: ....... aaggga ggga aactgcagct gggggcagg ggacaagcat tcgtcctata tgaaaaggtg     60

accctcaccc cagtgtgctc aactcaccct tcaggttaaa aataaccgag gtaaaggcca tggtggggcg ggggaggtgg    140

tggagaaggt cctgtcttcc cactatctgc ccatcagcgc tctggagggg cggaatgtgc tcaaggacta aaaaaaggcc    220

      AvaI
ctggagcccg aggggctggg gcagcagacc tttcatgggc aaatctgggg gccctgccgt cctcttgtca cctccagagc    300

caagggatca agggaggagg agccaggcag ggagagaggt gggagggggt ccctccggaa gggctccaaa tttaggcaag    380

                               PstI                   ↓    PstI
ggggtggggg aagtggcata taaaggagct gcagcgttga gggcagccag ACTCCTGCAG CCCAGgtaag tgtggtttgt    460

        PvuII
gcaggggcac tgtgcagctg tgcacaggga ccctgccttc ctgggagtga gctgcagggc aggcgggggc ccacgggttc    540

tgctggaagc gcccttctcc agccctgtgg caggtccatg aggagcgcga ctcctgtctc ctgtctctct ggttccctgc    620

ccggttctca gcacttgctc cctcctttgg gatggcttcc tgcccgttgg ccctgtgctg ctttgcctcc ctgtcttggt    700

cagcacttgt cccagtcctc aatctccgtc tccctccttc cttgccctcc cttcgtgcct gtgtgttctt ttctccctgc    780

ctttctctgt ttctgtctct gcttccacac ccctgccctc ccacattcac tcactctgca tcccttcctt tcttcctact    860

ccctgccggt ccaccctgct ggtgtctgtc ggctctctgg gctgggcctc ctcgtcccat cacttgtctc catgattctc    940

ttctctgcac tttgtggctc ctgctcctcc agTGGTTCAG ATTCTCCGGG ATTCCCTGTG AGGAGAGAGT CAGgtgagcc   1020

        PstI
gcacccagtc tcctctgcag acagcagagc aggctttctc cgggagccga gtccaccccc tggcaacccc aggcactcct   1100

                                                PvuII
tccccccatg ccgacctctt acaaagatcc cccctcagcc ccagctgaat ggggtgggtg gccccaggca gatcctttca   1180

ctgggtaatg agaggagagc aataggtcct ggaaaggagg tggtggggca aggaatggga atgggaaaga cagggtcacc   1260

                  AvaI                                       Met Asp Asp Ser
cacgaaagcc tgagtctgcc tcgggagtca caggaccccc tgtcttggcc ccagGGGAGA GGCAAG ATG GAT GAC TCC  1338
Rabbit HCβ: ........ gacagatggg aggaaccagc ttcttccgct cactacagGT ACAGCC ATG GGG GAC GCG    58
                                                                 Gly     Ala
                                                           XbaI
Gln Met Ala Asp Phe Gly Ala Ala Ala Gln Tyr Leu Arg Lys Ser Glu Lys Glu Arg Leu Glu Ala
CAG ATG GCC GAC TTT GGG GCG GCA GCC CAG TAC CTC CGC AAG TCC GAG AAA GAG CGT CTA GAG GCC   1404
GAG ATG GCC GTG TTT GGG GCT GCC GCC CCC TAC CTG CGC AAG TCA GAG AAG GAA CGG CTG GAA GCC    124
Glu         Val                     Pro

Gln Thr Arg Pro Phe Asp Ile Arg Thr Glu Arg Phe Val Pro Asp Asp Lys Glu Glu Phe Val Glu
CAG ACA CGG CCT TTC GAC ATC CGC ACT GAG CGC TTT GTG CCC GAT GAC AAG GAG GAG TTC GTC GAG   1470
CAG ACC AGG CCT TTT GAC CTC AAG AAA GAT GTC TTT GTG CCT GAT GAC AAA GAG GAG TTC GTC AAG    190
                        Leu Lys Lys Asp Val                                         Lys

                    AvaI
Ala Lys Ile Val Ser Arg Glu Gly Gly Lys Val Thr Ala Glu Thr Glu Asn Gly Lys
GCC AAG ATC GTG TCT CGG GAG GGA GGC AAG GTC ACC GCC GAA ACC GAG AAC GGC AAG gtg tgt ...   1533
GCC AAG ATC ATG TCT CGA GAG GGT GGC AAA GTC ACT GCT GAG ACC GAG CAT GGC AAG gtg ggt ...    253
            Met                                         His
                    AvaI
```

FIGURE 2. Sequence of the 5' regions of the rabbit α-and β-myosin HC genes. The exons are in upper case type, while the introns and non-coding regions are in lower case. The arrow denotes the putative cap site of the HCα gene.

determine if the distant location of the putative promoter region was due to the presence of intervening gene sequences, S1 nuclease mapping experiments were performed using single-stranded probes for the 5' regions of the genes (Figure 3).

From the HCα gene (Figure 3A), we isolated a 360-nt probe, which was end-labeled at the XbaI site 73 nts downstream from the initiation codon. The probe was hybridized with RNA from ventricular and atrial muscles, and the hybrids were treated with S1 nuclease (31). As shown

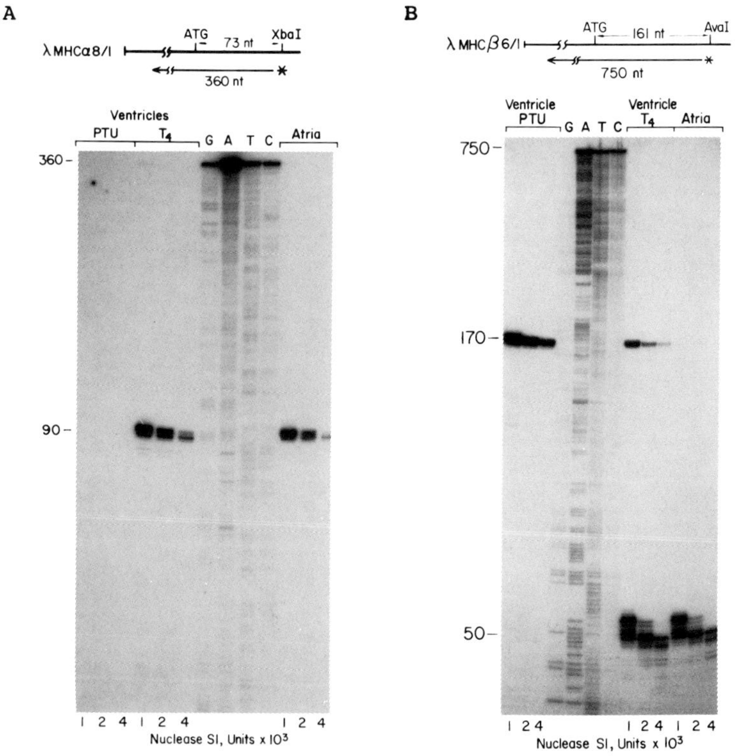

FIGURE 3. Mapping of the 5' boundaries of the N-terminal exons. At the top are schematic representations of the single-stranded probes, and at the bottom, autoradiograms of the S1-resistant labeled fragments fractionated on 6% polyacrylamide-urea sequencing gels. The RNAs are designated as indicated in the text. Lanes marked G,A,T,C are the Maxam-Gilbert sequencing ladders (32) of the respective probes.

in Figure 3A, a 90-nt fragment was protected following hybridization with RNAs from both thyrotoxic (T_4) rabbit ventricles containing predominantly HCα mRNA and from atrial muscle, but not with RNA from hypothyroid (PTU) rabbit ventricles. The results indicate that the 5' boundary for the N-terminal exon of the HCα gene is approximately 17 nts upstream of the initiation codon. Similarly, hybridization of the probe from the HCβ gene (Figure 3B) with RNA from hypothyroid rabbit ventricles resulted in the protection of a 170-nt fragment. Since the probe is end-labeled at the AvaI site 161 nts from the initiator ATG (Figure 2 and 3B), the N-terminal exon of the HCβ gene contains an additional 9 bps of untranslated sequences.

Examination of the genomic sequences (Figure 2) at the 5' portions of these exons in both genes reveals an AG dinucleotide and polypyrimidine sequence characteristic of 3' splice junctions (29). The presence of 3' consensus splice sites located 12 and 8 bps upstream of the initiation codons of the α- and β-HC genes, respectively, and the absence of CAT and TATA boxes within 100 bps of the N-terminal exons indicate that there are additional 5' coding regions. The slight difference in the location of the splice junction delineated by S1 nuclease mapping of the HCα gene probably results from the upstream untranslated sequences being identical to part of the 3' splice site.

Figure 3B also shows that hybridization of the HCβ probe to RNA from thyrotoxic rabbit ventricles protects a fragment of 170 nts and multiple bands around 50 nts. The 50-nt fragments result from hybridization of the probe to HCα mRNA and cleavage of the probe at a region of sequence divergence 18 amino acids upstream from the AvaI site (Figure 2). On the other hand, the protection of a 170-nt fragment probably results from the presence of HCβ mRNA in this particular RNA sample, since a protected fragment of similar size is not observed with RNA from atrial muscles, which contain only HCα mRNA.

To characterize the 5' untranslated sequences of the α- and β-HC genes, we synthesized cDNAs for the corresponding regions by primer extension. End-labeled, 21-nt primers encompassing the second to eighth amino acids of HCα or the first 7 amino acids of HCβ (Figure 2) were annealed to ventricular RNAs, and the primers were extended with reverse transcriptase (33). As shown in Figure 4, the HCα-gene primer was elongated to 92 nts with ventricular RNA from T_4-treated rabbits (HCα mRNA), and remained unchanged with RNA from PTU-treated animals (HCβ mRNA). On the other hand,

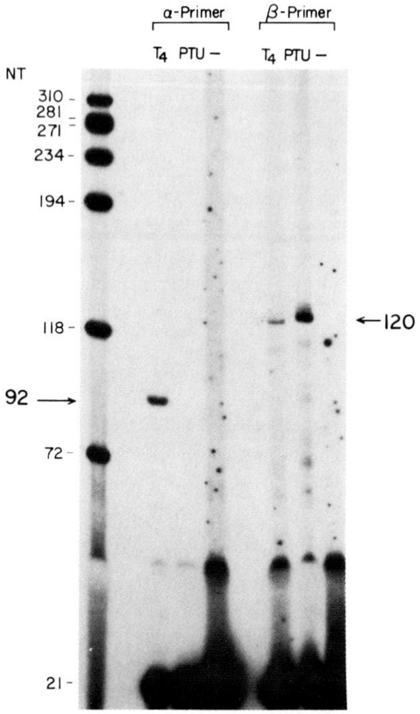

FIGURE 4. Primer extension analysis of the α- and β-HC mRNAs. Autoradiogram of the cDNAs transcribed from RNAs from thyrotoxic and hypothyroid rabbit ventricles fractionated on a 6% polyacrylamide-urea sequencing gel. RNAs are designated as indicated in the text. Lanes marked (-) show the fractionation of the respective primers alone.

extension of the HCβ-gene primer resulted in a 120-nt cDNA with both RNAs. As shown by the S1 mapping results (Figure 3B), the synthesis of a 120-nt cDNA with RNA from thyrotoxic rabbit ventricles probably results from residual HCβ mRNA in this sample. Thus, the cDNA synthesized with the HCβ-gene primer is specific for the respective mRNA. The data indicate that the 5' untranslated regions of the cardiac myosin HCα and HCβ mRNAs encode approximately 68 and 99 nts, respectively. Moreover, the homogeneous size of the primer-extended cDNAs strongly suggests that these genes are not transcribed from multiple initiation sites, nor alternately spliced in the 5' region.

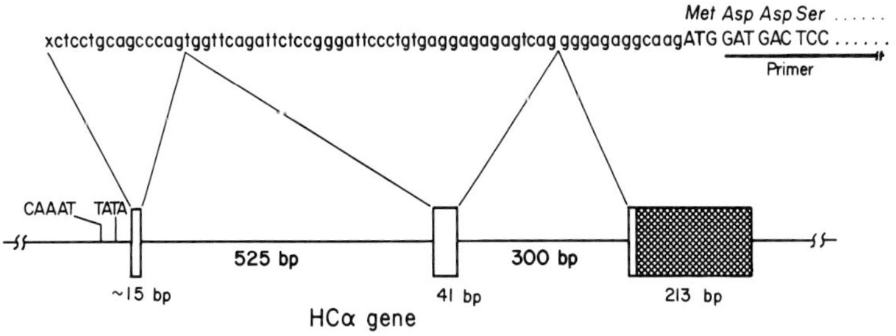

FIGURE 5. Sequence and genomic organization of the 5' untranslated sequences of the cardiac myosin HCα mRNA. The underlined portion of the mRNA sequence corresponds to the primer. The locations of the exons (boxes) in the 5' region of the myosin HCα gene are shown at the bottom.

Figure 5 shows the sequence of the 5' end of the HCα mRNA obtained by sequencing of the primer-extended cDNA. The sequence is complete to within 1 or 2 nts of the 5' end. Sequence identity of the cDNA and the 15 bps of genomic sequences upstream from the primer rigorously demonstrates that the primer-extended cDNA corresponds to HCα mRNA. Comparison of the cDNA and genomic sequences (Figures 2 and 5) identified two additional exons encoding the untranslated region of the mRNA. The penultimate exon of 41 bps is 313 bps from the N-terminus, and the first exon of 15 bps is an additional 525 bps further upstream. Thus, the 5' end of the HCα gene is located 26 and 59 bps, respectively, downstream from the consensus TATA and CAT sequences previously identified. The cap site of the gene probably corresponds to the A residue designated by the arrow in Figure 2, since transcription generally begins with this base (34) and because the untranslated sequences (Figure 5) are only missing 1 or 2 nts at the 5' terminus.

The presence of two introns in the 5' untranslated sequences of the myosin HCα gene is unique among the eukaryotic genes characterized thus far. The unusual organization of the 5' sequences may reflect a functional role for either the untranslated sequences of the second exon, or sequence elements in the 5' introns in regulating the expression of myosin HCα.

Comparison of the rabbit myosin HCα gene sequences with those of the homologous rat gene (18) reveals striking conservation of non-coding sequences (Figure 6). In both genes, a 400-bp region immediately upstream of the cap site is conserved not only in its sequence (>80% homology) but also in its location relative to the translational initiation site. Downstream, the sequences are divergent with only short regions of homology, some of which may be involved in maintaining the splice junctions of the second exon. Significantly, the latter regions do not correspond to the two untranslated exons. The sequence homology in the 400-bp region is comparable to that observed in the protein coding sequences (35) and strongly suggests that this region contains additional sequence elements (such as enhancers) that are important in regulating the expression of the HCα gene.

Comparisons with other eukaryotic genes revealed significant sequence homologies with only a few genes. Of interest is the rat growth hormone gene (36), where the sequence GTGTGGTTTG (underlined in Figure 2) is exactly conserved, while the sequence denoted in Figure 6 is identical in 20 of 21 nts. The former has similarities to a core enhancer sequence (37), while the latter is repeated imperfectly in the HCα gene (brackets, Figure 6), and is also homologous to sequences in the rat HCα gene. Both conserved sequences are

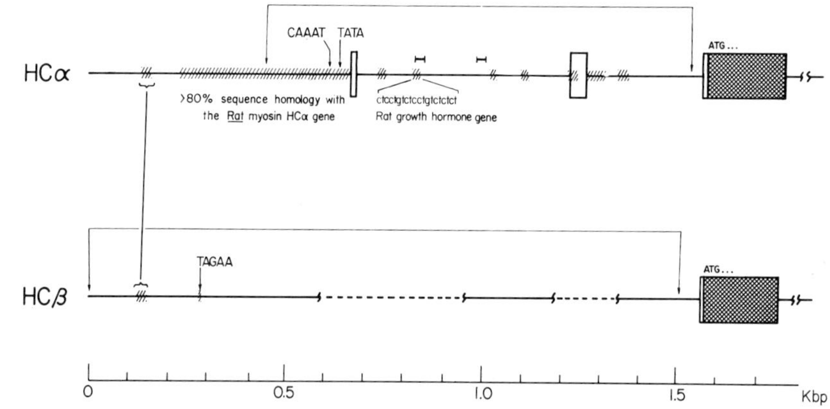

FIGURE 6. 5' Flanking regions of the rabbit α- and β-ventricular myosin HC genes. The exons are represented by boxes with the shaded and unshaded portions corresponding to coding and untranslated sequences respectively. Regions of homology between the rabbit and rat myosin HCα genes are indicated by the hatched areas. Unsequenced portions of the HCβ gene are denoted by the dashed lines.

located in the first introns of the respective genes. Since growth hormone gene expression is regulated by thyroid hormone, the sequence homology may reflect a role for these regions in thyroid hormone regulation of myosin HCα gene expression.

The upstream regions of the HCβ gene sequenced so far (Figure 6) show only limited sequence homology with the HCα gene. The comparisons, however, identify a region that may correspond to a TATA box with a variant consensus sequence of TAGAA. The overall non-homology of upstream sequences is consistent with the differential regulation of α- and β-myosin HC gene expression.

Transcriptional Regulation of the α- and β-Myosin HC Genes.

The divergent 5' sequences of the cardiac myosin HC gene provide effective hybridization probes for individual nuclear transcripts. Thus, to determine if the pre-translational regulation of myosin HC expression (12) by thyroid hormone results from either transcriptional control, or post-transcriptional processing, we have measured directly the transcription of these genes in isolated nuclei.

Previous studies show that the administration of thyroid hormone to hypothyroid rabbits results in a rapid increase in the relative level of HCα mRNA from <5% in the hypothyroid state to >95% within 3 days (12). Therefore, cardiac myocyte nuclei were isolated and partially purified (38) from hypothyroid rabbits and from hypothyroid animals treated for 5 days with 3,5,3'-triiodothyronine (T_3). The nuclei were incubated in vitro in the presence of α^{32}P-UTP, which labeled already initiated RNA transcripts (14,39). In these experiments, RNA synthesis continued for approximately 10 min. and was 50% inhibited by α-amanitin (1μg/ml). The HCα- and HC-β transcripts were then quantitated by filter hybridization to M13 subclones containing 5' regions (bracketed by arrows in Figure 6) of the respective genes. As shown in Figure 7, the transcription of the cardiac myosin HC genes is significantly altered by thyroid hormone. In hypothyroid rabbits (PTU), transcription of the HCβ gene is 4-5 fold higher than that of the HCα gene. Following thyroid hormone treatment (T_3), however, there is not only an increase in the relative rate of transcription of the HCα gene, but also a decrease in the specific transcription of the HCβ gene. The large change in transcripion clearly demonstrates that the effect of thyroid hormone on myosin HC

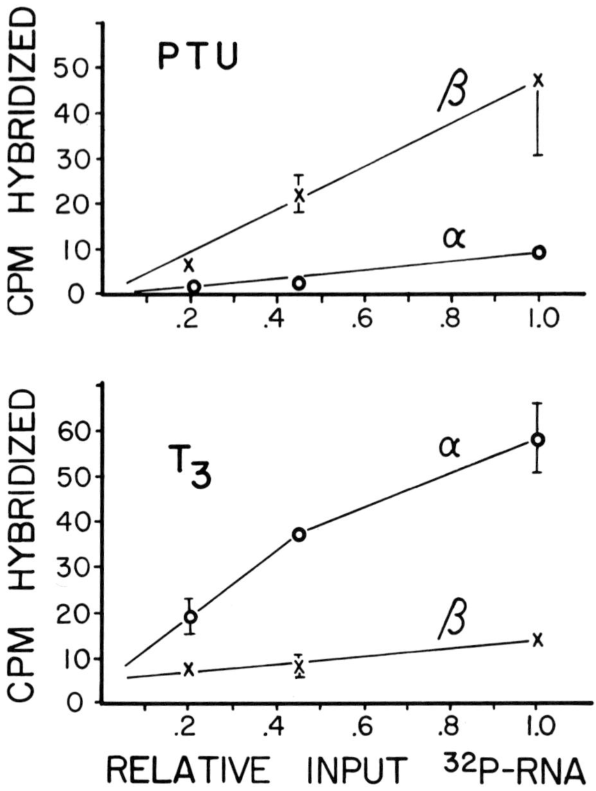

FIGURE 7. Relative transcription of α- and β-myosin HC genes. Heart nuclei isolated from hypothyroid rabbits (PTU) before and after treatment with T_3 were incubated in vitro in the presence of ^{32}P-UTP for 10 min. at 29° C. Myosin HC transcripts were measured by filter hybridization to the HCα- (o---o), and HCβ- (x---x) gene specific probes.

expression is primarily regulated at the level of transcriptional initiation.

Thus, the structural analysis of the 5' coding and flanking regions of the cardiac myosin HC genes not only identified a unique genomic organization and putative regulatory regions, but also enabled the direct demonstration of transcriptional regulation in response to thyroid hormone. Further studies of nuclear run-off transcription should provide information on whether the thyroid hormone is directly involved in controling myosin HC gene transcription, and if additional post-transcriptional processes modulate expression. Furthermore, the structural information now allows us to define regulatory DNA sequences by in vitro transcription and DNA-mediated gene transfer experiments to obtain insights into the molecular mechanisms that regulate the expression of this gene family.

REFERENCES

1. Whalen, R (1980). Contractile protein isozymes in muscle development: the embryonic phenotype. In Pette D (ed): "Plasticity of Muscle," Berlin, de Gruyter, p 177.
2. Gauthier GF, Lowey S (1979). Distribution of myosin isoenzymes among skeletal muscle fiber types. J Cell Biol 81:10.
3. Barany M (1967). ATPase activity of myosin correlated with speed of muscle shortening. J Gen Physiol 50:197.
4. Hoh JFY, Yeoh GPS, Thomas MAW, Higginbottom L (1979). Structural differences in the heavy chains of rat ventricular myosin isozymes. FEBS Lett 97:330.
5. Flink IL, Rader JH, Morkin E (1979). Thyroid hormone stimulates synthesis of a cardiac myosin isozyme. J Biol Chem 254:3105.
6. Chizzonite RA, Everett AW, Clark WA, Jakovcic S, Rabinowitz M, Zak R (1982). Isolation and characterization of two molecular variants of myosin heavy chain from rabbit ventricle. J Biol Chem 257:2056.
7. Sartore S, Pierobon-Bormioli S, Schiaffino S (1978). Immunohistochemical evidence for myosin polymorphism in the chicken heart. Nature 274:82.
8. Everett AW, Clark WA, Chizzonite RA, Zak R (1983). Change in synthesis rates of α- and β-myosin heavy chains in rabbit heart after treatment with thyroid hormone. J Biol Chem 258:2421.
9. Lompre AM, Mercadier JJ, Wisnewsky C, Bouveret P, Pantaloni C, d'Albis A, Schwartz K (1981). Species and age-dependent changes in the relative amounts of cardiac myosin isoenzymes in mammals. Develop Biol 84:286.
10. Lompre AM, Schwartz K, d'Albis A, Lacombe B, Thiem NV, Swynghedauw B (1979). Myosin isoenzyme redistribution in chronic heart overload. Nature 282:105.
11. Gorza L, Pauletto P, Pessina AC, Sartore S, Schiaffino S (1981). Isomyosin distribution in normal and pressure-overloaded rat ventricular myocardium. Circ Res 49:1003.

12. Everett AW, Sinha AM, Umeda PK, Jakovcic S, Rabinowitz,M, Zak R (1984). Regulation of myosin synthesis by thyroid hormone: Relative change in the α- and β-myosin heavy chain mRNA levels in rabbit heart. Biochemistry 23:1596.
13. Spindler SR, Mellon SH, Baxter JD (1982). Growth hormone gene transcription is regulated by thyroid and glucocorticoid hormones in cultured rat pituitary tumor cells. J Biol Chem 257:11627.
14. Yaffe B, Samuels HH (1984). Hormonal regultion of the growth hormone gene. J Biol Chem 259:6284.
15. Towle HC, Mariash CN, Oppenheimer JH (1980). Changes in the hepatic levels of messenger ribonucleic acid for malic enzyme during induction by thyroid hormone or diet. Biochemistry 19:579.
16. Friedman DJ, Umeda PK, Sinha AM, Hsu H-J, Jakovcic S, Rabinowitz M (1983). Isolation and characterization of genomic clones specifying rabbit α and β ventricular myosin heavy chains. Proc Natl Acad Sci USA 81:3044.
17. Sinha AM, Friedman DJ, Jakovcic S, Rabinowitz M, Umeda PK (1983). Expression of rabbit ventricular α-myosin heavy chain mRNA sequenes in atrial muscle. J Biol Chem 259:6674.
18. Mahdavi V, Chambers AP, Nadal-Ginard B (1984). Cardiac α- and β-myosin heavy chain genes are organized in tandem. Proc Natl Acad Sci USA 81:2626.
19. Strehler EF, Mahdavi V, Periasamy M, Nadal-Ginard B (1984). Intron positions are conserved in the 5' end region of myosin heavy-chain genes. J Biol Chem 260:468.
20. Efstratiadis A, Posakony JW, Maniatis T, Lawn RM, O'Connell C, Spritz RA, DeRiel JK, Forget BG, Weissman SM, Slightom JL, Blechl AE, Smithies O, Baralle FE, Shoulders CC, Proudfoot NJ (1980). The structure and evolution of the human β-globin gene family. Cell 21:653.
21. Wahli W, Dawid IB, Wyler T, Weber R, Ryffel GU (1980). Comparative analysis of the structural organization of two closely related vitellogenin genes in X. laevis. Cell 20:107.
22. Karn J, Brenner S, Barnett L (1983). Protein structural domains in the C. elegans unc-54 myosn heavy chain gene are not separated by introns. Proc Natl Acad Sci USA 80:4253.
23. Wydro RM, Nguyen HT, Gubits RM, Nadal-Ginard B (1983). Characterization of sarcomeric myosin heavy chain genes. J Biol Chem 258:670.
24. Bernstein SI, Mogami K, Donady JJ, Emerson, CP (1983). Drosophila muscle myosin heavy chain encoded by a single gene in a cluster of muscle mutations. Nature 302:393.

25. Rozek CE, Davidson N (1983). Drosophila has one myosin heavy-chain gene with three developmentally regulated transcripts. Cell 32:23.
26. Robbins J, Freyer GA, Chisholm D, Gilliam TC (1982). Isolation of multiple genomic sequences coding for chicken myosin heavy chain protein. J Biol Chem 257:549.
27. Leinwand LA, Saez L, McNally E, Nadal-Ginard B (1983). Isolation and characterization of human myosin heavy chain genes. Proc Natl Acad Sci USA 80:3716.
28. Tong SW, Elzinga M (1983). The sequence of the NH_2-terminal 204-residue fragment of the heavy chain of rabbit skeletal muscle myosin. J Biol Chem 258:13100.
29. Mount S (1982). A catalogue of splice junction sequences. Nuc Acids Res 10:459.
30. Breathnach R, Chambon P (1981). Organization and expression of eukaryotic split genes coding for protein Ann Rev Biochem 50:349.
31. Orkin SH, Goff, SC (1981). The duplicated human α-globin genes: their relative expression as measured by RNA analysis. Cell 24:345.
32. Maxam AM, Gilbert, W (1980). Sequencing end-labeled DNA with base-specific chemical cleavages. In Grossman L, Moldave K (eds): "Methods in Enzymology, Vol 65," New York, Academic Press, p 499.
33. Gosh PK, Reddy VD, Lebowitz P, Weissmann SM (1980). Determination of RNA sequences by primed directed synthesis and sequencing of their cDNA transcripts. Ibid., p 580.
34. Corden J, Wasylyk B, Buchwalder A, Sassone-Corsi P, Kedinger C, Chambon P (1980). Promoter sequences of eukaryotic protein-coding genes. Science 209:1406
35. Kavinsky CJ, Umeda PK, Levin JE, Sinha AM, Nigro JM, Jakovcic S, Rabinowitz M (1983). Analysis of cloned mRNA sequences encoding subfragment 2 and part of subfragment 1 of α and β myosin heavy chains of rabbit heart. J Biol Chem 259:2775.
36. Barta A, Richards RI, Baxter JD, Shine J (1981). Primary structure and evolution of rat growth hormone gene. Proc Natl Acad Sci USA 78:4867.
37. Weiher H, Konig M, Gruss P (1983). Multiple point mutations affecting simian virus 40 enhancer. Science 219:626.
38. Jackowski G, Liew C-C (1980). Fractionaton of rat ventricular nuclei. Biochem J 188:363.
39. McKinght GS, Palmiter RD (1979). Trascriptional regulation of the ovalbumin and conalbumin genes by steroid hormones in chick oviduct. J Biol Chem 254:9050.

Molecular Biology of Muscle Development, pages 825–839

ANALYSIS OF MYOSIN HEAVY CHAIN ISOFORMS IN THE AVIAN HEART

Barry Zadeh, Naoji Toyota, Arlene Gonzalez-Sanchez, Mary Ann Matheson, Kimi Hamilton, and David Bader[1]

Department of Cell Biology, Cornell University Medical College, New York, New York

ABSTRACT Myosin heavy chain heterogeneity is found in many striated muscle types. To characterize the isoform expression in the avian heart, monoclonal antibodies have been generated which react specifically with the various cardiac muscle types. In this study, these antibodies were used to examine the distribution of myosin heavy chains in the adult, during development and in vitro. The results indicate that 1) at least one distinct isoform can be identified in each cell type, 2) the pattern of distribution differs from other species studied, 3) specific isoform accumulation is determined at or near the onset of atrial and ventricular differentiation while the expression of at least one conductive system isoform is not detected until late in embryonic life, and 4) regulation of heavy chain expression in vitro is more permissive than in the intact heart.

INTRODUCTION

Study of myosin isoform expression to date has identified polymorphism among the various muscle types examined (11,16,22,26,32,34). These myosins may differ in either heavy or light chain subunit

[1]This work was supported by NIH grant #HL-06850-10, AM31305 and by the New York Heart Association.

composition or both. In the adult heart there are three types of striated muscle present; atrial and ventricular working myocardium and specialized cells of the conductive system. Heterogeneity of the myosin subunits of the different muscle types in the heart has been reported in several species(2-4,15,17,20+21,27+28,33). In the adult chicken, the atrial and ventricular myosins have been shown to be biochemically and immunochemically distinct(5,26). However , difficulty in obtaining purified conductive system myosin has made it impractical to study the expression of myosin subunit isoforms in conductive system cells using biochemical methods. Similarly, the expression of myosin subunit isoforms in the developing heart and in cultured cardiac tissues has been limited(2,13+14,25,30,36).

In this study, immunochemical methods have been used to help define the expression of tissue specific isoforms of the MHC subunit in cardiac tissues. A series of monoclonal antibodies, generated against adult avian tissues, have been produced which react specifically in the heart with a MHC present in either atrial, vetricular or conductive system cells. With these antibodies we have been able to characterize the distribution of these isoforms in atrial, ventricular and conductive system myocytes during deveolpment and in cell culture environment. The present study examines 1) the appearance and distribution of three tissue specific MHC isoforms during cardiac myogenesis, and 2) the effect of altering the cellular environment on expression of MHC isoforms in the developing heart by neural ablation and monolayer cell culture.

RESULTS

The antibodies used in these studies were produced against adult avian skeletal muscle myosins as described by Fischman and Masaki(8) according to Galfre et al(10) as modified by Gefter et al(12). They react in the heart with specificities as seen in Table 1. Western blot analysis confirms that each of these McAbs is reactive with the MHC only(1,13+14).

The MHCs identified by these antibodies in

TABLE 1

Antibody	Ventricle	Atria	Conductive System
MF-20	+	+	+
ALD-19	+	-	-
B-1	-	+	-
ALD-58	-	-	+

the adult heart were analyzed by affinity chromatography and peptide analysis. Radiolabelled myosins of the atria and ventricles were prepared and fractionated on affinity columns comprised of the McAbs coupled indirectly to Sepharose 4B. These affinity purified myosins were digested with S. Aureus protease and the resulting peptide maps compared. As shown in Figure 2,the atrial and ventricular myosins present in the working myocardium are clearly distinct, whereas the myosins of the conductive system, reactive with ALD-58, are indistinguishable in the atria and ventricle. Indirect immunocytochemical staining of whole atria and ventricle sections were done to identify topographic patterns of isoform distribution. They confirm that a population of cells exist in both chambers which are reactive with ALD-58. These cells which are located primarily in the subendocarial connective tissues and in perivascular areas, have been determined previously to be conductive system myocytes(26).

To determine if the MHC epitopes specific to the adult heart were shared in the developing heart chambers, myosins were prepared from different structures present in early and late stage embryo hearts. RIA were performed after correction for MHC content by PAGE dye elution standardization(1,7). These assays demonstrate that the timing of

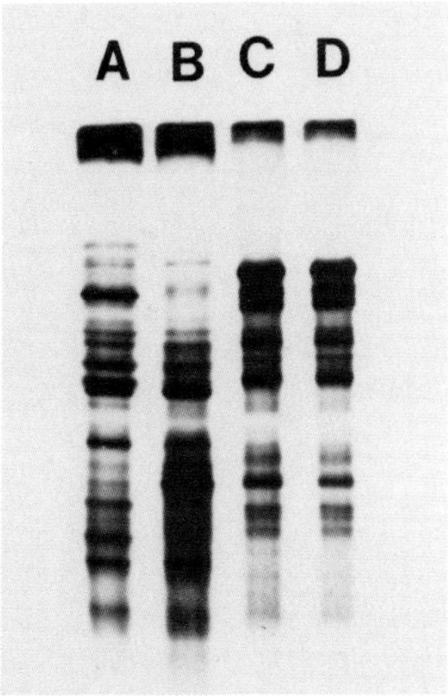

FIGURE 1. Peptide maps of myosins from ventricles and atria bound by ALD-19(A) and B-1(B), respectively and by ALD-58, ventricle(C) and atria(D).

appearance differs for each epitope. The ventricle specific epitope, reactive with ALD-19, first appears at stage 10; the atrial specific epitope, reactive with B-1, at stage 15; and the conductive system specific epitope, reactive with ALD-58, is not detectible before stage 36(Table 2). Sections were prepared from the heart at several stages in early and late embryos and reacted with our McAbs for examination by phase and epiflorescent microscopy. Reactivity with each antibody is first localized to the embryologic structure destined to become the adult tissue type for which they are specific. Once each isoform appeared in the respective tissue,e.g. the presumptive ventricle, that MHC was maintained throughout development in that tissue. Additionally, while no cross reactivity was ever observed, the presence of a MHC which was reactive with MF-20 and none of the tissue specific McAbs could not be excluded by RIA in the various structures in the early embryo.

To establish the importance of tissue related factors in the regulation of MHC

TABLE 2

ANTIBODY	STAGE at APPEARANCE
ALD-19	10
B-1	15
ALD-58	36

expression, we have used two experimental systems. The possible involvement of direct neural interaction on the expression of the MHCs in the heart during development was evaluated using the aneural heart preparation developed by Kirby et al (19). In this model, selective extirpation of segments of the neural crest results in anatomic and functional denervation of the avian heart. At various times after neural crest ablation in the early embryo, the hearts were removed for assay of atrial, ventricular and conductive system MHCs. Myosins were purified and the MHCs identified by RIA. Tissue distribution was again confirmed with immunocytochemical localization. As shown in figure 2, the pattern of MHC reactivity as compared with age matched, sham operated controls was unchanged.

The other system used to investigate the possible contribution of tissue factors to MHC isoform regulation is primary monolayer cell culture. In this model, atria and ventricles are seperated by microdissection, the cells are released by cyclic enzyme digestion, the myocyte population is enriched by differential adherence to substrate and the cell populations are characterized immunocytochemically prior to plating(36+37). These cells are then grown at confluence in media which promotes myocyte contractility while inhibiting non-myocyte proliferation. To assess MHC accumulation in these cells, various time points in culture were selected

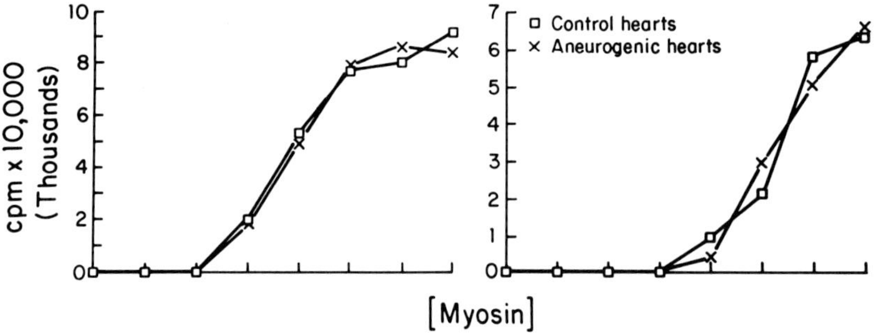

A. B.

FIGURE 2. Radioimmuneassay analysis of ventricle(A) and atria(B) myosins with ALD-58 comparing aneurogenic and control hearts.

and the myosins isolated for RIA or the cells characterized by direct and indirect immunocytochemistry. Several modulations of epitope accumulation were identified that were not seen in age matched tissue controls or in aneural hearts . Atrial myocytes have decreased amounts of B-1 reactive MHC at several stages assayed while developing cross reactivity to ALD-19 at another stage (figure 3a). In contrast, ventricular myocytes maintained expression of the tissue specific epitope, reactive with ALD-19, at one stage while having decreased accumulation at various other stages assayed. Ventricular myocytes never developed cross reactivity to McAbs B-1 or ALD-58(Figure 3b). Conductive system type MHC could never be identified in cultured myocytes at any stage from either atrial or ventricle.

Total sarcomeric type MHC was assessed in all cultures by reactivity to McAb MF-20. Both RIA and immunocytochemistry confirm that myosin accumulation was uneffected throughout the culture period in the setting of decreased reactivity with the tissue specific MHC antibodies. This finding implies that some of the MHCs acculated in these cells do not share the same epitopes with any of the MHCs present in the adult heart.

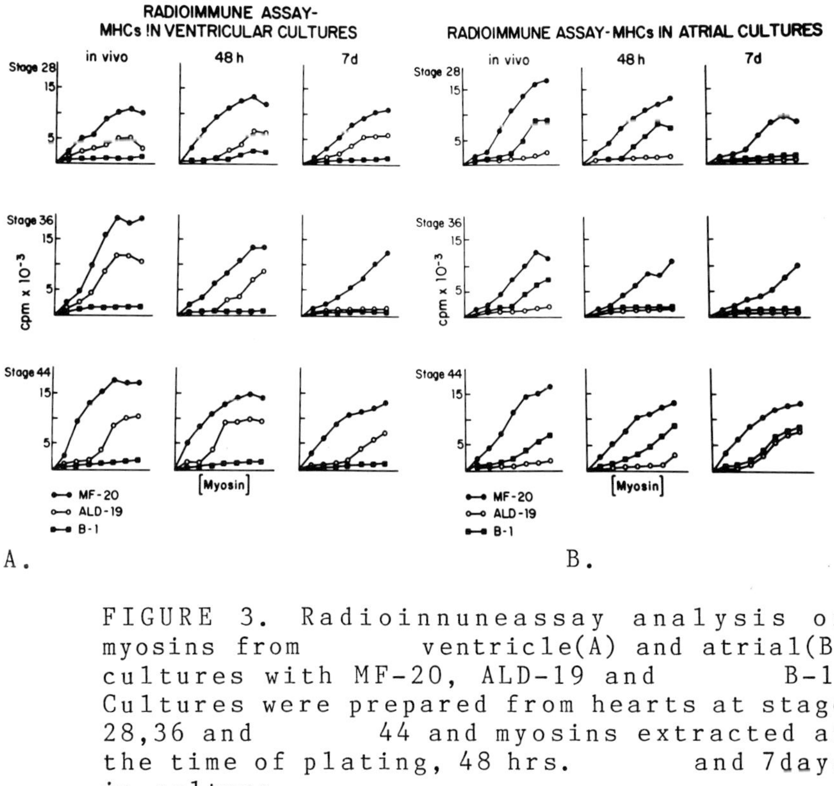

FIGURE 3. Radioinnuneassay analysis of myosins from ventricle(A) and atrial(B) cultures with MF-20, ALD-19 and B-1. Cultures were prepared from hearts at stage 28,36 and 44 and myosins extracted at the time of plating, 48 hrs. and 7days in culture.

DISCUSSION

The production of a panel of McAbs in our laboratory which react specifically with the various MHC isoforms has facilitated characterization of MHC heterogeneity. In this study, we have used these McAbs and a variety of immunochemical techniques to characterize MHC isoform expression in the avian heart. In the adult, we have demonstrated that the atrial and ventricular cells of the working myocardium express distinct myosin isoforms, whereas the conductive system cells of both chambers express a myosin which is very similar if not identical. During development, the appearance of each specific MHC varies temporally in the atria, ventricle and conductive system cells but once expressed that MHC is maintained throught development. We have also shown that the MHC present in the embryo in each cell type is immunochemically similar to the MHC of the respective adult tissue. Additionally, we have begun studies of the effect which cellular environment has on MHC expression. If direct neural-myocyte contact is abolished in the heart during development in vivo the expression of MHCs is uneffected. However, cell isolation in monolayer culture conditions has a profound effect on isoform accumulation in a manner which is stage dependent and tissue specific. These results, when compared with MHC expression in small mammals, indicate that the patterns of isoform expression in similar tissues can vary markedly among different species. The functional significance of this variability has not yet been clarified.

To study MHCs in the adult heart, four antibodies were selected from our panel based on specific reactivities with those MHC isoforms expressed. The adult heart contains three striated muscle types; ventricle, atria and conductive system myocytes. The MHCs they produce react specifically with McAbs ALD-19, B-1 and ALD-58 respectively and share the epitope recognized by McAb MF-20. Using affinity colunm chromatography, we have prepared highly purified isoforms for biochemical analysis. Peptide map analysis of these isoforms indicates that different MHCs exist in the atrial and ventricular working myocytes. This finding confirms reports by Sartore et al (26), that the MHCs in atria and ventricle are immunochemically distinct and by Obinata et al.

(23) that a single ventricular myosin is present in the chick.

In contrast to the working myocardium, our data suggest that the conductive system myocytes of both chambers in the adult chick express at least one common MHC. Further, this MHC is not present in the working myocardium of the atria or ventricles. These findings are consistent with Sartore et al(26) and Saito et al(24), who demonstrated that the myosins of the conductive system can be distict in either their heavy or light chain content. Whalen et al (35) however, have demonstrated a similar MHC in the working ventricle and the conductive cells of the bovine heart. The diversity in these studies may represent interspecies variability. Alternatively, multiple isoforms may be coexpressed in these cells and remain undetected with current methods.

To determine the appearance and distribution of the MHCs found in the developing embryo, we have also used immunocheical techniques. Our data demonstrate that each isoform appears at a different embryologic stage. The ventricular specific MHC appears at stage 10, at the onset of contractile activity , and can be related to the appearance of sarcomeric structures in the presumptive ventricle. The atrial MHC does not appear until stage 15. The atria forms later in development and the appearance of a MHC reactive with McAb B-1 corresponds to the regional differentiation of the caudal regions of the developing heart. The conductive system MHC is further delayed in appearance, until late in development at stage 36. This finding does not imply that conductive system myocytes are not present or functional earlier in development as in the atrium. Indeed, DeHaan (6) and Forsgren et al(9) have demonstated the presence of Purkinje type cells in the heart from early stages in embryogenesis. Our results indicate that the appearance of the adult type MHC recognized by ALD-58 occurs as secondary phenomenon in these cells but further data on conductive system specific MHCs is necessary to prove this assumption.

It is clear from out data, that the

distribution of each isoform is limited to those embryologic structures destined to become the adult cell types that express that isoform. Comparison of peptide maps derived from affinity purified myosins of the embryo and adult demonstrate similarities between the myosins of each cell type(13+14). Nevertheless, it is not possible to state unequivocally that these myosins are identical or if they simply share common epitopes. Also, the presence of any coexpressed MHC cannot be excluded. Furthermore, the presence of a primordial type myosin, present in each cell type before the appearance of the specific isoforms has been postulated by Sweeney et al(29).

Once established in the respective tissues, the expression of each tissue specific isoform is maintained throughout development and no other isoforms have been identified whose appearance occurs later in development. Additionally, no neonatal MHCs have been detected in the avian heart. These findings indicate that after differential expression of MHCs has been established in the avian heart, no modulation of alpha and beta like isoforms occurs and each tissue maintains the accumulation of at least one unique isoform throught life. This pattern of expression differs markedly from several other species previously studied, specifically the mammal group. These animals undergo developmentally regulated transition in the ventricle specific MHCs from the embryonic to the adult isoform (3,16,18,20,28) and may express immunochemically similar isoforms in different cardiac cell types.

The effect of changing the cellular environment of developing myocytes on MHC expression has been studied in two unrelated experiments. To determine if a direct neural contact is necessary for the orderly appearance and maintainance of the tissue specific isoforms, a model has been developed which utilizes segmental neural crest ablation and is termed "aneurogenic heart"(19). Immunochemical analysis of the MHCs prepared from the various chambers of these hearts has failed to identify any differences from normal age matched controls. A direct neural/myocyte interaction does not appear to be essential during the development and maintainance of tissue specific MHC expression in the avian embryo. This finding does not indicate that indirect neural stimulation is not of importance in the control of cardiac MHC expression, because circulating levels of these agents have not been defined in this model. In addition Toyota and Shimada (30) have demonstrated, using an in vitro model, that certain indirect neural factors are important in the regulation of specific isoforms of troponin in the developing avian heart. Further study of the intact heart during development is necessary to define if MHC accumulation is also independent of these factors.

A second technique to alter the cellular mileau of developing myocytes from the heart is enzymatic isolation and growth in a monolayer tissue culture. We have prepared cultures from various stages during development, both before and after the appearance of the specific MHCs. Atrial and ventricular cultures were highly enriched for striated myocytes and assayed at a variety of times in culture by both immunocytochemistry and RIA. In contrast to the intact heart and the aneurogenic model, modulation of MHC immunoreactivity was observed in these cultures. The alterations which were observed, however, are not uniform in nature but are stage dependent and tissue specific. While culture conditions were tightly controlled, the pattern of tissue specific MHC accumulation differed in the atria and ventricle cultures. The MHC reactive with ALD-58 was never identified in cultures from either atria or ventricles. These findings imply that embryonic chick cardiomyocytes undergo a transition in MHC accumulation in vitro which never occurs in the intact heart. Similar alterations in protein accumulation in vitro have been described for MHC isoforms in small mammals(2,31) and for troponins(30). While the mechanisms responsible for these changes are unknown, it is evident that the intracellular controls which regulate tissue specific differentiation in vivo are not adequate to maintain isoform specificity in vitro in the absence of tissue related factors. The permissiveness of MHC expression in this model may result in the re-expression or accumulation of a non-specific primordial type MHC under selected conditions.

In conclusion, we have used immunochemical methods to characterize heterogeneity in MHCs in those cardiac tissues not amenable to analysis with biochemical techniques alone. Our findings demonstrate that the pattern of MHC distribution in similiar tissues has marked interspecies variability. In addition, in the avian model, this variability in tissue specific isoforms allows us to use MHCs as markers of myocyte type and provides a powerful tool with which to study the development of different striated muscle cells in the heart during embryogenesis.

REFERENCE

1. Bader DM, Masaki T, Fischman DA (1982). Immunochemical analysis of myosin heavy chain during avian myogenesis in vivo and in vitro. J Cell Bio 95:763.
2. Cantini M, Sartore S, Schiaffino S (1980). Myosin types in cultured muscle cells. J Cell Bio 85:903.
3. Chizzonite RA, Everett AW, Clark WA, Jakovcic S, Rubinowitz M, Zak R (1982). Isolation and characterization of two molecular variants of myosin heavy chain from rabbit ventricle. J Biol Chem 257:2056.
4. Clarke WA, Chizzonite RA, Everett AW, Rabinowitz M, Zak R (1982). Species correlation between cardiac isomyosins. J Biol Chem 257:5449.
5. Dalla-Libera LS, Sartore S, Schiaffino S (1979). Comparative analysis of chicken atrial and ventricular myosins. Biochem Biophys Acta 581:283.
6. DeHaan RL (1961). Differentiation of the atrioventricular conductive system of the heart. Circulation 24:458.
7. Fenner C, Trout RR, Mason DT, Wilkman-Coffelt

J (1975). Anal Biochem 63:595.
8. Fischman DA, and Masaki T (1982). Immunochemical analysis of myosin with monoclonal antibodies. In Anderson W, Sadler W (eds):"Perspectives in differentiation and hypertrophy," p.279.
9. Forsgren SE, Strehler E, Thornell LE (1982). Differentiation of Purkinje fibers and ordinary ventricular and atrial myocytes in the bovine heart: an immuno and enzyme histochemical study. Histochem J 14:929.
10. Galfre G, Howe SC, Milstein C, Butcher GW, Howard JC (1977). Antibodies to major histocompatibility antigens produced by hybrid cell lines. Nature 266:550.
11. Gautier GF, Lowey S, Hobbs AS (1978). Fast and slow myosin in developing muscle fibers. Nature 274:25.
12. Gefter ML, Margulies DH, Scharff MD (1977). A simple method for polyethylene glycol promoted fusion of mouse myeloma cells. Somat Cell Genet 3:231.
13. Gonzalez-Sanchez A, Bader DM (1984) Immunochemical analysis of myosin heavy chains in the developing chicken heart. Devel Bio 103:151.
14. Gonzaleaz-Sanchez A, Bader DM (1985). Characterization of a myosin heavy chain in the conductive system of the adult and developing chicken heart.
15. Gorza L, Sartore S, Schiaffino S (1982). Myosin types and fiber types in cardiac muscle. 11. Atrial myocardium. J Cell Biol. 95:838.
16. Hoh JF, McGrath PA, Hale PT (1977). Electrophoretic pattern of multiple forms of rat cardiac myosin:Effects of hypophysectomy and thyrosine replacement. J Mol Cell Cardiol 10:1053.
17. Hoh JF, Yeoh CPS, Thomas MAW, Higginbuttom G (1979). Structural differences in the heavy chains of rat ventricular myosin isoenzymes. FEBS Lett 97:330.
18. Kavinsky CJ, Umeda PK, Levin JE, Sinha AM, Nigro JM, Jakovcic S, Rabinowitz M (1984).

Analysis of cloned mRNA sequences encoding subfragment 2 and part of subfragment 1 of alpha and beta myosin heavy chains of rabbit heart. J Biol Chem 259:2775.
19. Kirby ML, Stewart DF (1983). Neural crest origin of cardiac ganglion cells in the chick embryo: identification and extirpation. Dev Biol 97:433.
20. Lompre AM, Mercadier JJ, Wisnewsky C, Bouveret P, Pantaloni C, D'Albis A, Schwartz K (1981). Species and age dependent changes in the relative amounts of cardiac myosin isozymes in mammals. Dev Biol 84:286.
21. Maldavi U, Perisamy M, Nadal-Ginard B (1982). Molecular characterization of two myosin heavy chain genes expressed in the adult heart. Nature 297:659.
22. Masaki T, Yoshizaki C (1974). Differentiation of myosin in chick embryos. J.Biochem 76:123.
23. Obiata T, Masaki T, Takano-Ohmuro H, Suzuki T, Mikawa T (1984). In Nora JJ, Takao A (eds):"Etiology and morphogenesis of congenital heart disease,"New York:Futura Co.
24. Saito K, Tamina Y, Saito M, Matsumara K, Niki T, Mori H (1981). Comparison of the subunit compositions and ATPase activities of myosin in the myocardium and the conduction system. J Mol Cell Cardiol 13:311.
25. Samuel JL, Rappaport L, Mercadier JJ, Lompre AM, Sartore S, Triban C, Schiaffino S, Schwartz K (1983). Distribution of myosin isozymes within single cardiac cells. An immunohistochemical study. Circ Res 52:200.
26. Sartore S, Pierobon-Bormisli S, Schiaffino S (1978). Immunohistochemical evidence for myosin polymorphism in the chicken heart. Nature 274:82.
27. Sartore S, Dalla Libra L, Schiaffino S (1979). Fractionation of rabbit ventricle myosins by affinity chromatography with insolubilized antimyosin antibodies. FEBS Lett 106:197.
28. Sinha AM, Umeda PK, Kavinsky CJ, Rajamanickam C, Jakovcic H, Rabinowitz M (1982). Molecular cloning of mRNA sequences for cardiac alpha and beta myosin heavy chains: expression in

ventricles of normal,hypothyroid and thyrotoxic rabbits. Proc Nat Acad Sci USA 79:5847.
29. Sweeney LJ, Clark WA, Zak R, Manasek FJ (1983). Myosin heavy chain expression during cardiac embryogenesis in the chick. J Cell Biol 97:51a. Toyota N, Shimada Y (1983). Isoform variants of troponin in skeletal and cardiac muscle cells cultured with and without nerves. Cell 33:297.
30. Watras J (1981). Changes in rat cardiac myosin during development and in culture. J Mol Cell Cardiol 13:1011.
31. Whalen RG, Schwartz K, Bouveret P, Sell SM, Gros F (1979). Contractile protein isozymes in muscle development:Identification of an embryonic form of myosin heavy chain. Proc Nat Acad Sci USA 76:5197.
32. Whalen RG, Sell SM (1980). Myosin from fetal hearts contains the skeletal muscle embryonic light chain. Nature 286:731.
33. Whalen RG, SellSM, Butler-Brown GS, Schwartz S, Bouveret P, Pinset-Harstrom I (1981). Three myosin heavy chain isozymes appear sequentially in rat muscle development. Nature 292:805.
34. Whalen RG, Sell SM, Eriksson A, Thornell LE (1982). Myosin subunit types in skeletal and cardiac tissues and their developmental distribution. Dev Biol 91:478.
35. Zadeh BJ, Fischman DA, Bader DM (1984). Immunochemical identification of chick atrial and ventricular myocytes in vitro. In Vitro 20:245.
36. Zadeh BJ, Fischman DA, Bader DM (in Review). Myosin heavy chain expression during avian cardiac myogenesis in vitro. Dev Biol.

Molecular Biology of Muscle Development, pages 841–852

EVOLVING PHENOTYPE OF THE HEART[1]

Asish Nag,[2] Lauren Sweeney,[3], Brenda Eisenberg[4], Francis Manasek[3], and Radovan Zak

Department of Medicine and Pharmacological and Physiological Sciences, The University of Chicago, Chicago, Illinois 60637

ABSTRACT

The change in isomyosin profile of avian and mammalian hearts was examined during the embryonic period in primary culture and in serial sections of the entire ventricular wall of the adult heart. Two methods have been used to classify myocytes: immunofluorescent staining with McAbs specific for heavy chains of V1 and V3 isomyosins, and cytochemical staining for Ca^{2+} activated myosin ATPase.

The chicken ventricle showed strong reactivity with anti-V3 Ab throughout all stages of embryonic and adult development. The atrium also reacted with anti-V3 Ab from its initial stage of incorporation into the heart (st. 12) through the 4th day of incubation (st. 24). The atrial reactivity then declined so that from the 6th day (st. 29) through the remaining embryonic and adult life there was no significant atrial reactivity with anti-V3 Ab. These data are consistent with our previous observation that the

1 This work was supported in part by National Institute of Health grants HL-20592 and HL-16637 and the Muscular Dystrophy Association of America.
2 Department of Biological Sciences, Oakland University, Rochester, Michigan 48063.
3 Department of Anatomy, The University of Chicago, Chicago, Illinois 60637.
4 Department of Physiology, Rush Medical College, Chicago, Illinois 60612.

V3 isomyosin is the first to appear in all cross-striated muscles.

In contrast to birds, the mammalian heart contains a mixture of V1 and V3 isomyosins within both chambers. Using cultured cells we have examined whether myocytes are synchronized with respect to myosin expression. We have examined the isomyosin profile of cells derived from 16 day old rat embryonic ventricles grown in a synthetic medium in the absence of presence of physiological concentrations of thyroid hormone (T4). Myocytes grown in T4 free media for 7 days contained a majority of cells showing bright fluorescence after staining with anti-V3 Ab as well as weak ATPase activity. A small population of cells, however, reacted strongly with anti-V1 Ab and had a high ATPase activity. Myocytes grown in T4 supplemented cultures converted after 7 days to a predominance of V1 containing cells. Nevertheless, a small population of these cells remained which was strongly reactive for V3 isomyosin. These data are consistent with the existence of two populations of myocytes with different responsiveness to T4.

The heterogeneity in the myocyte population was examined in the ventricular wall of adult (1 year) rabbit. The isomyosin population varies through the epi- to endocardium of the ventricular wall with V1 predominating in the outer epicardial third of the wall and the lowest amount in the middle third of the wall. A mixture of stained and unstained myocytes is seen in the endo- and subendocardial regions. Myocytes with similar staining properties appear in large clusters, long strand and major anatomical regions. However, heterogeneous myocytes are also seen connected by an intercalated disc. The heterogeneity between two adjacent cells expressing different isomyosin might depend on a stage of genetic reprogramming reflecting the different populations of myocytes.

INTRODUCTION

Two classes of myosin have been detected in the heart: ventricular and atrial. These two classes can be further divided into V1, V2, and V3 in the ventricles and A1 and A2 in the atria. This can be accomplished through electrophoretic analysis in the native state where the migration rate reflects the combined properties of both the heavy (HC) and light (LC) chains. The multiplicity of ventricular isomyosins results from the association between HC α and HC β which form α and β homodimers (V1 and V3) and a heterodimer (V2). The primary structure of HC alpha and beta differs as shown by the analysis of both proteins[1] and of cDNA corresponding to their respective messages[2,3]. The two pairs of light chains of ventricular isomyosin, however, are identical[4].

The atrial myosins are indistinguishable from the HC alpha of the ventricle[5,6]. The two pair light chains (LCs) of atrial myosin differ from those present in the ventricular isomyosins. This can be determined by the exchange of LCs present in atrial isomyosins with those of the ventricle. The cross-hybridized atrial myosin now migrates as a single band only which has electrophoretic mobility identical to that of V1 isomyosin[5]. The different mobility of A1 and A2 isomyosin is not understood at this time but it may result from differential phosphorylation.

The proportions of the individual molecular variants of myosin in the heart is specific for a given animal but it does not remain constant. Instead, it changes both during the normal development and the altered functional, hormonal and nutritional state of the animal. The mechanisms by which multiple factors determine cardiac phenotype, however, remains unknown. In this study we report on the isomyosin transitions during embryonic development in primary cultures of embryonic myocytes and in the ventricular wall of adult animals.

METHOD

The isomyosin content of the heart has been evaluated by indirect immunofluorescence with HC specific monoclonal antibodies[7], by cytochemical assay of ATPase activity after alkaline and acid preincubation to differentiate the V1 and V3 activities[7], by electrophoretic analysis of myosin in the native state[8] and by assay of Ca-activated ATPase activity using bands separated on pyrophorate gels[9]. The SDS-denatured myosin was treated with chymotrypsin and then peptide mapping of the partial proteolytic cleavage products was performed. The reaction was terminated by 3 minutes of heating in the presence of 2% SDS and the peptides were then analysed by SDS-polyacrylamide gel electrophoresis[5].

RESULTS

Expression of Myosin HCs During Cardiac Embryogenesis:

We examined the class of myosin HCs present in the heart during chicken embryogenesis by using antibodies specific for ventricular myosin which is the pure V3 form in chicken. This antibody reacted strongly with the ventricle throughout all stages of embryonic (beginning at 1-1/2 days, stage 12) and post-hatching development. The developing atria were also reactive but only during the initial stage, from incorporation into the heart (stage 12) through the 4th day (stage 25). In contrast to the ventricular myocardium, the atrial reactivity subsequently declined so that from the 6th day (stage 29) throughout the remaining embryonic period there was no significant reactivity noticable. The staining of the post-hatched chicken was also negative. These data thus indicate that the initial stage of atrial development of the chick is characterized by the expression of heavy chains which are identical to that present in the ventricle. During later stages of development this myosin is replaced by the other type which has high ATPase typical of the adult atria.

Phenotypic Transitions in Adult Heart

The effect of thyroid hormone on the expression of atrial myosin. A variety of factors have been shown to influence the expression of cardiac myosin. But thyroid hormone has been shown in numerous studies of the ventricles to be the most potent one. Little information, however, is available about thyroid hormones impact on the atria.

Figure 1[5] indicates the SDS gel electrophoretic profiles of peptides generated from T4 and PTU treated animals (A_T and A_{PTU}) by partial chymotryptic cleavage. The peptide maps remain identical to that of ventricular V1 isomyosin (V_{T4}). The map of V1 myosin is different from V3 isomyosin (V_N) which is the species present in PTU treated animals. (In Figure 1 the arrows indicate peptides typical of V1 myosin while the asterisks indicate those of V3. LC indicates ventricular and atrial light chains). The atrial HCs are indistinguishable from those of ventricular HCα.

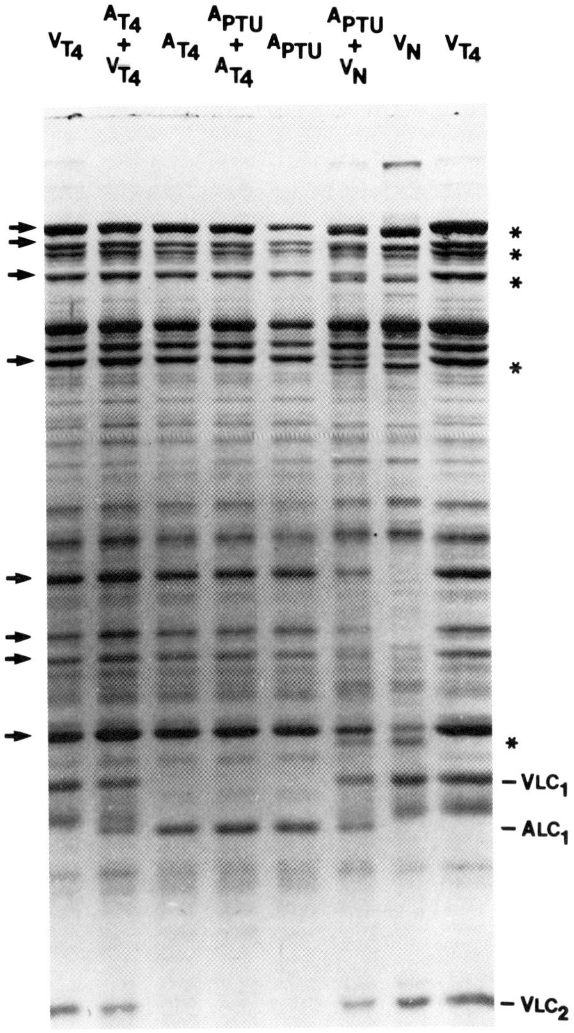

Figure 1.

Effect of treatment with 6(OH)-dopamine on isomyosin profile of the heart. Because of the close association between the thyroid and adrenergic systems especially during the early period following birth, we have sought to determine whether peripheral sympatectomy causes a shift in ventricular isomyosins. Rats were given 40mg/kg of 6(OH)-dopamine twice within 24 hours for two consecutive weeks. 21 days after the last treatment the N-epinephrine content of the heart and isomyosin profile were determined.

Figure 2 shows that treatment resulted in a 90% depletion of N-epi stores and in a moderate (30%) decrease in the content of V1 isomyosin.

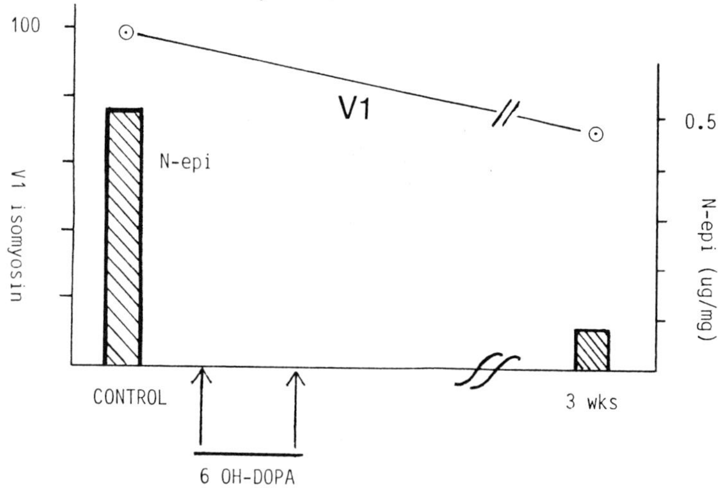

Figure 2. The Effect of Peripheral Sympatectomy on Isomyosin Composition of Rat Ventricle.

Transmural distribution of ventricular isomyosins. A possible clue for identifying the factors which control the expression of heavy chains can be obtained by following the isomyosin distribution within the myocytes of the ventricular wall. Using rabbit heart tissue excised from the region midway between the apex and the base of the left ventricle the segments were oriented so that a transmural surface was in the plane of the section. The segments were frozen in liquid nitrogen precooled with isopentane and sectioned 6 um thick. In the adult rabbit we found a marked transmural gradient with immunofluorescence and ATPase staining[7] as seen in Figure 3.

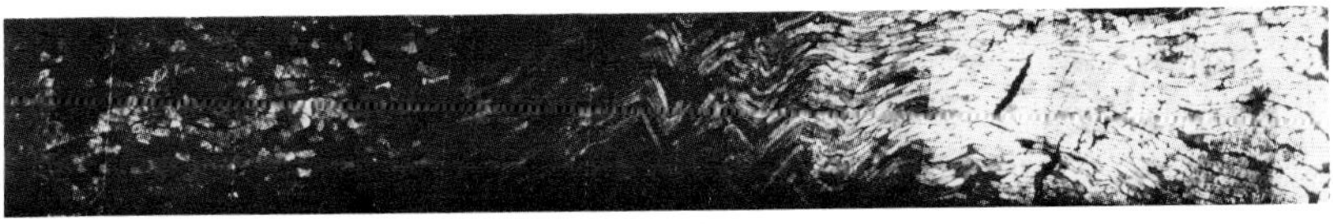

Figure 3.

The V1 isomyosin was found to predominate in the epicardial zone (right) while the endocardial regions remaining mostly non-reactive with anti-V1 antibodies which indicates the presence of V3 isomyosin (left). The cells with similar staining properties appeared in large clusters or strands. However, stained and non-stained myocytes were also frequently found directly connected by an intercalated disc.

Expression of ventricular myosin in cultured myocytes.

Since the transmural distribution of isomyosin indicates cellular heterogeneity, we sought to determine how individual cells respond to manipulation of the thyroid hormone level. Primary cultures of myocytes derived from ventricles of rat embryos (16 days of gestation) were grown in a serum-free defined medium in the absence or presence of thyroid hormone. The culture medium consisted of 99% F12 Ham's nutrient mixture without glutamine, 0.024% fetuin, 1% BSA, 0.02 mg/ml ascorbic acid, 50 μg/ml endothelial cell growth supplement, 10 mg/ml epidermal growth hormone and 0.1 ml/ml insulin-transferrin-selenium mixture. When the effect of thyroid hormone was examined, T4 was added to the above medium to obtain 10nM concentration. The cells were fixed at selected time intervals and the type of myosin was determined by indirect immunofluorescence and by cytochemical assay of ATPase activity[9].

The myocytes grown in the absence of T4 converted into cells strongly reactive with anti-V3 antibody and had low ATPase activity. A small population of cells, however, failed to display any V3 activity. These cells stained strongly with V1 antibody and had high ATPase activity. Seven days of growth in T4 supplemented medium converted most the cells to V1. But, again, a small population failed to respond to the thyroid hormone. Identical results were obtained where the type of myosin was determined by native gel electrophoresis as shown in Figure 4.

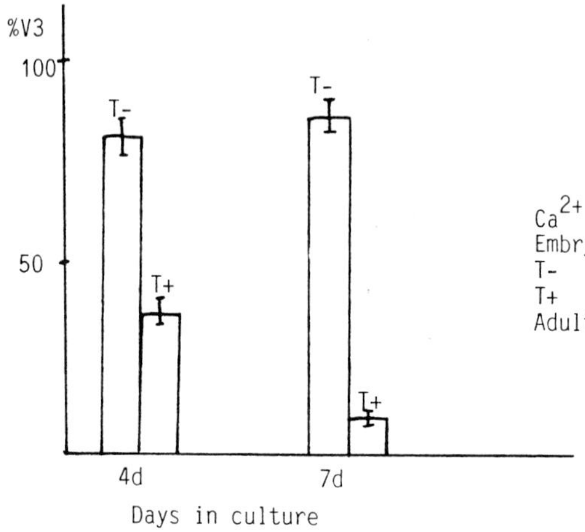

Ca^{2+} - ATPase activity		
Embryo	(18 days)	0.19 ± 0.05
T-	(7 days)	0.17 ± 0.03
T+	(7 days)	0.67 ± 0.15
Adult	(2 mon.)	0.75 ± 0.17

Figure 4. Isomyosin Shifts in Primary Culture of Rat Embryonic Myocytes.

DISCUSSION

The isomyosin profile of the heart differs between animal species[8]. In hearts of small animals, e.g. mouse, HCα predominates in both ventricles and atria. With an increase in body size and a consequent decline in basal metabolic rate, the HCβ gradually predominates in the ventricle and also appears in the atria. Although the myosin HC phenotype is typical of a given animal species it readily changes as a result of various experimental interventions as well as during normal development, embryonic and post-natal. Some of the factors which induce the expression of either V1 or V3 isomyosin are listed in Figure 5[12]. It must be noted that all of these studies utilized ventricles only.

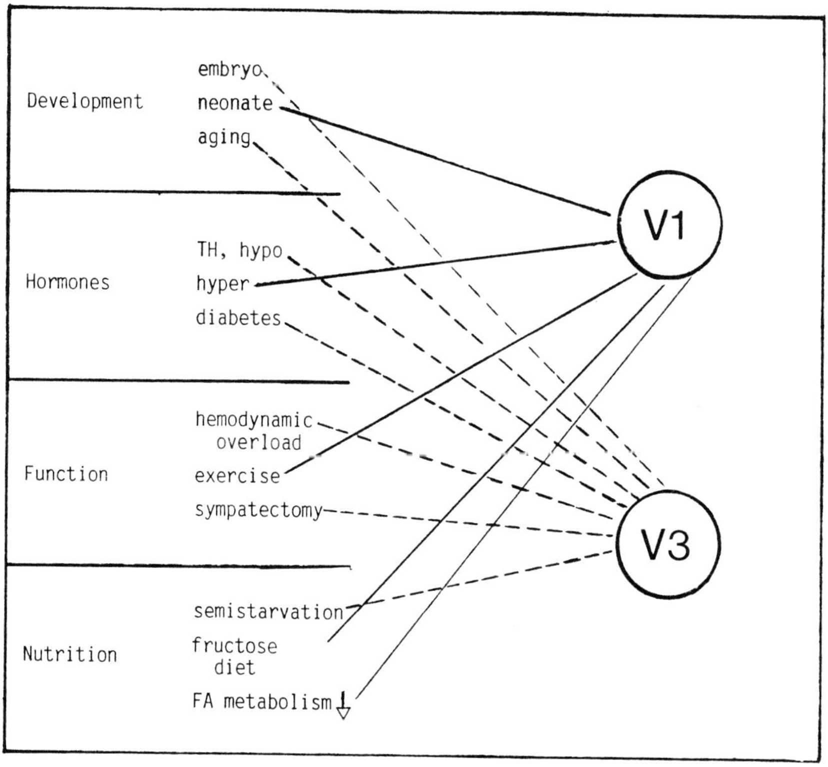

Figure 5. Changes in Ventricular Isomyosin Complement.

Only a few measurements are available that address the atria. Manipulation of the thyroid state did not result in any change in the atrial myosin of rabbit. The dynamism of atrial phenotype, however, can be seen during embryonic development. In our study the first isomyosin detected during embryogenesis is immunologically indistinguishable from that expressed in the adult ventricle. This form is replaced during development with another one which has V1-like characteristics (this study and 13). A change in atrial phenotype can also be detected in the adult. For example, a recent study[14] indicated a pronounced shift toward HCβ in hypertrophied human atria with chronic pressure overload.

As far as the determinants of cardiac phenotype is concerned, the observed heterogeneity seen in the ventricular wall is of particular interest. Whether the uneven distribution of myocytes with high (V1) and low (V3) ATPase activities have a functional consequence is difficult to assess at this time. The following table shows regional gradients of some properties of the heart.

Property	Portion of the Wall: Epicardial	Endocardial
Phylogenetic origin	new	old
Predominant myocytes	V 1	V 3
Electrical		
Length of AP	short	long
Mechanical		
Length of contraction	short	long
Onset of force generation	late	early
Morphological		
Sarcomere length	short	long
Cellular hypertrophy (pressure overload)	large	small
Metabolic		
Onset of ischemia (pH)	late	early
Mitochondrial respiration (P/O)	high	low

Table 1. Transmural Gradients

The fact that myocytes of different isomyosin content were found in cells directly connected by an intercalated disc suggests that factors different from regional work load or cellular microenvironment are responsible for the uncoordinated change in the cellular phenotype seen during maturation. For example, expression of different amounts of a give isomyosin might require genetic reprogramming. This possibility is suggested by the study of cultured embryonic myocytes. Although most of the cells do respond in synchrony to the manipulation of the thyroid state as in vivo cells do, a small population of cells is present which is unresponsive. This lack of responsiveness might depend on the stage of myocyte differentiation.

REFERENCES

1. Chizzonite, R.A., Everett, A.W., Clark, W.A., Jakovcic, S., Rabinowitz, M. and Zak, R. (1982). Isolation and characterization of two molecular variants of myosin heavy chain from rabbit ventricles. J. Biol. Chem. 257:2056-2065.

2. Sinha, A.M., Umeda, P.K., Kavinsky, C.J., Rajamanickam, C., Hsu, H-J., Jakovcic, S., and Rabinowitz, M. (1982). Molecular cloning of mRNA sequences for cardiac α and β-form myosin heavy chains: expression in ventricles of normal, hypothyroid and thyrotoxic rabbits. Proc. Natl. Acad. Sci. 79:5847.

3. Mahlavi, V., Periasamy, and Nadal-Ginard, B. (1982). Molecular characterization of two myowin heavy chain genes expressed in the adult heart. Nature 297:659.

4. Wilkman-Coffelt, J. and Srivastava, S. (1979). Differences in atrial and ventricular myosin light chains. FEBS letters, 106(1):207.

5. Chizzonite, R.A., Everett, A.W., Prior, G. and Zak, R. (1984). Comparison of myosin ehavy chains in atria and ventricles from hyperthyroid, hypothyroid and euthyroid rabbits. J. Biol. Chem. 259:15564.

6. Sinha, A.M., Friedman, D.J., Nigro, J.M., Jakovcic, S., Rabinowitz, M. and Umeda, P.K. (1984). Expression of rababit ventricular α-myosin heavy chain messenger RNA sequences in atrial muscle. J. Biol. Chem. 259:6674.

7. Eisenberg, B.R., Edwards, J.A. and Zak, R. (1985). Transmural distribution of isomyosin in rabbit ventricle during maturation examined by immunofluorescence and staining for Ca-activated ATPase. Circ. Res. (in press).

8. Clark, W.A., Chizzonaite, R.A., Everett, A.W., Rabinowitz, M. and Zak, R. (1982). Species correlations between cardiac isomyosins. J. Biol. Chem. 257:5449.

9. Nag, A.C., Cheng, M., and Zak, R. (1985). Distribution of isomyosin in cultured cardiac myocytes as determined by monoclonal antibodies and adenosine triphosphatase activity. Exp. Cell Res. (in press).

10. Nag, A.C., and Cheng, M. (1984). Expression of myosin isoenzymes in cardiac-muscle cells in culture. Biochem. J. 221:21.

11. Lompre, A.M., Mercadier, J.J., Wisnewsky, C., Bourveret, P., Pantaloni, C., d'Albis, A., and Schwartz, K. (1981). Species- and age- dependent changes in the relative amounts of cardiac myosin isoenzymes in mammmals. Develop. Bio. 84:286.

12. Zak, R. (1985). Biological mechanisms of hypertrophy.

In Fozzard, H.M., Haber, E., Jennings, R.B., Katz, A.M., and Morgan, H.E. (eds): "Handbook of Experimental Cardiology," New York, Raven Press.

13. Gonzalez-Sanchez, A., and Bader, D. (1984). Immunochemical analysis of myosin heavy chains in the developing chicken heart. Devel. Biol. 103:151.

14. Gorza, L., Mercadier, J.J., Schwartz, K., Thornell, L.E., Saverio, S., and Schiaffino, S. (1984). Myosin types in human heart: an immunofluorescence study of normal and hypertrophied atrial and ventricular myocardium. Circ. Res. 54:694.

Molecular Biology of Muscle Development, pages 853–863

REGULATION OF CARDIAC MYOSIN ISOENZYMES BY THYROID HORMONE[1]

Eugene Morkin, Donald Sheer[2], Thomas A. Gustafson, Irwin L. Flink, and Bruce E. Markham

Departments of Internal Medicine and Pharmacology
University of Arizona
Tucson, Arizona 85724

ABSTRACT Expression of the genes coding for myosin heavy chains in the heart are controlled by thyroid hormone, but also can be modified by a high carbohydrate diet and glucocorticoids. In skeletal muscle, thyroid hormone has different effects on these genes and longer treatment is required. The possible physiological significance of changes in myosin isoforms is discussed in terms of current molecular models of contraction and effects on myocardial performance.

Thyroid hormone regulates the expression of cardiac myosin isoenzymes (1). In heart, these proteins are hexamers containing two heavy peptide chains, and four light chains per molecule. Apparently, only the expression of the myosin heavy chains (MHCs) is hormonally regulated. This has been ascribed to increased transcription of the gene coding for α-MHC and inhibition of β-MHC mRNA synthesis (2,3). In the rat, seven forms have been identified, their respective cDNAs cloned and completely or partly sequenced (4-6). In addition to the

[1]This work was supported by research grants from the Arizona Affiliate of the American Heart Association, the National Institutes of Health (HL-20984), and the Gustavus and Louise Pfeiffer Research Foundation.
[2]Present address: Clinical Endocrinology Branch, National Institutes of Health, Bethesda, Maryland.

two mRNA types found in ventricular tissue, separate forms have been identified in fast skeletal muscles, as well as in embryonic and fetal muscles. MHCs in slow-twitch skeletal muscles are products of the same gene coding for ventricular muscle β-MHCs (3). Efforts are underway in several laboratories to understand the control of this complex multigene family. Here a brief overview of this emerging area of research is presented, and some comments are made related to the physiological significance of myosin isoenzyme changes within the heart.

Ventricular myosin isoenzyme composition is dependent upon thyroid status and animal species. The three ventricular myosin isoenzymes, designated V_1, V_2, and V_3, result from combinations of α- and β-MHCs. Functionally, these isoenzymes have different rates of ATP hydrolysis. Thus, ATP hydrolysis is most rapid in the V_1 form and is slowest in V_3. Although the V_2 form has not been well characterized it is thought to be intermediate in activity.

In euthyroid rat ventricle, the V_1 form, which contains two α-MHCs, is the major myosin constituent. This isoenzyme also predominates in the hyperthyroid state. The V_3 isoenzyme, comprised of two β-MHCs, is the major form in the hypothyroid state. By contrast, in rabbit ventricle V_3 is the major myosin isoform. Administration of thyroid hormone stimulates synthesis of α-MHCs, increasing the amount of the V_1 form. Thus, ventricular myosin isoenzyme composition varies depending upon animal species and thyroid hormone status.

The question naturally arises as to whether thyroid hormone regulates the expression of MHC genes in other muscle types. Recently, the effects of thyroid hormone on MHC mRNAs in ventricular and soleus muscles of hypothyroid rats have been compared (7), since in this condition, both muscles have a predominance of β-MHC mRNA. To analyze the changes in myosin gene expression, we have developed a rapid, simple dot-blot assay based upon hybridization of total cellular RNA to synthetic ^{32}P-labeled oligonucleotides. Probes 20 bases in length were constructed complementary to the unique 3' flanking sequences of six types of myosin mRNA. Although the coding regions of the myosin genes show a high degree of homology, the 3' untranslated sequences are not as closely conserved and are specific for each heavy chain mRNA type.

In these experiments, administration of triiodothyronine, T_3, (2 μg/100 g) resulted in a rapid increase in α-MHC mRNA and a corresponding decrease in β-MHC. The increase in α-MHC mRNA was evident within 6 h after treatment and was nearly maximal at 24 h. At 48 h, binding of the radioactive α-MHC probe was increased by 800% over the control value. During this time, the amount of radioactive β-MHC probe bound to the mRNA spot decreased to about 25% of the untreated value. After 7 wks of treatment at the same dosage of T_3, hybridization of the radioactive α-MHC probe to ventricular muscle total RNA decreased to about 50% of the 48 h value. Interestingly, ventricular RNA showed binding to the probe for fetal mRNA, which declined after T_3 treatment to about 60% of its initial value.

Comparison of the changes in ventricular tissue with those in soleus revealed some important differences. During the initial 48 h after injection of T_3 there were no significant differences in hybridization of any of the six synthetic probes to total RNA from the soleus muscle. To test the effects of longer treatment, the same dose of T_3 was given on a daily basis for periods up to 9 wks. After 7 wks, the amount of binding of the labeled probe for a fast muscle mRNA, Fast II mRNA (4), increased by 250%. Nevertheless, the amount of β-MHC probe bound per mg RNA remained constant. Thus, in contrast to its effects on ventricular muscle, thyroid hormone appeared to stimulate expression of Fast II, without appreciably decreasing the amount of β-MHC mRNA per mg RNA. Radioactivity associated with hybridization of the probe for embryonic message declined to about 50% of its control value after 7 wks of treatment. Analysis of hybridization to cellular RNA from soleus muscle by other ^{32}P-labeled MHC probes revealed only minor changes.

The changes in MHC mRNAs observed in these experiments probably reflect changes in the rates of gene transcription. The markedly longer time needed to stimulate synthesis of Fast II mRNA in soleus virtually precludes an explanation based on differences in mRNA half-lives. Also, β-MHC gene expression in soleus was much less affected than in ventricular muscle. Possibly, the longer time needed to stimulate transcription in soleus of a fast-type MHC mRNA is related to the slower accumulation of an intermediate in the steps between binding of T_3 to specific nuclear receptors and

TABLE 1
CHARACTERISTICS OF DOSE-RESPONSE CURVES FOR THYROID HORMONE ANALOGS

	ED_{50} Estimates		E_{max} Estimates[b]	
Compound	CO_2 production from glucose[a]	%V_1 myosin	CO_2 production from glucose	%V_1 myosin
T_3	0.18±0.03	0.46±.02	1.00	1.00
dT_3	0.31±0.02	0.56±.04	0.83	0.99
Triac	0.58±0.04	0.78±.13	0.54	0.71
T_2	0.62±0.03	0.75±.21	0.43	0.33
rT_3		0.86±.13		0.08

[a]Values are dpm X 10^4/mg/h.
[b]Efficacy relative to T_3.

stimulation of Fast II gene transcription. Alternatively, there may be different mechanisms for thyroid hormone control of gene expression in these tissues. Interestingly, the turnover of myosin is markedly longer in skeletal muscles than heart. The estimated half-life for myosin in ventricle is about 6 days whereas half-lives in excess of 150 days have been reported for skeletal muscle myosins (8).

The efficacy of thyroid hormone analogs in switching myosin isoforms relative to their effect on myocardial oxygen consumption (CO_2 production) is among the many aspects of regulation of MHC expression in need of further clarification. This information should be useful in evaluating the biological activity of intracellular T_3 metabolites as well as in identifying compounds which might be potentially useful as positive inotropic agents in hypertrophied and failing hearts.

The important characteristics of dose-response curves for the thyroid hormone analogs 3,5,3'-L-triiodothyronine (T_3), 3,5,3'-D-triiodothyronine (dT_3),

3,3'5'-L-triiodothyronine (rT_3), 3,5,3'-L-triiodo-thyroacetic acid (Triac), and 3,5-L-diiodothyronine (T_2) obtained in thyroidectomized rats (9) are shown in Table 1. $^{14}CO_2$ production from labeled glucose in isolated atria was compared with the change in ventricular V_1 content. The relative efficacy of the analogs tested was as follows: T_3>dT_3>Triac>T_2>rT_3. Furthermore, the ratio of the T_3 analog dose (E_{max}) required to maximally elevate atrial CO_2 production to that needed to stimulate synthesis of the V_1 isoform was calculated, since it should be predictive of the separation of these two biological activities. None of the analogs exhibited a ratio greater than 1.5, which probably indicates they do not possess any significant advantage over T_3.

A number of factors are reputed to modify the actions of thyroid hormone on other enzymes, such as β-adrenergic agonists and antagonists, glucocorticoids and a high carbohydrate diet. The results obtained in a study of these factors (9) have been summarized in Figure 1A-C.

Isoproterenol, a β-adrenergic agonist, had no appreciable influence on stimulating expression of the V_1 isoform in thyroidectomized (THX) rats (Figure 1A). Also, treatment with propranolol, a β-adrenergic antagonist, did not alter the dose-response curve for stimulation of V_1 expression by T_3 in hypothyroid animals.

As shown in Figure 1B, adrenalectomy (ADX) in euthyroid rats lead to an average decrease of about 33% in the V_1 form and a reciprocal increase in the V_3 form (not shown). These effects were reversed by administration of hydrocortisone.

A number of enzymes which are influenced by thyroid hormone, including malic enzyme, the two hexose monophosphate shunt dehydrogenases and mitochondrial glycerol-phosphate dehydrogenase, can be induced in thyroid-deficient rats by feeding a high carbohydrate diet (10). It occurred to us that cardiac myosin isoenzymes might be subject to similar regulation. When hypophysectomized (HPX) rats were tube fed with a diet containing 76% of the calories as glucose or fructose the V_1 form increased from about 12% to 36% of total myosin. Furthermore, administration of a partial replacement dose of T_3 together with the high carbohydrate diet had synergistic actions.

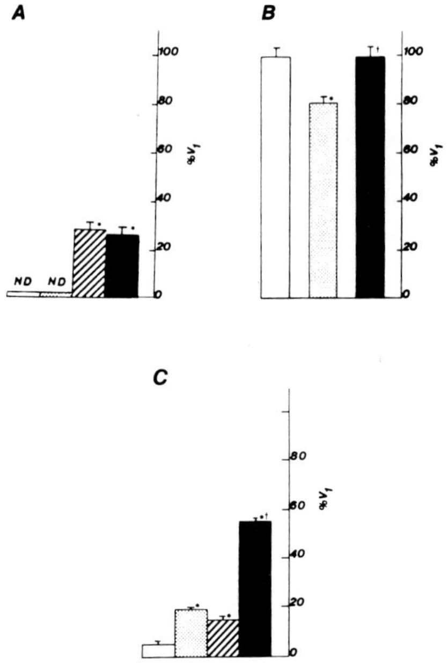

Figure 1. A) Effects of isoproterenol alone and in combination with partial T_3 replacement on V_1 content in THX rats. (□) THX, (▩) THX plus isoproterenol (0.5 mg/100 g daily), (▨) THX treated with T_3 (0.4 μg/100 g daily), (■) THX plus isoproterenol and T_3. ND, not detectable; *P < .001 compared with control group; †P < .001 compared with T_3-treated group; N = 5.

B) Influence of ADX in euthyroid rats and replacement with hydrocortisone on V_1 content. (□) Intact, (▩) ADX, (■) ADX plus hydrocortisone acetate (0.2 mg/100 g daily). *P < .001 compared with intact animals; †P < .001 compared with ADX animals; N = 3.

C) Effects of high carbohydrate diet and partial T_3 replacement on V_1 content in HPX rats. (□) HPX, (▩) HPX fed fructose diet, (▨) HPX plus T_3 (0.3 μg/100 g daily), (■) HPX fed high carbohydrate diet plus T_3. *P < .001 compared with control group; †P < .001 compared with T_3-treated group; N = 5.

It is now believed that T_3 stimulates protein synthesis by binding to nuclear receptors which regulate the concentration of specific proteins. Increases in mRNAs coding for a number of proteins, including α- and β-MHC have been demonstrated, as mentioned above. It seems likely, therefore, that interactions between regulatory signals produced by T_3, carbohydrates, and perhaps other factors, occur at a pretranslational level.

Our experiments are consistent with the hypothesis that MHC mRNA expression may be regulated by a primary carbohydrate-generated signal which is modulated by thyroid hormone, glucocorticoids and other factors. Alternatively, there may be multiple, independent mechanisms for control of MHC mRNA synthesis. Definitive support for a single vs. multiple regulatory mechanisms must await further studies.

The possible physiological significance of changes in myosin isoenzymes, such as those induced by alterations in thyroid status, also have been the subject of older and more recent studies. Direct inotropic effects of thyroid hormone have been shown by measurements of the mechanical performance of isolated papillary muscles from thyrotoxic cat hearts (11). These findings occurred in the presence and absence of intact norepinephrine stores and over a wide range of temperature and contraction frequency. An increase in the rate of isometric tension developments also has been found both in intact and glycerinated muscles from thyrotoxic rabbits, which is consistent with a direct effect (12).

The molecular basis for these actions may be related, in part, to changes in the myosin isoenzyme content, since the myosin ATPase activity of a muscle is closely correlated to its intrinsic speed of contraction. The relative proportions of ventricular myosin isoenzymes has been correlated with the velocity of shortening in papillary muscles and small segments of ventricular muscle (13). Papillary muscles that contained 100% V_1 shortened, under zero load, about six times faster than muscles that contained 100% V_3 myosin.

Histochemical studies have shown increased staining for myosin ATPase in virtually all ventricular fibers after thyroid hormone administration (14,15). Using antibodies that are specific for myosin isoforms it has been possible to show that individual fibers may reaction with antibodies for both V_1 and V_3 myosins (16).

A major question that arises from this evidence is how the different myosin isoforms are able to interact without the slowest ATP cleaving form limiting the speed of contraction. To relate the chemistry of actin-myosin interaction to mechanical events during muscle contraction Lymn and Taylor (17) proposed a four-step

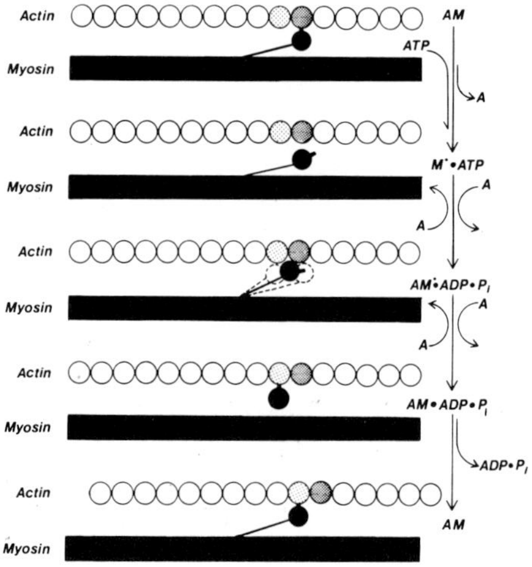

Figure 2. A current model of muscle contraction and actomyosin ATPase.

model of contraction: 1) binding of ATP and rapid dissociation of actomyosin, 2) hydrolysis of ATP on the free myosin head, 3) recombination of actin with the myosin products complex, and 4) release of myosin products. This mechanism gained wide attention because it could be related easily to the cycling of myosin cross-bridges during contraction. Subsequent studies, however, indicated that the model had to be modified in several important ways to be realistic (18). First, rather than a single hydrolysis step, several intermediate states in hydrolysis of ATP by myosin heads have been identified, which represent different conformational states of myosin. Other evidence suggested that the reaction products of ATP, ADP and P_i, may be released sequentially from the active site. Furthermore, kinetic studies revealed that association-dissociation between actin and the myosin products complex occurs at least 1,000 times more rapidly than hydrolysis of ATP. This would imply that cross-bridges (myosin heads) cycle on and off the thin filament many times during a single turn of the ATP cleavage cycle. Only after myosin heads have undergone the conformation changes associated with ATP hydrolysis would the complexes formed with actin contribute to force development. With a mechanochemical model of this type, myosin isoforms possessing greatly differing activities could coexist and work together without the lower activity forms limiting contraction speed.

Although there is a strong correlation between the steady state hydrolysis of MgATP and the unloaded shortening velocity, the cleavage step is too slow to directly limit the velocity of unloaded contraction. Recent measurements of ADP dissociation from bovine cardiac actomyosin at physiological temperatures suggested that the rate constant for this step (550-800 sec^{-1} at 38°C) is the correct order of magnitude to limit the rate of myosin dissociation from actin and is likely to be the rate-limiting step of cross-bridge dissociation (19). Thus, the rate constant of ADP dissociation may be the molecular step that limits the unloaded shortening velocity in cardiac muscle. Since the rate constant for ADP release seems to vary in proportion to the steady state ATPase rate in several striated muscle myosins that have been studied, the rate of ADP dissociation for the predominant myosin isoform may limit, in general, the velocity of unloaded shortening.

Finally, how do the biochemical, anatomical and physiological actions we have described relate to the performance of the intact heart? To gain a clearer description of the effects of the hormone on the intact heart, left ventricular (LV) mechanical performance has been assessed in calves with implanted sonomicrometer crystals and pressure transducers (20). As would be anticipated, heart rate, LV systolic pressure and the rate of rise in LV pressure (dP/dt) are increased significantly after treatment with thyroid hormone. LV internal diameter is increased in both systole and diastole, however, fractional shortening is unchanged. The increases in ventricular dimensions and systolic blood pressure combine to increase peak wall stress. Despite this increase in afterload, mean velocity of circumferential fiber shortening remains normal. Thus, in the intact heart at rest the direct inotropic effect of thyroid hormone actually does not produce an increase in speed of shortening, rather speed of shortening remains normal despite the greater afterload.

ACKNOWLEDGMENTS

We thank Mrs. Evelyn English for her assistance with the preparation of this manuscript.

REFERENCES

1. Morkin E, Flink IL, Goldman S (1983). Prog Cardiovasc Dis 25:435.
2. Everett AW, Sinha AM, Umeda PK, Jakovcic S, Rabinowitz M, Zak R (1984). Biochemistry 23:1596.
3. Lompre A-M, Nadal-Ginard B, Mahdavi J (1984) J Biol Chem 259:6437.
4. Nadal-Ginard B, et al. (1982). Structure and regulation of mammalian sarcomeric myosin heavy-chain gene. In Pearson ML, Epstein HF (eds): "Muscle Development: Molecular and Cellular Control," Cold Spring Harbor Laboratory, p 143.
5. Mahdavi V, Chambers AP, Nadal-Ginard B (1984). Proc Natl Acad Sci USA 81:2626.
6. Periasamy M, Wieczorek DF, Nadal-Ginard B (1984). J Biol Chem 259:13573.
7. Gustafson TA, Markham BE, Morkin E (1985). Submitted for publication.
8. Kimata S, Morkin E (1971). Am J Physiol 221:1706.
9. Sheer D, Morkin E (1984). J Pharmacol Exp Ther 229:872.
10. Tepperman HM, Tepperman J (1964). Am J Physiol 206:357.
11. Buccino RA, Spann JF, Pool PE, (1967). J Clin Invest 46:1669.
12. Skelton CL, Su JY, Pool E (1976). Cardiovasc Res 10:380.

13. Pagani ED, Julian FS (1984). Circ Res 54:586.
14. Morkin E, Banerjee SK, Stern LZ (1977). FEBS Lett. 79:357.
15. Weisberg A, Winegrad S, Tucker M, McClellan G (1982) Circ Res 51:802.
16. Sartore S, Gorza L, Bormioli SP, Libera LD, Schiaffino S (1981). J Cell Biol 88:226.
17. Lymn RW, Taylor EW (1971). Biochemistry 10:4617
18. Stein LA, Chock PB, Eisenberg E (1981). Proc Natl Acad Sci USA 78:1346.
19. Siemankowski RF, White HD (1984). J Biol Chem 259:5045.
20. Goldman S, Olajos M, Friedman H, Roeske WR, Morkin E (1981). Am J Physiol 242:H113.

Molecular Biology of Muscle Development, pages 865–874

MOLECULAR ANALYSIS OF THE DNA SURROUNDING THE DUCHENNE MUSCULAR DYSTROPHY LOCUS[1]

Louis M. Kunkel, Anthony P. Monaco, William Middlesworth, Corlee Bertelson[2] and Kenneth Fischbeck[3]

Division of Genetics, Children's Hospital, Boston, MA 02115

ABSTRACT A phenol accelerated competitive DNA reassociation method (1) has been utilized to specifically clone DNA fragments from within and around the Duchenne muscular dystrophy locus. Sheared DNA isolated from a patient (BB) who was diagnosed as having Duchenne muscular dystrophy, chronic granulomatous disease, retinitis pigmentosa and a minute X chromosome interstitial deletion (2) was combined in a 200-fold excess with MboI cleaved DNA isolated from a 49,XXXXY lymphoid cell line. The mixture was subjected to a phenol enhanced reassociation technique (3), and those reassociated molecules maintaining MboI termini were cloned in a recombinant plasmid vector. Analysis of recombinant clones from the resulting library has identified 7 human DNA fragments absent from the patients' DNA. All seven absent clones have been localized to the Xp21 region of the human X chromosome and are each potential linkage markers for the diseases affecting this boy. One of the seven absent clones has been shown to be absent from the DNA of a second unrelated boy with Duchenne muscular dystrophy but no apparent

[1]This work was supported by the Muscular Dystrophy Association and the National Institutes of Health (Program Project HD18658)
[2]Brandeis University, Waltham, Massachusetts 02254
[3]Department of Neurology, University of Pennsylvania, Philadelphia, Pennsylvania 19104

cytological deletion and no associated chronic granulomatous disease or retinitis pigmentosa. The absent clones described here are each potential startpoints in the eventual cloning of three X-linked disease loci; Duchenne muscular dystrophy, chronic granulomatous disease and retinitis pigmentosa.

INTRODUCTION

Duchenne muscular dystrophy (DMD) represents one of the numerous X-linked loci for which an aberrent genetic product has yet to be identified (4). The DMD locus is apparently disrupted in various X-autosome translocations (5,6) with resultant phenotype of DMD in females. Human X-chromosome specific DNA fragments which detect restriction fragment length polymorphisms (RFLPs; 7) have established the location of DMD in the Xp21 region of the X chromosome (8) which was implicated by the various translocations (5,6). One male with 3 X-linked diseases: DMD, retinitis pigmentosa (RP), chronic granulomatous disease (CGD) and the McLeod red blood cell phenotype has been shown to have an interstitial deletion of part of the Xp21 region of the human X chromosome (2). It is assumed that this patient has lost DNA sequences encoding for the DMD genetic product (2).

We describe here the rationale and method by which the DNA of the patient with DMD, RP and CGD can be utilized to obtain molecular clones of human DNA which will serve as startpoints in the characterization of the DMD locus. The first step was to obtain a library of human DNA highly enriched for segments absent from the patient. This was accomplished by differential reassociation between sheared DNA isolated from the patient (2) and MboI cleaved 49,XXXXY DNA followed by cloning of reassociated molecules maintaining MboI termini (1). We describe the characterization of the enriched library and our first attempts at defining those absent clones which are closest to the DMD locus.

RESULTS

DNA (250ug), isolated from the male patient (BB) with a minute interstitial deletion of part of band Xp21, DMD, RP and CGD (2) was sheared by sonication to a mean size of

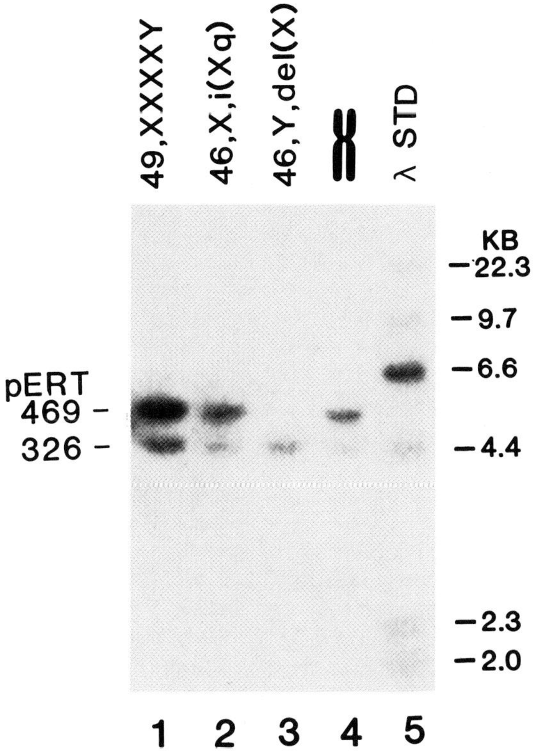

FIGURE 1. Hybridization analysis of pERT clones 469 and 326 was accomplished as previously described (1). One ug of the following Hind III cleaved DNA samples was loaded to a horizontal agarose gel slot: lane 1, a 49,XXXXY lymphoid cell line (GM1202) DNA sample; lane 2, a 46,X,i(Xq) lymphoid cell line DNA sample with 3 copies of Xq and one copy Xp (1); lane 3, the 46,Y,del(X) DNA sample (2); lane 4, a hamster-human hybrid cell DNA sample with an intact human X chromosome among rodent chromosomes (9); lane 5, Hind III cleaved λphage DNA. Radiolabeling of inserts and the processing of nitrocellulose filters was as previously described (1).

1000 base pairs. The sheared DNA was mixed as previously described (1) with MboI cleaved DNA (1.25 ug) isolated from a 49,XXXXY lymphoid cell line. The mixture was heated at 100° for 5' and then subjected to a phenol enhanced reassociation technique (PERT; 1,3). The reassociated molecules were mixed with BamHI cleaved dephosphorylated plasmid DNA (pBR322) and T4 DNA ligase. The ligation mixture was used to transform bacteria (MC1061) and resulted in more than 3,000 recombinant colonies from 5 ug of reassociated DNA. Analysis of recombinant clones was performed as previously described (1) and involved the isolation of individual DNA inserts and testing for their hybridization with DNA samples isolated from normals but not with DNA from the patient (e.g. Figure 1). The clone designated pERT469 hybridizes to a 5.5 kb Hind III fragment in two of the human DNA samples but not to the third (the DNA isolated from the 46,Y,del(X) patient) utilized in the competition reassociation. Another clone (pERT326) hybridizes to a 4.4 kb HindIII fragment in all DNA samples and is clearly present in the patient's DNA. The clone (pERT469) hybridizes with an intensity depending on the X chromosome content of the human DNA used, and it hybridizes to the same sized Hind III fragment in a rodent human hybrid cell DNA sample retaining an intact human X chromosome. Thus this sequence (pERT469) is tentatively localized to the human X chromosome on normal X chromosomes and it is completely absent from DNA of the patient (BB) with the deletion in the X chromosome short arm.

Hybridization analysis for 312 independently isolated inserts from the "PERT library" was undertaken as indicated in figure 1, and the results are summarized in Table 1. Of the 176 clones yielding a simple hybridization pattern, 7 (4%) are completely absent from the DNA of the male patient (BB) with the Xp deletion. Each of the absent clones hybridizes to a different sized Hind III fragment in the genome of normals, and each is localized to the human X chromosome. As previously published (1), the remaining clones hybridized in an expected manner, with some highly repeated sequences (general lane smear) and many moderately repeated sequences which hybridized very strongly to a few Hind III fragments. Certain sequences failed to hybridize to DNA on the nitrocellulose filter; it is assumed that these represented extremely small inserts (the average size of all inserts tested was approximately 160 bp). The majority of clones were judged as unique in the genome and were derived from autosomes (Table 1). One alteration from expected was the apparent enrichment for clones

TABLE 1

SUMMARY OF PERT CLONE HYBRIDIZATION RESULTS

Number Hybridized*	312	Percent
No Signal	44	14.1
Repeated	92	29.5
Unique	176	56.4
From X chromosome	52	29.5**
Absent in 46,Y(del(X))	7	4.0**

*The number hybridized refers to the total clones radiolabeled with the breakdown of their hybridization pattern given below. Hybridization analysis was performed as described in Figure 1 legend.

**The percentage of X clones and absent clones was calculated as a proportion of unique sequence clones.

bearing inserts derived from the X chromosome (29%). The trace 49,XXXXY DNA used in the competition reassociation is expected to have approximately 10% DNA sequences derived from the X chromosome.

Having established the X chromosome origin of the seven cloned DNA inserts absent in the DNA of the 46,Y,del(X) patient (BB), mapping panels were utilized to localize the sequences more precisely. Four of the clones have previously been published (1) as mapping to 3 regions in and around Xp21. The remaining unpublished 3 clones each hybridize to one of these same three regions (data not presented). Thus, all seven clones localize to the Xp21 region of the human X chromosome thought to be deleted in the 46,Y,del(X) patient (2).

In an attempt to subdivide those clones which might reside closest to DMD, the seven absent PERT clones and the probe 754 (also absent from the 46,Y,del(X) patient's DNA; 2) were hybridized to a panel of TaqI cleaved DNA isolated from 25 unrelated DMD males. An example of the hybridization for 2 PERT clones to the same 15 DNA samples is given in Figure 2. The cloned fragment pERT379 hybridizes to a 2.5 kb TaqI fragment in all DMD male DNA samples. In contrast is the fragment pERT87 which fails to hybridize to a 2.6 kb TaqI fragment in one of the 15 DNA samples (indicated by an arrow in Figure 2). The failure of hybridization of pERT87 to this DNA samples has been confirmed by hybridization of pERT87 to 4 different

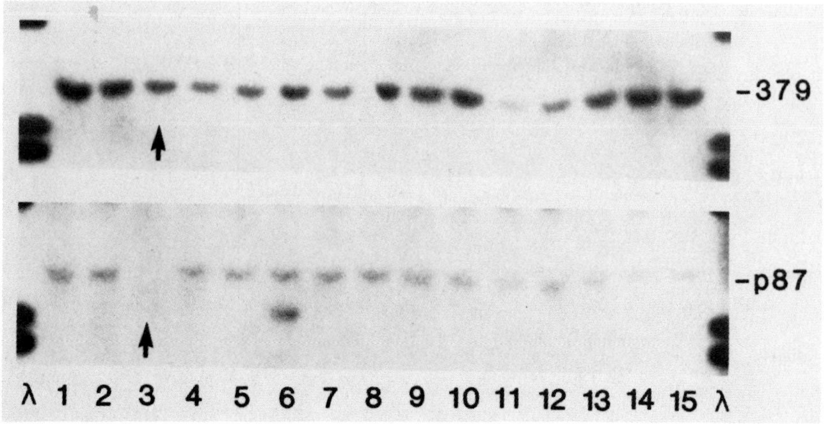

FIGURE 2. Hybridization of pERT379 and pERT87 to identical nitrocellulose filters bearing the immobilized TaqI digested DNA of 15 unrelated boys with DMD. Electrophoresis, radiolabeling and hybridization were as previously described (1). The TaqI fragment hybridizing with pERT379 and 87 is indicated to the side of the figure, and the lane with an absent TaqI fragment is indicated by an arrow. Molecular weight markers are indicated as and were Hind III cleaved lambda phage DNA.

restriction enzyme digestions of the patient's DNA (data not shown). Of the 25 DMD male DNA samples tested, only this one male sample (figure 2) exhibited absence of any DNA sequences. The pERT87 insert was only absent from this one DNA sample, and none of the remaining pERT clones or 754 were absent from any of the other DMD male DNA samples.

DISCUSSION

The rationale of the experiments presented here has been to obtain molecular probes for the DMD locus which can be eventually used to clone the DNA region altered in diseased individuals. The overall approach is outlined in figure 3. The first step of the analysis is to obtain

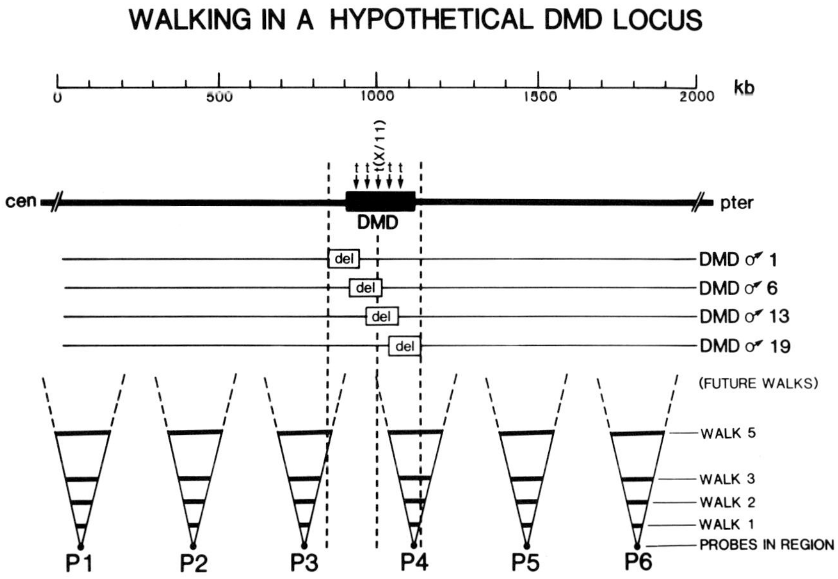

FIGURE 3. A possible schematic representation of the DNA region surrounding the DMD genetic locus. The DMD locus is indicated by a dark block, and the various X; autosome translocation breakpoints are indicated by a t and an arrow. A kb scale is given above. Four boys with presumed deletions are indicated, and 6 cloned probes are indicated at the lower portion of the figure. Sequential chromosome "walks" are indicated as walk 1, etc., and the proportional amount of DNA obtained at each walk is indicated in the cone diagram.

sufficient numbers of cloned DNA fragments in and around the DMD locus. Each of these cloned fragments of DNA can be used in chromosome "walking" (10) experiments to obtain more DNA from each region; towards a goal of identifying segments which are deleted in a subset of affected individuals. Eventually a segment of cloned DNA will be identified which is disrupted by a chromosomal translocation which gives rise to DMD in females (5,6). By a combination of deletion detection and translocation breakpoint mapping the region of DNA encoding the DMD genetic product should be identified. We chose a competition reassociation strategy to obtain the necessary startpoints (as indicated at the bottom of figure 3). A phenol-enhanced reassociation technique (3) allowed extremely high Cot (M/sec) values to be obtained. Such Cot values were necessary to allow sufficient competition of trace DNA (49,XXXXY) with MboI termini and still allow those unique MboI fragments with no complementary sequence in the deletion DNA sample to self-hybridize and be cloned as MboI fragments. Cloning of the reassociated molecules has led to the construction of a library which is highly enriched for unique DNA sequences absent from a patient with a deletion of the X chromosome short arm. In principle, the methods utilized here are applicable to any situation where one DNA contains a group of MboI fragments which are missing from another DNA sample. The acquisition of seven cloned DNA fragments which are absent from the 46,Y,del(X) patient's DNA sample used in competition reassociation was a result well above that expected from chance alone ($P<0.0001$). It is assumed that the high Cot values predicted for the reassociation reaction (3) were reached and the competition reassociation was successful. A three-fold enrichment for X chromosome DNA sequences in the PERT library (Table 1), relative to that expected for 49,XXXXY DNA, is consistent with the expectations of the competition reassociation. An effective 4 fold amplification of X DNA sequences in the trace 49,XXXXY DNA vs. the driver 46,Y,del(X) DNA increases the probability that a unique X chromosome DNA sequence will escape competition when compared with an autosomal sequence.

The seven absent clones presented here coupled with the published probe 754 (2) represent the necessary startpoints (indicated in figure 3) towards characterization of the entire Xp21 region in which the DMD locus resides. The results presented in figure 2 indicate that at least a short stretch of DNA complementary to pERT87 is missing from the DNA of a boy affected with only the disease DMD. This clone is assumed to represent a chromosome "walking" startpoint which has detected a presumed deletion of all or a portion of the DMD locus (potentially the probe indicated in figure 3 as P4). Further experiments are necessary to characterize the extent of the deleted material in this single DMD male and the relevance of the deletion to the DMD phenotype. In addition, analysis of this region for RFLP alleles detected by pERT87 clone (7) is in progress to precisely localize the proximity of this region to DMD. A combination of chromosome "walking" with pERT87 and the acquisition of new fragments from the enriched library described here should result in the eventual identification of the DNA segment encoding the DMD defective product.

ACKNOWLEDGMENTS

We would like to thank Drs. U. Francke and H. Ochs for provision of the 46,Y,del(X) cell line, Dr. P. Pearson for the use of the cloned probe 754 and Dr. S. Latt for useful discussion and critical reading of this manuscript.

REFERENCES

1. Kunkel LM, Monaco AP, Middlesworth W, Ochs HD, Latt SA (1985) Specific cloning of DNA fragments absent from the DNA of a male patient with an X chromosome deletion. Proc Natl Acad Sci USA, Submitted.
2. Francke U, Ochs HD, de Martinville B, Giacalone J, Lindgren V, Disteche C, Pagon RA, Hofker MH, van Ommen G-JB, Pearson PL, Wedgwood RJ (1985) Minor Xp21 chromosome deletion in a male associated with expression of Duchenne muscular dystrophy, chronic granulomatous disease, retinitis pigmentosa and McLeod syndrome. Am J Hum Genet, In press.
3. Kohne DE, Levinson SA, Byers MJ (1977) Room temperature method for increasing the rate of DNA reassociation by many thousandfold: the phenol emulsion reassociation technique. Biochemistry 16:5329-5341.
4. McKusick V (1983) Mendelian Inheritance in Man Baltimore: Johns Hopkins.
5. Jacobs PA, Hunt PA, Bart RD (1981) Duchenne muscular dystrophy (DMD) in a female with an X/autosome translocation: further evidence that the DMD locus is at Xp21. Am J Hum Genet 33:531.
6. Worton RG, Duff C, Sylvester JE, Schmickel RD, Willard HF (1984) Duchenne muscular dystrophy involving translocation of the DMD gene next to ribosomal RNA genes. Science 224:1447-1449.

7. Botstein D, White RL, Skolnick M, Davis RW (1980) Construction of a genetic linkage map in man using restriction fragment length polymorphisms. Am J Hum Genet 32:314-331.
8. Davies KE, Pearson PL, Harper PS, Murray JM, O'Brien T (1983) Linkage analysis of two cloned DNA sequences flanking the Duchenne muscular dystrophy locus on the short arm of the human X chromosome. Nucl Acids Res 11:2303-2312.
9. Bruns, GAP, Gusella JF, Keys C, Leary AC, Housman D, Gerald PS (1982) Isolation of X-chromosome DNA sequences. Adv Exp Med Biol 154:60-72.
10. Bender W, Arkam M, Karch F, Beachy PA, Peifer M, Spierer P, Lewis EB, Hogness DS (1983) Molecular genetics of the bithorax complex in Drosphila melanogaster. Science 221:23-29.

Molecular Biology of Muscle Development, pages 875–886

DEVELOPING GENETIC MARKERS BASED ON RESTRICTION FRAGMENT LENGTH POLYMORPHISMS NEAR THE LOCUS FOR X-LINKED MUSCULAR DYSTROPHY

Uta Francke

Department of Human Genetics, Yale University
New Haven, Connecticut 06510

ABSTRACT With knowledge of the primary molecular defect lacking, there is indirect evidence that the gene for X-linked muscular dystrophy (Duchenne type, DMD) is located in band Xp21 on the short arm of the human X chromosome. Attempting to develop DNA based genetic markers near the DMD locus, we have identified two patients with minor interstitial deletions of part of band Xp21 and have isolated their abnormal X chromosomes in somatic cell hybrids on a rodent cell background. In addition, rodent x human cell hybrids were constructed with human cells in which parts of Xp are translocated to other chromosomes. The resulting panel of somatic cell hybrids allows dissection of the short arm of the X chromosome into 7 distinct regions, 4 of which are within band Xp21. When the hybrids were screened, by Southern blot analysis, with 20 single-copy cloned sequences previously mapped to the X short arm, 6 of those were found to map to one of 3 regions within band Xp21. Four of these recognise DNA polymorphisms and will be useful for linkage studies of DMD families. A male patient who suffered from DMD and from two other X-linked disorders and who has a very minor cytologically detectable deletion of Xp21 was

This work was supported by NIH research grants GM 26105 and GM 32156 and by a Basic Research Grant from the Muscular Dystrophy Association of America.

missing one of the 20 DNA sequences tested. DNA from this patient will be useful for subtraction cloning strategies with the goal of isolating more probes from this small region and of eventually identifying the sequences involved in these disorders.

INTRODUCTION

Duchenne muscular dystrophy (DMD) is inherited as an X-linked recessive genetically lethal disorder. The incidence is approximately 1 in 4,000 males and the mutation rate is higher than for any other known genetic disorder in man [1]. The primary biochemical lesion leading to this disorder has not yet been established. The diagnosis of affected males and heterozygous females is based on secondary manifestations, e.g., clinical criteria, electromyography and serum enzyme levels.

In families with multiple affected individuals, detection of the mutant DMD allele could be accomplished by genetic studies if closely linked polymorphic markers were available. Until a few years ago, the location of the DMD gene on the X chromosome was not known, and linkage studies with any of the available X-linked markers revealed no measurable linkage. This situation has changed recently with two major breakthroughs.

First, the position of the DMD locus is now well-established on the short arm of the X chromosome in band Xp21. This localization is based on the finding of more than 10 unrelated females with clinical findings of pseudohypertrophic muscular dystrophy consistent with classical DMD by all diagnostic criteria, except for the fact that the patients are female. All of these females were found to have de novo balanced X/autosome translocations with a breakpoint in Xp21 and involvement of different autosomal sites in each case. X chromosome replication studies have demonstrated that the normal X is exclusively or predominantly late replicating. The conclusion has been generally accepted that disruption of the continuity (or a small deletion) of DNA sequences in the vicinity of the breakpoints in Xp21 produces the DMD phenotype by inactivating or deleting the normal allele [2-4].

Second, a new class of genetic markers has been discovered, based on DNA sequence variation that results in detectable differences in restriction fragment sizes. These are detected by cleavage of total human DNA with restriction endonucleases and Southern blot analysis using single-copy DNA fragments as hybridization probes. A number of so-called "restriction fragment length polymorphisms" (RFLP) have been detected on the short arm of the human X chromosome. Two that have been used widely in linkage studies with X-linked muscular dystrophy are RC8 (locus designation DXS9) [5] and L1.28 (DXS7) [6]. On the linkage map, these RFLP loci seem to be equidistant from the DMD locus, approximately 17 cM away. DXS9 maps distal and DXS7 maps proximal to DMD. On the cytological map, both of these loci are outside of band Xp21. The evidence provided by linkage studies with these two RFLP markers is consistent with location of the DMD gene in or near Xp21. Similar recombinational distances have been obtained between both these RFLP loci and the locus for Becker's muscular dystrophy (BMD) suggesting that the BMD locus may be near the DMD locus, or the two disorders could possibly be allelic [7]. In the less common Becker type the dystrophy progresses more slowly and does not prevent reproduction. In addition, sporadic cases exist with intermediate severity that are difficult to classify as either DMD or BMD. Thus, X-linked muscular dystrophy (XLMD) encompasses a clinical spectrum with a range of severity. How many genes there are on the human X chromosome that can give rise to an XLMD phenotype, when mutated or deleted, is yet unknown. Genetic linkage analyses of disease phenotypes with closely linked marker loci will be required to approach this question. Use of the two available RFLP markers has improved the power of genetic prediction for male fetuses and female first degree relatives. However, while the frequency of heterozygosity at DXS7 seems to be around 30%, heterozygotes at DXS9 are less frequent than originally assumed (equal or less than 20%). Even in families that are informative at both loci, single crossovers can occur simultaneously in both intervals since the markers are relatively far away from the XLMD locus. To find DNA sequences within band Xp21 and closer to the gene(s) for X-linked muscular dystrophy we have used cytogenetic and somatic cell genetic approaches.

RESULTS

Interstitial deletions of band Xp21.

Tools for the development of RFLP markers near XLMD are provided by the discovery of individuals with small interstitial deletions of part of band Xp21. We have recently reported two such deletion patients. KC was a young woman with normal gonadal development and mild mental retardation. She was found to have a small de novo interstitial deletion of most of band Xp21, karyotype designation 46,X,del(X) (pter→p21.3::p21.1→qter) [8]. This small deletion was associated with apparently random X inactivation. Replication studies on lymphocytes and skin fibroblasts revealed that in 45% of her cells, the normal X was late replicating. This patient had partial ornithine transcarbamylase (OTC) deficiency as documented by elevated urinary orotic acid excretion and increased serum glutamine after a protein load [8]. These findings confirm the mapping of the structural gene for OTC to this deletion which we had accomplished using a cloned OTC probe for in situ hybridization to human chromosomes and for Southern blot analysis of somatic cell hybrid DNA [9]. Testing of neutrophil function revealed that the patient is heterozygous for chronic granulomatous disease (CGD), suggesting that a gene for CGD also maps within her deletion [8].

The second patient (BB) was a male affected with several clinical conditions that are known to exist as X-linked recessive Mendelian traits: Chronic granulomatous disease (CGD) associated with cytochrome b deficiency and with the McLeod phenotype in the Kell red cell antigen system, retinitis pigmentosa (RP) and DMD. When his karyotype was analysed, no major X chromosome rearrangement was detected, but the central dark band on the short arm (Xp21) appeared slightly reduced in size. The possibility that the material missing from Xp had been inserted at a different chromosomal site could not be excluded by high-resolution chromosome banding studies. Molecular analyses were necessary to characterize his deletion [10].

X-autosome translocations.

Cells from two females with clinical manifestations of muscular dystrophy and de novo translocations were utilized. Fibroblast strain GM 1695 carries a 46,X,t(X;11)(p21.1;q13) reciprocal translocation [11]. Lymphoblastoid cells GM 6007, also obtained from the Human Genetic Mutant Cell Repository at the IMR, Camden NJ, have the translocation 46,X,t(X;9)(p21;p22) [3]. Individuals MN-3 and MN-4 were phenotypically normal carriers of a balanced reciprocal translocation with a breakpoint on the short arm of the X chromosome distal to band Xp21: t(X;14)(p22.31;q21) [12].

Production of somatic cell hybrids containing different regions of the human X chromosome short arm.

An established pseudo-diploid Chinese hamster cell line V79/380-6, deficient in the activity of hypoxanthine guanine phosphoribosyltransferase (HGPRT) was fused to cells with the Xp deletions and translocations described above. Hybrids were selected in HAT medium for retention of the active human X chromosome, or for that portion of it that carried the human HGPRT gene. Selection in medium containing toxic guanine analogues yielded subclones that had lost the HGPRT bearing portion of the human X chromosome. Details of the production and characterization, by chromosome banding, by isozyme and DNA analyses, of these series of hybrids have been reported in the original papers: Series XII and XIII [12]; series XVIII [8]; series 26 and 28 [13]; and series 29 [10]. From members of these 6 different series of Chinese hamster x human cell hybrids a mapping panel was constructed that allows dissection of Xp into 7 distinct regions. Hybrids with a normal human X as the only human chromosome and cells with a 48,XXXX chromosome constitution were used to confirm the X chromosomal assignment of any DNA fragment to be regionally mapped.

Mapping of 20 single copy DNA sequences to regions of Xp.

DNA extracted from the regional mapping panel, control hybrids and parental cell lines was cleaved with the appropriate restriction enzymes, electrophoresed and blotted as described elsewhere [10,13]. The filters were hybridised, sequentially and sometimes simultaneously, with ^{32}P-labelled probes that had been isolated in 7 different laboratories and were known to recognize sequences on the short arm of the X chromosome [10,13]. The results are summarized in Figure 1.

The anonymous sequences are identified by their locus designations. According to established convention [16] D stands for DNA, X for the chromosomal assignment, S for single copy fragment and the numeral is a consecutive ID number. Information about the origin and characteristics of these cloned sequences can be found in the tables published in the Human Gene Mapping Reports [16,17]. Probes that recognise restriction fragment length polymorphisms (RFLP) are indicated by (+), but the search for RFLP has not been extensive for some of the others and polymorphisms may still be discovered.

Identification of RFLP within band Xp21 near the DMD gene.

We have identified 6 cloned DNA fragments that map within band Xp21, 3 of them proximal and 3 of them distal to the DMD locus as defined by the breakpoints in two X/autosome translocation females. Five of these are random DNA segments and one is of known function (OTC). At least 3 of them recognize restriction fragment length polymorphisms that are highly informative. OTC detects an MspI polymorphism with four haplotypes and a heterozygote frequency of 60-70% [18]. 754 recognizes a PstI polymorphism with two alleles and a heterozygote frequency of about 60% [19]. Probe C7 [20] which maps distal to DMD within band Xp21 (Figure 1) recognises an EcoRV polymorphism with two alleles (Wrögemann K and Mandel JL, unpublished).

Family linkage studies.

A multi-center collaborative study is under way to determine the recombination distances between XLMD phenotypes and the RFLPs detected with probes OTC, 754 and C7. Preliminary data on informative meioses in females heterozygous at 3 of these loci are consistent with the gene order derived from cytological mapping: Xpter - C7 - DMD - 754 - OTC - Xcen (Rozen, Harper, Fox, Nussbaum, Fischbeck, Spiro, Blattner, Horwich, Pearson, Mandel, Francke, unpublished).

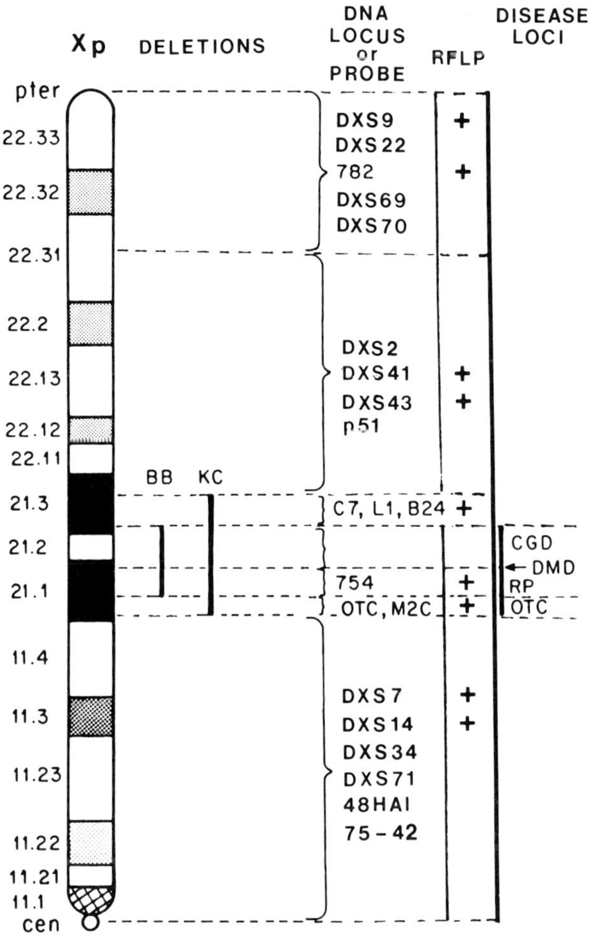

Figure 1. Diagram of human X chromosome from short arm terminus (pter) to centromere (cen) with band designations according to ISCN 1981 nomenclature [14], relative distances are based on band measurements [15]. The deletions in patients BB and KC are indicated by brackets. Dotted lines demarkate the 7 regions to which genes and DNA fragments have been assigned using our somatic cell hybrid panel. RFLP + means that restriction fragment length polymorphisms have been identified, the + symbols

within regions of band Xp21 refer to probes C7, B24, 754 and OTC. Assignments of disease loci are based on phenotypic manifestations in the deletion patients. BB suffered from CGD, DMD and RP but not from OTC deficiency; KC, the female, was a proven heterozygote for CGD and for OTC deficiency.

DISCUSSION

The finding that the DNA sequence recognized by X-specific probe 754 was absent in the male patient BB provides strong evidence for the presence of a deletion rather than a balanced insertion with the chromosomal material missing from Xp21 being inserted elsewhere in the genome. This result seems surprising since until recently no males with cytologically visible deletions of the X, who must be nullisomic for the genes present in the deletion, have been known to exist. With the exception of mild mental retardation, BB did not manifest nonspecific developmental consequences of this nullisomy.

The extent of the deletion can be inferred from the molecular mapping studies summarized in Figure 1. Sequences recognized by probes C7, L1, B24 (distal) and OTC, M2C (proximal) are flanking the BB deletion but are missing in the KC deletion. Therefore, the BB deletion must be entirely contained within the KC deletion. The XLMD locus, as defined by the breakpoints in the X/autosome translocation chromosomes from two females with DMD (dotted line intersecting sub-band Xp21.1) clearly falls within the KC deletion since there are flanking DNA markers. We assume that sequences responsible for the XLMD phenotype map within the BB deletion because (1) he suffered from the disorder, (2) his proximal deletion breakpoint is separated from the XLMD locus by the 754 sequence and there is genetic recombination between 754 and DMD, and (3) the distal breakpoint leading to the BB deletion has to be in sub-band Xp21.3 or at the p21.2/p21.3 border because band Xp21.2 appears to be absent cytologically [10].

Based on these considerations, a feasible strategy for the isolation of sequences close to and flanking the XLMD locus would involve subtraction cloning [21] in which human DNA enriched for X-linked sequences or an X-specific gene library were hybridized with an excess of BB DNA in an attempt to isolate sequences not present in BB. Such sequences could then be mapped to specific sub-regions of band Xp21 with the hybrid mapping panel described above. Sequences mapping within the BB deletion would be used to search for RFLP in human populations. In addition, such sequences will be used to screen DNA from patients with DMD in order to detect others with small sub-microscopic deletions.

In order to identify the "gene" for DMD, i.e., DNA sequences that, when mutated, deleted or disrupted by a translocation event, give rise to an XLMD phenotype, the spontaneous de novo translocations found in females with the disorder will be of tremendous importance. It is anticipated that the translocation events may have involved different areas on the molecular map of the region and, therefore, characterization of each of these events on the molecular level will help to define the extent of the region that is important. Somatic cell hybrids that contain one of the derivative parts of the X chromosome in the absence of the normal X will allow the identification of sequences that fall between the breakpoints present in different X/autosome translocations and, thus, into the region of interest.

In analogy to the homeotic gene mutations in Drosophila, the approach discussed here utilizes cytogenetic observations of chromosome abnormalities for the localization and eventual cloning of genes that are of developmental or clinical importance.

ACKNOWLEDGEMENTS

N. Oliver, W. Woolf, P. Tetri, B. de Martinville, V. Lindgren, J. Giacalone, B. Foellmer, and J. Cowan carried out parts of the research summarized here. H.D. Ochs and C. Disteche referred patient BB. L.M. Kunkel, G. Bruns, J.L. Mandel, P. Pearson, P. Szabo, K. Davies, and A. Horwich made available cloned sequences from the human X chromosome.

REFERENCES

1. Moser H (1983). Duchenne muscular dystrophy. Pathogenetic aspects and genetic prevention. Hum Genet 66:17.

2. Jacobs PA, Hunt PA, Mayer M, Bart RD (1981). Duchenne muscular dystrophy (DMD) in a female with an X/autosome translocation: Further evidence that the DMD locus is at Xp21. Am J Hum Genet 33:513.

3. Emanuel BS, Zackai EH, Tucker SH (1983). Further evidence for Xp21 location of Duchenne muscular dystrophy (DMD) locus: X;9 translocation in a female with DMD. J Med Genet 20:461.

4. Verellen-Dumoulin C, Freund M, de Meyer R, Laterre CH, Frederic J, Thompson MW, Markovic VD, Worton RG (1984). Expression of an X-linked muscular dystrophy in a female due to translocation involving Xp21 and non-random inactivation of the normal X chromosome. Hum Genet 67:115.

5. Murray JM, Davies KE, Harper PS, Meredith L, Mueller, CR, Williamson R (1982). Linkage relationship of a cloned DNA sequence on the short arm of the X chromosome to Duchenne muscular dystrophy. Nature 300:69.

6. Davies KE, Pearson PL, Harper PS, Murray JM, O'Brien T, Williamson R (1983). Linkage analysis of two cloned DNA sequences flanking the Duchenne muscular dystrophy locus on the short arm of the human X chromosome. Nucleic Acids Res 11:2303.

7. Kingston HM, Sarfarazi M, Thomas NST, Harper PS (1984). Localisation of the Becker muscular dystrophy gene on the short arm of the X chromosome by linkage to cloned DNA sequences. Hum Genet 67:6.

8. Francke U (1984). Random X-inactivation resulting in mosaic nullisomy of region Xp21.1→21.3 associated with heterozygosity for ornithine transcarbamylase deficiency and for chronic granulomatous disease. Cytogenet Cell Genet 38:298.

9. Lindgren V, de Martinville B, Horwich AL, Rosenberg LE, Francke U (1984). Human ornithine transcarbamylase locus mapped to band Xp21, near the Duchenne muscular dystrophy locus. Science 226:698.

10. Francke U, Ochs HD, de Martinville B, Giacalone J, Lindgren V, Disteche C, Pagon RA, Hofker MH, van Ommen G-J, Pearson PL, Wedgwood RJ (1985). Minor Xp21 chromosome deletion in a male associated with expression of Duchenne muscular dystrophy, chronic granulomatous disease, retinitis pigmentosa and McLeod syndrome. Amer J Hum Genet 37:250.

11. Greenstein RM, Reardon MP, Chan TS (1979). An (X;11) translocation in a girl with Duchenne muscular dystrophy. Cytogenet Cell Genet 27:389.

12. Francke U, Busby N, Shaw D, Hansen S, Brown MG (1976). Intrachromosomal gene mapping in man: assignment of nucleoside phosphorylase to region 14cen→14q21 by interspecific hybridization of cells with a t(X;14)(p22;p21) translocation. Somat Cell Genet 2:27.

13. De Martinville B, Kunkel L, Bruns G, Morle F, Koenig M, Mandel JL, Horwich A, Latt SA, Gusella JF, Housman D, Francke U (1985). Localization of DNA sequences in region Xp21 of the human X chromosome: Search for molecular markers close to the Duchenne muscular dystrophy locus. Amer J Hum Genet 37:235.

14. ISCN (1981), an international system for human cytogenetic nomenclature - High-resolution banding (1981). Cytogenet Cell Genet 31:1.

15. Francke U (1981). High-resolution ideograms of trypsin-Giemsa banded human chromosomes. Cytogenet. Cell Genet. 31:25.

16. Skolnick MH, Francke U (1982). Report of the committee on human gene mapping by recombinant DNA techniques. Cytogenet Cell Genet 31:194.

17. Skolnick MH, Willard, HF, Menlove LA (1984). Report of the committee on human gene mapping by recombinant DNA techniques. Cytogenet Cell Genet 37:210.

18. Rozen R, Fox J, Fenton WA, Horwich AL, Rosenberg LE (1985). Gene deletion and restriction fragment length polymorphisms at the human ornithine transcarbamylase locus. Nature 313:815.

19. Hofker MH, Wapenaar M, Goor N, Bakker E, van Ommen G-JB, Pearson PL (1985). Isolation of probes detecting restriction fragment length polymorphisms from chromosome specific libraries: Potential use for diagnosis of Duchenne muscular dystrophy. Hum Genet, in press.

20. Camerino G, Koenig M, Moisan JP, Oberle I, Morle F, Jaye M, de la Salle H, Weil D, Hellkuhl B, Grzeschik KH, Mandel JL (1984). Regional mapping of coagulation factor IX gene and of several unique DNA sequences on the human X chromosome. Cytogenet Cell Genet 37:431.

21. Lamar EE, Palmer (1984). Y-encoded, species-specific DNA in mice: Evidence that the Y chromosome exists in two polymorphic forms in inbred strains. Cell 37:171.

Molecular Biology of Muscle Development, pages 887–901

APPROACHING THE DUCHENNE MUSCULAR DYSTROPHY GENE THROUGH A TRANSLOCATION INVOLVING RIBOSOMAL RNA GENES

R.G. Worton, C. Duff, C. Logan, P.N. Ray, V. Kean, M.W. Thompson, J.E. Sylvester[1] and R.D. Schmickel[1]

Department of Genetics and Research Institute, Hospital for Sick Children, Toronto, and Departments of Medical Genetics, Medical Biophysics and Pediatrics, University of Toronto

ABSTRACT Duchenne muscular dystrophy (DMD) is an X-linked genetic disorder normally affecting males. Several females have been described whose disease is the result of an X;autosome translocation with the exchange point in band p21 at or near the site of the DMD locus. In one of these patients the autosome involved is 21 with an exchange point through a block of tandemly repeated genes encoding 18S and 28S rRNA. In the translocated chromosomes "rDNA" must be adjacent to the DMD region of the X, allowing subclones of rDNA to be used as probes in screening for a junction fragment. We have isolated the two translocation chromosomes in somatic cell hybrids, free of the other rDNA-containing human chromosomes, and have identified the putative junction near the 3' end of the 28S rRNA gene. Libraries of gel purified fragments as well as partial digest libraries of hybrid cell DNA have been constructed and screened. Several experiments indicate that the junction fragments are not present in the libraries suggesting that the region is refractory to cloning.

[1]Department of Human Genetics, University of Pennsylvania, Philadelphia

INTRODUCTION

Duchenne muscular dystrophy (DMD) is the most common (frequency of 1/3300 males) and the most severe (lethal by age 20) of the several muscular dystrophies. The gene responsible is X-linked. About one third of all affected males are the result of new mutation while two-thirds inherit their disease from a heterozygous mother. Offspring of carriers are at high genetic risk; male children have a 50% chance of having the disease, female children have a 50% chance of being carriers. To date the disease is poorly understood, the basic biochemical defect is unknown, and there is no definitive prenatal or carrier test.

In this report we describe a new approach to studying the disease utilizing a patient with a chromosomal rearrangement that might allow access to DNA at the site of the Duchenne gene.

DMD Resulting from X-Autosome Translocation

Although X-linked disorders normally affect only males, seven females have been reported with this form of muscular dystrophy (Table 1). In all seven the patient also has an X-autosome translocation (1-7). Although the autosome involved is different in each of the seven, the exchange point in the X chromosome is at band Xp21 (in the middle of the short arm) in each case. This has led to the suggestion that the gene responsible for DMD is at band Xp21, and that the translocation has disrupted or inactivated the normal gene function on the translocated chromosome. This conclusion is further strengthened by the fact that in no case were the parents found to carry the translocation nor was the mother a proven carrier of the Duchenne Gene. Thus, the *de novo* appearance of the defective gene correlates with the *de novo* appearance of a chromosomal rearrangement at band p21 on the X chromosome. Furthermore in females who carry an X-autosome translocation the X-inactivation process is non-random resulting in preferential inactivation of the normal X chromosome. These girls are no exception and therefore their normal allele is presumed to be inactive in most cells allowing the full expression of the disease.

It should also be noted that quite independently of these translocation cases the gene responsible for Duchenne dystrophy in males has been mapped to the middle of the X chromosome short arm at Xp21 by linkage studies in families segregating restriction fragment length polymorphisms that map to Xp (8,9).

On the basis of this evidence it is reasonable to conclude that the exchange point in some of the translocation cases might be within the DMD gene, in others it might be close to the DMD gene separating it from a cis-acting regulatory element, while in others it might be somewhat removed from the gene, but still close enough for a disruption in chromatin structure to affect the activity of the gene. Cloning of DNA from the X;autosome junction region in such a patient would therefore represent a major advance in identifying sequences in the vicinity of the DMD gene.

TABLE 1
A SUMMARY OF SEVEN DMD TRANSLOCATION CASES

Case[a] (Ref)	Pheno-type	Exchange Pts. X	Exchange Pts. Auto	Trans. Inher.	%Inactive Normal X[b]
1	DMD	Xp21	11q13	de novo	"most"
2	DMD	Xp21[d]	1p34	de novo	100%
3	DMD[c]	Xp21	3q13	de novo	93%
4	DMD	Xp21	5q35	de novo	98%
5	DMD	Xp21	6q21	de novo[e]	98%
6	DMD	Xp21	9p22	de novo	98%
7	DMD	Xp21	21p12	de novo	93%

[a]Case number is also reference number
[b]% of lymphocytes or fibroblasts in which the normal X is late-replicating
[c]Also mentally retarded and dysmorphic
[d]A complex rearrangement with two breaks in the X at Xp21 and Xp11
[e]Mother's karyotype was normal but father was not available

The t(X;21) Translocation Case

One of the seven patients, described by Verellen-Dumoulin et al(7), has a unique potential for cloning DNA from the region of the translocation junction. The important feature of this patient's karyotype is that her translocation involves a break in the tandem array of ribosomal RNA (rRNA) genes on the short arm of chromosome 21. This allows the opportunity to use ribosomal gene sequences (rDNA) as molecular probes to identify cloned fragments that span the translocation junction. Figure 1 shows a schematic of the translocation showing the origin of the two translocation derived chromosomes called der(21) and der(X).

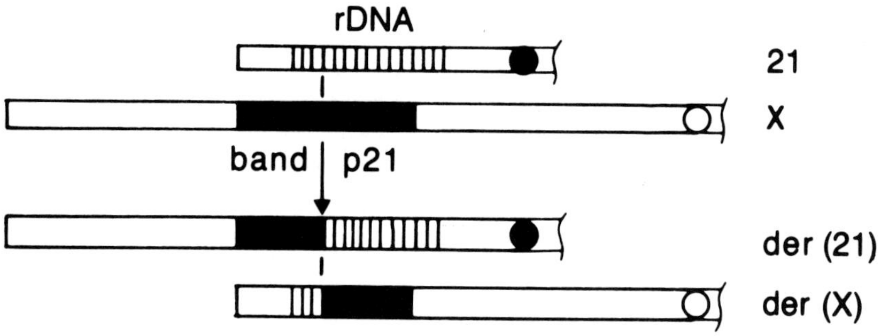

FIGURE 1 Schematic of the short arms of chromosome 21, chromosome X, and the two translocation derived chromosomes der(21) and der(X). (ref.7)

RESULTS

The approach of using rDNA probes to screen for clones containing the junction region is somewhat complicated by the fact that the ribosomal RNA genes are present in 300-400 copies (10) distributed over the 5 pairs of acrocentric chromosomes 13,14,15,21,22. Also, the rDNA repeat unit is 43kb in size (11) so that any particular segment of the rDNA repeat unit used as probe has a 50% chance of being more that 20kb from the translocation junction making it unsuitable for screening a phage library of 15-20kb fragments. For these reasons we first separated the translocation chromosomes from the acrocentric chromosomes by segregation in somatic cell hybrids, and then mapped the location of the translocation junction within the rDNA repeat unit by Southern blot analysis of these segregated hybrids.

Isolation of Somatic Cell Hybrids

Hybrids were generated by Sendai virus-induced fusion of the patient's fibroblasts with mouse A9 cells deficient in hypoxanthine phosphoribosyltransferase (HPRT). Hybrids were selected in HAT medium (alpha medium containing 10ug of hypoxanthine per ml., 10ug of thymidine per ml., and 1 uM methotrexate) containing 2uM ouabain (a dose sufficient to kill human cells but not A9 cells or hybrids)(12).

After culture for about 1 year in HAT medium, all hybrid lines retained the der(X) chromosome since the only active HPRT gene is on the long arm of this chromosome. Three independent hybrid clones were selected for further study. Hybrids A2 and F1 each contained only the der(X) translocation product. Hybrid C2 contained both der(X) and der(21) plus one or two other non-acrocentric human chromosomes. Subclones A2-4 and F1-3 were generated from A2 and F1 directly; subclones A2-T4, F1-T5, and C2-T10 were generated by plating in medium containing 6-thioguanine to select for subclones that had lost the der(X) chromosome. Clones A2-4 and F1-3 contained the der(X); A2-T4 and F1-T5 had no human chromosomes; and C2-T10 contained only the der(21) and in some cells a human chromosome 5(13).

Distribution of rDNA on the Translocation Chromosomes

To examine the der(X) and der(21) chromosomes for the presence of rDNA, human specific rDNA probes were used on Southern blots of DNA from the hybrid cell lines described above. A map of the rDNA repeat unit, and the segments used as probes are shown in Figure 2. The data obtained with pBES as a probe are shown in Figure 3.

The pBES probe revealed the expected 6.3-kb band in DNA from the patient and her father (Fig. 3). The band was present in hybrids A2 and F1 and in subclones A2-4 and F1-3, but not in subclones A2-T4 and F1-T5 selected in 6-thioguanine. Since the subclones A2-T4 and F1-T5 have lost their der(X) chromosome, this result shows that the rDNA sequences present in the hybrids are carried on the der(X) chromosome and are not a result of a minor

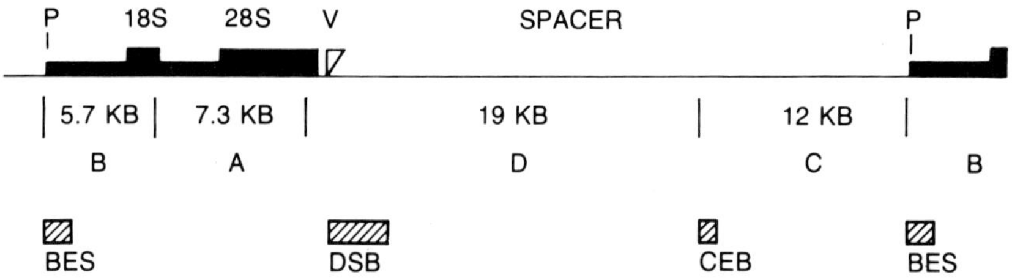

FIGURE 2. The ribosomal RNA gene complex. The 43kb unit is tandemly repeated. Transcription begins at the promotor (P) and generates a 14kb transcript from which the 18S and 28S rRNA molecules are processed. The repeat unit has been defined in terms of its EcoRI restriction sites (11) which generate fragments of size shown. A variable region (V) exists within the D region. The subcloned segments used as probes are BES, an EcoRI/SalI fragment of the B region (14), DSB, a SalI/BamHI fragment of the D region and CEB, an EcoRI/BamHI fragment of the C region. All are spacer region probes which show little or no cross homology to mouse ribosomal gene sequences.

population of cells carrying one or more human acrocentric chromosomes. The intensity of the band from hybrid cells was less than that from 100-fold less human DNA but greater than that from 400-fold less human DNA. Assuming that there are 300-400 copies of the rDNA repeat unit in human diploid cells (10,11) and correcting for the hybrids having about 20 percent more DNA per cell, we estimate that the der(X) chromosome carries three to five copies of the rDNA repeat unit.

Similar considerations for the hybrid C2-T10, which carries the der(21) suggest approximately 40 to 60 copies of rDNA per der(21) chromosome (last lane of Figure 3).

These results show that the hybrids A2 and F1 are ideal sources of DNA from which to attempt cloning of a DNA segment spanning the translocation site.

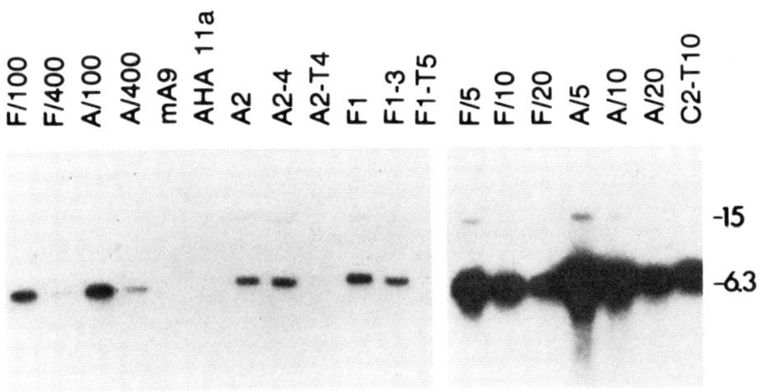

FIGURE 3. Autoradiogram of Southern blot(15) with labelled pBES as probe. DNA was prepared(16) from hybrid clones A2 and F1, subclones A2-4 and F1-3, thioguanine selected subclones A2-T4 and F1-T5 and C2-T10, mouse A9, AHA-11a (a hybrid containing a single human X chromosome and no other human chromosome)(17), and lymphoblast cultures from the patient(A) and her father(F). In most lanes 5ug of BamHI-digested DNA was applied; in lanes marked F/n or A/n 5ug of DNA (F or A) was diluted 1:n before loading. Electrophoresis and Southern blotting were performed as previously described(13).

Orientation of rDNA on the Chromosome

In hybrid lines A2 or F1, the der(X) chromosome should contain 3-5 copies of rDNA flanked on one side by X chromosome sequences and on the other by sequences from the distal end of chromosome 21, the so-called "satellite" region. There must be two junctions, the naturally occuring satellite junction and the translocation-specific X junction. Visualization of these junctions should be possible through Southern blot analysis of hybrid cell DNA, since restriction fragments spanning either junction are expected to be of a different size from the normal repeat fragment, and of an intensity 1/3 to 1/5 of the main rDNA band.

Such a unique band of reduced intensity is barely visible in the BamHI digest of Figure 3. This 15kb band is found in hybrids A2 and F1 at about 1/5 the intensity of the 6.3kb band. This fragment is unlikely to be from the X junction because a similar fragment is seen in the patient's father and he does not carry the translocation chromosome (Figure 3). Two pieces of evidence suggest that it might represent the satellite junction. First, the signal strength from the patient and her father is consistent with about 10 copies per cell, the expected result if all five pairs of acrocentric chromosomes had the same junction region. Secondly, the 15kb fragment is not present in hybrid C2-T10 carrying the der(21) chromosome since this chromosome lacks the satellite end of the rDNA complex (Figure 3; last lane).

This interpretation has been further strengthened by additional observations. First, the presence of a minor band, in addition to the expected major band, is seen not only with BamHI but also with six other enzymes demonstrating that there is a discontinuity in the rDNA map and not simply a BamHI polymorphism (data not presented). Furthermore, a series of double digests with several enzymes plus EcoRI (which cuts at the 5' edge of the BES probe) reveals no minor bands whereas in similar double digests with the same enzymes plus KpnI (which cuts to the 3' side of the BES probe) four of the six double digests reveal a minor band in addition to the main band. This confirms that the unique sites responsible for the minor bands are on the 5' side of BES.

Until these results were obtained the orientation of the rDNA complex on the acrocentric chromosomes was unknown. If our observed discontinuity to the 5' side of BES is indeed the satellite junction, then the satellite sequences must be to the 5' side of the gene and the 5' to 3' orientation of the complex on the chromosome must be telomere to centromere. A map of the putative satellite junction is shown in Figure 4. The 15kb BamHI junction fragment has recently been isolated from a library of 14-16kb fragments from a BamHI digest of human DNA. Two clones analyzed in detail have the same restriction map as that revealed by Southern blotting (Figure 4). Subclones from the satellite end will be mapped by in situ hybridization and by hybrid panel mapping to verify that the cloned segments are from the vicinity of the satellite junction.

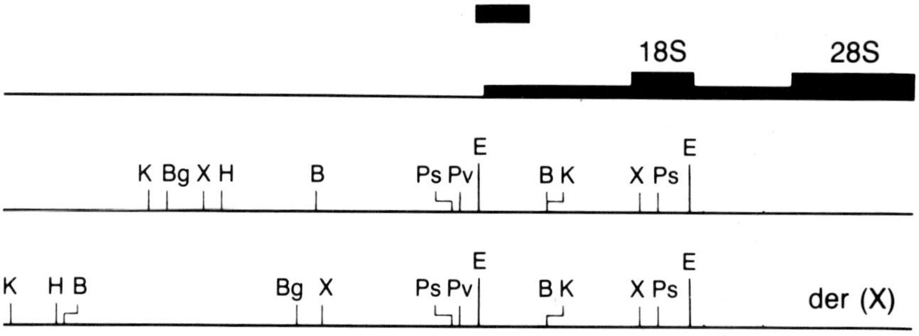

FIGURE 4. Restriction map of the normal rDNA repeat unit and the putative satellite junction region. The line below the schematic is the normal rDNA restriction map. The bottom line is the map obtained from the putative junction fragments visualized as minor bands on Southern blots. The junction region is to the left of the PstI and PvuII sites on the 5' side of BES. The BES position is indicated by the block above the figure.

Location of the X Junction within the rDNA Repeat Unit

From the orientation of the rDNA complex depicted in Figure 4 we predict that the X chromosome sequences must be to the 3' side of any probe used to reveal the junction. With this in mind we have utilized a set of four human-specific probes to search for the location of the X junction in hybrids A2 and F1. The first three probes tested were pBES, pDSB and pCEB (Figure 2) and they failed to reveal any minor band characteristic of a junction fragment. By utilizing each probe on several restriction digests we were able to eliminate approximately 38 of the 43kb as a possible site of the 21:X junction. The only region we were not able to carefully examine with these probes was about 5kb surrounding the 3' end of the 28S rRNA gene. This region was difficult to study because to the left lies the 18S, 5.8S (not shown) and 28S genes whose sequences strongly hybridize with the 300 copies of mouse rDNA obscurring the pattern from the human genes.

The fourth probe examined was an XbaI/KpnI fragment from the internal transcribed spacer region near the 3' end of the 18S gene. This probe called pAXK shows

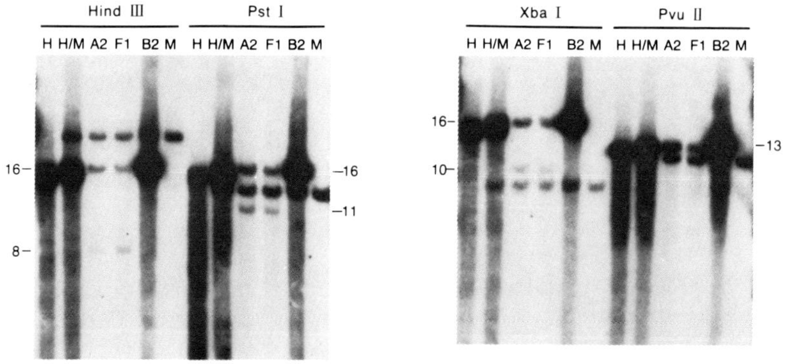

FIGURE 5. Autoradiogram of Southern blot with labelled pAXK as probe. Lane H contains 0.5ug of human DNA, M is 5ug of mouse A9 DNA, and H/M is 5ug mouse A9 plus 0.5ug human DNA to mimic a hybrid with a single human acrocentric chromosome. Lanes A2 and F1 contain 5ug of hybrid cell DNA containing sequences from the der(X) chromosome. Lane B2 is 5ug of DNA from a subclone of the C2-T10 line carrying the der(21) chromosome.

some cross hybridization to mouse ribosomal genes due to a 55 bp homologous region at the 3' end of the mouse 18S gene. Hybrid cell DNA was digested with XbaI, PstI, PvuII and HindIII, electrophoresed and blotted to nitrocellulose. The results of probing with pAXK are presented in Figure 5. Three enzymes, XbaI, PstI and HindIII all show a minor human band, in addition to the normal human and mouse bands. These are 8kb(HindIII), 11kb(Pst) and 10kb(XbaI).

Because the pAXK probe sequences are near the 5' end of the restriction fragments revealed in Figure 5, the minor bands are indicative of new sites on the 3' side of the probe. The putative X junction map based on these minor restriction fragments is presented in Figure 6. Since the PvuII digest did not reveal a minor band we believe the X;21 junction must be in the region between the PvuII site in the 28S gene and the new HindIII site presumed to be on the X chromosome side of the junction.

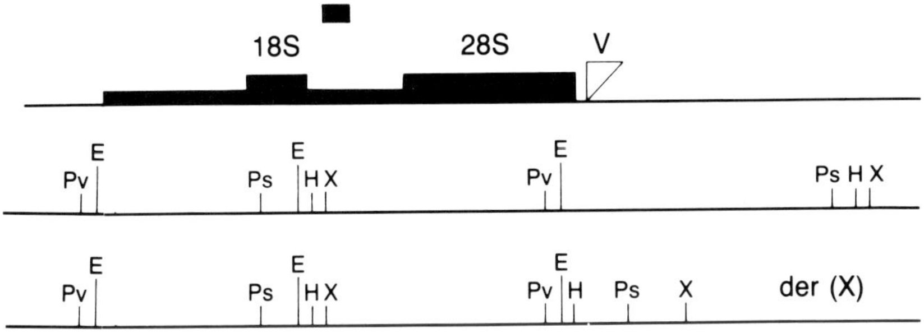

FIGURE 6. Restriction map of the normal rDNA repeat unit and the putative X junction region. The line below the schematic is the normal rDNA restriction map. The bottom line is the map obtained from the putative junction fragments visualized as minor bands on Southern blots. The junction region is to the right of the PvuII and EcoRI sites in the 28S gene. The AXK probe position is indicated by the block above the figure.

Attempts to Clone the Junction Region

Numerous attempts have been made to identify clones containing the putative X-junction region. It has become abundantly clear that the junction region has properties that render it difficult to clone. Our strategies have included the following:

First, a Sau 3A partial digest of hybrid A2-4 DNA was size fractionated on sucrose gradients and fragments of size 15-20kb were cloned into the vector EMBL-3B. A total of 4 x 10^6 clones were screened with the pAXK probe. Of approximately 80 positive clones analysed, all but one were mouse rDNA as determined by restriction mapping and hybridization to mouse repetitive sequences. The remaining one was a human clone derived from a normal rDNA repeat unit. This library was subsequently re-screened with a smaller probe pAAA, an AvaI/AvaI fragment of pAXK, that does not overlap the 18S gene and shows no cross homology to mouse. No positive clones were identified.

Second, the 10kb XbaI junction fragment, (Figure 5) was partially purified by cutting a band from an agarose gel, and cloning into the XbaI sites of a modified Charon 4A vector. A total of 4x10^5 plaques were screened with pAXK. The 6 positive clones were all of mouse origin. A subsequent re-screening of the library with pAAA revealed no positive signals.

Third, the 8.5kb XbaI-PstI fragment (Figure 6) was gel purified and cloned into plasmids pUC13 and pSP65 in a RecA host(18). A total of 1.2x10^5 clones from three independently constructed libraries were screened with probes pAXK and pAAA. No positive signals were found with pAAA and only mouse clones were picked up by pAXK.

Finally, phage cloning in a fully Rec$^-$ host was attempted based on the observation of Wyman _et al_.(19) that unclonable sequences from human chromosome 14 bearing short inverted repeats were clonable in a _recA recBC sbcB_ host. Again the 10kb XbaI fragment was purified on an agarose gel, fragments were ligated to purified left and right arms of Charon 35 phage (20) and the library was plated on the _recA recBC sbcB_ host obtained from Arlene Wyman. A total of 2x10^6 plaques were screened with pAAA and no positive plaques were revealed. In all library screens described above, positive controls were included containing the A fragment of human rDNA, and in all cases these gave strong positive signals by colony or plaque hybridization.

Clearly, the putative junction fragments that are revealed so clearly on Southern blots of hybrid cell DNA are not found in DNA libraries of sizes calculated to contain many copies. We are forced to conclude that some part of this region is refractory to cloning. It is interesting to note that the EcoRI fragment A of rDNA was itself difficult to clone. It appeared in phage libraries at a frequency 25-fold reduced from that of the B fragment, despite their equal representation in the genome. Furthermore, in one vector host system, the only A region clones obtained had deletions from either the A region insert or from one of the phage arms (21) suggesting a lethal interaction between phage and insert sequences.

Despite the difficulties encountered, we are optimistic that junction clones will eventually be obtained. In recent experiments we have identified new restriction enzymes that will aid in junction fragment purification. We have also isolated new human-specific probes from within the 28S gene that will allow the cloning of fragments containing shorter stretches of rDNA. Once a junction fragment has been isolated, we can leave ribosomal genes behind and begin to concentrate on the X chromosome sequences. It is anticipated that this will provide a new approach to the DMD gene itself, and at the very least should provide new probes tightly linked to the DMD locus for use in carrier identification and prenatal diagnosis.

REFERENCES

1. Greenstein RM, Reardon MP, Chan TS (1977). An X/autosome translocation in a girl with Duchenne muscular dystrophy (DMD): Evidence for DMD gene localization. Pediatr Res 11:457 (abstract).
2. Lindenbaum RH, Clarke G, Patel C, Moncrieff C, Hughes JT (1979). Muscular dystrophy in an X;1 translocation female suggests that Duchenne locus is on X chromosome short arm. J Med Genet 16:389-393.
3. Canki N, Dutrillaux B, Tivadar I (1979). Dystrophie musculaire de Duchenne chez une petite fille porteuse d'une translocation t(X;3)(p21;q13) de novo. Ann Genet (Paris) 22:35-39.
4. Jacobs PA, Hunt PA, Mayer M, Bart RD (1981). Duchenne muscular dystrophy (DMD) in a female with an X/autosome translocation:Further evidence that the DMD locus is at Xp21. Am J Hum Genet 33:513-518.
5. Zatz M, Vianna-Morgante AM, Campos P, Diament AJ (1981). Translocation (X;6) in a female with Duchenne muscular dystrophy: Implications for the localisation of the DMD locus. J Med Genet 18:442-447.
6. Emanuel BS, Zackai EH, Tucker SH (1983). Further evidence for Xp21 location of Duchenne muscular dystrophy (DMD) locus: X;9 translocation in a female with DMD. J Med Genet 20 (6):461-463.
7. Verellen-Dumoulin Ch., Freund M, De Meyer R, Laterre Ch., Frederic J, Thompson MW, Markovic VD, Worton RG (1984). Expression of an X-linked muscular dystrophy in a female due to translocation involving Xp21 and non-random inactivation of the normal X chromosome. Hum Genet 67:115-119.
8. Murray JM, Davies KE, Harper PS, Meredith L, Mueller CR, Williamson R (1982). Linkage relationship of a cloned DNA sequence on the short arm of the X chromosome to Duchenne muscular dystrophy. Nature 300:69-71.
9. Davies KE, Pearson PL, Harper PS, Murray JM, O'Brien TO, Sarfarazi M, Williamson R (1983). Linkage analysis of two cloned DNA sequences flanking the Duchenne muscular dystrophy locus on the short arm of the human X chromosome. Nucleic Acid Res 8:2303-2312.

10. Schmickel RD, Knoller M (1977). Characterization and localization of the Human Genes for Ribosomal Ribonucleic Acid. Pediatr Res 11:929.
11. Wellauer PK, Dawid IB (1979). Isolation and Sequence Organization of Human Ribosomal DNA. J Mol Biol 128:289.
12. Farrell SA, Worton RG (1977). Chromosome loss is responsible for segregation at the HPRT locus in Chinese Hamster cell intraspecific hybrids. Somat Cell Genet 3:539.
13. Worton RG, Duff C, Sylvester JE, Schmickel RD, Willard HF (1984). Duchenne Muscular Dystrophy Involving Translocation of the dmd Gene Next to Ribosomal RNA Genes. Science 224:1447.
14. Erickson JM, Rushford CL, Dorney DJ, Wilson GH, Schmickel RD (1981). Structure and variation of human ribosomal DNA:molecular analysis of cloned fragments. Gene 16:1.
15. Southern EM (1975). Detection of Specific Sequences Among DNA Fragments Separated by Gel Electrophoresis. J Mol Biol 98:503.
16. Willard HF Smith KD, Sutherland J (1983). Isolation and Characterization of a major tandem repeat family from the human X Chromosome. Nucl Acids Res 11:2017.
17. Dorman BP, Shimizu N, Ruddle FH (1978). Genetic analysis of the human cell surface: Antigenic markers for the human X chromosome in human-mouse hybrids. Proc Natl Acad Sci U.S.A. 75:2363.
18. Hanahan D (1983). Studies on Transformation of Escherichia coli with Plasmids. J Mol Biol 166:557.
19. Wyman A, Wolfe L, Botstein D (1984). Isolation of the highly polymorphic locus D14S1 and other lethal/unstable human sequences in a new cloning host. Am J Hum Genet 36(suppl.):159S (abstract).
20. Loenen WAM, Blattner FR (1983). Lambda Charon vectors (Ch32,33,34 and 35) adapted for DNA cloning in recombination-deficient hosts. Gene 26:171.
21. Respess J, Erickson J, Rushford C, Jackson D, Schmickel R (1983). Molecular Analysis of Rearrangements in Human Ribosomal RNA Gene Clones. Gene 24:331.

Molecular Biology of Muscle Development, pages 903–909

THE GENETIC MAP OF THE HUMAN X CHROMOSOME AND ITS USE IN THE STUDY OF X-LINKED MUSCULAR DYSTROPHIES[1]

Dennis Drayna and Ray White[2]

Howard Hughes Medical Institute & Dept. of Human Genetics
University of Utah School of Medicine
Salt Lake City, Utah 84132

ABSTRACT We have constructed a genetic linkage map of the human X chromosome by studying the inheritance of 21 DNA polymorphisms in a series of 38 normal families. The map reveals that the total length of the human X is 195 recombination units. The markers are sufficiently closely spaced to show useful linkage to any X-linked disease. As an example, we have chosen to study the genetic location of two muscular dystrophies, Duchenne and Becker. Our studies demonstrate that the linkage information derived from the normal linkage map can greatly assist in mapping, and thus in predicting, these diseases.

INTRODUCTION

DNA probes which reveal useful restriction fragment length polymorphisms (RFLP's) on the human X chromosome have been derived from both arbitrary sequences and known genes. The polymorphic probes used in our study are listed in Table 1 (1-13). The physical locations of these sequences have been determined by *in situ* hybridization and by hybridization to DNA of somatic cell hybrids which

[1]This work was supported by the Howard Hughes Medical Institute and by NIH grant number 5R01GM29789-05.
[2]Present address: Howard Hughes Medical Institute, University of Utah Medical Center, Salt Lake City, Utah 84132.

contain different portions of the human X chromosome (14,15). The markers span the entire length of the X chromosome, from Xp22.3 to Xq28.

TABLE 1
POLYMORPHIC PROBES

Probe	Physical Location	Polymorphism	Frequency	Reference
St 14	Xq28	Taq I	.33/.20/.15 .12/.10/.08 .02	(1)
		Msp I	.30/.20/.20 .15/.15	
FVIII	Xq28	Bcl I	.30/.70	(2)
DX13	Xq28	Bgl II	.65/.35	(3)
F IX	Xq27-28	Taq I	.30/.70	(4)
52A	Xq27	Taq I	.50/.50	(3)
43-15	Xq24-qter	Bgl II	.81/.19	(5)
HPRT	Xq26	BamH I	.70/.22/.08	(6)
22-33	Xq24-qter	Taq I	.84/.16	(5)
S21	Xq22	Taq I	.65/.35	(3)
		Msp I	.90/.10	
19-2	Xq13-q22	Msp I	.80/.20	(5)
DXYS1	Xq13	Taq I	.52/.48	(7)
58	Xcen-p21	Msp I	.65/.35	(5)
L1.28	Xp1	Taq I	.65/.35	(8)
OTCase	Xp21	Msp I	.30/.25/.25 .20	(9)
		BamH I	.65/.35	
754	Xp21	Pst I	.60/.40	(10)
99-6	Xp21-p22	Pst I	.71/.29	(5)
RC8	Xp21	Taq I	.85/.15	(11)
D2	Xp21-p22	Pvu II	.29/.71	(5)
dic56	Xp22.3	Bcl I	.45/.55	(12)
Xg^a	Xp22.3	protein polymorphism	.35/.65	(13)

RESULTS

We have studied the inheritance of RFLP markers on the X chromosome in a series of normal Utah families, each characterized by the presence of large sibships and all four grandparents. The availability of the maternal grandfather allows determination of the phase of the two maternal chromosomes. This makes it possible to simply count recombinant and nonrecombinant X chromosomes in the children, and to calculate an exact value for the recombination fraction observed between any two markers.

In order to construct a useful genetic map, it is necessary to determine gene order as well as recombination fraction (*i.e.*, genetic distance). Information about gene order can be obtained from simple pairwise crosses, but the variance on pairwise recombination fractions is typically too large to allow reliance on such information alone. Better information on marker order can be obtained from multifactor crosses, which can be scored when a family is informative for three or more linked markers. Computational methods which apply the method of maximum likelihood to the problem of gene order have been developed. We have employed two of these, named ILINK and LINKMAP (16), to quantify our confidence in various orders based on our data set with these markers. In practice, we have analysed data on five loci at a time and derived our map by simultaneously considering all of the two-factor, three-factor, four-factor, and five-factor crosses present between the five markers. The entire length of the chromosome was analysed with this technique, and information from each set of five loci was included in the analysis of the next, partially overlapping, set of five loci. The map obtained by this procedure represents a reasonable estimate of the most likely map of the entire chromosome, and gives a most likely gene order and distance for all 21 loci. The current genetic map of the X chromosome is shown in figure 1. The total length of the chromosome is approximately 190 recombination units, with the short arm displaying approximately the same genetic length as the long arm.

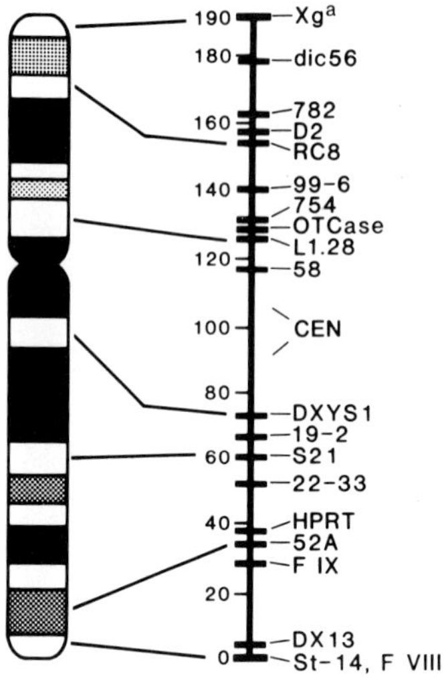

FIGURE 1. Map of chromosome X. The Xg locus codes for a red blood cell antigen; HPRT codes for hypoxanthine phosphoribosyl transferase; F IX codes for blood clotting factor 9. Others are arbitrary loci mapped with DNA probes. The scale of the recombination map is in centimorgans.

Since the majority of markers on the X chromosome are not highly polymorphic, they are not informative enough to provide large numbers of meioses which can be scored for recombination. This results in relatively large variances on gene distance measurements when markers are taken two at a time. When we take markers five at a time, however, substantially more information is available for mapping. Most of the variances obtained using this technique are modest, usually plus or minus half of the intermarker distance or less.

The problem of small numbers is acute in the case of many genetic diseases. Most genetic diseases are rare, and analysis of linkage between a DNA marker and the disease will necessarily be based on only a few meioses. Several muscular dystrophies, including X-linked Emery-Dreyfuss dystrophy, fall into this category. The variance on any linkage based on just a few meioses will be large, and it will be impossible to tell is a given marker is closely linked to the disease or not. A potential solution exists, however, in the information available from the normal genetic map.

If the genetic distances between the DNA markers are known with a high degree of confidence, then the variance on the location of a disease gene will be constrained (16).

We chose Duchenne and Becker muscular dystrophies to test the ability of the normal map to refine the location of genetic diseases. We identified six Duchenne families and two Becker families in Utah, all of them having either three or four living generations. We then examined the linkage of four DNA markers in the Xp21 region to both of these diseases. In the case of Duchenne, the physical order of the DNA markers with respect to the disease is known: the DMD locus lies distal to 754 and proximal to 99-6 (12, 17). Thus, the problem is to determine the linkage distances of these two markers to the disease. Since most of our DNA markers are simple two-allele systems, many matings were not informative and the number of scorable opportunities for recombination between a DNA marker and the disease was small, usually around 10. The information available from the disease families alone permitted us to map the Duchenne locus between 754 and 99-6, 18 recombination units from each. However, the variances on these distances were large (± 15 recombination units in both cases).

In contrast, the normal map shows 11 recombination units between 754 and 99-6, with a variance of ± 6 recombination units (Figure 1). This information adds significantly to the limited information from the disease families; when we combine the two data sets, normal and Duchenne, analysis places both 754 and 99-6 substantially closer to DMD, with a greatly reduced variance. Results obtained with Becker dystrophy demonstrate that the normal map provides a similar improvement in resolution of linkage distances,and support the idea that the Duchenne and Becker dystrophy loci are very close together.

CONCLUSION

Our results demonstrate that the normal linkage map of the human X chromosome can significantly improve resolution of genetic map location in cases of rare diseases. This ability makes the normal genetic map an essential tool for studying the inheritance of many muscular dystrophies.

ACKNOWLEDGMENTS

D.D. gratefully acknowledges the support of the William C. Gibson postdoctoral fellowship of the Muscular Dystrophy Association. R.W. is an Investigator of the Howard Hughes Medical Institute. We thank Mark Hoff and Patricia Callahan for technical assistance, and Barbara Ogden for assistance in sampling muscular dystrophy families.

REFERENCES

1. Oberle I, Drayna D, Camerino G, White R, Mandel J-L (1985). The telomere of the human X chromosome: presence of a highly polymorphic DNA marker and analysis of recombination frequency. Proc Natl Acad Sci USA: in press.
2. Gitschier J, Drayna D, Tuddenham T, White R, Lawn R (1985). A BclI polymorphism in the factor VIII gene allows genetic mapping and diagnosis of haemophilia A. Nature: in press.
3. Drayna D, Davies K, Hartley D, Mandel J-L, Camerino G, Williamson R, White R (1984). Genetic mapping of the human X chromosome by using restriction fragment length polymorphisms. Proc Natl Acad Sci USA 81:2836.
4. Camerino G, Grzeshik K, Jaye M, delaSalle H, Tolstoshev T,Lecocq J, Heilig R, Mandel J-L (1984). Regional localization on the human X chromosome and polymorphism of the coagulation factor IX gene (hemophilia B locus). Proc Natl Acad Sci USA 81:498.
5. Aldridge J, Kunkel L, Bruns G, Tantrahavi U, Lalande M, Brewster T, Moreau E, Wilson M, Bromley W, Roderick T, Latt S (1984). A strategy to reveal high frequency RFLP's along the human X chromosome. Am J. Hum Genet 32:314.
6. Nussbaum R, Crowder W, Nyhan W, Caskey C (1983). A three-allele restriction fragment length polymorphism at the hypoxanthine phosphoribosyltransferase locus in man. Proc Natl Acad Sci USA 80:403.
7 Page D, de Martinville B, Barker D, Wyman A, White R, Francke U, Botstein D (1982). Single copy sequence hybridizes to polymorphic and homologous loci on human X and Y chromosomes. Proc Natl Acad Sci USA 79:5352.

8. Davies K, Pearson P, Harper P, Murray J, O'Brien T, Sarfarzi M, Williamson R (1983). Linkage analysis of two cloned DNA segments flanking the Duchenne muscular dystrophy locus on the short arm of the human X chromosome. Nucl Acids Res 11:2302.
9. Rozen R, Fox J, Fenton W, Horwich A, Rosenberg L (1985). Gene deletion and restriction fragment length polymorphisms at the human ornithine transcarbamylase locus. Nature 313:815.
10. Pearson P, Bakker B (1984). Personal communication.
11. Murray J., Davies K, Harper P, Meredeth L, Mueller C, Williamson R (1982). Linkage relationship of a cloned DNA sequence on the short arm of the X chromosome to Duchenne muscular dystrophy. Nature 300:69.
12. Kunkel L (1985). Personal communication.
13. Race R, Sanger R (1975). "Blood Groups in Man", 6th ed. Oxford:Blackwell, p 610.
14. Hartley D, Davies K, Drayna D, White R, Williamson R (1984). A cytological map of the human X chromosome--evidence for nonrandom recombination. Nucl Acids Res 12:5277.
15. Kunkel L, Tantrahavi U, Eisenhard M, Latt S (1982). Regional localization on the human X of DNA segments cloned from flow sorted libraries. Nucl Acids Res 10:1557.
16. Lathrop M, Lalouel JM, Julier C, Ott J (1984). Strategies for multilocus analysis in humans. Proc Natl Acad Sci USA 81:3443.
17. Francke U (1985). Personal communication.

Molecular Biology of Muscle Development, pages 911–919

LINKAGE ANALYSIS OF DUCHENNE PHENOTYPE AND G6PD ISOTYPE IN MYOBLASTS FROM DOUBLY HETEROZYGOUS CARRIERS*

Cecelia Webster[1], Giorgio Filippi[2], Antoniettina Rinaldi[3], Camillo Mastropaolo[4], Massimo Tondi[4], Marcello Siniscalco[5], and Helen M. Blau[1]

[1]Department of Pharmacology, Stanford University School of Medicine, Stanford, California 94305; [2]Department of Medical Genetics, University of Trieste, Trieste, Italy; [3]Department of Biology, University of Cagliari, Cagliari, Italy; [4]Department of Child Neurology, University of Sassari, Sassari, Italy; [5]Sloan-Kettering Institute for Cancer Research, 425 East 68th Street, New York, New York 10021.

ABSTRACT We previously determined that myoblasts, but not fibroblasts, from muscle of DMD patients are decreased in number, have reduced proliferative capacity, increased doubling times, and altered morphology in culture. The aim of the studies presented here was to determine whether this myoblast phenotype is a primary expression of the DMD locus. Toward this end, we examined whether the DMD phenotype consistently segregated as if it were genetically linked to another marker on the X-chromosome, glucose-6-phosphate dehydrogenase (G6PD). Muscle samples were obtained from five doubly heterozygous carriers of DMD and a polymorphism of the enzyme G6PD, and compared with those of five controls who were heterozygous for G6PD only. Dissociated cells were plated at limiting dilution and 1,237 individual clones were analyzed for both G6PD isotype and DMD phenotype. All assays were performed at the single cell level. This was critical

*This work was supported by a grant to H.M.B. from the Task Force on Genetics of the Muscular Dystrophy Association of America (MDA) and a grant to M.S. from the MDA ; C.W. is supported by an NIH training grant (CA 09302). H.M.B. is the recipient of an NIH Research Career Development Award.

to the analysis, since the clones of interest are those capable of less than 15 doublings in culture. The results show that the DMD phenotype and a given G6PD isotype are not linked in the cells of any of the five carriers. We conclude that the DMD phenotype observed in culture is unlikely to be a primary expression of the defective gene.

INTRODUCTION

The heritable muscular dystrophies are a genetically determined group of diseases characterized by progressive muscular weakness. The primary defect is not known for any of the dystrophies and classification has relied on such clinical characteristics as age of onset, mode of inheritance, muscle groups involved, rate of progression, and clinical course (1). The most common and best defined clinically, is X-linked Duchenne muscular dystrophy (DMD). Due to lack of information concerning the primary defect, there is currently no treatment for the disease.

The study of dystrophic muscle has been hindered by the fact that both biopsy and cell culture material are heterogeneous and contain a large and variable number of adipocytes and fibroblasts. To overcome this problem, we defined procedures for routinely isolating pure populations of human muscle cells from children and adults in large quantity, which could be stored frozen and used in multiple well-controlled studies of human muscle differentiation in vitro (2).

We applied these methods to muscle samples from a number of patients with DMD and age-matched controls and found no significant differences in several important parameters of differentiation in culture: cell fusion, synthesis of several contractile proteins, accumulation and distribution of acetylcholine receptors, and creatine kinase isozyme composition (3). In contrast, a clonal analysis of the cells obtained from DMD biopsies revealed that both the yield and proliferative capacity of the undifferentiated myoblasts were markedly reduced (4,5). Only 5% of the normal number of myoblasts was obtained and a large proportion of the clones to which these gave rise exhibited an altered morphology and growth capacity limited to less than 15 doublings (10^4 cells) as compared to the 40 doublings (10^{12} cells) observed with controls, or a difference in yield of eight orders of magnitude. Given these results, we proposed a hypothesis for the etiology of DMD based on a primary growth defect intrinsic to the muscle cell: Dividing myoblasts are required for muscle growth and maintenance and the limited capacity of DMD myoblasts to grow is directly related to the progressive muscle degeneration characteristic of the disease (4).

The experiments described here were designed to determine whether the observed defect was a primary or secondary expression of the DMD gene. Individual clones of myoblasts from DMD carriers who were heterozygous for another X-linked function, glucose-6-phosphate dehydrogenase (G6PD), were analyzed. In accordance with the Lyon hypothesis, myoblasts carrying the DMD locus on an active X-chromosome should express the defective phenotype; conversely, myoblasts carrying the DMD locus on an inactive X-chromosome should exhibit normal growth potential and morphology. Therefore, the DMD phenotype should segregate consistently with one G6PD isotype. It was the availability of DMD carriers heterozygous for the Mediterranean variant, $G6PD^{M}$, which made this study possible. The small clones which exhibited the DMD phenotype had insufficient G6PD activity for the electrophoretic or kinetic analyses required to distinguish many polymorphisms, whereas $G6PD^{M}$ could be assayed histochemically at the single cell level. The identity of the cells as muscle was ensured by the use of a monoclonal antibody specific to a muscle cell surface antigen, 5.1H11. The results described below demonstrate a random association of the DMD phenotype with G6PD isotype in the myoblasts of each of the five carriers analyzed.

RESULTS

Muscle Samples.

Muscle biopsies were obtained from the biceps of five obligate carriers of DMD who were also heterozygous for G6PD B (normal) and M (mediterranean; a heat labile variant, 6), and from five age and sex matched controls with normal muscle. Three controls were heterozygous for $G6PD^{M/B}$ and two were heterozygous for the Seattle variant of G6PD (7, Table 1). These ten samples were shipped from Sardinia in F10 medium on ice (C II, C III, C IV and the five controls) or at room temperature (C I and C V) and spent between 2 and 7 days in transit. A sixth control, obtained at Stanford University Medical Center, was used to determine the effects of transport on cell behavior in culture. This sample was divided into three parts: (1) dissociated and plated immediately, (2) placed on ice in F10 for seven days prior to dissociation, and (3) kept at room temperature in F10 for seven days prior to dissociation.

Cell Cultures.

Upon arrival, muscle samples were dissociated, an aliquot of cells plated as previously described (2), and the remainder frozen for later use in replicate experiments. Individual clones were established by plating cells at limiting dilution in 96- or 24-well plates. Clones incapable of growing beyond 10 doublings, or 1,000 cells, were counted and assayed for G6PD isotype in situ in these wells. Clones which grew beyond this number were sequentially passaged, assayed for cell number and G6PD isotype, and frozen for future use.

Identification of Myogenic Clones.

Clones of small size and especially those which express the DMD phenotype are frequently not capable of fusion into myotubes. To determine that these small clones were myogenic we used a monoclonal antibody which recognizes a human muscle-specific cell surface antigen, 5.1H11 (8) assayed by immunofluorescence (9).

Histochemical Assay for G6PD Isotype.

The Mediterranean variant of $G6PD^M$, which is prevalent in Sardinia (10), permitted in situ single cell analysis of G6PD isotype and direct correlation with the DMD phenotype in small clones with poor growth. Individual cells of a clone were tested for G6PD activity using a modification of the histochemical technique of Rosenstraus and Chasin (6,11). By this method, the final product of a coupled enzyme reaction is NBT-formazan which stains cells with active G6PD dark blue. We found that although differentiated myotubes in culture have only trace amounts of G6PD activity, undifferentiated myoblasts express G6PD at a high

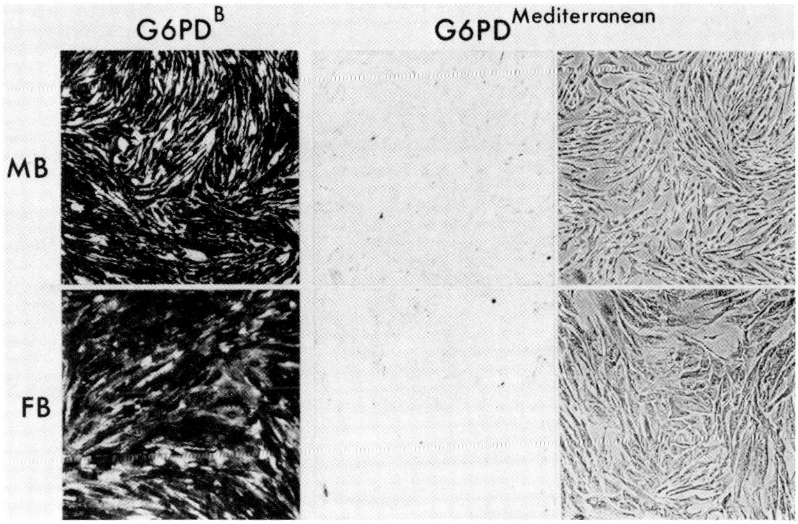

Figure 1. In situ staining of myoblasts (top) and fibroblasts (bottom) for G6PD activity after heat inactivation for 5 minutes at 54°C. Under brightfield illumination, cells expressing the normal isozyme stain darkly (left), whereas cells which express the Mediterranean variant have no enzyme activity (center). G6PD^{M} clones are also shown in phase contrast (right). Mag. 40x.

level, equivalent to that detected in fibroblast cultures (Figure 1). By visual assessment, clones at 37°C had similar enzyme activity regardless of whether they expressed G6PD^{M} or G6PD^{B}. However, if the clones were exposed to 54° for 5 min prior to the assay, enzyme activity was essentially abolished in G6PD^{M} clones, but was not affected in G6PD^{B} clones.

Cell Yield.

The cell yield was determined from the efficiency of plating of dissociated cells. The transport control showed that the yield and proliferative capacity of myoblasts was

TABLE 1
MYOBLAST CLONES WITH LIMITED PROLIFERATIVE CAPACITY

Sample Donor	Age (Yrs)	$G6PD^M$ %D/M*	$G6PD^B$ %D/M*
Normal			
N I	30	--	--
N II	32	0	8
N III	42	--	--
N IV	42	0	7
N V	48	14	13
Carrier			
C I	28	19	17
C II	38	26	31
C III	42	11	5
C IV	44	14	14
C V	52	19	20

* %D/M is the percentage of myoblast clones which expressed the DMD phenotype: altered morphology and capable of fewer than 15 population doublings.

The total number of myoblast clones analyzed for G6PD isotype was 203 for normal controls and 1034 for carriers.

N I and N III are $G6PD^{B/Seattle}$, a low activity variant which cannot be assayed *in situ*.

not affected by the shipping conditions from Sardinia. Furthermore, the average yield of myoblasts from the five carrier samples did not differ significantly from that of the five control samples, and was approximately 3,000 cells per 0.1 g tissue, in both cases. The results are in good agreement with the myoblast yields which we previously reported for normal individuals (2).

Segregation of Growth Properties and G6PD Isotype.

Using the assays described above, 1,237 individual clones from the $G6PD^{M/B}$ heterozygotes were scored for proliferative capacity and G6PD isotype. The results are summarized in Table 1. The percent of $G6PD^M$ and $G6PD^B$ clones which were limited in proliferative capacity to less than 15 doublings, or 30,000 cells, is shown for the five carriers and three controls. The proportion of total clones with limited growth was increased more than two-fold, for carriers relative to controls. However, these clones, which exhibited the dystrophic phenotype, did not segregate with a single G6PD isotype in any given DMD carrier.

DISCUSSION

Linkage of the DMD phenotype with a specific G6PD isotype was not observed in the clones of any of the carrier or control samples analyzed. These results support the hypothesis that the limited proliferative potential of DMD myoblasts in culture is secondary: a consequence of the demand placed upon the cells to replicate *in vivo* in response to fiber degeneration and regeneration in the course of the disease. An alternative explanation for our results is that they are due to mitotic recombination within the region of the DMD locus. In this case, the absence of linkage with a particular G6PD isotype would not rule out the possibility that the DMD phenotype is a primary expression of the gene. Although theoretically possible, there is currently no evidence to support this hypothesis.

The total yield of myoblasts from the biopsies of DMD carriers was not reduced; however, the proportion of the myoblasts which gave rise to clones with a DMD phenotype was significantly increased. Thus, in the muscle of DMD carriers, no significant loss of satellite cells has occurred. This is in contrast with the muscle from children with DMD, in which the yield of clonable myoblasts is reduced 20-fold (4). Furthermore, there was no evidence for selection against one particular myoblast population *in vivo* or *in vitro*. This was determined by an analysis of the $G6PD^{M/B}$ mosaicism in the myoblasts of these carriers (6). However, the proportion of all clones which was capable of less than 15 doublings in culture was significantly increased for carriers. Both the reduction in number of myoblasts from muscle of DMD patients and the reduction in proliferative capacity in an increased proportion of clones from DMD carriers could result from proliferative demands on the cells to provide the additional myoblasts required to replenish degenerating tissue. In further support of this conclusion, we have observed that the proportion of clones with a dystrophic phenotype increases with the age of the DMD patient (12).

From these and previous results utilizing pure populations of myogenic cells (3-5), it would appear that either the DMD locus is not expressed in myoblasts or myotubes in culture or has not yet been detected by the assays used. Alternatively, since DMD appears to be developmentally regulated, the degree of myotube maturation currently achieved in culture may not be sufficient to recapitulate its expression in culture. At present fetal isoforms are typically observed in muscle cells in vitro (13). It is possible that co-culture of muscle with nerve (14,15) or hormonal modulation could promote maturation of myotubes to a point at which they express proteins characteristic of later stages of development. Experiments to examine this possibility are currently in progress in our laboratory.

In the course of the studies described here, cell stocks of myoblasts which are genetically identical but phenotypically distinct in their X-chromosome expression have been generated. They are being maintained frozen in a cell bank. These cells provide a potential resource for experiments designed to test the possibilities described above or to assay the expression and elucidate the function in muscle cells of a putative DMD gene.

REFERENCES

1. Rowland LP (1985). Clinical perspective: Phenotypic expression in muscular dystrophy. In Strohman RC and Wolf S (eds): "Gene Expression in Muscle," Adv. Exp. Med. Biol. 182:3.

2. Blau HM and Webster C (1981). Isolation and characterization of human muscle cells. Proc. Natl. Acad. Sci. USA 78:5623.

3. Blau HM, Webster C, Chiu C-P, Guttman, S, and Chandler F (1983). Differentiation properties of pure populations of human dystrophic muscle cells. Exp. Cell Res. 144:495.

4. Blau HM, Webster C, and Pavlath GK (1983). Defective myoblasts identified in Duchenne muscular dystrophy. Proc. Natl. Acad. Sci. USA 80:4856.

5. Blau HM, Webster C, Pavlath GK, and Chiu C-P (1985). Evidence for defective myoblasts in Duchenne muscular dystrophy. In Strohman RC and Wolf S (eds): "Gene Expression in Muscle," Adv. Exp. Med. Biol. 182:85.
6. Webster C, Filippi G, Rinaldi A, Mastropaolo C, Tondi M, Siniscalco M, and Blau HM. Characterization of the myoblast defect in Duchenne muscular dystrophy (DMD): Linkage analysis of DMD and G6PD phenotypes in myoblasts from double heterozygotes (in preparation).
7. Lenzerini L, Meera Khan P, Filippi G, Rattazzi MC, Ray AK, and Siniscalco M (1969). Characterization of glucose-6-phosphate dehydrogenase variants. I. Occurrence of a G6PD Seattle-like variant in Sardinia and its interaction with the G6PD Mediterranean variant. Amer. J. Hum. Genet. 21: 142.
8. Walsh FS and Ritter MA (1981). Surface antigen differentiation during human myogenesis in culture. Nature 289: 60.
9. Chiu C-P and Blau HM (1985). 5-Azacytidine permits gene activation in a previously noninducible cell type. Cell 40:417.
10. Siniscalco M, Bernini LF, Filippi G, Latte B, Meera Khan P, Piomelli S, and Rattazzi MC (1966). Population genetics of haemoglobin variants, thalassaemia and glucose-6-phosphate dehydrogenase deficiency, with particular reference to the malaria hypothesis. Bull. W.H.O. 34:379.
11. Rosenstraus M and Chasin LA (1975). Isolation of mammalian cell mutants deficient in glucose-6-phosphate dehydrogenase activity: Linkage to hypoxanthine phosphoribosyl transferase. Proc. Natl. Acad. Sci. USA 72:493.
12. Webster C and Blau HM. Age related decline in the proliferative potential of myoblast clones from Duchenne muscular dystrophy patients (in preparation).
13. Strohman RC and Wolf S (1985). Ontologic perspective: terminal differentiation and maturation as differently regulated stages in muscle development. In Strohman RC and Wolf S (eds): "Gene Expression in Muscle," Adv. Exp. Med. Biol. 182:15.
14. Ecob MS, Butler-Browne GS, and Whalen RG (1983). The adult fast isozyme of myosin is present in a nerve-muscle tissue culture system. Differentiation 25:84.
15. Toyota N and Shimada Y (1983). Isoform variants of troponin in skeletal and cardiac muscle cells cultured with and without nerves. Cell 33:297.

Molecular Biology of Muscle Development, pages 921–928

PROLIFERATIVE CAPACITY OF DUCHENNE AND WILD-TYPE MYOBLASTS DERIVED FROM A DMD-G6PD DOUBLE HETEROZYGOTE[1]

Orest Hurko, Linda McKee, Judith GEM Zuurveld
Frank S. Walsh[2] and Herbert M. Swick[3]

Department of Neurology
The Johns Hopkins University School of Medicine
Baltimore, Maryland 21205

ABSTRACT We compared the proliferative capacity of two sets of myoblasts which were autosomally identical but which differed only in the expression of an X chromosome bearing the mutant Duchenne muscular dystrophy gene (DMD+) or its wild type allele (DMD wt). These cells were obtained from an obligate DMD heterozygote who was also heterozygous for the X-encoded marker enzyme, glucose-6-phosphate dehydrogenase (G6PD). A stock muscle culture was repeatedly passaged, and aliquots of cells were taken for analysis at multiple times, from early passage until senescence. The relative proportion of G6PD A and G6PD B cells was determined at each of these times, after separation of myoblasts and fibroblasts by one of the following independent methods: single cell cloning, fibroblast preplating, or fluorescence activated cell sorting after antibody staining for the human muscle specific surface antigen, 5.1H11. At all stages there were equal numbers of myoblasts expressing G6PD A or G6PD B, indicating no proliferataive disadvantage to cells carrying the DMD+ mutation on their active X chromosome.

[1]This work was supported by the Muscular Dystrophy Association
[2]Institute of Neurology, London, England
[3]Medical College of Wisconsin, Milwaukee, Wisconsin

INTRODUCTION

Duchenne muscular dystrophy is an X-linked human disorder which results in degeneration of mature striated muscle in hemizygote boys (1). Although the DMD gene has been localized with great precision (2-4), neither its primary gene product nor the cell(s) in which it is expressed is known (5). It has been reported that mononuclear myoblasts derived from muscle biopsies of hemizygote DMD boys divide more slowly and have a limited capacity for doubling in vitro compared to those derived from age-matched normal controls (6). This observation prompted the hypothesis that the DMD gene is expressed in myoblasts and acts directly in those cells to limit their proliferative capacity (6). Alternatively, this impaired growth of myoblasts might be a secondary phenomenon, perhaps resulting from repeated cycles of degeneration and regeneration in vivo, partially exhausting the proliferative capacity of these cells before cultures were established (7).

We undertook to distinguish between these alternatives, by comparing the growth of DMD+ and DMD wt myoblasts derived from an obligate DMD heterozygote woman, whose maternal and paternal X-chromosomes were marked by genes encoding electrophoretically distinguishable variants of G6PD, an enzyme expressed in mononuclear myoblasts (8) and regenerating muscle (9), though not in mature skeletal muscle (9). Because of the stability of lyonization (10) and the extreme rarity of mitotic recombination (11), measurement of the relative frequencies of G6PD A and G6PD B cells after various times in culture permitted the assessment of the relative frequency of DMD+ and DMD wt cells both in mixed cultures, and those in which either fibroblasts or myoblasts had been eliminated.

METHODS

Primary muscle cultures were established according to Yasin (13) and a stock culture was repeatedly subcultured at low density, with aliquots taken at each passage for fractionation and analysis. Fibroblast- and myoblast-enriched fractions were prepared by the method of Richler and Yaffe (14); cloning was done by limiting dilution in 24 well Linbro plates. Fluorescence-activated cell sorting (FACS) was performed on dissociated mononuclear cultures which had been stained with monoclonal antibody 5.1H11 and rhodaminated sheep anti-mouse antibody (Cappell Laboratories), using a Becton-Dickinson FACS II sorter. Electrophoresis for G6PD analysis was done on cellulose acetate (Helena) and scanned using a dual wavelength TLC scanner (Shimadsu CS-930).

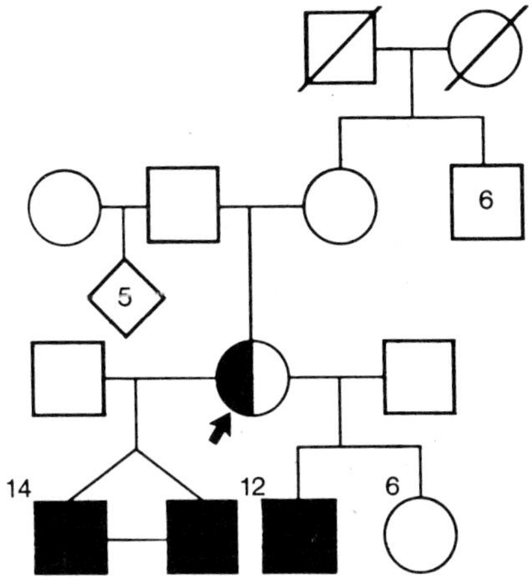

FIGURE 1 - Pedigree of YVL (arrow), an obligate carrier of DM who is also a G6PD heterozygote.

RESULTS

Clinical Information.

YVL is a 30-year-old obligate DMD heterozygote whose pedigree is shown in Figure 1. After being identified as a G6PD heterozygote in a screening program of black DMD heterozygotes conducted in our laboratory, she consentd to donate a muscle biopsy specimen for this study, in full compliance with the regulations of the Human Subjects Committees of the Johns Hopkins Hospital and the Medical College of Wisconsin. A portion of this muscle specimen, taken from the left vastus lateralis, was used for tissue culture and the remainder was examined by standard histopathologic and electron microscopic techniques (12). This demonstrated minor myopathic changes: central nucleation, variation of fiber size and type I predominance, but no delta lesions, necrosis, degeneration or active regeneration.

Clonal Analysis.

Limited dilution cloning was performed at passages 4, 6, 8 and 10 of the parent culture, at which time the cells had undergone an estimated 10, 12, 16 or 19 doublings in vitro, respectively. Cloning efficiencies (grossly visible colonies per plated cell, three weeks after plating) for the first three experiments ranged between 8% and 32%. Roughly equal numbers of myotube forming and fibroblast colonies were found (Table I). Each of these groups was made up of similar numbers of G6PD A and G6PD B colonies. Passage 10 cells failed to give any colonies of sufficient size for assay (approximately 10^4 cells, an additional 14 doublings, starting from a single cell). This indicated that by this late stage, both DMD+ and DMD wt cells had senesced.

TABLE 1
CLONAL ANALYSIS

Doublings*	Fibroblast Colonies	Muscle Colonies	Total Colonies
24	4A, 10B	9A, 13B	13A, 23B
26	-	-	7A, 7B
30	8A, 6B	6A, 7B	14A, 13B

*Estimated doublings in vitro at the time of electrophoresis.

Fibroblast Preplating.

In order to augment the number of cells that we could assay, we also undertook to separate large numbers of myoblasts from fibroblasts at various stages of the parent culture, and then to assess the relative proportion of G6PD A and B in each fraction. Preplating of mixed muscle cultures on uncoated tissue culture plastic has been reported to allow enrichment for either fibroblasts (early-attaching) or myoblasts (late-attaching) (14). At passages 5 through 11, equal intensities of G6PD A and G6PD B signals were found in unfractionated mixed cultures, fibroblast-enriched and myoblast-enriched fractions.

FACS Sorting.

In order to improve the separation between myoblasts and fibroblasts, we developed a novel separation technique based on fluorescent antibody staining with McAb 5.1H11 which recognizes a surface antigen of human myoblasts and myotubes, but does not recognize fibroblasts (15). Nonenzymatic dissociation of muscle cultures results in a viable suspension of mononuclear cells, which retain the 5.1H11 surface antigen. Fluorescence-activated cell sorting allows separation of viable antigen negative and antigen positive cells. The validity of this procedure was verified by (A) differences in cell morphology of sorted antigen positive and negative fractions, both in growth medium and after attempted induction of myoblast fusion in differentiation medium; (B) a two-fold increase in the specific activity of creatine phosphokinase in antigen positive cultures, compared to an unsorted control population; (C) the absence of a bright fluorescent fraction in skin fibroblast preparation stained with McAb 5.1H11 or of concurrent muscle cultures exposed only to fluorescent second antibody.

Aliquots for FACS sorting were diverted from the stock culture at passages 3, 11, 13 and 15; allowed to multiply further to an estimated 14, 19, 21 and 26 doublings _in vitro_; and then stained and sorted. In the three earlier sorts, viable antigen positive and antigen negative fractions were recovered and were found to generate G6PD A and G6PD B signals of equal intensity. In the last sorting experiment we could not identify a significant 5.1H11 antigen-positive fraction, indicating the virtual absence of identifiable myoblasts.

DISCUSSION

Myoblasts derived from hemizygote DMD patients have been reported to have a decreased proliferative capacity (10 doublings in vitro) compared to those from controls (30-40 doublings in vitro) (6). It was proposed that this reflects a direct action of the DMD gene in myoblasts (6). Alternatively, this may be a secondary phenomenon, perhaps the result of multiple cycles of degeneration and regeneration in vivo (7). These experiments were undertaken to distinguish between these two possibilities by culturing cells from a heterozygote, who, by virtue of lyonization, is a mosaic of cells either expressing the DMD+ gene or its allele.

With the possible exception of genes near Xpter, each somatic cell in early female embryos randomly inactivates either an entire paternal or maternal X-chromosome, and stably transmits this pattern of inactivation to all its progeny (9). Clonal analysis of heterozygotes from a variety of human X-linked disorders has confirmed the Lyon hypothesis (16-18) and, in studies using a syntenic G6PD marker (16), confirmed the rarity of mitotic recombination. Several such studies have demonstrated an in vivo tissue-specific selective advantage for cells expressing either the mutant gene (adrenoleukodystrophy (16)) or the wild type allele (Lesch-Nyhan (17); Wiskott-Aldrich (18)).

This study uses a similar analysis, relying on a syntenic G6PD marker to distinguish DMD+ and DMD wt cells, finding them present in equal numbers both in fibroblast and myoblast fractions. DMD+ and DMD wt myoblasts were shown by clonal analysis to have undergone an estimated 30 doublings in vitro, considerably in excess of the 10 doubling limit reported as the maximum for most myoblasts from DMD hemizygotes (6). Had there been a mere 3% difference in doubling rate between DMD+ and DMD wt cells, after twenty doublings the DMD+ cells would have constituted only 25% of the total population. Such a difference could have been detected by our assays, but was not observed.

As the pedigree of our obligate heterozygote does not show DMD in her siblings or in previous generations, it is possible that the mutant DMD gene is restricted to her germ line. In view of the mild myopathic changes in her muscle biopsy, we consider this unlikely. Furthermore, it is also possible that a proliferative disadvantage conferred by the DMD was masked by a "hitchhiker" gene elsewhere on one of her X chromosomes, exactly countering this effect (19). Therefore, we will extend these experiments to double heterozygotes from other families.

Our present data suggest that the Duchenne gene does not act directly in myoblasts to interfere with their growth. These experiments do not exclude the possibility that the gene is expressed in myoblasts but leads to some other as yet unidentified phenotype. Unfortunately, the possibility also remains that the DMD gene may not be expressed in tissue-cultured myoblasts, instead finding expression only in more mature muscle or some other cell type.

ACKNOWLEDGEMENTS

I would like to thank Dr. Pamela Talalay for critical reading of this manuscript and Ms. Diane Emerson for its skillful preparation.

REFERENCES

1. Cullen JJ, Mastaglia FL (1980). Morphologic changes in dystrophic muscle. Brit Med Bull 36:145.
2. Worton RG, Duff C, Sylvester JE, Schmickel RD, Willard HF (1984). Duchenne muscular dystrophy involving translocation of the dmd gene next to ribosomal RNA genes. Science 224:1447.
3. Kunkel L (1985). Genetic analysis of X-linked neuromuscular disease. These proceedings.
4. Davies KE, Pearson PL, Harper PS, Murray JM, O'Brien T, Saffrazi M, Williamson R (1983). Linkage analysis of two cloned DNA sequences flanking the Duchenne muscular dystrophy locus on the short arm of the human X chromosome. Nucleic Acids Res 11:2302.
5. Lucy JA (1980). Is there a membrane defect in muscle and other cells? Brit Med Bull 36:187.
6. Blau, HM, Webster C, Pavlath GK (1983). Defective myoblasts identified in Duchenne muscular dystrophy. Proc Nat Acad Sci USA 80:4856.
7. Wright WE (1985). Myoblast senescence in muscular dystrophy. Experimental Cell Research (in press).

8. Quinn CA, Goodfellow PN, Povey S, Walsh FS (1981). Human-rat muscle somatic cell hybrids form myotubes and express human muscle gene products. Proc Nat Acad Sci USA 78:5031.
9. Wagner KR, Kaufman FC, Max SR (1978). The pentose phosphate pathway in regenerating skeletal muscle. Biochem J 170:17.
10. Lyon, MF (1972). X-chromosome inactivation and developmental patterns in mammals. Biol Rev 47:1.
11. Lewin, B (1980). "Gene expression 2." New York: Wiley, p 139.
12. Carpenter S, Karpati G (1984). "Pathology of skeletal muscle." New York: Churchill Livingstone, p 63.
13. Yasin R, Van Beers G, Nurse KCE, Al-ani S, Landon DN, Thompson EJ (1977). A quantitative technique for growing human adult skeletal muscle in culture starting from mononucleated cells. J Neurol Sci 32:347.
14. Richler C, Yaffe D (1970). The in vitro cultivation and differentiation capacities of myogenic cell lines. Devel Biol 23:1.
15. Walsh FS, Ritter MA (1981). Surface antigen differentiation during myogenesis in culture. Nature 289:60.
16. Migeon BR, Moser HW, Moser AB, Axelman J, Sillence D, Norum RA (1981). Adrenoleukodystrophy: Evidence for X-linkage, inactivation, and selection favoring the mutant allele in heterozygous cells. Proc Nat Acad Sci USA 78:5066.
17. Migeon BR, Der Kaloustian VM, Nyhan WL, Young WJ, Childs B (1968). X-linked hypoxanthine guanine phosphoribosyl transferase deficiency: heterozygote has two clonal populations. Science 160:425.
18. Prchal JT, Carroll AJ, Prchal JF, Crist WM, Skalka HW, Gealy WJ, Harley J, Malluh A (1980). Wiskott-Aldrich syndrome: Cellular impairments and their implications for carrier detection. Blood 56:1048.
19. Migeon BR (1978). Clonal analysis of development X-inactivation and cell communication as determinants of female phenotype. In Subtelny S, Sussex IM (eds): "The Clonal Basis of Development," New York: Academic Press, p 205.

Index